THE CITY READER

…e second edition of *The City Reader* brings together the very best on the city. Classic writings …uch authors as Robert Park, Lewis Mumford, Raymond Unwin, Jane Jacobs, Le Corbusier, … Kevin Lynch meet the best contemporary writings of, among others, Peter Hall, Mike Davis, …ia Sassen, Dolores Hayden, and Manuel Castells.

…five generous selections are included: a combination of thirty extracts from the first edition …twenty-five entirely new selections. Each piece is introduced with a brief intellectual …aphy and a review of the author's writings and related literature, an explanation of …he piece fits into the broader context of urban history and practice, competing ideological …ctives on the city, and the major current debates concerning race and gender, global …turing, the impact of technology, and postmodernism.

…**d T. LeGates** is Professor of Urban Studies at San Francisco State University. **Frederic** …s Lecturer in Urban Studies at Stanford University.

Comments on the first edition

"An excellent, wide-ranging, stimulating reader; attractively presented and easy to read."
Brian Whalley, *Department of Built Environment, De Montfort University*

"A comprehensive mapping of the terrain of Urban Studies, old and new."
Jamie Peck, *Department of Geography, University of Manchester*

"An excellent overview, real breadth of coverage. Particularly valuable as a collection of key contributions which give a real flavour for the temporal development of Urban Studies."
David Valler, *Department of Town and Regional Planning, University of Sheffield*

"An excellent comprehensive overview of urban development and source material."
Allan Bryce, *School of Architecture, University of Dundee*

"This volume is a most welcome collection, without precedent in range and quality."
Alan Simpson, *Urban Design Associates*

"An excellent bringing together of the most important papers and ideas that are relevant to the study of the urban environment."
K. J. Bussey, *Department of Land Economy, Paisley University*

"A real achievement. This book brings together 99 percent of the prominent names in Urban Studies."
Ian Robert Douglas, *Watson Institute for International Studies, Brown University, USA*

"An excellent range of texts. *The City Reader* gathers together some central classics of urban theory, with a few surprises and a number of other pieces which can be difficult to acquire. Editors' comments are consistently illuminating."
Nick Freeman, *Department of English, University of Bristol*

"This is an essential reader for teaching about the cities and Urban Planning in developing countries."
Horng-Chang Hsieh, *Urban Planning Department, Taiwan University*

"I think this is a splendid selection of writings which illustrate the development of modern thinking on urban problems. This is by far the best book of its type."
Dr Tom Begg, *Queen Margaret College, Edinburgh*

"Provides an international overview of urban design issues and an historical perspective on visionary planners who have shaped thinking about development."
Andrew McCafferty, *Department of Built Environment, Northumbria University*

THE CITY READER

Second edition

edited by

Richard T. LeGates
and
Frederic Stout

London and New York

First published 1996
by Routledge
11 New Fetter Lane, London EC4P 4EE

Simultaneously published in the USA and Canada
by Routledge
29 West 35th Street, New York, NY 10001

Second edition published 2000
Reprinted 2000, 2001

Routledge is an imprint of the Taylor & Francis Group

Typeset in Sabon by Solius (Bristol) Ltd, Bristol

Printed and bound in Great Britain by
St Edmundsbury Press, Bury St Edmunds, Suffolk

British Library Cataloguing in Publication Data
A catalogue record for this book is available from the British Library

Library of Congress Cataloging in Publication Data
The city reader / edited by Richard T. LeGates & Frederic Stout. – 2nd ed.
p. cm.
Includes bibliographical references and index.
1. Urban policy. 2. Cities and towns. 3. City planning. I. LeGates, Richard T. II. Stout, Frederic.
HT151.C586 1999 99-31896
307.76–dc21

ISBN 0–415–19070–3 (hbk)
ISBN 0–415–19071–1 (pbk)

To Courtney Elizabeth LeGates
and Amy Catherine Stout

They are the future

CONTENTS

2 URBAN CULTURE AND SOCIETY

3 URBAN SPACE

PLATES

ACKNOWLEDGEMENTS

Many people contributed to this anthology. We owe a particular debt of gratitude to Sarah Lloyd, our editor at Routledge, for her enthusiasm, encouragement, and insightful comments on the manuscript at every stage in its development. Sarah Carty at Routledge ably assisted Ms. Lloyd and solved innumerable problems along the way. Tristan Palmer, our original Routledge editor, deserves much of the credit for the success of the first edition. His helpful advice and editorial suggestions continue to exert a strong influence on this edition.

The contents and underlying pedagogy of the anthology were shaped by the extremely helpful comments of a distinguished panel of advisors consisting of Eugenie L. Birch, Professor of City and Regional Planning at the University of Pennsylvania; Karen Christiansen, Assistant Professor of City and Regional Planning at the University of California, Berkeley; Sir Peter Hall, Bartlett Professor of City Planning at University College, London; Robin Hambleton, Associate Dean of the Faculty of the Built Environment, University of the West of England; Nancy Kleniewski, Dean of the College of Humanities and Social Sciences, University of Massachusetts, Lowell; Judith Martin, Professor of Urban Studies and Director of the Urban Studies Program, University of Minnesota; Leonie Sandercock, Professor of Planning, Policy and Landscape and Chair of the Department of Planning, Policy, and Landscape, Royal Melbourne Institute of Technology; Steven V. Ward, Professor of Planning, Oxford Brookes University; and David Wilmoth, Vice-Chancellor of the Royal Melbourne Institute of Technology.

We received constant encouragement and many valuable suggestions, both for selections to include and for approaches to critical commentary, from our colleagues. We wish particularly to thank Rufus Browning, Roger Crawford, Rich DeLeon, William Issel, Deborah LeVeen, Christopher McGee, Raquel Pinderhughes, Norman Schneider, Genie Stowers, and David Tabb at San Francisco State University; Paul Turner, Leonard Ortolano, and James L. Gibbs, Jr. at Stanford University; Cheyney Ryan at the University of Oregon, Chester Hartman at the Poverty and Race Research and Action Council, John Mollenkopf at the Graduate Center of the City University of New York, and Julia van Haaften of the New York Public Library. Jay Vance of the University of California, Berkeley, Alexander Garvin of Yale University, Peter Calthorpe of Calthorpe Associates, and Chris McGee of San Francisco State University were generous in sharing their insights about what visual images to include and their own copies of images they had assembled over the years. Many others, too numerous to mention, made helpful suggestions. All errors and infelicities are, of course, ours.

A number of people helped us with technical support throughout the writing and editing process. Particularly helpful were Alex Keller, LaVonne Jacobsen, Thoreau Lovell, and Andrew Roderick at San Francisco State University and Stephanie Bazirjian of the Stanford University Center for Teaching and Learning. Tiffany Haas, also of CTL, was especially helpful, supportive, and resourceful in many moments of deadline-crashing crisis.

For unfailing courtesy and helpfulness we thank the staffs of the Cecil H. Green Library and the Henry Meyer Memorial Library at Stanford, the J. Paul Leonard Library at San Francisco State University, the San Francisco Public Library, and the University of California, Berkeley, College of Environmental Design, Institute of Governmental Studies, Bancroft, and Doe libraries. We also wish to thank the staffs of the Museum of the City of New York, the Museum of Modern Art in New York, and the George Eastman House.

As always, Joanne Fraser, Courtney LeGates, and Lisa Ryan provided moral support and were unfailingly patient, often in difficult circumstances.

Finally, this project grew out of years of classroom teaching, and we would be severely amiss if we did not thank the many students of the Stanford Program on Urban Studies, the San Francisco State University Urban Studies Program, and the University of California, Berkeley, Department of City and Regional Planning who read and commented on the selections included as well as on many that did not meet their high standards. Students are the intended audience of this book, and they were the ultimate judges of what readings made the final cut as "essential."

INTRODUCTION

During the last thirty years our students in urban studies and city planning courses at Stanford University, San Francisco State University, and the University of California, Berkeley have often asked us what is the best writing on a given topic or what single new writing will let them know what is happening in a given area right now. Since there was no one source to which we could refer them, we both accumulated photocopies of what we consider to be *essential* writings and bibliographic references to many more. As time passed our colleagues began to come to us for suggested course readings, and we in turn added to our list other selections that they have found most useful. We realized that a systematic organization of the best writings we use to meet both requests would make a good anthology to introduce students of urban studies and city planning to the field and to supplement course texts used in these and other courses concerned with cities. Accordingly we set to work on *The City Reader*, which was published by Routledge in 1996. The selections in the anthology represent both kinds of essential readings – enduring writings we use consistently and the most exciting new writings to which we point our students. We have been very pleased with the reception of *The City Reader*. It has become required reading in many courses related to cities throughout the world.

This second edition retains the basic concept of the first edition, but adds more material, including new writings published in the intervening three years. Based both on our own experience of using the first edition and the most helpful feedback we received from a distinguished group of reviewers, we added many new selections so that this second edition has far more material than the first edition. Entirely new photo sections add visual material to the anthology. This edition also has important changes in emphasis. We have created a new section on urban space and pay more attention to geographical material throughout. We have reorganized and supplemented the material on urban planning. The twenty selections on urban planning in Parts 5, 6, and 7 of this edition could serve as the basis for a course in urban planning. Throughout this edition of *The City Reader* there is a new emphasis on and substantial new material related to sustainable urban development.

The book focuses on *essential* writings. We picked enduring issues in urban studies and planning across different cultures and times. In our courses we have found that H. D. F. Kitto's "The Polis" raises fundamental questions about individuals' relations to their communities that are as relevant today as they were 2,400 years ago; that Louis Wirth's sixty-year-old essay on "Urbanism as a Way of Life" speaks to our students trying to understand contemporary urban violence, economic dislocation, homelessness, and anomie; and that our students are excited by William Julius Wilson's theories on the Black underclass and Manuel Castells' reflections on the rise of the network society. Most writings in this edition of *The City Reader* are from twentieth-century writers, and almost half were written very recently.

This is an *international* anthology. In an increasingly global world, students must learn from writers beyond the borders of their country of origin. In addition to writers from the United States, the second edition now contains writings by scholars from Austria, Australia, Belgium, Canada, England, France, Germany, Greece, Scotland, and Spain. Some of the writers included are world citizens whose countries of birth, academic training, and current residence are all different and whose perspective is truly global. Space limitations precluded including material whose primary focus is on African, Asian, or South American cities, but many of the urban realities and urban processes are applicable everywhere precisely because they have become so internationalized.

This is an *interdisciplinary* anthology. The disciplines represented in the anthology include anthropology, architecture, archaeology, city planning, classics, culture studies, demography, economics, geography, history, landscape architecture, photography, political science, and sociology. But many of the writers blend insights from more than one discipline. And some of the best writing in the anthology, such as Castells' writings about the informational city and Dolores Hayden's writings on non-sexist cities, don't fit in conventional disciplinary boxes at all.

The writings in this anthology seek to combine *theory and practice*. "Urban studies" is the term commonly used to refer to the academic study of cities. Knowledge about cities generated by social scientists and others is sometimes taught in a single program, sometimes dispersed among academic departments. The goal of these courses is primarily to teach students to *understand* cities, only secondarily to empower them to change cities. On the other hand, professional city planning, town planning, and regional planning courses explicitly train students to work as city planners. Often planning courses are taught as part of graduate or undergraduate professional degree programs; sometimes as part of geography, architecture, or departments in the social sciences. This anthology blends both the goal of understanding cities and the goal of planning them. We feel planning should be informed by understanding and that understanding can be enhanced by studying planning.

This anthology includes material on *race and gender* issues in cities, both in the selection of writings and introductions. Diversity characterizes many cities throughout the world and writings need to include the situation, contributions, and perspective of women and people of color as an *integral* part of the writing. To produce balanced coverage of issues of race and gender we have included essential writings by and about women and people of color. We also include consideration of diverse groups in our introductions to the writings.

Any anthology of essential writings on cities should have a *flexible organization*. There is no one best way to organize material on cities. The content of urban studies and city planning courses varies widely and courses are organized in as many different ways as there are courses. This dictates a flexible structure for the book. Readings are grouped into eight broad categories: The Evolution of Cities; Urban Culture and Society; Urban Space; Urban Politics, Governance, and Economics; Urban Planning History and Visions; Urban Planning Theory and Practice; Perspectives on Urban Design; and The Future of the City.

Part 1, The Evolution of Cities, is chronological and works as a unit in the sequence in which the selections are presented. Some teachers may pick and choose selections from this section or use some selections as part of courses that do not have a chronological evolution section. The three sections on urban planning and design contain twenty selections in all and could form the core readings for an entire course in urban planning. The other groupings work well if selections are read in the order in which they are presented, but different sequencing may work better in the context of a given course. Professors experiment; students enjoy!

One goal in picking the selections was to expose students to *great scholarship*. Almost everything written on the emergence of cities acknowledges a debt to the meticulous empirical research and creative theory building of Australian archaeologist V. Gordon Childe, on the Greek cities to H. D. F. Kitto's delightfully written interpretation of the polis, or on medieval cities to Belgian historian Henri

Pirenne's provocative theories on the relationship between the revival of trade and the emergence of medieval cities. Students can learn a great deal from the way Childe, Kitto, and Pirenne think and write beyond the substantive content of the work.

The anthology begins with The Evolution of Cities. We have found that even our brightest and most experienced students bring time- and place-specific cultural concepts to their study of cities. We warn them, and other readers of this anthology, against too quickly assuming what a "suburb," a "slum," an "ancient city" or any other urban settlement is. Some of our San Francisco Bay Area students who grew up in the affluent, predominantly white, residential suburb of Palo Alto think of Palo Alto when they think of a suburb. Material on the evolution on cities enriches their understanding that there were suburbs in medieval European cities composed of traders free from the medieval guilds, suburbs where Manchester's bourgeoisie fled the pollution and stench of industrial Manchester in the 1840s, and a proliferation of streetcar suburbs as nineteenth century technology, entrepreneurship, and public tastes made them possible. These students may be surprised to realize there are a great range of suburbs today – including working-class suburbs, Black suburbs, and technoburbs. Study of the evolution of cities sharpens awareness of these differences. And understanding is the key to successful city planning.

Part 2, Urban Culture and Society, reflects our own interest in the relationship between urban history and urban culture studies. It is placed near the beginning of the anthology because we feel these materials lay a good foundation for all that follows. If there was a single most important insight that the great American intellectual Lewis Mumford emphasized in sixty years of tireless polemics against determinists of every stripe, it was that a city is an expression of the human spirit; not just a physical entity or a locus of economic activity. Accordingly we begin Part 2 with Lewis Mumford's essay "What is a City?" Other writings in this section review important contributions sociologists and anthropologists have made to our understanding of urban culture, explore the culture of poverty and underclass debates, and juxtapose liberal and conservative views on how to respond to urban poverty. We close Part 2 with Sharon Zukin's "Whose Culture? Whose City?" and an original essay by Frederic Stout on the city and the emergence of modern visual culture.

Part 3 deals with urban space, an essential concern of urban geographers. Writings about urban space come from many disciplines and are essential to understanding cities. The physical form of cities is fundamental to urban planners, whose work involves the shaping of urban space. Understanding urban space is important for architects, landscape architects, urban designers, and anyone interested in cities. The great scholar of vernacular landscape J. B. Jackson spent a lifetime decoding the meaning of ordinary landscapes. We include his description of the logic underlying the physical organization of a hypothetical, ordinary town in the Western United States both for its intrinsic interest and to illustrate the kind of intellectual inquiry Jackson pioneered. We have included in the section on urban space other writings by sociologists, architects, and planners, from Witold Rybczynski's description of shopping malls to Mike Davis's rant against the oppressive design of "Fortress L.A." and to Saskia Sassen's work on cities in a global economy.

Part 4 is titled Urban Politics, Governance, and Economics. Several selections in this section explore the way in which political scientists, sociologists, and others now think about urban politics and governance. We particularly emphasize the debate between pluralists and structuralists, Marxist and non-Marxist theories, and regime theory. Other selections in Part 4 emphasize the impact of global restructuring on cities and interconnections between urban economics, politics, and society.

Part 5, Urban Planning: History and Visions, begins with an overview of urban planning history adapted from our introductory essay for a nine-volume set of reprints titled *Early Urban Planning 1870–1940* (Routledge/

Thoemmes, 1996). We tell our beginning students that the best city planning is utopian, or at least idealistic, and encourage them to envision urban futures that they would like to see. Part 5 contains powerful visions of urban futures by Frederick Law Olmsted, Ebenezer Howard, Patrick Geddes, Le Corbusier, Frank Lloyd Wright, and Peter Calthorpe which should challenge students to think through their own values about the importance of aesthetics, community, the relation of man to nature, efficiency, and other enduring issues. We have added photographs and line drawings illustrating a number of the classic planning visions. We hope our students won't lose the power to dream as they tackle research methodology, computers, finance, computer-assisted design and other necessary professional coursework. Rather, we hope that they will see these technical skills as necessary knowledge to translate their own visions into reality.

Part 6 is titled Urban Planning Theory and Practice. Peter Hall's masterful review, "The City of Theory," provides a fine introduction to urban planning theory. Newly added selections by Alexander Garvin from *The American City: What Works, What Doesn't* and by Edward J. Kaiser and David R. Godschalk on twentieth-century land use planning greatly strengthen the material on the practice of urban planning in the first edition. Consistent with the second edition's emphasis on sustainable urban development, we include in the section on urban planning theory and practice a selection by Stephen Wheeler on planning sustainable and livable cities.

The seventh section covers urban design. In addition to classic writings by Kevin Lynch, William Whyte, Allan Jacobs and Donald Appleyard, and Dolores Hayden that students enjoyed from the first edition, they will find included in this edition a chapter from Camillo Sitte's classic book *The Art of Building Cities* and a chapter on the principles of ecological design from Sym Van Der Ryn and Stuart Cowan's important new book *An Introduction to Ecological Design*.

The final section of the anthology, The Future of The City, picks up evolutionary themes from Part 1 and extends the visionary section of Part 5. We open this section with Melvin Webber's prescient 1968 essay "The Post-City Age." In this compact little piece, written when Bill Gates was in junior high school, Webber anticipated many of the debates now occurring about the future of cities in the information age. We include other writings predicting what alternative urban futures *might* be like and describing the authors' normative views about what they *should* be like. Whether the focus is technology, demographics or ecological concerns, all these possible futures constitute the urban destiny of humankind.

We have said a good deal about the role of visions in urban studies and planning. We close with our own vision of how this anthology will be used. It is aimed primarily at students who will be encountering many of the writers and writings for the first time. We hope the writings touch responsive chords and inspire students to think more deeply and read more widely. To that end, for each selection we point the way to other related writings by the same authors and other writers on the same subject. We hope this second edition of *The City Reader* will prove to be a book that professionally oriented students, professors, and practitioners will keep and periodically reread. One test to which we put each of the essential writings included is that it should still be relevant to reread and enjoy in twenty-five years.

Richard LeGates
Frederic Stout
San Francisco, February 1999

Lisa Ryan 1998

*P*rologue

KINGSLEY DAVIS

"The Urbanization of the Human Population"

Scientific American (1965)

Editors' introduction Demography – from the Greek *demos*: "people" – is the study of human populations. Kingsley Davis (1908–1996) pioneered the study of historical urban demography and was particularly fascinated by the history of world urbanization; that is, the increase over time of the proportion of the total human population that is urban as opposed to rural.

The following selection synthesizes Davis's conclusions about how urbanization has occurred throughout the world during all of human history. He raises fundamental issues and lays out a clear framework for understanding population dynamics and urban growth. Davis's careful distinctions of possible sources of urbanization are fundamental. He concludes that urbanization is caused by rural–urban migration, not because of other possible factors such as differential birth and mortality rates.

Davis's extraordinary data on how tiny European urban settlements were after the fall of Rome, and how slowly they grew throughout the Middle Ages and early modern period, provides the demographic backdrop for historian Henri Pirenne's account of the nature of medieval cities (p. 37). During the long period of medieval urbanization the proportion of the population that was urban as opposed to rural changed very slowly. In sharp contrast, Davis concludes that as the Industrial Revolution occurred in England, rapid population growth combined with rural–urban shifts changed both the proportion of the population living in cities and absolute city size very quickly. Friedrich Engels (p. 46) describes in horrifying detail what this revolution in urban demography meant to the impoverished urban proletariat of Manchester and other nineteenth-century industrial cities. His analysis is extremely relevant in assessing prospects for the twenty-first century as the advanced industrial societies and eventually the world reach what some environmental analysts regard as the full "carrying capacity" of the globe.

Davis argues that urbanization follows an attenuated S curve in which pre-industrial cities urbanize very slowly at the long bottom of the S, shoot up at the middle of the S as they industrialize, and then level off at the top of the S. He observes that advanced industrialized countries are now reaching the top part of an S curve, many rapidly urbanizing Third World countries are at the steep middle of the S, and other emerging countries are still moving along the long, slowly rising bottom of the S.

The developing countries of Asia, South America, and Africa already have many huge and rapidly growing cities. As the twenty-first century progresses it appears likely that the human population will increasingly live in "mega cities" of 10 million inhabitants and more often flow together in vast urban connurbations.

Davis concludes that there will be an end to *urbanization* – but not necessarily to absolute population growth, the physical size of cities or the absolute number of people cities contain. He found that the rural population in Third World countries today continues to grow as these countries urbanize, unlike European cities in the nineteenth century where industrialization led to depopulation of rural areas. His vision of Third World societies unable to sustain their populations helps to explain Saskia Sassen's description of growing poverty and inequality worldwide and the growth of large, poorly paid immigrant labor forces in the largest cities in the developed world (p. 208).

Research and scholarly debate continues on the nature and causes of world urbanization. Historians continue to shed light on the growth of cities, but because the records from which they work are often fragmentary and incomplete not everyone agrees with Davis or any other standard account. Debate continues on the relative importance of war, plague, medical advances, trade, technology, religion, and ideology on urban growth. And debate is even more intense in the normative area – about what, if anything, governments should do about population growth and urbanization.

Davis stresses the impact of overall population growth (which he sees as a real danger) on world urbanization and implies that family planning is essential if cities are to meet human needs. But many governments reject family planning on religious or policy grounds, and some European countries now face declining populations and are currently debating the desirability of enacting family-friendly policies to reward child-bearing.

Davis's other writings include many articles and studies on demographics and natural resources as well as two anthologies: *Cities: Their Origin, Growth and Human Impact* (San Francisco, W. H. Freeman, 1973) and, with Mikhail S. Bernstram, *Resources, Environment and Population: Present Knowledge, Future Options* (New York: Population Council, and Oxford: Oxford University Press, 1991).

Data on world urbanization is contained in Tertius Chandler and Gerald Fox, *3000 Years of Urban Growth* (New York: Academic Press, 1974). Further insight on demography and urbanization can be found in World Bank, *World Development Report: Infrastructure for Development* (Oxford: Oxford University Press, 1994), and Ad van der Woude, Akira Hayami, and Jan de Vries (eds.), *Urbanization in History: A Process of Dynamic Interaction* (Oxford: Oxford University Press, 1990).

For recent developments in Third World urbanization, see Alan Gilbert (ed.), *The Mega-City in Latin America* (New York: United Nations University Press, 1996), Carole Rakodi (ed.), *The Urban Challenge in Africa* (New York: United Nations University Press, 1997), and Fu-chen Lo and Yue-man Yeung (eds.), *Emerging World Cities in Pacific Asia* (New York: United Nations University Press, 1996).

For an environmental view of world urbanization, consult Cedric Pugh (ed.), *Sustainabiliy, the Environment and Urbanization* (London: Earthscan, 1996). Lester R. Brown and Jodi L. Jacobson provide a summary of recent world population studies and reflections on the future in *The Future of Urbanization: Facing the Ecological and Economic Constraints* (New York: Worldwatch Paper No. 77, 1987).

Urbanized societies, in which a majority of the people live crowded together in towns and cities, represent a new and fundamental step in man's social evolution. Although cities themselves first appeared some 5,500 years ago, they were small and surrounded by an overwhelming majority of rural people; moreover, they relapsed easily to village or small-town status. The urbanized societies of today, in contrast, not only have urban agglomerations of a size never before attained but also have a high proportion of their population concentrated in such agglomerations. In 1960, for example, nearly 52 million Americans lived in only 16 urbanized areas. Together these areas covered less land than one of the smaller counties (Cochise) of Arizona. According to one definition used by the U.S. Bureau of the Census, 96 million people – 53 percent of the nation's population – were concentrated in 213 urbanized areas that together occupied only 0.7 percent of the nation's land. Another definition used by the bureau puts the urban population at about 70 percent. The large and dense agglomerations comprising the urban population involve a degree of human contact and of social complexity never before known. They exceed in size the communities of any other large animal; they suggest the behavior of communal insects rather than of mammals.

Neither the recency nor the speed of this evolutionary development is widely appreciated. Before 1850 no society could be described as predominantly urbanized, and by 1900 only one – Great Britain – could be so regarded. Today, only 65 years later, all industrial nations are highly urbanized, and in the world as a whole the process of urbanization is accelerating rapidly.

Some years ago my associates and I at Columbia University undertook to document the progress of urbanization by compiling data on the world's cities and the proportion of human beings living in them; in recent years the work has been continued in our center – International Population and Urban Research – at the University of California at Berkeley. The data obtained in these investigations . . . show the historical trend in terms of one index of urbanization: the proportion of the population living in cities of 100,000 or larger. Statistics of this kind are only approximations of reality, but they are accurate enough to demonstrate how urbanization has accelerated. Between 1850 and 1950 the index changed at a much higher rate than from 1800 to 1850, but the rate of change from 1950 to 1960 was twice that of the preceding 50 years! If the pace of increase that obtained between 1950 and 1960 were to remain the same, by 1990 the fraction of the world's people living in cities of 100,000 or larger would be more than half. Using another index of urbanization – the proportion of the world's population living in urban places of all sizes – we found that by 1960 the figure had already reached 33 percent.

Clearly the world as a whole is not fully urbanized, but it soon will be. This change in human life is so recent that even the most urbanized countries still exhibit the rural origins of their institutions. Its full implications for man's organic and social evolution can only be surmised.

In discussing the trend – and its implications insofar as they can be perceived – I shall use the term "urbanization" in a particular way. It refers here to the proportion of the total population concentrated in urban settlements, or else to a rise in this proportion. A common mistake is to think of urbanization as simply the growth of cities. Since the total population is composed of both the urban population and the rural, however, the "proportion urban" is a function of both of them. Accordingly, cities can grow without any urbanization, provided that the rural population grows at an equal or a greater rate.

Historically, urbanization and the growth of cities have occurred together, which accounts for the confusion. As the reader will soon see, it is necessary to distinguish the two trends. In the most advanced countries today, for example, urban populations are still growing, but their proportion of the total population is tending to remain stable or to diminish. In other words, the process of urbanization – the switch from a spread-out pattern of human settlement to one of concentration in urban centers – is a change that has a beginning and an end, but the growth of cities has no inherent limit. Such growth could continue even after everyone was living in cities, through sheer excess of births over deaths.

The difference between a rural village and an urban community is of course one of degree; a precise operational distinction is somewhat arbitrary, and it varies from one nation to another. Since data are available for communities of various sizes, a dividing line can be chosen at will. One convenient index of urbanization, for example, is the proportion of people living in places of 100,000 or more. In the following analysis I shall depend on two indexes: the one just mentioned and the proportion of population classed as "urban" in the official statistics of each country. In practice the two indexes are highly correlated; therefore either one can be used as an index of urbanization.

Actually the hardest problem is not that of determining the "floor" of the urban category but of ascertaining the boundary of places that are clearly urban by any definition. How far east is the boundary of Los Angeles? Where along the

Hooghly River does Calcutta leave off and the countryside begin? In the past the population of cities and towns has usually been given as the number of people living within the political boundaries. Thus the population of New York is frequently given as around eight million, this being the population of the city proper. The error in such a figure was not large before World War I, but since then, particularly in the advanced countries, urban populations have been spilling over the narrow political boundaries at a tremendous rate. In 1960 the New York–Northeastern New Jersey urbanized area, as delineated by the Bureau of the Census, had more than 14 million people. That delineation showed it to be the largest city in the world and nearly twice as large as New York City proper.

As a result of the outward spread of urbanites, counts made on the basis of political boundaries alone underestimate the city populations and exaggerate the rural. For this reason our office delineated the metropolitan areas of as many countries as possible for dates around 1950. These areas included the central, or political, cities and the zones around them that are receiving the spillover.

This reassessment raised the estimated proportion of the world's population in cities of 100,000 or larger from 15.1 percent to 16.7 percent. As of 1960 we have used wherever possible the "urban agglomeration" data now furnished to the United Nations by many countries. The U.S., for example, provides data for "urbanized areas," meaning cities of 50,000 or larger and the built-up agglomerations around them.

. . . My concern is with the degree of urbanization in whole societies. It is curious that thousands of years elapsed between the first appearance of small cities and the emergence of urbanized societies in the nineteenth century. It is also curious that the region where urbanized societies arose – northwestern Europe – was not the one that had given rise to the major cities of the past; on the contrary, it was a region where urbanization had been at an extremely low ebb. Indeed, the societies of northwestern Europe in medieval times were so rural that it is hard for modern minds to comprehend them. Perhaps it was the nonurban character of these societies that erased the parasitic nature of towns and eventually provided a new basis for a revolutionary degree of urbanization.

At any rate, two seemingly adverse conditions may have presaged the age to come: one the low productivity of medieval agriculture in both per-acre and per-man terms, the other the feudal social system. The first meant that towns could not prosper on the basis of local agriculture alone but had to trade and to manufacture something to trade. The second meant that they could not gain political dominance over their hinterlands and thus become warring city-states. Hence they specialized in commerce and manufacture and evolved local institutions suited to this role. Craftsmen were housed in the towns, because there the merchants could regulate quality and cost. Competition among towns stimulated specialization and technological innovation. The need for literacy, accounting skills and geographical knowledge caused the towns to invest in secular education.

Although the medieval towns remained small and never embraced more than a minor fraction of each region's population, the close connection between industry and commerce that they fostered, together with their emphasis on technique, set the stage for the ultimate breakthrough in urbanization. This breakthrough came only with the enormous growth in productivity caused by the use of inanimate energy and machinery. How difficult it was to achieve the transition is agonizingly apparent from statistics showing that even with the conquest of the New World the growth of urbanization during three postmedieval centuries in Europe was barely perceptible. I have assembled population estimates at two or more dates for 33 towns and cities in the sixteenth century, 46 in the seventeenth and 61 in the eighteenth. The average rate of growth during the three centuries was less than 0.6 percent per year. Estimates of the growth of Europe's population as a whole between 1650 and 1800 work out to slightly more than 0.4 percent. The advantage of the towns was evidently very slight. Taking only the cities of 100,000 or more inhabitants, one finds that in 1600 their combined population was 1.6 percent of the estimated population of Europe; in 1700, 1.9 percent; and in 1800, 2.2 percent. On the eve of the industrial revolution Europe was still an overwhelmingly agrarian region.

With industrialization, however, the transformation was striking. By 1801 nearly a tenth of the people of England and Wales were living in cities of 100,000 or larger. This proportion doubled in 40 years and doubled again in another 60 years. By 1900 Britain was an urbanized society. In general, the later each country became industrialized, the faster was its urbanization. The change from a population with 10 percent of its members in cities of 100,000 or larger to one in which 30 percent lived in such cities took about 79 years in England and Wales, 66 in the U.S., 48 in Germany, 36 in Japan and 26 in Australia. The close association between economic development and urbanization has persisted: . . . in 199 countries around 1960 the proportion of the population living in cities varied sharply with per capita income.

Clearly, modern urbanization is best understood in terms of its connection with economic growth, and its implications are best perceived in its latest manifestations in advanced countries. What becomes apparent as one examines the trend in these countries is that urbanization is a finite process, a cycle through which nations go in their transition from agrarian to industrial society. The intensive urbanization of most of the advanced countries began within the past hundred years; in the underdeveloped countries it got under way more recently. In some of the advanced countries its end is now in sight. The fact that it will end, however, does not mean that either economic development or the growth of cities will necessarily end.

The typical cycle of urbanization can be represented by a curve in the shape of an attenuated S. Starting from the bottom of the S, the first bend tends to come early and to be followed by a long attenuation. In the United Kingdom, for instance, the swiftest rise in the proportion of people living in cities of 100,000 or larger occurred from 1811 to 1851. In the U.S. it occurred from 1820 to 1890, in Greece from 1879 to 1921. As the proportion climbs above 50 percent the curve begins to flatten out; it falters, or even declines, when the proportion urban has reached about 75 percent. In the United Kingdom, one of the world's most urban countries, the proportion was slightly higher in 1926 (78.7 percent) than in 1961 (78.3 percent).

At the end of the curve some ambiguity appears. As a society becomes advanced enough to be highly urbanized it can also afford considerable suburbanization and fringe development. In a sense the slowing down of urbanization is thus more apparent than real: an increasing proportion of urbanites simply live in the country and are classified as rural. Many countries now try to compensate for this ambiguity by enlarging the boundaries of urban places; they did so in numerous censuses taken around 1960. Whether in these cases the old classification of urban or the new one is erroneous depends on how one looks at it; at a very advanced stage the entire concept of urbanization becomes ambiguous.

The end of urbanization cannot be unraveled without going into the ways in which economic development governs urbanization. Here the first question is: where do the urbanites come from? The possible answers are few: the proportion of people in cities can rise because rural settlements grow larger and are reclassified as towns or cities; because the excess of births over deaths is greater in the city than in the country, or because people move from the country to the city.

The first factor has usually had only slight influence. The second has apparently never been the case. Indeed, a chief obstacle to the growth of cities in the past has been their excessive mortality. London's water in the middle of the nineteenth century came mainly from wells and rivers that drained cesspools, graveyards and tidal areas. The city was regularly ravaged by cholera. Tables for 1841 show an expectation of life of about 36 years for London and 26 for Liverpool and Manchester, as compared to 41 for England and Wales as a whole. After 1850, mainly as a result of sanitary measures and some improvement in nutrition and housing, city health improved, but as late as the period 1901–1910 the death rate of the urban counties in England and Wales, as modified to make the age structure comparable, was 33 percent higher than the death rate of the rural counties. As Bernard Benjamin, a chief statistician of the British General Register Office, has remarked: "Living in the town involved not only a higher risk of epidemic and crowd diseases . . . but also a higher risk of degenerative disease – the harder wear and tear of factory employment and urban discomfort." By 1950, however, virtually the entire differential had been wiped out.

As for birth rates, during rapid urbanization in the past they were notably lower in cities than in rural areas. In fact, the gap tended to widen somewhat as urbanization proceeded in the latter half of the nineteenth century and the first quarter of the twentieth. In 1800 urban women in the U.S. had 36 percent fewer children than rural women did; in 1840, 38 percent and in 1930, 41 percent. Thereafter the difference diminished.

With mortality in the cities higher and birth rates lower, and with reclassification a minor factor, the only real source for the growth in the proportion of people in urban areas during the industrial transition was rural–urban migration. This source had to be plentiful enough not only to overcome the substantial disadvantage of the cities in natural increase but also, above that, to furnish a big margin of growth in their populations. If, for example, the cities had a death rate a third higher and a birth rate a third lower than the rural rates (as was typical in the latter half of the nineteenth century), they would require each year perhaps 40 to 45 migrants from elsewhere per 1,000 of their population to maintain a growth rate of 3 percent per year. Such a rate of migration could easily be maintained as long as the rural portion of the population was large, but when this condition ceased to obtain, the maintenance of the same urban rate meant an increasing drain on the countryside.

Why did the rural–urban migration occur? The reason was that the rise in technological enhancement of human productivity, together with certain constant factors, rewarded urban concentration. One of the constant factors was that agriculture uses land as its prime instrument of production and hence spreads out people who are engaged in it, whereas manufacturing, commerce and services use land only as a site. Moreover, the demand for agricultural products is less elastic than the demand for services and manufactures. As productivity grows, services and manufactures can absorb more manpower by paying higher wages. Since nonagricultural activities can use land simply as a site, they can locate near one another (in towns and cities) and thus minimize the fraction of space inevitably involved in the division of labor. At the same time, as agricultural technology is improved, capital costs in farming rise and manpower becomes not only less needed but also economically more burdensome. A substantial portion of the agricultural population is therefore sufficiently disadvantaged, in relative terms, to be attracted by higher wages in other sectors.

In this light one sees why a large *flow* of people from farms to cities was generated in every country that passed through the industrial revolution. One also sees why, with an even higher proportion of people already in cities and with the inability of city people to replace themselves by reproduction, the drain eventually became so heavy that in many nations the rural population began to decline in absolute as well as relative terms. In Sweden it declined after 1920, in England and Wales after 1861, in Belgium after 1910.

Realizing that urbanization is transitional and finite, one comes on another fact – a fact that throws light on the circumstances in which urbanization comes to an end. A basic feature of the transition is the profound switch from agricultural to nonagricultural employment. This change is associated with urbanization but not identical with it. The difference emerges particularly in the later stages. Then the availability of automobiles, radios, motion pictures and electricity, as well as the reduction of the workweek and the workday, mitigate the disadvantages of living in the country. Concurrently the expanding size of cities makes them more difficult to live in. The population classed as "rural" is accordingly enlarged, both from cities and from true farms. For these reasons the "rural" population in some industrial countries never did fall in absolute size. In all the industrial countries, however, the population dependent on agriculture – which the reader will recognize as a more functional definition of the nonurban population than mere rural residence – decreased in absolute as well as relative terms. In the U.S., for example, the net migration from farms totaled more than 27 million between 1920 and 1959 and thus averaged approximately 700,000 a year. As a result the farm population declined from 32.5 million in 1916 to 20.5 million in 1960, in spite of the large excess of births in farm families. In 1964, by a stricter American definition classifying as "farm families" only those families actually earning their living from agriculture, the farm population was down to 12.9 million. This number represented 6.8 percent of the nation's population; the comparable figure for 1880 was 44 percent. In Great Britain the number of males occupied in agriculture was, at its peak, 1.8 million, in 1851; by 1961 it had fallen to 0.5 million.

In the later stages of the cycle, then, urbanization in the industrial countries tends to cease. Hence the connection between economic development and the growth of cities also ceases. The change is explained by two circumstances. First, there is no longer enough farm population to furnish a significant migration to the cities. (What can 12.9 million American farmers contribute to the growth of the 100 million people already in urbanized areas?) Second, the rural nonfarm population, nourished by refugees from the expanding cities, begins to increase as fast as the city population. The effort of census bureaus to count fringe residents as urban simply pushes the definition of "urban" away from the notion of dense settlement and in the direction of the term "nonfarm." As the urban population becomes more "rural," which is to say less densely settled, the advanced industrial peoples are for a time able to enjoy the amenities of urban life without the excessive crowding of the past.

Here, however, one again encounters the fact that a cessation of urbanization does not necessarily mean a cessation of city growth. An example is provided by New Zealand. Between 1945 and 1961 the proportion of New Zealand's population classed as urban – that is, the ratio between urban and rural residents – changed hardly at all (from 61.3 percent to 63.6 percent) but the urban population increased by 50 percent. In Japan between 1940 and 1950 urbanization actually decreased slightly, but the urban population increased by 13 percent.

The point to be kept in mind is that once urbanization ceases, city growth becomes a function of general population growth. Enough farm-to-city migration may still occur to redress the difference in natural increase. The reproductive rate of urbanites tends, however, to increase when they live at lower densities, and the reproductive rate of "urbanized" farmers tends to decrease; hence little migration is required to make the urban increase equal the national increase.

I now turn to the currently underdeveloped countries. With the advanced nations having slackened their rate of urbanization, it is the others – representing three-fourths of humanity – that are mainly responsible for the rapid urbanization now characterizing the world as a whole. In fact, between 1950 and 1960 the proportion of the population in cities of 100,000 or more rose about a third faster in the underdeveloped regions than in the developed ones. Among the underdeveloped regions the pace was slow in eastern and southern Europe but in the rest of the underdeveloped world the proportion in cities rose twice as fast as it did in the industrialized countries, even though the latter countries in many cases broadened their definitions of urban places to include more suburban and fringe residents.

Because of the characteristic pattern of urbanization, the current rates of urbanization in underdeveloped countries could be expected to exceed those now existing in countries far advanced in the cycle. On discovering that this is the case one is tempted to say that the underdeveloped regions are now in the typical stage of urbanization associated with early economic development. This notion, however, is erroneous. In their urbanization the underdeveloped countries are definitely not recreating past history. Indeed, the best grasp of their present situation comes from analyzing how their course differs from the previous pattern of development.

The first thing to note is that today's underdeveloped countries are urbanizing not only more rapidly than the industrial nations are now but also more rapidly than the industrial nations did in the heyday of their urban growth. The difference, however, is not large. In 40 underdeveloped countries for which we have data in recent decades, the average gain in the proportion of the population urban was 20 percent per decade; in 16 industrial countries, during the decades of their most rapid urbanization (mainly in the nineteenth century), the average gain per decade was 15 percent.

This finding that urbanization is proceeding only a little faster in underdeveloped countries than it did historically in the advanced nations may be questioned by the reader. It seemingly belies the widespread impression that cities throughout the nonindustrial parts of the world are bursting with people. There is, however, no contradiction. One must recall the basic distinction between a change in the proportion of the population urban, which is a ratio, and the absolute growth of cities. The popular impression is correct: the cities in underdeveloped areas are growing at a disconcerting rate. They are far outstripping the city boom of the industrializing era in the nineteenth century. If they continue their recent rate of growth, they will double their population every 15 years.

In 34 underdeveloped countries for which we have data relating to the 1940s and 1950s, the average annual gain in the urban population was 4.5 percent. The figure is remarkably similar for the various regions: 4.7 percent in seven countries of Africa, 4.7 percent in 15 countries of Asia and 4.3 percent in 12 countries of Latin America. In contrast, in nine European countries during their period of fastest urban population growth (mostly in the latter half of the nineteenth century) the average gain per year was 2.1 percent. Even the frontier industrial countries – the U.S., Australia–New Zealand, Canada and Argentina – which received huge numbers of immigrants had a smaller population growth in towns and cities: 4.2 percent per year. In Japan and the U.S.S.R. the rate was respectively 5.4 and 4.3 percent per year, but their economic growth began only recently.

How is it possible that the contrast in growth between today's underdeveloped countries and yesterday's industrializing countries is sharper with respect to the absolute urban population than with respect to the urban share of the total population? The answer lies in another profound difference between the two sets of countries – a difference in total population growth, rural as well as urban. Contemporary underdeveloped populations have been growing since 1940 more than twice as fast as industrialized populations, and their increase far exceeds the growth of the latter at the peak of their expansion. The only rivals in an earlier day were the frontier nations, which had the help of great streams of immigrants. Today the underdeveloped nations – already densely settled, tragically impoverished and with gloomy economic prospects – are multiplying their people by sheer biological increase at a rate that is unprecedented. It is this population boom that is overwhelmingly responsible for the rapid inflation of city populations in such countries. Contrary to popular opinion both inside and outside those countries, the main factor is not rural– urban migration.

This point can be demonstrated easily by a calculation that has the effect of eliminating the influence of general population growth on urban growth. The calculation involves assuming that the total population of a given country remained constant over a period of time but that the percentage urban changed as it did historically. In this manner one obtains the growth of the absolute urban population that would have occurred if rural–urban migration were the only factor affecting it. As an example, Costa Rica had in 1927 a total population of 471,500, of which 88,600, or 18.8 percent, was urban. By 1963 the country's total population was 1,325,200 and the urban population was 456,600, or 34.5 percent. If the total population had remained at 471,500 but the percentage urban had still risen from 18.8 to 34.5, the absolute urban population in 1963 would have been only 162,700. That is the growth that would have occurred in the urban population if rural–urban migration had been the only factor. In actuality the urban population rose to 456,600. In other words, only 20 percent of the rapid growth of Costa Rica's towns and cities was attributable to urbanization per se; 44 percent was attributable solely to the country's general population increase, the remainder to the joint operation of both factors. Similarly, in Mexico between 1940 and 1960, 50 percent of the urban population increase was attributable to national multiplication alone and only 22 percent to urbanization alone.

The past performance of the advanced countries presents a sharp contrast. In Switzerland between 1850 and 1888, when the proportion urban resembled that in Costa Rica recently, general population growth alone accounted for only 19 percent of the increase of town and city people, and rural–urban migration alone accounted for 69 percent. In France between 1846 and 1911 only 21 percent of the growth in the absolute urban population was due to general growth alone.

The conclusion to which this contrast points is that one anxiety of governments in the underdeveloped nations is misplaced. Impressed by the mushrooming in their cities of shantytowns filled with ragged peasants, they attribute the fantastically fast city growth to rural–urban migration. Actually this migration now does little more than make up for the small difference in the birth rate between city and countryside. In the history of the industrial nations, as we have seen, the sizable difference between urban and rural birth rates and death rates required that cities, if they were to grow, had to have an enormous influx of people from farms and villages. Today in the underdeveloped countries the towns and cities have only a slight disadvantage in fertility, and their old disadvantage in mortality not only has been wiped out but also in many cases has been reversed. During the nineteenth century the urbanizing nations were learning how to keep crowded populations in cities from dying like flies. Now the lesson has been learned, and it is being applied to cities even in countries just emerging from tribalism. In fact, a disproportionate share of public health funds goes into cities. As a result, throughout the nonindustrial world people in cities are multiplying as never before, and rural–urban migration is playing a much lesser role.

The trends just described have an important implication for the rural population. Given the explosive overall population growth in underdeveloped countries, it follows that if the rural population is not to pile up on the land and reach an economically absurd density, a high rate of rural–urban migration must be maintained. Indeed, the exodus from rural areas should be higher than in the past. But this high rate of internal movement is not taking place, and there is some doubt that it could conceivably do so.

To elaborate, I shall return to my earlier point that in the evolution of industrialized countries the rural citizenry often declined in absolute as well as relative terms. The rural population of France – 26.8 million in 1846 – was down to 20.8 million by 1926 and 17.2 million by 1962, notwithstanding a gain in the nation's total population during this period. Sweden's rural population dropped from 4.3 million in 1910 to 3.5 million in 1960. Since the category "rural" includes an increasing portion of urbanites living in fringe areas, the historical drop was more drastic and consistent specifically in the farm population. In the U.S., although the "rural" population never quite ceased to grow, the farm contingent began its long descent shortly after the turn of the century; today it is less than two-fifths of what it was in 1910.

This transformation is not occurring in contemporary underdeveloped countries. In spite of the enormous growth of their cities, their rural populations – and their more narrowly defined agricultural populations – are growing at a rate that in many cases exceeds the rise of even the urban population during the evolution of the now advanced countries. The poor countries thus confront a grave dilemma. If they do not substantially step up the exodus from rural areas, these areas will be swamped with underemployed farmers. If they do step up the exodus, the cities will grow at a disastrous rate.

The rapid growth of cities in the advanced countries, painful though it was, had the effect of solving a problem – the problem of the rural population. The growth of cities enabled agricultural holdings to be consolidated, allowed increased capitalization and in general resulted in greater efficiency. Now, however, the underdeveloped countries are experiencing an even more rapid urban growth – and are suffering from urban problems – but urbanization is not solving their rural ills.

A case in point is Venezuela. Its capital, Caracas, jumped from a population of 359,000 in 1941 to 1,507,000 in 1963; other Venezuelan towns and cities equaled or exceeded this growth. Is this rapid rise denuding the countryside of people? No, the Venezuelan farm population increased in the decade 1951–1961 by 11 percent. The only thing that declined was the amount of cultivated land. As a result the agricultural population density became worse. In 1950 there were some 64 males engaged in agriculture per square mile of cultivated land; in 1961 there were 78. (Compare this with 4.8 males occupied in agriculture per square mile of cultivated land in Canada, 6.8 in the U.S. and 15.6 in Argentina.) With each male occupied in agriculture there are of course dependants. Approximately 225 persons in Venezuela are trying to live from each square mile of cultivated land. Most of the growth of cities in Venezuela is attributable to overall population growth. If the general population had not grown at all, and internal migration had been large enough to produce the actual shift in the proportion in cities, the increase in urban population would have been only 28 percent of what it was and the rural population would have been reduced by 57 percent.

The story of Venezuela is being repeated virtually everywhere in the underdeveloped world. It is not only Caracas that has thousands of squatters living in self-constructed junk houses on land that does not belong to them. By whatever name they are called, the squatters are to be found in all major cities in the poorer countries. They live in broad gullies beneath the main plain in San Salvador and on the hillsides of Rio de Janeiro and Bogotá. They tend to occupy with implacable determination parks, school grounds and vacant lots. Amman, the capital of Jordan, grew from 12,000 in 1958 to 247,000 in 1961. A good part of it is slums, and urban amenities are lacking most of the time for most of the people. Greater Baghdad now has an estimated 850,000 people; its slums, like those in many other underdeveloped countries, are in two zones: the central part of the city and the outlying areas. Here are the *sarifa* areas, characterized by self-built reed huts; these areas account for about 45 percent of the housing in the entire city and are devoid of amenities, including even latrines. In addition to such urban problems, all the countries struggling for higher living levels find their rural population growing too and piling up on already crowded land. I have characterized urbanization as a transformation that, unlike economic development, is finally accomplished and comes to an end. At the 1950–1960 rate the term "urbanized world" will be applicable well before the end of the century. One should scarcely expect, however, that mankind will complete its urbanization without major complications. One sign of trouble ahead turns on the distinction I made at the start between urbanization and city growth *per se*. Around the globe today city growth is disproportionate to urbanization. The discrepancy is paradoxical in the industrial nations and worse than paradoxical in the nonindustrial.

It is in this respect that the nonindustrial nations, which still make up the great majority of nations, are far from repeating past history. In the nineteenth and early twentieth centuries the growth of cities arose from and contributed to economic advancement. Cities took surplus manpower from the countryside and put it to work producing goods and services that in turn helped to modernize agriculture. But today in underdeveloped countries, as in present-day advanced nations, city growth has become increasingly unhinged from economic development and hence from rural–urban migration. It derives in greater degree from overall population growth, and this growth in nonindustrial lands has become unprecedented because of modern health techniques combined with high birth rates.

The speed of world population growth is twice what it was before 1940, and the swiftest increase has shifted from the advanced to the backward nations. In the latter countries, consequently, it is virtually impossible to create city services fast enough to take care of the huge, never-ending cohorts of babies and peasants swelling the urban masses. It is even harder to expand agricultural land and capital fast enough to accommodate the enormous natural increase on farms. The problem is not urbanization, not rural–urban migration, but human multiplication. It is a problem that is new in both its scale and its setting, and runaway city growth is only one of its painful expressions.

As long as the human population expands, cities will expand too, regardless of whether urbanization increases or declines. This means that some individual cities will reach a size that will make nineteenth-century metropolises look like small towns. If the New York urbanized area should continue to grow only as fast as the nation's population (according to medium projections of the latter by the Bureau of the Census), it would reach 21 million by 1985 and 30 million by 2010. I have calculated that if India's population should grow as the U.N. projections indicate it will, the largest city in India in the year 2000 will have between 36 and 66 million inhabitants.

What is the implication of such giant agglomerations for human density? In 1950 the New York–Northeastern New Jersey urbanized area had an average density of 9,810 persons per square mile. With 30 million people in the year 2010, the density would be 24,000 per square mile. Although this level is exceeded now in parts of New York City (which averages about 25,000 per square mile) and many other cities, it is a high density to be spread over such a big area; it would cover, remember, the suburban areas to which people moved to escape high density. Actually, however, the density of the New York urbanized region is dropping, not increasing, as the population grows. The reason is that the territory covered by the urban agglomeration is growing faster than the population: it grew by 51 percent from 1950 to 1960, whereas the population rose by 15 percent.

If, then, one projects the rise in population and the rise in territory for the New York urbanized region one finds the density problem solved. It is not solved for long, though, because New York is not the only city in the region that is expanding. So are Philadelphia, Trenton, Hartford, New Haven and so on. By 1960 a huge stretch of territory about 600 miles long and 30 to 100 miles wide along the eastern seaboard contained some 37 million people. (I am speaking of a longer section of the seaboard than the Boston-to-Washington conurbation referred to by some other authors.) Since the whole area is becoming one big polynucleated city, its population cannot long expand without a rise in density. Thus persistent human multiplication promises to frustrate the ceaseless search for space – for ample residential lots, wide-open suburban school grounds, sprawling shopping centers, one-floor factories, broad freeways.

How people feel about giant agglomerations is best indicated by their headlong effort to escape them. The bigger the city, the higher the cost of space; yet the more the level of living rises, the more people are willing to pay for low-density living. Nevertheless, as urbanized areas expand and collide, it seems probable that life in low-density surroundings will become too dear for the great majority.

One can of course imagine that cities may cease to grow and may even shrink in size while the population in general continues to multiply. Even this dream, however, would not permanently solve the problem of space. It would eventually obliterate the distinction between urban and rural, but at the expense of the rural.

It seems plain that the only way to stop urban crowding and to solve most of the urban problems besetting both the developed and the underdeveloped nations is to reduce the overall rate of population growth. Policies designed to do this have as yet little intelligence and power behind them. Urban planners continue to treat population growth as something to be planned for, not something to be itself planned. Any talk about applying brakes to city growth is therefore purely speculative, overshadowed as it is by the reality of uncontrolled population increase.

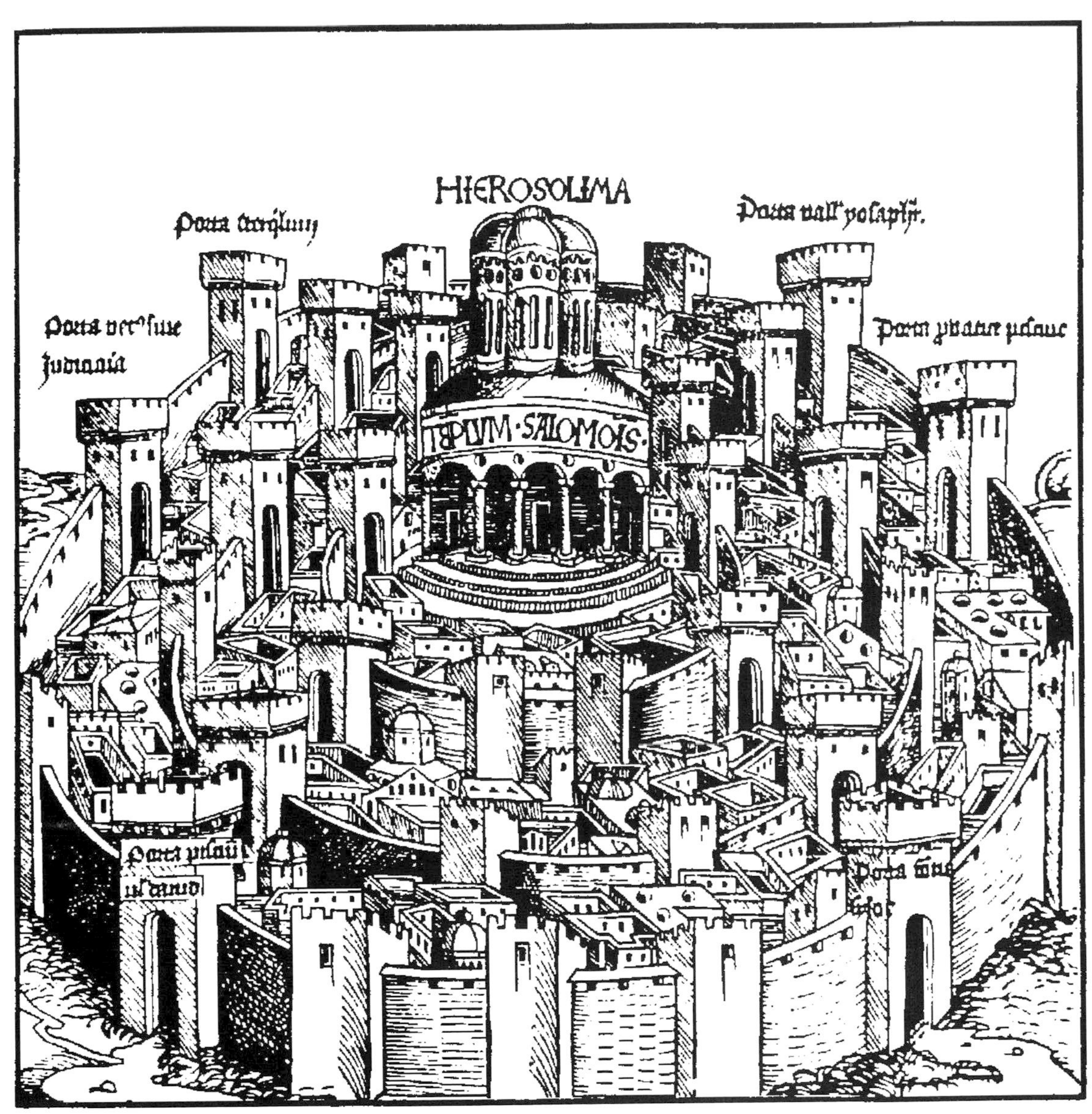

PART 1

The Evolution of Cities

PART ONE

INTRODUCTION

Cities are civilization. Humankind's rise to urban civilization took tens of thousands of years, but ever since the first true cities arose in Mesopotamia sometime between 4000 and 3000 BCE, the influence of city-based cultures and the steady spread and increase of urban populations around the world have been the central facts of human history. As Kingsley Davis points out (p. 3) "urbanization" and "the growth of cities" are not the same thing. "Urbanization," as Davis defines it, is the *increase in the proportion* of a population that is urban as opposed to rural. That such an increase could take place without the growth of cities *per se* (for example, by the death of vast numbers of the rural population) or that city populations could grow without an increase in urbanization (as when the total population, urban and rural, increases at a similar rate) are important concepts that underlie the history of urban life. This definition of urbanization helps to explain, for example, how immigration from the countryside to the city has repeatedly been the key factor in the history of urban development, as it continues to be today.

The first stage of urban history is what the Australian archaeologist V. Gordon Childe called "the Urban Revolution," that momentous shift from simple tribal communities and village-based agricultural production to the complex social, economic, and political systems that characterized the earliest cities of Mesopotamia, Egypt, and the Indus Valley. True, the earliest cities, in the ancient Near East and elsewhere, grew out of accumulated neolithic knowledge, and certain extensive neolithic communities (such as Catal Hüyük in Anatolia) pre-date the Mesopotamian cities by several millennia and may be regarded as at least proto-urban. For Childe, however, the development of writing was a crucial cultural element of true urbanism, and the emergence of the cities of the ancient Near East, where writing began, constituted the second of a series of massive transformations that gave shape to the whole of human evolutionary development. Although the successive stages overlapped, each of Childe's three "revolutions" (the agricultural, the urban, and the industrial) totally changed the world as it had been before.

In certain important respects, all the ancient cities are remarkably similar. Most are walled – except in Egypt, where the surrounding deserts may have been regarded as sufficient defenses – and all contain a distinct citadel precinct, separately walled, encompassing a temple, a palace, and the central granary. Most of the earliest cities also boasted some sort of pyramid or ziggurat. And, as Karl Wittfogel pointed out in *Oriental Despotism* (1957), almost all were located along major rivers and based their power (and that of their rulers) on the control of massive irrigation systems serving the surrounding countryside.

Thus, both the physical structure and socio-economic complexity of the earliest cities are unlike anything that had come before. Whereas the neolithic village had been ruled by a council of elders, the cities were mostly ruled by totalitarian god-kings and their attendant priests who formed a class totally apart from the rest of the citizenry. And whereas neolithic communities may have built earthen enclosures as ceremonial centers for ritual pageantry and hill forts for refuge and defense, the ancient cities – from Ur and Babylon on the Euphrates to Teotihuacan and Tenochtitlan in the Valley of Mexico – transformed these institutions into elaborate structures so massive that their remains are still visible today. Many of the ancient cities elsewhere – in Shang and Chou China, sub-Saharan Africa, Southeast Asia, and Mesoamerica – arose quite independently of the cities of the ancient Near East. Still, what is remarkable is how similar ancient cities everywhere were in terms of social structure, economic function, political order, and architectural monumentality.

The cities of ancient Greece, on the other hand, developed on a very different model. Perhaps because they arose in narrow mountain valleys rather than on broad alluvial plains, the Greek cities that emerged around 1200 BCE and developed into an astonishing cultural efflorescence by 500 BCE, were small (sometimes with a population of only a few thousand), economically self-contained, and almost village-like in their social and political institutions. It was the concept of urban citizenship and democratic self-government that was the distinctive contribution of the Greeks to the evolution of urban civilization. Greek democracy was by no means perfect and hardly inclusive. Women, slaves, and foreigners were all excluded from the rights and responsibilities of citizenship. But the cultural, artistic, and intellectual consequences of the democratic principle were extraordinary. "Within a couple of centuries," writes Lewis Mumford in *The City in History* (1961), "the Greeks discovered more about the nature and potentialities of man than the Egyptians or the Sumerians seem to have discovered in as many millennia."

If cities are civilization, they are also the cultural instrumentality by which humanity has attempted, since neolithic times, to achieve a higher, more inclusive concept of community. At the core of the Greek contribution to the history of urban civilization was the concept of the "polis." Sometimes translated as "city-state," at other times identified as the collective citizenry of a Greek city, the polis, as described so brilliantly by H. D. F. Kitto in *The Greeks* (1951), was both a community and a *sense* of community that helped to define the Greek citizen's relationship to his city and his fellow citizens, to the world at large, and to himself. In *The Politics*, Aristotle called man the "*zoon politikon*" (the "political animal" or, more properly, "the animal that belongs to a polis") and described the ideal city-state as one small enough so that a single citizen's voice could be heard by all his assembled fellow citizens. For the Greek citizen, social life was lived in the agora or public marketplace, and contact with rural nature was immediately at hand. In that sense, the polis was a reincarnation, in an urban context, of the face-to-face human relationships that characterized the pre-urban community of the neolithic village.

Marking another sharp break in the history of urban life, the city of Rome began as a cluster of villages along the Tiber in central Italy, emerged as a powerful republic similar to the earlier Greek cities, but then exploded into a giant metropolis and a city of world empire. Rome's contributions to civilization were considerable. Its roads, aqueducts, and sewers set new standards of engineering excellence. Its systems of military and colonial administration spread a common law, and established a common peace, throughout a large and populous area that extended from Persia to the borders of Scotland. Roman imperial expansion also spread Roman literature, philosophy, and art, establishing the basis for a widespread cultural hegemony. And Rome planted colonial towns wherever its legions marched, often leaving traces of an original *castrum* laid out along the cardinal points of the compass at the center of later medieval cities.

But if the administrative and infrastructural accomplishments of the Romans were impressive, their record in the field of social development is more problematic. In the place of the Greek conception of community and participation in the life of the polis, the Romans erected a citizenship of imperial privilege rooted in a rigid social hierarchy of patricians, clients, and plebeians. Beginning with Augustus, the Roman emperors proclaimed themselves gods, staged extravagant spectacles to awe the cowed populace, and ruled by "bread and circuses." Rome, at a population of one million, became a parasite on the entire Mediterranean world, and both city and empire eventually fell of their own weight.

For much of the medieval period that followed the fall of the Roman Empire in the West, Europe was a cultural backwater. In the early Middle Ages, self-contained monastery communities kept the larger world at bay, many provincial towns retreated inside the walls of the Roman amphitheaters, and the population of Rome itself dwindled to a few thousand. During this same period Chang'an in China reached one million and Teotihuacan in the Valley of Mexico reached 200,000. Raided by Vikings from the north and invaded by North African Arabs on its southern flank, most of Europe reverted to rural conditions, and serfdom became widespread under a system of warlord feudalism. Meanwhile, the cities of Islam – Samara and Baghdad and Moorish Cordova – were the real centers of power in what had been the Roman Empire. And other urban centers – the Khmer civilization at Angkor, Luoyang in Sung and Ch'in China, Great Zimbabwe in Africa – often surpassed Europe's cities in wealth and power.

After about the year CE 1000, however, Europe began to revive, and the late medieval cities became true centers of commerce, culture, and community. As Henri Pirenne argues in *Medieval Cities* (1925), it was the economic function of the great trading towns that led inevitably to their growing power and political independence. Having used their wealth to win from the barons the right to self-government, the medieval towns became islands of freedom in a sea of feudal obligation.

The defensive walls of the medieval city provided a clear demarcation line between the urban and the rural, and the small size of most towns allowed for an easy reciprocity between urban industry and commerce, on the one hand, and agricultural pursuits on the other. Within the town walls, the guilds provided for the organization of economic and social life, while the church saw to the citizens' spiritual needs and established a framework for social ritual and communal unity. Cathedrals, guildhouses, charitable institutions, universities, and colorful marketplaces were all characteristic medieval institutions. Together, they established the perfect stage for what Lewis Mumford called "the urban drama," but as soon as "the unity of this social order was broken [with the advent of nation-statism and capitalist industrialization] everything about it was set in confusion . . . and the city became a battleground for conflicting cultures, dissonant ways of life."

The slow decay of medieval urban unity began with the Renaissance and the rise of absolutist monarchies. The new national rulers built their royal palaces, such as Louis XIV's Versailles, outside of the traditional urban centers. Their interventions into the existing urban fabric included building broad boulevards and open squares fit for the display of baroque pomp and power. The Enlightenment and the Age of Revolutions brought down the divine right of kings and reestablished the political power of the urban commercial interests. And in the end, it was capitalism and the new industrial economic order that destroyed the last vestiges of the medieval city by separating the church from its social role and reducing the marketplace to its purely economic functions. The capitalist city, especially the city of the Industrial Revolution, created an entirely new urban paradigm and established the physical, social, economic, and political preconditions of all that was to follow. With the Industrial Revolution, we see the emergence of the modern city.

The political and economic consequences of the Renaissance had helped to spread European domination worldwide through extensive projects of exploration, discovery, and imperialist expansion. The forces of industrialization helped to complete that process of world domination by dividing the world between the advanced industrialized nations (originally Europe and North America) and the underdeveloped, nonindustrialized nations. It also created a new social order based on property-owning capitalists and propertyless proletarians. And the cities, especially the new industrial centers, became dismal conurbations of factories and slums such as the world had never before seen.

One of the earliest and most acute observers of the new urban-industrial order was Friedrich Engels, himself the son of a leading German industrialist. In *The Condition of the Working Class in England in 1844* (1845), Engels detailed the unrelenting squalor and misery that characterized the working-class districts of Manchester and the strategies employed by the capitalist bourgeoisie to protect themselves from the physical and social horror that was the source of their wealth. There were many responses to these horrifying conditions – the introduction of urban parks, systems of public hygiene, agitation for poor relief and model housing, even utopian visions of perfect societies – and all of these contributed to the development of modern urban planning, as described in Parts 5 and 6 of this volume. Yet another compelling strategy was middle-class flight to the suburbs. Suburbanization, with its consequent segregation by social class, became one of the continuing features of the modern city and one of the sources of its ongoing social disharmony and class conflict.

In multiethnic societies, racial divisions compounded class distinctions to create an even further crisis of community in the form of racially segregated neighborhoods that have remained as symbols of inequality and oppression. *The Philadelphia Negro* (1899) by W. E. B. Du Bois specifically describes the African-American district of Philadelphia, Pennsylvania, as it developed in the years following the American Civil War, but the social and cultural dynamic of housing segregation and racial discrimination in the workplace that Du Bois describes can be applied to ghetto and barrio experiences throughout the United States and to Third World immigrant communities worldwide. In the years since Du Bois first surveyed the life of the racially segregated ghetto community, conditions have in some ways grown worse: so much so that the persistence of racial segregation, and the emergence of an "underclass" population radically disconnected from the rest of the urban community, threaten the social stability of some of the largest, wealthiest cities in the world.

Meanwhile, the model of middle-class suburbia has grown in size and influence to the point where it is no longer just an appendage to the central city. Instead, it now actually defines many cities, leaving the old inner cores to the poorest elements of the urban population and in need of massive efforts at renewal and redevelopment. Although the first modern suburbs were built along interurban railroad lines, the newer suburbs, especially those developments built after World War II, were automobile-based and created the "sprawl" that was first seen in North America and that characterizes more and more cities worldwide. The new tract-home developments have spawned a vast literature, much of it criticizing suburbia as a cultural wasteland and a segregated sanctuary of class privilege. But one of the best analyses of contemporary suburbia, Herbert Gans's *The Levittowners*, is also one of the most sympathetic. Gans experienced the Long Island, New York, tract-home suburbia built by developer Arthur Levitt at first hand and describes a family-oriented community of skilled workers and mid-level managers – that is, a true *middle* class, not an upper-middle-class elite.

Although the Levittowns on the East Coast of the United States were clearly suburbs within the New York metropolitan region, a new city arose in California during the early decades of the twentieth century that signaled a new phase in the history of urbanism worldwide. Sometimes dismissed as a mere conurbation of suburbs in search of a city and derided as the ultimate in mindless, chaotic sprawl, Los Angeles did indeed break all the existing rules and natural boundaries of urban development but emerged finally as a new, radically decentralized urban paradigm: the contemporary multi-nucleated metropolis poised on the edge of postmodernity. In *The Urban Wilderness* (1972), Sam Bass Warner, Jr., analyzes how Los Angeles grew from a city of less than 600,000 in 1920 to over 10 million today and how that growth was fueled by clear historical, economic, and demographic realities.

The essential characteristics of Los Angeles were present almost from the beginning, particularly its sense of "spatial freedom" and its preference for the middle-class single-family dwelling as an "expression of its design for living." For good or ill, these characteristics were further emphasized by a grid pattern of freeways and a reliance – many would say over-reliance – on the automobile that replaced a once-extensive network of streetcars and created a metropolis without a single downtown. Today, Los Angeles is a true world city, and its products – both industrial and cultural – are influential around the globe.

In the nineteenth century, middle-class suburbs developed outside major urban centers, spaced along commuter rail lines. In the twentieth century, the influence of the automobile turned once-attractive small-scale suburbs into an endless, congested sprawl. These first two stages in the development of suburbia depended on the existence of a vital central city, both as a center for production and employment and for cultural amenities. With the emergence of Los Angeles, however, that pattern began to change, and today the new "Edge City" suburban ring is clearly different from the earlier suburban developments in size, complexity, and even function. This is where most of the new houses, most of the new jobs, and even most of the new cultural centers are located. Increasingly, the major commute pattern is not from suburb to central city, but from suburb to suburb. Indeed, as Robert Fishman argues in *Bourgeois Utopias: The Rise and Fall of Suburbia* (1987), the new Edge City suburbs are not suburbs at all, but a fundamentally new kind of decentralized city that he calls "technoburbs."

What the future holds for urban civilization is infinitely debatable. Will central cities disappear? Will "edge cities" take over the primary urban functions? Will the urbanization process itself reverse direction (as Melvin Webber predicted in 1968 (p. 535) and lead to counter-urbanization and a general dispersal of the human population? Or will certain urban centers become worldwide command and control centers, internally characterized by the uneasy side-by-side coexistence of corporate power and service-sector marginality, as Saskia Sassen argues in "A New Geography of Centers and Margins" (p. 208). No one knows for certain, but it increasingly appears that urban history is on the cusp of a major new transformation. Looming on the horizon is a new paradigm, a new discontinuity in urban history, tentatively described as postmodernism, that will be characterized by telecommunications technologies, global systems of economic exchange, and new ecological constraints. Perhaps part-Los Angeles/part-technoburb, the new urban paradigm that will emerge promises to be a major new stage in the history of cities, in the history of civilization, and in the evolution of the human community.

V. GORDON CHILDE

"The Urban Revolution"

Town Planning Review (1950)

Editors' introduction V. Gordon Childe (1892–1957) is arguably the single most influential archaeologist of the twentieth century. Born in Australia, Childe won a scholarship to Queens College, Oxford, returned to Australia where he briefly pursued a career in left-wing politics, then returned to Great Britain as Professor of Archaeology at the University of Edinburgh and, later, Director of the Institute of Archaeology at the University of London.

Childe's most important book, the one that revolutionized the world of archaeological research by laying out an entirely new theoretical framework for understanding the phases of human development throughout history and pre-history, was *Man Makes Himself* (1936). In that pioneering work, Childe threw out the "three age system" (Stone Age, Bronze Age, Iron Age) that had been left over from nineteenth-century conceptions of human historical development. In its place he proposed a series of four stages (paleolithic, neolithic, urban, industrial) punctuated by three "revolutions" (or, as we might term them today, "paradigm shifts"). According to Childe, the first revolution – from Old Stone Age hunter-gatherer cultures to settled agriculture – was the Neolithic Revolution. The second – the movement from neolithic agriculture to complex, hierarchical systems of manufacturing and trade that began during the fourth and third millennia BCE – was the Urban Revolution. And the third major shift in the record of human cultural and historical development – the only truly new development since the rise of cities – was the Industrial Revolution of the eighteenth and nineteenth centuries.

Childe is best known for his writings on the first cities, which arose in Mesopotamia (present day Iraq) beginning about 4000 BCE. These cities sprang up in the area bounded by the Tigris and Euphrates Rivers – often referred to as "the Fertile Crescent." Plate 1 "The Palace of Sargon II of Khorsabad" illustrates the form these first cities took. Monumental gates, massive mud-brick walls, courtyards, residences for priest-kings and a ziggurat.

Childe's work continues to figure prominently in ongoing debates about when, where, and why the first cities arose and in the antecedent debate about what a city is. Not everyone has accepted Childe's notion that the shift from neolithic to urban was a total break with the past. Evidence of ancient earthworks, wells, irrigation systems, and even continental trade networks have been traced back as far as 10,000 years in a number of areas in both the Old World and the New. Archaeologist James Mellaart has argued that evidence from the great neolithic communities of Catal Hüyük and Hacilar in ancient Turkey, which predate the earliest Mesopotamian cities by some thousands of years, calls the entire Childe theory into question. Still, it is clear that in most locations agriculture generally predated the rise of the first cities by not just thousands, but tens of thousands of years, and that the full elaboration of those cultural institutions we associate with urban life only emerged with the rise of the Mesopotamian cities.

Not everyone agrees with Childe's definition of a city. Archaeologists excavating older, smaller, less culturally advanced settlements than the Mesopotamian cities Childe studied often argue that these settlements were urban enough to qualify as true cities. Scholars working in South and Central America point out that many of the cultural features Childe believed essential to the definition of a city (including the wheel, writing, and the plow) did not exist in large and culturally advanced Amerindian settlements, which appear truly urban in other respects.

In the selection from *Town Planning Review* reprinted here, Childe details the constituent elements of the Urban Revolution that accompanied the initial rise of complex civilizations in Mesopotamia and elsewhere in the ancient Near East. Childe felt that the major factors motivating the transformation were rooted in the material base of the society: its means of production and its available physical and technological resources. Thus, the economic division of labor, the elaboration of socio-political hierarchies, and even the emergence of basic religious and intellectual patterns of thought characteristic of urban civilizations all rested on the underlying need to increase food production through massive irrigation systems and to protect the communities themselves through the erection of massive walls and fortifications.

Many modern scholars question the deterministic Marxist categories Childe employed. Although he stresses the importance of writing as an element of any truly urban society, Childe has been faulted for his apparent disregard of the primacy of non-material aspects of culture. His system has very little room for what Lewis Mumford (p. 92) called "the urban drama" or what Jane Jacobs (p. 106) called the "street ballet." Still, no one has ever called Childe's vision limited or ideologically cramped. On the contrary, he provided an expansive macro-historical foundation upon which generations of others have built.

A tireless researcher and writer, Childe produced a veritable stream of books, many of which are still classics. Among the most notable are *The Dawn of European Civilization* (London: Routledge and Kegan Paul, 1925), *The Most Ancient East* (New York: Grove Press, 1928), *What Happened In History* (Harmondsworth: Penguin, 1942), *Social Evolution* (London: Watts, 1951), and many more.

Other books on Mesopotamian cities include Nicholas Postgate and J. N. Postgate, *Early Mesopotamia: Society and Economy at the Dawn of History* (London and New York: Routledge, 1994), Georges Roux, *Ancient Iraq* (New York: Penguin, 1993), and C. Leonard Wooley's classics, *The Sumerians* (Oxford: Oxford University Press, 1928) and *Ur of the Chaldees* (Oxford: Oxford University Press, 1929).

Books on the rise of the earliest cities elsewhere in the world include Mortimer Wheeler, *Civilizations of the Indus Valley and Beyond* (London: Thames and Hudson, 1966), Karl Wittfogel, *Oriental Despotism* (New Haven, CT: Yale University Press, 1957), Basil Davidson, *The Lost Cities of Africa* (Boston: Little, Brown, 1959), Richard E. W. Adams, *Prehistoric Mesoamerica* (Norman: University of Oklahoma Press, 1991), Sylvanus G. Morely and George W. Brainerd, *The Ancient Maya* (Stanford: Stanford University Press, 1956), Jacques Soustelle, *The Daily Life of the Aztecs* (New York: Macmillan, 1962), James Mellaart, *Earliest Civilizations of the Near East* (New York: McGraw-Hill, 1965) and *Catal Hüyük* (New York: McGraw-Hill, 1967), and Paul Wheatley, *The Pivot of the Four Quarters: A Preliminary Inquiry into the Origins and Character of the Ancient Chinese City* (Chicago: Aldine, 1971).

The concept of 'city' is notoriously hard to define. The aim of the present essay is to present the city historically – or rather prehistorically – as the resultant and symbol of a 'revolution' that initiated a new economic stage in the evolution of society. The word 'revolution' must not of course be taken as denoting a sudden violent catastrophe; it is here used for the culmination of a progressive change in the economic structure and social organization of communities that caused, or was accompanied by, a dramatic increase in the population affected – an increase that would appear as an obvious bend in the population graph were vital statistics available. Just such a bend is observable at the time of the Industrial Revolution in England. Though not demonstrable statistically, comparable changes of direction must have occurred at two earlier points in the demographic history of Britain and other regions. Though perhaps less sharp and less durable, these too should indicate equally revolutionary changes in economy. They may then be regarded likewise as marking transitions between stages in economic and social development.

Sociologists and ethnographers last century classified existing pre-industrial societies in a hierarchy of three evolutionary stages, denominated respectively 'savagery,' 'barbarism' and 'civilization.' If they be defined by suitably selected criteria, the logical hierarchy of stages can be transformed into a temporal sequence of ages, proved archaeologically to follow one another in the same order wherever they occur. Savagery and barbarism are conveniently recognized and appropriately defined by the methods adopted for procuring food. Savages live exclusively on wild food obtained by collecting, hunting or fishing. Barbarians on the contrary at least supplement these natural resources by cultivating edible plants and – in the Old World north of the Tropics – also by breeding animals for food.

Throughout the Pleistocene Period – the Palaeolithic Age of archaeologists – all known human societies were savage in the foregoing sense, and a few savage tribes have survived in out of the way parts to the present day. In the archaeological record barbarism began less than ten thousand years ago with the Neolithic Age of archaeologists. It thus represents a later, as well as a higher stage, than savagery. Civilization cannot be defined in quite such simple terms. Etymologically the word is connected with 'city,' and sure enough life in cities begins with this stage. But 'city' is itself ambiguous so archaeologists like to use 'writing' as a criterion of civilization; it should be easily recognizable and proves to be a reliable index to more profound characters. Note, however, that, because a people is said to be civilized or literate, it does not follow that all its members can read and write, nor that they all lived in cities. Now there is no recorded instance of a community of savages civilizing themselves, adopting urban life or inventing a script. Wherever cities have been built, villages of preliterate farmers existed previously (save perhaps where an already civilized people have colonized uninhabited tracts). So civilization, wherever and whenever it arose, succeeded barbarism.

We have seen that a revolution as here defined should be reflected in the population statistics. In the case of the Urban Revolution the increase was mainly accounted for by the multiplication of the numbers of persons living together, i.e., in a single built-up area. The first cities represented settlement units of hitherto unprecedented size. Of course it was not just their size that constituted their distinctive character. We shall find that by modern standards they appeared ridiculously small and we might meet agglomerations of population today to which the name city would have to be refused. Yet a certain size of settlement and density of population is an essential feature of civilization.

Now the density of population is determined by the food supply which in turn is limited by natural resources, the techniques for their exploitation and the means of transport and food-preservation available. The last factors have proved to be variables in the course of human history, and the technique of obtaining food has already been used to distinguish the consecutive stages termed savagery and barbarism. Under the gathering economy of savagery population was always exceedingly sparse. In aboriginal America the carrying capacity of normal unimproved land seems to have been from .05 to .10 per square mile. Only under exceptionally favourable conditions did the fishing tribes of the Northwest Pacific coast attain densities of over one human to the square mile. As far as we can guess from the extant remains, population densities in Palaeolithic and pre-neolithic Europe were less than the normal American. Moreover such hunters and collectors usually live in small roving bands. At best several bands may come together for quite brief periods on ceremonial occasions such as the Australian corroborees. Only in exceptionally favoured regions can fishing tribes establish anything like villages. Some settlements on the Pacific coasts comprised thirty or so substantial and durable houses, accommodating groups of several hundred persons. But even these villages were only occupied during the winter; for the rest of the year their inhabitants disposed in smaller groups. Nothing comparable has been found in pre-neolithic times in the Old World.

The Neolithic Revolution certainly allowed an expansion of population and enormously increased the carrying capacity of suitable land. On the Pacific Islands neolithic societies today attain a density of 30 or more persons to the square mile. In pre-Columbian North America, however, where the land is not obviously restricted by surrounding seas, the maximum density recorded is just under 2 to the square mile.

Neolithic farmers could of course, and certainly did, live together in permanent villages, though, owing to the extravagant rural economy generally practised, unless the crops were watered by irrigation, the villages had to be shifted at least every twenty years. But on the whole the growth of population was not reflected so much in the enlargement of the settlement unit as in a multiplication of settlements. In ethnography neolithic villages can boast only a few hundred inhabitants (a couple of 'pueblos' in New Mexico house over a thousand, but perhaps they cannot be regarded as neolithic). In prehistoric Europe the largest neolithic village yet known, Barkaer in Jutland, comprised 52 small, one-roomed dwellings, but 16 to 30 houses was a more normal figure; so the average, local group in neolithic times would average 200 to 400 members.

These low figures are of course the result of technical limitations. In the absence of wheeled vehicles and roads for the transport of bulky crops men had to live within easy walking distance of their cultivations. At the same time the normal rural economy of the Neolithic Age, what is now termed slash-and-burn or *jhumming*, condemns much more than half the arable land to lie fallow so that large areas were required. As soon as the population of a settlement rose above the numbers that could be supported from the accessible land, the excess had to hive off and found a new settlement.

The Neolithic Revolution had other consequences beside increasing the population, and their exploitation might in the end help to provide for the surplus increase. The new economy allowed, and indeed required, the farmer to produce every year more food than was needed to keep him and his family alive. In other words it made possible the regular production of a social surplus. Owing to the low efficiency of neolithic technique, the surplus produced was insignificant at first, but it could be increased till it demanded a reorganization of society.

Now in any Stone Age society, palaeolithic or neolithic, savage or barbarian, everybody can at least in theory make at home the few indispensable tools, the modest cloths and the simple ornaments everyone requires. But every member of the local community, not disqualified by age, must contribute actively to the communal food supply by personally collecting, hunting, fishing, gardening or herding. As long as this holds good, there can be no full-time specialists, no persons nor class of persons who depend for their livelihood on food produced by others and secured in exchange for material or immaterial goods or services.

We find indeed today among Stone Age barbarians and even savages expert craftsmen (for instance flint-knappers among the Ona of Tierra del Fuego), men who claim to be experts in magic, and even chiefs. In Palaeolithic Europe too there is some evidence for magicians and indications of chieftainship in pre-neolithic times. But on closer observation we discover that today these experts are not full-time specialists. The Ona flintworker must spend most of his time hunting; he only adds to his diet and his prestige by making arrowheads for clients who reward him with presents. Similarly a pre-Columbian chief, though entitled to customary gifts and services from his followers, must still personally lead hunting and fishing expeditions and indeed could only maintain his authority by his industry and prowess in these pursuits. The same holds good of barbarian societies that are still in the neolithic stage, like the Polynesians where industry in gardening takes the place of prowess in hunting. The reason is that there simply will not be enough food to go round unless every member of the group contributes to the supply. The social surplus is not big enough to feed idle mouths.

Social division of labour, save those rudiments imposed by age and sex, is thus impossible. On the contrary community of employment, the common absorption in obtaining food by similar devices guarantees a certain solidarity to the group. For co-operation is essential to secure food and shelter and for defence against foes, human and subhuman. This identity of economic interests and pursuits is echoed and magnified by identity of language, custom and belief; rigid conformity is enforced as effectively as industry in the common quest for food. But conformity and industrious co-operation need no State organization to maintain them. The local group usually consists either of a single clan (persons who believe themselves descended from a common ancestor or who have earned a mystical claim to such descent by ceremonial adoption) or a group of clans related by habitual intermarriage. And the sentiment of kinship is reinforced or supplemented by common rites focused on some ancestral shrine or sacred place. Archaeology can provide no evidence for kinship organization, but shrines occupied the central place in preliterate villages in Mesopotamia, and the long barrow, a collective tomb that overlooks the presumed site of most neolithic villages in Britain, may well have been also the ancestral shrine on which converged the emotions and ceremonial activities of the villagers below. However, the solidarity thus idealized and concretely symbolized, is really based on the same principles as that of a pack of wolves or a herd of sheep; Durkheim has called it 'mechanical.'

Now among some advanced barbarians (for instance tattooers or woodcarvers among the Maori) still technologically neolithic we find expert craftsmen tending towards the status of full-time professionals, but only at the cost of breaking away from the local community. If no single village can produce a surplus large enough to feed a full-time specialist all the year round, each should produce enough to keep him a week or so. By going round from village to village an expert might thus live entirely from his craft. Such itinerants will lose their membership of the sedentary kinship group. They may in the end form an analogous organization of their own – a craft clan, which, if it remain hereditary, may become a caste, or, if it recruit its members mainly by adoption (apprenticeship throughout Antiquity and the Middle Age was just temporary adoption), may turn into a guild. But such specialists by emancipation from kinship ties, have also forfeited the protection of the kinship organization which alone under barbarism, guaranteed to its members security of person and property. Society must be reorganized to accommodate and protect them.

In pre-history specialization of labour presumably began with similar itinerant experts. Archaeological proof is hardly to be expected, but in ethnography metal-workers are nearly always full time specialists. And in Europe at the beginning of the Bronze Age metal seems to have been worked and purveyed by perambulating smiths who seem to have functioned like tinkers and other itinerants of much more recent times. Though there is no such positive evidence, the same probably happened in Asia at the beginning of metallurgy. There must of course have been in addition other specialist craftsmen whom, as the Polynesian example warns us, archaeologists could not recognize because they worked in perishable materials. One result of the Urban Revolution will be to rescue such specialists from nomadism and to guarantee them security in a new social organization.

About 5,000 years ago irrigation cultivation (combined with stockbreeding and fishing) in the valleys of the Nile, the Tigris-Euphrates and the Indus had begun to yield a social surplus, large enough to support a number of resident specialists who were themselves released from food-production. Water-transport, supplemented in Mesopotamia and the Indus valley by wheeled vehicles and even in Egypt by pack animals, made it easy to gather food stuffs at a few centres. At the same time dependence on river water for the irrigation of the crops restricted the cultivable areas while the necessity of canalizing the waters and protecting habitations against annual floods encouraged the aggregation of population. Thus arose the first cities – units of settlement ten times as great as any known neolithic village. It can be argued that all cities in the old world are offshoots of those of Egypt, Mesopotamia and the Indus basin. So the latter need not be taken into account if a minimum definition of civilization is to be inferred from a comparison of its independent manifestations.

But some three millennia later cities arose in Central America, and it is impossible to prove that the Mayas owed anything directly to the urban civilizations of the Old World. Their achievements must therefore be taken into account in our comparison, and their inclusion seriously complicates the task of defining the essential preconditions for the Urban Revolution. In the Old World the rural economy which yielded the surplus was based on the cultivation of cereals combined with stock-breeding. But this economy had been made more efficient as a result of the adoption of irrigation (allowing cultivation without prolonged fallow periods) and of important inventions and discoveries – metallurgy, the plough, the sailing boat and the wheel. None of these devices was known to the Maya; they bred no animals for milk or meat; though they cultivated the cereal maize, they used the same sort of slash-and-burn method as neolithic farmers in prehistoric Europe or in the Pacific Islands today. Hence the minimum definition of a city, the greatest factor common to the Old World and the New will be substantially reduced and impoverished by the inclusion of the Maya. Nevertheless ten rather abstract criteria, all deducible from archaeological data, serve to distinguish even the

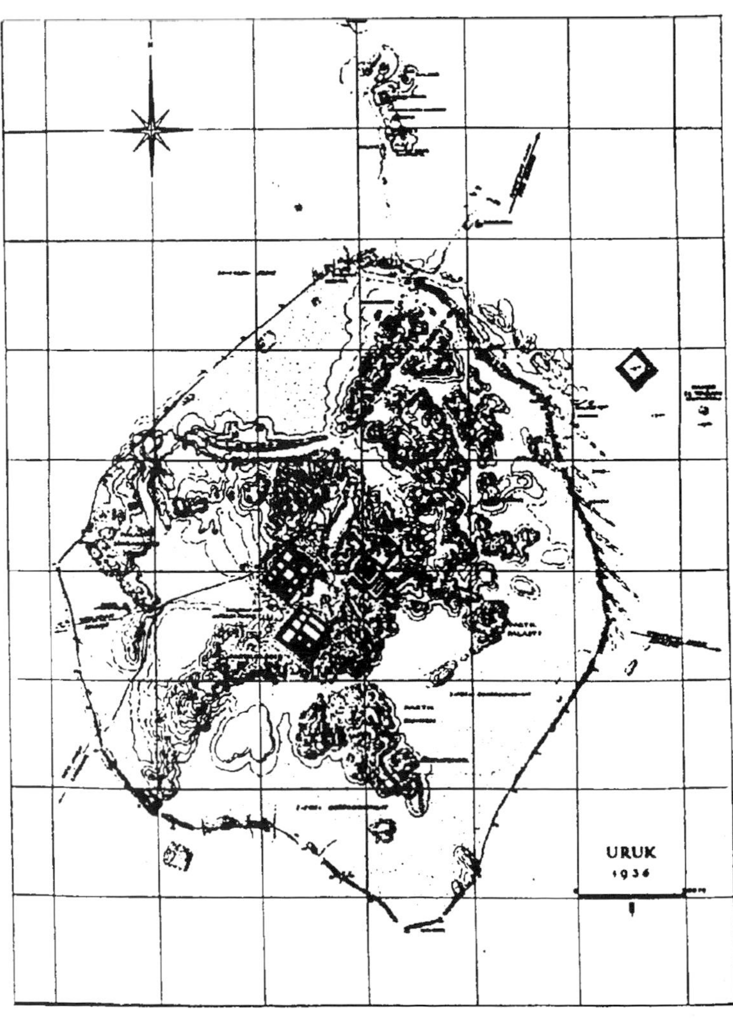

Figure 1 Plan of the city of Erek (Uruk)

earliest cities from any older or contemporary village.

(1) In point of size the first cities must have been more extensive and more densely populated than any previous settlements, although considerably smaller than many villages today. It is indeed only in Mesopotamia and India that the first urban populations can be estimated with any confidence or precision. There excavation has been sufficiently extensive and intensive to reveal both the total area and the density of building in sample quarters and in both respects has disclosed significant agreement with the less industrialized Oriental cities today. The population of Sumerian cities, thus calculated, ranged between 7,000 and 20,000; Harappa and Mohenjo-daro in the Indus valley must have approximated to the higher figure. We can only infer that Egyptian and Maya cities were of comparable magnitude from the scale of public works, presumably executed by urban populations.

(2) In composition and function the urban population already differed from that of any village. Very likely indeed most citizens were still also peasants, harvesting the lands and waters adjacent to the city. But all cities must have accommodated in addition classes who did not themselves procure their own food by agriculture, stock-breeding, fishing or collecting – full-time specialist craftsmen, transport workers, merchants, officials and priests. All these were of course supported by the surplus produced by the peasants living in the city and in dependent villages, but they did not secure their share directly by exchanging their products or services for grains or fish with individual peasants.

(3) Each primary producer paid over the tiny surplus he could wring from the soil with his still very limited technical equipment as tithe or tax to an imaginary deity or a divine king who thus concentrated the surplus. Without this concentration, owing to the low productivity of the rural economy, no effective capital would have been available.

(4) Truly monumental public buildings not only distinguish each known city from any village but also symbolize the concentration of the social surplus. Every Sumerian city was from the first dominated by one or more stately temples, centrally situated on a brick platform raised above the surrounding dwellings and usually connected with an artificial mountain, the staged tower or ziggurat. But attached to the temples were workshops and magazines, and an important appurtenance of each principal temple was a great granary. Harappa, in the Indus basin, was dominated by an artificial citadel, girt with a massive rampart of kiln-baked bricks, containing presumably a palace and immediately overlooking an enormous granary and the barracks of artisans. No early temples nor palaces have been excavated in Egypt, but the whole Nile valley was dominated by the gigantic tombs of the divine pharaohs while royal granaries are attested from the literary record. Finally the Maya cities are known almost exclusively from the temples and pyramids of sculptured stone round which they grew up.

Hence in Sumer the social surplus was first effectively concentrated in the hands of a god and stored in his granary. That was probably true in Central America while in Egypt the pharaoh (king) was himself a god. But of course the imaginary deities were served by quite real priests who, besides celebrating elaborate and often sanguinary rites in their honour, administered their divine masters' earthly estates. In Sumer indeed the god very soon, if not even before the revolution, shared his wealth and power with a mortal viceregent, the 'City-King,' who acted as civil ruler and leader in war. The divine pharaoh was naturally assisted by a whole hierarchy of officials.

(5) All those not engaged in food-production were of course supported in the first instance by the surplus accumulated in temple or royal granaries and were thus dependent on temple or court. But naturally priests, civil and military leaders and officials absorbed a major share of the concentrated surplus and thus formed a 'ruling class.' Unlike a palaeolithic magician or a neolithic chief, they were, as an Egyptian scribe actually put it, 'exempt from all manual tasks.' On the other hand, the lower classes were not only guaranteed peace and security, but were relieved from intellectual tasks which many find more irksome than any physical labour. Besides reassuring the masses that the sun was going to rise next day and the river would flood again next year (people who have not five thousand years of recorded experience of natural uniformities behind them are really worried about such matters!), the ruling classes did

confer substantial benefits upon their subjects in the way of planning and organization.

(6) They were in fact compelled to invent systems of recording and exact, but practically useful, sciences. The mere administration of the vast revenues of a Sumerian temple or an Egyptian pharaoh by a perpetual corporation of priests or officials obliged its members to devise conventional methods of recording that should be intelligible to all their colleagues and successors, that is, to invent systems of writing and numeral notation. Writing is thus a significant, as well as a convenient, mark of civilization. But while writing is a trait common to Egypt, Mesopotamia, the Indus valley and Central America, the characters themselves were different in each region and so were the normal writing materials – papyrus in Egypt, clay in Mesopotamia. The engraved seals or stelae that provide the sole extant evidence for early Indus and Maya writing no more represent the normal vehicles for the scripts than do the comparable documents from Egypt and Sumer.

(7) The invention of writing – or shall we say the inventions of scripts – enabled the leisured clerks to proceed to the elaboration of exact and predictive sciences – arithmetic, geometry and astronomy. Obviously beneficial and explicitly attested by the Egyptian and Maya documents was the correct determination of the tropic year and the creation of a calendar. For it enabled the rulers to regulate successfully the cycle of agricultural operations. But once more the Egyptian, Maya and Babylonian calendars were as different as any systems based on a single natural unit could be. Calendrical and mathematical sciences are common features of the earliest civilizations and they too are corollaries of the archaeologists' criterion, writing.

(8) Other specialists, supported by the concentrated social surplus, gave a new direction to artistic expression. Savages even in palaeolithic times had tried, sometimes with astonishing success, to depict animals and even men as they saw them – concretely and naturalistically. Neolithic peasants never did that; they hardly ever tried to represent natural objects, but preferred to symbolize them by abstract geometrical patterns which at most may suggest by a few traits a fantastical man or beast or plant. But Egyptian, Sumerian, Indus and Maya artist-craftsmen – full-time sculptors, painters, or seal-engravers – began once more to carve, model or draw likenesses of persons or things, but no longer with the naive naturalism of the hunter, but according to conceptualized and sophisticated styles which differ in each of the four urban centres.

(9) A further part of the concentrated social surplus was used to pay for the importation of raw materials, needed for industry or cult and not available locally. Regular 'foreign' trade over quite long distances was a feature of all early civilizations and, though common enough among barbarians later, is not certainly attested in the Old World before 3000 B.C. nor in the New before the Maya 'empire.' Thereafter regular trade extended from Egypt at least as far as Byblos on the Syrian coast while Mesopotamia was related by commerce with the Indus valley. While the objects of international trade were at first mainly 'luxuries,' they already included industrial materials, in the Old World notably metal the place of which in the New was perhaps taken by obsidian. To this extent the first cities were dependent for vital materials on long distance trade as no neolithic village ever was.

(10) So in the city, specialist craftsmen were both provided with raw materials needed for the employment of their skill and also guaranteed security in a State organization based now on residence rather than kinship. Itinerancy was no longer obligatory. The city was a community to which a craftsman could belong politically as well as economically.

Yet in return for security they became dependent on temple or court and were relegated to the lower classes. The peasant masses gained even less material advantages; in Egypt for instance metal did not replace the old stone and wood tools for agricultural work. Yet, however imperfectly, even the earliest urban communities must have been held together by a sort of solidarity missing from any neolithic village. Peasants, craftsmen, priests and rulers form a community, not only by reason of identity of language and belief, but also because each performs mutually complementary functions, needed for the well-being (as redefined under civilization) of the whole. In fact the earliest cities illustrate a first approximation to an organic solidarity based upon a functional complementarity and interdependence between all its members such as subsist between the constituent cells of an organism. Of course this was only a very distant approximation. However necessary the concentration of the surplus really were with the existing forces of production, there seemed a glaring conflict on economic interests between the tiny ruling class, who annexed the bulk of the social surplus, and the vast majority who were left with a bare subsistence and effectively excluded from the spiritual benefits of civilization. So solidarity had still to be maintained by the ideological devices appropriate to the mechanical solidarity of barbarism as expressed in the pre-eminence of the temple or the sepulchral shrine, and now supplemented by the force of the new State organization. There could be no room for sceptics or sectaries in the oldest cities.

These ten traits exhaust the factors common to the oldest cities that archaeology, at best helped out with fragmentary and often ambiguous written sources, can detect. No specific elements of town planning for example can be proved characteristic of all such cities; for on the one hand the Egyptian and Maya cities have not yet been excavated; on the other neolithic villages were often walled, an elaborate system of sewers drained the Orcadian hamlet of Skara Brae; two-storeyed houses were built in pre-Columbian pueblos, and so on.

The common factors are quite abstract. Concretely Egyptian, Sumerian, Indus and Maya civilizations were as different as the plans of their temples, the signs of their scripts and their artistic conventions. In view of this divergence and because there is so far no evidence for a temporal priority of one Old World centre (for instance, Egypt) over the rest nor yet for contact between Central America and any other urban centre, the four revolutions just considered may be regarded as mutually independent. On the contrary, all later civilizations in the Old World may in a sense be regarded as lineal descendants of those of Egypt, Mesopotamia or the Indus.

But this was not a case of like producing like. The maritime civilizations of Bronze Age Crete or classical Greece for example, to say nothing of our own, differ more from their reputed ancestors than these did among themselves. But the urban revolutions that gave them birth did not start from scratch. They could and probably did draw upon the capital accumulated in the three allegedly primary centres. That is most obvious in the case of cultural capital. Even today we use the Egyptians' calendar and the Sumerians' divisions of the day and the hour. Our European ancestors did not have to invent for themselves these divisions of time nor repeat the observations on which they are based; they took over – and very slightly improved – systems elaborated 5,000 years ago! But the same is in a sense true of material capital as well. The Egyptians, the Sumerians and the Indus people had accumulated vast reserves of surplus food. At the same time they had to import from abroad necessary raw materials like metals and building timber as well as 'luxuries.' Communities controlling these natural resources could in exchange claim a slice of the urban surplus. They could use it as capital to support full-time specialists – craftsmen or rulers – until the latters' achievement in technique and organization had so enriched barbarian economics that they too could produce a substantial surplus in their turn.

H. D. F. KITTO

"The Polis"

from *The Greeks* (1951)

Editors' introduction At its peak Ancient Athens had only about as many residents as Peoria, Illinois (1990 population 113,504) – not a city which leaps out as a great center of world civilization. But British classicist Humphrey Davy Findley Kitto (1897–1982) reminds us not to commit the vulgar error of confusing size with significance. During its golden age, Athens and the 700 or so other tiny settlements of ancient Greece made a monumental contribution to human culture. What the Greeks achieved in philosophy, literature, drama, poetry, art, logic, mathematics, sculpture, and architecture has exercised a profound influence on Western civilization.

A Greek invention of enduring interest to urbanists is the polis. Since we have not got the thing that the Greeks called "the polis," Kitto notes, we do not possess an equivalent word. "City-state" or, perhaps, "self-governing community" come closest.

The classical Greek polis came of age in the fifth century BCE, about half way between the emergence of the great Mesopotamian cities Childe describes and the present time. The physical form of the polis stressed public space. Private houses were low and turned away from the street. In contrast the Greeks emphasized public temples, stadiums, the agora (a combined marketplace and public forum) and theaters like Athens' magnificent Theater of Dionysus illustrated in Plate 2. In the larger polei, like Athens, these public buildings were spacious and often beautifully constructed of marble. Even in the smaller ones the community devoted many of its resources to them.

If the physical form of the polis was often stunning, it is the social organization of the polis that remains of particular fascination. The polis represents a form of community that has exerted a powerful fascination for more than two millennia.

In the following selection Kitto describes how the polis made it possible for each citizen to realize his spiritual, moral, and intellectual capacities. The polis was a living community; almost an extended family. While the Greeks were very private in many ways, Kitto notes that their public life was essentially communistic. The polis as a social institution defined the very nature of being human for its citizens.

Not that the polis supported development of every resident: women and slaves were not citizens and did not participate in much of the life of the polis. Foreigners could attend plays in the Greek theater, but were barred from many institutions reserved to the (free, non-foreign, male) citizens. Sir Peter Hall in *Cities in Civilization* further questions the extent to which many citizens actually participated in public affairs. He hypothesizes that only a small percentage of those eligible to participate in public decision-making actually did so. He also notes that while farmers and other of the least educated and least articulate citizens of the Greek polis may have been physically present and possessed the same voting rights as educated upper-class Athenians, it is unlikely that they participated very effectively compared with the higher classes. For the most part, Hall believes, they were passive spectators rather than active participants in the public affairs.

While a balanced view of the polis must acknowledge the existence of slavery, exclusion of women from civic life, limitations on the rights of foreigners, and the influence of education and class on social relations, Athens and the other Greek poleis were astonishingly democractic compared with any other urban civilization that preceded them. It is easy to dismiss Kitto as a hopeless romantic and his depiction of the Greek polis as an ivory tower depiction of a Camelot that never was. But that may be too harsh. The Greek polis as a social institution did represent a remarkable advance over social relations in any previous society. And the values it represented for its citizens are of enduring importance in an imperfect world.

In the debate about why the polis arose in Greece when it did, Kitto rejects deterministic answers. Geographical and economic determinists argue that the mountainous terrain required little, separate city-states. Kitto attributes the rise of the polis to the *character of the Greeks* rather than these external factors.

Kitto expresses nostalgia for human qualities of life in the polis which appear threatened today. Compare the vision of polis as a supportive, humanistic, structure for human fulfillment with the vision of large modern cities as centers of alienation and anomie depicted by Louis Wirth (p. 97), or ghettos housing the black underclass as described by William Julius Wilson (p. 112). Note the connections between humanistic values Kitto felt that the polis nurtured and J. B. Jackson's "almost perfect town" (p. 162) and the return to human-scale community Peter Calthorpe advocates in "Pedestrian Pockets" (p. 350). Compare the positive values Kitto describes in the polis as well as the status and gender barriers with Dolores Hayden's vision of a non-sexist city (p. 503).

Other books helpful in understanding the polis and its significance are Christian Meier, *Athens: A Portrait of the City in its Golden Age* (New York: Metropolitan Books, 1998), Cecil Maurice Bowra, *The Greek Experience* (London: Weidenfeld and Nicolson, 1957), and a new edition of classic writings by a great classicist – Jacob Burckhardt, *The Greeks and Greek Civilization* (New York: St. Martins, 1998).

For accounts of Greek city planning see Richard Ernest Wycherley, *How the Greeks Built Cities*, 2nd edn. (London: Macmillan, 1963) and *The Stones of Athens* (Princeton, NJ: Princeton University Press, 1978). Spiro Kostof, "Polis and Akropolis," Chapter 7 of *A History of Architecture* (New York: Oxford University Press, 1980) is also helpful.

Two masterful accounts of the role of cities in civilization give particular emphasis to the contribution of the Greek polis. See Lewis Mumford's chapters on "The Emergence of the Polis" and "Citizen Versus Ideal City" in *The City in History* (New York: Harcourt Brace Jovanovich, 1961), and Chapter 2 of Sir Peter Hall's *Cities in Civilization* "The Fountainhead" (New York: Pantheon Books, 1998).

"Polis" is the Greek word which we translate "city-state". It is a bad translation, because the normal polis was not much like a city, and was very much more than a state. But translation, like politics, is the art of the possible; since we have not got the thing which the Greeks called "the polis", we do not possess an equivalent word. From now on, we will avoid the misleading term "city-state", and use the Greek word instead ... We will first inquire how this political system arose, then we will try to reconstitute the word "polis" and recover its real meaning by watching it in action. It may be a long task, but all the time we shall be improving our acquaintance with the Greeks. Without a clear conception what the polis was, and what it meant to the Greek, it is quite impossible to understand properly Greek history, the Greek mind, or the Greek achievement.

First then, what was the polis? ...

. . . In Crete . . . we find over fifty quite independent poleis, fifty small "states" . . . What is true of Crete is true of Greece in general, or at least of those parts which play any considerable part in Greek history . . .

It is important to realize their size. The modern reader picks up a translation of Plato's *Republic* or Aristotle's *Politics*; he finds Plato ordaining that his ideal city shall have 5,000 citizens, and Aristotle that each citizen should be able to know all the others by sight; and he smiles, perhaps, at such philosophic fantasies. But Plato and Aristotle are not fantasts. Plato is imagining a polis on the normal Hellenic scale; indeed he implies that many existing Greek poleis are too small – for many had less than 5,000 citizens. Aristotle says, in his amusing way . . . that a polis of ten citizens would be impossible, because it could not be self-sufficient, and that a polis of a hundred thousand would be absurd, because it could not govern itself properly . . . Aristotle speaks of a hundred thousand citizens; if we allow each to have a wife and four children, and then add a liberal number of slaves and resident aliens, we shall arrive at something like a million – the population of Birmingham; and to Aristotle an independent "state" as populous as Birmingham is a lecture-room joke . . .

In fact, only three poleis had more than 20,000 citizens: Syracuse and Acragas (Girgenti) in Sicily, and Athens. At the outbreak of the Peloponnesian War the population of Attica was probably about 350,000, half Athenian (men, women and children), about a tenth resident aliens, and the rest slaves. Sparta, or Lacedaemon, had a much smaller citizen-body, though it was larger in area. The Spartans had conquered and annexed Messenia, and possessed 3,200 square miles of territory. By Greek standards this was an enormous area: it would take a good walker two days to cross it. The important commercial city of Corinth had a territory of 330 square miles . . . The island of Ceos, which is about as big as Bute, was divided into four poleis. It had therefore four armies, four governments, possibly four different calendars, and, it may be, four different currencies and systems of measures – though this is less likely. Mycenae was in historical times a shrunken relic of Agamemnon's capital, but still independent. She sent an army to help the Greek cause against Persia at the battle of Plataea; the army consisted of eighty men. Even by Greek standards this was small, but we do not hear that any jokes were made about an army sharing a cab.

To think on this scale is difficult for us, who regard a state of ten million as small, and are accustomed to states which, like the U.S.A. and the U.S.S.R., are so big that they have to be referred to by their initials; but when the adjustable reader has become accustomed to the scale, he will not commit the vulgar error of confusing size with significance . . .

But before we deal with the nature of the polis, the reader might like to know how it happened that the relatively spacious pattern of pre-Dorian Greece became such a mosaic of small fragments. The Classical scholar too would like to know; there are no records, so that all we can do is to suggest plausible reasons. There are historical, geographical and economic reasons; and when these have been duly set forth, we may conclude perhaps that the most important reason of all is simply that this is the way in which the Greeks preferred to live.

[Here Kitto describes the evolution of the Greek acropolis from a fortified hilltop strong-point built for protection against Dorian invaders to a place of assembly, religion, and commerce.]

At this point we may invoke the very sociable habits of the Greeks, ancient or modern. The English farmer likes to build his house on his land, and to come into town when he has to. What little leisure he has he likes to spend on the very satisfying occupation of looking over a gate. The Greek prefers to live in the town or village, to walk out to his work, and to spend his rather ampler leisure talking in the town or village square. Therefore the market becomes a market-town, naturally beneath the acropolis. This became the center of the communal life of the people – and we shall see presently how important that was.

But why did not such towns form larger units? This is the important question.

There is an economic point. The physical barriers which Greece has so abundantly made the transport of goods difficult, except by sea, and the sea was not yet used with any confidence. Moreover, the variety of which we spoke earlier enabled quite a small area to be reasonably self-sufficient for a people who made such small material demands on life as the Greek.

Both of these facts tend in the same direction; there was in Greece no great economic interdependence, no reciprocal pull between the different parts of the country, strong enough to counteract the desire of the Greek to live in small communities.

There is a geographical point. It is sometimes asserted that this system of independent poleis was imposed on Greece by the physical character of the country. The theory is attractive, especially to those who like to have one majestic explanation of any phenomenon, but it does not seem to be true. It is of course obvious that the physical subdivision of the country helped; the system could not have existed, for example, in Egypt, a country which depends entirely on the proper management of the Nile flood, and therefore must have a central government. But there are countries cut up quite as much as Greece – Scotland, for instance – which have never developed the polis-system; and conversely there were in Greece many neighbouring poleis, such as Corinth and Sicyon, which remained independent of each other although between them there was no physical barrier that would seriously incommode a modern cyclist. Moreover, it was precisely the most mountainous parts of Greece that never developed poleis, or not until later days – Arcadia and Aetolia, for example, which had something like a canton-system. The polis flourished in those parts where communications were relatively easy. So that we are still looking for our explanation.

Economics and geography helped, but the real explanation is the character of the Greeks . . . As it will take some time to deal with this, we may first clear out of the way an important historical point. How did it come about that so preposterous a system was able to last for more than twenty minutes?

The ironies of history are many and bitter, but at least this must be put to the credit of the gods, that they arranged for the Greeks to have the Eastern Mediterranean almost to themselves long enough to work out what was almost a laboratory-experiment to test how far, and in what conditions, human nature is capable of creating and sustaining a civilization . . . this lively and intelligent Greek people was for some centuries allowed to live under the apparently absurd system which suited and developed its genius instead of becoming absorbed in the dull mass of a large empire, which would have smothered its spiritual growth . . . no history of Greece can be intelligible until one has understood what the polis meant to the Greek; and when we have understood that, we shall also understand why the Greeks developed it, and so obstinately tried to maintain it. Let us then examine the word in action.

It meant at first that which was later called the Acropolis, the stronghold of the whole community and the centre of its public life . . . "polis" very soon meant either the citadel or the whole people which, as it were, "used" this citadel. So we read in Thucydides, "Epidamnus is a polis on the right as you sail into the Ionian gulf." This is not like saying "Bristol is a city on the right as you sail up the Bristol Channel", for Bristol is not an independent state which might be at war with Gloucester, but only an urban area with a purely local administration. Thucydides' words imply that there is a town – though possibly a very small one – called Epidamnus, which is the political centre of the Epidamnians, who live in the territory of which the town is the centre – not the "capital" – and are Epidamnians whether they live in the town or in one of the villages in this territory.

Sometimes the territory and the town have different names. Thus, Attica is the territory occupied by the Athenian people; it comprised Athens – the "polis" in the narrower sense – the Piraeus, and many villages; but the people collectively were Athenians, not Attics, and a citizen was an Athenian in whatever part of Attica he might live.

In this sense "polis" is our "state" . . . The actual business of governing might be entrusted to a monarch, acting in the name of all according to traditional usages, or to the heads of certain noble families, or to a council of citizens owning so much property, or to all the citizens. All these and many modifications of them, were natural forms of "polity"; all were sharply distinguished by the Greek from Oriental monarchy, in which the monarch is irresponsible, not holding his powers in trust by the grace of god, but being himself a god. If there were irresponsible government there was no polis . . .

. . . [T]he size of the polis made it possible for a member to appeal to all his fellow citizens in person, and this he naturally did if he thought that another member of the polis had injured him. It was the common assumption of the Greeks that the polis took its origin in the desire for Justice. Individuals are lawless, but the polis will see to it that wrongs are redressed. But not by an elaborate machinery of state-justice, for such a machine could not be operated except by individuals, who may be as unjust as the original wrongdoer. The injured party will be sure of obtaining Justice only if he can declare his wrongs to the whole polis. The word therefore now means "people" in actual distinction from state.

[. . .]

. . . Demosthenes the orator talks of a man who, literally, "avoids the city" – a translation which might lead the unwary to suppose that he lived in something corresponding to the Lake District, or Purley. But the phrase "avoids the polis" tells us nothing about his domicile; it means that he took no part in public life – and was therefore something of an oddity. The affairs of the community did not interest him.

We have now learned enough about the word polis to realize that there is no possible English rendering of such a common phrase as, "It is everyone's duty to help the polis." We cannot say "help the state", for that arouses no enthusiasm; it is "the state" that takes half our incomes from us. Not "the community", for with us "the community" is too big and too various to be grasped except theoretically. One's village, one's trade union, one's class, are entities that mean something to us at once, but "work for the community", though an admirable sentiment, is to most of us vague and flabby. In the years before the war, what did most parts of Great Britain know about the depressed areas? How much do bankers, miners and farmworkers understand each other? But the "polis" every Greek knew; there it was, complete, before his eyes. He could see the fields which gave it its sustenance – or did not, if the crops failed; he could see how agriculture, trade and industry dovetailed into one another; he knew the frontiers, where they were strong and where weak; if any malcontents were planning a *coup*, it was difficult for them to conceal the fact. The entire life of the polis, and the relation between its parts, were much easier to grasp, because of the small scale of things. Therefore to say "It is everyone's duty to help the polis" was not to express a fine sentiment but to speak the plainest and most urgent common sense. Public affairs had an immediacy and a concreteness which they cannot possibly have for us.

[. . .]

Pericles' Funeral Speech, recorded or recreated by Thucydides, will illustrate this immediacy, and will also take our conception of the polis a little further. Each year, Thucydides tells us, if citizens had died in war – and they had, more often than not – a funeral oration was delivered by "a man chosen by the polis". Today, that would be someone nominated by the Prime Minister, or the British Academy, or the BBC [British Broadcasting Corporation]. In Athens it meant that someone was chosen by the Assembly who had often spoken to that Assembly; and on this occasion Pericles spoke from a specially high platform, that his voice might reach as many as possible. Let us consider two phrases that Pericles used in that speech.

He is comparing the Athenian polis with the Spartan, and makes the point that the Spartans admit foreign visitors only grudgingly, and from time to time expel all strangers, "while we make our polis common to all". "Polis" here is not the political unit; there is no question of naturalizing foreigners – which the Greeks did rarely, simply because the polis was so intimate a union. Pericles means here: "We throw open to all our common cultural life", as is shown by the words that follow, difficult though they are to translate: "nor do we deny them any instruction or spectacle" – words that are almost meaningless until we realize that the drama, tragic and comic, the performance of choral hymns, public recitals of Homer, games, were all necessary and normal parts of "political" life. This is the sort of thing Pericles has in mind when he speaks of "instruction and spectacle", and of "making the polis open to all".

But we must go further than this. A perusal of the speech will show that in praising the Athenian polis Pericles is praising more than a state, a nation, or a people: he is praising a way of life; he means no less when, a little later, he calls Athens the "school of Hellas". – And what of that? Do not we praise "the English way of life"? The difference is this; we expect our State to be quite indifferent to "the English way of life" – indeed, the idea that the State should actively try to promote it would fill most of us with alarm. The Greeks thought of the polis as an active, formative thing, training the minds and characters of the citizens; we think of it as a piece of machinery for the production of safety and convenience. The training in virtue, which the medieval state left to the Church, and the polis made its own concern, the modern state leaves to God knows what.

"Polis", then, originally "citadel", may mean as much as "the whole communal life of the people, political, cultural, moral" – even "economic", for how else are we to understand another phrase in this same speech, "the produce of the whole world comes to us, because of the magnitude of our polis"? This must mean "our national wealth".

Religion too was bound up with the polis – though not every form of religion. The Olympian gods were indeed worshipped by Greeks everywhere, but each polis had, if not its own gods, at least its own particular cults of these gods . . . But beyond tese Olympians, each polis had its minor local deities, "heroes" and nymphs, each worshipped with his immemorial rite, and scarcely imagined to exist outside the particular locality where the rite was performed. So . . . there is a sense in which it is true to say that the polis is an independent religious, as well as political, unit . . .

[. . .]

. . . Aristotle made a remark which we most inadequately translate "Man is a political animal." What Aristotle really said is "Man is a creature who lives in a polis"; and what he goes on to demonstrate, in his *Politics*, is that the polis is the only framework within which man can fully realize his spiritual, moral and intellectual capacities.

Such are some of the implications of this word . . . The polis was a living community, based on kinship, real or assumed – a kind of extended family, turning as much as possible of life into family life, and of course having its family quarrels, which were the more bitter because they were family quarrels.

This it is that explains not only the polis but also much of what the Greek made and thought, that he was essentially social. In the winning of his livelihood he was essentially individualist: in the filling of his life he was essentially "communist". Religion, art, games, the discussion of things – all these were needs of life that could be fully satisfied only through the polis – not, as with us, through voluntary associations of like-minded people, or through entrepreneurs appealing to individuals. (This partly explains the difference between Greek drama and the modern cinema.) Moreover, he wanted to play his own part in running the affairs of the community. When we realize how many of the necessary, interesting and exciting activities of life the Greek enjoyed through the polis, all of them in the open air, within sight of the same acropolis, with the same ring of mountains or of sea visibly enclosing the life of every member of the state – then it becomes possible to understand Greek history, to understand that in spite of the promptings of common sense the Greek could not bring himself to sacrifice the polis, with its vivid and comprehensive life, to a wider but less interesting unity . . .

[. . .]

HENRI PIRENNE

"City Origins" and "Cities and European Civilization"

from *Medieval Cities* (1925)

Editors' introduction The Greek polis influenced Roman ideas of city building and society. The Greek model influenced monumental marble public buildings, orthogonal streets, inward-turning private residences, and theaters and stadiums of Roman cities. But Rome achieved a population of a million people, a size so large that Aristotle had given it as an absurd example of a size inconceivable to imagine for a polis. Unlike the society of fragmented city-states Kitto described, Roman traders carried goods all over the Roman Empire from present day Iran to Scotland. Both the social structure and the physical form of Roman cities took on an imperial character as centers of military and political power, which increasingly differentiated them from the Greek polis.

Between the death of Roman Emperor Justinian in CE 565 and the Renaissance of the eleventh century, European cities withered in size to tiny shadows of their former selves. Then, beginning in the eleventh century, they began to grow in size and change in function. Their populations began the slow ascent of the shallow part of Kingsley Davis's urbanization S curve.

Exactly what happened and why during the period of medieval cities has provoked much scholarly debate. Even more interesting and controversial is the re-emergence of cities and their contribution to European history from the eleventh century onward. Why did cities begin to re-emerge in the eleventh century?

In the following selection Belgian historian Henri Pirenne (1862–1935) emphasizes the role of trade in both the decline of cities at the end of the Roman Empire and their subsequent re-emergence in the eleventh century. Pirenne argues that the barbarian invaders were absorbed into the Roman culture they overthrew, often without physically destroying Roman cities or even Roman social institutions. Generally the barbarians wanted to enjoy, not destroy, the Roman cities. Far more damaging to the Roman system of cities, according to Pirenne, was the Islamic conquest of the Mediterranean, which choked off long-distance trade routes. As trade stagnated, cities in Europe lost their economic reason for existing and withered. Regions became self-contained. By the time of Charlemagne (CE 742–814) the largest settlements established by Rome had shrunk to tiny religious places. Defense and administrative functions were added somewhat later so that medieval cities blended religion, defense, and administration.

Just as lack of trade atrophied and transmuted post-Roman cities, Pirenne argues that it was trade that revived cities during the eleventh century. Merchants emerged as a separate class – a group independent of the clergy, the landed aristocracy, or the vast submerged population of serfs. By the eleventh century many European cities had bustling marketplaces like the one in Monpazier, France, illustrated in Plate 3.

The merchants driving the re-emergence of European cities often lived and traded in suburbs below the walls of medieval cities built on hills. The word suburb itself is derived from the Latin "below the town." More important the emerging merchant class was free of many of the political, legal, and social restrictions that kept medieval society so changeless.

As the merchants grew in numbers and influence, Pirenne argues, they revolutionized the social structure of cities. Cities took on new life. The old stagnant class structure loosened up. The cities produced and exchanged new goods, and new distinctively urban forms of thought and culture emerged. And urban culture in turn revolutionized social relations and thought throughout the countryside.

Contrast Childe's views on the role of agricultural production in the origins of Mesopotamian cities (p. 22) and Kitto's emphasis on the importance of defense and religion to the emergence of the Greek polis with what Pirenne has to say about economics and trade in the re-emergence of cities in Europe. Compare Pirenne's account of the positive contributions of capitalism to European culture with Engels's devastating description of Manchester, England, during the full flowering of mid-nineteenth-century capitalism (p. 46). Compare Pirenne's views on the essentially economic function of cities to Mumford's view of the city as stage for human culture (p. 92).

Pirenne's thesis is fully developed in *Medieval Cities* (Princeton, NJ: Princeton University Press, 1925). Other notable books by Pirenne available in English translation include *Economic and Social History of Medieval Europe* (London: Routledge & Kegan Paul, 1958), *A History of Europe: From the Invasions to the XVI Century* (New Hyde Park, NY: University Books, 1965) and *Mohammed and Charlemagne* (London: Unwin University Books, 1974).

For overviews of European medieval cities see Paul M. Hohenberg and Lynn Hollen Lees, *The Making of Urban Europe 1000–1950* (Cambridge, MA: Harvard University Press, 1985), Charles Tilly and William P. Blockmans (eds.), *Cities and the Rise of States in Europe, A.D. 1000 to 1800* (Boulder, CO: Westview, 1994), and David Nicholas, *The Growth of the Medieval City: From Antiquity to the Early Fourteenth Century* (New York: Addison-Wesley, 1997). Chapters 4 and 5 on "Medieval Towns" and "The Renaissance: Italy Sets a Pattern" in A. E. J. Morris, *History of Urban Form Before the Industrial Revolution* (Harlow, UK: Longman Scientific & Technical, 1994) are also helpful.

Fernand Braudel's three volume history of capitalism, *The Structures of Everyday Life, The Wheels of Commerce,* and *The Perspective of the World* (Berkeley: University of California Press, 1992), contains an enormous amount of information on the relationship between cities and capitalism.

CITY ORIGINS

An interesting question is whether or not cities existed in the midst of that essentially agricultural civilization into which western Europe had developed in the course of the ninth century. The answer depends on the meaning given to the word "city." If by it is meant a locality the population of which, instead of living by working the soil, devotes itself to commercial activity, the answer will have to be "No." The answer will also be in the negative if we understand by "city" a community endowed with legal personality and possessing laws and institutions peculiar to itself. On the other hand, if we think of a city as a center of administration and as a fortress, it is clear that the Carolingian period knew nearly as many cities as the centuries which followed it must have known. That is merely another way of saying that the cities which were then to be found were without two of the fundamental attributes of the cities of the Middle Ages and of modern times: a middle-class population and a communal organization.

Primitive though it may be, every stable society feels the need of providing its members with centers of assembly, or meeting places. Observance of religious rites, maintenance of markets, and political and judicial gatherings necessarily bring about the designation of localities intended for the assembly of those who wish to or who must participate therein.

Military needs have a still more positive effect. Populations have to prepare refuges where will be found momentary protection from the enemy in case of invasion. War is as old as humanity, and the construction of fortresses almost as old as war. The first buildings erected by man seem, indeed, to have been protecting walls ... Their plan and their construction depended naturally upon the conformation of the terrain and upon the building materials at hand. But the general arrangement of them was everywhere the same. It consisted of a space, square or circular in shape, surrounded by ramparts made of trunks of trees, or mud or blocks of stone, protected by a moat and entered by gates. In short, it was an enclosure. And it is an interesting fact that the words which in modern English and in modern Russian (town and *gorod*) designate a city, originally designated an enclosure.

In ordinary times, these enclosures remained empty. The people resorted to them only on the occasion of religious or civic ceremonies, or when war constrained them to seek refuge there with their herds. But, little by little with the march of civilization, their intermittent animation became a continuous animation. Temples arose; magistrates or chieftains established their residence; merchants and artisans came to settle. What first had been only an occasional center of assembly became a city, the administrative, religious, political, and economic center of all the territory of the tribe whose name it customarily took.

This explains why, in many societies and particularly in classic antiquity, the political life of the cities was not restricted to the circumference of their walls. The city, indeed, had been built for the tribe, and every man in it, whether dwelling within or without the walls, was equally a citizen thereof. Neither Greece nor Rome knew anything analogous to the strictly local and particularist bourgeoisie of the Middle Ages. The life of the city was blended with the national life. The law of the city was, like the religion itself of the city, common to all the people whose capital it was and who constituted with it a single autonomous republic.

The municipal system, then, was identified in antiquity with the constitutional system. And when Rome extended her dominion over all the Mediterranean world, she made it the basis of the administrative system of her Empire. This system withstood, in western Europe, the Germanic invasions. Vestigial but thoroughly definite relics of it were still to be found in Gaul, in Spain, in Africa, and in Italy, long after the fifth century. Little by little, however, the increasing weakness of social organization did away with most of its characteristic features ... At the same time the thrust of Islam in the Mediterranean, in making impossible the commerce which up to now had still sustained a certain activity in the cities, condemned them to an inevitable decline. But it did not condemn them to death. Curtailed and weakened though they were, they survived. Their social function did not altogether disappear. In the agricultural social order of the time, they retained in spite of everything a fundamental importance. It is necessary to take full count of the role they played, in order to understand what was to befall them later.

As has been stated above, the Church had based its diocesan boundaries on the boundaries of the Roman cities. Held in respect by the barbarians, it therefore continued to maintain, after their occupation of the provinces of the Empire, the municipal system upon which it had been based. The dying out of trade and the exodus of foreign merchants had no influence on the ecclesiastical organization. The cities where the bishops resided became poorer and less populous without the bishops themselves feeling the effects. On the contrary, the more that general prosperity declined, the more their power and their influence had a chance to assert itself. Endowed with a prestige which was the greater because the State had disappeared, sustained by donations from their congregations, and partners with the Carolingians in the governing of society, they were in a commanding position by virtue of, at one and the same time, their moral authority, their economic power, and their political activity.

When the Empire of Charlemagne foundered, their status, far from being adversely affected, was made still more secure. The feudal princes, who had ruined the power of the Monarchy, did not touch that of the Church, for its divine origin protected it from their attacks. They feared the bishops, who could fling at them the terrible weapon of excommunication. They revered them as the supernatural guardians of order and justice. In the midst of the anarchy of the tenth and eleventh centuries the ascendancy of the Church remained, therefore, unimpaired . . .

This prestige of the bishops naturally lent to their places of residence – that is to say, to the old Roman cities – considerable importance. It is highly probable that this was what saved them. In the economy of the ninth century they no longer had any excuse for existence. In ceasing to be commercial centers they must have lost, quite evidently, the greatest part of their population. The merchants who once frequented them, or dwelt there, disappeared and with them disappeared the urban character which they had still preserved during the Merovingian era. Lay society no longer had the least use for them. Round about them the great demesnes lived their own life. There is no evidence that the State, itself constituted on a purely agricultural basis, had any cause to be interested in their fate. It is quite characteristic, and quite illuminating, that the palaces (*palatia*) of the Carolingian princes were not located in the towns. They were, without exception, in the country . . .

[. . .]

The State, on its part, in exercising administrative powers could contribute in no way to the continued existence of the Roman cities. The countries which formed the political districts of the Empire were without their chief-towns, just as the Empire itself was without a capital. The counts, to whom the supervision of them was entrusted, did not settle down in any fixed spot. They were constantly traveling about their districts in order to preside over judicial assemblies, to levy taxes, and to raise troops . . .

On the contrary, the immobility which ecclesiastical discipline enforced upon a bishop permanently held him to the city where was established the see of his particular diocese . . . Each diocese comprised the territory about the city which contained its cathedral and kept in constant touch with it . . .

[. . .]

During the last days of the Lower Empire, and still more during the Merovingian era, the power of the bishops over the city populace consistently increased. They had profited by the growing disorganization of civil society to accept, or to arrogate to themselves, an authority which the inhabitants did not take pains to dispute with them, and which the State had no interest in and, moreover, no means of denying them . . .

When the disappearance of trade, in the ninth century, annihilated the last vestiges of city life and put an end to what still remained of a municipal population, the influence of the bishops, already so extensive, became unrivalled. Henceforward the towns were entirely under their control. In them were to be found, in fact, practically only inhabitants dependent more or less directly upon the Church.

Though no precise information is available, it is, nevertheless, possible to conjecture as to the nature of this population. It was composed of the clerics of the cathedral church and of the other churches grouped nearby; of the monks of the monasteries which, especially after the ninth century, came to be established, sometimes in great numbers, in the see of the diocese; of the teachers and the students of the ecclesiastical schools; and finally, of servitors and artisans, free or serf, who were indispensable to the needs of the religious group and to the daily existence of the clerical agglomeration. Almost always there was to be found in the town a weekly market whither the peasants from round about brought their produce. Sometimes, even, an annual fair was held there. At the gates a market toll was levied on everything that came in or went out. A mint was in operation within the walls. There were also to be found there a number of keeps occupied by vassals of the bishop, by his advocate or by his castellan. To all of this must be added, finally, the granaries and the storehouses where were stored the harvests from the monastical demesnes brought in, at stated periods, by the tenant-farmers. At the great yearly festivals the congregation of the diocese poured into the town and gave it, for several days, the animation of unaccustomed bustle and stir.

All this little world accepted the bishop as both its spiritual and temporal head. Religious and secular authority were united or, to put it better, were blended in his person . . . [T]here was no longer any field in the administration of the town wherein, whether by law or by prerogative, he did not intervene as the guardian of order, peace, and the common weal. A theocratic form of government had completely replaced the municipal regimen of antiquity . . .

[. . .]

These towns were fortresses as well as episcopal residences. In the last days of the Roman Empire they had been enclosed by walls as a protection against the barbarians. These walls were still in existence almost everywhere and the bishops busied themselves with keeping them up or with restoring them with the greater zeal in that the incursions of the Saracens and the Norsemen had given increasingly impressive proof, during the ninth century, of the need of protection. The old Roman *enceintes* continued, therefore, to protect the towns against new perils.

Their form remained, under Charlemagne, what it had been under Constantine. As a general rule, it took the shape of a rectangle surrounded by ramparts flanked by towers and communicating with the outside by gates, customarily to the number of four. The space so enclosed was very restricted and the length of its sides rarely exceeded four to five hundred yards. Moreover, it was far from being entirely built up; between the houses cultivated fields and gardens were to be found. The outskirts (*suburbium*), which in the Merovingian era still extended beyond the walls, had disappeared . . .

[. . .]

In the midst of the insecurity and the disorders which imparted so lugubrious a character to the second half of the ninth century, it therefore fell to the towns to fulfill a true mission of protection. They were, in every sense of the word, the ramparts of a society invaded, under tribute, and terrorized. Soon, from another cause, they were not to be alone in filling that role.

[. . .]

[Pirenne describes the disintegration of the Frankish state into territories controlled by princes. The princes established burgs (fortresses) which complemented the towns as centers for defense against invaders, but had none of the towns' other characteristics.]

It is therefore a safe conclusion that the period which opened with the Carolingian era knew cities neither in the social sense, nor in the economic sense, nor in the legal sense of that word. The towns and burgs were merely fortified places and headquarters of administration. Their inhabitants enjoyed neither special laws nor institutions of their own, and their manner of living did not distinguish them in any way from the rest of society.

Commercial and industrial activity were completely foreign to them. In no respect were they out of key with the agricultural civilization of their times. The groups they formed were, after all, of trifling importance. It is not possible, in the lack of reliable information, to give an exact figure, but everything indicates that the population of the burgs never consisted of more than a few hundred men and that of the towns probably did not pass the figure of two to three thousand souls.

The towns and burgs played, however, an essential role in the history of cities. They were, so to speak, the stepping-stones thereto. Round about their walls cities were to take shape after the economic renaissance, whose first symptoms appeared in the course of the tenth century, had made itself manifest.

CITIES AND EUROPEAN CIVILIZATION

The birth of cities marked the beginning of a new era in the internal history of Western Europe. Until then, society had recognized only two active orders: the clergy and the nobility. In taking its place beside them, the middle class rounded the social order out or, rather, gave the finishing touch thereto. Thenceforth its composition was not to change; it had all its constituent elements, and the modifications which it was to undergo in the course of centuries were, strictly speaking, nothing more than different combinations in the alloy. Like the clergy and like the nobility, the middle class was itself a privileged order. It formed a distinct legal group and the special law it enjoyed isolated it from the mass of the rural inhabitants which continued to make up the immense majority of the population. Indeed, as has already been seen, it was obliged to preserve intact its exceptional status and to reserve to itself the benefits arising therefrom. Freedom, as the middle class conceived it, was a monopoly. Nothing was less liberal than the caste idea, which was the cause of its strength until it became, at the end of the Middle Ages, a cause of weakness. Nevertheless, to that middle class was reserved the mission of spreading the idea of liberty far and wide and of becoming, without having consciously desired to be, the means of the gradual enfranchisement of the rural classes. The sole fact of its existence was due, indeed, to have an immediate effect upon these latter and, little by little, to attenuate the contrast which at the start separated them from it. In vain it strove to keep them under its influence, to refuse them a share in its privileges, to exclude them from engaging in trade and industry. It had not the power to arrest an evolution of which it was the cause and which it could not suppress save by itself vanishing.

For the formation of the city groups disturbed at once the economic organization of the country districts. Production, as it was there carried on, had served until then merely to support the life of the peasant and supply the prestations due to his seigneur. Upon the suspension of commerce, nothing impelled him to ask of the soil a surplus which it would have been impossible for him to get rid of, since he no longer had outside markets to call upon. He was content to provide for his daily bread, certain of the morrow and longing for no amelioration of his lot, since he could not conceive the possibility of it. The small markets of the towns and the burgs were too insignificant and their demand was too regular to rouse him enough to get out of his rut and intensify his labor. But suddenly these markets sprang into new life. The number of buyers was multiplied, and all at once he had the assurance of being able to sell the produce he brought to them. It was only natural for him to have profited from an opportunity as favorable as this. It depended on himself alone to sell, if he produced enough, and forthwith he began to till the land which hitherto he had let lie fallow. His work took on a new significance; it brought him profits, the chance of thrift and of an existence which became more comfortable as it became more active. The situation was still more favorable in that the surplus revenues from the soil belonged to him in his own right. The claims of the seigneur were fixed by demesnial custom at an immutable rate, so that the increase in the income from the land benefited only the tenant.

But the seigneur himself had a chance to profit from the new situation wherein the development of the cities placed the country districts. He had enormous reserves of uncultivated land: woods, heaths, marshes, and fens. Nothing could be simpler than to put them under cultivation and through them to profit from these new outlets which were becoming more and more exigent and remunerative as the towns grew in size and multiplied in number. The increase in population would furnish the necessary hands for the work of clearing and draining. It was enough to call for men; they would not fail to show up.

By the end of the eleventh century the movement was already manifest in its full force. Monasteries and local princes thenceforth were busy transforming the sterile parts of their demesnes into revenue-producing land. The area of cultivated land which, since the end of the Roman Empire, had not been increased, kept growing continually greater . . .

Meanwhile, on all sides, the seigneurs, both lay and ecclesiastic, were founding "new" towns. So was called a village established on virgin soil, occupants of which received plots of land in return for an annual rental. But these new towns, the number of which continued to grow in the course of the twelfth century, were at the same time towns. For in order to attract the farmers the seigneur promised them exemption from the taxes which bore down upon the serfs. In general, he reserved to himself only jurisdiction over them; he abolished in their favor the old claims which still existed in the demesnial organization . . .

Thus a new type of peasant appeared, quite different from the old. The latter had serfdom as a characteristic; the former enjoyed freedom. And this freedom, the cause of which was the economic disturbance communicated by the towns to the organization of the country districts, was itself copied after that of the cities. The inhabitants of the new towns were, strictly speaking, rural burghers. They even bore, in a good number of charters, the name of *burgenses*. They received a legal constitution and a local autonomy which was manifestly borrowed from city institutions, so much so that it may be said that the latter went beyond the circumference of their walls in order to reach the country districts and acquaint them with liberty.

And this new freedom, as it progressed, was not long in making headway even in the old demesnes, whose archaic constitution could not be maintained in the midst of a reorganized social order. Either by voluntary emancipation, or by prescription or usurpation, the seigneurs permitted it to be gradually substituted for the serfdom which had so long been the normal condition of their tenants. The form of government of the people was there changed at the same time as the form of government of the land, since both were consequences of an economic situation on the way to disappearing. Commerce now supplied all the necessaries which the demesnes had hitherto been obliged to obtain by their own efforts. It was no longer essential for each of them to produce all the commodities for which it had use. It sufficed to go get them at some nearby city . . .

. . . Trade, which was becoming more and more active, necessarily favored agricultural production, broke down the limits which had hitherto bounded it, drew it towards the towns, modernized it, and at the same time set it free. Man was therefore detached from the soil to which he had so long been enthralled, and free labor was substituted more and more generally for serf labor . . .

The emancipation of the rural classes was only one of the consequences provoked by the economic revival of which the towns were both the result and the instrument. It coincided with the increasing importance of liquid capital. During the demesnial era of the Middle Ages, there was no other form of wealth than that which lay in real estate. It ensured to the holder both personal liberty and social prestige. It was the guaranty of the privileged status of the clergy and the nobility. Exclusive holders of the land, they lived by the labor of their tenants whom they protected and whom they ruled. The serfdom of the masses was the necessary consequence of such a social organization. There was no alternative save to own the land and be lord, or to till it for another and be a serf.

But with the origin of the middle class there took its place in the sun a class of men whose existence was in flagrant contradiction to this traditional order of things. The land upon which they settled they not only did not cultivate but did not even own. They demonstrated and made increasingly clear the possibility of living and growing rich by the sole act of selling, or producing exchange values.

Landed capital had been everything, and now by the side of it was made plain the power of liquid capital. Heretofore money had been sterile. The great lay or ecclesiastic proprietors in whose hands was concentrated the very scant stock of currency in circulation, by means of either the land taxes which they levied upon their tenants or the alms which the congregations brought to the church, normally had no way of making it bear fruit . . . As a general rule cash was hoarded by its possessors and most often changed into vessels or ornaments for the church, which might be melted down in case of need. Trade, naturally, released this captive money and restored its proper function. Thanks to this, it became again the instrument of exchange and the measure of values, and since the towns were the centers of trade it necessarily flowed towards them. In circulating, its power was multiplied by the number of transactions in which it served. Its use, at the same time, became more general; payments in kind gave way more and more to payments in money.

A new motion of wealth made its appearance: that of mercantile wealth, consisting no longer in land but in money or commodities of trade measurable in money. During the course of the eleventh century, true capitalists already existed in a number of cities . . . These city capitalists soon formed the habit of putting a part of their profits into land. The best means of consolidating their fortune and their credit was, in fact, the buying up of land. They devoted a part of their gains to the purchase of real estate, first of all in the same town where they dwelt and later in the country. But they changed themselves, especially, into money-lenders. The economic crisis provoked by the irruption of trade into the life of society had caused the ruin of, or at least trouble to, the landed proprietors who had not been able to adapt themselves to it. For in speeding up the circulation of money a natural result was the decreasing of its value and by that very fact the raising of all prices. The period contemporary with the formation of the cities was a period of high cost of living, as favorable to the businessmen and artisans of the middle class as it was painful to the holders of the land who did not succeed in increasing their revenues. By the end of the eleventh century many of them were obliged to have recourse to the capital of the merchants in order to keep going . . .

But more important operations were already current at this era. There was no lack of merchants rich enough to agree to loans of considerable amount . . . The kings themselves had recourse, in the course of the twelfth century, to the good services of the city financiers . . .

[. . .]

The power of liquid capital, concentrated in the cities, not only gave them an economic ascendancy but contributed also towards making them take part in political life. For as long as society had known no other power than that which derived from the possession of the land, the clergy and the nobility alone had had a share in the government . . .

But as soon as the economic revival enabled [the prince] to augment his revenues, and cash, thanks to it, began to flow to his coffers, he took immediate advantage of circumstances . . . Identical economic causes had changed simultaneously the organization of the land and the governing of the people. Just as they enabled the peasants to free themselves, and the proprietors to substitute the quit-rent for the demesnial mansus, so they enabled the princes, thanks to their salaried agents, to lay hold of the direct government of their territories. This political innovation, like the social innovations with which it was contemporary, implied the diffusion of ready cash and the circulation of money . . .

The connections which were necessarily established between the princes and the burghers also had political consequences of the greatest import. It was necessary to take heed of those cities whose increasing wealth gave them a steadily increasing importance, and which could put on the field, in case of need, thousands of well-equipped men . . .

[. . .]

. . . [The cities'] natural tendency led them to become municipal republics. There is but little doubt but that, if they had had the power, they would have everywhere become States within the State. But they did not succeed in realizing this ideal save where the power of the State was impotent to counterbalance their efforts.

[. . .]

[The territorial government] did not treat them as mere subjects. It had too much need of them not to have regard for their interests. Its finances rested in great part upon them, and to the extent that they augmented the power of the State and therewith its expenses, it felt more and more frequently the need of going to the pocketbooks of the burghers . . . Little by little the princes formed the habit of calling the burghers into the councils of prelates and nobles with whom they conferred upon their affairs. The instances of such convocations were still rare in the twelfth century; they multiplied in the thirteenth; and in the fourteenth century the custom was definitely legalized by the institution of the Estates in which the cities obtained, after the clergy and the nobility, a place which soon became, although the third in dignity, the first in importance.

Although the middle classes, as we have just seen, had an influence of very vast import upon the social, economic, and political changes which were manifest in Western Europe in the course of the twelfth century, it does not seem at first glance that they played much of a role in the intellectual movement. It was not, in fact, until the fourteenth century that a literature and an art was brought forth from the bosom of the middle classes, animated with their spirit. Until then, science remained the exclusive monopoly of the clergy and employed no other tongue than the Latin. What literature was written in the vernacular had to do solely with the nobility, or at least expressed only the ideas and the sentiments which pertained to the nobility as a class. Architecture and sculpture produced their masterpieces only in the construction and ornamentation of the churches. The market and belfries, of which the oldest specimens date back to the beginning of the thirteenth century . . .

remained still faithful to the architectural style of the great religious edifices.

Upon closer inspection, however, it does not take long to discover that city life really did make its own contribution to the moral spirit of the Middle Ages. To be sure, its intellectual culture was dominated by practical considerations which, before the period of the Renaissance, kept it from putting forth any independent effort. But from the very first it showed that characteristic of being an exclusively lay culture. By the middle of the twelfth century the municipal councils were busy founding schools for the children of the burghers, which were the first lay schools since the end of antiquity. By means of them, instruction ceased to be furnished exclusively for the benefit of the novices of the monasteries and the future parish priests. Knowledge of reading and writing, being indispensable to the practice of commerce, ceased to be reserved for the members of the clergy alone. The burgher was initiated into them long before the noble, because what was for the noble only an intellectual luxury was for him a daily need . . .

However, the teaching in these communal schools was limited, until the period of the Renaissance, to elementary instruction. All who wished to have more were obliged to turn to the clerical establishments. It was from these latter that came the "clerks" who, starting at the end of the twelfth century, were charged with the correspondence and the accounts of the city, as well as the publication of the manifold Acts necessitated by commercial life. All these "clerks" were, furthermore, laymen, the cities having never taken into their service, in contradistinction to the princes, members of the clergy who by virtue of the privileges they enjoyed would have escaped their jurisdiction.

The language which the municipal scribes employed was naturally, at first, Latin. But after the first years of the thirteenth century they adopted more and more generally the use of national idioms. It was by the cities that the common tongue was introduced for the first time into administrative usage. Thereby they showed an initiative which corresponded perfectly to that lay spirit of which they were the preeminent representatives in the civilization of the Middle Ages.

This lay spirit, moreover, was allied with the most intense religious fervor. If the burghers were very frequently in conflict with the ecclesiastic authorities, if the bishops thundered fulsomely against them with sentences of excommunication, and if, by way of counterattack, they sometimes gave way to decidedly pronounced anti-clerical tendencies, they were, for all of that, none the less animated by a profound and ardent faith . . .

Both lay and mystic at the same time, the burghers of the Middle Ages were thus singularly well prepared for the role which they were to play in the two great future movements of ideas: the Renaissance, the child of the lay mind, and the Reformation, towards which religious mysticism was leading.

FRIEDRICH ENGELS

"The Great Towns"

from *The Condition of the Working Class in England in 1844* (1845)

Editors' introduction It was the peculiar fate of Friedrich Engels (1820–1895) to live most of his adult life in the shadow of his better-known friend and partner Karl Marx and to be remembered as a fiercely bearded icon of international communism. It was, however, a more humanly accessible Engels who, full of youthful idealism at the age of only 24, came face to face with the social horrors of the Industrial Revolution. Young Engels was sent by his industrialist father to learn business management in the factories of Manchester in the English Midlands. The unintended consequence of that particular paternal decision was *The Condition of the Working Class in England in 1844* (1845), a book that ranks as one of the earliest masterpieces of urban sociology.

By 1844 the Industrial Revolution had transformed conditions in many English cities. Manchester, England, which Engels observed in detail, was emblematic of the new industrial cities. Plate 4, from Augustus Welby Pugin's *Contrasts* (1841), contrasts the skyline of a fifteenth-century city dominated by church steeples with the same town in 1840. In the second view mills, factories, and a huge prison dominate the scene.

In the selection "The Great Towns" reprinted here, Engels employs a peripatetic method. Although he summarizes the socialist theory of the origin and historic role of the industrial working class, and although he quotes from many contemporaneous sources to bolster his analysis, Engels constructs the bulk of his argument by merely walking around the city and reporting what he sees. Quickly growing impatient with *telling* his readers about the social misery of working-class life, Engels begins *showing* them the horrors of industrial urbanism by conducting them on a tour of Manchester's working-class districts. As in Dante's *Inferno*, the tour descends deeper and deeper into the filth, misery, and despair that constitute the greater part of the Manchester conurbation.

Engels wrote just as the first daguerreotypes of cities were being produced, but unfortunately he did not illustrate his book with actual pictures of the conditions that he described. Later in the nineteenth century Jacob Riis, Lewis Hine, and other photographers were to document slum conditions. The response of these and other photographers to the new reality of the nineteenth-centuiry city are discussed in the selection by Frederic Stout, "Visions of a New Reality" (p. 143).

No one can read *The Condition of the Working Class* without acknowledging that Engels had come to know the various neighborhoods of proletarian Manchester – Old Town, Irish Town, Long Millgate, and Salford – intimately and that his observations were acute and objective. Of particular interest are his descriptions of the public health consequences (in terms of air and water pollution) of unrestrained overbuilding. Engels also observed that the façades of the main thoroughfares mask the horrors that lie beyond from the eyes of the factory owners and the middle-class managers who commute into the city from outlying suburbs. Plate 5, from *Frank Leslie's Illustrated Newspaper* in 1887, contrasts the belching factories and slums that comprise the bulk of the city with a spacious main thoroughfare in the city.

The entire tradition of twentieth-century urban planning, capitalist and socialist alike, owes an enormous debt to Engels. The connection he draws between the physical decrepitude of the urban infrastructure and the alienation and despair of the urban poor remains valid to the present day. The urban parks movement and the construction of ideal company towns – Saltaire and Port Sunlight in the United Kingdom, Lowell and Pullman in the United States – as well as more recent attempts at inner-city redevelopment, all address issues first identified by Engels.

The conditions described by Engels form the basis for the social realist tradition in literature, a tradition that begins with Charles Dickens and Mrs. Gaskell in England and is continued in the works of Upton Sinclair and Theodore Dreiser in the United States.

See Chapter 10 of Sir Peter Hall's *Cities in Civilization* (New York: Pantheon, 1998) "Manchester 1760–1840", for more on early Manchester. Hall's chapter contains a bibliography of source material on economic and social conditions in early Manchester.

Other significant investigations of urban poverty in England include Henry Mayhew's *London Labour and the London Poor* (4 volumes, 1851–62), Charles Booth's *Conditions and Occupation of the People in East London and Hackney* (*Journal of the Royal Historical Society*, 1887), Jack London's *People of the Abyss* (New York: Macmillan, 1903), and George Orwell's *Down and Out in Paris and London* (London: Secker and Warburg, 1933).

In America, key works include Jacob Riis, *How the Other Half Lives* (New York: Scribners, 1903) and a whole series of reports on conditions in the African-American ghettos such as W. E. B. Du Bois's *The Philadelphia Negro* (p. 56), St. Clair Drake and Horace Cayton's *Black Metropolis* (Chicago: University of Chicago Press, 1945), and William Julius Wilson's *When Work Disappears* (p. 112).

For a sampling of Engels most important writings, as well as those of his lifelong friend Karl Marx, see Robert C. Tucker, *The Marx–Engels Reader* (New York: W. W. Norton, 1978).

For an excellent summary of nineteenth-century urban poverty conditions, consult "The City of Dreadful Night" in Peter Hall's *Cities of Tomorrow* (London: Basil Blackwell, 1988). For further information on Engels's Manchester, and its connections to an emerging social realism in fiction, consult literary historian Steven Marcus's magisterial *Engels, Manchester, and the Working Class* (New York: Random House, 1974). Also of interest is Robert Roberts's extraordinary *The Classic Slum* (Manchester: University of Manchester Press, 1971), a first-person account of growing up in Salford during the early years of the twentieth century.

A town, such as London, where a man may wander for hours together without reaching the beginning of the end, without meeting the slightest hint which could lead to the inference that there is open country within reach, is a strange thing. This colossal centralization, this heaping together of two and a half millions of human beings at one point, has multiplied the power of this two and a half millions a hundred-fold; has raised London to the commercial capital of the world, created the giant docks and assembled the thousand vessels that continually cover the Thames. I know nothing more imposing than the view which the Thames offers during the ascent from the sea to London Bridge. The masses of buildings, the wharves on both sides, especially from Woolwich upwards, the countless ships along both shores, crowding ever closer and closer together, until, at last, only a narrow passage remains in the middle of the river, a passage through which hundreds of steamers shoot by one another; all this is so vast, so impressive, that a man cannot collect himself, but is lost in the marvel of England's greatness before he sets foot upon English soil.

But the sacrifices which all this has cost become apparent later. After roaming the streets of the capital a day or two, making headway with difficulty through the human turmoil and the endless lines of vehicles, after visiting the slums of the metropolis, one realizes for the first time that these Londoners have been forced to sacrifice the best qualities of their human nature, to bring to pass all the marvels of civilization which crowd their city; that a hundred powers which slumbered within them have remained inactive, have been suppressed in order that a few might be developed more fully and multiply through union with those of others. The very turmoil of the streets has something repulsive, something against which human nature rebels. The hundreds of thousands of all classes and ranks crowding past each other, are they not all human beings with the same qualities and powers, and with the same interest in being happy? And have they not, in the end, to seek happiness in the same way, by the same means? And still they crowd by one another as though they had nothing in common, nothing to do with one another, and their only agreement is the tacit one, that each keep to his own side of the pavement, so as not to delay the opposing streams of the crowd, while it occurs to no man to honour another with so much as a glance. The brutal indifference, the unfeeling isolation of each in his private interest becomes the more repellent and offensive, the more these individuals are crowded together, within a limited space. And, however much one may be aware that this isolation of the individual, this narrow self-seeking is the fundamental principle of our society everywhere, it is nowhere so shamelessly barefaced, so self-conscious as just here in the crowding of the great city. The dissolution of mankind into monads, of which each one has a separate principle and a separate purpose, the world of atoms, is here carried out to its utmost extreme.

Hence it comes, too, that the social war, the war of each against all, is here openly declared. . . . , people regard each other only as useful objects; each exploits the other, and the end of it all is, that the stronger treads the weaker under foot, and that the powerful few, the capitalists, seize everything for themselves, while to the weak many, the poor, scarcely a bare existence remains.

What is true of London, is true of Manchester, Birmingham, Leeds, is true of all great towns. Everywhere barbarous indifference, hard egotism on one hand, and nameless misery on the other, everywhere social warfare, every man's house in a state of siege, everywhere reciprocal plundering under the protection of the law, and all so shameless, so openly avowed that one shrinks before the consequences of our social state as they manifest themselves here undisguised, and can only wonder that the whole crazy fabric still hangs together.

Since capital, the direct or indirect control of the means of subsistence and production, is the weapon with which this social warfare is carried on, it is clear that all the disadvantages of such a state must fall upon the poor. For him no man has the slightest concern. Cast into the whirlpool, he must struggle through as well as he can. If he is so happy as to find work, i.e. if the bourgeoisie does him the favour to enrich itself by means of him, wages await him which scarcely suffice to keep body and soul together; if he can get no work he may steal, if he is not afraid of the police, or starve, in which case the police will take care that he does so in a quiet and inoffensive manner. During my residence in England, at least twenty or thirty persons have died of simple starvation under the most revolting circumstances, and a jury has rarely been found possessed of the courage to speak the plain truth in the matter. Let the testimony of the witnesses be never so clear and unequivocal, the bourgeoisie, from which the jury is selected, always finds some backdoor through which to escape the frightful verdict, death from starvation. The bourgeoisie dare not speak the truth in these cases. for it would speak its own condemnation. But indirectly, far more than directly, many have died of starvation, where long continued want of proper nourishment has called forth fatal illness, when it has produced such debility that causes which might otherwise have remained inoperative, brought on severe illness and death. The English working-men call this social murder, and accuse our whole society of perpetrating this crime perpetually. Are they wrong?

True, it is only individuals who starve, but what security has the working-man that it may not be his turn tomorrow? Who assures him employment, who vouches for it that, if for any reason or no reason his lord and master discharges him tomorrow, he can struggle along with those dependent upon him, until he may find some one else "to give him bread"? Who guarantees that willingness to work shall suffice to obtain work, that uprightness, industry, thrift, and the rest of the virtues recommended by the bourgeoisie, are really his road to happiness? No one. He knows that he has something today, and that it does not depend upon himself whether he shall have something tomorrow. He knows that every breeze that blows, every whim of his employer, every bad turn of trade may hurl him back into the fierce whirlpool from which he has temporarily saved himself, and in which it is hard and often impossible to keep his head above water. He knows that, though he may have the means of living today, it is very uncertain whether he shall tomorrow.

[. . .]

Manchester lies at the foot of the southern slope of a range of hills, which stretch hither from Oldham, their last peak, Kersallmoor, being at once the racecourse and the Mons Sacer of Manchester. Manchester proper lies on the left bank of the Irwell, between that stream and the two smaller ones, the Irk and the Medlock, which here empty into the Irwell. On the right bank of the Irwell, bounded by a sharp curve of the river, lies Salford, and farther westward Pendleton; northward from the Irwell lie Upper and Lower Broughton; northward of the Irk, Cheetham Hill; south of the Medlock lies Hulme; farther east Chorlton on Medlock; still farther, pretty well to the east of Manchester, Ardwick. The whole assemblage of buildings is commonly called Manchester, and contains about four hundred thousand inhabitants, rather more than less. The town itself is peculiarly built, so that a person may live in it for years, and go in and out daily without coming into contact with a working-people's quarter or even with workers; that is, so long as he confines himself to his business or to pleasure walks. This arises chiefly from the fact, that by unconscious tacit agreement, as well as with outspoken conscious determination, the working-people's quarters are sharply separated from the sections of the city reserved for the middle class; or, if this does not succeed, they are concealed with the cloak of charity. Manchester contains, at its heart, a rather extended commercial district, perhaps half a mile long and about as broad, and consisting almost wholly of offices and warehouses. Nearly the whole district is abandoned by dwellers, and is lonely and deserted at night; only watchmen and policemen traverse its narrow lanes with their dark lanterns. This district is cut through by certain main thoroughfares upon which the vast traffic concentrates, and in which the ground level is lined with brilliant shops. In these streets the upper floors are occupied, here and there, and there is a good deal of life upon them until late at night. With the exception of this commercial district, all Manchester proper, all Salford and Hulme, a great part of Pendleton and Chorlton, two-thirds of Ardwick, and single stretches of Cheetham Hill and Broughton are all unmixed working-people's quarters, stretching like a girdle, averaging a mile and a half in breadth, around the commercial district. Outside, beyond this girdle, lives the upper and middle bourgeoisie, the middle bourgeoisie in regularly laid out streets in the vicinity of the working quarters, especially in Chorlton and the lower-lying portions of Cheetham Hill; the upper bourgeoisie in remoter villas with gardens in Chorlton and Ardwick or on the breezy heights of Cheetham Hill, Broughton and Pendleton, in free, wholesome country air, in fine, comfortable homes, passed once every half or quarter hour by omnibuses going into the city. And the finest part of the arrangement is this, that the members of this money aristocracy can take the shortest road through the middle of all the labouring districts to their places of business, without ever seeing that they are in the midst of the grimy misery that lurks to the right and the left. For the thoroughfares leading from the Exchange in all directions out of the city are lined, on both sides, with an almost unbroken series of shops, and are so kept in the hands of the middle and lower bourgeoisie, which, out of self-interest, cares for a decent and cleanly external appearance and *can* care for it. True, these shops bear some relation to the districts which lie behind them, and are more elegant in the commercial and residential quarters than when they hide grimy working-men's dwellings; but they suffice to conceal from the eyes of the wealthy men and women of strong stomachs and weak nerves the misery and grime which form the complement of their wealth. So, for instance, Deansgate, which leads from the Old Church directly southward, is lined first with mills and warehouses, then with second-rate shops and alehouses; farther south, when it leaves the commercial district, with less inviting shops, which grow dirtier and more interrupted by beerhouses and gin palaces the farther one goes, until at the southern end the appearance of the shops leaves no doubt that workers and workers only are their customers. So Market Street running south east from the Exchange; at first brilliant shops of the best sort, with counting-houses or warehouses above; in the continuation, Piccadilly, immense hotels and warehouses; in the farther continuation, London Road, in the neighbourhood of the Medlock, factories, beerhouses, shops for the humbler bourgeoisie and the working population; and from this point onward, large gardens and villas of the wealthier merchants and manufacturers. In this way any one who knows Manchester can infer the adjoining districts, from the appearance of the thoroughfare, but one is seldom in a position to catch from the street a glimpse of the real labouring districts. I know very well that this hypocritical plan is more or less common to all great cities; I know, too, that the retail dealers are forced by the nature of their business to take possession of the great highways; I know that there are more good buildings than bad ones upon such streets everywhere, and that the value of land is greater near them than in remoter districts; but at the same time I have never seen so systematic a shutting out of the working-class from the thoroughfares, so tender a concealment of everything which might affront the eye and the nerves of the bourgeoisie, as in Manchester. And yet, in other respects, Manchester is less built according to a plan, after official regulations, is more an outgrowth of accident, than any other city; and when I consider in this connection the eager assurances of the middle-class, that the working-class is doing famously, I cannot help feeling that the liberal manufacturers, the "Big Wigs" of Manchester, are not so innocent after all, in the matter of this sensitive method of construction.

I may mention just here that the mills almost all adjoin the rivers or the different canals that ramify throughout the city, before I proceed at once to describe the labouring quarters. First of all, there is the Old Town of Manchester [Figure 1], which lies between the northern boundary of the commercial district and the Irk. Here the streets, even the better ones, are narrow and winding, as Todd Street, Long Millgate, Withy Grove, and Shude Hill, the houses dirty, old, and tumble-down, and the construction of the side streets utterly horrible. Going from the Old Church to Long Millgate, the stroller has at once a row of old-fashioned houses at the right, of which not one has kept its original level; these are remnants of the old pre-manufacturing Manchester, whose former inhabitants have removed with their descendants into better-built districts, and have left the houses, which were not good enough for them, to a working-class population strongly mixed with Irish blood. Here one is in an almost undisguised working-men's quarter, for even the shops and beerhouses hardly take the trouble to exhibit a trifling degree of cleanliness. But all this is nothing in comparison with the courts and lanes which lie behind, to which access can be gained only through covered passages in which no two human beings can pass at the same time. Of the irregular cramming together of dwellings in ways which defy all rational plan, of the tangle in which they

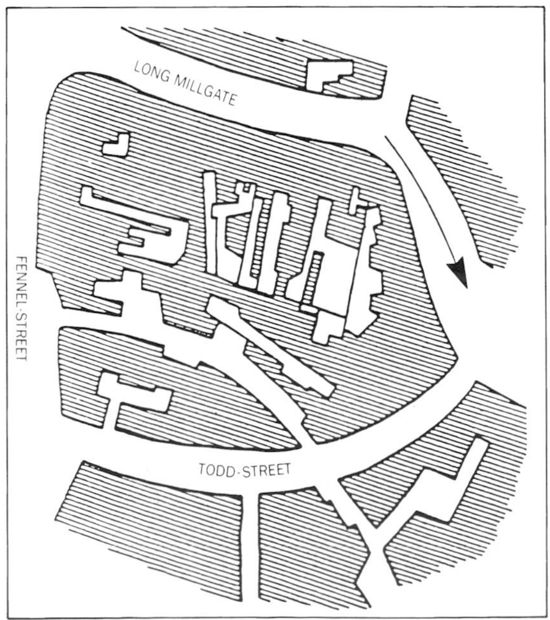

Figure 1

are crowded literally one upon the other, it is impossible to convey an idea. And it is not the buildings surviving from the old times of Manchester which are to blame for this; the confusion has only recently reached its height when every scrap of space left by the old way of building has been filled up and patched over until not a foot of land is left to be further occupied.

[. . .]

The south bank of the Irk is here very steep and between fifteen and thirty feet high. On this declivitous hillside there are planted three rows of houses, of which the lowest rise directly out of the river, while the front walls of the highest stand on the crest of the hill in Long Millgate. Among them are mills on the river; in short, the method of construction is as crowded and disorderly here as in the lower part of Long Millgate. Right and left a multitude of covered passages lead from the main street into numerous courts, and he who turns in thither gets into a filth and disgusting grime the equal of which is not to be found – especially in the courts which lead down to the Irk and which contain unqualifiedly the most horrible dwellings which I have yet beheld. In one of these courts there stands directly at the entrance, at the end of the covered passage, a privy without a door, so dirty that the inhabitants can pass into and out of the court only by passing through foul pools of stagnant urine and excrement. This is the first court on the Irk above Ducie Bridge – in case any one should care to look into it. Below it on the river there are several tanneries which fill the whole neighbourhood with the stench of animal putrefaction. Below Ducie Bridge the only entrance to most of the houses is by means of narrow dirty stairs and over heaps of refuse and filth. The first court below Ducie Bridge, known as Allen's Court, was in such a state at the time of the cholera that the sanitary police ordered it evacuated, swept, and disinfected with chloride of lime. Dr. Kay gives a terrible description of the state of this court at that time. Since then it seems to have been partially torn away and rebuilt; at least looking down from Ducie Bridge, the passer-by sees several ruined walls and heaps of debris with some newer houses. The view from this bridge, mercifully concealed from mortals of small stature by a parapet as high as a man, is characteristic for the whole district. At the bottom flows, or rather stagnates, the Irk, a narrow, coal-black, foul-smelling stream full of debris and refuse, which it deposits on the shallower right bank. In dry weather, a long string of the most disgusting, blackish-green, slime pools are left standing on this bank, from the depths of which bubbles of miasmatic gas constantly arise and give forth a stench unendurable even on the bridge forty or fifty feet above the surface of the stream. But besides this, the stream itself is checked every few paces by high weirs, behind which slime and refuse accumulate and rot in thick masses. Above the bridge are tanneries, bonemills, and gasworks, from which all drains and refuse find their way into the Irk, which receives further the contents of all the neighbouring sewers and privies. It may be easily imagined, therefore, what sort of residue the stream deposits. Below the bridge you look upon the piles of debris, the refuse, filth, and offal from the courts on the steep left bank; here each house is packed close behind its neighbour and a piece of each is visible, all black, smoky, crumbling, ancient, with broken panes and window-frames. The background is furnished by old barrack-like factory buildings. On the lower right bank stands a long row of houses and mills; the second house being a ruin without a roof, piled with debris; the third stands so low that the lowest floor is uninhabitable, and therefore without windows or doors. Here the background embraces the pauper burial-ground, the station of the Liverpool and Leeds railway, and, in the rear of this, the Workhouse, the "Poor-Law-Bastille" of Manchester, which, like a citadel, looks threateningly down from behind its high walls and parapets on the hilltop, upon the working people's quarter below.

Above Ducie Bridge, the left bank grows more flat and the right bank steeper, but the condition of the dwellings on both banks grows worse rather than better. He who turns to the left here from the main street, Long Millgate, is lost; he wanders from one court to another, turns countless corners, passes nothing but narrow, filthy nooks and alleys, until after a few minutes he has lost all clue, and knows not whither to turn. Everywhere half or wholly ruined buildings, some of them actually uninhabited, which means a great deal here; rarely a wooden or stone floor to be seen in the houses, almost uniformly broken, ill-fitting windows and doors, and a state of filth! Everywhere heaps of debris, refuse, and offal; standing pools for gutters, and a stench which alone would make it impossible for a human being in any degree civilized to live in such a district. The newly-built extension of the Leeds railway, which crosses the Irk here, has swept away some of these courts and lanes, laying others completely open to view. Immediately under the railway bridge there stands a court, the filth and horrors of which surpass all the others by far, just because it was hitherto so shut off, so secluded that the way to it could not be found without a good deal of trouble. I should never have discovered it myself, without the breaks made by the railway, though I thought I knew this whole region thoroughly. Passing along a rough bank, among stakes and washing-lines, one penetrates into this chaos of small one-storeyed, one-roomed huts, in most of which there is no artificial floor; kitchen, living and sleeping room all in one. In such a hole, scarcely five feet long by six broad, I found two beds – and such bedsteads and beds! – which, with a staircase and chimney-place, exactly filled the room. In several others I found absolutely nothing, while the door stood open, and the inhabitants leaned against it. Everywhere before the doors refuse and offal; that any sort of pavement lay underneath could not be seen but only felt, here and there, with the feet. This whole collection of cattle-sheds for human beings was surrounded on two sides by houses and a factory, and on the third by the river, and besides the narrow stair up the bank, a narrow doorway alone led out into another almost equally ill-built, ill-kept labyrinth of dwellings.

Enough! The whole side of the Irk is built in this way, a planless, knotted chaos of houses, more or less on the verge of uninhabitableness, whose unclean interiors fully correspond with their filthy external surroundings. And how could the people be clean with no proper opportunity for satisfying the most natural and ordinary wants? Privies are so rare here that they are either filled up every day, or are too remote for most of the inhabitants to use. How can people wash when they have only the dirty Irk water at hand, while pumps and water pipes can be found in decent parts of the city alone? In truth, it cannot be charged to the account of these helots of modern society if their dwellings are not more clean than the pig-sties which are here and there to be seen among them. The landlords are not ashamed to let dwellings like the six or seven cellars on the quay directly below Scotland Bridge, the floors of which stand at least two feet below the low-water level of the Irk that flows not six feet away from them; or like the upper floor of the corner-house on the opposite shore directly above the bridge, where the ground-floor, utterly uninhabitable, stands deprived of all fittings for doors and windows, a case by no means rare in this region, when this open ground-floor is used as a privy by the whole neighbourhood for want of other facilities!

If we leave the Irk and penetrate once more on the opposite side from Long Millgate into the midst of the working-men's dwellings, we shall come into a somewhat newer quarter, which stretches from St. Michael's Church to Withy Grove and Shude Hill. Here there is somewhat better order. In place of the chaos of buildings, we find at least long straight lanes and alleys or courts, built according to a plan and usually square. But if, in the former case, every house was built according to caprice, here each lane and court is so built, without reference to the situation of the adjoining ones. The lanes run now in this direction, now in that, while every two minutes the wanderer gets into a blind alley, or, on turning a corner, finds himself back where he started from; certainly no one who has not lived a considerable time in this labyrinth can find his way through it.

If I may use the word at all in speaking of this district, the ventilation of these streets and courts is, in consequence of this confusion, quite as imperfect as in the Irk region; and if this quarter may, nevertheless, be said to have some advantage over that of the Irk, the houses being newer and the streets occasionally having gutters, nearly every house has, on the other hand, a cellar dwelling, which is rarely found in the Irk district, by reason of the greater age and more careless construction of the houses. As for the rest the filth, debris, and offal heaps, and the pools in the streets are common to both quarters, and in the district now under discussion, another feature most injurious to the cleanliness of the inhabitants, is the multitude of pigs walking about in all the alleys, rooting into the offal heaps, or kept imprisoned in small pens. Here, as in most of the working-men's quarters of Manchester, the pork-raisers rent the courts and build pigpens in them. In almost every court one or even several such pens may be found into which the inhabitants of the court throw all refuse and offal, whence the swine grow fat; and the atmosphere, confined on all four sides, is utterly corrupted by putrefying animal and vegetable substances. Through this quarter, a broad and measurably decent street has been cut, Millers Street, and the background has been pretty successfully concealed. But if any one should be led by curiosity to pass through one of the numerous passages which lead into the courts, he will find this piggery repeated at every twenty paces.

Such is the Old Town of Manchester, and on re-reading my description, I am forced to admit that instead of being exaggerated, it is far from black enough to convey a true impression of the filth, ruin, and uninhabitableness, the defiance of all considerations of cleanliness, ventilation, and health which characterize the construction of this single district, containing at least twenty to thirty thousand inhabitants. And such a district exists in the heart of the second city of England, the first manufacturing city of the world. If any one wishes to see in how little space a human being can move, how little air – and *such* air! – he can breathe, how little of civilization he may share and yet live, it is only necessary to travel hither. True, this is the *Old* Town, and the people of Manchester emphasize the fact whenever any one mentions to them the frightful condition of this Hell upon Earth; but what does that prove? Everything which here arouses horror and indignation is of recent origin, belongs to the *industrial* epoch. The couple of hundred houses, which belong to old Manchester, have been long since abandoned by their original inhabitants; the industrial epoch alone has crammed into them the swarms of workers whom they now shelter; the industrial epoch alone has built up every spot between these old houses to win a covering for the masses whom it has conjured hither from the agricultural districts and from Ireland; the industrial epoch alone enables the owners of these cattlesheds to rent them for high prices to human beings, to plunder the poverty of the workers, to undermine the health of thousands, in order that they *alone*, the owners, may grow rich. In the industrial epoch alone has it become possible that the worker scarcely freed from feudal servitude could be used as mere material, a mere chattel; that he must let himself be crowded into a dwelling too bad for every other, which he for his hard-earned wages buys the right to let go utterly to ruin. This manufacture has achieved, which, without these workers, this poverty, this slavery could not have lived. True, the original construction of this quarter was bad, little good could have been made out of it; but, have the landowners, has the municipality done anything to improve it when rebuilding? On the contrary, wherever a nook or corner was free, a house has been run up; where a superfluous passage remained, it has been built up; the value of land rose with the blossoming out of manufacture, and the more it rose, the more madly was the work of building up carried on, without reference to the health or comfort of the inhabitants, with sole reference to the highest possible profit on the principle that *no hole is so bad but that some poor creature must take it who can pay for nothing better*.

[. . .]

It may not be out of place to make some general observations just here as to the customary construction of working-men's quarters in Manchester. We have seen how in the Old Town pure accident determined the grouping of the houses in general. Every house is built without reference to any other, and the scraps of space between them are called courts for want of another name. In the somewhat newer portions of the same quarter, and in other working- men's quarters, dating from the early days of industrial activity, a somewhat more orderly arrangement may be found. The space between two streets is divided into more regular, usually square courts.

These courts were built in this way from the beginning, and communicate with the streets by means of covered passages. If the totally planless construction is injurious to the health of the workers by preventing ventilation, this method of shutting them up in courts surrounded on all sides by buildings is far more so. The air simply cannot escape; the chimneys of the houses are the sole drains for the imprisoned atmosphere of the courts, and they serve the purpose only so long as fire is kept burning. Moreover, the houses surrounding such courts are usually built back to back, having the rear wall in common; and this alone suffices to prevent any sufficient through ventilation. And, as the police charged with care of the streets does not trouble itself about the condition of these courts, as everything quietly lies where it is thrown, there is no cause for wonder at the filth and heaps of ashes and offal to be found here. I have been in courts, in Millers Street, at least half a foot below the level of the thoroughfare, and without the slightest drainage for the water that accumulates in them in rainy weather! More recently another different method of building was adopted, and has now become general. Working-men's cottages are almost never built singly, but always by the dozen or score; a single contractor building up one or two streets at a time. These are then arranged as follows: One front is formed of cottages of the best class, so fortunate as to possess a back door and small court, and these command the highest rent. In the rear of these cottages runs a narrow alley, the back street, built up at both ends, into which either a narrow roadway or a covered passage leads from one side. The cottages which face this back street command least rent, and are most neglected. These have their rear walls in common with the third row of cottages which face a second street, and command less rent than the first row and more than the second. The streets are laid out somewhat as in [Figure 2].

By this method of construction, comparatively good ventilation can be obtained for the first row of cottages, and the third row is no worse off than in the former method. The middle row, on the other hand, is at least as badly ventilated as the houses in the courts, and the back street is always in the same filthy, disgusting condition as they. The contractors prefer this method because it saves them space, and furnishes the means of fleecing better-paid workers through the higher rents of the cottages in the first and third rows. These three different forms of cottage building are found all over Manchester and throughout Lancashire and Yorkshire, often mixed up together, but usually separate enough to indicate the relative age of parts of towns. The third system, that of the back alleys, prevails largely in the great working-men's district east of St. George's Road and Ancoats Street, and is the one most often found in the other working-men's quarters of Manchester and its suburbs.

[. . .]

Such are the various working-people's quarters of Manchester as I had occasion to observe them personally during twenty months. If we briefly formulate the result of our wanderings, we must admit that 350,000 working-people of Manchester and its environs live, almost all of them, in wretched, damp, filthy cottages, that the streets which surround them are usually in the most miserable and filthy condition, laid out without the slightest reference to ventilation, with reference solely to the profit secured by the contractor. In a word, we must confess that in the working-men's dwellings of Manchester, no cleanliness, no convenience, and consequently no comfortable family life is possible; that in such dwellings only a physically degenerate race, robbed of all humanity, degraded, reduced morally and physically to bestiality, could feel comfortable and at home.

[. . .]

To sum up briefly the facts thus far cited. The great towns are chiefly inhabited by working-people, since in the best case there is one bourgeois for two workers, often for three, here and there for four; these workers have no property whatsoever of their own, and live wholly upon wages, which usually go from hand to mouth. Society, composed wholly of atoms,

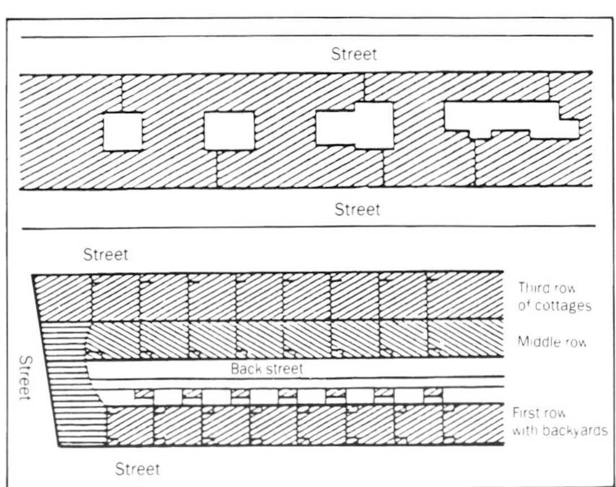

Figure 2

does not trouble itself about them; leaves them to care for themselves and their families, yet supplies them no means of doing this in an efficient and permanent manner. Every working-man, even the best, is therefore constantly exposed to loss of work and food, that is, to death by starvation, and many perish in this way. The dwellings of the workers are everywhere badly planned, badly built, and kept in the worst condition, badly ventilated, damp, and unwholesome. The inhabitants are confined to the smallest possible space, and at least one family usually sleeps in each room. The interior arrangement of the dwellings is poverty-stricken in various degrees, down to the utter absence of even the most necessary furniture. The clothing of the workers, too, is generally scanty, and that of great multitudes is in rags. The food is, in general, bad; often almost unfit for use, and in many cases, at least at times, insufficient in quantity, so that, in extreme cases, death by starvation results. Thus the working-class of the great cities offers a graduated scale of conditions in life, in the best cases a temporarily endurable existence for hard work and good wages, good and endurable, that is, from the worker's standpoint; in the worst cases, bitter want, reaching even homelessness and death by starvation. The average is much nearer the worst case than the best. And this series does not fall into fixed classes, so that one can say, this fraction of the working-class is well off, has always been so, and remains so. If that is the case here and there, if single branches of work have in general an advantage over others, yet the condition of the workers in each branch is subject to such great fluctuations that a single working-man may be so placed as to pass through the whole range from comparative comfort to the extremest need, even to death by starvation, while almost every English working-man can tell a tale of marked changes of fortune.

W. E. B. DU BOIS

"The Negro Problems of Philadelphia," "The Question of Earning a Living" and "Color Prejudice"

from *The Philadelphia Negro* (1899)

Editors' introduction William Edward Burghardt Du Bois (1868–1963) was one of the preeminent intellectuals of his generation. As a professor, editor, author, novelist, playwright, and politician he made notable contributions in history, sociology, ethnic studies, literature, politics, and other fields. A brilliant student, Du Bois excelled at Fisk University in Nashville, Tennessee, the University of Berlin, where he studied with the great sociologist Max Weber, and at Harvard University, where in 1895 he obtained the first Ph.D. degree Harvard awarded to an African-American.

Du Bois defies easy classification. He was always an independent and critical thinker. During his long and varied career he was a pan-Africanist who advocated solidarity among Black Africans and Blacks elsewhere in the world; a radical pacifist who was indicted, tried, and acquitted as an unregistered foreign agent during the McCarthy era for circulating the Stockholm peace plan; a humanist who wrote novels and plays and published many of the writers of the "Harlem Renaissance"; a civil rights leader who founded the NAACP's publication *Crisis* in 1910 and served as its influential editor until 1934; a writer of children's books which taught Black pride; and a world political figure who urged United Nations protection for Black Americans as a nation within a nation. Du Bois joined the Communist Party at age 93 and became a Ghanaian citizen just before his death in 1963.

At the time that Du Bois completed his education, Philadelphia had the largest and oldest settlement of African-Americans in the northern United States. The settlement house movement was underway, and some well-intentioned Philadelphians were concerned to understand "the Negro problem" and to help the many poor Blacks in the city. Two wealthy leaders of Philadelphia society suggested a study of Negroes in the Seventh Ward, the city's Black ghetto.

Du Bois was given a one-year appointment as an assistant instructor in the Sociology Department at the University of Pennsylvania. Living with his bride of three months in one room over a cafeteria in the worst part of Philadelphia's worst Black ghetto, with no contact with students and little with faculty, Du Bois wrote *The Philadelphia Negro* from which the following selection is taken. He was only 31 when his monumental study was published.

While Du Bois found many problems in Philadelphia's African-American community in the 1890s, there was work available for able bodied laborers, no evidence of drug use, substantial homeownership, middle and upper income craftspeople, businessmen, and professionals to serve the community and act as role models, and little Black-on-Black violent crime. This is in marked contrast to William Julius Wilson's description of poor Black ghetto areas of Chicago in the 1980s (p. 112). Wilson describes "underclass" ghettos in Chicago consisting almost entirely of renters (many in public housing), with very few employed residents, extremely high concentrations of single-parent families, welfare dependency, drug use, and violent crime.

Ethnographic studies by sociologists and anthropologists often shed light on variations within communities that are viewed as homogenous by outsiders. While white Philadelphians who never visited the Seventh Ward tended to view the area as homogenous and all African-Americans as similar, Du Bois found a physical and social structure within the neighborhood – alleys peopled by criminals, loafers, and prostitutes separate from streets of the working poor and still other streets where an established group of Black middle class homeowners lived.

In addition to *The Philadelphia Negro* (Philadelphia: University of Pennsylvania Press, 1899) from which the following selection is taken, Du Bois's writings include *Suppression of the Slave Trade to the United States of America* (New York: Longmans, Green and Co., 1896), *Souls of Black Folk* (Chicago: A. C. McClurg and Co., 1903), *The Negro* (New York: Henry Holt and Co., 1915), *Black Reconstruction* (New York: Harcourt, Brace, 1935), *The World and Africa* (New York: The Viking Press, 1947), and many other works.

For more by and about W. E. B. Du Bois see *The Autobiography of W. E. B. Du Bois* (New York: International Publishers, 1968), Francis L. Broderick, *W. E. B. Du Bois* (Palo Alto, CA: Stanford University Press, 1959), Walter Wilson (ed.), *The Selected Writings of W. E. B. Du Bois* (New York: New American Library, 1970), Henry Lee Moon, *The Emerging Thought of W. E. B. Du Bois* (New York: Simon and Schuster, 1972), Marable Manning, *W. E. B. DuBois, Black Radical Democrat* (Boston: Twayne Publishers, 1986), and Patricia and Fredrick McKissack, *W. E. B. Du Bois* (New York: Franklin Watts, 1990).

4. THE NEGRO PROBLEMS OF PHILADELPHIA

In Philadelphia, as elsewhere in the United States, the existence of certain peculiar social problems affecting the Negro people are plainly manifest. Here is a large group of people – perhaps forty- five thousand, a city within a city – who do not form an integral part of the larger social group. This in itself is not altogether unusual; there are other unassimilated groups: Jews, Italians, even Americans; and yet in the case of the Negroes the segregation is more conspicuous, more patent to the eye, and so intertwined with a long historic evolution, with peculiarly pressing social problems of poverty, ignorance, crime and labor, that the Negro problem far surpasses in scientific interest and social gravity most of the other race or class questions.

The student of these questions must first ask, What is the real condition of this group of human beings? Of whom is it composed, what sub-groups and classes exist, what sort of individuals are being considered? Further, the student must clearly recognize that a complete study must not confine itself to the group, but must specially notice the environment; the physical environment of city, sections and houses, the far mightier social environment – the surrounding world of custom, wish, whim and thought which envelops this group and powerfully influences its social development.

[. . .]

The Seventh Ward starts from the historic center of Negro settlement in the city, South Seventh street and Lombard, and includes the long narrow strip, beginning at South Seventh and extending west, with South and Spruce streets as boundaries, as far as the Schuylkill River. The colored population of this ward numbered 3,621 in 1860, 4,616 in 1870, and 8,861 in 1890. It is a thickly populated district of varying character; north of it is the residence and business section of the city; south of it a middle class and workingmen's residence section; at the east end it joins Negro, Italian and Jewish slums; at the west end, the wharves of the river and an industrial section separating it from the grounds of the University of Pennsylvania and the residence section of West Philadelphia.

Starting at Seventh street and walking along Lombard, let us glance at the general character of the ward. Pausing a moment at the corner of Seventh and Lombard, we can at a glance view the worst Negro slums of the city. The houses are mostly brick, some wood, not very old, and in general uncared for rather than dilapidated. The blocks between Eighth, Pine, Sixth, and South have for many decades been the center of Negro population. Here the riots of the thirties took place, and here once was a depth of poverty and degradation almost unbelievable. Even today there are many evidences of degradation . . . The alleys near, as Ratcliffe street, Middle alley, Brown's court, Barclay street, etc., are haunts of noted criminals, male and female, of gamblers and prostitutes, and at the same time of many poverty-stricken people, decent but not energetic. There is an abundance of political clubs, and nearly all the houses are practically lodging houses, with a miscellaneous and shifting population. The corners, night and day, are filled with Negro loafers – able-bodied young men and women, all cheerful, some with good natured, open faces, some with traces of crime and excess, a few pinched with poverty. They are mostly gamblers, thieves and prostitutes, and few have fixed and steady occupation of any kind. Some are stevedores, porters, laborers and laundresses. On its face this slum is noisy and dissipated, but not brutal, although now and then highway robberies and murderous assaults in other parts of the city are traced to its denizens. Nevertheless a stranger can usually walk about here day and night with little fear of being molested if he be not too inquisitive.

Passing up Lombard, beyond Eighth, the atmosphere suddenly changes, because these next two blocks have few alleys and the residences are good-sized and pleasant. Here some of the best Negro families of the ward live. Some are wealthy in a small way, nearly all are Philadelphia born, and they represent an early wave of emigration from the old slum section . . .

[. . .]

21. THE QUESTION OF EARNING A LIVING

For a group of freedmen the question of economic survival is the most pressing of all questions; the problem as to how, under the circumstances of modern life, any group of people can earn a decent living, so as to maintain their standard of life, is not always easy to answer. But when the question is complicated by the fact that the group has a low degree of efficiency on account of previous training; is in competition with well-trained, eager and often ruthless competitors; is more or less handicapped by a somewhat wide-reaching discrimination; and finally is seeking not merely to maintain a standard of living but steadily to raise it to a higher plane – such a situation presents baffling problems to the sociologist and philanthropist.

Of the men 21 years of age and over, there were in gainful occupations, the following:

47. COLOR PREJUDICE

Incidentally throughout this study the prejudice against the Negro has been again and again mentioned. It is time now to reduce this somewhat indefinite term to something tangible. Everybody speaks of the matter, everybody knows that it exists, but in just what form it shows itself or how influential it is few agree. In the Negro's mind, color prejudice in Philadelphia is that widespread feeling of dislike for his blood, which keeps him and his children out of decent employment, from certain public conveniences and amusements, from hiring houses in many sections, and in general, from being recognized as a man. Negroes regard this prejudice as the chief cause of their present unfortunate condition. On the other hand most white people are quite unconscious of any such powerful and vindictive feeling; they regard color prejudice as the easily explicable feeling that intimate social intercourse with a lower race is not only undesirable but impractical if our present standards of culture are to be maintained, and although they are aware that some people feel the aversion more intensely than others, they cannot see how such a feeling has much influence on the real situation or alters the social condition of the mass of Negroes.

In the learned professions		61	2.0	per cent
Conducting business on their own account		207	6.5	
In the skilled trades		236	7.0	
Clerks, etc.		159	5.0	
Laborers, better class	602			
Laborers, common class	852	1454	45.0	
Servants		1079	34.0	
Miscellaneous		11	0.5	
		3207	100	per cent
Total male population 21 and over		3850		

Taking the occupations of women 21 years of age and over, we have:

Domestic servants	1262	37.0	per cent
Housewives and day laborers	937	27.0	
Housewives	568	17.0	
Day laborers, maids, etc.	297	9.0	
In skilled trades	221	6.0	
Conducting businesses	63	2.0	
Clerks, etc.	40	1.0	
Learned professions	37	1.0	
	3425	100	per cent
Total female population 21 and over	3740		

As a matter of fact, color prejudice in this city is something between these two extreme views: it is not today responsible for all, or perhaps the greater part of the Negro problems, or of the disabilities under which the race labors; on the other hand it is a far more powerful social force than most Philadelphians realize. The practical results of the attitude of most of the inhabitants of Philadelphia towards persons of Negro descent are as follows:

1. As to getting work:

   No matter how well trained a Negro may be, or how fitted for work of any kind, he cannot in the ordinary course of competition hope to be much more than a menial servant.

   He cannot get clerical or supervisory work to do save in exceptional cases.

   He cannot teach save in a few of the remaining Negro schools.

   He cannot become a mechanic except for small transient jobs, and cannot join a trades union.

   A Negro woman has but three careers open to her in this city: domestic service, sewing, or married life.

2. As to keeping work:

   The Negro suffers in competition more severely than white men.

   Change in fashion is causing him to be replaced by whites in the better-paid positions of domestic service.

   Whim and accident will cause him to lose a hard-earned place more quickly than the same things would affect a white man.

   Being few in number compared with the whites the crime or carelessness of a few of his race is easily imputed to all, and the reputation of the good, industrious, and reliable suffer thereby.

   Because Negro workmen may not often work side by side with white workmen, the individual black workman is rated not only by his own efficiency, but by the efficiency of a whole group of black fellow workmen which may often be low.

   Because of these difficulties which virtually increase competition in his case, he is forced to take lower wages for the same work than white workmen.

3. As to entering new lines of work:

   Men are used to seeing Negroes in inferior positions; when, therefore, by any chance a Negro gets in a better position, most men immediately conclude that he is not fitted for it, even before he has a chance to show his fitness.

   If, therefore, he set up a store, men will not patronize him.

   If he is put into public position men will complain.

   If he gain a position in the commercial world, men will quietly secure his dismissal or see that a white man succeeds him.

4. As to his expenditure:

The comparative smallness of the patronage of the Negro, and the dislike of other customers, makes it usual to increase the charges or difficulties in certain directions in which a Negro must spend money.

He must pay more house-rent for worse houses than most white people pay.

He is sometimes liable to insult or reluctant service in some restaurants, hotels and stores, at public resorts, theatres and places of recreation; and at nearly all barber shops.

5. As to his children:

The Negro finds it extremely difficult to rear children in such an atmosphere and not have them either cringing or impudent: if he impresses upon them patience with their lot, they may grow up satisfied with their condition; if he inspires them with ambition to rise, they may grow to despise their own people, hate the whites, and become embittered with the world.

His children are discriminated against, often in public schools.

They are advised when seeking employment to become waiters and maids.

They are liable to species of insult and temptation peculiarly trying to children.

6. As to social intercourse:

In all walks of life the Negro is liable to meet some objection to his presence or some discourteous treatment; and the ties of friendship or memory seldom are strong enough to hold across the color line.

If an invitation is issued to the public for any occasion, the Negro can never know whether he would be welcomed or not; if he goes he is liable to have his feelings hurt and get into unpleasant altercation; if he stays away, he is blamed for indifference.

If he meet a lifelong white friend on the street, he is in a dilemma; if he does not greet the friend he is put down as boorish and impolite; if he does greet the friend he is liable to be flatly snubbed.

If by chance he is introduced to a white woman or man, he expects to be ignored on the next meeting, and usually is.

White friends may call on him, but he is scarcely expected to call on them, save for strictly business matters.

If he gain the affections of a white woman and marry her he may invariably expect that slurs will be thrown on her reputation and on his, and that both his and her race will shun their company. When he dies he cannot be buried beside white corpses.

7. The result:

Any one of these things happening now and then would not be remarkable or call for especial comment; but when one group of people suffer all these little differences of treatment and discriminations and insults continually, the result is either discouragement, or bitterness, or over-sensitiveness, or recklessness. And a people feeling thus cannot do their best.

Presumably the first impulse of the average Philadelphian would be emphatically to deny any such marked and blighting discrimination as the above against a group of citizens in this metropolis. Every one knows that in the past color prejudice in the city was deep and passionate; living men can remember when a Negro could not sit in a street car or walk many streets in peace. These times have passed, however, and many imagine discrimination against the Negro has passed with them. Careful inquiry will convince any such one of his error. To be sure a colored man to-day can walk the streets of Philadelphia without personal insult; he can go to theatres, parks and some places of amusement without meeting more than stares and discourtesy; he can be accommodated at most hotels and restaurants, although his treatment in some would not be pleasant. All this is a vast advance and augurs much for the future. And yet all that has been said of the remaining discrimination is but too true.

During the investigation of 1896 there was collected a number of actual cases, which may illustrate the discriminations spoken of. So far as possible these have been sifted and only those which seem undoubtedly true have been selected:

I. As to getting work

It is hardly necessary to dwell upon the situation of the Negro in regard to work in the higher walks of life: the white boy may start in the lawyer's office and work himself into a lucrative practice; he may serve a physician as office boy or enter a hospital in a minor position, and have his talent alone between him and affluence and fame; if he is bright in school, he may make his mark in a university, become a tutor with some time and much inspiration for study, and eventually fill a professor's chair. All these careers are at the very outset closed to the Negro on account of his color; what lawyer would give even a minor case to a Negro assistant? What university would appoint a promising young Negro as tutor? Thus the young white man starts in life knowing that within some limits and barring accidents, talent and application will tell. The young Negro starts knowing that on all sides his advance is made difficult if not wholly shut off by his color. Let us come, however, to ordinary occupations which concern more nearly the mass of Negroes. Philadelphia is a great industrial and business center with thousands of foremen, managers and clerks – the lieutenants of industry who direct its progress. They are paid for thinking and for skill to direct, and naturally such positions are coveted because they are well paid, well thought-of and carry some authority. To such positions Negro boys and girls may not aspire no matter what their qualifications. Even as teachers and ordinary clerks and stenographers they find almost no openings. Let us note some actual instances:

A young woman who graduated with credit from the Girls Normal School in 1892 has taught in the kindergarten, acted as substitute, and waited in vain for a permanent position. Once she was allowed to substitute in a school with white teachers; the principal commended her work, but when the permanent appointment was made a white woman got it.

A girl who graduated from a Pennsylvania high school and from a business college sought work in the city as a stenographer and type-writer. A prominent lawyer undertook to find her a position; he went to friends and said, "Here is a girl that does excellent work and is of good character; can you not give her work?" Several immediately answered yes. "But," said the lawyer, "I will be perfectly frank with you and tell you she is colored;" and not in the whole city could he find a man willing to employ her. It happened, however, that the girl was so light in complexion that few not knowing would have suspected her descent. The lawyer therefore gave her temporary work in his own office until she found a position outside the city. "But," said he, "to this day I have not dared to tell my clerks that they worked beside a Negress." Another woman graduated from the high school and the Palmer College of Shorthand, but all over the city has met with nothing but refusal of work.

Several graduates in pharmacy have sought three years' required apprenticeship in the city and in only one case did one succeed, although they offered to work for nothing. One young pharmacist came from Massachusetts and for weeks sought in vain for work here at any price; "I wouldn't have a darky to clean out my store, much less to stand behind the counter," answered one druggist.

A colored man answered an advertisement for a clerk in the suburbs. "What do you suppose we'd want of a nigger?" was the plain answer. A graduate of the University of Pennsylvania in mechanical engineering, well recommended, obtained work in the city, through an advertisement, on account of his excellent record. He worked a few hours and then was discharged because he was found to be colored. He is now a waiter at the University Club, where his white fellow graduates dine. Another young man attended Spring Garden Institute and studied drawing for lithography. He had good references from the institute and elsewhere, but application at the five largest establishments in the city could secure him no work. A telegraph operator has hunted in vain for an opening, and two graduates of the Central High School have sunk to menial labor. "What's the use of an education?" asked one. Mr. A—— has elsewhere been employed as a traveling salesman. He applied for a position here by letter and was told he could have one. When they saw him they had no work for him.

Such cases could be multiplied indefinitely. But that is not necessary; one has but to note that, notwithstanding the acknowledged ability of many colored men, the Negro is conspicuously absent from all places of honor, trust, emolument, as well as from those of respectable grade in commerce and industry.

Even in the world of skilled labor the Negro is largely excluded. Many would explain the absence of Negroes from higher vocations by saying that while a few may now and then be found competent, the great mass are not fitted for that sort of work and are destined for some time to form a laboring class. In the matter of the trades, however, there can be raised no serious question of ability; for years the Negroes filled satisfactorily the trades of the city, and to-day in many parts of the South they are still prominent. And yet in Philadelphia a determined prejudice, aided by public opinion, has succeeded nearly in driving them from the field:

A——, who works at a bookbinding establishment on Front street, has learned to bind books and often does so for his friends. He is not allowed to work at the trade in the shop, however, but must remain a porter at a porter's wages.

B—— is a brushmaker; he has applied at several establishments, but they would not even examine his testimonials. They simply said: "We do not employ colored people."

C—— is a shoemaker; he tried to get work in some of the large department stores. They "had no place" for him.

D—— was a bricklayer, but experienced so much trouble in getting work that he is now a messenger.

E—— is a painter, but has found it impossible to get work because he is colored.

F—— is a telegraph line man, who formerly worked in Richmond, Va. When he applied here he was told that Negroes were not employed.

G—— is an iron puddler, who belonged to a Pittsburgh union. Here he was not recognized as a union man and could not get work except as a stevedore.

H—— was a cooper, but could get no work trials, and is now a common laborer.

I—— is a candy-maker, but has never been able to find employment in the city; he was always told the white help would not work with him.

J—— is a carpenter; he can only secure odd jobs or work where only Negroes are employed.

K—— was an upholsterer, but could get no work save in the few colored shops which had workmen; he is now a waiter on a dining car.

L—— was a first-class baker; he applied for work some time ago near Green street and was told shortly, "We don't work no niggers here."

[. . .]

HERBERT J. GANS

"Levittown and America"

from *The Levittowners* (1967)

Editors' introduction Herbert Gans is the author of two of the most fascinating and influential books of urban sociology ever published. *The Urban Villagers* is a brilliant study of the Italian-American immigrant community of Boston's North End neighborhood. *The Levittowners*, from which the following selection is taken, is Gans's analysis of post-World War II tract-home suburbia. Both are examples of the participant-observer methodology at its best. While never losing sight of objective scholarship, Gans lets the reader see and experience urban communities from the inside out.

Robert Fishman (p. 77) describes affluent suburbs surrounding mid-nineteenth-century Manchester, England, and Sam Bass Warner, Jr. (p. 69) documents how suburban the entire Los Angeles region had become by the 1920s. However, it was the period immediately following World War II that saw the massive middle-class suburbanization that established many of the suburbs we know today.

East Coast developer William Levitt capitalized on the enormous pent up demand for entry-level single-family suburban homes for returning servicemen and their new families. He combined mass production techniques and fantasy to create massive "Levittowns" in New York and Pennsylvania, which defined the post-war version of the American dream. Levittowns were built on vast tracts of exurban land. They were filled with spanking new, cute, affordable mass-produced imitation Cape Cod colonial homes like those in Plate 6's 1947 streetscape of Levittown, New York. In the late 1940s and 1950s young families lined up to buy into Levitt's version of the American dream and other developers borrowed freely from Levitt's ideas. But architects, social critics, and city planners were less enthusiastic about what this new physical and social form represented.

The suburban developments of the 1940s, 1950s, and 1960s, in America and elsewhere, gave birth to a massive literature, much of it critical. Damned as automobile-dependent and socially/racially segregated, the post-World War II suburbs were called "sprawl" and stigmatized as "anti-cities" (to use Lewis Mumford's term) contributing to a stifling social conformity and cultural mediocrity. Titles such as John Keats's *The Crack in the Picture Window* (1956), Richard Gordon's *The Split-Level Trap* (1961), Kenneth Jackson's *Crabgrass Frontier* (1985), and Mark Baldassare's *Trouble in Paradise* (1986) capture the tone of much of the commentary on the suburban way of life. One prominent feminist critic – Dolores Hayden in *Redesigning the American Dream* (1984) – has charged that the first Levittown was built specifically for "the returning veteran, the beribboned male war hero who wanted his wife to stay home."

Given the overwhelming anti-suburban bias of most of these analyses, Gans's view of Levittown is remarkable in that it steadfastly rejects the notion that an easily definable "suburban way of life" even exists, much less that suburbia represents a distinctly new kind of socio-cultural place. On the contrary, Gans found the Levittowners to be much like hard-working middle-class people anywhere in the world engaged in the process of adapting their needs to new social and environmental situations. To be sure,

Levittown had its problems. Gans analyzes them in some detail and makes clear that Levittown is no utopia. But neither is it a spiritual wasteland.

As a classic analysis of life in a contemporary suburban development, Gans's *The Levittowners* should be compared and contrasted with other views of suburbia (see above) and to such anti-suburban celebrators of inner-city neighborhoods as Jane Jacobs (p. 106). Gans was prophetic in 1967 about the probability that "yet another ring of suburban communities will spring up around American cities" in the (then) near future. Such prescience points directly to the new communities described by Joel Garreau in *Edge City* (New York: Anchor, 1992) and to Robert Fishman's "technoburbs" (p. 77).

This selection is from *The Levittowners* (New York: Pantheon, 1967). Gans's other great work of urban sociology is *The Urban Villagers* (New York: Free Press, 1962). Related books by Gans include *People, Plans, and Policies: Essays on Poverty, Racism, and other National Urban Problems* (New York: Columbia University Press; Russell Sage Foundation, 1993), *The War Against the Poor: The Underclass and Antipoverty Policy* (New York: Basic Books, 1995), and *Middle American Individualism: The Future of Liberal Democracy* (New York: Free Press, 1988).

For more on Levittown see Barbara M. Kelley, *Expanding the American Dream: Building and Rebuilding Levittown* (Buffalo: State University of New York Press, 1993) and W. D. Wetherell, *The Man Who Loved Levittown* (Pittsburgh: University of Pittsburgh Press, 1985).

Other notable books on suburbia include Kenneth T. Jackson, *Crabgrass Frontier: The Suburbanization of the United States* (London and New York: Oxford University Press, 1985), John R. Stilgoe, *Borderland: Origins of the American Suburb, 1820–1939* (New Haven, CT: Yale University Press, 1990), Barbara M. Kelley, *Suburbia Re-examined* (New York: Greenwood Press, 1989), and Philip Langdon, *A Better Place to Live: Reshaping the American Suburb* (Amherst: University of Massachusetts Press, 1994).

CONFLICT, PLURALISM, AND COMMUNITY

Although a part of my study was concerned with the possibilities of change and innovation, I do not mean to suggest that Levit town is badly in need of either. The community may displease the professional city planner and the intellectual defender of cosmopolitan culture, but perhaps more than any other type of community, Levittown permits most of its residents to be what they want to be – to center their lives around the home and the family, to be among neighbors whom they can trust, to find friends to share leisure hours, and to participate in organizations that provide sociability and the opportunity to be of service to others.

That Levittown has its faults and problems is undeniable . . . physical and social isolation, familial and governmental financial problems, insufficient public transportation, less than perfect provision of public services, inadequate decision-making and feedback processes, lack of representation for minorities and over-representation for the builder, and the entire array of familial and individual problems common to any population. Many of them can be traced back to three basic shortcomings, none distinctive to Levittown or the Levittowners.

One is the difficulty of coping with conflict. Like the rest of the country, Levittown is beset with conflict: class conflict between the lower middle class group and the smaller working and upper middle class groups; generational conflict between adults, children, adolescents, and the elderly. The existence of conflict is no drawback, but the way conflict is handled leaves much to be desired. Levittowners, like other Americans, do not really accept the inevitability of conflict. Insisting that a consensus is possible, they only exacerbate the conflict, for each group demands that the other conform to its values and accept its priorities. When power is a valuable prize and resources are scarce, such a perspective is understandable, but in Levittown the exercise of power is not an end in itself for most people; they want it mainly to control the allocation of resources. Since resources are not so scarce, however, the classes and age groups could resolve their conflicts more constructively than they do, giving each group at least some of what it wants. If the inevitability of conflicting interests were accepted, differences might be less threatening, and this would make it easier to reach the needed compromises. I am not sanguine that this will happen, for if people think resources are scarce, they act as if they are scarce, and will not pay an extra $20 a year in taxes to implement minority demands. Even so, conditions to make viable compromises happen are more favorable in Levittown than in larger or poorer communities.

The second shortcoming, closely related to the first, is the inability to deal with pluralism. People have not recognized the diversity of American society, and they are not able to accept other life styles. Indeed, they cannot handle conflict because they cannot accept pluralism. Adults are unwilling to tolerate adolescent culture, and vice versa. Lower middle class people oppose the ways of the working class and upper middle class, and each of these groups is hostile to the other two. Perhaps the inability to cope with pluralism is greater in Levittown than elsewhere because it is a community of young families who are raising children. Children are essentially asocial and unacculturated beings, easily influenced by new ideas. As a result, their parents feel an intense need to defend familial values; to make sure that their children grow up according to parental norms and not by those of their playmates from another class. The need to shield the children from what are considered harmful influences begins on the block, but it is translated into the conflict over the school, the definitional struggles within the voluntary associations whose programs affect the socialization of children, and, ultimately, into political conflicts. Each group wants to put its cultural stamp on the organizations and institutions that are the community, for otherwise the family and its culture are not safe. In a society in which extended families are unimportant and the nuclear family cannot provide the full panoply of personnel and activities to hold children in the family culture, parents must use community institutions for this purpose, and every portion of the community therefore becomes a battleground for the defense of familial values.

This thesis must not be exaggerated, for much of the conflict is, as it has always been, between the haves and the have-nots. Even if Levittown's median income is considerably above the national average even for white families, no one feels affluent enough to let other people determine how their own income should be spent. Most of the political conflict in the community rages over how much of the family income should be given over to the community, and then, how it should be used. In fact consensus about municipal policies and expenditures exists only about the house. Because many Levittowners are first-time homeowners, they are especially eager to protect that home against loss of value, both as property and as status image. But every class has its own status image and its own status fears. Working class people do not want to be joined by lower class neighbors or to be forced to adopt middle class styles. Lower middle class people do not want more working class neighbors or to be forced to adopt cosmopolitan styles, and upper middle class people want neither group to dominate them. These fears are not, as commonly thought, attributes of status-seeking, for few Levittowners are seeking higher status: they are fears about self-image. When people reject pluralism, they do so because accepting the viability of other ways of living suggests that their own is not as absolute as they need to believe. The outcome is the constant search for compatible people and the rejection of those who are different.

When the three class groups – not to mention their subgroupings and yet other groups with different values – must live together and share a common government, every group tries to make sure that the institutions and facilities which serve the entire community maintain its own status and culture, and no one is happy when another group wins. If working class groups can persuade the Township Committee to allocate funds for a firehouse, middle class groups unite in a temporary coalition to guarantee that a library is also established. When the upper middle class group attempts to influence school policy to shape education to its standard, lower middle class residents raise the specter of Levittown aping Brookline and Scarsdale, while working class people become fearful that the schools will neglect discipline or that taxes will rise further. Consequently, each group seeks power to prevent others from shaping the institutions that must be shared. They do not seek power as an end in itself, but only to guarantee that their priorities will be met by the community. Similarly, they do not demand lower taxes simply for economic reasons (except for those few really hard pressed) but in order to be sure that community institutions are responsible to their familial values and status needs. Obviously, power sought for these ends is hard to share, and decisions for levying and allocating public funds are difficult to compromise.

The third shortcoming of the community, then, is the failure to establish a meaningful relationship between home and community and to reconcile class-cultural diversity with government and the provision of public services. Levittowners, like other Americans, not only see government as a parasite and public services as a useless expenditure of funds better spent privately, but they do not allow government to adapt these services to the diversity among the residents. Government is committed to the establishment of a single (and limited) set of public services, and its freedom to do otherwise is restricted by legislation and, of course, by American tradition.

Government has always been a minor supplier of services basic to everyday life, and an enemy whose encroachment on private life must be resisted. The primary source of this conception is the historic American prejudice against public services, which stems in part from the rural tradition of the individual and his family as a self-sufficient unit, but which is perpetuated by contemporary cultural values and made possible by the alliance which enables at least middle class families to live with only minimal dependence on local government. The bias against public services does not interfere with their use, however, but only with their financing and their extension and proliferation. Nor does it lead Levittowners to reject government outright, but only to channel it into a few limited functions. Among these, the primary one is the protection of the home against diversity.

Government thus becomes a defense agency, to be taken over by one group to defend itself against others in and out of the community. The idea that it could have positive functions, such as the provision of facilities to make life richer and more comfortable, is resisted, for every new governmental function is seen first as an attempt by one community group to increase its dominance over others. Of course, these attempts are rarely manifest, for the political dialogue deals mainly with substantive matters, but when Levittowners spoke against a proposal, they were reacting principally against those who proposed it rather than against its substance.

Until government can tailor its actions to the community's diversity, and until people can accept the inevitability of conflict and pluralism in order to give government that responsibility, they will prefer to spend their money for privately and commercially supplied services. Unlike city hall, the marketplace is sensitive to diversities among the customers and does not require them to engage in political conflict to get what they want. Of course not all people can choose the marketplace over city hall, but Levittowners are affluent enough to do so. Moreover, until parents have steered their children safely into their own class and culture – or have given up trying – they are likely to seek out relatively homogeneous communities and small ones, so that they have some control over government's inroads against personal and familial autonomy. This not only maintains the sovereignty of hundreds of small local governments but also contributes to the desire to own a house and a free-standing one.

LEVITTOWN AS AMERICA

The strengths and weaknesses of Levittown are those of many American communities, and the Levittowners closely resemble other young middle class Americans. They are not America, for they are not a numerical majority of the population, but they represent the major constituency of the latest and most powerful economic and political institutions in American society – the favored customers and voters whom these seek to attract and satisfy. Upper middle class

Americans may spend more per capita and join more groups, but they are fewer in number than the lower middle classes. Working and lower class people are more numerous but they have less money and power; and people over 40, who still outnumber young adults, are already committed to most of the goods, affiliations, and ideas they will need in their lifetime.

Even so, Levittowners are not really members of the national society, or for that matter, of a mass society. They are not apathetic conformists ripe for takeover by a totalitarian elite or corporate merchandiser; they are not conspicuous consumers and slaves to sudden whims of cultural and political fashion; they are not even organization men or particularly other-directed personalities. Clearly, inner-directed strivers are a minority in Levittown, and tradition-directed people would not think of moving to a new community of strangers, but most people maintain a balance between inner personal goals and the social adjustment necessary to live with neighbors and friends that, I suspect, is prevalent all over lower middle class America . . . Although ethnic, religious, and regional differences are eroding, the never-ending conflicts over other differences are good evidence that Levittowners are far from becoming mass men.

Although they are citizens of a national polity and their lives are shaped by national economic, social, and political forces, Levittowners deceive themselves into thinking that the community, or rather the home, is the single most influential unit in their lives. Of course, in one way they are right; it is the place where they can be most influential, for if they cannot persuade the decision-makers, they can influence family members. Home is also the site of maximal freedom, for within its walls people can do what they want more easily than anywhere else. But because they are free and influential only at home, their dependence on the national society ought to be obvious to them. This not being the case, the real problem is that Levittowners have not yet become aware of how much they are a part of the national society and economy.

In viewing their homes as the center of life, Levittowners are still using a societal model that fit the rural America of self-sufficient farmers and the feudal Europe of self-isolating extended families. Yet the critics who argue about the individual versus mass society are also anachronistic: they are still thinking of the individual artist or intellectual who must shield himself from a society which either rejects him or coopts him to produce popular culture. Both Levittowners and critics have to learn that they live in a national society characterized by pluralism and bureaucracy, and that the basic conflict is not between individual (or family) and society, but between the classes (and other interest groups) who live together in a bureaucratized political and cultural democracy. The prime challenge is how to live with bureaucracy; how to use it rather than be used by it; how to obtain individual freedom and social resources from it through political action.

Yet even though Levittowners and other lower middle class Americans continue to be home-centered, they are much more "in the world" than their parents and grandparents were. Those coming out of ethnic working class backgrounds have rejected the "amoral familism" which pits every family against every other in the struggle to survive and the ethnocentrism which made other cultures and even other neighborhoods bitter enemies. This generation trusts its neighbors, participates with them in social and civic activities and no longer sees government as inevitably corrupt. Even working class Levittowners have begun to give up the suspicion that isolated their ancestors from all but family and childhood friends. Similarly, the descendants of rural Protestant America have given up the xenophobia that turned previous generations against the Catholic and Jewish immigrants, they have almost forgotten the intolerant Puritanism which triggered attacks against pleasure and enjoyment, and they no longer fully accept the doctrine of laissez faire that justifies the defense of all individual rights and privileges against others' needs.

These and other changes have come about not because people are now better or more tolerant human beings, but because they are affluent. For the Levittowners, life is not a fight for survival any more; they have been able to move into a community in which income and status are equitably enough distributed so that neighbors are no longer treated as enemies, even if they are still criticized for social and cultural deviance. By any yardstick one chooses, Levittowners treat their fellow residents more ethically and more democratically than did their parents and grandparents.

They also live a "fuller" and "richer" life. Their culture may be less subtle and sophisticated than that of the intellectual, their family life less healthy than that advocated by psychiatrists, and their politics less thoughtful and democratic than the political philosophers' – yet all of these are superior to what prevailed among the working and lower middle classes of past generations.

But beyond these changes, it is striking how little American culture among the Levittowners differs from what de Tocqueville reported in his travels through small-town middle class America a century ago. Of course, he was here before the economy needed an industrial proletariat, but the equality of men and women, the power of the child over his parents, the importance of the voluntary association, the social functions of the church, and the rejection of high culture seem to be holdovers from his time, and so is the adherence to the traditional virtues: individual honesty, thrift, religiously inspired morality, Franklinesque individualism and Victorian prudery. Some Levittowners have retained the values of rural ancestors; some have only begun to practice them as affluence enabled them to give up the values of a survival-centered culture. Still other eternal verities remain: class conflict is as alive as ever, even if the struggle is milder and the have-nots in Levittown have much more than the truly poor. Working class culture continues to flourish, even though its rough edges are wearing smooth and its extended family and public institutions are not brought to the suburbs. Affluence and better education have made a difference, but they have not made the factory worker middle class, any more than college attendance has made lower middle class people cosmopolitan.

What seems to have happened is that improvements and innovations are added to old culture patterns, giving affluent Americans a foot in several worlds. They have more knowledge and a broader outlook than their ancestors, and they enjoy the advantages of technology, but these are superimposed on old ways. While conservative critics rail about technology's dehumanization of modern man, the Levittowners who spend their days programming computers come home at night to practice the very homely and old-fashioned virtues these critics defend. For example, they have television sets, but they watch much the same popular comedies and melodramas their ancestors saw on the nineteenth century stage. The melodramas are less crude and vaudeville is more respectable; the girls dance with covered bosoms, but Ed Sullivan's program is pure vaudeville and *The Jackie Gleason Show* even retains traces of the working class music hall. The overlay of old and new is not all good, of course: the new technology has created methods of war and destruction which the old insularity allows Americans to unleash without much shame or guilt, and some Levittowners may find work less satisfying than their ancestors. But only some, for the majority's parents slaved in exhausting jobs which made them too tired to enjoy the advantages of suburbia even if they could have afforded them. On the whole, however, the Levittowners have only benefitted from the changes in society and economy that have occurred in this century, and if they were not given to outmoded models of social reality, they might feel freer about extending these benefits to less fortunate sectors of American society. But whether people's models are anachronistic or avant-garde, they are rarely willing to surrender their own powers and privileges to others.

SAM BASS WARNER, JR.

"The Megalopolis: 1920–"

from *The Urban Wilderness: A History of the American City* (1972)

Editors' introduction In *The Urban Wilderness*, Sam Bass Warner traces the history of American cities beginning with traditional colonial planning in New England and continuing down to the emergence of Los Angeles as a new kind of metropolitan complex. He writes movingly about what he calls "the neglect of everyday life" in American urban policy, and provides strong support for the Great Society social programs of the 1960s.

The Urban Wilderness is particularly interesting for its balanced treatment of Los Angeles. In the following selection Warner describes how early Los Angeles had become dependent on the automobile. Plate 7 shows developers in the Westlake district of Los Angeles in front of a sign, which proudly proclaims "50 Foot Boulevard Here." They are typical of thousands of small and medium-sized developers who built the auto-dependent communities in the Los Angeles metropolitan region during the period Warner describes.

In *The City in History* (1960), Lewis Mumford had denounced Los Angeles as an automobile-dependent "anti-city" suffering from "metropolitan elephantiasis" and "sprawling gigantism." Later observers of Los Angeles such as Mike Davis in *City of Quartz* (p. 193), Edward Soja in Postmodern Geographies (p. 180), and Norman Klein in *The History of Forgetting: Los Angeles and the Erasure of Memory* (1997) denounced Los Angeles as an obscene nightmare of fabulous wealth side-by-side with hopeless poverty and racist oppression. And literary views of Los Angeles – from Nathaniel West's *Day of the Locust* (1939) to Ridley Scott's film *Blade Runner* (1982) – have been apocalyptic. Warner does not blink at the obvious problems that so many others have pointed out. He finds the racial and class divisions a challenge that cannot be ignored if Americans are to realize the promise of an open, inclusive urban society. But Warner sees in Los Angeles and other cities of the West Coast a "new freedom" that suggests a potential hitherto unknown in the history of American urbanization.

Warner has devoted his life to studying and writing the history of the American city. This selection is from *The Urban Wilderness: A History of the American City* (Berkeley: University of California Press, 1995, originally published by Harper & Row in 1972), a book that many regard as the single best one-volume history of American urban society. Warner's first book, *Streetcar Suburbs: The Process of Growth In Boston, 1870–1900* (Cambridge, MA: Harvard University Press, 1962), examined the growth of the Boston metropolitan region at the end of the nineteenth century. This was followed by *The Private City: Philadelphia in Three Periods of its Growth* (Philadelphia: University of Pennsylvania Press, 1968), a study of Philadelphia's development from colonial outpost to modern metropolis. This body of work placed Warner in a pantheon of great American urban historians that included Arthur Schlesinger, Sr., author of *The Rise of the City, 1878–1898* (1933), Carl Bridenbaugh, author of *Cities in the Wilderness* (London and New York: Oxford University Press, 1971) and *Cities in Revolt* (London and New York: Oxford University Press, 1971), and Richard Wade, author of *The Urban Frontier* (Urbana: University of Illinois Press, 1996).

For more on Los Angeles, consult R. A. Nadeau's *Los Angeles: From Mission to Modern City* (New York: Longman, 1960) and R. Marchand's *The Emergence of Los Angeles: Population and Housing in the City of Dreams* (London: Pion, 1986). Interesting architectural and cultural analyses of Los Angeles can be found in Rayner Banham's *Los Angeles: The Architecture of Four Ecologies* (London: Penguin, 1971) and Dolores Hayden's *The Power of Place: Urban Landscapes as Public History* (Cambridge, MA: MIT Press, 1995).

For analyses of the troubled race relations of Los Angeles, see Mauricio Mazon, *The Zoot-Suit Riots* (Austin: University of Texas Press, 1984), Raphael J. Sonenshein, *Politics in Black and White* (Princeton, NJ: Princeton University Press, 1993), and Mark Baldassare (ed.), *The Los Angeles Riots: Lessons for the Urban Future* (Boulder, CO: Westview, 1994).

Los Angeles, city of war material, swimming pools, and smog, wonderfully exemplifies the urban consequences stemming from the change in structure of the national economy and its institutions. It is par excellence a city of the past half-century. In 1920 the city proper had grown to be the tenth largest in the nation, about the same size as Pittsburgh (Los Angeles, 577,000; Pittsburgh, 588,000), and its metropolitan population had reached almost a million. Thanks largely to the prosperity and land rush of the twenties, the metropolitan area sustained a population of 2,785,000 on the eve of World War II, and the wartime infusion of business and workers raised this figure to 9,475,000 in 1970. Today Los Angeles is the second largest cluster of population and the third largest manufacturing center in the United States (Chicago Consolidated Statistical Area 7,612,000; New York CSA 16,179,000). It has now become the economic capital of the Pacific and the Southwest, the heart of the fast-growing San Diego-San Francisco megalopolis.

Like all great American cities, Los Angeles grew not by accretion of economic functions taken from other cities but by being geographically located in the center of new developments. Chicago rose with the settlement of the Midwest, Los Angeles with the waves of migration to California and the Southwest. Moreover, new resources and industries fired its growth. The railroads, the prairie farms, the forests of Michigan and Wisconsin, and Lake Superior ore had made Chicago a center for transportation, food processing, lumber, steel, and machinery. Similarly oil, a warm sunny climate, and the airlines made Los Angeles the capital of petroleum refining, of the national distribution of fruit and vegetables, and of movies, as well as the focal point of the nation's aircraft, aerospace, and war-research industries. Migrants added banks, stores, and residentiary industries of all kinds, and the local specialties encouraged complementary industries, until by the fifties the city was functioning as a widely diversified metropolis given over to manufacturing, commerce, service, and war production.

During the twenties and thirties, irrigated agriculture, oil discoveries, the motion-picture boom, and waves of Midwestern and Texan migrants seeking a pleasant place to live and work swelled the city's size. Oil revenues in part financed the construction of an ocean port at Long Beach, and favorable rail connections to the Southwest and the East (the city lies a few miles closer to Chicago than San Francisco does) made Los Angeles a preferred site for warehouses and branch plants of national corporations. During the twenties, for example, both Ford and Goodyear built Pacific plants there, and many other firms followed suit. But the city was handicapped by the circumstance that its factories were some two thousand miles from the western edge of the Midwestern manufacturing belt at St. Louis. Therefore the sectors of the metropolitan economy devoted to general machinery and metalworking – sectors of vital importance to a fully elaborated industrial region – did not develop during these years. Los Angeles in 1940, for all its impressive size, was not yet committed to manufacturing.

Thirty years of almost continuous hot and

cold wars ended this anomaly. Tremendous aircraft orders from the federal government not only caused that particular industry to shoot ahead, but federal sponsorship of aerospace research and all sorts of war material fostered supportive manufacturing, until today the city is the only fully diversified manufacturing region outside the belts of the Northeast and Midwest. To be sure, Los Angeles still has its agricultural, aircraft, electronic, and movie specialties, just as Chicago still concentrates on steel, machinery, and printing and New York on garments, leather, printing, electrical equipment, and national offices, but since the 1950s a full range of complementarities has been available in Los Angeles to boost further expansion.

Three special characteristics of the Los Angeles metropolis stand out by comparison with the earlier examples of New York and Chicago: its high degree of spatial freedom, its potential for a more equitable and inclusive class and racial society, and its growth in response to deliberate federal programs.

The land-use and transportation structure of Los Angeles gives glimpses of a more humane environment than we have yet enjoyed. The special factor of the city's social geography is its low density of settlement, the ease and scope of movement of the overwhelming proportion of its citizens, and its comparative lack of domination by a single downtown area. It has thus escaped the rigid core, sector, and ring structure of business and residential occupation that tyrannized the industrial metropolis and from which older cities are only now beginning to extricate themselves. Los Angeles is an amorphous metropolis, and vast tracts of it have a rather uniform low-density settlement of five to twenty-four persons per acre. Along the Pacific in the Santa Monica and Long Beach areas and in a crescent of housing from Santa Monica through Beverly Hills to the old core city there are apartment houses and multiple-dwelling neighborhoods that resemble those in Chicago or New York. Also scattered through the metropolis, especially along its shopping strips, stand many of the motel-like courts that are the contemporary slum-tenement style of the American city; but the single-family dwelling has long been the glory of Los Angeles and the expression of its design for living. Sixty-four percent of all its occupied housing in 1967 was given over to single-family dwelling units.

The plan for a metropolis composed of single-family houses did not emerge from the drawing boards of freeway engineers; their constructions followed an already entrenched preference of the Angelenos. During the twenties three social factors had converged to establish the Los Angeles plan: the cultural preference of Americans for detached private homes, the need to supply water for burgeoning land development, and the sheer pleasure and freedom bestowed by the automobile.

Three out of four of the army of migrants who came into southern California during the early part of the twentieth century were white native-born Americans from the cities to the east and from the farms and small towns of the Midwest. City dwellers and farmers alike brought with them an ingrained tradition of the single-family house as the measure of a home and of Main Street or suburbia as the measure of satisfactory living. The American has often had to share his housing with others – in rented rooms, in two-family houses, in tenements of three, six, or more flats – but given the opportunity he has customarily sought a house of his own. Moreover, just as in the case of Chicago's middle class, in the Los Angeles region no thirst for the big-city life of skyscrapers, restaurants, and theaters has tempted him to sacrifice the privacy of a tree-shaded lawn and garden for apartment luxury or for the urban habits of the nineteenth-century inhabitants of European or American industrial cities. In sum, people came to southern California seeking a warm, sunlit, home-town city.

To provide these amenities in the first decades of the twentieth century in the face of the aridity of Los Angeles, residential property carried high land-preparation costs and so had to be developed in tracts of considerable size. An expensive water supply had to be meshed with public transportation to a degree unheard of in the modest subdivisions common in Eastern cities. Speculators capitalized on the situation by building a wide-range complex of electric interurban streetcars. They hoped that their initial investment in lines that stretched from twenty to thirty-five miles out from the downtown area would be justified by massive profits from future land development and ensuing heavy traffic on their routes. Los Angeles in 1920, compared to other cities of the period, was extraordinarily extended into large-scale suburban development.

The automotive boom of the twenties carried these trends toward diffusion to modern proportions. The general ownership of automobiles and their use in commuting allowed developers to open up smaller tracts beyond walking distance from the interurbans. When these lines began to lose money from competition with automobiles and when traffic jams in downtown Los Angeles became intolerable, the municipality called for the construction of a rapid-transit system to alleviate traffic, revitalize the street railways, and save the downtown. Other cities had voted for subways in the twenties, but Los Angeles did not follow the precedents; its citizens voted down the proposals. Their city was so new and open that they had no image before them of a desirable downtown, and they had no habit of listening to the appeals of downtown business leaders. Then, too, their tradition told them that happiness lay in another style of life.

During the Great Depression the public transportation system cut back its service, and after World War II the interurban lines closed down; today the public bus lines handle a small volume of passengers – 400,000 fewer than in 1939. Without the discipline of street railways, commerce drained from the core city to spread out along strips of land like Wilshire Boulevard. Suburban towns like Glendale and Pasadena established their own downtowns, and suburban shopping centers and office clusters have sprung up to form a multicentered metropolis.

The key decision in the determination of the spatial freedom of its residents came in 1939, when the Los Angeles Freeway plans were settled into a multicentered pattern. A failing public transit, serious traffic jams, and a new state statute that permitted construction of limited-access highways had prompted the City of Los Angeles to commission still another study of its transportation problems. The Works Progress Administration of the federal government carried out a traffic census, and on the basis of these findings the City of Los Angeles Transportation and Engineering Board made its recommendations; freeways would be the solution to the region's traffic difficulties. The unusual multicity and multicounty membership of the advisory board may have accounted for its metropolitan orientation. Besides Los Angeles officials, representatives from such scattered places as Glendale, Beverly Hills, Redondo Beach, Huntington Beach, Whittier, Pasadena, and the San Fernando and San Gabriel valleys sat on the board. The 1939 plan called for limited-access express highways to be laid out in the form of a giant grid, which would be capable of carrying automobile traffic both into and out of the overcrowded Los Angeles central business district and would guide it across the city without the necessity of its going through the downtown area. The principal justification for this plan was the board's recognition of the already highly dispersed character of the region. Many of its alignments were to follow the much-traveled state highways that crisscrossed the area. The report may well have reflected too the politics of the Transportation Board itself. Although no record survives of its discussions, it seems highly probable that the members from the more distant areas would not have accepted the conventional hub-and-wheel design that was at this time being proposed for old American single-core cities, since such a plan would have drawn business away from their own centers. The board did expect, however, that Los Angeles would eventually grow to be a conventional single-centered city and that in time commuter railroads and subways would be required.

The 1939 proposal for freeways derives its historical importance from the fact that it was subsequently adopted and adapted in a succession of plans and projects. The first undertakings, begun in 1940 with the Pasadena Freeway, all converged on the downtown (the Hollywood, San Bernardino, Santa Ana, and Harbor Freeways), thereby beginning a radial scheme for serving the downtown, but wartime and postwar planning studies continued to repeat the basic grid strategy of the 1939 report. Then in 1956, when the federal government passed its Interstate Highway law, the California legislature set up a committee to establish routes for the state. The routes that were then adopted incorporated the 1939 proposal for a grid system of metropolitan freeways, and many such highways have been built in the ensuing years. The grid is still being extended at its outer margins in order to keep up with the spread of the metropolis, and it is being added to at the center to relieve traffic congestion further, but for many years now Los Angeles has enjoyed a transportation system that permits its residents to move swiftly from subcenter to subcenter over the entire region without having to go through the downtown or any lesser center.

The social consequences of the multicentered, low-density metropolitan region are manifold and are important to our urban future. First and foremost was the increase in the job choices offered the urban resident. In 1967 there were three automobiles for every seven persons in the Los Angeles region; 41 percent of the households had access to one vehicle, 44 percent had access to two or more vehicles, and only 15 percent of the households lacked a car. Such a distribution of automobiles and freeways gives the Los Angeles employee the widest choice of job opportunities ever possible in an American city. An hour's drive from any point in the region makes hundreds of possible employers accessible. A man can live in the San Fernando Valley and work in the old industrial sector of southeast Los Angeles, or he can commute from the old core-city neighborhoods to the new steel mills at Fontana. These are extreme commuting distances, to be sure, but recent studies show that blue-collar workers in particular are crisscrossing the whole area in search of the best jobs. In the past urban workers were pinned down to living next to their mills, or they purchased their economic freedom by commuting from crowded core-city working-class and slum districts along the radial lines of the streetcar system. These journeys were long, the cars crowded and slow, and transfers and waits in the cold and wet often necessary. Such trials cannot be compared with the ease of an hour's run today in a private car or car pool. In Los Angeles most commuters go directly from their home to the company parking lot, and the majority travel alone; more than 70 percent of all weekday automobile trips are taken by a single driver.

To this economic freedom must be added the social advantages that have accrued to the Los Angeles public. The greater number of car trips are not work-related at all but are undertaken for social or recreational purposes or for shopping. Thus scattered friends and relatives, long shopping strips and outlying shopping centers, the Pacific beaches, and national parks are all within easy reach of most families. At the same time the city is able to grow by continuing to build in its low-density popular single-family or low-rise apartment-court style.

Like all American cities, the Los Angeles automobile metropolis does not extend its amenities equally to all its residents but reinforces sharp differences governed by class and race. Boyle Heights, an early twentieth-century neighborhood in East Los Angeles, offers some of the attractions of small-town life to its Mexican-American residents. Watts and the black ghetto have many of the same features. Yet the poor, lacking cars and perhaps also fearful of their reception in other parts of the metropolis, do not travel over large sections of the region as residents of the San Fernando Valley might do. Women and old people especially suffer from a dearth of cars. The women thus lack access to jobs that would boost family income, and to essential shopping and health services. For the men, transportation doesn't seem to be the problem. Once they find employment they seem to be able to purchase cars on time.

The solution to the disadvantaged position of those without cars is neither difficult nor expensive, but like all our cities Los Angeles remains heedless of the needs and suffering of its poor, its blacks, its Mexican Americans, and its old people. After the Watts riot the transportation problems of the poor were made plain, and the state of California, with federal funding, set up a demonstration project in southeast Los Angeles to attempt to deal with the problem. A social science team intervened on behalf of the poor in several ways. They became the spokesmen of the neighborhoods before the bus companies; they waged campaigns to have route maps printed and distributed and to persuade the companies to post the schedules at the bus stops. They discovered that the last bus might leave a factory's gates a few minutes before the men finished their daily shifts; they discovered routes that could be instituted or altered to reflect the commuting habits of the residents. It seems clear from the findings of this team that the cost cutting of private and public bus companies should be monitored in every city by a political agency representing the transportation interests of those outside the circle of the automobile world. The altering of existing public transportation, however, was not sufficient. The women needed short runs along shopping strips and to shopping centers and they needed long trips in every direction to reach scattered job sites. To this end the experiment rented buses for some new regular routes and also used station wagons and cars for a kind of metropolitan taxi service to jobs at such widely dispersed destinations as Lockheed in Burbank, Douglas Aircraft at Long Beach, Torrance, and Santa Monica, American Electric at La Mirada, and Sergent-Fletcher in El Monte. The cost of these services was high, but could have been much reduced if established permanently. The demonstration project was a success, but in 1971, when the funds were gone, predictably the grant by the federal government was not renewed and the service closed down.

Judging from this experience, a few millions of dollars a year spent in maintaining an agency to represent the carless and offer its own flexible small-bus service would redress the worst of the access problems attendant on Los Angeles' unequal distribution of family income. To be sure, those without cars are the ones who suffer most from low wages, from job discrimination against women, old people, blacks, and the ill-educated, and from regional and national callousness toward the unemployed and the underemployed, but even a modest effort would be far more useful to the poor than the currently fashionable enthusiasm for rail lines. Railroads by their very nature offer high-volume service only along their own narrow strip. Los Angeles is neither a linear nor a radial city; it is a multicentered city that calls for multidestination public transportation.

[. . .]

Since all of the metropolises and the three megalopolises of the United States are growing into the form of Los Angeles, it is important that we understand the social potential of this diffuse layout of building. The open character of Los Angeles has resulted in a land-use structure more favorable for the achievement of racial and class justice than any that has so far existed in any large American city. This structure is only a potential for justice, however, and remains far from realization.

Los Angeles is no better than most cities when it comes to racial discrimination, segregation, and disadvantages for its poor . . . The isolation of blacks is almost as extreme as in Chicago. The Mexican Americans are less rigidly segregated from the Anglo Americans than are the blacks, but nonetheless they are highly segregated. When the frustration of the black community exploded with the Watts riots of 1965, a national chain reaction was set in motion, and since that time there have been a number of Mexican-American riots expressing the conflicts of another disadvantaged community. All the hostility and failings of our society, so fully documented in 1968 by the Kerner Commission and in other reports, are part of Los Angeles as well. But if it wishes to build a more inclusive society, it has a special advantage in the availability of land for redevelopment and in abundant fringe land for new construction. The prerequisite for making good use of the land in terms of its society is a commitment by the national government and the Los Angeles public and its officials to make a decent house and neighborhood open to every resident. With such a commitment, new housing can become the framework for an inclusive urban society, rejecting the city's present mechanism for class, racial, and ethnic segregation.

The original low density of Los Angeles' construction has saved it from mortgaging its future. Its built-up areas are not so crammed with structures that new construction in old areas brings serious dislocation of present residents, as is the case in Chicago or New York. There is vacant land all over the Los Angeles region. This land is in the form of weedy lots and unused spaces, some large enough for development in their own right, some requiring some demolition of adjacent houses. If public agencies were to build on five-acre tracts they would find land, either vacant or sparsely occupied, by the hundreds of parcels all over the city. Such land is ideal for the design and construction of low-rise apartments, two-family housing, and town houses. In other words, existing styles of housing could become the basis for a massive public housing program that would not disrupt the physical fabric of the city. Federal programs for land clearance, housing, and rent subsidies allow such housing to be offered today to the fourth of the Los Angeles population who cannot afford decent housing at present prices. Such federal programs, however, would have to be so funded as to allow for a large-scale and sustained undertakings.

At least three important social consequences

would be derived from such a small-parcel program. First, the chronic shortage of low-income housing would be relieved. Second, the abundance of scattered sites would enable blacks, Mexican Americans, and the poor to choose to settle either near their present neighborhoods or in new distant sections of the city. The degree of social pioneering undertaken by any family would be according to their own choice and would not depend on the decision of public authorities. Third, the five-acre tracts would have to be designed to fit into the styles prevailing in their environs or else face strong local opposition. Perhaps this requirement that design attain at least the levels of popular taste would save the projects from the degradation of some of the philanthropic architectural styles that now stigmatize public housing.

At its fringes the Los Angeles freeway brings large tracts of vacant land within the reach of commuters. This same situation prevails in the Eastern and Midwestern megalopolis and all metropolises. Here publicly sponsored or managed new towns of 50,000 to 200,000 residents could be used as a device for increasing the residential options of black and low-income families. The key advantage of the new-town concept, or entire-city building scheme, over the conventional suburban subdivision lies in its coordination of public facilities with employment and housing. The new town can be built around industries to provide jobs for a range of skills and classes. The schools and health services can be built in the beginning so that the residents do not create unnecessary conflict by overloading small local community services. Finally, because these would be large projects, the social engineering of building at all class levels would become feasible. Los Angeles is currently building a beautiful new town at Irvine. Its extent is 53,000 acres, and it will ultimately hold 430,000 inhabitants. It has jobs, recreation, an airport, and community facilities laid out in such a way that the native California land form is undisturbed. It is a brilliant example of the advantage of large-scale new-town development over the spread of subdivisions. Despite all this, Irvine is a social scandal – an all-white, upper-middle-class enclave. It is exactly the sort of project that at once shows the high potential of the new metropolis and its bigoted, class-bound failure to realize that potential.

Finally, Los Angeles should be understood as an outstanding example of regional and national planning. Its port at Long Beach, its interstate water-supply system, and its national parks and forests are excellent examples of well-designed and coordinated large-scale planning. Its war-stimulated growth is of interest precisely because it demonstrates an important but as yet little exercised capability of the federal government to influence the prosperity of a metropolitan region. The Los Angeles experience shows that we have depressed cities and depressed regions only because we have chosen to let them go unattended.

Several principles governing a successful national urban policy can be derived from the federal sponsorship in Los Angeles. First, federal orders for products and services, direct intervention in the construction of plants, and long-term aid in the building of water and electric facilities encouraged and sustained its growth. In the case of huge aircraft orders and the subsequent erection of aircraft plants with federal funds, the investment helped an industry that had already taken root to settle in more firmly. The additional contracts in the fields of aerospace and war material were instrumental in bringing the blossoming electronics industry into the city from the East.

Second, the magnitude of orders and assistance exceeded any later federal attempts to foster business in depressed areas through favored purchasing, small-business loans, and Office of Economic Opportunity programs. Aside from the money spent for war material, tools were being forged here for a national policy directed to urban growth and regional employment guidance, but our low-key use of these tools and our compulsion to tie them to the Congressional pork barrel have rendered them ineffective.

Third, the federal infusion of capital into the area continued over a long period – at least thirty years – without serious interruption, so that the war specialties of Los Angeles have been consistently nourished.

Fourth, the Los Angeles case showed that special institutions could be used to upgrade a regional labor force that had been neither particularly skilled nor outstandingly intelligent. The government sponsored here the foundation of nonprofit research and development corporations, such as RAND and the Systems

Development Corporation, to supplement the educational and research capabilities of the California Institute of Technology, the University of Southern California, and the University of California. These additional institutions, by attracting scientists and engineers to the region, did much to accelerate the city's movement into higher levels of technology, and at a pace more rapid than would have prevailed if the city had depended only upon its universities and the technical staffs of resident aircraft and electronic concerns. In the Boston metropolitan region, similar paramilitary institutions helped raise the level of a management and labor force formerly connected to the dragging textile industry, and it seems probable that the Houston Manned Spacecraft Center will have the same effect in Texas. These instances suggest that the federal government could create research and service institutions for civilians as a device to upgrade the industries and labor forces of such places as Atlanta, Chicago, Detroit, and New York.

Fifth and finally, the federal investment in Los Angeles, moved by an unconscious planning wisdom, supplemented and stimulated an existing trend. There was no effort to reverse the path of popular migration or to go against the locational trends of the economy as might have been the case in such places as Appalachia or East St. Louis, Illinois. It is obvious that to roll with the economy, to relieve temporary distress, and to help people to move is a prerequisite for national urban programs.

What the federal effort in Los Angeles lacked was self-conscious dedication. The United States has the tools to plan its regional and national urban growth through the twin job-awarding devices of the giant corporation and the federal budget. Since World War II the magnitude of the corporate and federal effort has in large measure determined the national location of jobs; it has designated the industries and places where jobs would be plentiful and where they would be sparse. The depressed condition of the core of New York City is a casualty of the federal and corporate concentration on war just as surely as the prosperity of Los Angeles is its hero. If we are to mitigate the appalling human waste and suffering of our cities, both institutions must be made to act as self-conscious agents of urban reordering. The federal government must follow European examples by adopting a carefully considered, politically viable policy for urban growth so that the modernization of our cities and our economy can go forward more humanely; the corporations must be socialized to the point at which they can be held accountable for the public consequences of their behavior. Like the legitimate demands for decent work, comfortable housing, and an open inclusive society, the promise of today's economy and its cities will not be realized unless the American public demands that government and business serve the goals of a humane society.

ROBERT FISHMAN

"Beyond Suburbia: The Rise of the Technoburb"

from *Bourgeois Utopias: The Rise and Fall of Suburbia* (1987)

Editors' introduction Robert Fishman is a professor of history at Rutgers University who established his academic reputation with his first book, the magisterial *Urban Utopias in the Twentieth Century* (1977), a study of the work of Ebenezer Howard, Le Corbusier, and Frank Lloyd Wright. For his second book, he decided to address a totally prosaic, nonvisionary subject – the history of suburbia – only to discover that "the suburban ideal" was, in the final analysis, yet another form of utopia, the utopia of the middle class.

The real focus of *Bourgeois Utopias* is the suburban ideal, more than suburbia itself, and the logic of Fishman's analysis leads him to many surprising insights and conclusions. In the medieval period and up through the eighteenth century, suburbs were clusters of houses inhabited by poor and/or disreputable people on the outskirts of towns. When suburbs were first established for the upper and middle classes, the ideal was to create a perfect synthesis of urban sophistication and rural virtue. Here was a conception as utopian as that of any visionary social reformer but with an important difference: "Where other modern utopias have been collectivist," writes Fishman, "suburbia has built its vision of community on the primacy of private property and the individual family."

What the suburbia that Herbert Gans described in *The Levittowners* (p. 63) has evolved into today is "technoburbia," a dominant new urban reality that can no longer be considered suburbia in the traditional sense. In Redmond, Washington (Plate 8), Microsoft Corporation's corporate headquarters mix with residential neighborhoods, retail centers, and even bands of open space to make up a new urban form where city and suburb, urbanized and unurbanized areas, high-tech and conventional development flow together.

To describe this new reality Fishman has coined two new terms, "technoburb" and "technocity." Fishman defines technoburbs as peripheral zones, perhaps as large as a county, that have emerged as viable socio-economic units. The new technoburbs are spread out along highway growth corridors. Along the highways of metropolitan regions shopping malls, industrial parks, campus-like office complexes, hospitals, schools, and a whole range of housing types succeed each other. Silicon Valley in the San Francisco Bay Area is an example of such a technoburb.

By "technocity" Fishman means the whole metropolitan region, which has been transformed by the coming of the technoburb. In Fishman's view we may still refer to the New York Metropolitan region as "New York City," but increasingly by "New York City" we mean the entire New York City region. And much of the economic and cultural life of the region no longer resides just in the core city. The old central cities have become increasingly marginal, while the technoburb has emerged as the focus of American life. In Fishman's view, the new technoburbs surrounding the old urban cores do not represent "the suburbanization of the United States," as Kenneth Jackson would have it, but "the end of suburbia in its traditional sense and the creation of a new kind of decentralized city." That suburbia has become the city itself is, perhaps, the final irony of modern urbanism.

Fishman lays out a strong indictment of what is wrong with technoburbs. They consist of an unplanned jumble of discordant elements – housing, industry, commerce, even agriculture – with little coherent pattern or structure. They waste land. Technoburbs are dependent on highway systems, yet their highway systems are in a state of chronic chaos. They have no proper boundaries, but consist of a crazy quilt of separate and overlapping political jurisdictions which make meaningful regionwide planning virtually impossible.

Yet Fishman notes that all new urban forms appeared chaotic in their early stages. Even the most "organic" cityscapes of the past evolved slowly after much chaos and trial and error. For example, it took planners of genius like Frederick Law Olmsted (p. 314) and Ebenezer Howard (p. 321) to create orderly parks and garden suburbs out of the chaos of the nineteenth-century city or to imagine and actually build garden cities.

Fishman acknowledges that there is a functional logic to sprawl. Perhaps, he speculates, if sprawl is better understood and better managed it might prove to be a positive rather than a negative development. Fishman looks to Frank Lloyd Wrights' Broadacre City vision (p. 344) as an example of how inspired planners may yet devise an aesthetic to tame technoburbia.

The technocity, Fishman concludes, is still under construction both physically and culturally. How technocities will evolve is unclear. The jury is still out on whether technoburbia will ultimately be judged as an advance over earlier urban forms.

This selection is from Robert Fishman, *Bourgeois Utopias: The Rise and Fall of Suburbia* (New York: Basic Books, 1987). Fishman's other major book on cities is *Urban Utopias in the Twentieth Century* (New York: Basic Books, 1977).

For other views of emerging postmodern suburbia, see journalist Joel Garreau's *Edge City* (New York: Anchor, 1992), Manuel Castells, *The Rise of The Network Society* (Cambridge, MA: Blackwell, 1996), Manuel Castells and Peter Hall, *Technopoles of the World* (London and New York: Routledge, 1994), and Edward Soja's studies of the Los Angeles region cited in the selection "Taking Los Angeles Apart" (p. 180).

If the nineteenth century could be called the Age of Great Cities, post-1945 America would appear to be the Age of Great Suburbs. As central cities stagnated or declined in both population and industry, growth was channeled almost exclusively to the peripheries. Between 1950 and 1970 American central cities grew by 10 million people, their suburbs by 85 million. Suburbs, moreover, accounted for at least three-quarters of all new manufacturing and retail jobs generated during that period. By 1970 the percentage of Americans living in suburbs was almost exactly double what it had been in 1940, and more Americans lived in suburban areas (37.6 percent) than in central cities (31.4 percent) or in rural areas (31 percent). In the 1970s central cities experienced a net out-migration of 13 million people, combined with an unprecedented deindustrialization, increasing poverty levels, and housing decay.

[. . .]

From its origins in eighteenth-century London, suburbia has served as a specialized portion of the expanding metropolis. Whether it was inside or outside the political borders of the central city, it was always functionally dependent on the urban core. Conversely, the growth of suburbia always meant a strengthening of the specialized services at the core.

In my view, the most important feature of

postwar American development has been the almost simultaneous decentralization of housing, industry, specialized services, and office jobs; the consequent breakaway of the urban periphery from a central city it no longer needs; and the creation of a decentralized environment that nevertheless possesses all the economic and technological dynamism we associate with the city. This phenomenon, as remarkable as it is unique, is not suburbanization but a new city.

Unfortunately, we lack a convenient name for this new city, which has taken shape on the outskirts of all our major urban centers. Some have used the terms "exurbia" or "outer city." I suggest (with apologies) two neologisms: the "technoburb" and the "techno-city." By "technoburb" I mean a peripheral zone, perhaps as large as a county, that has emerged as a viable socioeconomic unit. Spread out along its highway growth corridors are shopping malls, industrial parks, campuslike office complexes, hospitals, schools, and a full range of housing types. Its residents look to their immediate surroundings rather than to the city for their jobs and other needs; and its industries find not only the employees they need but also the specialized services.

The new city is a technoburb not only because high tech industries have found their most congenial homes in such archetypal technoburbs as Silicon Valley in northern California and Route 128 in Massachusetts. In most technoburbs such industries make up only a small minority of jobs, but the very existence of the decentralized city is made possible only through the advanced communications technology which has so completely superseded the face-to-face contact of the traditional city. The technoburb has generated urban diversity without traditional urban concentration.

By "techno-city" I mean the whole metropolitan region that has been transformed by the coming of the technoburb. The techno-city usually still bears the name of its principal city, for example, "the New York metropolitan area"; its sports teams bear that city's name (even if they no longer play within the boundaries of the central city); and its television stations appear to broadcast from the central city. But the economic and social life of the region increasingly bypasses its supposed core. The techno-city is truly multicentered, along the pattern that Los Angeles first created. The technoburbs, which might stretch over seventy miles from the core in all directions, are often in more direct communication with one another – or with other techno-cities across the country – than they are with the core. The techno-city's real structure is aptly expressed by the circular superhighways or beltways that serve so well to define the perimeters of the new city. The beltways put every part of the urban periphery in contact with every other part without passing through the central city at all.

[. . .]

The old central cities have become increasingly marginal, while the technoburb has emerged as the focus of American life. The traditional suburbanite – commuting at ever-increasing cost to a center where the available resources barely duplicate those available much closer to home – becomes increasingly rare. In this transformed urban ecology the history of suburbia comes to an end.

PROPHETS OF THE TECHNO-CITY

Like all new urban forms, the techno-city and its technoburbs emerged not only unpredicted but unobserved. We are still seeing this new city through the intellectual categories of the old metropolis. Only two prophets, I believe, perceived the underlying forces that would lead to the techno-city at the time of their first emergence. Their thoughts are therefore particularly valuable in understanding the new city.

At the turn of the twentieth century, when the power and attraction of the great city was at its peak, H. G. Wells daringly asserted that the technological forces that had created the industrial metropolis were now moving to destroy it. In his 1900 essay "The Probable Diffusion of Great Cities," Wells argued that the seemingly inexorable concentration of people and resources in the largest cities would soon be reversed. In the course of the twentieth century, he prophesied, the metropolis would see its own resources drain away to decentralized "urban regions" so vast that the very concept of "the city" would become, in his phrase, "as obsolete as 'mailcoach.'"

Wells based his prediction on a penetrating analysis of the emerging networks of transportation and communication. Throughout the nineteenth century, rail transportation had been a relatively simple system favoring direct access to large centers. With the spread of branchlines and electric tramways, however, a complex rail network had been created that could serve as the basis for a decentralized region. (As Wells wrote, Henry E. Huntington was proving the truth of his propositions for the Los Angeles region.)

Wells pictured the "urban region" of the year 2000 as a series of villages with small homes and factories set in the open fields, yet connected by high speed rail transportation to any other point in the region. (It was a vision not very different from those who saw Los Angeles developing into just such a network of villages.) The old cities would not completely disappear, but they would lose both their financial and their industrial functions, surviving simply because of an inherent human love of crowds. The "post-urban" city, Wells predicted, will be "essentially a bazaar, a great gallery of shops and places of concourse and rendezvous, a pedestrian place, its pathways reinforced by lifts and moving platforms, and shielded from the weather, and altogether a very spacious, brilliant, and entertaining agglomeration." In short, the great metropolis will dwindle to what we would today call a massive shopping mall, while the productive life of the society would take place in the decentralized urban region.

Wells's prediction was taken up in the late 1920s and early 1930s by Frank Lloyd Wright, who moved from similar assumptions to an even more radical view. Wright had actually seen the beginnings of the automobile and truck era; he was, perhaps not coincidentally, living mostly in Los Angeles in the late 1910s and early 1920s. Wright, like Wells, argued that "the great city was no longer modern" and that it was destined to be replaced by a decentralized society.

He called this new society Broadacre City. It has often been confused with a kind of universal suburbanization, but for Wright "Broadacres" was the exact opposite of the suburbia he despised. He saw correctly that suburbia represented the essential extension of the city into the countryside, whereas Broadacres represented the disappearance of all previously existing cities.

As Wright envisioned it, Broadacres was based on universal automobile ownership combined with a network of superhighways, which removed the need for population to cluster in a particular spot. Indeed, any such clustering was necessarily inefficient, a point of congestion rather than of communication. The city would thus spread out over the countryside at densities low enough to permit each family to have its own homestead and even to engage in part-time agriculture. Yet these homesteads would not be isolated; their access to the superhighway grid would put them within easy reach of as many jobs and specialized services as any nineteenth century urbanite. Traveling at more than sixty miles an hour, each citizen would create his own city within the hundreds of square miles he could reach in an hour's drive.

Like Wells, Wright saw industrial production inevitably leaving the cities for the space and convenience of rural sites. But Wright went one step further in his attempt to envision the way that a radically decentralized environment could generate that diversity and excitement which only cities had possessed.

He saw that even in the most scattered environment, the crossing of major highways would possess a certain special status. These intersections would be the natural sites of what he called the roadside market, a remarkable anticipation of the shopping center: "great spacious roadside pleasure places these markets, rising high and handsome like some flexible form of pavilion – designed as places of co-operative exchange, not only of commodities but of cultural facilities." To the roadside markets he added a range of highly civilized yet small scale institutions: schools, a modern cathedral, a center for festivities, and the like. In such an environment, even the entertainment functions of the city would disappear. Soon, Wright devoutly wished, the centralized city itself would disappear.

Taken together, Wells's and Wright's prophecies constitute a remarkable insight into the decentralizing tendencies of modern technology and society. Both were presented in utopian form, an image of the future presented as somehow "inevitable" yet without any sustained attention to how it would actually be achieved. Nevertheless, something like the transformation that Wells and Wright foresaw has taken place in the United States, a transformation all the more remarkable in that it occurred without a clear recognition that it was happening. While diverse groups were engaged in what they believed was "the suburbanization" of America, they were in fact creating a new city.

[. . .]

TECHNOBURB/TECHNO-CITY: THE STRUCTURE OF THE NEW METROPOLIS

To claim that there is a pattern or structure in the new American city is to contradict what appears to be overwhelming evidence. One might sum up the structure of the technoburb by saying that it goes against every rule of planning. It is based on two extravagances that have always aroused the ire of planners: the waste of land inherent in a single family house with its own yard, and the waste of energy inherent in the use of the personal automobile. The new city is absolutely dependent on its road system, yet that system is almost always in a state of chaos and congestion. The landscape of the technoburb is a hopeless jumble of housing, industry, commerce, and even agricultural uses. Finally, the technoburb has no proper boundaries; however defined, it is divided into a crazy quilt of separate and overlapping political jurisdictions, which make any kind of coordinated planning virtually impossible.

Yet the technoburb has become the real locus of growth and innovation in our society. And there is a real structure in what appears to be wasteful sprawl, which provides enough logic and efficiency for the technoburb to fulfill at least some of its promises.

If there is a single basic principle in the structure of the technoburb, it is the renewed linkage of work and residence. The suburb had separated the two into distinct environments; its logic was that of the massive commute, in which workers from the periphery traveled each morning to a single core and then dispersed each evening. The technoburb, however, contains both work and residence within a single decentralized environment.

By the standards of a preindustrial city where people often lived and worked under the same roof, or even of the turn of the century industrial zones where factories were an integral part of working class neighborhoods, the linkage between work and residence in the technoburb is hardly close. A recent study of New Jersey shows that most workers along the state's growth corridors now live in the same county in which they work. But this relative dispersion must be contrasted to the former pattern of commuting into urban cores like Newark or New York. In most cases traveling time to work diminishes, even when the distances traveled are still substantial; as the 1980 census indicates, the average journey to work appears to be diminishing both in distance and, more importantly, in time.

For commuting within the technoburb is multidirectional, following the great grid of highways and secondary roads that, as Frank Lloyd Wright understood, defines the community. This multiplicity of destinations makes public transportation highly inefficient, but it does remove that terrible bottleneck which necessarily occurred when work was concentrated at a single core within the region. Each house in a technoburb is within a reasonable driving time of a truly "urban" array of jobs and services, just as each workplace along the highways can draw upon an "urban" pool of workers.

Those who believed that the energy crisis of the 1970s would cripple the technoburb failed to realize that the new city had evolved its own pattern of transportation in which a multitude of relatively short automobile journeys in a multitude of different directions substitutes for that great tidal wash in and out of a single urban core which had previously defined commuting. With housing, jobs, and services all on the periphery, this sprawl develops its own form of relative efficiency. The truly inefficient form would be any attempted revival of the former pattern of long distance mass transit commuting into a core area. To account for the new linkage of work and residence in the technoburb, we must first confront this paradox: the new city required a massive and coordinated relocation of housing, industry, and other "core" functions to the periphery; yet there were no coordinators directing the process. Indeed, the technoburb emerged in spite of, not because of, the conscious purposes motivating the main actors. The postwar housing boom was an attempt to escape from urban conditions; the new highways sought to channel traffic into the cities; planners attempted to limit peripheral growth; the government programs that did the most to destroy the hegemony of the old industrial metropolis were precisely those designed to save it.

This paradox can be seen clearly in the area of transportation policy. Wright had grasped the basic point in his Broadacre City plan: a fully developed highway grid eliminates the primacy of a central business district. It creates a whole series of highway crossings, which can serve as business centers while promoting the multi-directional travel that prevents any single center from attaining unique importance. Yet, from the time of Robert Moses to the present, highway planners have imagined that the new roads, like the older rail transportation, would enhance the importance of the old centers by funneling cars and trucks into the downtown area and the surrounding industrial belt. At most, the highways were to serve traditional suburbanization; in other words, the movement from the periphery to the core during morning rush hours and the reverse movement in the afternoon. The beltways, those crucial "Main Streets" of the technoburb, were designed simply to allow interstate traffic to avoid going through the central cities.

The history of the technoburb, therefore, is the history of those deeper structural features of modern society first described by Wells and Wright taking precedence over conscious intentions. For purposes of clarity I shall now divide this discussion of the making of the techno-city into two interrelated topics: housing and job location.

Housing

The great American postwar housing boom was perhaps the purest example of the suburban dream in action, yet its ultimate consequence was to render suburbia obsolete. Between 1950 and 1970, on the average, 1.2 million housing units were built each year, the vast majority as suburban single family dwellings; the nation's housing stock increased by 21 million units or over 50 percent. In the 1970s the boom continued even more strongly: twenty million more new units were added, almost as many as in the previous two decades. It was precisely this vast production of new residences that shifted the center of gravity in the United States from the urban core to the periphery and thus ensured that these vital and expanding areas could no longer remain simply bedroom communities.

This great building boom, which seems so characteristic of post-1945 conditions, in fact had its origins early in the twentieth century in the first attempts to universalize suburbia throughout the United States. It can be seen essentially as a continuation of the 1920s building boom, which had been cut off for two decades by the Depression and the war. As George Sternlieb reminds us, the American automobile industry in 1929 was producing as many cars per capita as it did in the 1980s, and real estate developers had already plotted out subdivisions in out-lying areas that were only built up in the 1960s and 1970s.

[. . .]

Even the late 1970s combination of stagnant real income with high interest rates, gasoline prices, and land values did not diminish the desirability of the new single family house. In 1981 a median American family earned only 70 percent of what was needed to make the payments on the median priced house; by 1986, the median family could once again afford the median house. Single family houses still constitute 67 percent of all occupied units, down only 2 percent since 1970 despite the increase in costs; moreover, a survey of potential home buyers in 1986 showed that 85 percent intended to purchase a detached, single family suburban house, while only 15 percent were looking at condominium apartments or townhouses. The "single," as builders call it, is still alive and well on the urban periphery.

This continuing appeal of the single should not, however, obscure the crucial changes that have transformed the meaning and context of the house. The new suburban house of the 1950s, like its predecessors for more than a century, existed precisely to isolate women and the family from urban economic life; it defined an exclusive zone of residence between city and country. Now a new house might adjoin a landscaped office park with more square feet of new office space than in a downtown building, or might be just down the highway from an enclosed shopping mall with a sales volume that exceeds those of the downtown department stores, or might overlook a high tech research laboratory making products that are exported around the world. No longer a refuge, the single family detached house on the periphery is preferred as a convenient base from which both spouses can rapidly reach their jobs.

Without the simultaneous movement of jobs along with housing, the great "suburban" boom would surely have exhausted itself in ever longer journeys to workplaces in a crowded core on overburdened highways and mass transit facilities. And the new peripheral communities would have been in reality the "isolation wards" for women that critics have called them, instead of becoming the setting for the reintegration of middle-class women into the work force as they have. The unchanging image of the suburban house and the suburban bedroom community has obscured the crucial importance of this transformation in work location, the subject of the next section.

Job location

As those who have tried to plan the process have painfully learned, job location has its own autonomous rules. The movement of factories away from the urban core after 1945 took place independently of the housing boom and probably would have occurred without it. Nevertheless, the simultaneous movement of housing and jobs in the 1950s and 1960s created an unforeseen "critical mass" of entrepreneurship and expertise on the perimeters, which allowed the technoburb to challenge successfully the two century long economic dominance of the central city.

[. . .]

At the same time, the growing importance of trucking meant that factories were no longer as dependent on the confluence of rail lines which existed only in the old factory zones. Workers had their automobiles, so factories could scatter along the periphery without concern about the absence of mass transit. (The scattering of aircraft plants and other factories in Los Angeles in the 1930s prefigured this trend.) The process gained momentum as a result of thousands of uncoordinated decisions in which managers allowed their inner city plants to run down and directed new investment toward the outskirts . . .

These changes in job location during the 1950s and 1960s were, however, only a prelude to the real triumph of the technoburb: the luring of both managerial office employment and advanced technological laboratories and production facilities from the core to the peripheries. This process may be divided into three parts. First came the establishment of "high tech" growth corridors in such diverse locations as Silicon Valley, California; Silicon Prairie, between Dallas and Forth Worth; the Atlanta Beltway; Route 1 between Princeton and New Brunswick, New Jersey; Westchester County, New York; Route 202 near Valley Forge, Pennsylvania; and Route 128 outside Boston. The second step was the movement of office bureaucracies, especially the so-called back office, from center city high-rises to technoburb office parks; and the final phase was the movement of production-service employment – banks, accountants, lawyers, advertising agencies, skilled technicians, and the like – to locations within the technoburb, thus creating that vital base of support personnel for larger firms.

Indeed, this dramatic surge toward the technoburb has been so sweeping that we must now ask whether Wright's ultimate prophecy will be fulfilled: the disappearance of the old urban centers. Is the present-day boom in downtown office construction and inner city gentrification simply a last hurrah for the old city before deeper trends in decentralization lead to its ultimate decay?

In my view, the final diffusion that Wells and Wright predicted is unlikely, if only because both underestimated the forces of economic and political centralization that continue to exist in the late twentieth century. If physical decentralization had indeed meant economic decentralization, then the urban cores would by now be ghost towns. But large and powerful organizations still seek out a central location that validates their importance, and the historic core of great cities still meets that need better than the office complexes on the outskirts. Moreover, the corporate and government headquarters in the core still attract a wide variety of specialized support services – law firms, advertising, publishing, media, restaurants, entertainment centers, museums, and more – that continue to make the center cities viable.

The old factory zones around the core have also survived, but only in the painfully anomalous sense of housing those too poor to earn admission to the new city of prosperity at the periphery. The big city, therefore, will not disappear in the foreseeable future, and residents of the technoburbs will continue to confront uneasily both the economic power and elite culture of the urban core and its poverty. Nevertheless, the technoburb has become the true center of American society.

THE MEANING OF THE NEW CITY

Beyond the structure of the techno-city and its technoburbs, there is the larger question: what is the impact of this decentralized environment on our culture? Can anyone say of the technoburb, as Olmsted said of the suburb a century ago, that it represents "the most attractive, the most refined, and the most soundly wholesome forms of domestic life, and the best application of the arts of civilization to which mankind has yet attained"? Most planners in fact say the exact opposite. Their indictment can be divided into two parts. First, decentralization has been a social and economic disaster for the old city and for the poor, who have been increasingly relegated to its crowded, decayed zones. It has resegregated American society into an affluent outer city and an indigent inner city, while erecting ever higher barriers that prevent the poor from sharing in the jobs and housing of the technoburbs.

Second, decentralization has been seen as a cultural disaster. While the rich and diverse architectural heritage of the cities decays, the technoburb has been built up as a standardized and simplified sprawl, consuming time and space, destroying the natural landscape. The wealth that postindustrial America has generated has been used to create an ugly and wasteful pseudocity, too spread out to be efficient, too superficial to create a true culture. The truth of both indictments is impossible to deny, yet it must be rescued from the polemical overstatements that seem to afflict anyone who deals with these topics. The first charge is the more fundamental, for it points to a genuine structural discontinuity in post-1945 decentralization. By detaching itself physically, socially, and economically from the city, the technoburb is profoundly antiurban as suburbia never had been. Suburbanization strengthened the central core as the cultural and economic heart of an expanding region; by excluding industry, suburbia left intact and even augmented the urban factory districts.

Technoburb development, however, completely undermines the factory district and potentially threatens even the commercial core. The competition from new sites on the outskirts renders obsolete the whole complex of housing and factory sites that had been built up in the years 1890 to 1930 and provides alternatives to the core for even the most specialized shopping and administrative services.

This competition, moreover, has occurred in the context of a massive migration of southern blacks to northern cities. Blacks, Hispanics, and other recent migrants could afford housing only in the old factory districts, which were being abandoned by both employers and the white working class. The result was a twentieth century version of Disraeli's "two nations." Now, however, the outer reaches of affluence include both the middle class and the better-off working class – a majority of the population; while the largely black and Hispanic minority are forced into decaying neighborhoods, which lack not only decent housing but jobs.

This bleak picture has been modified somewhat by the continued ability of the traditional urban cores to retain certain key areas of white collar and professional employment; and by the choice of some highly paid core workers to live in high-rise or recently renovated housing around the core. Compared both to the decaying factory zones and to peripheral expansion, the "gentrification" phenomenon has been highly visible yet statistically insignificant. It has done as much to displace low income city dwellers as to benefit them. The late twentieth century American environment thus shows all the signs of the two nations syndrome: one caught in an environment of poverty, cut off from the majority culture, speaking its own languages and dialects; the other an increasingly homogenized culture of affluence, more and more remote from an urban environment it finds dangerous.

[. . .]

The case against the technoburb can easily be

summarized. Compared even to the traditional suburb, it at first appears impossible to comprehend. It has no clear boundaries; it includes discordant rural, urban, and suburban elements; and it can best be measured in counties rather than in city blocks. Consequently the new city lacks any recognizable center to give meaning to the whole. Major civic institutions seem scattered at random over an undifferentiated landscape.

Even planned developments – however harmonious they might appear from the inside – can be no more than fragments in a fragmented environment. A single house, a single street, even a cluster of streets and houses can be and frequently are well designed. But true public space is lacking or totally commercialized. Only the remaining pockets of undeveloped farmland maintain real openness, and these pockets are inevitably developed, precipitating further flight and further sprawl.

The case for the techno-city can only be made hesitantly and conditionally. Nevertheless, we can hope that its deficiencies are in large part the early awkwardness of a new urban type. All new city forms appear in their early stages to be chaotic. "There were a hundred thousand shapes and substances of incompleteness, wildly mingled out of their places, upside down, burrowing in the earth, aspiring in the earth, moldering in the water, and unintelligible as any dream." This was Charles Dickens describing London in 1848, in his novel *Dombey and Son* (Chapter 6). As I have indicated, sprawl has a functional logic that may not be apparent to those accustomed to more traditional cities. If that logic is understood imaginatively, as Wells and especially Wright attempted to do, then perhaps a matching aesthetic can be devised.

We must remember that even the most "organic" cityscapes of the past evolved slowly after much chaos and trial and error. The classic late nineteenth century railroad suburb – the standard against which critics judge today's sprawl – evolved out of the disorder of nineteenth century metropolitan growth. First, planners of genius like John Nash and Frederick Law Olmsted comprehended the process and devised aesthetic formulas to guide it. These formulas were then communicated – slowly and incompletely – to speculative builders, who nevertheless managed to capture the basic idea. Finally, individual property owners constantly upgraded their holdings to eliminate discordant elements and bring their community closer to the ideal.

We might hope that a similar process is now at work in the postsuburban outer city. As a starting point for a technoburb aesthetic, there are Wright's Broadacre City plans and drawings, which still repay study for anyone seeking a vision of a modern yet organic American landscape. More useful still is the American New Town tradition, starting from Radburn, New Jersey, with its careful designs intended to reconcile decentralization with older ideas of community. Already, New Town designs have been adopted by speculative builders, not only in a highly publicized project like James Rouse's Columbia, Maryland, but in hundreds of smaller planned communities, which are beginning to leave their mark on the landscape.

At the level of civic architecture there is Wright's Marin County Civic Center to serve as a model for public monuments in a decentralized environment. The multilevel, enclosed shopping mall has attained a spaciousness not unworthy of the great urban shopping districts of the past, while newly built college campuses and campuslike office complexes and research centers contribute significantly to the environment. Some commercial highway strips have been rescued from cacophony and have managed to achieve a liveliness that is not tawdry. (This evolution parallels the evolution of the nineteenth century urban core, originally a remarkably ugly cluster of small buildings and large signs, which was transformed into a reasonably dignified center for commerce by the turn of the century.)

Most importantly, there is a growing sense that open land must be preserved as an integral part of the landscape, through regional land use plans, purchases for parklands, and tax abatements for working farms. These governmental measures, combined with thousands of small scale efforts by individuals, could create a fitting environment for the new city. These efforts, moreover, could provide the starting point for a more profound diversification of the outer city. An increased understanding and respect for the landscape of each region could lead to a growing rejection of a mass culture that erases all such distinctions. The techno-city, therefore, is still under construction, both physically and culturally. Its economic and social successes are undeniable, as are its costs. Most importantly, the new pattern of decentralization has fundamentally altered the urban form on which suburbia had depended for its function and meaning. Whatever the fate of the new city, suburbia in its traditional sense now belongs to the past.

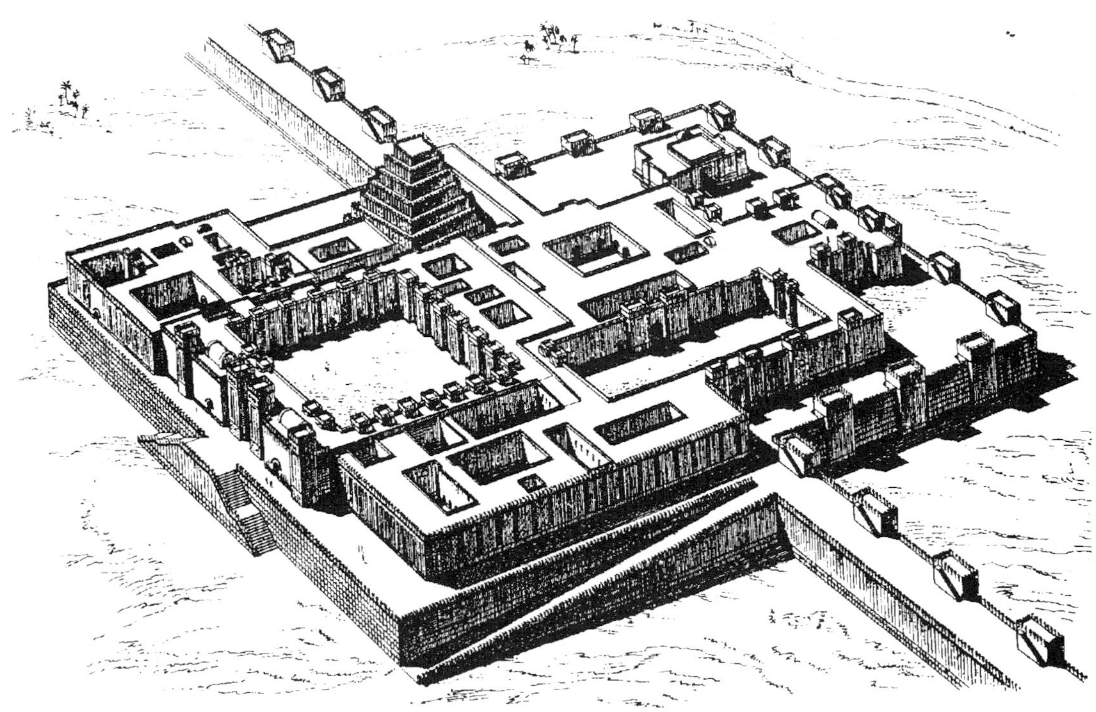

Plate 1 Palace of Sargon II of Khorsabad. The first cities arose about 4000 BCE in Mesopotamia between the Tigris and Euphrates Rivers during what V. Gordon Childe termed "the urban revolution." Note the mud brick walls, monumental gate, and ziggurat in this very substantial Mesopotamian palace.

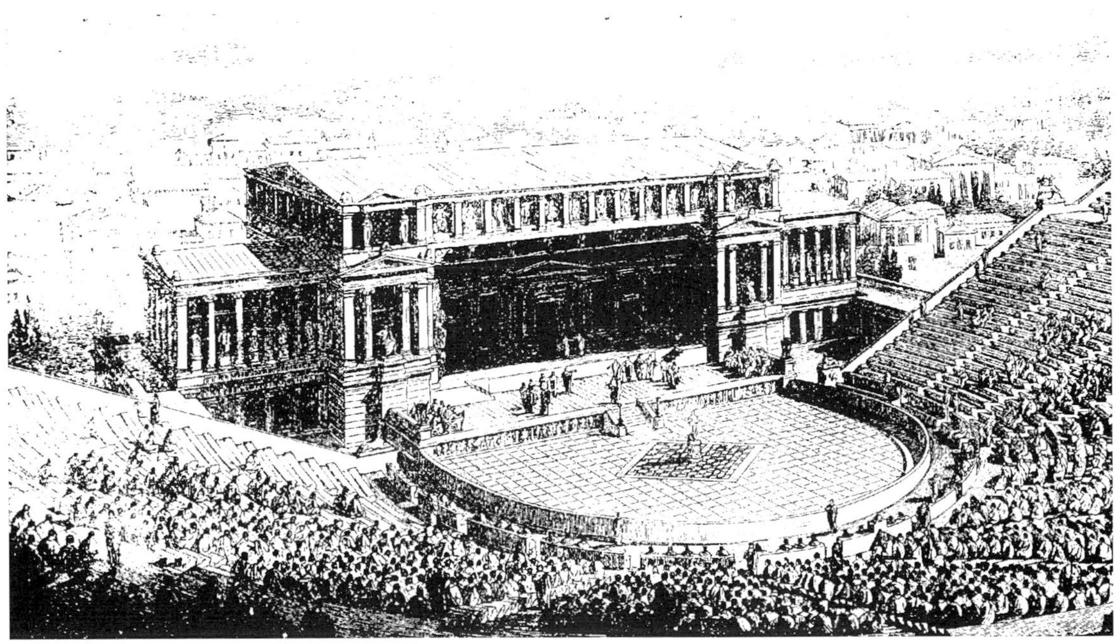

Plate 2 Theater of Dionysus, Athens, Greece, fifth Century BCE. The Greek polis stressed public over private life. Citizens conversed, shopped, and settled disputes in public agora, exercised and competed in public stadiums and gymnasia, and shared in the political and cultural life of the community in large open-air structures like the magnificent Theater of Dionysus.

Plate 3 Medieval marketplace, Monpazier, France. As Europe began to revive after the period of disorganization and strife that followed the fall of the Roman Empire, cities like Monpazier, France, fostered trade, economic expansion, and free institutions. (© Jay Vance.)

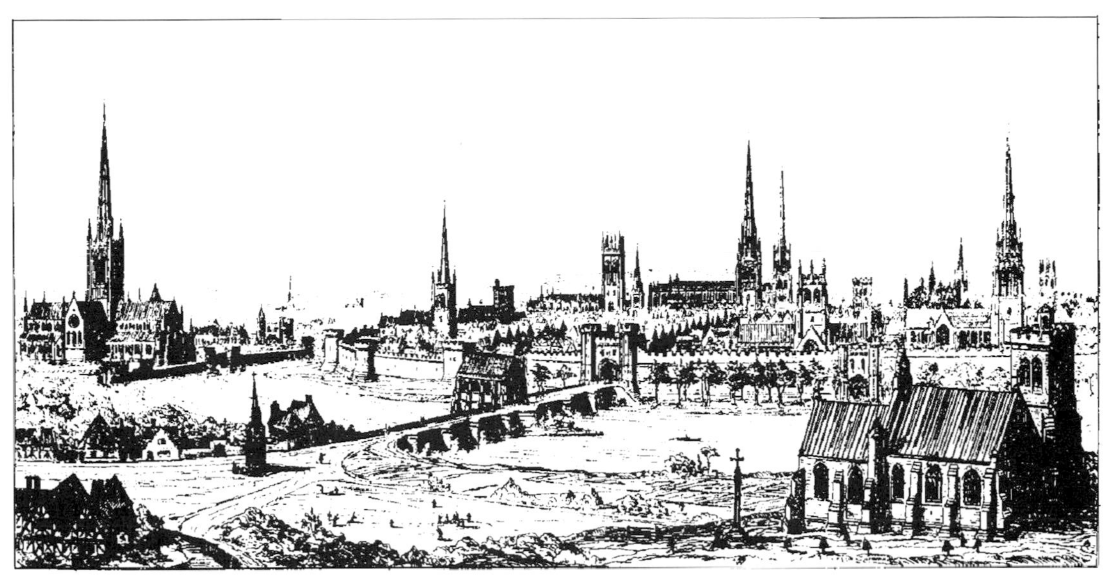

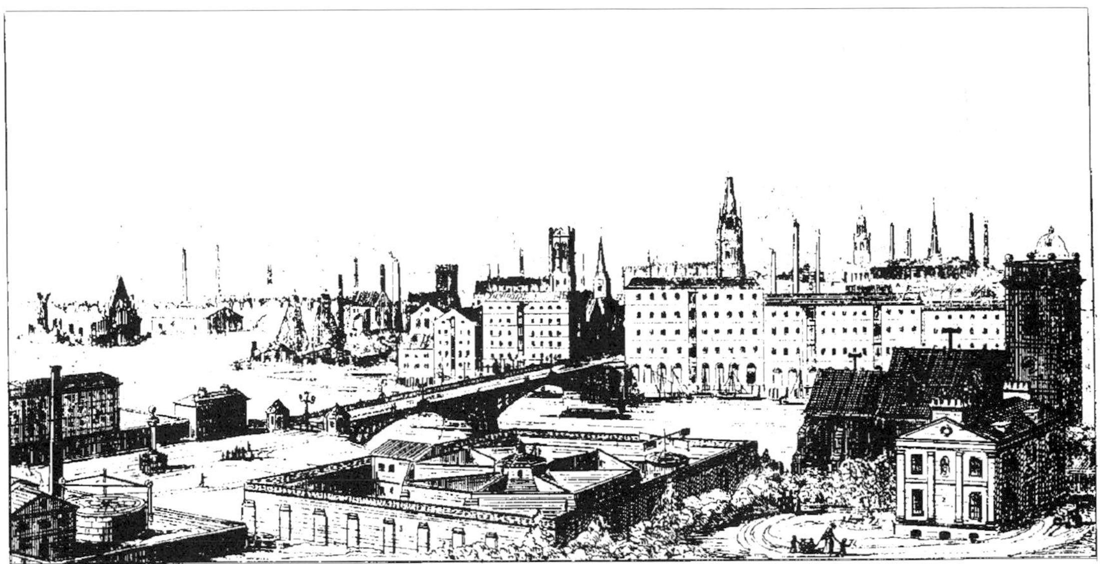

Plate 4 The nineteenth-century industrial city. During the first part of the nineteenth century new industrial cities based on steam-powered machinery sprang up in Europe. In these prints titled "A Catholic town in 1440" and "The same town in 1840," Augustus Welby Pugin, a contemporary observer, contrasts the same city before and after the Industrial Revolution. The first print shows a city where church spires are the dominant architectural element, the land surrounding the medieval city walls is largely empty, and the air and water are clean. In the second print, factory smokestacks have largely replaced steeples, the air is filled with smoke, development has sprawled to the once-empty land, and the foreground is dominated by a massive panopticon prison. Pugin subtitled his work: "A Parallel Between the Noble Edifices of the Middle Ages and the Corresponding Buildings of the Present Day Shewing The Present Decay of Taste."

Plate 5 Birds-eye view of La Crosse, Wisconsin, showing the principal business street. One of the features that Friedrick Engels noted in mid-nineteenth-century Manchester, England, was the appearance of large streets where the upper classes could travel without coming into contact with the squalid living conditions of the residential slums which comprised the greater part of the city. This illustration from a popular American illustrated weekly depicted with pride the principal business street in a Midwestern industrial city. The belching factories and packed residential slums, which then comprised much of La Crosse, are not mentioned in the caption.

Plate 6 Levittown, New York, 1947. While suburbs have a long and varied history, it was during the period after World War II that many of the suburbs surrounding US cities arose. Levittown, New York (and its counterpart Levittown, Pennsylvania), provided entirely new communities of affordable, cute, single-family houses on individual lots to returning GIs and other first-time (white) homebuyers. (© Levittown, New York Public Library.)

Plate 7 The auto-centered metropolis. Developers show off the location of a proposed new "50 Foot Boulevard" in the Westwood Village area of Santa Monica in 1922. Capitalizing on the mass production of Henry Ford's Model-T, they and thousands of their counterparts built in Southern California communities that required a car. Security Pacific Collection. Los Angeles Public Library. (© Los Angeles Public Library.)

Plate 8 Technoburbia. Robert Fishman uses the term technoburbia to describe the form of urban development that jumbles business, residential, commercial, and other uses together in the area surrounding older core cities. Here, Microsoft Corporate Headquarters in Redmond, Washington, forms part of a pattern that includes residential areas, open space, shopping centers, and other uses. (© Microsoft Corporation.)

PART 2

Urban Culture and Society

PART TWO

INTRODUCTION

As Shakespeare wrote – and as urbanists ever since have never tired of quoting – "the people are the city." Urban history expresses the progressive evolution of the city as an institution. Urban form and design describe the physical appearance and infrastructural layout of cities, but it is the people of the city – their individual aspirations and collective struggles, their day-to-day lives and their moments of heightened awareness – that constitute the core subject of urban studies and the final purpose of city planning.

In turning to the people of the cities themselves, we move to a consideration of the subtle and ever-shifting interplay between society, community, and culture. This section addresses how urban society affects urban culture and how culture affects the daily lives and the life prospects of city dwellers. It asks what culture is in an urban context and how it expresses itself in different social contexts – either as high culture or popular culture. Finally, it analyzes what community is in an urban context and speculates about what it could be.

In studying the people of the city, the key methodology is sociology, that "science of society" that arose alongside the emergence of the modern industrial city itself. Urban sociology has always been allied with urban anthropology, and in recent years new formulations – social studies, social theory, social relations, culture studies – have joined the discipline either as adjuncts or rivals. But the basic sociological vision remains central to all investigations of people in cities.

There is no better person with whom to begin a section on "Urban Culture and Society" than Lewis Mumford, one of the great public intellectuals of the twentieth century. Mumford never lost sight of the human dimension of cities. For over sixty years he sparred with those who argued that cities arose and prospered for purely economic reasons or that cities were best defined in terms of size and density. Not so!, thundered Mumford: cities are expressions of the human spirit and cities exist to contribute to the ever-evolving human personality. This perspective comes through loud and clear in "What Is a City?" To Mumford, defining a city in terms of population size, or density, or attributes of the built environment is inadequate. Rather, the human side of cities is their very essence, and city streets are a stage on which life's drama is played out. Like William Whyte (p. 483), Jane Jacobs (p. 106), and Sharon Zukin (p. 131), Mumford takes real delight in city life. For him, cities reflect and enlarge the human spirit, and he argues that the project of creating better, more human cities will enrich civilization itself.

Mumford, of course, was not alone in focusing on the connections between urban life and the human personality. In "Urbanism as a Way of Life," Louis Wirth asked the fundamental question "What does it mean to be urban?" and concluded that an urban "way of life" resulted in an "urban type" of character and personality. Wirth was one of a gifted group of sociologists at the University of Chicago who, in the 1920s and 1930s, developed a pioneering body of urban sociological theory that still shapes the field of urban sociology today. Studying rural migrants to Chicago from the peasant societies of Southern and Eastern Europe, Wirth perceived that the whole way of life in modern cities was fundamentally different from the way of life in rural cultures. In "Urbanism as a Way of Life," he attempts to abstract the essential characteristics of urban as opposed to rural life and to find the sources of the widely perceived urban characteristics of brusqueness and impersonality. As the face-to-face transactions of static rural village life are replaced by the distanced and mediated transactions of a large city, human personalities are transformed, and the new urbanites respond to each other and to society as a whole in entirely different ways than they did in their rural folk communities.

While Wirth's is a *theoretical* study, it is important to remember that his theories were generated by empirical observations that he and his colleagues conducted in Chicago during the early decades of the twentieth century. Architectural critic and urban community activist Jane Jacobs did her own kind of street-level social observation in writing *The Death and Life of Great American Cities* (1961), a book that shook the complacent world of establishment planning by re-introducing the values of community to the design of urban spaces. Architecture and urban design may not *determine* human behavior, but bad design can numb the human spirit and good design can have powerful, positive influences on human beings. Of the many values designers seek to build into their designs perhaps none is more important than fostering community and human interaction. To Jacobs, traffic engineering should be only one consideration in designing a street. In "The Uses of Sidewalks: Safety," she argues that a street designed so that people can see their children from house windows and will want to congregate on the front door stoops – one very much like her own Hudson Street in Greenwich Village – will be much more user-friendly. It will also be much safer than one that moves traffic efficiently but is inhospitable to neighborhood life and insensitive to the potential for street life to reduce urban crime. Jacobs stresses the importance of designing streets to promote safety, particularly for women. A safer environment, she argues, is essential to the creation and preservation of community.

The Black ghettos of the United States have been the subject of an enormous body of sociological research that both celebrates the distinctive culture and analyzes the social pathology of segregated, poverty-ridden inner-city communities. Beginning with W. E. B. Du Bois's study of *The Philadelphia Negro* (1899) (p. 56) and continuing through the work of E. Franklin Frazier (*The Negro Family in the United States*, 1939), St. Clair Drake and Horace Cayton (*Black Metropolis*, 1945), and Kenneth B. Clark (*Dark Ghetto*, 1965), African-American scholars have taken the lead in examining the social and cultural dynamics of ghetto communities in America's northern cities. More recently, an important debate, called the "underclass" debate, arose concerning the plight of the mostly Black residents of American inner-city ghettos. Among the principals in that debate are Charles Murray, a white libertarian conservative, and William Julius Wilson, an African-American sociologist.

Charles Murray's "Choosing a Future" from *Losing Ground* (1984) here follows Wilson's "From Institutional to Jobless Ghettos" from *When Work Disappears* (1996). Comparing Murray and Wilson illustrates the role of ideology in shaping urban theory. Both agree that the situation of the poorest urban Blacks in the United States has grown worse during the last generation and that poor ghetto Blacks today, especially youth, are in deep trouble. But they part company on why this is so and what to do about it. Wilson stresses the loss of jobs accessible to unskilled Black youth. It is this loss of jobs, Wilson argues, that has destroyed ghetto family structure and lies at the root of crime, substance abuse, and other ghetto ills. Murray looks at many of the same facts and reaches different conclusions. As he sees it, many young Black males are not "marriageable" because they lack minimum job skills, have substance abuse problems, or are in prison. Without "marriageable" males, many young Black women cannot form two-parent nuclear families, and teen pregnancies, out-of-wedlock births, and female-headed, welfare-dependent, single-parent households result.

Wilson would like to see government intervene with a universal, not race-specific, full employment program. If the poor of all races are employed, he reasons, family stability will return, and substance abuse and criminal behavior will drop. Charles Murray disagrees. According to him, the last thing the poor need is more government assistance. After government poverty programs increased, he notes, the condition of inner-city Blacks only got worse. Murray argues that patronizing government programs sapped initiative and created perverse incentives to stay out of the labor market. Murray's remedy: cut government welfare and social service programs. When the previously unemployed know they are making it on their own by working long hours at hard jobs, without paternalistic government programs, Murray argues, they will regain self respect and cut down on drugs and criminality.

Culture, of course, is not just the product of an identifiable artist or intellectual class. Working-class neighborhoods and inner-city ghettos also produce formal poetry and the linguistic inventiveness of street talk, rhapsodies and jazz music, paintings and graffiti. These are surely elements of "the urban drama," but who will dominate a city's culture? Who will be hegemonic in the socially "contested city"? In "Whose Culture? Whose City?" from *The Cultures of Cities* (1995), sociologist and cultural analyst Sharon Zukin explores the ways that authentic popular culture is being increasingly privatized and commercialized in a contested urban environment. As private corporations take over public parks in the name of civility and as the fashion industry appropriates the attitudes and demeanors of the street culture of urban youth, an authentic and democratic public culture may, she fears, be irrevocably lost. In the final analysis, the urban cultural drama may not be just about the abstractions of class and socio-economic dynamics. Instead, it must encompass the real, everyday experience of all citizens, rich and poor alike, in the public places of their cities. Respecting and making provision for these experiences will be the ultimate building blocks of the urban community.

In exploring urban society and culture, sociologists are trained to investigate a range of measurable conditions and observable social behaviors and to analyze their findings in terms of class, race, ethnicity, and socio-economic status. But urban culture has other dimensions that do not lend themselves so easily to quantitative analysis or narrative description. Among these are the expressions of the creative arts and the other cultural productions that emerge from urban communities everywhere, indeed from the very conditions of modern urban life. In "Visions of a New Reality: The City and the Emergence of Modern Visual Culture," the image-essay that ends Part 3, Frederic Stout argues that popular illustrated journalism, photography, and cinema are among the most characteristic of urban cultural genres and that visual culture generally is an artifact of modern urban society.

LEWIS MUMFORD

"What Is a City?"

Architectural Record (1937)

Editors' introduction Lewis Mumford (1895–1990) has been called America's last great public intellectual. Beginning with his first book in 1922 and continuing throughout a career that saw the publication of some twenty-five influential volumes, Mumford made signal contributions to social philosophy, American literary and cultural history, the history of technology and, preeminently, the history of cities and urban planning practice.

Mumford saw the urban experience as an integral component in the development of human culture and the human personality. He consistently argued that the physical design of cities and their economic functions were secondary to their relationship to the natural environment and to the spiritual values of human community. Mumford applied these principles to his architectural criticism for *The New Yorker* magazine in the 1920s, his work with the Regional Planning Association of America, his campaign against plans to build a highway through Washington Square in New York's Greenwich Village in the 1950s, and his lifelong championing of the Garden City ideals of Ebenezer Howard.

In "What Is a City?" Mumford lays out his fundamental propositions about city planning and the human potential, both individual and social, of urban life. The city, he writes, is "a theater of social action," and everything else – art, politics, education, commerce – only serves to make the "social drama . . . more richly significant, as a stage-set, well-designed, intensifies and underlines the gestures of the actors and the action of the play." It was a theme and an image to which Mumford would return over and over again. In his chapter on "The Nature of the Ancient City" in *The City in History* (1961), he wrote that the city is "above all things a theater" and, as if commenting on the cultural conformity of the 1950s, warned that an urban civilization that has lost its sense of dramatic dialogue "is bound to have a fatal last act."

Mumford's influence on modern urban planning theory can hardly be overstated. His "urban drama" idea clearly resonates with an entire line of urban cultural analysts. Jane Jacobs, for example, talks about "street ballet" (p. 106). William Whyte (p. 483) says that a good urban plaza should function like a stage. Allan Jacobs and Donald Appleyard (p. 491) urge planners to fulfill human needs for "fantasy and exoticism." The city, they write, "has always been a place of excitement; it is a theater, a stage upon which citizens can display themselves and be seen by others."

As a historian, Mumford is the antithesis of Henri Pirenne (p. 37), whom Mumford considered too much of an economic determinist despite his "excellent basic scholarship." Mumford's emphasis on community values and the city's role in enlarging the potential of the human personality connects him with a long line of urban theorists that includes Louis Wirth (p. 97) and many others.

The City in History (New York: Harcourt Brace, 1961) is undoubtedly Mumford's masterpiece, but an earlier version of the same material, *The Culture of Cities* (New York: Harcourt Brace, 1938), is still of interest. *The Urban Prospect* (New York: Harcourt Brace, 1968) is an outstanding collection of his essays on urban planning and culture, and *The Myth of the Machine* (New York: Harcourt Brace, 1967) and *The Pentagon of Power* (New York: Harcourt Brace, 1970) are excellent analyses of the influence of technology on human culture. The magisterial *The Transformations of Man* (New York: Harper, 1956) invites comparison with V. Gordon Childe's theory of the urban revolution (p. 22). A sampling of Mumford's writings are included in Donald L. Miller (ed.), *The Lewis Mumford Reader* (Athens: University of Georgia Press, 1995).

Mumford's illuminating correspondence with Patrick Geddes (p. 330) is contained in Frank G. Novak, *Lewis Mumford and Patrick Geddes: The Correspondence* (London: Routledge, 1995).

Mumford is being rediscovered by the current generation of environmental planners. Examples of recent books applying his perspective to current ecological issues are Mark Luccarelli Lewis, *Mumford and the Ecological Region: The Politics of Planning* (New York: Guilford Press, 1997) and Robert Wojtowicz, *Lewis Mumford and American Modernism: Eutopian Theories for Architecture and Urban Planning* (Cambridge: Cambridge University Press, 1998).

Biographies of Lewis Mumford are Donald L. Miller's *Lewis Mumford: A Life* (New York: Weidenfeld & Nicolson, 1989), Thomas P. Hughes and Agatha C. Hughes (eds.), *Lewis Mumford: Public Intellectual* (Oxford: Oxford University Press, 1990), and Frank G. Novak, *Lewis Mumford* (New York: Twayne Publishers, 1998).

A bibliography of Mumford's writings is Elmer S. Newman, *Lewis Mumford: A Bibliography, 1914–1970* (New York: Harcourt Brace Jovanovich, 1971).

Most of our housing and city planning has been handicapped because those who have undertaken the work have had no clear notion of the social functions of the city. They sought to derive these functions from a cursory survey of the activities and interests of the contemporary urban scene. And they did not, apparently, suspect that there might be gross deficiencies, misdirected efforts, mistaken expenditures here that would not be set straight by merely building sanitary tenements or straightening out and widening irregular streets.

The city as a purely physical fact has been subject to numerous investigations. But what is the city as a social institution? The earlier answers to these questions, in Aristotle, Plato, and the Utopian writers from Sir Thomas More to Robert Owen, have been on the whole more satisfactory than those of the more systematic sociologists: most contemporary treatises on "urban sociology" in America throw no important light upon the problem. One of the soundest definitions of the city was that framed by John Stow, an honest observer of Elizabethan London, who said:

> Men are congregated into cities and commonwealths for honesty and utility's sake, these shortly be the commodities that do come by cities, commonalties and corporations. First, men by this nearness of conversation are withdrawn from barbarous fixity and force, to certain mildness of manners, and to humanity and justice . . . Good behavior is yet called urbanitas because it is rather found in cities than elsewhere. In sum, by often hearing, men be better persuaded in religion, and for that they live in the eyes of others, they be by example the more easily trained to justice, and by shamefastness restrained from injury.
>
> And whereas commonwealths and kingdoms cannot have, next after God, any surer foundation than the love and good will of one man towards another, that also is closely bred and maintained in cities, where men by mutual society and companying together, do grow to alliances, commonalties, and corporations.

It is with no hope of adding much to the essential insight of this description of the urban process that I would sum up the sociological concept of the city in the following terms:

The city is a related collection of primary groups and purposive associations: the first, like family and neighborhood, are common to all communities, while the second are especially characteristic of city life. These varied groups support themselves through economic organizations that are likewise of a more or less corporate, or at least publicly regulated, character; and they are all housed in permanent structures, within a relatively limited area. The essential physical means of a city's existence are the fixed site, the durable shelter, the permanent facilities for assembly, interchange, and storage; the essential social means are the social division of labor, which serves not merely the economic life but the cultural processes. The city in its complete sense, then, is a geographic plexus, an economic organization, an institutional process, a theater of social action, and an aesthetic symbol of collective unity. The city fosters art and is art; the city creates the theater and is the theater. It is in the city, the city as theater, that man's more purposive activities are focused, and work out, through conflicting and cooperating personalities, events, groups, into more significant culminations.

Without the social drama that comes into existence through the focusing and intensification of group activity there is not a single function performed in the city that could not be performed – and has not in fact been performed – in the open country. The physical organization of the city may deflate this drama or make it frustrate; or it may, through the deliberate efforts of art, politics, and education, make the drama more richly significant, as a stage-set, well-designed, intensifies and underlines the gestures of the actors and the action of the play. It is not for nothing that men have dwelt so often on the beauty or the ugliness of cities: these attributes qualify men's social activities. And if there is a deep reluctance on the part of the true city dweller to leave his cramped quarters for the physically more benign environment of a suburb – even a model garden suburb! – his instincts are usually justified: in its various and many-sided life, in its very opportunities for social disharmony and conflict, the city creates drama; the suburb lacks it.

One may describe the city, in its social aspect, as a special framework directed toward the creation of differentiated opportunities for a common life and a significant collective drama. As indirect forms of association, with the aid of signs and symbols and specialized organizations, supplement direct face-to-face intercourse, the personalities of the citizens themselves become many-faceted: they reflect their specialized interests, their more intensively trained aptitudes, their finer discriminations and selections: the personality no longer presents a more or less unbroken traditional face to reality as a whole. Here lies the possibility of personal disintegration; and here lies the need for reintegration through wider participation in a concrete and visible collective whole. What men cannot imagine as a vague formless society, they can live through and experience as citizens in a city. Their unified plans and buildings become a symbol of their social relatedness; and when the physical environment itself becomes disordered and incoherent, the social functions that it harbors become more difficult to express.

One further conclusion follows from this concept of the city: social facts are primary, and the physical organization of a city, its industries and its markets, its lines of communication and traffic, must be subservient to its social needs. Whereas in the development of the city during the last century we expanded the physical plant recklessly and treated the essential social nucleus, the organs of government and education and social service, as mere afterthought, today we must treat the social nucleus as the essential element in every valid city plan: the spotting and inter-relationship of schools, libraries, theaters, community centers is the first task in defining the urban neighborhood and laying down the outlines of an integrated city.

In giving this sociological answer to the question: What is a City? one has likewise provided the clue to a number of important other questions. Above all, one has the criterion for a clear decision as to what is the desirable size of a city – or may a city perhaps continue to grow until a single continuous urban area might cover half the American continent, with the rest of the world tributary to this mass? From the standpoint of the purely physical organization of urban utilities – which is almost the only matter upon which metropolitan planners in the past have concentrated – this latter process might indeed go on indefinitely. But if the city is a theater of social activity, and if its needs are defined by the opportunities it offers to differentiated social groups, acting through a specific nucleus of civic institutes and associations, definite limitations on size follow from this fact.

In one of Le Corbusier's early schemes for an ideal city, he chose three million as the number to be accommodated: the number was roughly the size of the urban aggregate of Paris, but that hardly explains why it should have been taken as a norm for a more rational type of city development. If the size of an urban unit, however, is a function of its productive organization and its opportunities for active social intercourse and culture, certain definite facts emerge as to adequate ratio of population to the process to be served. Thus, at the present level of culture in America, a million people are needed to support a university. Many factors may enter which will change the size of both the university and the population base; nevertheless one can say provisionally that if a million people are needed to provide a sufficient number of students for a university, then two million people should have two universities. One can also say that, other things being equal, five million people will not provide a more effective university than one million people would. The alternative to recognizing these ratios is to keep on overcrowding and overbuilding a few existing institutions, thereby limiting, rather than expanding, their genuine educational facilities.

What is important is not an absolute figure as to population or area: although in certain aspects of life, such as the size of city that is capable of reproducing itself through natural fertility, one can already lay down such figures. What is more important is to express size *always as a function of the social relationships to be served* . . . There is an optimum numerical size, beyond which each further increment of inhabitants creates difficulties out of all proportion to the benefits. There is also an optimum area of expansion, beyond which further urban growth tends to paralyze rather than to further important social relationships. Rapid means of transportation have given a regional area with a radius of from forty to a hundred miles, the unity that London and Hampstead had before the coming of the underground railroad. But the activities of small children are still bounded by a walking distance of about a quarter of a mile; and for men to congregate freely and frequently in neighborhoods the maximum distance means nothing, although it may properly define the area served for a selective minority by a university, a central reference library, or a completely equipped hospital. The area of potential urban settlement has been vastly increased by the motor car and the airplane; but the necessity for solid contiguous growth, for the purposes of intercourse, has in turn been lessened by the telephone and the radio. In the Middle Ages a distance of less than a half a mile from the city's center usually defined its utmost limits. The block-by-block accretion of the big city, along its corridor avenues, is in all important respects a denial of the vastly improved type of urban grouping that our fresh inventions have brought in. For all occasional types of intercourse, the region is the unit of social life but the region cannot function effectively, as a well-knit unit, if the entire area is densely filled with people – since their very presence will clog its arteries of traffic and congest its social facilities.

Limitations on size, density, and area are absolutely necessary to effective social intercourse; and they are therefore the most important instruments of rational economic and civic planning. The unwillingness in the past to establish such limits has been due mainly to two facts: the assumption that all upward changes in magnitude were signs of progress and automatically "good for business," and the belief that such limitations were essentially arbitrary, in that they proposed to "decrease economic opportunity" – that is, opportunity for profiting by congestion – and to halt the inevitable course of change. Both these objections are superstitious.

Limitations on height are now common in American cities; drastic limitations on density are the rule in all municipal housing estates in England: that which could not be done has been done. Such limitations do not obviously limit the population itself: they merely give the planner and administrator the opportunity to multiply the number of centers in which the population is housed, instead of permitting a few existing centers to aggrandize themselves on a monopolistic pattern. These limitations are necessary to break up the functionless, hypertrophied urban masses of the past. Under this mode of planning, the planner proposes to replace the "mononucleated city," as Professor Warren Thompson has called it, with a new type of "polynucleated city," in which a cluster of communities, adequately spaced and bounded, shall do duty for the badly organized mass city. Twenty such cities, in a region whose environment and whose resources were adequately planned, would have all the benefits of a metropolis that held a million people, without its ponderous disabilities: its capital frozen into unprofitable utilities, and its land values congealed at levels that stand in the way of effective adaptation to new needs.

Mark the change that is in process today. The emerging sources of power, transport, and communication do not follow the old highway network at all. Giant power strides over the hills, ignoring the limitations of wheeled vehicles; the airplane, even more liberated, flies over swamps and mountains, and terminates its journey, not on an avenue, but in a field. Even the highway for fast motor transportation abandons the pattern of the horse-and-buggy era. The new highways, like those of New Jersey and Westchester, to mention only examples drawn locally, are based more or less on a system definitively formulated by Benton MacKaye in his various papers on the Townless Highway. The most complete plans form an independent highway network, isolated both from the adjacent countryside and the towns that they bypass: as free from communal encroachments as the railroad system. In such a network no single center will, like the metropolis of old, become the focal point of all regional advantages: on the contrary, the "whole region" becomes open for settlement.

Even without intelligent public control, the likelihood is that within the next generation this dissociation and decentralization of urban facilities will go even farther. The Townless Highway begets the Highwayless Town in which the needs of close and continuous human association on all levels will be uppermost. This is just the opposite of the earlier mechanocentric picture of Roadtown, as pictured by Edgar Chambless and the Spanish projectors of the Linear City. For the highwayless town is based upon the notion of effective zoning of functions through initial public design, rather than by blind legal ordinances. It is a town in which the various functional parts of the structure are isolated topographically as urban islands, appropriately designed for their specific use with no attempt to provide a uniform plan of the same general pattern for the industrial, the commercial, the domestic, and the civic parts.

The first systematic sketch of this type of town was made by Messrs. Wright and Stein in their design for Radburn in 1929; a new type of plan that was repeated on a limited scale – and apparently in complete independence – by planners in Köln and Hamburg at about the same time. Because of restrictions on design that favored a conventional type of suburban house and stale architectural forms, the implications of this new type of planning were not carried very far in Radburn. But in outline the main relationships are clear: the differentiation of foot traffic from wheeled traffic in independent systems, the insulation of residence quarters from through roads; the discontinuous street pattern; the polarization of social life in specially spotted civic nuclei, beginning in the neighborhood with the school and the playground and the swimming pool. This type of planning was carried to a logical conclusion in perhaps the most functional and most socially intelligent of all Le Corbusier's many urban plans: that for Nemours in North Africa, in 1934.

Through these convergent efforts, the principles of the polynucleated city have been well established. Such plans must result in a fuller opportunity for the primary group, with all its habits of frequent direct meeting and face-to-face intercourse: they must also result in a more complicated pattern and a more comprehensive life for the region, for this geographic area can only now, for the first time, be treated as an instantaneous whole for all the functions of social existence. Instead of trusting to the mere massing of population to produce the necessary social concentration and social drama, we must now seek these results through deliberate local nucleation and a finer regional articulation. The words are jargon; but the importance of their meaning should not be missed. To embody these new possibilities in city life, which come to us not merely through better technical organization but through acuter sociological understanding, and to dramatize the activities themselves in appropriate individual and urban structures, forms the task of the coming generation.

LOUIS WIRTH

"Urbanism as a Way of Life"

American Journal of Sociology (1938)

Editors' introduction Louis Wirth (1897–1952) was a member of the famed "Chicago School" of urban sociology that included such academic luminaries as Ernest W. Burgess (author of "The Growth of the City," p. 153), Robert E. Park, St. Clair Drake, and Horace Cayton. Together, these scholars at the University of Chicago set out virtually to reinvent modern sociology by taking academic research to the streets and by using the city of Chicago itself as a "living laboratory" for the study of urban problems and social processes.

Wirth's major contribution to urban sociology was the formulation of nothing less fundamental than a meaningful and logically coherent "sociological definition" of urban life. As he lays it out in the magnificent synthesis that is his 1938 essay "Urbanism as a Way of Life," a "sociologically significant definition of the city" looks beyond the mere physical structure of the city, or its economic product, or its characteristic cultural institutions – however important all these may be – to discover those underlying "elements of urbanism which mark it as a distinctive mode of human group life."

Wirth argues that three key characteristics of cities – large population size, social heterogeneity, and population density – contribute to the development of a peculiarly "urban way of life" and, indeed, a distinct "urban personality." For centuries, at least as far back as Aesop's fable of the city mouse and the country mouse, casual observers have noted sharp personality differences between urban and rural people and between nature-based and machine-based styles of living. Wirth attempts to explain those differences in terms of the functional responses of urban dwellers to the characteristic environmental conditions of modern urban society. If, for example, city people are regarded as rather more socially tolerant than rural people – and, at the same time, more impersonal and seemingly less friendly – these are merely adaptations to the experience of living in large, dense, socially diverse urban environments.

Although some see Wirth's explanation of the sociology of urban life as nothing more than the social scientific verification of the obvious, others have argued that there is actually no such thing as an "urban personality" or an "urban way of life." Herbert Gans, for example, argues that both inner-city "urban villagers" and suburbanites tend to maintain their pre-existing cultures and personalities (p. 63), and Oscar Lewis's work on "the culture of poverty" suggests that culture and personality types differ widely with socio-economic class. Wirth's work, however, led to the development of a whole school of urban social ecology, and Wirth's basic ideas about personality and adaptation to urban conditions inform the full range of more recent urban planning theories and the planning practitioners who attempt to create and nurture a sense of community in the urban environment.

Other books by Louis Wirth include *Contemporary Social Problems* (Chicago: University of Chicago Press, 1940), *The Effect of War on American Minorities* (New York: Social Science Research Council, 1943), *Community Life and Social Policy* (Chicago: University of Chicago Press, 1956), and *The Ghetto* (Chicago: University of Chicago Press, 1956). For other important analyses of the relationship between urban life and the human personality, see Erving Goffman, *The Presentation of Self in Everyday Life* (Garden City, NY: Doubleday, 1959) and Richard Sennett, *The Uses of Disorder: Personal Identity and City Life* (New York: W. W. Norton, 1970).

THE CITY AND CONTEMPORARY CIVILIZATION

Just as the beginning of Western civilization is marked by the permanent settlement of formerly nomadic peoples in the Mediterranean basin, so the beginning of what is distinctively modern in our civilization is best signalized by the growth of great cities. Nowhere has mankind been farther removed from organic nature than under the conditions of life characteristic of great cities . . . The city and the country may be regarded as two poles in reference to one or the other of which all human settlements tend to arrange themselves. In viewing urban-industrial and rural-folk society as ideal types of communities, we may obtain a perspective for the analysis of the basic models of human association as they appear in contemporary civilization.

A SOCIOLOGICAL DEFINITION OF THE CITY

Despite the preponderant significance of the city in our civilization, however, our knowledge of the nature of urbanism and the process of urbanization is meager. Many attempts have indeed been made to isolate the distinguishing characteristics of urban life. Geographers, historians, economists, and political scientists have incorporated the points of view of their respective disciplines into diverse definitions of the city. While it is in no sense intended to supersede these, the formulation of a sociological approach to the city may incidentally serve to call attention to the interrelations between them by emphasizing the peculiar characteristics of the city as a particular form of human association. A sociologically significant definition of the city seeks to select those elements of urbanism which mark it as a distinctive mode of human group life.

[. . .]

While urbanism, or that complex of traits which makes up the characteristic mode of life in cities, and urbanization, which denotes the development and extensions of these factors, are thus not exclusively found in settlements which are cities in the physical and demographic sense, they do, nevertheless, find their most pronounced expression in such areas, especially in metropolitan cities. In formulating a definition of the city it is necessary to exercise caution in order to avoid identifying urbanism as a way of life with any specific locally or historically conditioned cultural influences which, while they may significantly affect the specific character of the community, are not the essential determinants of its character as a city.

It is particularly important to call attention to the danger of confusing urbanism with industrialism and modern capitalism. The rise of cities in the modern world is undoubtedly not independent of the emergence of modern power-driven machine technology, mass production, and capitalistic enterprise. But different as the cities of earlier epochs may have been by virtue of their development in a preindustrial and precapitalistic order from the great cities of today, they were, nevertheless, cities.

For sociological purposes a city may be defined as a relatively large, dense, and permanent settlement of socially heterogeneous individuals. On the basis of the postulates which this minimal definition suggests, a theory of urbanism may be formulated in the light of existing knowledge concerning social groups.

A THEORY OF URBANISM

In the rich literature on the city we look in vain for a theory of urbanism presenting in a systematic fashion the available knowledge concerning the city as a social entity. We do indeed have excellent formulations of theories on such special problems as the growth of the city viewed as a historical trend and as a recurrent process, and we have a wealth of literature presenting insights of sociological relevance and empirical studies offering detailed information on a variety of particular aspects of urban life. But despite the multiplication of research and textbooks on the city, we do not as yet have a comprehensive body of competent hypotheses which may be derived from a set of postulates implicitly contained in a sociological definition of the city, and from our general sociological knowledge which may be substantiated through empirical research. The closest approximations to a systematic theory of urbanism that we have are to be found in a penetrating essay, "Die Stadt," by Max Weber, and a memorable paper by Robert E. Park titled "The City: Suggestions for the Investigation of Human Behavior in the Urban Environment." But even these excellent contributions are far from constituting an ordered and coherent framework of theory upon which research might profitably proceed.

In the pages that follow, we shall seek to set forth a limited number of identifying characteristics of the city. Given these characteristics we shall then indicate what consequences or further characteristics follow from them in the light of general sociological theory and empirical research. We hope in this manner to arrive at the essential propositions comprising a theory of urbanism. Some of these propositions can be supported by a considerable body of already available research materials; others may be accepted as hypotheses for which a certain amount of presumptive evidence exists, but for which more ample and exact verification would be required. At least such a procedure will, it is hoped, show what in the way of systematic knowledge of the city we now have and what are the crucial and fruitful hypotheses for future research.

[. . .]

There are a number of sociological propositions concerning the relationship between (a) numbers of population, (b) density of settlement, (c) heterogeneity of inhabitants and group life, which can be formulated on the basis of observation and research.

SIZE OF THE POPULATION AGGREGATE

Ever since Aristotle's *Politics*, it has been recognized that increasing the number of inhabitants in a settlement beyond a certain limit will affect the relationships between them and the character of the city. Large numbers involve, as has been pointed out, a greater range of individual variation. Furthermore, the greater the number of individuals participating in a process of interaction, the greater is the potential differentiation between them. The personal traits, the occupations, the cultural life, and the ideas of the members of an urban community may, therefore, be expected to range between more widely separated poles than those of rural inhabitants.

That such variations should give rise to the spatial segregation of individuals according to color, ethnic heritage, economic and social status, tastes and preferences, may readily be inferred. The bonds of kinship, of neighborliness, and the sentiments arising out of living together for generations under a common folk tradition are likely to be absent or, at best, relatively weak in an aggregate the members of which have such diverse origins and backgrounds. Under such circumstances competition and formal control mechanisms furnish the substitutes for the bonds of solidarity that are relied upon to hold a folk society together.

[. . .]

The multiplication of persons in a state of interaction under conditions which make their contact as full personalities impossible produces that segmentalization of human relationships which has sometimes been seized upon by students of the mental life of the cities as an explanation for the "schizoid" character of urban personality. This is not to say that the urban inhabitants have fewer acquaintances than rural inhabitants, for the reverse may actually be true; it means rather that in relation to the number of people whom they see and with whom they rub elbows in the course of daily life, they know a smaller proportion, and of these they have less intensive knowledge.

Characteristically, urbanites meet one another in highly segmental roles. They are, to be sure, dependent upon more people for the satisfactions of their life-needs than are rural people and thus are associated with a greater number of organized groups, but they are less dependent upon particular persons, and their dependence upon others is confined to a highly fractionalized aspect of the other's round of activity. This is essentially what is meant by saying that the city is characterized by secondary rather than primary contacts. The contacts of the city may indeed be face to face, but they are nevertheless impersonal, superficial, transitory, and segmental. The reserve, the indifference, and the blasé outlook which urbanites manifest in their relationships may thus be regarded as devices for immunizing themselves against the personal claims and expectations of others.

The superficiality, the anonymity, and the transitory character of urban social relations make intelligible, also, the sophistication and the rationality generally ascribed to city-dwellers. Our acquaintances tend to stand in a relationship of utility to us in the sense that the role which each one plays in our life is overwhelmingly regarded as a means for the achievement of our own ends. Whereas, therefore, the individual gains, on the one hand, a certain degree of emancipation or freedom from the personal and emotional controls of intimate groups, he loses, on the other hand, the spontaneous self-expression, the morale, and the sense of participation that comes with living in an integrated society. This constitutes essentially the state of anomie or the social void to which Durkheim alludes in attempting to account for the various forms of social disorganization in technological society.

The segmental character and utilitarian accent of interpersonal relations in the city find their institutional expression in the proliferation of specialized tasks which we see in their most developed form in the professions. The operations of the pecuniary nexus lead to predatory relationships, which tend to obstruct the efficient functioning of the social order unless checked by professional codes and occupational etiquette. The premium put upon utility and efficiency suggests the adaptability of the corporate device for the organization of enterprises in which individuals can engage only in groups. The advantage that the corporation has over the individual entrepreneur and the partnership in the urban-industrial world derives not only from the possibility it affords of centralizing the resources of thousands of individuals or from the legal privilege of limited liability and perpetual succession, but from the fact that the corporation has no soul.

[. . .]

DENSITY

As in the case of numbers, so in the case of concentration in limited space certain consequences of relevance in sociological analysis of the city emerge. Of these only a few can be indicated.

As Darwin pointed out for flora and fauna and as Durkheim noted in the case of human societies, an increase in numbers when area is held constant (i.e. an increase in density) tends to produce differentiation and specialization, since only in this way can the area support increased numbers. Density thus reinforces the effect of numbers in diversifying men and their activities and in increasing the complexity of the social structure.

On the subjective side, as Simmel has suggested, the close physical contact of numerous individuals necessarily produces a shift in the mediums through which we orient ourselves to the urban milieu, especially to our fellow-men. Typically, our physical contacts are close but our social contacts are distant. The urban world puts a premium on visual recognition. We see the uniform which denotes the role of the functionaries and are oblivious to the personal eccentricities that are hidden behind the uniform. We tend to acquire and develop a sensitivity to a world of artifacts and become progressively farther removed from the world of nature.

We are exposed to glaring contrasts between splendor and squalor, between riches and poverty, intelligence and ignorance, order and chaos. The competition for space is great, so that each area generally tends to be put to the use which yields the greatest economic return. Place of work tends to become dissociated from place of residence, for the proximity of industrial and commercial establishments makes an area both economically and socially undesirable for residential purposes.

Density, land values, rentals, accessibility, healthfulness, prestige, aesthetic consideration, absence of nuisances such as noise, smoke, and dirt determine the desirability of various areas of the city as places of settlement for different sections of the population . . . The different parts of the city thus acquire specialized functions. The city consequently tends to resemble a mosaic of social worlds in which the transition from one to the other is abrupt. The juxtaposition of divergent personalities and modes of life tends to produce a relativistic perspective and a sense of toleration of differences which may be regarded as prerequisites for rationality and which lead toward the secularization of life.

The close living together and working together of individuals who have no sentimental and emotional ties foster a spirit of competition, aggrandizement, and mutual exploitation. To counteract irresponsibility and potential disorder, formal controls tend to be resorted to. Without rigid adherence to predictable routines a large, compact society would scarcely be able to maintain itself. The clock and the traffic signal are symbolic of the basis of our social order in the urban world. Frequent close physical contact, coupled with great social distance, accentuates the reserve of unattached individuals toward one another and, unless compensated for by other opportunities for response, gives rise to loneliness. The necessary frequent movement of great numbers of individuals in a congested habitat gives occasion to friction and irritation. Nervous tensions which derive from such personal frustrations are accentuated by the rapid tempo and the complicated technology under which life in dense areas must be lived.

HETEROGENEITY

The social interaction among such a variety of personality types in the urban milieu tends to break down the rigidity of caste lines and to complicate the class structure, and thus induces a more ramified and differentiated framework of social stratification than is found in more integrated societies. The heightened mobility of the individual, which brings him within the range of stimulation by a great number of diverse individuals and subjects him to fluctuating status in the differentiated social groups that compose the social structure of the city, tends toward the acceptance of instability and insecurity in the world at large as a norm. This fact helps to account, too, for the sophistication and cosmopolitanism of the urbanite. No single group has the undivided allegiance of the individual. The groups with which he is affiliated do not lend themselves readily to a simple hierarchical arrangement. By virtue of his different interests arising out of different aspects of social life, the individual acquires membership in widely divergent groups, each of which functions only with reference to a single segment of his personality. Nor do these groups easily permit of a concentric arrangement so that the narrower ones fall within the circumference of the more inclusive ones, as is more likely to be the case in the rural community or in primitive societies. Rather the groups with which the person typically is affiliated are tangential to each other or intersect in highly variable fashion.

Partly as a result of the physical footlooseness of the population and partly as a result of their social mobility, the turnover in group membership generally is rapid. Place of residence, place and character of employment, income and interests fluctuate, and the task of holding organizations together and maintaining and promoting intimate and lasting acquaintanceship between the members is difficult. This applies strikingly to the local areas within the city into which persons become segregated more by virtue of differences in race, language, income, and social status, than through choice or positive attraction to people like themselves. Overwhelmingly the city-dweller is not a homeowner, and since a transitory habitat does not generate binding traditions and sentiments, only rarely is he truly a neighbor. There is little opportunity for the individual to obtain a conception of the city as a whole or to survey his place in the total scheme. Consequently he finds it difficult to determine what is to his own "best interests" and to decide between the issues and leaders presented to him by the agencies of mass suggestion. Individuals who are thus detached from the organized bodies which integrate society comprise the fluid masses that make collective behavior in the urban community so unpredictable and hence so problematical.

Although the city, through the recruitment of variant types to perform its diverse tasks and the accentuation of their uniqueness through competition and the premium upon eccentricity, novelty, efficient performance, and inventiveness, produces a highly differentiated population, it also exercises a leveling influence. Wherever large numbers of differently constituted individuals congregate, the process of depersonalization also enters . . . Individuality under these circumstances must be replaced by categories. When large numbers have to make common use of facilities and institutions, an arrangement must be made to adjust the facilities and institutions to the needs of the average person rather than to those of particular individuals. The services of the public utilities, of the recreational, educational, and cultural institutions, must be adjusted to mass requirements. Similarly, the cultural institutions, such as the schools, the movies, the radio, and the newspapers, by virtue of their mass clientele, must necessarily operate as leveling influences. The political process as it appears in urban life could not be understood without taking account of the mass appeals made through modern propaganda techniques. If the individual would participate at all in the social, political, and economic life of the city, he must subordinate some of his individuality to the demands of the larger community and in that measure immerse himself in mass movements.

THE RELATION BETWEEN A THEORY OF URBANISM AND SOCIOLOGICAL RESEARCH

By means of a body of theory such as that illustratively sketched above, the complicated and many-sided phenomena of urbanism may be analyzed in terms of a limited number of basic categories. The sociological approach to the city thus acquires an essential unity and coherence enabling the empirical investigator not merely to focus more distinctly upon the problems and processes that properly fall in his province but also to treat his subject matter in a more integrated and systematic fashion. A few typical findings of empirical research in the field of urbanism, with special reference to the United States, may be indicated to substantiate the theoretical propositions set forth in the preceding pages, and some of the crucial problems for further study may be outlined.

On the basis of the three variables, number, density of settlement, and degree of heterogeneity, of the urban population, it appears possible to explain the characteristics of urban life and to account for the differences between cities of various sizes and types.

Urbanism as a characteristic mode of life may be approached empirically from three interrelated perspectives: (1) as a physical structure comprising a population base, a technology, and an ecological order; (2) as a system of social organization involving a characteristic social structure, a series of social institutions, and a typical pattern of social relationships; and (3) as a set of attitudes and ideas, and a constellation of personalities engaging in typical forms of collective behavior and subject to characteristic mechanisms of social control

URBANISM IN ECOLOGICAL PERSPECTIVE

Since in the case of physical structure and ecological processes we are able to operate with fairly objective indices, it becomes possible to arrive at quite precise and generally quantitative results. The dominance of the city over its hinterland becomes explicable through the functional characteristics of the city which derive in large measure from the effect of numbers and density. Many of the technical facilities and the skills and organizations to which urban life gives rise can grow and prosper only in cities where the demand is sufficiently great. The nature and scope of the services rendered by these organizations and institutions and the advantage which they enjoy over the less developed facilities of smaller towns enhances the dominance of the city and the dependence of ever wider regions upon the central metropolis.

The urban population composition shows the operation of selective and differentiating factors. Cities contain a larger proportion of persons in the prime of life than rural areas which contain more old and very young people. In this, as in so many other respects, the larger the city the more this specific characteristic of urbanism is apparent. With the exception of the largest cities, which have attracted the bulk of the foreign-born males, and a few other special types of cities, women predominate numerically over men. The heterogeneity of the urban population is further indicated along racial and ethnic lines. The foreign born and their children constitute nearly two-thirds of all the inhabitants of cities of one million and over. Their proportion in the urban population declines as the size of the city decreases, until in the rural areas they comprise only about one-sixth of the total population. The larger cities similarly have attracted more Negroes and other racial groups than have the smaller communities. Considering that age, sex, race, and ethnic origin are associated with other factors such as occupation and interest, it becomes clear that one major characteristic of the urban-dweller is his dissimilarity from his fellows. Never before have such large masses of people of diverse traits as we find in our cities been thrown together into such close physical contact as in the great cities of America. Cities generally, and American cities in particular, comprise a motley of peoples and cultures, of highly differentiated modes of life between which there often is only the faintest communication, the greatest indifference and the broadest tolerance, occasionally bitter strife, but always the sharpest contrast.

The failure of the urban population to reproduce itself appears to be a biological consequence of a combination of factors in the complex of urban life, and the decline in the birth-rate generally may be regarded as one of the most significant signs of the urbanization of the Western world. While the proportion of deaths in cities is slightly greater than in the country, the outstanding difference between the failure of present-day cities to maintain their population and that of cities of the past is that in former times it was due to the exceedingly high death-rates in cities, whereas today, since cities have become more livable from a health standpoint, it is due to low birth-rates. These biological characteristics of the urban population are significant sociologically, not merely because they reflect the urban mode of existence but also because they condition the growth and future dominance of cities and their basic social organization. Since cities are the consumers rather than the producers of men, the value of human life and the social estimation of the personality will not be unaffected by the balance between births and deaths. The pattern of land use, of land values, rentals, and ownership, the nature and functioning of the physical structures, of housing, of transportation and communication facilities, of public utilities – these and many other phases of the physical mechanism of the city are not isolated phenomena unrelated to the city as a social entity, but are affected by and affect the urban mode of life

URBANISM AS A FORM OF SOCIAL ORGANIZATION

The distinctive features of the urban mode of life have often been described sociologically as consisting of the substitution of secondary for primary contacts, the weakening of bonds of kinship, and the declining social significance of the family, the disappearance of the neighborhood, and the undermining of the traditional basis of social solidarity. All these phenomena can be substantially verified through objective indices. Thus, for instance, the low and declining urban reproduction rates suggest that the city is not conducive to the traditional type of family life, including the rearing of children and the maintenance of the home as the locus of a whole round of vital activities. The transfer of industrial, educational, and recreational activities to specialized institutions outside the home has deprived the family of some of its most characteristic historical functions. In cities mothers are more likely to be employed, lodgers are more frequently part of the household, marriage tends to be postponed, and the proportion of single and unattached people is greater. Families are smaller and more frequently without children than in the country. The family as a unit of social life is emancipated from the larger kinship group characteristic of the country, and the individual members pursue their own diverging interests in their vocational, educational, religious, recreational, and political life.

[. . .]

On the whole, the city discourages an economic life in which the individual in time of crisis has a basis of subsistence to fall back upon, and it discourages self-employment. While incomes of city people are on the average higher than those of country people, the cost of living seems to be higher in the larger cities. Home ownership involves greater burdens and is rarer. Rents are higher and absorb a large proportion of the income. Although the urban-dweller has the benefit of many communal services, he spends a large proportion of his income for such items as recreation and advancement and a smaller proportion for food. What the communal services do not furnish the urbanite must purchase, and there is virtually no human need which has remained unexploited by commercialism. Catering to thrills and furnishing means of escape from drudgery, monotony, and routine thus become one of the major functions of urban recreation, which at its best furnishes means for creative self-expression and spontaneous group association, but which more typically in the urban world results in passive spectatorism on the one hand, or sensational record-smashing feats on the other.

Being reduced to a stage of virtual impotence as an individual, the urbanite is bound to exert himself by joining with others of similar interest into organized groups to obtain his ends. This results in the enormous multiplication of voluntary organizations directed toward as great a variety of objectives as there are human needs and interests. While on the one hand the traditional ties of human association are weakened, urban existence involves a much greater degree of interdependence between man and man and a more complicated, fragile, and volatile form of mutual interrelations over many phases of which the individual as such can exert scarcely any control. Frequently there is only the most tenuous relationship between the economic position or other basic factors that determine the individual's existence in the urban world and the voluntary groups with which he is affiliated. While in a primitive and in a rural society it is generally possible to predict on the basis of a few known factors who will belong to what and who will associate with whom in almost every relationship of life, in the city we can only project the general pattern of group formation and affiliation, and this pattern will display many incongruities and contradictions.

URBAN PERSONALITY AND COLLECTIVE BEHAVIOR

It is largely through the activities of the voluntary groups, be their objectives economic, political, educational, religious, recreational, or cultural, that the urbanite expresses and develops his personality, acquires status, and is able to carry on the round of activities that constitute his life-career. It may easily be inferred, however, that the organizational framework which these highly differentiated functions call into being does not of itself insure the consistency and integrity of the personalities whose interests it enlists. Personal disorganization, mental breakdown, suicide, delinquency, crime, corruption, and disorder might be expected under these circumstances to be more prevalent in the urban than in the rural community. This has been confirmed in so far as comparable indices are available; but the mechanisms underlying these phenomena require further analysis.

Since for most group purposes it is impossible in the city to appeal individually to the large number of discrete and differentiated individuals, and since it is only through the organizations to which men belong that their interests and resources can be enlisted for a collective cause, it may be inferred that social control in the city should typically proceed through formally organized groups. It follows, too, that the masses of men in the city are subject to manipulation by symbols and stereotypes managed by individuals working from afar or operating invisibly behind the scenes through their control of the instruments of communication. Self-government either in the economic, the political, or the cultural realm is under these circumstances reduced to a mere figure of speech or, at best, is subject to the unstable equilibrium of pressure groups. In view of the ineffectiveness of actual kinship ties we create fictional kinship groups. In the face of the disappearance of the territorial unit as a basis of social solidarity we create interest units. Meanwhile the city as a community resolves itself into a series of tenuous segmental relationships superimposed upon a territorial base with a definite center but without a definite periphery and upon a division of labor which far transcends the immediate locality and is world-wide in scope. The larger the number of persons in a state of interaction with one another the lower is the level of communication and the greater is the tendency for communication to proceed on an elementary level, i.e. on the basis of those things which are assumed to be common or to be of interest to all.

It is obviously, therefore, to the emerging trends in the communication system and to the production and distribution technology that has come into existence with modern civilization that we must look for the symptoms which will indicate the probable future development of urbanism as a mode of social life. The direction of the ongoing changes in urbanism will for good or ill transform not only the city but the world. Some of the more basic of these factors and processes and the possibilities of their direction and control invite further detailed study.

It is only insofar as the sociologist has a clear conception of the city as a social entity and a workable theory of urbanism that he can hope to develop a unified body of reliable knowledge, which what passes as "urban sociology" is certainly not at the present time. By taking his point of departure from a theory of urbanism such as that sketched in the foregoing pages to be elaborated, tested, and revised in the light of further analysis and empirical research, it is to be hoped that the criteria of relevance and validity of factual data can be determined. The miscellaneous assortment of disconnected information which has hitherto found its way into sociological treatises on the city may thus be sifted and incorporated into a coherent body of knowledge. Incidentally, only by means of some such theory will the sociologists escape the futile practice of voicing in the name of sociological science a variety of often unsupportable judgments concerning such problems as poverty, housing, city-planning, sanitation, municipal administration, policing, marketing, transportation, and other technical issues. While the sociologist cannot solve any of these practical problems – at least not by himself – he may, if he discovers his proper function, have an important contribution to make to their comprehension and solution. The prospects for doing this are brightest through a general, theoretical, rather than through an *ad hoc* approach.

JANE JACOBS

"The Uses of Sidewalks: Safety"

from *The Death and Life of Great American Cities* (1961)

Editors' introduction Jane Jacobs started writing about city life and urban planning as a neighborhood activist, not as a trained professional. Dismissed as the original "little old lady in tennis shoes" and derided as a political amateur more concerned about personal safety issues than state-of-the-art planning techniques, she nonetheless struck a responsive chord with a 1960s public eager to believe the worst about arrogant city planning technocrats and just as eager to rally behind movements for neighborhood control and community resistance to bulldozer redevelopment.

The Death and Life of Great American Cities hit the world of city planning like an earthquake when it appeared in 1961. The book was a frontal attack on the planning establishment. Jacobs derided urban renewal as a process that only served to create instant slums. She questioned universally accepted articles of faith – for example, that parks were good and that crowding was bad. Indeed she suggested that parks were often dangerous and that crowded neighborhood sidewalks were the safest places for children to play. Jacobs ridiculed the planning establishment's most revered historical traditions as "the Radiant Garden City Beautiful" – an artful phrase that not only airily dismissed the contributions of Le Corbusier (p. 336) , Ebenezer Howard (p. 321), and Daniel Burnham but lumped them together as well!

The selection from *The Death and Life of Great American Cities* reprinted here presents Jane Jacobs at her very best. In "The Uses of Sidewalks: Safety," she outlines her basic notions of what makes a neighborhood a community and what makes a city livable. Safety – particularly for women and children – comes from "eyes on the street," the kind of involved neighborhood surveillance of public space that modern planning practice in the Corbusian tradition had destroyed with its insistence on superblocks and skyscraper developments. A sense of personal belonging and social cohesiveness comes from well-defined neighborhoods and narrow, crowded, multi-use streets. And, finally, basic urban vitality comes from residents' participation in an intricate "street ballet," a diurnal pattern of observable and comprehensible human activity that is possible only in places like Jacobs's own Hudson Street in her beloved Greenwich Village.

It was this last quality, her unabashed love of cities and urban life, that is Jane Jacobs's most obvious and enduring characteristic. *The Death and Life of Great American Cities* was a scathing attack on the planning establishment – and, in many ways, it was a grassroots political call to arms – but it was also a loving invitation to experience the joys of city living that led many young, college-educated people to seek out neighborhoods like Greenwich Village as places to live, struggle, and raise families. In one sense, the book encouraged and justified middle-class gentrification of formerly working-class neighborhoods. In another, it found itself oddly reflected in the fantasy-nostalgia of "Sesame Street." But in all ways it was committedly urban, never suburban, at a time when inner-city communities were being increasingly abandoned to the forces of poverty, decay, and neglect.

Contrast Louis Wirth's theory of how population size, density, and heterogeneity in cities create a distinct urban personality (p. 97) with Jacobs's argument that these very same city characteristics may create neighborhood vitality, social cohesion, and the perception and reality of safety. Jacobs's notion of the "street ballet" invites comparison with Lewis Mumford's idea of the "urban drama" (p. 92) and William Whyte's emphasis on the importance of public plazas (p. 483). Jacobs's community activism in resistance to urban renewal places her within a long tradition that includes Paul Davidoff's "Advocacy and Pluralism in Planning" (p. 423), and Sherry Arnstein's "A Ladder of Citizen Participation" (p. 240).

Other important works by Jane Jacobs include *The Economy of Cities* (New York: Random House, 1969) and *Systems of Survival* (New York: Random House, 1992). In the former book Jacobs again turns conventional explanation on its head by arguing that the rise of cities may have preceded, and even accounted for, rural agricultural development. The latter is a Platonic dialogue on "the moral foundations of commerce and politics."

Streets in cities serve many purposes besides carrying vehicles, and city sidewalks – the pedestrian parts of the streets – serve many purposes besides carrying pedestrians. These uses are bound up with circulation but are not identical with it and in their own right they are at least as basic as circulation to the proper workings of cities.

A city sidewalk by itself is nothing. It is an abstraction. It means something only in conjunction with the buildings and other uses that border it, or border other sidewalks very near it. The same might be said of streets, in the sense that they serve other purposes besides carrying wheeled traffic in their middles. Streets and their sidewalks, the main public places of a city, are its most vital organs. Think of a city and what comes to mind? Its streets. If a city's streets look interesting, the city looks interesting; if they look dull, the city looks dull.

More than that, and here we get down to the first problem, if a city's streets are safe from barbarism and fear, the city is thereby tolerably safe from barbarism and fear. When people say that a city, or a part of it, is dangerous or is a jungle what they mean primarily is that they do not feel safe on the sidewalks. But sidewalks and those who use them are not passive beneficiaries of safety or helpless victims of danger. Sidewalks, their bordering uses, and their users, are active participants in the drama of civilization versus barbarism in cities. To keep the city safe is a fundamental task of a city's streets and its sidewalks.

This task is totally unlike any service that sidewalks and streets in little towns or true suburbs are called upon to do. Great cities are not like towns, only larger. They are not like suburbs, only denser. They differ from towns and suburbs in basic ways, and one of these is that cities are, by definition, full of strangers. To any one person, strangers are far more common in big cities than acquaintances. More common not just in places of public assembly, but more common at a man's own doorstep. Even residents who live near each other are strangers, and must be, because of the sheer number of people in small geographical compass.

The bedrock attribute of a successful city district is that a person must feel personally safe and secure on the street among all these strangers. He must not feel automatically menaced by them. A city district that fails in this respect also does badly in other ways and lays up for itself, and for its city at large, mountain on mountain of trouble.

Today barbarism has taken over many city streets, or people fear it has, which comes to much the same thing in the end. "I live in a lovely, quiet residential area," says a friend of mine who is hunting another place to live. "The only disturbing sound at night is the occasional scream of someone being mugged." It does not take many incidents of violence on a city street, or in a city district, to make people fear the streets ... And as they fear them, they use them less, which makes the streets still more unsafe.

To be sure, there are people with hobgoblins in their heads, and such people will never feel safe no matter what the objective circumstances are. But this is a different matter from the fear that besets normally prudent, tolerant and cheerful people who show nothing more than common sense in refusing to venture after dark – or in a few places, by day – into streets where they may well be assaulted, unseen or unrescued until too late. The barbarism and the real, not imagined, insecurity that gives rise to such fears cannot be tagged a problem of the slums. The problem is most serious, in fact, in genteel-looking "quiet residential areas" like that my friend was leaving.

It cannot be tagged as a problem of older parts of cities. The problem reaches its most baffling dimensions in some examples of rebuilt parts of cities, including supposedly the best examples of rebuilding, such as middle-income projects. The police precinct captain of a nationally admired project of this kind (admired by planners and lenders) has recently admonished residents not only about hanging around outdoors after dark but has urged them never to answer their doors without knowing the caller. Life here has much in common with life for the three little pigs or the seven little kids of the nursery thrillers. The problem of sidewalk and doorstep insecurity is as serious in cities which have made conscientious efforts at rebuilding as it is in those cities that have lagged. Nor is it illuminating to tag minority groups, or the poor, or the outcast with responsibility for city danger. There are immense variations in the degree of civilization and safety found among such groups and among the city areas where they live. Some of the safest sidewalks in New York City, for example, at any time of day or night, are those along which poor people or minority groups live. And some of the most dangerous are in streets occupied by the same kinds of people. All this can also be said of other cities.

[. . .]

The first thing to understand is that the public peace – the sidewalk and street peace – of cities is not kept primarily by the police, necessary as police are. It is kept primarily by an intricate, almost unconscious, network of voluntary controls and standards among the people themselves, and enforced by the people themselves. In some city areas – older public housing projects and streets with very high population turnover are often conspicuous examples – the keeping of public sidewalk law and order is left almost entirely to the police and special guards. Such places are jungles. No amount of police can enforce civilization where the normal, casual enforcement of it has broken down.

The second thing to understand is that the problem of insecurity cannot be solved by spreading people out more thinly, trading the characteristics of cities for the characteristics of suburbs. If this could solve danger on the city streets, then Los Angeles should be a safe city because superficially Los Angeles is almost all suburban. It has virtually no districts compact enough to qualify as dense city areas. Yet Los Angeles cannot, any more than any other great city, evade the truth that, being a city, it is composed of strangers not all of whom are nice. Los Angeles' crime figures are flabbergasting. Among the seventeen standard metropolitan areas with populations over a million, Los Angeles stands so pre-eminent in crime that it is in a category by itself. And this is markedly true of crimes associated with personal attack, the crimes that make people fear the streets.

[. . .]

This is something everyone already knows: A well-used city street is apt to be a safe street. A deserted city street is apt to be unsafe. But how does this work, really? And what makes a city street well used or shunned? . . . What about streets that are busy part of the time and then empty abruptly?

A city street equipped to handle strangers, and to make a safety asset, in itself, out of the presence of strangers, as the streets of successful city neighborhoods always do, must have three main qualities:

First, there must be a clear demarcation between what is public space and what is private space. Public and private spaces cannot ooze into each other as they do typically in suburban settings or in projects.

Second, there must be eyes upon the street, eyes belonging to those we might call the natural proprietors of the street. The buildings on a street equipped to handle strangers and to insure the safety of both residents and strangers must be oriented to the street. They cannot turn their backs or blank sides on it and leave it blind.

And third, the sidewalk must have users on it fairly continuously, both to add to the number of effective eyes on the street and to induce the people in buildings along the street to watch the sidewalks in sufficient numbers. Nobody enjoys sitting on a stoop or looking out a window at an empty street. Almost nobody does such a thing. Large numbers of people entertain themselves, off and on, by watching street activity.

In settlements that are smaller and simpler than big cities, controls on acceptable public behavior, if not on crime, seem to operate with greater or lesser success through a web of reputation, gossip, approval, disapproval and sanctions, all of which are powerful if people know each other and word travels. But a city's streets, which must control the behavior not only of the people of the city but also of visitors from suburbs and towns who want to have a big time away from the gossip and sanctions at home, have to operate by more direct, straightforward methods. It is a wonder cities have solved such an inherently difficult problem at all. And yet in many streets they do it magnificently.

It is futile to try to evade the issue of unsafe city streets by attempting to make some other features of a locality, say interior courtyards, or sheltered play spaces, safe instead. By definition again, the streets of a city must do most of the job of handling strangers, for this is where strangers come and go. The streets must not only defend the city against predatory strangers, they must protect the many, many peaceable and well-meaning strangers who use them, insuring their safety too as they pass through. Moreover, no normal person can spend his life in some artificial haven, and this includes children. Everyone must use the streets.

On the surface, we seem to have here some simple aims: to try to secure streets where the public space is unequivocally public, physically unmixed with private or with nothing-at-all space, so that the area needing surveillance has clear and practicable limits; and to see that these public street spaces have eyes on them as continuously as possible.

But it is not so simple to achieve these objects, especially the latter. You can't make people use streets they have no reason to use. You can't make people watch streets they do not want to watch. Safety on the streets by surveillance and mutual policing of one another sounds grim, but in real life it is not grim. The safety of the street works best, most casually, and with least frequent taint of hostility or suspicion precisely where people are using and most enjoying the city streets voluntarily and are least conscious, normally, that they are policing.

The basic requisite for such surveillance is a substantial quantity of stores and other public places sprinkled along the sidewalks of a district; enterprises and public places that are used by evening and night must be among them especially. Stores, bars and restaurants, as the chief examples, work in several different and complex ways to abet sidewalk safety.

First, they give people – both residents and strangers – concrete reasons for using the sidewalks on which these enterprises face.

Second, they draw people along the sidewalks past places which have no attractions to public use in themselves but which become traveled and peopled as routes to somewhere else; this influence does not carry very far geographically, so enterprises must be frequent in a city district if they are to populate with walkers those other stretches of street that lack public places along the sidewalk. Moreover, there should be many different kinds of enterprises, to give people reasons for crisscrossing paths.

Third, storekeepers and other small businessmen are typically strong proponents of peace and order themselves; they hate broken windows and holdups; they hate having customers made nervous about safety. They are great street watchers and sidewalk guardians if present in sufficient numbers.

Fourth, the activity generated by people on errands, or people aiming for food or drink, is itself an attraction to still other people.

This last point, that the sight of people attracts still other people, is something that city planners and city architectural designers seem to find incomprehensible. They operate on the premise that city people seek the sight of emptiness, obvious order and quiet. Nothing could be less true. People's love of watching activity and other people is constantly evident in cities everywhere.

[. . .]

Under the seeming disorder of the old city, wherever the old city is working successfully, is a marvelous order for maintaining the safety of the streets and the freedom of the city. It is a complex order. Its essence is intricacy of sidewalk use, bringing with it a constant succession of eyes. This order is all composed of movement and change, and although it is life, not art, we may fancifully call it the art form of the city and liken it to the dance – not to a simple-minded precision dance with everyone kicking up at the same time, twirling in unison and bowing off en masse, but to an intricate ballet in which the individual dancers and ensembles all have distinctive parts which miraculously reinforce each other and compose an orderly whole. The ballet of the good city sidewalk never repeats itself from place to place, and in any one place is always replete with new improvisations.

The stretch of Hudson Street where I live is each day the scene of an intricate sidewalk ballet. I make my own first entrance into it a little after eight when I put out the garbage can, surely a prosaic occupation, but I enjoy my part, my little clang, as the droves of junior high school students walk by the center of the stage dropping candy wrappers. (How do they eat so much candy so early in the morning?)

While I sweep up the wrappers I watch the other rituals of morning: Mr. Halpert unlocking the laundry's handcart from its mooring to a cellar door, Joe Cornacchia's son-in-law stacking out the empty crates from the delicatessen, the barber bringing out his sidewalk folding chair, Mr. Goldstein arranging the coils of wire which proclaim the hardware store is open, the wife of the tenement's superintendent depositing her chunky 3-year-old with a toy mandolin on the stoop, the vantage point from which he is learning the English his mother cannot speak. Now the primary children, heading for St. Luke's, dribble through to the south; the children for St. Veronica's cross, heading to the west, and the children for P.S. 41, heading toward the east. Two new entrances are being made from the wings: well-dressed and even elegant women and men with briefcases emerge from doorways and side streets . . . Most of these are heading for the bus and subways, but some hover on the curbs, stopping taxis which have miraculously appeared at the right moment, for the taxis are part of a wider morning ritual: having dropped passengers from midtown in the downtown financial district, they are now bringing downtowners up to midtown. Simultaneously, numbers of women in housedresses have emerged and as they crisscross with one another they pause for quick conversations that sound with either laughter or joint indignation; never, it seems, anything between. It is time for me to hurry to work too, and I exchange my ritual farewell with Mr. Lofaro, the short, thick-bodied, white-aproned fruit man who stands outside his doorway a little up the street, his arms folded, his feet planted, looking solid as earth itself. We nod; we each glance quickly up and down the street then look back to each other and smile. We have done this many a morning for more than ten years, and we both know what it means: All is well.

[. . .]

I know the deep night ballet and its seasons best from waking; long after midnight to tend a baby and, sitting in the dark, seeing the shadows and hearing the sounds of the sidewalk. Mostly it is a sound like infinitely pattering snatches of party conversation and, about three in the morning, singing, very good singing. Sometimes there is sharpness and anger or sad, sad weeping, or a flurry of search for a string of beads broken. One night, a young man came roaring along, bellowing terrible language at two girls whom he had apparently picked up and who were disappointing him. Doors opened; a wary semicircle formed around him, not too close, until the police came. Out came the heads, too, along Hudson Street, offering opinion, "Drunk . . . Crazy . . . A wild kid from the suburbs." (He turned out to be a wild kid from the suburbs. Sometimes, on Hudson Street, we are tempted to believe the suburbs must be a difficult place to bring up children.)

I have made the daily ballet of Hudson Street sound more frenetic than it is, because writing it telescopes it. In real life, it is not that way. In real life, to be sure, something is always going on, the ballet is never at a halt, but the general effect is peaceful and the general tenor even leisurely. People who know well such animated city streets will know how it is. I am afraid people who do not will always have it a little wrong in their heads like the old prints of rhinoceroses made from travelers' descriptions of rhinoceroses. On Hudson Street, the same as in the North End of Boston or in any other animated neighborhoods of great cities, we are not innately more competent at keeping the sidewalks safe than are the people who try to live off the hostile truce of Turf in a blind-eyed city. We are the lucky possessors of a city order that makes it relatively simple to keep the peace because there are plenty of eyes on the street. But there is nothing simple about that order itself, or the bewildering number of components that go into it. Most of those components are specialized in one way or another. They unite in their joint effect upon the sidewalk, which is not specialized in the least. That is its strength.

WILLIAM JULIUS WILSON

"From Institutional to Jobless Ghettos"

from *When Work Disappears: The World of the New Urban Poor* (1996)

Editors' introduction Harvard sociologist William Julius Wilson spent much of his career at the University of Chicago. Like the earlier Chicago School sociologists, Ernest W. Burgess (p. 153) and Louis Wirth (p. 97) writing in the l920s and 1930s, and St. Clair Drake and Horace Cayton writing in the 1940s, Wilson uses careful empirical studies of Chicago to generate important urban theory. An African-American, Wilson has been particularly concerned about the situation of poor Blacks in America's decaying central cities.

Wilson is critical of timid liberals who avoid confronting tough questions about race and poverty because they are afraid anything negative they say about Blacks will appear racist. He argues that there is an *urban underclass* (a term many liberals will not use) and that residents of poor Black ghettos today are socially isolated and caught in a tangle of pathology characterized by unemployment, crime (including violent Black-on-Black crime), teenage pregnancy, out-of-wedlock births, welfare dependency, and drug use. Wilson feels that the situation of Blacks, particularly poor urban Blacks, requires objective research and honest reportage. He is unwilling to let conservatives such as Charles Murray (p. 122) dominate theoretical discourse about the causes of and cures for Black poverty.

Wilson's research has convinced him that conditions for poor Black ghetto residents are far worse in many ways than a century ago when W. E. B. Du Bois studied *The Philadelphia Negro* (p. 56) or in 1945 when St. Clair Drake and Horace Cayton published *Black Metropolis*, their study of the African-American community of Chicago at that time. Both Du Bois's and Drake and Cayton's studies found more Blacks employed and role models of upward mobility present, a higher proportion of nuclear families, less Black-on-Black violence, and much less drug use than exist in the poorest Black ghettos today.

Wilson argues that the changed structure of the US economy is more responsible for the plight of poor Blacks today than is racism. A generation ago, Wilson argues, an able-bodied unskilled Black man could readily find work sufficient to support himself and a family – albeit often physically hard, racially segregated, and dirty work. But in the last generation unskilled manual urban jobs have largely disappeared. Without work, Black males cannot support a family. Hence, Wilson argues, Black ghettos today contain few "marriageable" Black males. A high incidence of out-of-wedlock births, family dissolution, and welfare dependent female-headed households follow directly from that fact. With little sense of self-worth, Wilson argues, unemployed Blacks naturally turn to drug dependence and crime.

One of Wilson's most controversial contentions is that race and racism are declining in importance as causes of Black distress. Paradoxically, he argues, less racial discrimination has made matters in Black ghettos worse. According to Wilson, as upwardly-mobile Blacks move out of Black ghetto areas, community leadership and positive role models disappear and pathology is concentrated.

Wilson is skeptical that *race-specific* policies like affirmative action will address problems as pervasive and profound as he describes. Rather he favors universal social policies aimed at improving the lot of all poor people regardless of race: new education, training, and particularly full employment policy.

This selection is from *When Work Disappears: The World of the New Urban Poor* (New York: Knopf, 1996). Wilson's two most influential prior books are *The Declining Significance of Race: Blacks and Changing American Institutions* 2nd edn. (Chicago: University of Chicago Press, 1980) and *The Truly Disadvantaged* (Chicago: University of Chicago Press, 1987).

For a recent overview of the literature on the underclass debate and writings by Latino scholars exploring the relevance and limitations of underclass theory for America's varied Latino communities see Joan Moore and Raquel Pinderhughes, *In the Barrios: Latinos and the Underclass Debate* (New York: Russell Sage Foundation, 1993). Conservative explanations of and suggested public policy regarding poverty and race include Edward Banfield, *The Unheavenly City Revisited* (Boston: Little Brown, 1974) and Charles Murray, *Losing Ground* (New York: Basic Books, 1984).

An elderly woman who has lived in one inner-city neighborhood on the South Side of Chicago for more than forty years reflected:

> I've been here since March 21, 1953. When I moved in, the neighborhood was intact. It was intact with homes, beautiful homes, mini mansions, with stores, laundromats, with cleaners, with Chinese [cleaners]. We had drugstores. We had hotels. We had doctors over on Thirty-ninth Street. We had doctors' offices in the neighborhood. We had the middle class and upper middle class. It has gone from affluent to where it is today. And I would like to see it come back, that we can have some of the things we had. Since I came in young, and I'm a senior citizen now, I would like to see some of the things come back so I can enjoy them like we did when we first came in.

[. . .]

A 91-year-old woman spoke of safety concerns: "It's not safe anymore because the streets aren't. When all the black businesses and shows closed down, the economy went to the dogs. The stores, the businesses, the shows, everywhere was lighted, the stores and businesses have disappeared."

The negative social forces triggered a decision by a concerned mother to send her son away.

> I have a 13-year-old. I sent him away when he was nine because the gangs was at him so tough, because he wouldn't join – he's a basketball player. That's all he ever cared about. They took his gym shoes off his feet. They took his clothes. Made him walk home from school. Jumped on him every day. Took his jacket off his back in subzero weather. You know, and we only live two blocks from the school . . . A boy pulled a gun to his head and told him, "If you don't join, next week you won't be here." I had to send him out of town. His father stayed out of town. He came here last week for a week. He said, "Mom, I want to come home so bad," I said no!

The social deterioration of ghetto neighborhoods is the central concern expressed in the testimony of these residents. As a representative from the media put it, the ghetto has gone "from bad to worse." Few observers of the urban scene in the late 1960s anticipated the extensive breakdown of social institutions and the sharp rise in rates of social dislocation that have since swept the ghettos and spread to other neighborhoods that were once stable. For example, in the neighborhood of Woodlawn, located on the South Side of Chicago, there were over eight hundred commercial and industrial establishments in 1950. Today, it is estimated that only about a hundred are left, many of them represented by "tiny catering places, barber shops, and thrift stores with no more than one or two employees." As Löic Wacquant, a member of the Urban Poverty and Family Life Study research team, put it:

> The once-lively streets – residents remember a time, not so long ago, when crowds were so dense at rush hour that one had to elbow one's way to the train station – now have the appearance of an empty, bombed-out war zone. The commercial strip has been reduced to a long tunnel of charred stores, vacant lots littered with broken glass and garbage, and dilapidated buildings left to rot in the shadow of the elevated train line. At the corner of Sixty-third Street and Cottage Grove Avenue, the handful of remaining establishments that struggle to survive are huddled behind wrought iron bars ... The only enterprises that seem to be thriving are liquor stores and currency exchanges, these "banks of the poor" where one can cash checks, pay bills and buy money orders for a fee.

[. . .]

In 1950, almost two-thirds of Woodlawn's population was white; by 1960 the white population had declined to just 10 percent. Despite the sudden white exodus, the number of residents in the neighborhood increased slightly during this period. After 1960, however, a sizable exodus of black residents followed, including a significant number of working- and middle-class families. The population of the neighborhood declined from over 80,000 in 1960 to 53,814 in 1970; it further slipped to 36,323 in 1980 and finally to 24,473 in 1990. The loss of residents was accompanied by a substantial reduction in the economic, social, and political resources that make a community vibrant. Woodlawn is only one of a growing number of poor black neighborhoods in Chicago plagued by depopulation and social and economic deterioration.

When the black respondents in our large ... survey were asked to rate their neighborhood as a place to live, only a third said that their area was a good or very good place to live and only 18 percent of those in the ghetto poverty census tracts felt that their neighborhood was a desirable place to live. (The Bureau of the Census defines a census tract as "a relatively homogeneous area with respect to population characteristics, economic status, and living conditions with an average population of 4,000." Poverty tracts are those in which at least 20 percent of the residents are poor, and ghetto poverty tracts are those in which at least 40 percent are poor.)

[. . .]

Many of the respondents described the negative effects of their neighborhood on their own personal outlook. An unmarried, employed clerical worker from a ghetto poverty census tract on the West Side stated:

> There is a more positive outlook if you come from an upwardly mobile neighborhood than you would here. In this type of neighborhood, all you hear is negative [things] and that can kind of bring you down when you're trying to make it. So your neighborhood definitely has something to do with it.

This view was shared by a 17-year-old college student and part-time worker from an impoverished West Side neighborhood.

> I'd say about 40 percent in my neighborhood ... I'd say 40 percent are alcoholics ... And ... only 5 percent of the alcoholics have homes. Then you got the other 35 percent who are in the street ... They probably live somewhere, but they in street, on the corner every day, same old thing, because they don't have no chance in life. They live based on today. [They say] "Oh, we gonna get high today." "Oh, whoopee!" "What you gonna do tomorrow, man?" "I don't know, man, I don't know." You can ask any of 'em: "What you gonna do tomorrow?" "I don't know, man. I know when it gets here." And I can really understand, you know, being in that state. If you around totally negative people, people who are not doing anything, that's the way you gonna be regardless.

The state of the inner-city public schools was another major concern expressed by our respondents. The complaints ranged from overcrowded conditions to unqualified and uncaring teachers. Sharply voicing her views on these subjects, a 25-year-old married mother of two children from a South Side census tract that just recently became poor stated: "My daughter ain't going to school here, she was going to a nursery school where I paid and of course they took the time and spent it with her, 'cause they was getting the money. But the public schools, no! They are overcrowded and the teachers don't care."

[. . .]

The respondents were also asked whether their neighborhoods had changed as a place to live over the years. Seventy-one percent of the African-American respondents felt that their neighborhoods had either stayed the same or had gotten worse.

An unemployed black man from a West Side housing project felt that the only thing that had changed in his neighborhood was that it was "going down instead of going back up." He further stated, "It ain't like it used to be. They laid off a lot of people. There used to be a time when you got a broken window, you call up housing and they send someone over to fix it, but it ain't like that no more."

Respondents frequently made statements about the increase in drug trafficking and drug consumption when discussing how their neighborhood had changed. "Well, OK, I realize there was drugs when I was growing up but they weren't as open as they are now," stated a divorced telephone dispatcher and mother of five children from a neighborhood that recently changed from a nonpoverty to a poverty area. "It's nothing to see a 10-year-old kid strung out or a 10-year-old kid selling drugs. I mean, when they were doing it back then they were sneaking around doing it. It's like an open thing now.

[. . .]

The feelings of many of the respondents in our study were summed up by a 33-year-old married mother of three from a very poor West Side neighborhood:

> If you live in an area in your neighborhood where you have people that don't work, don't have no means of support, you know, don't have no jobs, who're gonna break into your house to steal what you have, to sell to get them some money, then you can't live in a neighborhood and try to concentrate on tryin' to get ahead, then you get to work and you have to worry if somebody's breakin' into your house or not. So, you know, it's best to try to move in a decent area, to live in a community with people that works.

In 1959, less than one-third of the poverty population in the United States lived in metropolitan central cities. By 1991, the central cities included close to half of the nation's poor. Many of the most rapid increases in concentrated poverty have occurred in African-American neighborhoods. For example, in the ten community areas that represent the historic core of Chicago's Black Belt (see Figure 1), eight had rates of poverty in 1990 that exceeded 45 percent, including three with rates higher than 50 percent and three that surpassed 60 percent. Twenty-five years earlier, in 1970, only two of these neighborhoods had poverty rates above 40 percent.

In recent years, social scientists have paid particular attention to the increases in urban neighborhood poverty. "Defining an urban neighborhood for analytical purposes is no easy task." The community areas of Chicago referred to in Figure 1 include a number of adjacent census tracts. The seventy-seven community areas within the city of Chicago represent statistical units derived by urban sociologists at the University of Chicago for the 1930 census in their effort to analyze varying conditions within the city. These delineations were originally drawn up on the basis of settlement and history of the area, local identification and trade patterns, local institutions, and natural and artificial barriers. There have been major shifts in population and land use since then. But these units remain useful in tracing changes over time, and they continue to capture much of the contemporary reality of Chicago neighborhoods.

Other cities, however, do not have such convenient classifications of neighborhoods, which means that comparison across cities cannot be drawn using community areas. The measurable unit considered most appropriate to represent urban neighborhoods is the census tract. In attempts to examine this problem of ghetto poverty across the nation empirically, social scientists have tended to define ghetto neighborhoods as those located in the *ghetto poverty* census tracts. As indicated earlier, ghetto poverty census tracts are those in which at least 40 percent of the residents are poor. For example, Paul Jargowsky and Mary Jo Bane state: "Visits to various cities confirmed that the 40 percent criterion came very close to identifying areas that looked like ghettos in terms of their housing conditions. Moreover, the areas selected by the 40 percent criterion corresponded closely with the neighborhoods that city officials and local Census Bureau officials considered ghettos." The ghetto poor in Jargowsky and Bane's study are therefore designated as those among the poor who live in these ghetto poverty areas.

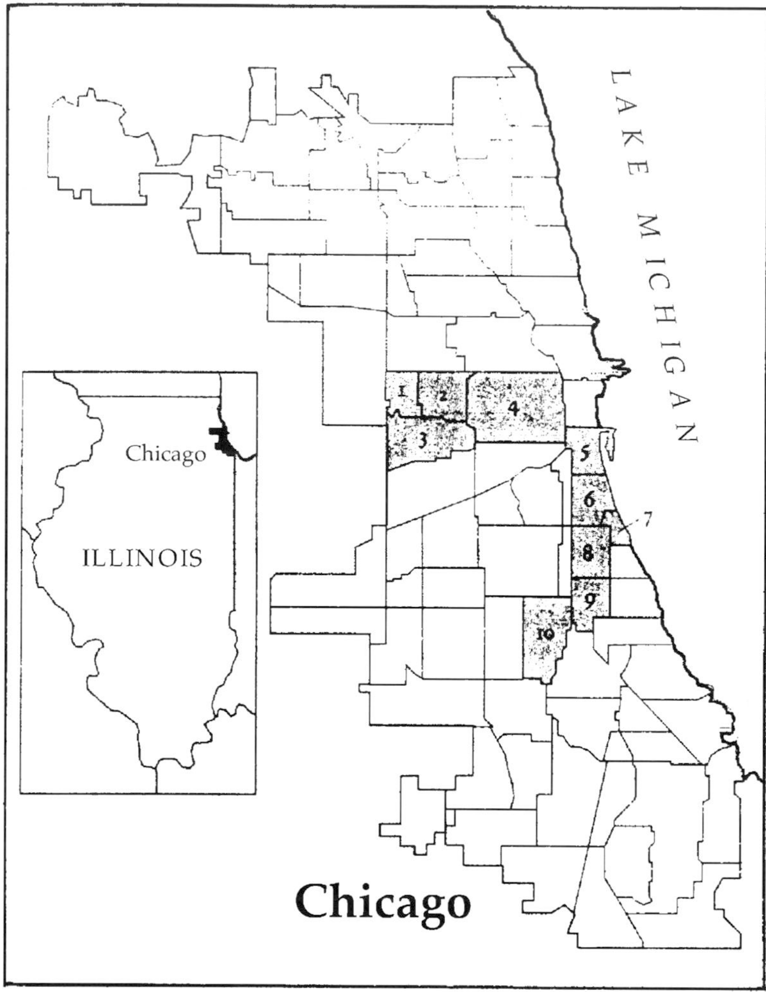

1. West Garfield Park
2. East Garfield Park
3. North Lawndale
4. Near West Side
5. Near South Side
6. Douglas
7. Oakland
8. Grand Boulevard
9. Washington Park
10. Englewood

Figure 1 Community areas in Chicago's Black Belt

Three-quarters of all the ghetto poor in metropolitan areas reside in one hundred of the nation's largest central cities; however, it is important to remember that the ghetto areas in these central cities also include a good many families and individuals who are not poor.

In the nation's one hundred largest central cities, nearly one in seven census tracts is at least 40 percent poor. The number of such tracts has more than doubled since 1970 – indeed, it is alarming that 579 tracts fell to ghetto poverty level in these cities between 1970 and 1980, and 624 additional tracts joined these ranks in the following decade.

Paul Jargowsky's research reveals that a vast majority of people (almost seven out of eight) living in metropolitan-area ghettos in 1990 were minority group members. The number of African-Americans in these ghettos grew by more than one-third from 1980 to 1990, reaching nearly 6 million. Most of this growth involved poor people. The proportion of metropolitan blacks who live in ghetto areas climbed from more than a third (37 percent) to almost half (45 percent). Indeed, the metropolitan black poor are becoming increasingly isolated. The poverty rate among metropolitan blacks who reside in ghettos increased while the rate among those who live in nonghettos decreased.

The increase in the *number* of ghetto blacks is related to the *geographical spread* of the ghetto. Jargowsky and Bane found that in the cities they studied (Philadelphia, Cleveland, Milwaukee, and Memphis) areas that had become ghettos by 1980 had been mixed-income tracts in 1970 – but tracts that were contiguous to areas identified as ghettos. The exodus of the nonpoor from mixed-income areas was a major factor in the spread of ghettos in these cities in the 1970s. Since 1980, ghetto census tracts have increased in a substantial majority of the metropolitan areas in the country, including those with fewer people living in them. Nine new ghetto census tracts were added in Philadelphia, even though it experienced one of the largest declines in the proportion of people living in ghetto tracts. In a number of other cities, including Baltimore, Boston, and Washington, D.C., a smaller percentage of poor blacks live in a larger number of ghetto census tracts. Chicago had a 61.5 percent increase in the number of ghetto census tracts from 1980 to 1990, even though the number of poor residing in those areas increased only slightly.

Jargowsky reflects on the significance of the substantial spread of ghetto areas:

> The geographic size of a city's ghetto has a large effect on the perception of the magnitude of the problem associated with ghetto poverty. How big an area of the city do you consider off limits? How far out of your way will you drive not to go through a dangerous area? Indeed, the lower density exacerbated the problem. More abandoned buildings mean more places for crack dens and criminal enterprises. Police trying to protect a given number of citizens have to be stretched over a wider number of square miles, making it less likely that criminals will be caught. Lower density also makes it harder for a sense of community to develop, or for people to feel that they can find safety in numbers. From the point of view of local political officials, the increase in the size of the ghetto is a disaster. Many of those leaving the ghetto settle in nonghetto areas outside the political jurisdiction of the central city. Thus, geographic size of the ghetto is expanding, cutting a wider swath through the hearts of our metropolitan areas.

In sum, the 1970s and 1980s witnessed a sharp growth in the number of census tracts classified as ghetto poverty areas, an increased concentration of the poor in these areas, and sharply divergent patterns of poverty concentration between racial minorities and whites. One of the legacies of historic racial and class subjugation in America is a unique and growing concentration of minority residents in the most impoverished areas of the nation's metropolises.

Some have argued that this concentration of poverty is not new but mirrors conditions prevalent in the 1930s. According to Douglas Massey and Nancy Denton, during the Depression poverty was just as concentrated in the ghettos of the 1930s as in those of the 1970s. The black communities of the 1930s and those of the 1970s shared a common experience: a high degree of racial segregation from the larger society. Massey and Denton argue that "concentrated poverty is created by a pernicious interaction between a group's overall rate of poverty and its degree of segregation in society. When a highly segregated group experiences a high or rising rate of poverty, geographically concentrated poverty is the inevitable result." However convincing the logic of that argument, it does not explain the following: In the ten neighborhoods that make up Chicago's Black Belt, the poverty rate increased almost 20 percent between 1970 and 1980 (from 32.5 to 50.4 percent) despite the fact that the overall black poverty rate for the city of Chicago increased only 7.5 percent during this same period (from 25.1 to 32.6 percent).

Concentrated poverty may be the inevitable result when a highly segregated group experiences an increase in its overall rate of poverty. But segregation does not explain why the concentration of poverty in *certain* neighborhoods of this segregated group should increase to nearly three times the group's *overall* rate of poverty increase. There is no doubt that the disproportionate concentration of poverty among African-Americans is one of the legacies of historic racial segregation. It is also true that segregation often compounds black vulnerability in the face of other changes in the society, including, as we shall soon see, economic changes. Nonetheless, to focus mainly on segregation to account for the growth of concentrated poverty is to overlook some of the dynamic aspects of the social and demographic changes occurring in cities like Chicago. Given the existence of segregation, we must consider the way in which other changes in society have interacted with segregation to produce the dramatic social transformation of inner-city neighborhoods, especially since 1970.

For example, the communities that make up the Black Belt in Chicago have been overwhelmingly black for the last four decades, yet they lost almost half their residents between 1970 and 1980. This rapid depopulation has had profound consequences for the social and economic deterioration of segregated Black Belt neighborhoods, including increases in concentrated poverty and joblessness. If comparisons are made strictly between the Depression years of the 1930s and the 1980s, rates of ghetto poverty and joblessness in these neighborhoods will indeed be similar. But such a comparison obscures significant changes that have occurred in these neighborhoods across the fifty-year span between those two points.

. . . Many of the gains made in inner-city neighborhoods following the Depression were wiped out after 1970. To maintain that concentrated black poverty in the 1970s or in the 1980s is equivalent in severity and pervasiveness to that which occurred during the Depression does not explain its dramatic rise since 1970; nor does it address a far more fundamental problem that is at the heart of the extraordinary increases in and spread of concentrated poverty – namely, the rapid growth of joblessness, which accelerated through these two decades. The problems reported by the residents of poor Chicago neighborhoods are not a consequence of poverty alone. Something far more devastating has happened that can only be attributed to the emergence of concentrated and persistent joblessness and its crippling effects on neighborhoods, families, and individuals. The city of Chicago epitomizes these changes.

[. . .]

The most fundamental difference between today's inner-city neighborhoods and those studied by Drake and Cayton [in the 1940s] is the much higher levels of joblessness. Indeed, there is a new poverty in our nation's metropolises that has consequences for a range of issues relating to the quality of life in urban areas, including race relations.

By "the new urban poverty," I mean poor, segregated neighborhoods in which a substantial majority of individual adults are either unemployed or have dropped out of the labor force altogether. For example, in 1990 only one in three adults aged 16 and over in the twelve Chicago community areas with ghetto poverty rates held a job in a typical week of the year. Each of these community areas, located on the South and West Sides of the city, is overwhelmingly black. We can add to these twelve high-jobless areas three additional predominantly black community areas, with rates approaching ghetto poverty, in which only 42 percent of the adult population were working in a typical week in 1990. Thus, in these fifteen black community areas – comprising a total population of 425,125 – only 37 percent of all the adults were gainfully employed in a typical week in 1990. By contrast, 54 percent of the adults in the seventeen other predominantly black community areas in Chicago – a total population of 545,408 – worked in a typical week in 1990. This was close to the citywide employment figure of 57 percent for all adults. Finally, except for one Asian community area with an employment rate of 46 percent, and one Latino community area with an employment rate of 49 percent, a majority of the adults held a job in a typical week in each of the remaining forty-five community areas of Chicago.

But Chicago is by no means the only city that features new poverty neighborhoods. In the ghetto census tracts of the nation's one hundred largest central cities, there were only 65.5 employed persons for every hundred adults who did not hold a job in a typical week in 1990. In contrast, the nonpoverty areas contained 182.3 employed persons for every hundred of those not working. In other words, the ratio of employed to jobless persons was three times greater in census tracts not marked by poverty.

Looking at Drake and Cayton's Bronzeville, I can illustrate the magnitude of the changes that have occurred in many, inner-city ghetto neighborhoods in recent years. A majority of adults held jobs in the three Bronzeville areas in 1950, but by 1990 only four in ten in Douglas worked in a typical week, one in three in Washington Park, and one in four in Grand Boulevard. In 1950, 69 percent of all males 14 and over who lived in the Bronzeville neighborhoods worked in a typical week, and in 1960, 64 percent of this group were so employed. However, by 1990 only 37 percent of all males 16 and over held jobs in a typical week in these three neighborhoods.

Upon the publication of the first edition of *Black Metropolis* in 1945, there was much greater class integration within the black community. As Drake and Cayton pointed out, Bronzeville residents had limited success in "sorting themselves out into broad community areas designated as 'lower class' and 'middle class' . . . Instead of middle class areas, Bronzeville tends to have middle-class buildings in all areas, or a few middle-class blocks here and there." Though they may have lived on different streets, blacks of all classes in inner-city areas such as Bronzeville lived in the same community and shopped at the same stores. Their children went to the same schools and played in the same parks. Although there was some class antagonism, their neighborhoods were more stable than the inner-city neighborhoods of today; in short, they featured higher levels of what social scientists call "social organization."

When I speak of social organization I am referring to the extent to which the residents of a neighborhood are able to maintain effective social control and realize their common goals. There are three major dimensions of neighborhood social organization: (1) the prevalence, strength, and interdependence of social networks; (2) the extent of collective supervision that the residents exercise and the degree of personal responsibility they assume in addressing neighborhood problems; and (3) the rate of resident participation in voluntary and formal organizations. Formal institutions (e.g., churches and political party organizations), voluntary associations (e.g., block clubs and parent-teacher organizations), and informal networks (e.g., neighborhood friends and acquaintances, coworkers, marital and parental ties) all reflect social organization.

Neighborhood social organization depends on the extent of local friendship ties, the degree of social cohesion, the level of resident participation in formal and informal voluntary associations, the density and stability of formal organizations, and the nature of informal social controls. Neighborhoods in which adults are able to interact in terms of obligations, expectations, and relationships are in a better position to supervise and control the activities and behavior of children. In neighborhoods with high levels of social organization, adults are empowered to act to improve the quality of neighborhood life – for example, by breaking up congregations of youths on street corners and by supervising the leisure activities of youngsters.

Neighborhoods plagued by high levels of joblessness are more likely to experience low levels of social organization: the two go hand in hand. High rates of joblessness trigger other neighborhood problems that undermine social organization, ranging from crime, gang violence, and drug trafficking to family breakups and problems in the organization of family life.

Consider, for example, the problems of drug trafficking and violent crime. As many studies have revealed, the decline in legitimate employment opportunities among inner-city residents has increased incentives to sell drugs. The distribution of crack in a neighborhood attracts individuals involved in violence and lawlessness. Between 1985 and 1992, there was a sharp increase in the murder rate among men under the age of 24; for men 18 years old and younger, murder rates doubled. Black males in particular have been involved in this upsurge in violence. For example, whereas the homicide rate for white males between 14 and 17 increased from 8 per 100,000 in 1984 to 14 in 1991, the rate for black males tripled during that time (from 32 per 100,000 to 112). This sharp rise in violent crime among younger males has accompanied the widespread outbreak of addiction to crack cocaine. The association is especially strong in inner-city ghetto neighborhoods plagued by joblessness and weak social organization.

Violent persons in the crack-cocaine marketplace have a powerful impact on the social organization of a neighborhood. Neighborhoods plagued by high levels of joblessness, insufficient economic opportunities, and high residential mobility are unable to control the volatile drug market and the violent crimes related to it. As informal controls weaken, the social processes that regulate behavior change.

As a result, the behavior and norms in the drug market are more likely to influence the action of others in the neighborhood, even those who are not involved in drug activity. Drug dealers cause the use and spread of guns in the neighborhood to escalate, which in turn raises the likelihood that others, particularly the youngsters, will come to view the possession of weapons as necessary or desirable for self-protection, settling disputes, and gaining respect from peers and other individuals.

Moreover, as Alfred Blumstein pointed out, the drug industry actively recruits teenagers in the neighborhood "partly because they will work more cheaply than adults, partly because they may be less vulnerable to the punishments imposed by the adult criminal justice system, partly because they tend to be daring and willing to take risks that more mature adults would eschew." Inner-city black youths with limited prospects for stable or attractive employment are easily lured into drug trafficking and therefore increasingly find themselves involved in the violent behavior that accompanies it.

A more direct relationship between joblessness and violent crime is revealed in recent research by Delbert Elliott of the University of Colorado, a study based on National Longitudinal Youth Survey data collected from 1976 to 1989, covering ages 11 to 30. As Elliott points out, the transition from adolescence to adulthood usually results in a sharp drop in most crimes as individuals take on new adult roles and responsibilities. "Participation in serious violent offending behavior (aggravated assault, forcible rape, and robbery) increases [for all males] from ages 11 and 12 to ages 15 and 16, then declines dramatically with advancing age." Although black and white males reveal similar age curves, "the negative slope of the age curve for blacks after age 20 is substantially less than that of whites."

The black-white differential in the proportion of males involved in serious violent crime, although almost even at age 11, increases to 3 : 2 over the remaining years of adolescence, and reaches a differential of nearly 4 : 1 during the late twenties. However, when Elliott compared only *employed* black and white males, he found no significant differences in violent behavior patterns among the two groups by age 21. Employed black males, like white males, experienced a precipitous decline in serious violent behavior following their adolescent period. Accordingly, a major reason for the racial gap in violent behavior after adolescence is joblessness; a large proportion of jobless black males do not assume adult roles and responsibilities, and their serious violent behavior is therefore more likely to extend into adulthood. The new poverty neighborhoods feature a high concentration of jobless males and, as a result, suffer rates of violent criminal behavior that exceed those in other urban neighborhoods.

. . . In 1990, 37 percent of Woodlawns 27,473 adults were employed and only 23 percent of Oakland's 4,935 adults were working. When asked how much of a problem unemployment was in their neighborhood, 73 percent of the residents in Woodlawn and 76 percent in Oakland identified it as a major problem. The responses to the survey also revealed the residents' concerns about a series of related problems, such as crime and drug abuse, that are symptomatic of severe problems of social organization. Indeed, crime was identified as a major problem by 66 percent of the residents in each neighborhood. Drug abuse was cited as a major problem by as many as 86 percent of the adult residents in Oakland and 79 percent of those in Woodlawn.

Although high-jobless neighborhoods also feature concentrated poverty, high rates of neighborhood poverty are less likely to trigger problems of social organization if the residents are working. This was the case in previous years when the working poor stood out in areas like Bronzeville. Today, the nonworking poor predominate in the highly segregated and impoverished neighborhoods.

The rise of new poverty neighborhoods represents a movement away from what the historian Allan Spear has called an institutional ghetto – whose structure and activities parallel those of the larger society, as portrayed in Drake and Cayton's description of Bronzeville – toward a jobless ghetto, which features a severe lack of basic opportunities and resources, and inadequate social controls.

What can account for the growing proportion of jobless adults and the corresponding increase in problems of social organization in innercity communities such as Bronzeville? An easy answer is racial segregation. However, a race-specific argument is not sufficient to explain recent changes in neighborhoods like Bronzeville.

After all, Bronzeville was just as *segregated by skin color in 1950* as it is today, yet the level of employment was much higher then.

Nonetheless, racial segregation does matter. If large segments of the African-American population had not been historically segregated in inner-city ghettos, we would not be talking about the new urban poverty. The segregated ghetto is not the result of voluntary or positive decisions on the part of the residents who live there. As Massey and Denton have carefully documented, the segregated ghetto is the product of systematic racial practices such as restrictive covenants, redlining by banks and insurance companies, zoning, panic peddling by real estate agents, and the creation of massive public housing projects in low-income areas.

Segregated ghettos are less conducive to employment and employment preparation than are other areas of the city. Segregation in ghettos exacerbates employment problems because it leads to weak informal employment networks and contributes to the social isolation of individuals and families, thereby reducing their chances of acquiring the human capital skills, including adequate educational training, that facilitate mobility in a society. Since no other group in society experiences the degree of segregation, isolation, and poverty concentration as do African-Americans, they are far more likely to be disadvantaged when they have to compete with other groups in society, including other despised groups, for resources and privileges.

To understand the new urban poverty, one has to account for the ways in which segregation interacts with other changes in society to produce the recent escalating rates of joblessness and problems of social organization in inner-city ghetto neighborhoods.

CHARLES MURRAY

"Choosing a Future"

from *Losing Ground: American Social Policy 1950–1980* (1984)

Editors' introduction Beginning with Margaret Thatcher's administration in the United Kingdom and Ronald Reagan's in the United States during the 1980s, conservative social policies that favor shrinking national government expenditures on social programs and privatizing government functions have been in vogue in many countries. Simultaneously, an influential new breed of conservative scholars undertook empirical research that emphasized the failures of government programs and constructed theories that argue in favor of less government generally, and particularly less national government meddling in issues of poverty and social inequality in cities.

Charles Murray's book *Losing Ground* was published in 1984, mid-way through Ronald Reagan's presidency. It quickly became a conservative icon. Murray juxtaposes data on massive increases in federal spending on welfare, education, job training, violence prevention, and other US social programs during the late 1960s and 1970s with data on *worsening* situations of poor people, particularly poor inner-city Blacks. By comparison, Murray argues, data for the 1950s and early 1960s show the condition of poor Blacks was improving with very little government spending on social programs.

Murray takes aim squarely at liberal social policies. He characterizes programs intended to support poor people, particularly poor Blacks, as paternalistic, a new form of unequal treatment which he feels undermines initiative and is as harmful as historic racial discrimination. Giving a teenage mother welfare to support an illegitimate child, Murray argues, undermines families and eliminates the incentive for the mother, father, or either's parents to act responsibly. It discourages the mother from taking necessary steps to lift herself out of dependency, condemning her to a life at the bottom of the social ladder. Unlike William Julius Wilson (p. 112), Murray does not see loss of unskilled jobs accessible to Blacks as the primary reason for the underclass. Murray blames the patronizing social programs that attempt to help poor Blacks through welfare. Accordingly he argues that welfare should be eliminated or changed to foster self-reliance. Welfare reform in the United States and other countries is heavily influenced by this view.

This selection is from *Losing Ground: American Social Policy 1950–1980* (New York: Basic Books, 1984). Murray's other books include an extremely controversial best-selling study of race and intelligence co-written by the late Richard Hernstein, *The Bell Curve* (New York: Free Press, 1994) and, with Louis A. Cox, *Beyond Probation: Juvenile Corrections and the Chronic Delinquent* (Beverly Hills, CA: Sage Publications, 1979).

For a scathing attack on the government programs for the poor during the US "war on poverty" see Daniel Patrick Moynihan, *Maximum Feasible Misunderstanding* (New York: Free Press, 1969). Other conservative critiques of urban social policy include Edward Banfield, *The Unheavenly City Revisited* (Boston: Little, Brown, 1974), and Emanuel S. Savas, *Privatizing the Public Sector* (Chatham, NJ: Chatham House, 1982) and *Privatization: The Key to Better Government* (Chatham, NJ: Chatham

House, 1987). Liberal alternatives include Francis Fox Piven and Richard A. Cloward, *Regulating the Poor: The Functions of Public Welfare* (New York: Pantheon, 1971), William Goldsmith and Edward J. Blakeley, *Separate Societies: Poverty and Inequality in U.S. Cities* (Philadelphia: Temple University Press, 1992), and Peter Marris, *Community Planning and Conceptions of Change* (London: Routledge & Kegan Paul, 1982).

. . . [I]f the behaviors of members of the underclass are founded on a rational appreciation of the rules of the game, and as long as the rules encourage dysfunctional values and behaviors, the future cannot look bright. Behaviors that work will tend to persist until they stop working. The rules will have to be changed. How might they be changed? I present three proposals: one for education, one for public welfare, and one for civil rights. The proposals of greatest theoretical interest involve education and public welfare . . . I begin . . . with the proposal for civil rights. It is simple, would cost no money to implement, and is urgently needed.

A PROPOSAL FOR SOCIAL POLICY AND RACE

Real reform of American social policy is out of the question until we settle the race issue. We have been dancing around it since 1964, wishing it would go away and at the same time letting it dominate, sub rosa, the formation of social policy.

The source of our difficulties has been the collision, with enormous attendant national anxiety and indecision, of two principles so much a part of the American ethos that hardly anyone, whatever his political position, can wholly embrace one and reject the other. The principles are equal treatment and a fair shake.

The principle of equal treatment demands that we all play by the same rules – which would seem to rule out any policy that gives preferential treatment to anyone. A fair shake demands that everyone have a reasonably equal chance at the brass ring – or at least a reasonably equal chance to get on the merry-go-round.

Thus hardly anyone, no matter how strictly noninterventionist, can watch with complete equanimity when a black child is deprived of a chance to develop his full potential for reasons that may be directly traced to a heritage of exploitation by whites. Neither can anyone, no matter how devoted to Affirmative Action, watch with complete equanimity when a white job applicant is turned down for a job in favor of a black who is less qualified. Something about it is fundamentally unfair – un-American – no matter how admirable the ultimate goal.

Until 1965, the principles of equal treatment and a fair shake did not compete. They created no tension. Their application to racial policy was simple: Make the nation color-blind. People were to be judged on their merits. But then the elite wisdom changed. Blacks were to be helped to catch up.

. . . In summarizing [the] results as they pertain to the poorest blacks, this harsh judgment is warranted: If an impartial observer from another country were shown the data on the black lower class from 1950 to 1980 but given no information about contemporaneous changes in society or public policy, that observer would infer that racial discrimination against the black poor increased drastically during the late 1960s and 1970s. No explanation except a surge in outright, virulent discrimination would as easily explain to a "blind" observer why things went so wrong.

Such an explanation is for practical purposes correct. Beginning in the last half of the 1960s, the black poor were subjected to new forms of racism with effects that outweighed the waning of the old forms of racism. Before the 1960s, we had a black underclass that was held down because blacks were systematically treated differently from whites, by whites. Now, we have a black underclass that is held down for the same generic reason – because blacks are systematically treated differently from whites, by whites.

The problem consists of a change in the nature of white condescension toward blacks. Historically, virtually all whites condescended toward virtually all blacks; there is nothing new in that. The condescension could be vicious in intent, in the form of "keeping niggers in their place." It could be benign, as in the excessive solicitousness with which whites who considered themselves enlightened tended to treat blacks.

These forms of condescension came under withering attack during the civil rights movement, to such an extent that certain manifestations of the condescension disappeared altogether in some circles. A variety of factors – among them, simply greater representation of blacks in the white professional world of work – made it easier for whites to develop relationships of authentic equality and respect with black colleagues. But from a policy standpoint, it became clear only shortly after the War on Poverty began that henceforth the black lower class was to be the object of a new condescension that would become intertwined with every aspect of social policy. Race is central to the problem of reforming social policy, not because it is intrinsically so but because the debate about what to do has been perverted by the underlying consciousness among whites that "they" – the people to be helped by social policy – are predominantly black, and blacks are owed a debt.

The result was that the intelligentsia and the policymakers, coincident with the revolution in social policy, began treating the black poor in ways that they would never consider treating people they respected. Is the black crime rate skyrocketing? Look at the black criminal's many grievances against society. Are black illegitimate birth rates five times those of whites? We must remember that blacks have a much broader view of the family than we do – aunts and grandmothers fill in. Did black labor force participation among the young plummet? We can hardly blame someone for having too much pride to work at a job sweeping floors. Are black high school graduates illiterate? The educational system is insensitive. Are their test scores a hundred points lower than others? The tests are biased. Do black youngsters lose jobs to white youngsters because their mannerisms and language make them incomprehensible to their prospective employers? The culture of the ghetto has its own validity.

That the condescension should be so deep and pervasive is monumentally ironic, for the injunction to respect the poor (after all, they are not to blame) was hammered home in the tracts of OEO [the Office of Economic Opportunity] and radical intellectuals. But condescension is the correct descriptor. Whites began to tolerate and make excuses for behavior among blacks that whites would disdain in themselves or their children.

The expression of this attitude in policy has been a few obvious steps – Affirmative Action, minority set-asides in government contracts, and the like – but the real effect was the one that I discussed in the history of the period. The white elite could not at one time cope with two reactions. They could not simultaneously feel compelled to make restitution for past wrongs to blacks and blame blacks for not taking advantage of their new opportunities. The system had to be blamed, and any deficiencies demonstrated by blacks had to be overlooked or covered up – by whites.

A central theme of this book [*Losing Ground*] has been that the consequences were disastrous for poor people of all races, but for poor blacks especially, and most emphatically for poor blacks in all-black communities – precisely that population that was the object of the most unremitting sympathy.

My proposal for dealing with the racial issue in social welfare is to repeal every bit of legislation and reverse every court decision that in any way requires, recommends, or awards differential treatment according to race, and thereby put us back onto the track that we left in 1965. We may argue about the appropriate limits of government intervention in trying to enforce the ideal, but at least it should be possible to identify the ideal: Race is not a morally admissible reason for treating one person differently from another. Period.

A PROPOSAL FOR EDUCATION

There is no such thing as an undeserving 5-year-old. Society, in the form of government intervention, is quite limited in what it can do to make up for many of the deficiencies of life that an unlucky 5-year-old experiences; it can, however, provide a good education and thereby give the child a chance at a different future.

The objective is a system that provides more effective education of the poor and disadvantaged without running afoul of the three laws of social programs. The objective is also to construct what is, in my view, a just system – one that does not sacrifice one student's interests to another's, and one that removes barriers in the way of those who want most badly to succeed and are prepared to make the greatest efforts to do so. So once again let us put ourselves in the position of bureaucrats of sweeping authority and large budgets. How shall we make things better?

We begin by installing a completely free educational system that goes from preschool to the loftiest graduate degrees, removing economic barriers entirely. Having done so, however, we find little change from the system that prevailed in 1980. Even then, kindergarten through high school were free to the student, and federal grants and loans worth $4.4 billion plus a very extensive system of private scholarships and loans were available for needy students who wanted to continue their education.

By making the system entirely free, we are not making more education newly accessible to large numbers of people, nor have we done anything about the quality of education.

We then make a second and much more powerful change. For many years, the notion of a voucher system for education has enjoyed a periodic vogue. In its pure form, it would give each parent of a child of school age a voucher that the parent could use to pay for schooling at any institution to which the child could gain admittance. The school would redeem the voucher for cash from the government. The proposals for voucher systems have generally foundered on accusations that they are a tool for the middle class and would leave the disadvantaged in the lurch. My proposition is rather different: a voucher system is the single most powerful method available to us to improve the education of the poor and disadvantaged. Vouchers thus become the second component of our educational reforms.

For one large segment of the population of poor and disadvantaged, the results are immediate, unequivocal, and dramatic. I refer to children whose parents take an active role in overseeing and encouraging their children's education. Such parents have been fighting one of the saddest of the battles of the poor – doing everything they can within the home environment, only to see their influence systematically undermined as soon as their children get out the door. When we give such parents vouchers, we find that they behave very much as their affluent counterparts behave when they are deciding upon a private school. They visit prospective schools, interview teachers, and place their children in schools that are demanding of the students and accountable to the parents for results. I suggest that when we give such parents vouchers, we will observe substantial convergence of black and white test scores in a single generation. All that such parents have ever needed is an educational system that operates on the same principles they do.

This is a sufficient improvement to justify the system, for we are in a no-lose situation with regard to the children whose parents do not play their part effectively. These children are sent to bad schools or no schools at all – just as they were in the past. How much worse can it be under the new system?

This defect in the voucher system leaves us, however, with a substantial number of students who are still getting no education through no fault of their own. Nor can we count on getting results if we round them up and dispatch them willy-nilly to the nearest accredited school. A school that can motivate and teach a child when there is backup from home cannot necessarily teach the children we are now discussing. Many of them are poor not only in money. Many have been developmentally impoverished as well, receiving very little of the early verbal and conceptual stimulation that happens as a matter of course when parents expect their children to be smart. Some arrive at the school door already believing themselves to be stupid, expecting to fail. We can be as angry as we wish at their parents, but we are still left with the job of devising a school that works for these children. What do we do – not in terms of a particular pedagogical program or curriculum, but in broad strokes?

First, whatever else, we decide to create a world that makes sense in the context of the society we want them to succeed in. The school is not an extension of the neighborhood. Within the confines of the school building and school day, we create a world that may seem as strange and irrelevant as Oz.

We do not do so with uniforms or elaborate rules or inspirational readings – the embellishments are left up to the school. Rather, we install one simple, inflexible procedure. Each course has an entrance test. Tenth-grade geometry has an entrance test; so does first-grade reading. Entrance tests for simple courses are simple; entrance tests for hard courses are hard. Their purpose is not to identify the best students, but to make sure that any student who gets in can, with an honest effort, complete the course work.

Our system does not carry with it any special teaching technique. It does, however, give the teacher full discretion over enforcing an orderly working environment. The teacher's only obligation is to teach those who want to learn.

The system is also infinitely forgiving. A student who has just flunked algebra three times running can enroll in that or any other math class for which he can pass the entrance test. He can enroll even if he has just been kicked out of three other classes for misbehavior. The question is never "What have you been in the past?" but always "What are you being as of now?"

The evolving outcomes of the system are complex. Some students begin by picking the easiest, least taxing courses, and approach them with as little motivation as their counterparts under the current system. Perhaps among this set of students are some who cannot or will not complete even the simplest courses. They drop by the wayside, failures of the system.

Among those who do complete courses, any courses, five things happen, all of them positive. First, the system is so constructed that to get into a course is in itself a small success ("I passed!"). Second, the students go into the course with a legitimate reason for believing that they can do the work; they passed a valid test that says they can. Third, they experience a success when they complete the course. Fourth, they experience – directly – a cause–effect relationship between their success in one course and their ability to get into the next course, no matter how small a step upward that next course may be. Fifth, all the while this is going on, they are likely to be observing other students no different from them – no richer, no smarter – who are moving upward faster than they are but using the same mechanism.

What of those who are disappointed, who try to get into a class and fail? Some will withdraw into themselves and be forever fearful of taking a chance on failure, as almost all do under the current system anyway. But there is a gradation to risk, and a peculiar sort of guarantee of success in our zero-transfer system. Whatever class a student finally takes, the student will have succeeded in gaining entrance to it. He will go into the classroom with official certification – based on reality – that he will be able to learn the material if he gives it an honest effort. The success–failure, cause–effect features of the system are indispensable for teaching some critical lessons:

- Effort is often rewarded with success.
- Effort is not always rewarded with success.
- Failure in one instance does not mean inability to succeed in anything else.
- Failure in one try does not mean perpetual failure.
- The better the preparation, the more likely the success.

None of these lessons is taught as well or as directly under the system prevailing in our current education of the disadvantaged. The central failing of the educational system for the poor and disadvantaged, and most especially poor and disadvantaged blacks, is not that it fails to provide meaningful ways for a student to succeed, though that is part of it. The central failing is not that ersatz success – fake curricula, fake grades, fake diplomas – sets the students up for failure when they leave the school, though that too is part of it. The central failing is that the system does not teach disadvantaged students, who see permanent failure all around them, how to fail. For students who are growing up expecting (whatever their dreams may be) ultimately to be a failure, with failure writ large, the first essential contravening lesson is that failure can come in small, digestible packages. Failure can be dealt with. It can be absorbed, analyzed, and converted to an asset.

We are now discussing a population of students – the children of what has become known as "the underclass" – that comes to the classroom with an array of disadvantages beyond simple economic poverty. I am not suggesting that, under our hypothetical system, all children of the underclass will become motivated students forthwith. Rather, some will. Perhaps it will be a small proportion; perhaps a large one. Certainly the effect interacts with the inherent abilities of the children involved. But some effect will be observed. Some children who are at the very bottom of the pile in the disadvantages they bear will act on the change in the reality of their environment. It will be an improvement over the situation in the system we have replaced, in which virtually none of them gets an education in anything except the futility of hoping.

A PROPOSAL FOR PUBLIC WELFARE

I begin with the proposition that it is within our resources to do enormous good for some people quickly. We have available to us a program that would convert a large proportion of the younger generation of hard-core unemployed into steady workers making a living wage. The same program would drastically reduce births to single teenage girls. It would reverse the trendline in the breakup of poor families. It would measurably increase the upward socioeconomic mobility of poor families. These improvements would affect some millions of persons.

All these are results that have eluded the efforts of the social programs installed since 1965, yet, from everything we know, there is no real question about whether they would occur under the program I propose. A wide variety of persuasive evidence from our own culture and around the world, from experimental data and longitudinal studies, from theory and practice, suggests that the program would achieve such results.

The proposed program, our final and most ambitious thought experiment, consists of scrapping the entire federal welfare and income-support structure for working-aged persons, including AFDC [Aid to Families with Dependent Children], Medicaid, Food Stamps, Unemployment Insurance, Worker's Compensation, subsidized housing, disability insurance, and the rest. It would leave the working-aged person with no recourse whatsoever except the job market, family members, friends, and public or private locally funded services. It is the Alexandrian solution: cut the knot, for there is no way to untie it.

It is difficult to examine such a proposal dispassionately. Those who dislike paying for welfare are for it without thinking. Others reflexively imagine bread lines and people starving in the streets. But as a means of gaining fresh perspective on the problem of effective reform, let us consider what this hypothetical society might look like.

A large majority of the population is unaffected. A surprising number of the huge American middle and working classes go from birth to grave without using any social welfare benefits until they receive their first Social Security check. Another portion of the population is technically affected, but the change in income is so small or so sporadic that it makes no difference in quality of life. A third group comprises persons who have to make new arrangements and behave in different ways. Sons and daughters who fail to find work continue to live with their parents or relatives or friends. Teenaged mothers have to rely on support from their parents or the father of the child and perhaps work as well. People laid off from work have to use their own savings or borrow from others to make do until the next job is found. All these changes involve great disruption in expectations and accustomed roles.

Along with the disruptions go other changes in behavior. Some parents do not want their young adult children continuing to live off their income, and become quite insistent about their children learning skills and getting jobs. This attitude is most prevalent among single mothers who have to depend most critically on the earning power of their offspring.

Parents tend to become upset at the prospect of a daughter's bringing home a baby that must be entirely supported on an already inadequate income. Some become so upset that they spend considerable parental energy avoiding such an eventuality. Potential fathers of such babies find themselves under more pressure not to cause such a problem, or to help with its solution if it occurs.

Adolescents who were not job-ready find they are job-ready after all. It turns out that they can work for low wages and accept the discipline of the workplace if the alternative is grim enough. After a few years, many – not all, but many – find that they have acquired salable skills, or that they are at the right place at the right time, or otherwise find that the original entry-level job has gradually been transformed into a secure job paying a decent wage. A few – not a lot, but a few – find that the process leads to affluence.

Perhaps the most rightful, deserved benefit goes to the much larger population of low-income families who have been doing things right all along and have been punished for it: the young man who has taken responsibility for his wife and child even though his friends with the same choice have called him a fool; the single mother who has worked full time and forfeited her right to welfare for very little extra money; the parents who have set an example for their children even as the rules of the game have taught their children that the example is outmoded. For these millions of people, the instantaneous result is that no one makes fun of them any longer. The longer-term result will be that they regain the status that is properly theirs. They will not only be the bedrock upon which the community is founded (which they always have been), they will be recognized as such. The process whereby they regain their position is not magical, but a matter of logic. When it becomes highly dysfunctional for a person to be dependent, status will accrue to being independent, and in fairly short order. Noneconomic rewards will once again reinforce the economic rewards of being a good parent and provider.

The prospective advantages are real and extremely plausible. In fact, if a government program of the traditional sort (one that would "do" something rather than simply get out of the way) could as plausibly promise these advantages, its passage would be a foregone conclusion. Congress, yearning for programs that are not retreads of failures, would be prepared to spend billions. Negative side-effects (as long as they were the traditionally acceptable negative side-effects) would be brushed aside as trivial in return for the benefits. For let me be quite clear: I am not suggesting that we dismantle income support for the working-aged to balance the budget or punish welfare cheats. I am hypothesizing, with the advantage of powerful collateral evidence, that the lives of large numbers of poor people would be radically changed for the better.

There is, however, a fourth segment of the population yet to be considered, those who are pauperized by the withdrawal of government supports and unable to make alternate arrangements: the teenaged mother who has no one to turn to; the incapacitated or the inept who are thrown out of the house; those to whom economic conditions have brought long periods in which there is no work to be had; those with illnesses not covered by insurance. What of these situations?

The first resort is the network of local services. Poor communities in our hypothetical society are still dotted with storefront health clinics, emergency relief agencies, employment services, legal services. They depend for support on local taxes or local philanthropy, and the local taxpayers and philanthropists tend to scrutinize them rather closely. But, by the same token, they also receive considerably more resources than they formerly did. The dismantling of the federal services has poured tens of billions of dollars back into the private economy. Some of that money no doubt has been spent on Mercedes and summer homes on the Cape. But some has been spent on capital investments that generate new jobs. And some has been spent on increased local services to the poor, voluntarily or as decreed by the municipality. In many cities, the coverage provided by this network of agencies is more generous, more humane, more wisely distributed, and more effective in its results than the services formerly subsidized by the federal government.

But we must expect that a large number of people will fall between the cracks. How might we go about trying to retain the advantages of a zero-level welfare system and still address the residual needs?

As we think about the nature of the population still in need, it becomes apparent that their basic problem in the vast majority of the cases is the lack of a job, and this problem is temporary. What they need is something to tide them over while finding a new place in the economy. So our first step is to re-install the Unemployment Insurance program in more or less its previous form. Properly administered, unemployment insurance makes sense. Even if it is restored with all the defects of current practice, the negative effects of unemployment insurance alone are relatively minor. Our objective is not to wipe out chicanery or to construct a theoretically unblemished system, but to meet legitimate human needs without doing more harm than good. Unemployment insurance is one of the least harmful ways of contributing to such ends. Thus the system has been amended to take care of the victims of short-term swings in the economy.

Who is left? We are now down to the hardest of the hard core of the welfare-dependent. They have no jobs. They have been unable to find jobs (or have not tried to find jobs) for a longer period of time than the unemployment benefits cover. They have no families who will help. They have no friends who will help. For some reason, they cannot get help from local services or private charities except for the soup kitchen and a bed in the Salvation Army hall.

What will be the size of this population? We have never tried a zero-level federal welfare system under conditions of late-twentieth-century national wealth, so we cannot do more than speculate. But we may speculate. Let us ask of whom the population might consist and how they might fare.

For any category of "needy" we may name, we find ourselves driven to one of two lines of thought. Either the person is in a category that is going to be at the top of the list of services that localities vote for themselves, and at the top of the list of private services, or the person is in a category where help really is not all that essential or desirable. The burden of the conclusion is not that every single person will be taken care of, but that the extent of resources to deal with needs is likely to be very great – not based on wishful thinking, but on extrapolations from reality.

To illustrate, let us consider the plight of the stereotypical welfare mother – never married, no skills, small children, no steady help from a man. It is safe to say that, now as in the 1950s, there is no one who has less sympathy from the white middle class, which is to be the source of most of the money for the private and local services we envision. Yet this same white middle class is a soft touch for people trying to make it on their own, and a soft touch for "deserving" needy mothers – AFDC was one of the most widely popular of the New Deal welfare measures, intended as it was for widows with small children. Thus we may envision two quite different scenarios.

In one scenario, the woman is presenting the local or private service with this proposition: "Help me find a job and day-care for my children, and I will take care of the rest." In effect, she puts herself into the same category as the widow and the deserted wife – identifies herself as one of the most obviously deserving of the deserving poor. Welfare mothers who want to get into the labor force are likely to find a wide range of help. In the other scenario, she asks for an outright and indefinite cash grant – in effect, a private or local version of AFDC – so that she can stay with the children and not hold a job. In the latter case, it is very easy to imagine situations in which she will not be able to find a local service or a private philanthropy to provide the help she seeks. The question we must now ask is: What's so bad about that? If children were always better off being with their mother all day and if, by the act of giving birth, a mother acquired the inalienable right to be with the child, then her situation would be unjust to her and injurious to her children. Neither assertion can be defended, however, especially not in the 1980s, when more mothers of all classes work away from the home than ever before, and even more especially not in view of the empirical record for the children growing up under the current welfare system. Why should the mother be exempted by the system from the pressures that must affect everyone else's decision to work?

As we survey these prospects, important questions remain unresolved. The first of these is why, if federal social transfers are treacherous, should locally mandated transfers be less so? Why should a municipality be permitted to legislate its own AFDC or Food Stamp program if their results are so inherently bad?

Part of the answer lies in conceptions of freedom. I have deliberately avoided raising them – the discussion is about how to help the disadvantaged, not about how to help the advantaged cut their taxes, to which arguments for personal freedom somehow always get diverted. Nonetheless, the point is valid: Local or even state systems leave much more room than a federal system for everyone, donors and recipients alike, to exercise freedom of choice about the kind of system they live under. Laws are more easily made and changed, and people who find them unacceptable have much more latitude in going somewhere more to their liking.

But the freedom of choice argument, while legitimate, is not necessary. We may put the advantages of local systems in terms of the Law of Imperfect Selection. A federal system must inherently employ very crude, inaccurate rules for deciding who gets what kind of help, and the results are as I outlined them in [another chapter]. At the opposite extreme – a neighbor helping a neighbor, a family member helping another family member – the law loses its validity nearly altogether. Very fine-grained judgments based on personal knowledge are being made about specific people and changing situations. In neighborhoods and small cities, the procedures can still bring much individualized information to bear on decisions. Even systems in large cities and states can do much better than a national system; a decaying industrial city in the Northeast and a booming sunbelt city of the same size can and probably should adopt much different rules about who gets what and how much.

A final and equally powerful argument for not impeding local systems is diversity. We know much more in the 1980s than we knew in the 1960s about what does not work. We have a lot to learn about what does work. Localities have been a rich source of experiments. Marva Collins in Chicago gives us an example of how a school can bring inner-city students up to national norms. Sister Falaka Fattah in Philadelphia shows us how homeless youths can be rescued from the streets. There are numberless such lessons waiting to be learned from the diversity of local efforts. By all means, let a hundred flowers bloom, and if the federal government can play a useful role in lending a hand and spreading the word of successes, so much the better.

The ultimate unresolved question about our proposal to abolish income maintenance for the working-aged is how many people will fall through the cracks. In whatever detail we try to foresee the consequences, the objection may always be raised: We cannot be sure that everyone will be taken care of in the degree to which we would wish. But this observation by no means settles the question. If one may point in objection to the child now fed by Food Stamps who would go hungry, one may also point with satisfaction to the child who would have an entirely different and better future. Hungry children should be fed; there is no argument about that. It is no less urgent that children be allowed to grow up in a system free of the forces that encourage them to remain poor and dependent. If a strategy reasonably promises to remove those forces, after so many attempts to "help the poor" have failed, it is worth thinking about.

But that rationale is too vague. Let me step outside the persona I have employed and put the issue in terms of one last intensely personal hypothetical example. Let us suppose that you, a parent, could know that tomorrow your own child would be made an orphan. You have a choice. You may put your child with an extremely poor family, so poor that your child will be badly clothed and will indeed sometimes be hungry. But you also know that the parents have worked hard all their lives, will make sure your child goes to school and studies, and will teach your child that independence is a primary value. Or you may put your child with a family with parents who have never worked, who will be incapable of overseeing your child's education – but who have plenty of food and good clothes, provided by others. If the choice about where one would put one's own child is as clear to you as it is to me, on what grounds does one justify support of a system that, indirectly but without doubt, makes the other choice for other children? The answer that "What we really want is a world where that choice is not forced upon us" is no answer. We have tried to have it that way. We failed. Everything we know about why we failed tells us that more of the same will not make the dilemma go away.

SHARON ZUKIN

"Whose Culture? Whose City?"

from *The Cultures of Cities* (1995)

Editors' introduction One way to look at urban society is through the prism of the arts and popular culture. Novels and short stories – as well as paintings, sculptures, musical compositions, and architectural designs – are all created out of specific urban contexts and all reflect, to one degree or another, the social, political, and economic conditions of the urban cultures that gave birth to them. And increasingly, movies, advertising, fashion, and expressions of the "street culture" of urban youth have supplanted high culture as the subjects of postmodern urban cultural analysis.

In "Whose Culture? Whose City?" sociologist Sharon Zukin looks at the way urban public space in New York is increasingly being appropriated and privatized by corporate and commercial forces, thus making a study of urban popular culture one part of a larger analysis of an ongoing cultural war on the streets of the city. "In recent years," she writes, "culture has . . . become a more explicit site of conflicts over social differences and urban fears." Many of these conflicts are between middle-class, mostly white citizens on the one hand and homeless, poor, and ethnic minority citizens on the other. In the case of Bryant Park (next to the New York Public Library and directly across the street from Zukin's office at the Graduate Center of the City University of New York), a private nonprofit corporation dominated by middle-class and corporate interests and inspired by William Whyte's ideas (p. 483) took over management of the space to rescue it from homeless panhandlers and drug dealers. Although many of the results seem quite positive – lively cafes and fashion shows brought many more users into the park than ever before – the underlying reality that disturbs Zukin is the gradual loss of meaningful *public* life under the control of inclusive, democratic forces that creeping privatization implies.

Like Jane Jacobs (p. 106), Zukin has a real love for what she proudly calls "my city," and, like Lewis Mumford (p. 92), she has a well-developed sensitivity to city life as a form of social theater.

This selection is excerpted from *The Cultures of Cities* (Oxford: Blackwell, 1995). Other books by Zukin on urban culture are *Landscapes of Power: From Detroit to Disney World* (Berkeley: University of California Press, 1991), and *Loft Living: Culture and Capital in Urban Change* (Baltimore, MD: Johns Hopkins University Press, 1982).

A consideration of urban culture opens the door to an entire body of analytical literature on the relationship of the arts – both high and popular – to society as a whole. For literature, Richard Lehan's *The City in Literature: An Intellectual and Cultural History* (Berkeley: University of California Press, 1998) and Burton Pike's *The Image of the City in Modern Literature* (Princeton, NJ: Princeton University Press, 1981) are excellent overviews. Raymond Williams's *The Country and the City* (London: Oxford University Press, 1973) is especially insightful on the relationship of Charles Dickens to nineteenth-century London.

For painting, consult Arnold Hauser's chapter on "Impressionism" in *The Social History of Art* (New York: Knopf, 1952) and T. J. Clark's *The Painting of Modern Life* (New York: Knopf, 1985). For an alternative view of privatization, Tyler Cowen's *In Praise of Commercial Culture* (Cambridge, MA: Harvard University Press, 1998) is a stimulating analysis that celebrates the role of the market in arts production and urban culture. For a specifically American view of the image of the city as a cultural force, see Morton and Lucia White's classic *The Intellectual versus the City: From Thomas Jefferson to Frank Lloyd Wright* (Cambridge, MA: Harvard University Press and MIT Press, 1962).

For the relationship of architecture to urban life – an enormous field – one can begin with Spiro Kostof's *A History of Architecture: Settings and Rituals* (New York: Oxford University Press, 1985), Craig Whitaker, *Architecture and the American Dream* (New York: Three Rivers Press, 1996), and Mark Girouard's *Cities and People: A Social and Architectural History* (New Haven, CT: Yale University Press, 1985). Also of interest is Henri Lefebvre's *The Production of Space* (London: Blackwell, 1991).

Other interesting perspectives on the general relationship of the arts to urban history and culture are Roland Barthes's "Semiology and the Urban" from M. Gottdiener and Alexandros Lagopoulos (eds.), *The City and the Sign* (New York: Columbia University Press, 1986), Vera Solberg's, *Constructing a Sociology of the Arts* (Cambridge: Cambridge University Press, 1990), Richard Sennett's *Flesh and Stone: The Body and the City in Western Civilization* (New York: W. W. Norton, 1994), and Ray Oldenburg's *The Great Good Place: Cafes, Coffee Shops, Community Centers, Beauty Parlors, General Stores, Bars, Hangouts and How They Get You Through the Day* (New York: Paragon House, 1989).

Cities are often criticized because they represent the basest instincts of human society. They are built versions of Leviathan and Mammon, mapping the power of the bureaucratic machine or the social pressures of money. We who live in cities like to think of "culture" as the antidote to this crass vision. The Acropolis of the urban art museum or concert hall, the trendy art gallery and cafe, restaurants that fuse ethnic traditions into culinary logos – cultural activities are supposed to lift us out of the mire of our everyday lives and into the sacred spaces of ritualized pleasures.

Yet culture is also a powerful means of controlling cities. As a source of images and memories, it symbolizes "who belongs" in specific places. As a set of architectural themes, it plays a leading role in urban redevelopment strategies based on historic preservation or local "heritage." With the disappearance of local manufacturing industries and periodic crises in government and finance, culture is more and more the business of cities – the basis of their tourist attractions and their unique, competitive edge. The growth of cultural consumption (of art, food, fashion, music, tourism) and the industries that cater to it fuels the city's symbolic economy, its visible ability to produce both symbols and space.

In recent years, culture has also become a more explicit site of conflicts over social differences and urban fears. Large numbers of new immigrants and ethnic minorities have put pressure on public institutions, from schools to political parties, to deal with their individual demands. Such high culture institutions as art museums and symphony orchestras have been driven to expand and diversify their offerings to appeal to a broader public. These pressures, broadly speaking, are both ethnic and aesthetic. By creating policies and ideologies of "multiculturalism," they have forced public institutions to change.

On a different level, city boosters increasingly compete for tourist dollars and financial investments by bolstering the city's image as a center of cultural innovation, including restaurants, avant garde performances, and architectural design. These cultural strategies of redevelopment have fewer critics than multiculturalism. But they often pit the self-interest of real estate developers, politicians, and expansion-minded cultural institutions against grassroots pressures from local communities.

At the same time, strangers mingling in public space and fears of violent crime have inspired the growth of private police forces, gated and barred communities, and a movement to design public spaces for maximum surveillance. These, too, are a source of contemporary urban culture. If one way of dealing with the material inequalities of city life has been to aestheticize diversity, another way has been to aestheticize fear.

Controlling the various cultures of cities suggests the possibility of controlling all sorts of urban ills, from violence and hate crime to economic decline. That this is an illusion has been amply shown by battles over multiculturalism and its warring factions – ethnic politics and urban riots. Yet the cultural power to create an image, to frame a vision, of the city has become more important as publics have become more mobile and diverse, and traditional institutions – both social classes and political parties – have become less relevant mechanisms of expressing identity. Those who create images stamp a collective identity. Whether they are media corporations like the Disney Company, art museums, or politicians, they are developing new spaces for public cultures. Significant public spaces of the late 19th and early 20th century – such as Central Park, the Broadway theater district, and the top of the Empire State Building – have been joined by Disney World, Bryant Park, and the entertainment-based retail shops of Sony Plaza. By accepting these spaces without questioning their representations of urban life, we risk succumbing to a visually seductive, privatized public culture.

THE SYMBOLIC ECONOMY

Anyone who walks through midtown Manhattan comes face to face with the symbolic economy. A significant number of new public spaces owe their particular shape and form to the intertwining of cultural symbols and entrepreneurial capital.

- The AT&T Building, whose Chippendale roof was a much criticized icon of postmodern architecture, has been sold to the Japanese entertainment giant Sony; the formerly open public areas at street level have been enclosed as retail stores and transformed into Sony-Plaza. Each store sells Sony products: video cameras in one shop, clothes and accessories related to performers under contract to Sony's music or film division in another . . .
- Two blocks away, Andre Emmerich, a leading contemporary art dealer, rented an empty storefront in a former bank branch to show three huge abstract canvases by the painter Al Held. Entitled *Harry, If I Told You, Would You Know?* the group of paintings was exhibited in raw space, amid falling plaster, peeling paint, exposed wires, and unfinished floors, and passersby viewed the exhibit from the street through large plate glass windows. The work of art was certainly for sale, yet it was displayed as if it were a free, public good; and it would never have been there had the storefront been rented by a more usual commercial tenant.
- On 42nd Street, across from my office, Bryant Park is considered one of the most successful public spaces to be created in New York City in recent years. After a period of decline, disuse, and daily occupation by vagrants and drug dealers, the park was taken over by a not-for-profit business association of local property owners and their major corporate tenants, called the Bryant Park Restoration Corporation. This group redesigned the park and organized daylong programs of cultural events; they renovated the kiosks and installed new food services; they hired a phalanx of private security guards. All this attracted nearby office workers, both women and men, who make the park a lively midday gathering place, as it had been prior to the mid 1970s – a public park under private control.

Building a city depends on how people combine the traditional economic factors of land, labor, and capital. But it also depends on how they manipulate symbolic languages of exclusion and entitlement. The look and feel of cities reflect decisions about what – and who – should be visible and what should not, on concepts of order and disorder, and on uses of aesthetic power. In this primal sense, the city has always had a symbolic economy. Modern cities also owe their existence to a second, more abstract symbolic economy devised by "place entrepreneurs," officials and investors whose ability to deal with the symbols of growth yields "real" results in real estate development, new businesses, and jobs.

Related to this entrepreneurial activity is a third, traditional symbolic economy of city advocates and business elites who, through a combination of philanthropy, civic pride, and desire to establish their identity as a patrician class, build the majestic art museums, parks, and architectural complexes that represent a world-class city. What is new about the symbolic economy since the 1970s is its symbiosis of image and product, the scope and scale of selling images on a national and even a global level, and the role of the symbolic economy in speaking for, or representing, the city.

[. . .]

The growth of the symbolic economy in finance, media, and entertainment may not change the way entrepreneurs do business. But it has already forced the growth of towns and cities, created a vast new work force, and changed the way consumers and employees think. In the early 1990s, employment in "entertainment and recreation" in the United States grew slightly more than in health care and six times more than in the auto industry. The facilities where these employees work – hotels, restaurants, expanses of new construction and undeveloped land – are more than just workplaces. They reshape geography and ecology; they are places of creation and transformation.

The Disney Company, for example, makes films and distributes them from Hollywood. It runs a television channel and sells commercial spinoffs, such as toys, books, and videos, from a national network of stores. Disney is also a real estate developer in Anaheim, Orlando, France, and Japan and the proposed developer of a theme park in Virginia and a hotel and theme park in Times Square. Moreover, as an employer, Disney has redefined work roles. Proposing a model for change in the emerging service economy, Disney has shifted from the white-collar worker described by C. Wright Mills in the 1950s to a new chameleon of "flexible" tasks. The planners at its corporate headquarters are "imagineers"; the costumed crowd-handlers at its theme parks are "cast members." Disney suggests that the symbolic economy is more than just the sum of the services it provides. The symbolic economy unifies material practices of finance, labor, art, performance, and design.

The prominence of culture industries also inspires a new language dealing with difference. It offers a coded means of discrimination, an undertone to the dominant discourse of democratization. Styles that develop on the streets are cycled through mass media, especially fashion and "urban music" magazines and MTV, where, divorced from their social context, they become images of cool. On urban billboards advertising designer perfumes or jeans, they are recycled to the streets, where they become a provocation, breeding imitation and even violence. The beachheads of designer stores, from Armani to A/X, from Ralph Lauren to Polo, are fiercely parodied for the "props" of fashion-conscious teenagers in inner city ghettos. The cacophony of demands for justice is translated into a coherent demand for jeans. Claims for public space by culture industries inspire the counterpolitics of display in late 20th century urban riots.

The symbolic economy recycles real estate as it does designer clothes. Visual display matters in American and European cities today, because the identities of places are established by sites of delectation. The sensual display of fruit at an urban farmers' market or gourmet food store puts a neighborhood "on the map" of visual delights and reclaims it for gentrification. A sidewalk cafe takes back the street from casual workers and homeless people. In Bryant Park, enormous white tents and a canopied walkway set the scene for spring and fall showings of New York fashion designers. Twice a year, the park is filled by the fashion media, paparazzi, store buyers, and supermodels doing the business of culture and reclaiming Bryant Park as a vital, important place. We New Yorkers become willing participants in the drama of the fashion business. As cultural consumers, we are drawn into the interrelated production of symbols and space.

Mass suburbanization since the 1950s has made it unreasonable to expect that most middle-class men and women will want to live in cities. But developing small places within the city as sites of visual delectation creates urban oases where everyone *appears* to be middle class. In the fronts of the restaurants or stores, at least, consumers are strolling, looking, eating, drinking, sometimes speaking English and sometimes not. In the back regions, an ethnic division of labor guarantees that immigrant workers are preparing food and cleaning up. This is not just a game of representations: developing the city's symbolic economy involves recycling workers, sorting people in housing markets, luring investment, and negotiating political claims for public goods and ethnic promotion. Cities from New York to Los Angeles and Miami seem to thrive by developing small districts around specific themes. Whether it is Times Square or el Calle Ocho, a commercial or an "ethnic" district, the narrative web spun by the symbolic economy around a specific place relies on a vision of cultural consumption and a social and an ethnic division of labor.

[. . .]

I also see public culture as socially constructed on the micro-level. It is produced by the many social encounters that make up daily life in the streets, shops, and parks – the spaces in which we experience public life in cities. The right to be in these spaces, to use them in certain ways, to invest them with a sense of our selves and our communities – to claim them as ours and to be claimed in turn by them – make up a constantly changing public culture. People with economic and political power have the greatest opportunity to shape public culture by controlling the building of the city's public spaces in stone and concrete. Yet public space is inherently democratic. The question of who can occupy public space, and so define an image of the city, is open-ended.

Talking about the cultures of cities in purely visual terms does not do justice to the material practices of politics and economics that create a symbolic economy. But neither does a strictly political-economic approach suggest the subtle powers of visual and spatial strategies of social differentiation. As I suggested in *Landscapes of Power* (1991), the rise of the cities' symbolic economy is rooted in two long-term changes – the economic decline of cities compared to suburban and nonurban spaces and the expansion of abstract financial speculation – and in such short-term factors, dating from the 1970s and 1980s, as new mass immigration, the growth of cultural consumption, and the marketing of identity politics. This is an inclusive, structural, and materialist view. If I am right, we cannot speak about cities today without understanding:

- how cities use culture as an economic base,
- how capitalizing on culture spills over into the privatization and militarization of public space, and
- how the power of culture is related to the aesthetics of fear.

CULTURE AS AN ECONOMIC BASE

Suppose we turn the old Marxist relation between a society's base and its superstructure on its head and think of culture as a way of producing basic goods. In fact, culture supplies the basic information – including symbols, patterns, and meaning – for nearly all the service industries. In our debased contemporary vocabulary, the word *culture* has become an abstraction for any economic activity that does not create material products like steel, cars, or computers. Stretching the term is a legacy of the advertising revolution of the early 20th century and the more recent escalation in political image making. Because culture is a system for producing symbols, every attempt to get people to buy a product becomes a culture industry. The sociologist Daniel Bell used to tell a joke about a circus employee whose job it was to follow the elephant and clean up after it; when asked, she said her job was in "the entertainment business." Today, she might say she was in "the culture industry." Culture is intertwined with capital and identity in the city's production systems.

From one point of view, cultural institutions establish a competitive advantage over other cities for attracting new businesses and corporate elites. Culture suggests the coherence and consistency of a brand name product. Like any commodity, "cultural" landscape has the possibility of generating other commodities. Historically, of course, the arrow of causality goes the other way. Only an economic surplus – sufficient to fund sacrifices for the temple, Michelangelos for the chapel, and bequests to art museums in the wills of robber barons – generates culture. But in American and European cities during the 1970s, culture became more of an instrument in the entrepreneurial strategies of local governments and business alliances. In the shift to a post-postwar economy, who could build the biggest modern art museum suggested the vitality of the financial sector. Who could turn the waterfront from docklands rubble to parks and marinas suggested the possibilities for expansion of the managerial and professional corps. This was probably as rational a response as any to the unbeatable isolationist challenge of suburban industrial parks and office campuses. The city, such planners and developers as James Rouse believed, would counter the visual homogeneity of the suburbs by playing the card of aesthetic diversity.

Yet culture also suggests a labor force that is well suited to the revolution of diminished expectations that began in the 1960s. In contrast to high-rolling rappers and rockers, "high" cultural producers are supposed to live on the margins; and the incomes of most visual artists, art curators, actors, writers, and musicians suggest they must be used to deprivation. A widespread appreciation of culture does not really temper the work force's demands. But, in contrast to workers in other industries, artists are flexible on job tasks and work hours, do not always join labor unions, and present a docile or even "cultured" persona. These qualities make them, like immigrants, desirable employees in service industries. Dissatisfaction with menial and dead-end jobs does not boil over into protest because their "real" identity comes from an activity outside the job.

[. . .]

Whether there is a singular, coherent vision no longer depends on the power of a single elite group. Constant political pressures by interest groups and complex interwoven networks of community groups, corporations, and public officials signal multiple visions. The ability to arrange these visions artfully, to orchestrate and choreograph images of diversity to speak for a larger whole, has been claimed by major nonprofit cultural institutions. This is especially true of art museums.

Since the 1980s, museums have fallen victim to their own market pressures. Reduced government funding and cutbacks in corporate support have made them more dependent than ever on attracting paying visitors ("gate"). They rely on their gift shops to contribute a larger share of their operating expenses. They try out new display techniques and seek crowd-pleasing exhibit ideas. In an attempt to reach a broader public, the Metropolitan Museum of Art and the Museum of Modern Art in New York have upgraded their restaurants and offer jazz performances on weekend evenings. Yet financial pressures have also led museums to capitalize on their visual holdings. By their marketing of cultural consumption, great art has become a public treasure, a tourist attraction, and a representation – divorced from the social context in which the art was produced – of public culture. Like Calvin Klein jeans on a bus stop billboard, the work of art and the museum itself have become icons of the city's symbolic economy.

[. . .]

PUBLIC SPACE

The fastest growing kind of public space in America is prisons. More jails are being built than housing, hospitals, or schools. No matter how well designed or brightly painted they may be, prisons are still closely guarded, built as cheaply as possible, and designed for surveillance. I can think of more pleasant public spaces, especially parks that I use in New York City. But is the Hudson River Park, near Battery Park City, or Bryant Park, on 42nd Street, less secure or exclusive than a prison? They share with the new wave of prison building several characteristics symptomatic of the times. Built or rebuilt as the city is in severe financial distress, they confirm the withdrawal of the public sector, and its replacement by the private sector, in defining public space. Reacting to previous failures of public space – due to crime, a perceived lower-class and minority-group presence, and disrepair – the new parks use design as an implicit code of inclusion and exclusion. Explicit rules of park use are posted in the parks and enforced by large numbers of sanitation workers and security guards, both public and private. By cleaning up public space, nearby property owners restore the attractiveness of their holdings and reconstruct the image of the city as well.

It is important to understand the histories of these symbolically central public spaces. The history of Central Park, for example, shows how, as definitions of who should have access to public space have changed, public cultures have steadily become more inclusive and democratic. From 1860 to 1880, the first uses of the park – for horseback riders and carriages – rapidly yielded to sports activities and promenades for the mainly immigrant working class. Over the next 100 years, continued democratization of access to the park developed together with a language of political equality. In the whole country, it became more difficult to enforce outright segregation by race, sex, or age.

[. . .]

In 1989, a private organization that manages Central Park, the Central Park Conservancy, demanded demolition of the Naumberg Bandshell, site of popular concerts from the 1930s to the 1950s, where homeless people gathered. Similarly, the Bryant Park Restoration Corporation started cleaning up the midtown business district by adopting the social design principles developed by William H. Whyte. Whyte's basic idea is that public spaces are made safe by attracting lots of "normal" users. The more normal users there are, the less space there will be for vagrants and criminals to maneuver. The Bryant Park Restoration Corporation intended their work to set a prototype for urban public space. They completely reorganized the landscape design of the park, opening it up to women, who tended to avoid the park even during daylight, and selling certain kinds of buffet food. They established a model of pacification by cappuccino.

Central Park, Bryant Park, and the Hudson River Park show how public spaces are becoming progressively less public: they are, in certain ways, more exclusive than at any time in the past 100 years. Each of these areas is governed, and largely or entirely financed, by a private organization, often working as a quasi-public authority. These private groups are much better funded than the corresponding public organization. Design in each park features a purposeful vision of urban leisure. A heightened concern for security inspires the most remarkable visible features: gates, private security guards, and eyes keeping the space under surveillance. The underlying assumption is that of a paying public, a public that values public space as an object of visual consumption. Yet it has become inconceivable in public discussions that control of the parks be left in public hands. When the *New York Times* praised plans to require developers to provide public access to the city's extensive waterfront, the newspaper said that only a public-private partnership could raise the funds to maintain such a significant public space.

A major reason for privatization of some public parks is that city governments cannot pay for taking care of them. Since the 1960s, while groups of all sorts have requested more use of the parks, the New York City Parks Department has been starved of government funds. Half the funding for Central Park is now raised privately by the Central Park Conservancy, which enjoys a corresponding influence on parks policy. Founded by private donors in 1980, the conservancy's original mission was to raise funds in the private sector to offset the park's physical deterioration. But it soon developed an authoritative cultural voice. The conservancy publicly defends the intentions of Olmsted and Vaux, the park's original designers, to create a "natural" landscape for contemplation. Most often, they beautify the park by restoring its 19th century buildings and bridges or setting up a nature program or skating facilities on one of its landscaped ponds. The conservancy has also become an arbiter between groups that want to use the park for sports or demonstrations, thus mediating between the homeless and the joggers, between athletes who come to the park from all over the city and those who come from low-income neighborhoods on the park's northern borders. The conservancy, moreover, has spoken loudly and often in favor of hiring nonunion labor . . .

In midtown, Bryant Park is an even more aggressive example of privatization. Declared a New York City landmark in 1975, the nine-acre park is essentially run by the Bryant Park Restoration Corporation, whose biggest corporate members are Home Box Office (HBO), a cable television network, and NYNEX, a regional telecommunications company. Like the Central Park Conservancy, the Bryant Park Restoration Corporation raises most of the park's budget, supervises maintenance, and decides on design and amenities.

The design of Bryant Park, in 1934, was based on an Olmstedian separation of a rural space of contemplation from the noisy city. By the late 1970s, this was determined to have the effect of walling off the park's intended public of office workers outside from drug dealers and loiterers inside. When the restoration corporation was formed, it took as its major challenge the development of a new design that would visually and spatially ensure security. The wall around the park was lowered, and the ground leveled to bring it closer to the surrounding streets. The restoration corporation bought movable chairs and painted them green, as in Parisian parks, responding to William H. Whyte's suggestion that park users like to create their own small spaces. Whyte recommended keeping "the undesirables" out by making a park *attractive.* Victorian kiosks selling cappuccino and sandwiches were built and painted, paths were repaved and covered with pebbles, a central lawn was opened up, and performers were enlisted to offer free entertainment in the afternoons. The restoration corporation hired its own security guards and pressured the New York City Police Department to supply uniformed officers. Four uniformed New York City police officers and four uniformed private security guards are on duty all day.

Plainclothes private security guards are also on patrol. A list posted at all entrances prohibits drug use, picking flowers, and drinking alcohol except for beverages bought at park concessions, which are limited to certain seating areas. It states the park's hours, 9 a.m. to 7 p.m., coinciding roughly with the business day. The rules specify that only homeless people connected to a particular shelter in the neighborhood have the right to rummage through the garbage cans for returnable bottles and cans. Unlike Parks Department workers, Bryant Park maintenance workers do not belong to a labor union. Starting salary for a maintenance worker is $6 an hour, half the starting rate of unionized workers in other city parks.

On a sunny summer day at noon, Bryant Park is full of office workers out to lunch – between 1,500 and 6,000 of them. The movable chairs and benches are filled; many people are sitting on the grass, on the edge of the fountain, even on the pebbled paths. Men and women eat picnic lunches singly, in couples, and in groups. Some traditional social hierarchies are subverted. Women feel free to glance at men passing by. Most men do not ogle the women. The dominant complexion of park users is white, with minority group members clustered outside the central green. Few people listen to the subsidized entertainment, an HBO comedian shouting into a microphone; no one notices when she finishes the show. A large sculpture by Alexander Calder stands in the middle of the lawn, on loan from an art gallery, both an icon and a benediction on the space. At sunset in the summer, HBO shows free movies from their stock of old films, a "take back the night" activity similar to those now being tried in other cities. This is a very deliberate exception to the rule of closing the park at night. During lunchtime, at least, the park visually represents an urban middle class: men and women who work in offices, jackets off, sleeves rolled up, mainly white. On the same day, at the same hour, another public space a block away – the tellers' line at Citibank – attracts a group that is not so well dressed, with more minority group members. The cultural strategies that have been chosen to revitalize Bryant Park carry with them the implication of controlling diversity while re-creating a consumable vision of civility.

[. . .]

Since its renovation, Bryant Park has changed character. It has become a place for people to be with others, to see others, a place of public sociability. John Berger once criticized New Yorkers for eating while walking alone on the street, alienating a social ritual from its proper context. Yet now, in the park, eating becomes a public ritual, a way of trusting strangers while maintaining private identities. Because of the police and security guards, the design, and the food, the park has become a visual and spatial representation of a middle-class public culture. The finishing touch will be a privately owned, expensive restaurant, whose rent payments will help finance the park's maintenance. This, however, is a degree of privatization that has stirred prolonged controversy. First envisioned in the 1980s, the restaurant remained the subject of public approvals processes until 1994.

The disadvantage of creating public space this way is that it owes so much to private-sector elites, both individual philanthropists and big corporations. This is especially the case for centrally located public spaces, the ones with the most potential for raising property values and with the greatest claim to be symbolic spaces for the city as a whole. Handing such spaces over to corporate executives and private investors means giving them carte blanche to remake public culture. It marks the erosion of public space in terms of its two basic principles: public stewardship and open access.

The Central Park Conservancy, a group of 30 private citizens who choose their own replacements, represents large corporations with headquarters in the city, major financial institutions, and public officials. The membership echoes both the new (the nonelected, tripartite Emergency Financial Control Board that has overseen New York City's budget since the fiscal crisis of 1975) and the old (the board of "gentlemen" trustees that originally guided the planning of Central Park in the 1860s) . . .

The Bryant Park Restoration Corporation, a subsidiary of the Bryant Park Business Improvement District, follows a fairly new model in New York State, and in smaller cities around the United States, that allows business and property owners in commercial districts to tax themselves voluntarily for maintenance and improvement of public areas and take these areas under their control. The concept originated in the 1970s as special assessment districts; in the 1980s, the name was changed to a more upbeat acronym, business improvement districts (BIDs). A BID can be incorporated in any commercial area. Because the city government has steadily reduced street cleaning and trash pickups in commercial streets since the fiscal crisis of 1975, there is a real incentive for business and property owners to take up the slack . . .

What kind of public culture is created under these conditions? Do urban BIDs create a Disney World in the streets, take the law into their own hands, and reward their entrepreneurial managers as richly as property values will allow? If elected public officials continue to urge the destruction of corrupt and bankrupt public institutions, I imagine a scenario of drastic privatization, with BIDs replacing the city government. As Republican Mayor Rudolph Giuliani said enthusiastically at the second annual NYC BIDs Association Conference in 1994, "This is a difficult time for the city and the country as we redefine ourselves. BIDs are one of the true success stories in the city. It's a tailor-made form of local government."

[. . .]

In their own way, under the guise of improving public spaces, BIDs nurture a visible social stratification. Like the Central Park Conservancy, they channel investment into a central space, a space with both real and symbolic meaning for elites as well as other groups. Like the Central Park Conservancy, the resources of the rich Manhattan BIDs far outstrip those even potentially available in other areas of the city, even if those areas set up BIDS. The rich BIDs' opportunity to exceed the constraints of the city's financial system confirms the fear that the prosperity of a few central spaces will stand in contrast to the impoverishment of the entire city.

BIDs can be equated with a return to civility, "an attempt to reclaim public space from the sense of menace that drives shoppers, and eventually store owners and citizens, to the suburbs." But rich BIDs can be criticized on the grounds of control, accountability, and vision. Public space that is no longer controlled by public agencies must inspire a liminal public culture open to all but governed by the private sector. Private management of public space does create some savings: saving money by hiring nonunion workers, saving time by removing design questions from the public arena. Because they choose an abstract aesthetic with no pretense of populism, private organizations avoid conflicts over representations of ethnic groups that public agencies encounter when they subsidize public art, including murals and statues.

Each area of the city gets a different form of visual consumption catering to a different constituency: culture functions as a mechanism of stratification. The public culture of midtown public space diffuses down through the poorer BIDs. It focuses on clean design, visible security, historic architectural features, and the sociability among strangers achieved by suburban shopping malls. Motifs of local identity are chosen by merchants and commercial property owners. Since most commercial property owners and merchants do not live in the area of their business or even in New York City, the sources of their vision of public culture may be eclectic: the nostalgically remembered city, European piazzas, suburban shopping malls, Disney World. In general, however, their vision of public space derives from commercial culture.

[. . .]

SECURITY, ETHNICITY, AND CULTURE

One of the most tangible threats to public culture comes from the politics of everyday fear. Physical assaults, random violence, hate crimes that target specific groups: the dangers of being in public spaces utterly destroy the principle of open access. Elderly men and women who live in cities commonly experience fear as a steady erosion of spaces and times available to them. An elderly Jewish politician who in the 1950s lived in Brownsville, a working-class Jewish neighborhood in Brooklyn where blacks began to move in in greater numbers as whites moved out, told me, "My wife used to be able to come out to meet me at night, after a political meeting, and leave the kids in our apartment with the door unlocked." A Jewish woman remembers about that same era, "I used to go to concerts in Manhattan wearing a fur coat and come home on the subway at 1 a.m." There may be some exaggeration in these memories, but the point is clear. And it is not altogether different from the message behind crimes against black men who venture into mainly white areas of the city at night or attacks on authority figures such as police officers and firefighters who try to exercise that authority against street gangs, drug dealers, and gun-toting kids. Cities are not safe enough for people to participate in a public culture.

"Getting tough" on crime by building more prisons and imposing the death penalty are all too common answers to the politics of fear. "Lock up the whole population," I heard a man say on the bus, at a stroke reducing the solution to its ridiculous extreme. Another answer is to privatize and militarize public space – making streets, parks, and even shops more secure but less free, or creating spaces, such as shopping malls and Disney World, that only *appear* to be public spaces because so many people use them for common purposes. It is not so easy, given a language of social equality, a tradition of civil rights, and a market economy, to enforce social distinctions in public space. The flight from "reality" that led to the privatization of public space in Disney World is an attempt to create a different, ultimately more menacing kind of public culture.

In *City of Quartz* (1990), Mike Davis describes the reshaping of public spaces in Los Angeles by surveillance and security procedures. Helicopters buzz the skies over ghetto neighborhoods, police hassle teenagers as putative gang members, homeowners buy into the type of armed defense they can afford . . . or have nerve enough to use. While Los Angeles may represent an extreme, high-tech example, I have also seen "Eyes on the Street" surveillance signs on lamp posts in small towns in Vermont and the design of Bryant Park gives evidence of a relatively low-tech but equally suggestive concern for public order. Indeed, Bryant Park may be a more typical public space than downtown Los Angeles because it has been "secured" within a democratic discourse of aestheticizing both cities and fear.

Gentrification, historic preservation, and other cultural strategies to enhance the visual appeal of urban spaces developed as major trends during the late 1960s and early 1970s. Yet these years were also a watershed in the institutionalization of urban fear. Voters and elites – a broadly conceived middle class in the United States – could have faced the choice of approving government policies to eliminate poverty, manage ethnic competition, and integrate everyone into common public institutions. Instead, they chose to buy protection, fueling the growth of the private security industry . . .

From the viewpoint of political economy, the withdrawal from public to private security employees is part of a general shift to privatization. Fiscal austerity limits government spending increases, even on the police. Yet private security cannot be free. The security costs borne by the private sector are passed on to the public by excluding potential criminal acts from segregated spaces, leaving the rest of the city to watch out for itself and be watched by the police. Crime, the criminal justice system, and private security forces absorb a high percentage of the unemployed, a "reserve army" in a more literal sense than Marx intended in his famous phrase about "the reserve army of the unemployed." While factory jobs disappear, urban workers, especially minority-group members, seek security jobs, and their mainly white colleagues in small, rural towns go to work in prisons.

[. . .]

In the past, those people who lived so close together they had to work out some etiquette for sharing, or dividing, public space were usually the poor. An exception that affected everyone was the system of racial segregation that worked by law in the south and by convention in many northern states until the 1960s, when – not surprisingly – perceptions of danger among whites increased. Like segregation, a traditional etiquette of public order of the urban poor involves dividing up territory by ethnic groups ... Among city dwellers today, innumerable informal etiquettes for survival in public spaces flourish. The "streetwise" scrutiny of passersby ... is one means for unarmed individuals to secure the streets. I think ethnicity – a cultural strategy for producing difference – is another, and it survives on the politics of fear by requiring people to keep their distance from certain aesthetic markers. These markers vary over time. Pants may be baggy or pegged, heads may be shaggy or shaved. Like fear itself, ethnicity becomes an aesthetic category.

[. . .]

In such a landscape, there are no safe places. The Los Angeles uprising of 1992 showed that, unlike in earlier riots, the powerless respect fewer geographical boundaries, except perhaps the neighborhoods where rich people live. Carjackings – the ultimate American violence – occur on the highway and in the parking lots of fast food restaurants ... Patrons of 24-hour automatic teller machines are robbed so often that the NYCE Network, with 10,000 machines in New York City, distributes a pamphlet of safety tips worthy of a military base: "As you approach an ATM, be aware of your surroundings. . . . When using an ATM at night, be sure it is located in a well lit area. And consider having someone accompany you." Someone, that is, other than the homeless man who stood by the door with an empty paper cup in his hand, until the New York City Council passed a law that forbade panhandlers to stand within 15 feet of an ATM. Or, as a Spanish-language subway advertisement cautions, "Mantengase alerta. Sus ojos, oidos y instinto son sus recursos naturales de seguridad en la ATM." In Chicago and Los Angeles, ATMs have been installed in police stations, so residents with bank accounts in the poorest neighborhoods will have a safe place to get cash.

For a brief moment in the late 1940s and early 1950s, working-class urban neighborhoods held the possibility of integrating white Americans and African-Americans in roughly the same social classes. This dream was laid to rest by movement to the suburbs, continued ethnic bias in employment, the decline of public services in expanding racial ghettos, criticism of integration movements for being associated with the Communist party, and fear of crime. Over the next 15 years, enough for a generation to grow up separate, the inner city developed its stereotyped image of "Otherness." The reality of minority groups' working-class life was demonized by a cultural view of the inner city "made up of four ideological domains: a physical environment of dilapidated houses, disused factories, and general dereliction; a romanticized notion of white working-class life with particular emphasis on the centrality of family life; a pathological image of black culture; and a stereotypical view of street culture."

By the 1980s, the development of a large black middle class with incomes more or less equal to white households' and the increase in immigrant groups raised a new possibility of developing ethnically and racially integrated cities. This time, however, there is a more explicit struggle over who will occupy the image of the city. Despite the real impoverishment of most urban populations, the larger issue is whether cities can again create an inclusive public culture. The forces of order have retreated into "small urban spaces," like privately managed public parks that can be refashioned to project an image of civility. Guardians of public institutions – teachers, cops – lack the time or inclination to *understand* the generalized ethnic Other . . .

Yet the groups that have inherited the city have a claim on its central symbolic spaces. Not only to the streets that serve as major parade routes, not only to the central parks, but also to the monumental spaces that confirm identity by offering visual testimony to a group's presence in history.

[. . .]

Many social critics have begun to write about new public spaces formed by the "transactional space" of telecommunications and computer technology, but my interest . . . is in public spaces as places that are physically *there*, as geographical and symbolic centers, as points of assembly where strangers mingle. Many Americans, born and raised in the suburbs, accept shopping centers as the pre-eminent public spaces of our time. Yet while shopping centers are undoubtedly gathering places, their private ownership has always raised questions about whether all the public has access to them and under what conditions. In the 1980s and 1990s, shopping centers became sites for hotels, post offices, and even schools, suggesting that public institutions can indeed function on private property . . .

When Disneyland recruited teenagers in South Central Los Angeles for summer jobs following the riots of 1992, it thrust into prominence a new confluence between the sources of contemporary public culture: a confluence between commercial culture and ethnic identity. Defining public culture in these terms recasts the way we view and describe the cultures of cities. Real cities are both material constructions, with human strengths and weaknesses, and symbolic projects developed by social representations, including affluence and technology, ethnicity and civility, local shopping streets and television news. Real cities are also macro-level struggles between major sources of change – global and local cultures, public stewardship and privatization, social diversity and homogeneity – and mid-level negotiations of power. Real cultures, for their part, are not torn by conflict between commercialism and ethnicity; they are made up of one-part corporate image selling and two-parts claims of group identity, and get their power from joining autobiography to hegemony – a powerful aesthetic fit with a collective lifestyle. This is the landscape of a symbolic economy . . . on sites as geographically and socially diverse as Disney World, the Massachusetts Museum of Contemporary Art, New York art worlds, Times Square, New York City restaurants, and ghetto shopping streets like 125th Street in Harlem. These are my sources; this is my "city."

How do we connect what we experience in public space with ideologies and rhetorics of public culture?

On the streets, the vernacular culture of the powerless provides a currency of economic exchange and a language of social revival. In other public spaces – grand plazas, waterfronts, and shopping streets reorganized by business improvement districts – another landscape incorporates vernacular culture or opposes it with its own image of identity and desire. Fear of reducing the distance between "us" and "them," between security guards and criminals, between elites and ethnic groups, makes culture a crucial weapon in reasserting order. Militant rhetoric belongs to the forces of order. "We will fight for every house in the city," said the New York City Police Commissioner in his inaugural address in 1993. "We will fight for every street. We will fight for every borough. And we will win." This Churchillian call echoes the appeal of right-wing journalist Patrick Buchanan, who explicitly identified cities and culture when he addressed the 1992 Republican National Convention in Houston: "And as those boys [in the National Guard] took back the streets of Los Angeles, block by block, my friends, we must take back our cities, and take back our culture, and take back our country."

But whose city? I ask. And whose culture?

FREDERIC STOUT

"Visions of a New Reality: The City and the Emergence of Modern Visual Culture"

(1999)

A fundamental precept of Marxist cultural analysis is that superstructures of thought and artistic expression rest upon and derive from a material base rooted in social and economic reality. Thus, each historical era creates characteristic forms of expression and explanatory discourse that reflect, indeed construct, the social reality of the period. Lewis Mumford spoke to this process when, in *The City in History* (1961), he wrote that in "the Book of Job, one beholds Jerusalem; in Plato, Sophocles, and Euripides, Athens; in Shakespeare and Marlowe . . . Elizabethan London."

For the cities of the Industrial Revolution, a number of forms of expression and modes of critical analysis arose to make sense of the dramatic and rapidly changing social reality. In literature, Balzac and Dickens pioneered a tradition of social realism particularly focused on the struggles of the urban poor which later attracted such followers as Emile Zola, Theodore Dreiser, Mrs. Gaskell, Upton Sinclair . . . and thousands more. In the realm of social and political analysis, the use of statistical evidence based on survey data compiled by government commissions and philanthropic organizations, the construction of great explanatory theories (such as those of Marx and Weber), and the development of social science methodologies helped observers of the urban milieu make sense of the revolutionary changes that were taking place around them. And in the realm of art, a whole new kind of visual culture emerged rooted in the observation of the new urban reality, both social and physical. It was a culture that began with nineteenth-century popular illustrated journalism and evolved to include mainstream traditions in the twentieth-century history of photography and cinema.

In the history of the fine arts, urban social realism plays a brief but highly visible role. Although much artistic social activism was largely confined to the satirical printmakers (except in those countries where state-sponsored socialist realism became the prescribed orthodoxy), works like William Hogarth's depictions of eighteenth-century London and Honoré Daumier's glimpses of nineteenth-century Paris are as influential as they are familiar. In painting, however, the realism of Courbet and Millet quickly give way to the impressionist celebration of light and the post-impressionist analysis of pure form. In America, the urban imagery of The Eight (dubbed "The Ashcan School" by a hostile critic) was revolutionary and shocking at their group show in 1907 but strictly old hat by the time modern abstraction was introduced at the Armory Show in 1912. One explanation for this is that The Eight stood not at the beginning but at the end of a fifty-year social realist tradition in the visual arts, a tradition that took place not in the salons of the easel painters but in the pages of the illustrated newspapers and magazines that flourished as a ubiquitous element of urban popular culture during the last half of the nineteenth century. Several of the Ashcan School painters – among them John Sloan, William Glackens, Everett Shinn, and Edward Hopper – had previously worked as journalistic illustrators, and it is to the pages of the popular newspapers that one must turn to see the earliest representations of the modern industrial city.

Illustrated journalism began in England as early as the 1820s, and in America in the 1840s, in response to a growing demand for information and entertainment on the part of a marginally literate but fully enfranchised working class and petit bourgeoisie. The first images of cities were almost always long-range or bird's-eye views (similar to the view of La Crosse, Wisconsin, from a 1887 issue of *Frank Leslie's Illustrated Newspaper* that appears as Plate 5). These static images were imitative of a received landscape tradition that had been commonly applied to rural and wild nature scenes, but they conveyed relatively little information about urban subject matter – note the inclusion of a street-scene cut-out in the La Crosse picture – and were soon replaced by more kinetic images and vignettes. "Contrast pictures" – both dual images such as "Music in the Street/Music in the Parlor" (Plate 9) from an 1868 issue of *The Illustrated News* and single images such as "The Hearth-stone of the Poor" (Plate 10) from an 1876 issue of *Harper's Weekly* by the exceptionally talented Sol Eytinge, Jr. – were particularly useful for comparing the lives of the urban rich and poor. Soon, a new kind of composite image emerged – "City Sketches" by the illustrator C. A. Barry from an 1855 issue of *Ballou's Pictorial Drawing-Room Companion* (Plate 11) is an interesting early example – that more accurately reflected the diverse, jumbled-together chaos that was the common experience of urban street life in the nineteenth century. Eventually, such composite pictures became a staple of popular illustrated journalism, and the step-by-step depiction of industrial processes became a favored topic, as in "Bicycles and Tricycles – How They Are Made" (Plate 12) from an 1887 issue of *Frank Leslie's Illustrated Newspaper*.

Handdrawn popular illustration died out around the turn of the century when a new technology of visual representation, photography, took over the journalistic duties of the newspaper illustrators. As a genre, popular illustration had accomplished much. It had created a visual culture embedded in the social reality of urban life and had urged visual art generally away from landscape toward cityscape, from stasis to kinesis. The contribution of photography would be even greater. Here was a technology of visual representation perfectly suited to its age. In the hands of journalists, it created a new and powerfully accurate kind of documentary record. In the hands of social activists, it created images of passion and outrage that could not be ignored. In the hands of the popular masses, it created snapshots and memories, both social and personal. And in the hands of artists – and who with a camera in his hands was not an artist? – it created art.

Photography is arguably the single most characteristic medium of expression and representation of the modern period, and a complete history of photography would include consideration of its use in science, medicine, criminology, education, and commerce as well as art. But as a medium of artistic expression, photography perfectly exemplified the spirit of the modern age, addressing every known genre and inventing new ones of its own. Nature and landscape, still life and the nude, staged heroic set pieces and candid genre scenes – all found new vitality in the hands of the masters of the new technology of visual representation. But it was urban subjects more than any others – the inescapable social and physical reality of the modern city – that captured photography's unique potential as an expressive medium and gave photography its historic and artistic *raison d'être*.

Although photography did not completely supplant handdrawn illustrations in journalism until the 1890s and early 1900s, the very first news photo may well have been "Burning Mills, Oswego, New York" (Plate 13), a daguerreotype by George N. Barnard that was printed in the *Oswego Daily Times* in 1853. While pictures of factory fires had been a staple of popular illustration, photography brought a new realism and intensity to the images presented to the public. Even today, dramatic pictures of big fires, even if they are from cities far away, have a special fascination and are often prominently placed in newspapers and the evening television news broadcasts.

The spectacle of the burning factory was a special case in a larger melodrama: the ongoing, day-to-day reality of modern urban life. Many of the very earliest photographs by Daguerre, Nadar, and Fox Talbot were simple street scenes made by the photographer simply turning his camera out the studio window. Like the "city sketches" type of popular illustration, these photographic street scenes conveyed a special sense of the vitality of urban life. Edward Anthony's "A Rainy Day on Broadway" (Plate 14) shows New York in 1859. The motion of those on the street and on the sidewalks is frozen in time – clearly Anthony's shutter speed was fast enough to eliminate most of the blurring that was so common in early photographs of moving subjects – but the image itself does not seem frozen. Rather, the image speaks directly to the viewer's own experience of street life, and the interaction of the viewed image and the viewer's imagination almost magically captures the hustle and bustle of New York street life.

Not all streets in the modern city were so pleasant. In "Bandits' Roost, 39½ Mulberry Street," of 1889 (Plate 15), Jacob Riis invites the viewer to look into one of New York's most notorious and dangerous back alleys. The author of *How the Other Half Lives* (1890), Riis was a crusading social reformer who had observed the dark side of the city as a police reporter for the local newspapers and he used his skills as a photographer to supplement his activism. Many middle-class viewers were undoubtedly repelled by images such as this one. It was the kind of view that one might catch fleetingly as one passed along a main thoroughfare, something to glance at quickly and then turn away. But Riis the reformer would have us not look away, and photography rivets the gaze on an unpleasant but inescapable social reality. In this, Riis's photographic work is directly linked to Friedrich Engels's descriptions of the Manchester slums in *The Condition of the Working Class in England in 1844*.

Jacob Riis was but one in a long and distinguished line of photographers who turned their art and their talents to the ends of political activism and social protest. One of the greatest of these visual documentors of the modern city was Lewis Hine. Hine's picture of a young girl tending a massive cloth-weaving machine in a North Carolina mill has become a nearly universal icon of industrial child labor, and his photographs of high-steel construction workers casually toiling hundreds of feet above the city are gut-wrenching images of working-class heroism. For a time, Hine worked at Ellis Island, the New York port of entry for immigrants to America. His portraits of the immigrants (Plate 16) are masterpieces of photographic humanism. Portraiture holds an important place in the history of photography. Whereas oil-painting portraiture had served the interests of the aristocracy and the haute bourgeoisie almost exclusively, the new medium allowed members of the broad urban middle class, and the new industrial working class as well, to record their images for posterity. And whereas the images of immigrants in popular illustration had tended toward racial and ethnic caricatures, Hine's portraits reveal the individuality and humanism behind the stereotypes. Another photograph depicting the immigrant condition – all the more powerful for its juxtaposition of social commentary and pure artistic composition – was Alfred Stieglitz's "The Steerage" of 1907 (Plate 17). Here, the interplay between the immigrants, mostly enveloped in shadows, and the physical structures of gangway and ladder combine simultaneously to fascinate the eye and touch the soul.

The Great Depression of the 1930s strengthened the documentary and social activist tendencies in photography that had been pioneered by Riis, Hine, and the photojournalists, especially in work that had been commissioned by the Works Progress Administration and other New Deal agencies. The photographs of Southern sharecroppers by Walker Evans and the images of Harlem streetlife by Gordon Parkes are notable examples of the intersection of art and social protest. Dorothea Lange was another activist photographer who, like Evans, was drawn to the documentation of rural poverty. Her pictures of dispossessed dust bowl emigres are among the most famous and most powerful images of the Depression era. And in her picture of a roadside sign (Plate 18), published in a 1939 Farm Security Administration report on rural migration, Lange captured both the bleakness of the Depression and the kinds of fears and insecurities that underlay Frank Lloyd Wright's "Broadacre City" proposal – a homestead of at least one acre per person – and other back-to-the-land schemes.

The people of the city and their problems were the dominant, but not the sole, subject of urban documentary photography. The very physical presence of the modern city was often evoked through images of architecture and infrastructure. Stieglitz's moody impression of the Flatiron Building cloaked in mist is one well-known example, and Margaret Bourke-White's formalist, monumental depiction of Hoover Dam that appeared on the cover of the first *Life* magazine is another. Charles Sheeler's "Ford Plant, Detroit" of 1927 (Plate 19) demonstrates the way in which pure architectural form – an assemblage of strong verticals and diagonals revealed through light and shadow – could be read both as pure cubist composition and powerful, if somewhat distanced, social documentation.

If the architecture of the city could present images of foreboding power or lyrical freedom, it was nonetheless the lives of the people of the city themselves that became the constantly recurring theme of photo-journalists and artists alike. The work of Weegee (Arthur Fertig) is a case in point. Although most of his pictures were made in the 1940s and 1950s and featured startling images of auto wrecks, transvestite bars, and gangland killings, the individual faces communicate directly to the viewer with a sense of direct, immediate reality that is the transcendent essence of urban life. "The Critic" (Plate 20), for example, was taken in 1943, but the people and the streetlife situation – so like the contrast pictures in popular illustration – are instantly recognizable as big-city icons of the painful urban conflation of luxury and poverty, indifference and despair.

The project of comprehending the modern city visually played, and continues to play, a central role in the history of art and consciousness. The great themes of the city – its kinetic activity, its juxtapositions and ironies, its massive forms and tiny details – provided the artist with a subject matter that could not be ignored and pioneered modes of visual perception and communication that were to fundamentally transform the nature of social life. Whereas popular illustrated journalism liberated visual images from the control of privileged elites, photography completed the democratization of the visual by placing it in the hands of the masses.

Eventually, the new modes of visual perception would be connected to narrative – particularly to the "urban narratives" of young rural innocents encountering the experience of the city for the first time, a plot line as old as Gilgamesh and given renewed meaning as a result of the millions of immigration stories that fueled the urbanization process during the nineteenth and twentieth centuries. And out of the interconnection of urban narrative and visual representation would come cinema, the ultimate realization of the kinetic imagery that urban life characterized and photography mirrored. In King Vidor's *The Crowd* of 1926 (Plate 21), we see the once-optimistic individual lost in the enveloping urban mass. In Fritz Lang's *Metropolis* of 1929 (Plate 22), we see the modern citadels of faceless power looming over the dehumanizing structures of class segregation and oppression. In both cases, the viewer confronts and subconsciously confirms the artist's perception of the reality of modern urban life. The confrontation may be discomfiting, but the ubiquity of modern visual culture makes the confrontations inescapable and memorable, perhaps more memorable even than actual encounters on the streets of the city. Writing about the historic city, Lewis Mumford found that they were characterized by an "urban drama" and that the dramatic dialogue was "one of the ultimate expressions of life in the city." In the modern city, it is the image – sometimes celebratory, sometimes haunting, always definitive in its explanatory value – that is paramount both as spectacle and revelation.

Plate 9 "Music in the Street, Music in the Parlor" (Unknown artist, 1868, *The Illustrated News*).

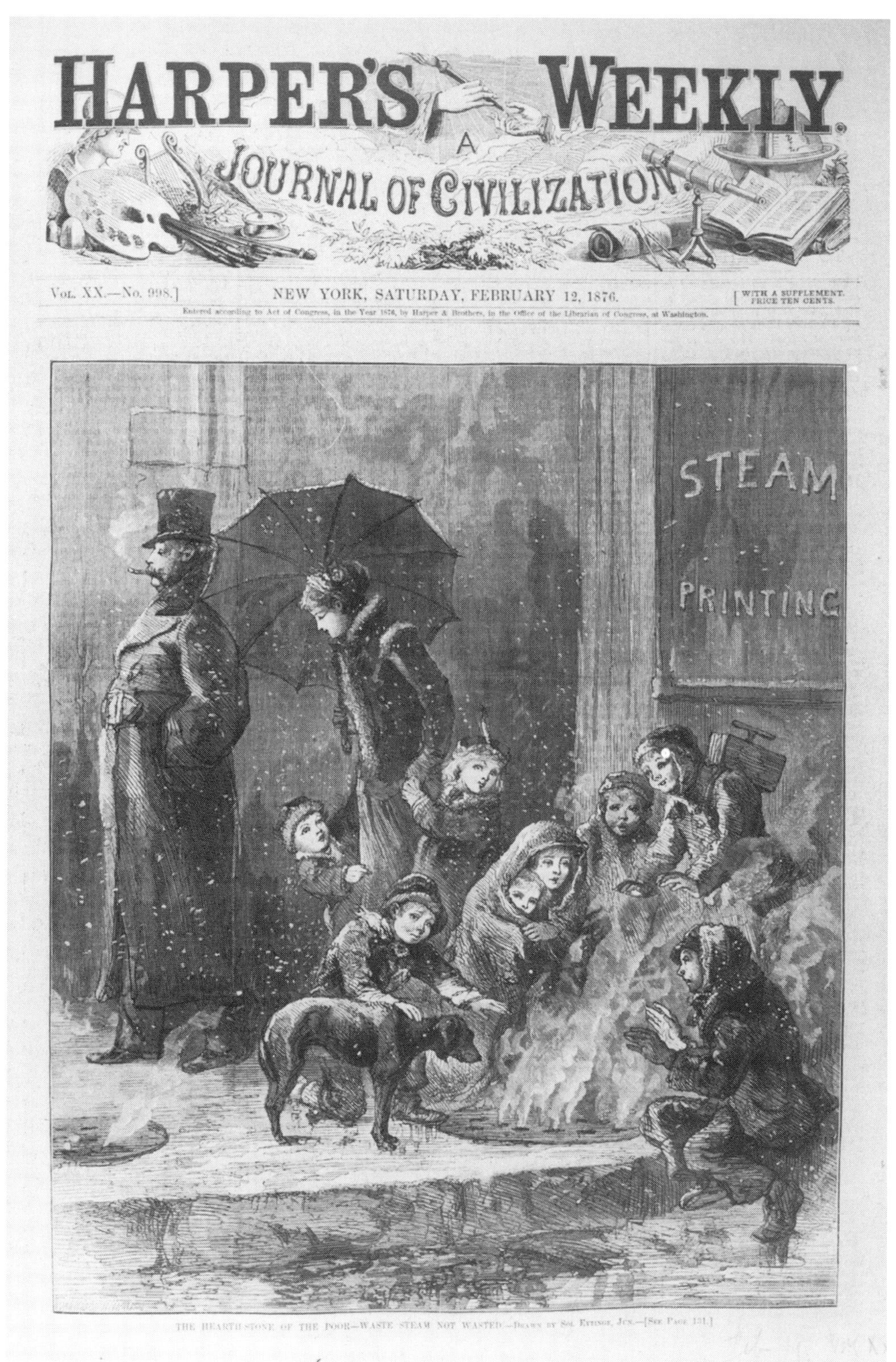

HARPER'S WEEKLY.

A JOURNAL OF CIVILIZATION

Vol. XX.—No. 998.] NEW YORK, SATURDAY, FEBRUARY 12, 1876. [WITH A SUPPLEMENT. PRICE TEN CENTS.

Entered according to Act of Congress, in the Year 1876, by Harper & Brothers, in the Office of the Librarian of Congress, at Washington.

THE HEARTH-STONE OF THE POOR—WASTE STEAM NOT WASTED.—Drawn by Sol Eytinge, Jun.—[See Page 131.]

Plate 10 "The Hearth-stone of the Poor" (Sol Eytinge, Jr., 1876, *Harper's Weekly*).

Plate 11 "City Sketches" (C. A. Barry, 1855, *Ballou's Pictorial Drawing-Room Companion*).

Plate 12 "Bicycles and Tricycles – How They Are Made" (Unknown artist, 1887, *Frank Leslie's Illustrated Newspaper*).

Plate 13 "Burning Mills, Oswego, New York" (George N. Barnard, 1853, *Oswego Daily Times*). (Courtesy George Eastman House.)

Plate 14 "A Rainy Day on Broadway" (Edward Anthony, 1859). (Courtesy George Eastman House.)

Plate 15 "Bandits' Roost, 39½ Mulberry Street" (Jacob Riis, New York, 1889). The Jacob A. Riis collection # 101. (Copyright © Museum of the City of New York.)

WELSH COAL MINER

SLAVIC STEEL-WORKER

Photos by Hine

ITALIAN LABORER

IRISH IRON WORKER

Plate 16 Ellis Island immigrant portraits (Lewis Hine, ca. 1910). In Grove S. Dow, *Social Problems of Today* (New York: Thomas Y. Crowell, 1925). Public domain.

Plate 17 "The Steerage" (Alfred Stieglitz, 1907). (Courtesy of the Museum of Modern Art, New York.)

Plate 18 Untitled (roadside sign) (Dorothea Lange, 1939). Farm Security Administration, US Department of Agriculture. Printed in C. E. Lively and Conrad Tauber, *Rural Migration in the United States* (Washington, DC: Works Progress Administration, 1939).

Plate 19 "Ford Plant, Detroit" (Charles Sheeler, 1927). (Courtesy of the Museum of Modern Art, New York.)

Plate 20 "The Critic" (Weegee (Arthur Fellig), 1943). (Copyright © 1994, International Center of Photography, New York. Bequest of Wilma Wilcox.)

Plate 21 Still from *The Crowd* (King Vidor, 1926). (Photograph courtesy of the Museum of Modern Art, New York, Film Stills Archive.)

Plate 22 Still from *Metropolis* (Fritz Lang, 1929). (Photograph courtesy of the Museum of Modern Art, New York, Film Stills Archive.)

PART 3
*U*rban Space

PART THREE

INTRODUCTION

The physical form of cities described in Part 1, on The Evolution Of Cities, ranged from Ur's mudbrick walls and ziggurat to the marble agora of the Greek polis; from the polluted slums of nineteenth-century Manchester to the high-tech research "campuses," malls, and residential areas of today's technoburbs. The size, density, spatial distribution of functions, and physical and cultural characteristics of cities described in that section display an astonishing variety over time.

This section is concerned with *urban geography*. It contains classic and contemporary writings on the physical and social structure of cities. Parts 5, 6 and 7 on Urban Planning History and Visions, Urban Planning Theory and Practice and Perspectives on Urban Design, return to this material, but extend the discussion to interventions to shape the form and design of cities and regions.

For many years geography focused on the study of physical space and the mapping of topographical and political features. Today academic geography has been reinvigorated by a new generation of scholars who define the discipline broadly and include analysis of social and cultural reality. Scholars from many disciplines have contributed to our understanding of urban space. In addition to geographers such as Edward Soja and David Harvey, this selection includes writings from the disciplines of architecture (Davis and Rybczynski), sociology (Burgess), landscape architecture (Jackson), and urban planning (Sassen). Writers from these and other disciplines throughout this book inform our understanding of urban geography.

Many cities have grown organically with little or no explicit overall plan or centrally controlled regulation. They contain vernacular architecture created by ordinary people without specialized training in the design professions. Other cities display a mix of organic growth and development planned at different periods in their history. The old crooked streets and irregular lots that constitute most of Boston, Massachusetts, contrast with the regularity of Back Bay. There a swampy area was developed during the latter part of the nineteenth century according to a plan by Frederick Law Olmsted. Disorderly districts of London that have evolved since the Middle Ages contrast with the area rebuilt after the great fire of 1666. Only a few entire cities like Australia's capital city Canberra and Brazil's capital city Brasilia have been built according to a consistent master plan.

It was sociologists, not geographers, who pioneered the modern field of urban geography. In addition to path-breaking work on the social structure of cities by Louis Wirth (p. 97), Robert Park, Ernest Burgess and others members of the Chicago School of urban sociology developed the first systematic theory about the physical form of cities. The first selection in this section, Ernest Burgess's "The Growth of the City," advances many provocative hypotheses. Burgess argued that there was an underlying logic to the physical form of cities. In his view they were organized in a series of concentric rings moving out from a central business district. Each ring had a distinct set of residents and functions. Physical form and human life were intimately linked. According to Burgess cities were not static. They were dynamic organisms with a constant flow of new residents coming into the inner rings and a flow from these rings outward over time. Burgess and other members of the Chicago School thought of cities as similar to natural ecological systems. Each part of an urban system had distinct characteristics and played a unique role in the total system. But each was related to the others and the whole system was in a state of constant flux.

Sociologists inspired by Burgess, land economists, and many geographers have studied city form during the last seventy-five years. Burgess's "concentric zone" theory is just one of a number of theories that seek to theoretically describe the internal structure of cities. It is often juxtaposed with theories developed by real estate economist Homer Hoyt in the 1930s, postulating an organization of cities in *sectors* radiating out along transportation corridors, and with an essay by geographers Chauncy Harris and Edward Ullman in the 1940s concluding that most cities have multiple nuclei rather than either concentric zone or sectoral organization. Even in the large postmodern Los Angeles region, geographer Edward Soja (p. 180) draws on this early theory and discerns a concentric and sectoral logic.

Burgess wrote about the way in which a single city is organized. A related topic in the study of urban form concerns systems of cities. Central place theory, developed by German geographer Walter Christaller, focuses on how cities within an entire region relate to each other. As computerized statistical analysis and geographical information systems for analyzing and mapping data rapidly increase our ability to understand urban form, the debates about the internal structure of cities and systems of cities continue.

The selections by landscape architect J. B. Jackson and architect Witold Rybczynski also note an underlying logic to the physical form, social structure and functions of cities. According to Jackson even small cities that evolved without city planners and architects display an underlying regularity and logic. Rybczynski, writing seventy-five years after Burgess and nearly half a century after Jackson, agrees with the premise of logic and regularity, but he sees a different logic underlying the new mall-based alternative downtown than existed at the time Burgess and Jackson wrote.

Despite the fact that until recently most cities have grown with little or no formal planning or professional design, little scholarly attention was devoted to studying vernacular urban form and popular design until Harvard and University of California, Berkeley-based landscape architecture professor J. B. Jackson pioneered the study of *vernacular* landscapes, the environments built by common people without trained architects, planners, or other professionals. Jackson was fascinated with the physical form of barns, fences, billboards, and grain silos; pioneer settlements with stumps in the field and muddy roads; dying former railroad towns where the train no longer stops; humble mobile homes in rural New Mexico; and gas stations and frozen custard stands along the principal artery across the American Southwest, Highway 66. Jackson was one of the first to argue that a study of vernacular architecture provides important understanding of the culture and values of the people who have built the built environment. According to Jackson a close look at Highway 66 in the 1950s reveals a great deal about the worldview of the highway builders and residents along the highway. Today, largely as a result of Jackson's influence, there is a large literature describing and analyzing the function and cultural meaning of Las Vegas casinos, White Tower hamburger stands, suburban tract homes, billboards, and other vernacular architecture.

Like Jackson, architect Witold Rybczynski begins his chapter on "The New Downtown" with a description of a small town, Plattsburgh, New York. Like Jackson's Optimo City in the 1950s Plattsburgh in the 1990s has seen better days. Much of what Jackson prophesized about the probable demise of Optimo had in fact occurred in Plattsburgh. Automobiles are a central factor. In fact by the early 1950s Optimo City was already in danger of being bypassed by a new interstate highway, and the town fathers were thinking of paving over the courthouse square as a parking lot. By the 1990s new roads and much higher automobile use had shifted most of Plattsburgh's downtown activities to the highways outside of town, leaving the main street of "old" downtown Plattsburgh with a thrift store, secondhand bookshops, a marginal movie theater, and vacant lots. Plattsburgh's customs house and 1,000-seat theater were gone, and its grand 500-room tourist hotel had been converted to a community college.

Rybczynski does not blame Plattsburgh's faded condition on the rise of global cities that Saskia Sassen describes (p. 208), on the deindustrialization and the loss of jobs that William Julius Wilson blames for the woes of Black ghettos in Chicago (p. 112), nor on the emergence of a threatening urban underclass that Mike Davis describes in Los Angeles (p. 193). Nor does he blame Plattsburgh's decline on migration of former residents to suburbs. Plattsburghians did not move to the suburbs. Most of them still live in large comfortable houses on quiet, tree-lined streets in the residential neighborhoods that surround the downtown, much as they did one hundred years ago. Rather, Rybczynski points to the rise of shopping centers – the new downtown – as the key force that undermined Plattsburgh's former economy. By the 1960s there were three new shopping centers all built on highways leading into Plattsburgh. "Downtown," Rybczynski concludes, "was finished." New technology – refrigerators that preserved food and personal mobility by automobile that made it possible for city residents to travel to stores – led to the creation of modern supermarkets. The freedom to move goods by truck contributed to the decline of downtown as small factories relocated to strip developments accessible by truck rather than train. Rybczynski cites J. B. Jackson with approval for the proposition that the commercial automobile, the truck, the pickup, the van, minivan, and jeep have been most effective in introducing a different spatial order to cities and regions.

While the first shopping center, the original Piggly Wiggly, opened in Memphis, Tennessee, in 1916, shopping centers were few and far between until the 1950s. According to Rybczynski there were only about 100 shopping centers in the entire United States in 1950, but there were 3,700 ten years later. What happened to Plattsburgh's downtown economy was repeated across the country as the "new downtown" of exurban shopping centers sapped the "old downtown." And today, Rybczynski argues, the shopping center has truly become the center for urban social life as well as economic activity.

University of California, Los Angeles, geography professor Edward Soja uses the Los Angeles metropolitan region as the basis for a discussion of the postmodern metropolis. He is bemused by the bewildering contrasts to be found in the Sixty-Mile Circle surrounding downtown Los Angeles. The pieces of the Los Angeles landscape Soja takes apart display huge contrasts: gigantic military bases and tiny eighteenth century missions; gated communities for the rich and squalid ghettos for the poor; buildings and whole regions engaged solely in post-industrial research and managerial activities and a thriving industrial sector employing low-wage immigrant labor. Soja emphasizes the fantastical imagery and theme park quality of the Los Angeles region. But underlying his irony is a serious message. Beneath the glitz and glitter, Los Angeles has a dark side. According to Soja, police stations, courts, and prisons are central features of downtown. The region's industry could not exist without poorly paid immigrant labor. Violent and depressing ghettos are threaded throughout the Los Angeles region. The natural ecology of this congested and smog-blanketed metropolis is in deep trouble.

In "Fortress L.A." social critic Mike Davis addresses the same issue – the dark side of the postmodern metropolis. Davis describes a built environment complete with surveillance cameras, barrel-shaped park benches designed to keep people from sleeping on them, overhead sprinklers to douse the homeless, windowless concrete hotel walls facing streets, and gated communities. All these artifacts provide a disturbing glimpse of Los Angeles and, by implication, postmodern cities emerging around the world.

Davis offers no way out of the apocalyptic world he describes, but geographer (and philosopher) David Harvey does. Drawing on both Marxist theory and postmodern critical theory, Harvey discusses social divisions and conflicting worldviews among the hugely varied inhabitants of the postmodern metropolis. Ultimately he concludes that justice is relative, largely determined by economic relations produced by a specific culture at a specific time in history. Nonetheless consciousness about the relativity of social justice and attention to the forms social oppression takes can sensitize policy-makers, if not resolve the underlying conflicts.

The final selection in this section, "A New Geography of Centers and Margins," by urban planner Saskia Sassen, moves from Soja, Davis, and Harvey's discussion of the contradictions in postmodern metropoles like Los Angeles and New York to an analysis of the underlying economic characteristics of these and other global cities. Based on her research on the economy of world cities, Sassen concludes that a truly global system of cities has emerged. New York, London, and Tokyo are the main bastions of international finance and control. Decisions made in these three cities affect the residents of cities around the world. Sassen concludes that information technology is the driving force behind the new geography, but rejects some facile stereotypes. Some writers argue that because the fax machine, phone, modem, and Internet make it possible to communicate easily and instantaneously anywhere in the world, place no longer matters and large cities will wither away. Sassen finds no evidence for this view. Rather she notes further growth of population and power in the commanding global cities. Sassen also dismisses facile divisions of the world into a developed center and an undeveloped margin. Rather, she argues, there are de facto Third World communities in developed global cities and outposts of global cities throughout the Third World.

ERNEST W. BURGESS

"The Growth of the City: An Introduction to a Research Project"

from Robert Park *et al.*, *The City* (1925)

Editors' introduction Ernest W. Burgess (1886–1966) was a member of the famed Sociology Department at the University of Chicago that included such luminaries as Louis Wirth, author of "Urbanism As a Way of Life" (p. 97), and former newspaper reporter and social reformer Robert E. Park. Together, these scholars set out to virtually reinvent modern sociology by taking academic research to the streets and by using the city of Chicago itself as a "living laboratory" for the study of urban problems and social dynamics.

Throughout a long and productive career, Burgess addressed a whole series of issues that connected the social dynamics of the city as a whole with the lives of its citizens. He wrote extensively on issues related to marriage and the family, the relation of personality to social groups, and, in the final decades of his life, problems of the elderly. However, his most famous contribution to the study of the city was the 1925 essay reprinted here: "The Growth of the City."

Subtitled "An Introduction to a Research Project," Burgess's seminal analysis of the interrelation of the social growth and the physical expansion of modern cities served generations of other urban sociologists, geographers, and planners as a kind of prolegomena to all future study of the city. Seeking to describe what he called "the pulse of the community," Burgess devised a theory that was thoroughly organic, dynamic, and developmental. "In the expansion of the city," he wrote, "a process of distribution takes place, which sifts and sorts and relocates individuals and groups by residence and occupation." And it was this dynamic process – "process" was one of Burgess's favorite words – that "gives form and character to the city."

Central to Burgess's analysis of urban growth was his famous model based on a series of concentric circles that divided the city into five zones. On one level, the concentric zone model was merely a map of contemporaneous Chicago with Zone I being the central business district area (known in Chicago as "The Loop"). On another level, the model was a theoretical diagram of a dynamic process Burgess called "succession," a term he borrowed from the science of plant ecology to describe "urban metabolism and mobility." By borrowing terminology from the natural sciences, and by drawing analogies between the urban and the natural worlds, Burgess established the study of "social ecology" as a distinct approach to understanding the underlying patterns of urban growth and development.

Following the publication of Burgess's essay, a number of urban theorists offered modifications and even refutations of the simple elegance of the concentric zone model. In 1939, real estate economist Homer Hoyt proposed a "sectoral model" for modern capitalist cities based on "wedges of activity" extending outward from the city center along transportation corridors. In 1945, geographers Chauncy Harris and Edward Ullman suggested a "multiple nuclei model," arguing that cities developed around several, not just one, center of economic activity.

But geographer Edward Soja, discussing postmodern Los Angeles (p. 180), notes that even in the polycentric archipelago of modern Los Angeles "population densities do mound up around the centres of cities" and that "there is also an accompanying concentric residential rhythm" to the region. Soja also notes "the emanation of fortuitous wedges or sectors from the centre," suggesting that Homer Hoyt's sectoral model still has some explanatory power even in postmodern Los Angeles. Soja notes that the Wilshire corridor, for example, "extends the citadels of the central city almost twenty miles westwards to the Pacific."

Burgess's influence was both widespread and longstanding, if not for the details of his model, then for the spirit of its basic insight. Burgess's view may be compared to J. B. Jackson's notion that there is an underlying logic to urban form that occurs even in the absence of formal planning (p. 162). The work of dozens of urban sociologists such as William Julius Wilson's study of the urban underclass (p. 112) owes a profound debt to Burgess. And just as importantly, Burgess's sense of urban dynamism and vitality captures the essence of those more lyric conceptions of the city such as Lewis Mumford's notion of "the urban drama" (p. 92) and Jane Jacobs's "ballet of the streets" (p. 106).

Computer technology, including statistical packages and geographic information systems (GIS) software, now makes it possible for present-day geographers and sociologists to summarize vast amounts of data and map the internal structure of cities in ever more sophisticated ways. Thus, understanding of the relationship between social groups and urban form pioneered by Burgess continues to advance by leaps and bounds. Robert E. Park, Ernest W. Burgess, and Roderick D. McKenzie, *The City* 3rd edn, revised (Chicago: University of Chicago Press, 1967), from which this selection is taken, is the best introductory collection of the works of the Chicago School of urban sociology. For the works of Burgess in particular, including a comprehensive bibliography, see Leonard S. Cottrell, Jr., *et al.* (eds.), *On Community, Family and Delinquency* (Chicago: University of Chicago Press, 1973). Homer Hoyt's sectoral theory is described in *The Structure and Growth of Residential Neighborhoods in American Cities* (Washington: Federal Housing Administration, 1939); Chauncy Harris and Edward Ullman's multiple nucleii model is presented in "The Nature of Cities," *Annals of the American Academy of Political and Social Science*, 242 (1945).

The outstanding fact of modern society is the growth of great cities. Nowhere else have the enormous changes which the machine industry has made in our social life registered themselves with such obviousness as in the cities. In the United States the transition from a rural to an urban civilization, though beginning later than in Europe, has taken place, if not more rapidly and completely, at any rate more logically in its most characteristic forms.

All the manifestations of modern life which are peculiarly urban – the skyscraper, the subway, the department store, the daily newspaper, and social work – are characteristically American. The more subtle changes in our social life, which in their cruder manifestations are termed "social problems," problems that alarm and bewilder us, such as divorce, delinquency, and social unrest, are to be found in their most acute forms in our largest American cities. The profound and "subversive" forces which have wrought these changes are measured in the physical growth and expansion of cities. That is the significance of the comparative statistics of Weber, Bucher, and other students.

These statistical studies, although dealing mainly with the effects of urban growth, brought out into clear relief certain distinctive characteristics of urban as compared with rural populations. The larger proportion of women to men in the cities than in the open country, the greater percentage of youth and middle-aged, the higher ratio of the foreign-born, the increased heterogeneity of occupation increase with the growth of the city and profoundly alter its social structure. These variations in the composition of population are indicative of all the changes going on in the social organization of the community. In fact, these changes are a part of the growth of the city and suggest the nature of the processes of growth.

The only aspect of growth adequately described by Bucher and Weber was the rather obvious process of the aggregation of urban population. Almost as overt a process, that of expansion, has been investigated from a different and very practical point of view by groups interested in city planning, zoning, and regional surveys. Even more significant than the increasing density of urban population is its correlative tendency to overflow, and so to extend over wider areas, and to incorporate these areas into a larger communal life. This paper, therefore, will treat first of the expansion of the city, and then of the less-known processes of urban metabolism and mobility which are closely related to expansion.

EXPANSION AS PHYSICAL GROWTH

The expansion of the city from the standpoint of the city plan, zoning, and regional surveys is thought of almost wholly in terms of its physical growth. Traction studies have dealt with the development of transportation in its relation to the distribution of population throughout the city. The surveys made by the Bell Telephone Company and other public utilities have attempted to forecast the direction and the rate of growth of the city in order to anticipate the future demands for the extension of their services. In the city plan the location of parks and boulevards, the widening of traffic streets, the provision for a civic center, are all in the interest of the future control of the physical development of the city.

This expansion in area of our largest cities is now being brought forcibly to our attention by the Plan for the Study of New York and Its Environs, and by the formation of the Chicago Regional Planning Association, which extends the metropolitan district of the city to a radius of 50 miles, embracing 4,000 square miles of territory. Both are attempting to measure expansion in order to deal with the changes that accompany city growth. In England, where more than one-half of the inhabitants live in cities having a population of 100,000 and over, the lively appreciation of the bearing of urban expansion on social organization is thus expressed by C. B. Fawcett:

> One of the most important and striking developments in the growth of the urban populations of the more advanced peoples of the world during the last few decades has been the appearance of a number of vast urban aggregates, or conur bations, far larger and more numerous than the great cities of any preceding age. These have usually been formed by the simultaneous expansion of a number of neighboring towns, which have grown out toward each other until they have reached a practical coalescence in one continuous urban area. Each such conurbation still has within it many nuclei of denser town growth, most of which represent the central areas of the various towns from which it has grown, and these nuclear patches are connected by the less densely urbanized areas which began as suburbs of these towns. The latter are still usually rather less continuously occupied by buildings, and often have many open spaces.
>
> These great aggregates of town dwellers are a new feature in the distribution of man over the earth. At the present day there are from thirty to forty of them, each containing more than a million people, whereas only a hundred years ago there were, outside the great centers of population on the waterways of China, not more than two or three. Such aggregations of people are phenomena of great geographical and social importance; they give rise to new problems in the organization of the life and well-being of their inhabitants and in their varied activities. Few of them have yet developed a social consciousness at all proportionate to their magnitude, or fully realized themselves as definite groupings of people with many common interests, emotions and thoughts.

In Europe and America the tendency of the great city to expand has been recognized in the term "the metropolitan area of the city," which far overruns its political limits, and, in the case of New York and Chicago, even state lines. The metropolitan area may be taken to include urban territory that is physically contiguous, but it is coming to be defined by that facility of transportation that enables a business man to live in a suburb of Chicago and to work in the loop, and his wife to shop at Marshall Field's and attend grand opera in the Auditorium.

EXPANSION AS A PROCESS

No study of expansion as a process has yet been made, although the materials for such a study and intimations of different aspects of the process are contained in city planning, zoning, and regional surveys. The typical processes of the expansion of the city can best be illustrated, perhaps, by a series of concentric circles, which may be numbered to designate both the successive zones of urban extension and the types of areas differentiated in the process of expansion [Figure 1].

[Figure 1] represents an ideal construction of the tendencies of any town or city to expand radially from its central business district – on the map "the Loop" (I). Encircling the downtown area there is normally an area in transition, which is being invaded by business and light manufacture (II). A third area (III) is inhabited by the workers in industries who have escaped from the area of deterioration (II) but who desire to live within easy access of their work. Beyond this zone is the "residential area" (IV) of high-class apartment buildings or of exclusive

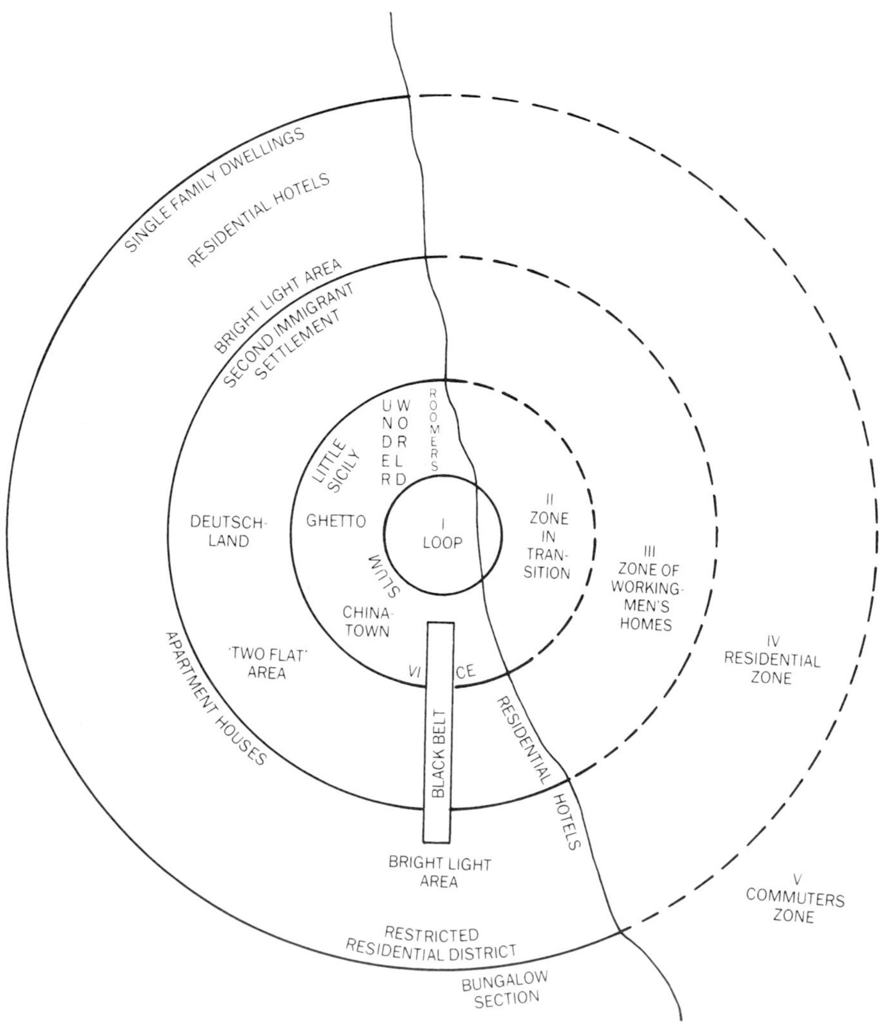

Figure 1

"restricted" districts of single family dwellings. Still farther, out beyond the city limits, is the commuters' zone: suburban areas, or satellite cities, within a thirty- to sixty-minute ride of the central business district.

This [figure] brings out clearly the main fact of expansion, namely, the tendency of each inner zone to extend its area by the invasion of the next outer zone. This aspect of expansion may be called *succession*, a process which has been studied in detail in plant ecology. If this [figure] is applied to Chicago, all four of these zones were in its early history included in the circumference of the inner zone, the present business district. The present boundaries of the area of deterioration were not many years ago those of the zone now inhabited by independent wage-earners, and within the memories of thousands of Chicagoans contained the residences of the "best families." It hardly needs to be added that neither Chicago nor any other city fits perfectly into this ideal scheme. Complications are introduced by the lake front, the Chicago River, railroad lines, historical factors in the location of industry, the relative degree of the resistance of communities to invasion, etc.

Besides extension and succession, the general process of expansion in urban growth involves the antagonistic and yet complementary processes of concentration and decentralization. In all cities there is the natural tendency for local and outside transportation to converge in the central business district. In the downtown section of every large city we expect to find the department stores, the skyscraper office buildings, the railroad stations, the great hotels, the theaters, the art museum, and the city hall. Quite naturally, almost inevitably, the economic, cultural, and political life centers here. The relation of centralization to the other processes of city life may be roughly gauged by the fact that over half a million people daily enter and leave Chicago's "loop." More recently sub-business centers have grown up in outlying zones. These "satellite loops" do not, it seems, represent the "hoped for" revival of the neighborhood, but rather a telescoping of several local communities into a larger economic unity. The Chicago of yesterday, an agglomeration of country towns and immigrant colonies, is undergoing a process of reorganization into a centralized decentralized system of local communities coalescing into sub-business areas visibly or invisibly dominated by the central business district. The actual processes of what may be called centralized decentralization are now being studied in the development of the chain store, which is only one illustration of the change in the basis of the urban organization.

Expansion, as we have seen, deals with the physical growth of the city, and with the extension of the technical services that have made city life not only livable, but comfortable, even luxurious. Certain of these basic necessities of urban life are possible only through tremendous development of communal existence. Three millions of people in Chicago are dependent upon one unified water system, one giant gas company, and one huge electric light plant. Yet, like most of the other aspects of our communal urban life, this economic co-operation is an example of co-operation without a shred of what the "spirit of co-operation" is commonly thought to signify. The great public utilities are a part of the mechanization of life in great cities, and have little or no other meaning for social organization.

Yet the processes of expansion, and especially the rate of expansion, may be studied not only in the physical growth and business development, but also in the consequent changes in the social organization and in personality types. How far is the growth of the city, in its physical and technical aspects, matched by a natural but adequate readjustment in the social organization? What, for a city, is a normal rate of expansion, a rate of expansion with which controlled changes in the social organization might successfully keep pace?

SOCIAL ORGANIZATION AND DISORGANIZATION AS PROCESSES OF METABOLISM

These questions may best be answered, perhaps, by thinking of urban growth as a resultant of organization and disorganization analogous to the anabolic and katabolic processes of metabolism in the body. In what way are individuals incorporated into the life of a city? By what process does a person become an organic part of his society? The natural process of acquiring culture is by birth. A person is born into a family already adjusted to a social environment – in this case the modern city. The natural rate of increase of population most favorable for assimilation may then be taken as the excess of the birth-rate over the death-rate, but is this the normal rate of city growth? Certainly, modern cities have increased and are increasing in population at a far higher rate. However, the natural rate of growth may be used to measure the disturbances of metabolism caused by any excessive increase, as those which followed the great influx of southern Negroes into northern cities since the war. In a similar way all cities show deviations in composition by age and sex from a standard population such as that of Sweden, unaffected in recent years by any great emigration or immigration. Here again, marked variations, as any great excess of males over females, or of females over males, or in the proportion of children, or of grown men or women, are symptomatic of abnormalities in social metabolism.

Normally the processes of disorganization and organization may be thought of as in reciprocal relationship to each other, and as co-operating in a moving equilibrium of social order toward an end vaguely or definitely regarded as progressive. So far as disorganization points to reorganization and makes for more efficient adjustment, disorganization must be conceived not as pathological, but as normal. Disorganization as preliminary to reorganization of attitudes and conduct is almost invariably the lot of the newcomer to the city, and the discarding of the habitual, and often of what has been to him the moral, is not infrequently accompanied by sharp mental conflict and sense of personal loss. Oftener, perhaps, the change gives sooner or later a feeling of emancipation and an urge toward new goals.

In the expansion of the city a process of distribution takes place which sifts and sorts and relocates individuals and groups by residence and occupation. The resulting differentiation of the cosmopolitan American city into areas is typically all from one pattern, with only interesting minor modifications. Within the central business district or on an adjoining street is the "main stem" of "hobohemia," the teeming Rialto of the homeless migratory man of the Middle West. In the zone of deterioration encircling the central business section are always to be found the so-called "slums" and "bad lands," with their submerged regions of poverty, degradation, and disease, and their underworlds of crime and vice. Within a deteriorating area are rooming-house districts, the purgatory of "lost souls." Nearby is the Latin Quarter, where creative and rebellious spirits resort. The slums are also crowded to over flowing with immigrant colonies – the Ghetto, Little Sicily, Greek town, Chinatown – fascinatingly combining old world heritages and American adaptations. Wedging out from here is the Black Belt with its free and disorderly life. The area of deterioration, while essentially one of decay, of stationary or declining population, is also one of regeneration, as witness the mission, the settlement, the artists' colony, radical centers – all obsessed with the vision of a new and better world.

The next zone is also inhabited predominatingly by factory and shop workers, but skilled and thrifty. This is an area of second immigrant settlement, generally of the second generation. It is the region of escape from the slum, the *Deutschland* of the aspiring Ghetto family. For *Deutschland* (literally "Germany") is the name given, half in envy, half in derision, to that region beyond the Ghetto where successful neighbors appear to be imitating German Jewish standards of living. But the inhabitant of this area in turn looks to the "Promised Land" beyond, to its residential hotels, its apartment-house region, its "satellite loops," and its "bright light" areas.

This differentiation into natural economic and cultural groupings gives form and character to the city. For segregation offers the group, and thereby the individuals who compose the group, a place and a role in the total organization of city life. Segregation limits development in certain directions, but releases it in others. These areas tend to accentuate certain traits, to attract and develop their kind of individuals, and so to become further differentiated.

The division of labor in the city likewise illustrates disorganization, reorganization and increasing differentiation. The immigrant from rural communities in Europe and America seldom brings with him economic skill of any great value in our industrial, commercial, or professional life. Yet interesting occupational selection has taken place by nationality, explainable more by racial temperament or circumstance than by old-world economic background as Irish policemen, Greek ice-cream parlors, Chinese laundries, Negro porters, Belgian janitors, etc.

The facts that in Chicago one million (996,589) individuals gainfully employed reported 509 occupations, and that over 1,000 men and women in *Who's Who* gave 116 different vocations give some notion of how in the city the minute differentiation of occupation "analyzes and sifts the population, separating and classifying the diverse elements." These figures also afford some intimation of the complexity and complication of the modern industrial mechanism and the intricate segregation and isolation of divergent economic groups. Interrelated with this economic division of labor is a corresponding division into social classes and into cultural and recreational groups. From this multiplicity of groups, with their different patterns of life, the person finds his congenial social world and – what is not feasible in the narrow confines of a village – may move and live in widely separated, and perchance conflicting, worlds. Personal disorganization may be but the failure to harmonize the canons of conduct of two divergent groups.

If the phenomena of expansion and metabolism indicate that a moderate degree of disorganization may and does facilitate social organization, they indicate as well that rapid urban expansion is accompanied by excessive increases in disease, crime, disorder, vice, insanity and suicide, rough indexes of social disorganization. But what are the indexes of the causes, rather than of the effects, of the disordered social metabolism of the city? The excess of the actual over the natural increase of population has already been suggested as a criterion. The significance of this increase consists in the immigration into a metropolitan city like New York and Chicago of tens of thousands of persons annually. Their invasion of the city has the effect of a tidal wave inundating first the immigrant colonies, the ports of first entry, dislodging thousands of inhabitants who overflow into the next zone, and so on and on until the momentum of the wave has spent its force on the last urban zone. The whole effect is to speed up expansion, to speed up industry, to speed up the "junking" process in the area of deterioration (II). These internal movements of the population become the more significant for study. What movement is going on in the city, and how may this movement be measured? It is easier, of course, to classify movement within the city than to measure it. There is the movement from residence to residence, change of occupation, labor turnover, movement to and from work, movement for recreation and adventure. This leads to the question: what is the significant aspect of movement for the study of the changes in city life? The answer to this question leads directly to the important distinction between movement and mobility.

MOBILITY AS THE PULSE OF THE COMMUNITY

Movement, per se, is not an evidence of change or of growth. In fact, movement may be a fixed and unchanging order of motion, designed to control a constant situation, as in routine movement. Movement that is significant for growth implies a change of movement in response to a new stimulus or situation. Change of movement of this type is called *mobility*. Movement of the nature of routine finds its typical expression in work. Change of movement, or mobility, is characteristically expressed in adventure. The great city, with its "bright lights," its emporiums of novelties and bargains, its palaces of amusement, its underworld of vice and crime, its risks of life and property from accident, robbery, and homicide, has become the region of the most intense degree of adventure and danger, excitement and thrill.

Mobility, it is evident, involves change, new experience, stimulation. Stimulation induces a response of the person to those objects in his environment which afford expression for his wishes. For the person, as for the physical organism, stimulation is essential to growth. Response to stimulation is wholesome so long as it is a correlated integral reaction of the entire personality. When the reaction is segmental, that is, detached from, and uncontrolled by, the organization of personality, it tends to become disorganizing or pathological. That is why stimulation for the sake of stimulation, as in the restless pursuit of pleasure, partakes of the nature of vice.

The mobility of city life, with its increase in the number and intensity of stimulations, tends inevitably to confuse and to demoralize the person. For an essential element in the mores and in personal morality is consistency, consistency of the type that is natural in the social control of the primary group. Where mobility is the greatest, and where in consequence primary controls break down completely, as in the zone of deterioration in the modern city, there develop areas of demoralization, of promiscuity, and of vice.

In our studies of the city it is found that areas of mobility are also the regions in which are found juvenile delinquency, boys' gangs, crime, poverty, wife desertion, divorce, abandoned infants, vice.

These concrete situations show why mobility is perhaps the best index of the state of metabolism of the city. Mobility may be thought of, in more than a fanciful sense, as the "pulse of the community." Like the pulse of the human body, it is a process which reflects and is indicative of all the changes that are taking place in the community, and which is susceptible of analysis into elements which may be stated numerically.

The elements entering into mobility may be classified under two main heads: (1) the state of mutability of the person, and (2) the number and kind of contacts or stimulations in his environment. The mutability of city populations varies with sex and age composition, and the degree of detachment of the person from the family and from other groups. All these factors may be expressed numerically. The new stimulations to which a population responds can be measured in terms of change of movement or of increasing contacts. Statistics on the movement of urban population may only measure routine, but an increase at a higher ratio than the increase of population measures mobility. In 1860 the horse-car lines of New York City carried about 50,000,000 passengers; in 1890 the trolley cars (and a few surviving horse-cars) transported about 500,000,000; in 1921, the elevated, subway, surface, and electric and steam suburban lines carried a total of more than 2,500,000,000 passengers. In Chicago the total annual rides per capita on the surface and elevated lines were 164 in 1890; 215 in 1900; 320 in 1910; and 338 in 1921. In addition, the rides per capita on steam and electric suburban lines almost doubled between 1916 (23) and 1921 (41), and the increasing use of the automobile must not be overlooked. For example, the number of automobiles in Illinois increased from 131,140 in 1915 to 833,920 in 1923.

Mobility may be measured not only by these changes of movement, but also by increase of contacts. While the increase of population of Chicago in 1912–22 was less than 25 percent (23.6 percent), the increase of letters delivered to Chicagoans was double that (49.6 percent) – from 693,048,196 to 1,038,007,854). In 1912 New York had 8.8 telephones; in 1922, 16.9 per 100 inhabitants. Boston had, in 1912, 10.1 telephones; ten years later, 19.5 telephones per 100 inhabitants. In the same decade the figures for Chicago increased from 12.3 to 21.6 per 100 population. But increase of the use of the telephone is probably more significant than increase in the number of telephones. The number of telephone calls in Chicago increased from 606,131,928 in 1914 to 944,010,586 in 1922, an increase of 55.7 percent, while the population increased only 13.4 percent.

Land values, since they reflect movement, afford one of the most sensitive indexes of mobility. The highest land values in Chicago are at the point of greatest mobility in the city, at the corner of State and Madison streets, in the Loop. A traffic count showed that at the rush period 31,000 people an hour, or 210,000 men and women in sixteen and one-half hours, passed the southwest corner. For over ten years land values in the Loop have been stationary but in the same time they have doubled, quadrupled and even sextupled in the strategic corners of the "satellite loops," an accurate index of the changes which have occurred. Our investigations so far seem to indicate that variations in land values, especially where correlated with differences in rents, offer perhaps the best single measure of mobility, and so of all the changes taking place in the expansion and growth of the city.

In general outline, I have attempted to present the point of view and methods of investigation which the department of sociology is employing in its studies in the growth of the city, namely, to describe urban expansion in terms of extension, succession, and concentration; to determine how expansion disturbs metabolism when disorganization is in excess of organization; and, finally, to define mobility and to propose it as a measure both of expansion and metabolism, susceptible to precise quantitative formulation, so that it may be regarded almost literally as the pulse of the community. In a way, this statement might serve as an introduction to any one of five or six research projects under way in the department. The project, however, in which I am directly engaged is an attempt to apply these methods of investigation to a cross-section of the city – to put this area, as it were, under the microscope, and so to study in more detail and with greater control and precision the processes which have been described here in the large. For this purpose the West Side Jewish community has been selected. This community includes the so-called "Ghetto," or area of first settlement, and Lawndale, the so-called "Deutschland," or area of second settlement. This area has certain obvious advantages for this study, from the standpoint of expansion, metabolism, and mobility. It exemplifies the tendency to expansion radially from the business center of the city. It is now relatively a homogeneous cultural group. Lawndale is itself an area in flux, with the tide of migrants still flowing in from the Ghetto and a constant egress to more desirable regions of the residential zone. In this area, too, it is also possible to study how the expected outcome of this high rate of mobility in social and personal disorganization is counteracted in large measure by the efficient communal organization of the Jewish community.

JOHN BRINCKERHOFF JACKSON

"The Almost Perfect Town"

Landscape (1952)

Editors' introduction John Brinckerhoff Jackson (1909–1997) can best be described as a historian of landscapes as physical, cultural, and conceptual artifacts. To Jackson, the landscape was neither wilderness nor rural nature "out there," but the totality of natural and built environments that simultaneously surround and infuse all forms of human activity.

Jackson taught for many years in schools of landscape architecture at Harvard and the University of California, Berkeley. He was the founder of the influential journal *Landscape* and its editor from 1951 to 1968. He may be regarded as an heir to the nineteenth-century French and English landscape gardening traditions and to the pioneering work of Frederick Law Olmsted (p. 314) and the American parks movement. What sets him clearly apart from those traditional roots, however, is that, for Jackson, the urban landscape is not just a city's parks, public gardens, official buildings, and treelined boulevards, but its highways, its shopping malls, its rundown warehouse districts, its standard-built two-bedroom houses, and its slums as well. Jackson sees these many elements of the built landscape not just as physical objects, but as social constructs full of meaning and moral implication. It is hardly surprising, then, that J. B. Jackson's approximate version of urban utopia – his *almost* perfect place – is the commonplace, vernacular, Main Street environment of a typical American small town.

In "The Almost Perfect Town" Jackson describes a semi-mythical Optimo City that is located in the American Southwest but could just as easily have been found almost anywhere in North America, Europe, Australia, and even parts of Asia and Africa beyond the margins of the great metropolitan regions. It is important that Optimo is a small place. Jackson observes that "the world of Optimo City is still complete" precisely because "the ties between country and town have not yet been broken." It is also important that Optimo has a history, however slender, that can be read in its architecture, in the layout of its streets, and in its traditional rivalry with Apache Center twenty miles away.

Jackson uses his loving, elegiac description of Optimo City as a way of criticizing developments in city planning after World War II that threatened to destroy local communities in the name of economic progress. He compares Optimo's unexceptional Courthouse Square to the Spanish plazas and the great public squares of the Baroque era in that they all serve as socially unifying communal centers. And he notes with dismay that some of Optimo's business leaders want to tear the courthouse down to build a parking lot and to replace it with the typical "bureaucrat modernism" of so many contemporary civic centers. Jackson would like to preserve positive features of small towns that Peter Calthorpe (p. 350) and other architects associated with the "new urbanism" seek to recapture in "pedestrian pockets" and related neo-traditional designs.

Jackson's Optimo City provides a revealing snapshot of small town America in the 1950s before freeway construction, mass auto-ownership, and the proliferation of highway-based shopping centers. It also represents a way of thinking about urban space that is timeless.

J. B. Jackson's "The Almost Perfect Town" and the following selection by Witold Rybczynski (p. 170) make a nice pair. Rybczynski picks up nearly half a century after Jackson with a description of the effects that the automobile and shopping centers have had on the downtown of another small American city. Review Jane Jacobs's critique of rigid, automobile-based planning (p. 106) for another defense of vernacular urban evolution.

Among the best of J. B. Jackson's booklength studies and collections of essays are *Landscapes* (Amherst: University of Massachusetts Press, 1970), *American Space* (New York: Norton, 1972), *Discovering the Vernacular Landscape* (New Haven, CT: Yale University Press, 1984), *The Necessity for Ruins, and Other Topics* (Amherst: University of Massachusetts Press, 1980), and *A Sense of Place, A Sense of Time* (New Haven, CT: Yale University Press, 1994). Also, see the journal *Landscape* for additional writings by J. B. Jackson himself and for a continuing outpouring of articles exploring the meaning of landscape by academics and the architects, urban planners, and landscape designers Jackson inspired.

Books by Jackson's student and his successor at Harvard John R. Stilgoe are *Common Landscape of America, 1580 to 1845* (New Haven, CT: Yale University Press, 1986) and *Outside Lies Magic: Regaining History and Awareness in Everyday Places* (New York: Walker and Co., 1998).

Other books on vernacular architecture include Robert Venturi, Denise Scott Brown, and Steven Izenour, *Learning from Las Vegas: The Forgotten Symbolism of Architectural Form* (Cambridge, MA: MIT Press, 1972), Dell Upton and John M. Vlach, *Common Places: Readings in American Vernacular Architecture* (Atlanta: University of Georgia Press, 1986), Ronald W. Haase, *Classic Cracker: Florida's Wood-Frame Vernacular Architecture* (Sarasota, FL: Pineapple Press, 1992), John A. Kouwenhoven, *Made in America: The Arts in Modern Civilization* (Garden City, NY: Doubleday, 1948), Thomas Carter (ed.), *Images of an American Land: Vernacular Architecture in the Western United States* (Albuquerque: University of New Mexico Press, 1997), Jim Heimann, *California Crazy: Roadside Vernacular Architecture* (San Francisco: Chronicle Books, 1981), Colleen Josephine Sheehy, *The Flamingo in the Garden: American Yard Art and the Vernacular Landscape* (London: Garland, 1998), Fred E. H. Schroeder, *Front Yard America: The Evolution and Meanings of a Vernacular Domestic Landscape* (Bowling Green, OH: Bowling Green State University, 1993), John Chase, *Glitter Stucco and Dumpster Diving: Reflections on Building Production in the Vernacular City* (London: Verso, 1998), Bernard Rudofsky, *Architecture Without Architects: A Short Introduction to Non-pedigreed Architecture* (Albuquerque: University of New Mexico Press, 1987), and Paul Oliver (ed.), *Encyclopedia of Vernacular Architecture of the World* (Cambridge: Cambridge University Press, 1998).

Optimo City (pop. 10,783, alt. 2,100 ft.), situated on a small rise overlooking the N. branch of the Apache River, is a farm and ranch center served by a spur of the S.P. County seat of Sheridan Co. Optimo City (originally established in 1843 as Ft. Gaffney). It was the scene of a bloody encounter with a party of marauding Indians in 1857. (See marker on courthouse lawn.) It is the location of a state Insane Asylum, of a sorghum processing plant and an overall factory. Annual County Fair and Cowboy Roundup Sept. 4. The highway now passes through a rolling countryside devoted to grain crops and cattle raising.

Thus would the state guide dispose of Optimo City and hasten on to a more spirited topic if Optimo City as such existed. Optimo City, however, is not one town, it is a hundred or more towns, all very much alike, scattered across the United States from the Alleghenies to the Pacific, most numerous west of the Mississippi and south of the Platte. When, for instance, you travel through Texas and Oklahoma and New Mexico and even parts of Kansas and Missouri, Optimo City is the blur of filling stations and motels you occasionally pass; the solitary traffic light, the glimpse up a side street of an elephantine courthouse surrounded by elms and sycamores, the brief congestion of mud-spattered pickup trucks that slows you down before you hit the open road once more. And fifty miles farther on Optimo City's identical twin appears on the horizon, and a half dozen more Optimos beyond that until at last, with some relief, you reach the metropolis with its new housing developments and factories and the cluster of downtown skyscrapers.

Optimo City, then, is actually a very familiar feature of the American landscape. But since you have never stopped there except to buy gas, it might be well to know it a little better. What is there to see? Not a great deal, yet more than you would at first suspect.

Optimo, being after all an imaginary average small town, has to have had an average small-town history, or at least a Western version of that average. The original Fort Gaffney (named after some inconspicuous worthy in the U.S. Army) was really little more than a stockade on a bluff overlooking a ford in the river; a few roads or trails in the old days straggled out into the plain (or desert as they called it then), lost heart and disappeared two or three miles from town. Occasionally even today someone digs up a fragment of the palisade or a bit of rust-eaten hardware in the backyards of the houses near the center of town, and the historical society possesses what it claims is the key to the principal gate. But on the whole, Optimo City is not much interested in its martial past. The fort as a military installation ceased to exist during the Civil War, and the last of the pioneers died a half century ago before anyone had the historical sense to take down his story. And when the county seat was located in the town the name was changed from Fort Gaffney with its frontier connotation to Optimo, which means (so the townspeople will tell you) "I hope for the best" in Latin.

What Optimo is really proud of even now is its identity as county seat. Sheridan County (and you will do well to remember that it was NOT named after the notorious Union general but after Horace Sheridan, an early member of the territorial legislature; Optimo still feels strongly about what it calls the War between the States) was organized in the 1870s and there ensued a brief but lively competition for the possession of the courthouse between Optimo and the next largest settlement, Apache Center, twenty miles away. Optimo City won, and Apache Center, a cowtown with one paved street, is not allowed to forget the fact. The football and basketball games between the Optimo Cougars and the Apache Braves are still characterized by a very special sort of rivalry. No matter how badly beaten Optimo City often is, it consoles itself by remembering that it is still the county seat, and that Apache Center, in spite of the brute cunning of its team, has still only one street paved. We shall presently come back to the meaning of that boast.

To get on with the history of Optimo.

THE INFLEXIBLE GRIDIRON

Aided by the state and Army engineers, the city fathers, back in the 1870s, surveyed and laid out the new metropolis. As a matter of course they located a square or public place in the center of the town and eventually they built their courthouse in the middle of the square; such having been the layout of every county seat these Western Americans had ever seen. Streets led from the center of each side of the square, being named Main Street North and South, and Sheridan Street East and West. Eventually these four streets and the square were surrounded by a gridiron pattern of streets and avenues – all numbered or lettered, and all of them totally oblivious of the topography of the town. Some streets charge up impossibly steep slopes, straight as an arrow; others lead off into the tangle of alders and cottonwoods near the river and get lost.

Strangely enough, this inflexibility in the plan has had some very pleasant results. South Main Street, which leads from the square down to the river, was too steep in the old days for heavily laden wagons to climb in wet weather, so at the foot of it on the flats near the river those merchants who dealt in farm produce and farm equipment built their stores and warehouses. The blacksmith and welder, the hay and grain supply, and finally the auction ring and the farmers' market found South Main the best location in town for their purpose – which purpose being primarily dealing with out-of-town farmers and ranchers. And when, after considerable pressure on the legislature and much resistance from Apache Center (which already had a railroad), the Southern Pacific built a spur to Optimo, the depot was naturally built at the foot of South Main. And of course the grain elevator and the stockyards were built near the railroad. The railroad spur was intended to make Optimo into a manufacturing city, and never did; all that ever came was a small overall factory and a plant for processing sorghum with a combined payroll of about 150. Most of the workers in the two establishments are Mexicans from south of the border – locally referred to in polite circles as "Latinos" or "Hispanos." They have built for themselves flimsy little houses under the cottonwoods and next to the river. "If ever we have an epidemic in Optimo," the men at the courthouse remark, "it will break out first of all in those Latino shacks." But they have done nothing as yet about providing them with better houses, and probably never will.

DOWNTOWN AND UPTOWN

Depot, market, factories, warehouses, slum – these features, combined with the fascination of the river bank and stockyards and the assorted public of railroaders and Latinos and occasional ranch hands – have all given South Main a very definite character: easy-going, loud, colorful, and perhaps during fair week or at shipping time a little disreputable. Boys on the Cougar football squad have specific orders to stay away from South Main, but they don't. Actually the whole of Optimo looks on the section with indulgence and pride; it makes the townspeople feel that they understand metropolitan problems when they can compare South Main with the New York waterfront.

North Main, up on the heights beyond the Courthouse Square and past the two or three blocks of retail stores, is (on the other hand) the very finest part of Optimo. The northwestern section of town, with its tree-shaded streets, its view over the river and the prairie, its summer breezes, has always been identified with wealth and fashion as Optimo understands them. Colonel Ephraim Powell (Confederate Army, Ret., owner of some of the best ranch country in the region) built his bride a handsome limestone house with a slate roof and a tower, and Walter Slymaker, proprietor of Slymaker's Mercantile and of the grain elevator, not to be outdone, built an even larger house farther up Main; so did Hooperson, first president of the bank. There are a dozen such houses in all, stone or Milwaukee brick with piazzas (or galleries, as the old timers still call them) and large, untidy gardens around them. It is worth noting, by the way, that the brightest claim to aristocratic heritage is this: grandfather came out West for his health. New England may have its "Mayflower" and "Arabella," east Texas its Three Hundred Founding Families, New Mexico its Conquistadores; but Optimo is loyal to the image of the delicate young college graduate who arrived by train with his law books, his set of Dickens, his taste for wine, and the custom of dressing for dinner. This legendary figure has about seen his day in the small talk of Optimo society, and the younger generation frankly doubts his having ever existed; but he (or his ghost) had a definite effect on local manners and ways of living. At all events, because of this memory Optimo looks down on those Western mining towns where Sarah Bernhardt and de Reszke and Oscar Wilde seem to have played so many one-night stands in now-vanished opera houses.

A WORLD IN ITSELF

Wickedness – or the suggestion of wickedness – at one end of Main, affluence and respectability at the other. How about Sheridan Street running East and West? That is where you'll find most of the stores; in the first four or five blocks on either side of the Courthouse Square. They form a rampart: narrow brick houses, most of them two stories high with elaborate cornices and long narrow windows; all of them devoid of modern commercial graces of chromium and black glass and artful window display, all of them ugly but all of them pretty uniform; and so you have on Sheridan Street something rarely seen in urban America: a harmonious and restful and dignified business section. Only eight or ten blocks of it in all, to be sure; turn any corner and you are at once in a residential area.

Here there is block after block of one-story frame houses with trees in front and picket fences or hedges; no sidewalk after the first block or so; a hideous church (without a cemetery of course); a small-time auto repair shop in someone's back yard; dirt roadway; and if you follow the road a few blocks more – say to 10th Street (after that there are no more signs) – you are likely to see a tractor turn into someone's drive with wisps of freshly cut alfalfa clinging to the vertical sickle bar. The countryside is that close to the heart of Optimo City, farmers are that much part of the town. And the glimpse of the tractor (like the glimpse of a deer or a fox driven out of the hills by a heavy winter) restores for a moment a feeling for an old kinship that seemed to have been destroyed forever. But this is what makes Optimo, the hundreds of Optimos throughout America, so valuable; the ties between country and town have not yet been broken. Limited though it may well be in many ways, the world of Optimo City is still complete.

The center of this world is Courthouse Square, with the courthouse, ponderous, barbaric, and imposing, in the center of that. The building and its woebegone little park not only interrupts the vistas of Main and Sheridan – it was intended to do this – it also interrupts the flow of traffic in all four directions. A sluggish eddy of vehicles and pedestrians is the result, Optimo's animate existence slowed and intensified. The houses on the four sides of the square are of the same vintage (and the same general architecture) as the monument in their midst: mid-nineteenth-century brick or stone; cornices like the brims of hats, fancy dripstones over the arched windows like eyebrows; painted blood-red or mustard-yellow or white; identical except for the six-story Gaffney Hotel and the classicism of the First National Bank.

Every house has a tin roof porch extending over the sidewalk, a sort of permanent awning which protects passersby and incidentally conceals the motley of store windows and signs. To walk around the square and down Sheridan Street under a succession of these galleries or metal awnings, crossing the strips of bright sunlight between the roofs of different height, is one of the delights of Optimo – one of its amenities in the English use of that word. You begin to understand why the Courthouse Square is such a popular part of town.

SATURDAY NIGHTS – BRIGHT LIGHTS

Saturday, of course, is the best day for seeing the full tide of human existence in Sheridan County. The rows of parked pickups are like cattle in a feed lot; the sidewalks in front of Slymaker's Mercantile, the Ranch Cafe, Sears, the drugstore, resound to the mincing steps of cowboy boots; farmers and ranchers, thumbs in their pants pockets, gather in groups to lament the drought (there is always a drought) and those men in Washington, while their wives go from store to movie house to store. Radios, jukeboxes, the bell in the courthouse tower; the teenagers doing "shave-and-a-haircut; bay rum" on the horns of their parents' cars as they drive round and round the square. The smell of hot coffee, beer, popcorn, exhaust, alfalfa, cow manure. A man is trying to sell a truckload of grapefruit so that he can buy a truckload of cinderblocks to sell somewhere else. Dogs; 10-year-old cowboys firing cap pistols at each other. The air is full of pigeons, floating candy wrappers, the flat strong accent erroneously called Texan.

All these people are here in the center of Optimo for many reasons – for sociability first of all, for news, for the spending and making of money; for relaxation. "Jim Guthrie and wife were in town last week, visiting friends and transacting business," is the way the *Sheridan Sentinel* describes it; and almost all of Jim Guthrie's business takes place in the square. That is one of the peculiarities of Optimo and one of the reasons why the square as an institution is so important. For it is around the square that the oldest and most essential urban (or county) services are established. Here are the firms under local control and ownership, those devoted almost exclusively to the interest of the surrounding countryside. Upstairs are the lawyers, doctors, dentists, insurance firms, the public stenographer, the Farm Bureau. Downstairs are the bank, the prescription drugstore, the newspaper office, and of course Slymaker's Mercantile and the Ranch Cafe.

INFLUENCE OF THE COURTHOUSE

Why have the chain stores not invaded this part of town in greater force? Some have already got a foothold, but most of them are at the far end of Sheridan or even out on the Federal Highway. The presence of the courthouse is partly responsible. The traditional services want to be as near the courthouse as they can, and real-estate values are high. The courthouse itself attracts so many out-of-town visitors that the problem of parking is acute. The only solution that occurs to the enlightened minds of the Chamber of Commerce is to tear the courthouse down, use the place for parking, and build a new one somewhere else. They have already had an architect draw a sketch of a new courthouse to go at the far end of Main Street: a chaste concrete cube with vertical motifs between the windows – a fine specimen of bureaucrat modernism. But the trouble is, where to get the money for a new courthouse when the old one is still quite evidently adequate and in constant use?

If you enter the courthouse you will be amazed by two things: the horrifying architecture of the place, and the variety of functions it fills. Courthouse means of course courtrooms, and there are two of those. Then there is the office of the County Treasurer, the Road Commissioner, the School Board, the Agricultural Agent, the Extension Agent, Sanitary Inspector, and usually a group of Federal agencies as well – PMA, Soil Conservation, FHA and so on. Finally the Red Cross, the Boy Scouts, and the District Nurse. No doubt many of these offices are tiresome examples of government interference in private matters; just the same, they are for better or worse part of almost every farmer's and rancher's business, and the courthouse, in spite of all the talk about county consolidation, is a more important place than ever.

As it is, the ugly old building has conferred upon Optimo a blessing which many larger and richer American towns can envy: a center for civic activity and a symbol for civic pride – something as different from the modern "civic center" as day is from night. Contrast the array of classic edifices, lost in the midst of waste space, the meaningless pomp of flagpoles and war memorials and dribbling fountains of any American city from San Francisco to Washington with the animation and harmony and the almost domestic intimacy of Optimo Courthouse Square, and you have a pretty good measure of what is wrong with much American city planning: civic consciousness has been divorced from everyday life, put in a special zone all by itself. Optimo City has its zones; but they are organically related to one another.

Doubtless the time will never come that the square is studied as a work of art. Why should it be? The craftsmanship in the details, the architecture of the building, the notions of urbanism in the layout of the square itself are all on a very countrified level. Still, such a square as this has dignity and even charm. The charm is perhaps antiquarian – a bit of rural America of seventy-five years ago; the dignity is something else again. It derives from the function of the courthouse and the square, and from its peculiarly national character.

COMMUNAL CENTER

The practice of erecting a public building in the center of an open place is in fact pretty well confined to America – more specifically to nineteenth-century America. The vast open areas favored by eighteenth-century European planners were usually kept free of construction, and public buildings – churches and palaces and law-courts – were located to face these squares; to command them, as it were. But they were not allowed to interfere with the original open effect. Even the plans of eighteenth-century American cities, such as Philadelphia and Reading and Savannah and Washington, always left the square or public place intact. Spanish America, of course, provides the best illustrations of all; the plaza, nine times out of ten, is surrounded by public buildings, but it is left free. Yet almost every American town laid out after (say) 1820 deliberately planted a public building in the center of its square. Sometimes it was a school, sometimes a city hall, more often a courthouse, and it was always approachable from all four sides and always as conspicuous as possible.

Why? Why did these pioneer city fathers go counter to the taste of the past in this matter?

One guess is as good as another. Perhaps they were so proud of their representative institutions that they wanted to give their public buildings the best location available. Perhaps frontier America was following an aesthetic movement, already at that date strong in Europe, that held that an open space was improved when it contained some prominent free-standing object – an obelisk or a statue or a triumphal arch. However that may have been, the pioneer Americans went Europe one better, and put the largest building in town right in the center of the square.

Thus the square ceased to be thought of in nineteenth-century America as a vacant space; it became a container or (if you prefer) a frame. A frame, so it happened, not merely for the courthouse, but for all activity of a communal sort. Few aesthetic experiments have ever produced such brilliantly practical results. A society which had long since ceased to rally around the individual leader and his residence and which was rapidly tiring of rallying around the meetinghouse or church all at once found a new symbol: local representative government, or the courthouse. A good deal of flagwaving resulted – as European travelers have always told us – and a good deal of very poor "representational" architecture; but Optimo acquired something to be proud of, something to moderate that American tendency to think of every town as existing entirely for money-making purposes.

SYMBOL OF INDEPENDENCE

At this juncture the protesting voice of the Chamber of Commerce is heard. "One moment. Before you finish with our courthouse you had better hear the other side of the question. If the courthouse were torn down we would not only have more parking space – sorely needed in Optimo – we would also get funds for widening Main Street into a four-lane highway. If Main Street were widened Optimo could attract many new businesses catering to tourists and other transients – restaurants and motels and garages and all sorts of drive-in establishments. In the last ten years" (continues the Chamber of Commerce) "Optimo has grown by twelve hundred. Twelve hundred! At that rate we'll still be a small town of less than twenty thousand in 1999. But if we had new businesses we'd grow fast and have better schools and a new hospital, and the young people wouldn't move to the cities. Or do you expect Optimo to go on depending on a few hundred tight-fisted farmers and ranchers for its livelihood?" The voice, now shaking with emotion, adds something about "eliminating" South Main by means of an embankment and a clover leaf and picnic grounds for tourists under the cottonwoods where the Latinos still reside.

These suggestions are very sensible ones on the whole. Translate them into more general terms and what they amount to is this: if we want to get ahead, the best thing to do is break with our own past, become as independent as possible of our immediate environment and at the same time become almost completely dependent for our well-being on some remote outside resource. Whatever you may think of such a program, you cannot very well deny that it has been successful for a large number of American towns. Think of the hayseed communities which have suddenly found themselves next to an oil field or a large factory or an Army installation, and which have cashed in on their good fortune by transforming themselves overnight, turning their backs on their former sources of income, and tripling their population in a few years! It is true that these towns put all their eggs in one basket, that they are totally at the mercy of some large enterprise quite beyond their control. But think of the freedom from local environment; think of the excitement and the money! Given the same circumstances – and the Southwest is full of surprises still – why should Optimo not do the same?

A COMMON DESTINY

Because there are many different kinds of towns just as there are many different kinds of men, a development which is good for one kind can be death on another. Apache Center (to use that abject community as an example), with its stockyards and its one paved street and its very limited responsibility to the county, as a community might well become a boom-town and no one would be worse off. Optimo seems to have a different destiny. For almost a hundred years – a long time in this part of the world – it has been identified with the surrounding landscape and been an essential part of it. Whatever wealth it possesses has come from the farms and ranches, not from the overall factory or from tourists. The bankers and merchants will tell you, of course, that without their ceaseless efforts and their vision the countryside could never have existed; the farmers and ranchers consider Optimo's prosperity and importance entirely their own creation. Both parties are right to the extent that the town is part of the landscape – one might even say part of every farm, since much farm business takes place in the town itself.

Now if Optimo suddenly became a year-round tourist resort, or the overall capital of the Southwest; what would happen to that relationship, do you suppose? It would vanish. The farmers and ranchers would soon find themselves crowded out, and would go elsewhere for those services and benefits which they now enjoy in Optimo. And as for Optimo itself, it would soon achieve the flow of traffic, the new store fronts, the housing developments, the payrolls and bank accounts it cannot help dreaming about; and in the same process achieve a total social and physical dislocation, and a loss of a sense of its own identity. County Seat of Sheridan County? Yes; but much more important: Southwestern branch of the "American Cloak and Garment Corporation"; or the LITTLE TOWN WITH THE BIG WELCOME – 300 tourist beds which, when empty for one night out of three, threaten bankruptcy to half the town.

As of the present, Optimo remains pretty much as it has been for the last generation. The Federal Highway still bypasses the center (what a roadblock, symbolical as well as actual, that courthouse is!); so if you want to see Optimo, you had better turn off at the top of the hill near the watertower of the lunatic asylum – now called Fairview State Rest Home, and with the hideous high fence around it torn down. The dirt road eventually becomes North Main. The old Slymaker place is still intact. The Powell mansion, galleries and all, belongs to the American Legion, and a funeral home has taken over the Hooperson house. Then comes downtown Optimo; and then the courthouse, huge and graceless, in detail and proportion more like a monstrous birdhouse than a monument. Stop here. You'll find nothing of interest in the stores, and no architectural gems down a side street. Even if there were, no one would be able to point them out. The historical society, largely in the hands of ladies, thinks of antiquity in terms of antiques, and art as anything that looks pretty on the mantelpiece.

The weather is likely to be scorching hot and dry, with a wild ineffectual breeze in the elms and sycamores. You'll find no restaurant in town with atmosphere – no chandeliers made out of wagon wheels, no wall decorations of famous brands, no bar disguised as the Hitching Rail or the Old Corral. Under a high ceiling with a two-bladed fan in the middle, you'll eat ham hock and beans, hot bread, iced tea without lemon, and like it or go without. But as compensation of sorts at the next table there will be two ranchers eating with their hats on, and discussing the affairs, public and private, of Optimo City. To hear them talk, you'd think they owned the town.

That's about all. There's the market at the foot of South Main, the Latino shacks around the overall factory, a grove of cottonwoods, and the Apache River (North Branch) trickling down a bed ten times too big; and then the open country. You may be glad to have left Optimo behind.

Or you may have liked it, and found it pleasantly old-fashioned. Perhaps it is; but it is in no danger of dying out quite yet. As we said to begin with, there is another Optimo City fifty miles farther on. The country is covered with them. Indeed they are so numerous that it sometimes seems as if Optimo and rural America were one and indivisible.

WITOLD RYBCZYNSKI

"The New Downtown"

from *City Life: Urban Expectations in a New World* (1995)

Editors' introduction When J. B. Jackson wrote "The Almost Perfect Town," Optimo's eight-or-ten block downtown made a "harmonious and restful and dignified business section." Yet by the early 1950s some members of the Optimo chamber of commerce were already thinking of tearing down the courthouse for more parking space and luring in federal funding to make Main Street into a four-lane highway.

Forty-three years later, Witold Rybczynski, a professor of urbanism at the University of Pennsylvania, describes the effects just such a strategy had on Plattsburgh, New York, and hundreds of other Optimos and Plattsburghs across the United States and around the world. Plattsburgh's once proper little downtown had by 1995 seen better days; it was suffused with a "general air of patient but unmistakable retreat." A number of downtown establishments had closed and others were just barely hanging on.

Before analyzing what caused Plattsburgh's downtown to wither, Rybczynski gives a nice snapshot of turn-of-the-century Plattsburgh's downtown. It had been shaped by forces that created many late nineteenth-century downtowns: railroad access, hotels, and retail stores and businesses. A trolley line with six and a half miles of track ran down one of the main streets, linking the downtown to residential neighborhoods, a fairground, the baseball park, army barracks, and a large resort hotel. Attractive window displays, clean sidewalks, and fashionably dressed shoppers thronged the downtown area.

Unlike many nineteenth-century mill towns whose economies collapsed when the mills closed, Plattsburgh did not suffer catastrophic loss of employment. William Julius Wilson (p. 112) and Mike Savage and Alan Warde (p. 264) correctly point to deindustrialization as a key source of urban economic and social problems. But Rybczynski does not blame deindustrialization for the decline of downtown Plattsburgh. True, a sewing machine factory and a luxury touring car plant closed, but a giant airbase and state college campus still provide employment for local residents. Depopulation and white flight may be the principal causes of distressed downtowns in other cities, but downtown Plattsburgh still has plenty of residents. Affluent white residents still live in the gracious old houses near downtown Plattsburgh just as they did a hundred years ago.

For Rybczynski changes in transportation technology – the rise of the automobile and truck – have much more explanatory power. So does the electric refrigerator, which made the modern supermarket possible and also allowed people to drive to a supermarket, shop, and then keep larger amounts of food fresh at home for longer periods of time. The decline of the railroads also hurt Plattsburgh. According to Rybczyinski it was the combination of personal household mobility by car and more flexible movement of goods by truck that doomed Plattsburgh's downtown.

Rybczynski traces shopping centers and supermarkets back to a few visionary projects built before World War I, but it was not until the 1950s and 1960s that they became a mass phenomenon. According to Rybczynski there were only eight large shopping centers in the entire United States in

1946. But during the decade of the 1960s more than 8,000 shopping centers were built. The pace of new shopping center construction has continued to accelerate since then.

Many shopping centers follow the model of the Southdale Shopping Center near Minneapolis. They consist of huge retail store buildings surrounded by parking lots. But shopping centers run the gamut from small indoor malls to the mammoth West Edmonton shopping mall in Alberta. Some shopping centers now have hotels, restaurants, banks, and even athletic clubs, classrooms, synagogues, and police stations in their malls. The West Edmonton Mall has an aviary, dolphin pool, the world's largest indoor water park (complete with surf) and a replica of the *Santa Maria*!

Are shopping centers and malls destined to be sterile, managed environments or do they offer the potential for a new form of urbanity? Rybczynski is less critical of shopping malls than some critics. He is hopeful that malls will offer a new urbanity. But like them or not, shopping centers and malls promise to be a feature of the landscape far into the future.

Witold Rybczynski has written extensively on architecture and urban planning issues. This selection is from *City Life: Urban Expectations in a New World* (New York: Touchstone Books, 1995). Other books by Rybczynski include *Home: A Short History of an Idea* (New York: Viking, 1986), *Looking Around: A Journey Through Architecure* (New York: Viking, 1993), and *A Place for Art* (Ottawa: National Gallery of Canada, 1993).

Bernard J. Frieden and Lynne B. Sagalyn, *Downtown, Inc.: How America Rebuilds Cities* (Cambridge, MA: MIT Press, 1989) describes the economics of locating new shopping centers and malls downtown. Other books about shopping centers and malls include Richard W. Longstreth, *City Center to Regional Mall: Architecture, the Automobile, and Retailing in Los Angeles, 1920–1950* (Cambridge, MA: MIT Press, 1998), Ira G. Zepp Jr., *The New Religious Image of Urban America: The Shopping Mall*, 2nd edn. (Niwot, CO: University of Colorado Press, 1997), Barry Maitland, *The New Architecture of the Retail Mall* (New York: Van Nostrand, Reinhold, 1997), and William Severini Kowinski, *The Malling Of America: An Inside Look at the Great Consumer Paradise*, (New York: W. Morrow, 1985).

Downtown Plattsburgh in upstate New York, where I used to go for lunch until the Metropole bar closed its dining room, has seen better days. It's not so much that things have been allowed to run down, although here and there boarded-up windows deface the staid Victorian storefronts. It is the general air of patient but unmistakable retreat. This is visible in the types of stores that line the main street – the secondhand bookshops, a thrift store, an outlet for used restaurant equipment – the sort of businesses whose survival depends on low overheads. There are no snappy or fashionable establishments; the signs tend to be home-made, the window displays unchanging and dusty. Merkel's, a department store that is descended from a tobacconist established by Isaac Merkel more than a hundred years ago, appears to be barely hanging on. A snack bar across the street keeps changing owners and menus; now it's an ice cream parlor, but next month, who knows? Down the street, the site of a fire remains an empty lot; there is not enough demand for commercial space. The movie theater, a proper one with a marquee, is still operating, although it doesn't show the kinds of movies that people line up for – my wife, Shirley, and I have sometimes been the only patrons. On the street there is evidence of sporadic attempts at civic beautification – benches and planters – and there is an attractively landscaped promenade alongside the Saranac River. But these improvements haven't had their desired effect. The streets are more or less empty; there is simply none of the bustle or activity normally associated with downtown life. The unbroken facades of the three-story brick buildings along the main street – stores below, offices and rooms above – remind me of an Edward Hopper painting, *Early Sunday Morning*, in which the artist portrays a row of small-town storefronts. The blank stares of the vacant windows and the still emptiness of the Plattsburgh street are Hopperesque, too.

Empty it may be today, but Plattsburgh had a proper little downtown once. The old Customs House is gone, as is the imposing Weed Building, a thousand-seat theater for drama and opera opened in 1893, as well as the Witherill, the biggest of five downtown hotels in operation at the turn of the century. Enough of the architectural heritage remains, however, to remind the visitor of what a handsome town this must have been. Still in operation is the Clinton County Courthouse, a grand building in stone and brick, surrounded by a bevy of lawyers' offices; not far away is an elegant obelisk designed by John Russell Pope, the architect of the National Gallery in Washington, DC. The monument commemorates a naval victory in the War of 1812. Further along the shore of Lake Champlain is the railroad depot, built in the Richardsonian Romanesque style in 1886, when Plattsburgh became an important stop on the Delaware & Hudson line between Montreal and New York City. The depot functioned as a transfer point to stagecoaches and to the paddle steamers that linked Plattsburgh to Burlington, Vermont, across the lake.

Plattsburgh's main street is called Margaret Street – the founders had a charming habit of naming streets after their wives and daughters. In a photograph of Margaret Street taken in 1918 one of the stores displays a telephone sign. Plattsburgh had had a telephone system as early as 1880, and an electric power company since 1889. There is a trolley car running down the center of the street. The Plattsburgh Traction Company operated six and a half miles of trolley line linking downtown to the residential neighborhoods, the fairgrounds, the baseball park, and the army barracks; a spur ran out to the Hotel Champlain, a rather magnificent 500-room resort hotel south of the city. The trolley car in the photograph is an open-air model, suggesting it is summer. There are also a horse-drawn buggy and several automobiles on the tidy, brick-paved street. On the sidewalk, a number of well-dressed men and women window-shop under striped canvas awnings. It is a distinctly urban scene.

What happened to this urbanity? To answer the question, one should note, first, what did not happen. Plattsburgh did not suffer the devastating loss of employment of many nineteenth-century mill towns in upstate New York and New England. True, the power supplied by the falls of the Saranac, which is what drew the founder Zephaniah Platt here in 1784, is no longer an industrial asset, and firms like the Williams Manufacturing Company (maker of the Helpmate sewing machine) and the Lozier Motor Company (of the Lozier touring car) no longer exist, but other industries have replaced them. The grand Hotel Champlain has been converted into a community college. The tourists still come, not by train and steamer, but by car and camper to nearby Adirondack State Park. Plattsburgh has a state college campus, and the army barracks have grown into a giant air force base, part of the Strategic Air Command. All in all, Plattsburgh has prospered.

The decline of downtown Plattsburgh has nothing to do with deindustrialization or with crime or with white reactions to black migration. Nor is it due to a drop in population. Since the middle of the nineteenth century, population growth has been steady, if unspectacular: in 1918 the population of the city of Plattsburgh was about 12,000 people; today it is nearly twice that size. The county has also grown and made the city an important regional center for shopping and entertainment. Plattsburghians did not move to the suburbs, and most of them still live in large comfortable houses on quiet, tree-lined streets in the residential neighborhoods that surround the downtown, much as they did one hundred years ago. What caused the downtown to change was neither urban decay nor suburban flight.

The first indication that change was in the air occurred on November 11, 1929, when the Plattsburgh Traction Company folded. This was less than two weeks after the stock market collapsed on Wall Street, but that was not the reason – the company had been losing money consistently for the previous nine years. Rising operating costs were part of the explanation, but the real problem was that fewer and fewer people were riding the trolley; they were driving cars. In 1900 there were only 8,000 private automobiles registered in the United States, but in the following decade this number grew to almost half a million, thanks in no small part to the introduction of the Ford Model T. By 1920, car ownership stood at eight million. Car ownership started in the large cities, but it spread quickly to smaller towns, judging from a report in the Plattsburgh *Daily Press*, which in 1928 reported traffic congestion on Margaret Street.

Downtown Plattsburgh was formed by the same forces as big cities: the railroad, hotels, and a concentration of stores and businesses. The railroad brought travelers, who in turn sustained the hotels, which offered civic amenities like dining rooms, bars, ballrooms, and evening entertainment. The stores and other businesses (including manufacturing) brought more people downtown, both shoppers and employees. But cars (and later planes) changed the way people traveled; when measured in terms of passenger miles, patronage of non-commuter passenger trains in the United States dropped 84 percent between 1945 and 1964. Local train service in Plattsburgh ceased in 1971 (the converted depot contains rental offices and a restaurant). Now only the New York–Montreal train stops twice a day to take on and discharge passengers, and in 1994, Amtrak announced that this train, too, would cease operation.

By the 1960s the Witherill was the only remaining first-class hotel in downtown Plattsburgh, and before the end of the decade it too had closed, unable to compete with the tourist cabins and motor courts that had sprung up along the main roads leading into the city. This also became the location for automotive needs: garages, car washes, showrooms, and used-car lots. This strip development did not affect other Plattsburgh downtown businesses until the shopping center showed up. By the 1960s, Plattsburgh had three shopping centers, all built next to highways, on the west, south, and north sides of the city: North Country Plaza, Plattsburgh Plaza, and the Skyway Shopping Center. Downtown was finished.

The success of shopping centers, in Plattsburgh and elsewhere, was predicated above all on the existence of the supermarket. In combination with the refrigerator and the automobile, the supermarket changed shopping habits. Since they could store food in refrigerators – and later freezers – housewives didn't have to shop every day; weekly shopping meant having to transport many bags of heavy groceries, which is where the car came in. In fact, the first supermarket pre-dated widespread car ownership and originated in an urban area. Piggly Wiggly, started in Memphis in 1916, was the first self-service grocery-store chain and the model for large supermarkets like King Kullen, which opened its first store in Queens in 1930. But large supermarkets were not really suited to downtown. Unlike department stores, supermarkets are spread out on one floor and, especially when parking is taken into account, require large building lots, which are more affordable on the edge of town.

Personal mobility was responsible for the shift from downtown to the strip, and personal mobility molded American cities and towns in a way that was impossible to imagine in Europe. Not only was automobile ownership higher, but American physical mobility was combined with a high degree of social mobility and the space to exploit the advantages of rapid, easy movement. Only in Canada, New Zealand, and Australia were these conditions duplicated, and it is no accident that urbanism in those countries took a similar course. (Eventually, private automobile ownership did increase in many European countries, and postwar European cities started to incorporate some of the features – strip development, shopping centers – previously seen only in North America and Australia.)

A key feature of this new American mobility was not only individual freedom, but also the freedom to move goods and services. Large long-distance trucks replaced the railroad. Since trucks arrived in the city on highways, the edge of town was the ideal location for distribution warehouses, the new railroad depots. Small industry and workshops, the kind that earlier would have been downtown, near the railroad tracks, also relocated; small trucking companies distributed services as well as goods around the city. They, too, settled beside the convenient highway. As the landscape historian J. B. Jackson wrote, "The automobile – especially the commercial automobile, the truck, the pickup, the van and minivan and jeep – has been most effective in introducing a different spatial order.

For what those vehicles contain (and distribute) is not only new attitudes toward work, new uses of time and space, new and more direct contacts with customers and consumers, but new techniques of problem solving." Jackson's valuable observation underlines the fact that the "automobile city" was also the "truck city," and that personal mobility affected not only *where* people worked, but also *how* they worked.

The strip spawned drive-in establishments like diners, dairy bars, and juke joints. By the 1960s, when teenagers could afford to buy cars, the strip also became a hangout and place for cruising. Despite the disappearance of the railroad and hotels – as well as grocery stores – downtown continued as a home for many traditional businesses; it was still where you bought a record album, a bouquet of flowers, or a hat, and it was where you went to a movie, got a prescription filled, or opened a bank account. Soon that changed, too. You could do all these things and more without setting foot in downtown because the shopping centers had arrived. The chief attraction of the early shopping centers, which usually included a supermarket, was the concentration of commercial establishments in one spot. This meant that shoppers could park their cars and walk from store to store, instead of driving up and down the strip.

From 1960 to 1970, more than 8,000 new centers opened in the United States, but the first shopping centers, sometimes referred to as "shopping villages," emerged much earlier, in the first decade of the century. The shopping village had three identifiable features: it consisted of a number of stores built and leased by a single developer; it provided plenty of free off-street parking; and it was usually located near the center of a planned suburb. According to the *Guinness Book of World Records,* the first shopping center opened at the turn of the century in Roland Park, an exclusive suburban enclave planned by the Olmsted brothers and George Kessler about five miles north of downtown Baltimore. With only six establishments, however, Roland Park barely qualifies as a shopping center.

A more impressive early example is Market Square, designed by Howard Van Doren Shaw in 1916 for the Chicago garden suburb of Lake Forest. Now on the National Register of Historic Places, the exquisite Arts and Crafts-style buildings house a combination of small stores and a Marshall Field department store. The buildings sit on three sides of a landscaped plaza across from the railroad station, and include two charming clock towers. These, as well as the intimate scale of the arcaded stores and the integration of apartments on the second floor, make Market Square not merely a shopping center but a true town center. Country Club Plaza, which opened in 1915, was the town center for Jesse Clyde Nichols's Country Club District outside Kansas City. Like Market Square, it is broken up into several buildings containing retail and commercial spaces – including a movie theater – and incorporating professional offices on the second stories. Shoppers can park in small lots discreetly concealed by low brick walls and walk through landscaped squares interspersed between the buildings. There is nothing discreet about the architecture, however, which is a flamboyant Spanish-Moorish concoction that includes a copy of Sevilla's Giralda tower. Nichols, the founder of the Urban Land Institute, a developers' trade association, was a tireless proselytizer for planned suburbs, and thanks to him, Country Club Plaza became well known.

Market Square and Country Club Plaza consciously recalled small-town shopping districts in the intimate, almost domestic scale of their architecture and in their layouts – the stores faced the street and the parking lots were in the rear. This was not accidental. The developers of the shopping village were also the developers of the surrounding residential areas, and retail areas were designed to fit into the overall master plan. One of the most attractive shopping complexes of this period was developed by the architect Addison Mizner in Palm Beach in 1924–25. Two picturesque pedestrian alleys lined with small shops – Via Mizner and Via Parigi – cut through the block; an arcade along Worth Avenue provided additional retail space. The pedestrian had the impression of walking through an old Spanish town, with crooked walls, wrought-iron balconies, worn steps, and clouds of falling bougainvillea, all artfully arranged by the architect. Mizner, whose scenographic approach to architecture is insufficiently appreciated, intended the complex to resemble a converted medieval castle. As he colorfully described it to a reporter, he wanted the shopping complex to appear as if "with the advent of more civilized times the armies were dismissed and commercially minded people converted the cellar-like rooms into small shops."

Not all the early shopping centers were part of planned suburbs. Farmers Markets, a California chain, were the 1930s equivalents of today's discount warehouses. The stores faced an inner, completely private pedestrian walkway, which presaged the inward-looking shopping centers of the future. To keep overheads low, Farmers Markets were built on cheap land on the edges of cities, and the peripheral location was no longer an inconvenience because most people drove to go shopping. This did not escape the attention of developers, who understood that with almost universal car ownership the pool of potential customers for any single shopping center – those who lived within a ten-minute drive, say, rather than within a ten-minute walk – had grown very large.

The spread of shopping centers was slowed by the depression and World War II, and in 1946 there were still only eight large shopping centers. The postwar period saw much new suburban construction, but just as the subdivision replaced the garden suburb, the shopping village was replaced by the regional shopping center. Probably the first such center was Northgate, which opened on the outskirts of Seattle on May 1, 1950. The architect John Graham, Jr., devised a long, open-air pedestrian way that was a sort of carless street lined with a department store and a number of smaller stores. The idea was that the department store – called Bon Marche – would attract people, who would then walk and shop along the way. In addition to stores and a supermarket, Northgate eventually acquired a gas station, a drive-in bank, a movie theater, and a bowling alley. Like all future suburban shopping centers, Northgate was built next to a highway. Unlike the earlier shopping villages, it was developed, literally, as a freestanding project – the inward-oriented building was surrounded on all sides by a 4,000-car parking lot that took up about three-quarters of the sixty-acre site.

During the 1950s the construction of shopping centers, in tandem with the construction of subdivisions, began in earnest; the total went from about 100 centers nationwide in 1950 to about 3,700 only a decade later. Not only were there more centers, but they were growing bigger. One of the largest was Northland in Detroit, which opened in 1954 and included more than a million square feet of rentable space and parking for 7,400 cars. The 250-acre site around the center was planned to accommodate a host of nonretail buildings, including offices, research laboratories, apartments, a hospital, and a hotel.

In 1956 a shopping center that was to become a model for the next three decades opened in Edina, a suburb of Minneapolis. Southdale (the early centers all seemed to be named after compass points) was not particularly large, just over half a million square feet, and it followed the usual pattern of buildings surrounded by parking lots. It incorporated one striking innovation, however: the public walking areas were indoors, air-conditioned in the summer and heated during the winter. The architect, Victor Gruen, a transplanted Viennese and a prolific designer of shopping centers (he had been the architect of Northland), cited the glass-roofed, nineteenth-century *gallerias* of Milan and Naples as his inspiration, even though the bland, modernistic interior of Southdale held no trace of its supposed Italian antecedents. Indoor shopping streets are attractive anywhere the climate is marked by hot, humid summers or harsh winters or a lot of rain; hence enclosed shopping malls appeared in the cold Midwest and Northeast, in the South and Southwest, in hot Southern California, and in the wet Northwest – that is to say, everywhere.

Starting in the sixties, most new regional shopping centers, following Southdale's success, were indoor shopping malls. To optimize the extra investment, malls were built on two or even three levels (an idea also introduced at Southdale). This made for shorter walking distances and more stores. In 1970, suburban Houston's Galleria, which did recall its Italian namesakes by providing a grand skylit promenade, opened with 1.5 million square feet of retail and commercial space; the following year Woodfield, outside Chicago, enclosed 1 million square feet of shopping under one roof. During the 1980s malls got even larger: the Del Amo Fashion Center in Torrance, California – 3 million square feet – and the mother of all malls, the Ghermezian brothers' West Edmonton Mall in

Alberta – 5.2 million square feet. In all, from 1970 to 1990, about 15,000 new shopping centers were built in the United States: during that period every seven hours, on average, a new center opened its parking lot to the public.

The Galleria in Houston, whose centerpiece is a year-round ice-skating rink, added yet another ingredient. The developer, Gerald Hines, incorporated a variety of nonretail uses within the mall itself. A hotel guest or an office worker could go out of the lobby and straight into the mall, a simple change with a great impact. Malls were no longer merely shopping centers; they were urban places. Although retail functions continued to dominate, mall developers started leasing space to a variety of clients, including health and athletic clubs, banks, brokerage houses, and medical centers. Malls now also house civic functions: with public libraries, in Saint John, New Brunswick, and Tucson, Arizona; a United Services Organizations (USO) outlet in Hampton, Virginia; a city hall branch office in Everett, Washington; and federal and state agencies elsewhere. The Sports Museum of New England, in East Cambridge, Massachusetts, is housed in a mall; so is a children's museum in Ogden, Utah. The Board of Education of Ottawa has been leasing space for a storefront classroom in a local shopping mall since November 1987, and a local high school recently opened a counseling center in the West Edmonton Mall, which also contains a small synagogue. Just as noncommercial spaces are showing up in shopping malls, malls are popping up in unexpected places: York University in suburban Toronto recently added a shopping mall to its campus, and Pittsburgh's new airport includes a mall with more than a hundred outlets, three food courts, and a chiropractor.

The introduction of noncommercial tenants into shopping malls, which made them more like traditional downtowns, raised the issue of public access. Were malls private property, as mall owners maintained, or had they become, as the American Civil Liberties Union argued, public places where principles of free speech applied? Mall owners were not keen on the idea of abortion groups arguing their cases in the food court, or of having their customers witness a violent altercation between the Ku Klux Klan and its opponents, which actually happened in a Connecticut mall in the mid-1980s. On the other hand, if large regional malls wanted to be a part of the community and attract a broad cross section of the population, it was in their interest to allow access to as many people as possible, including various community groups.

In 1976, the United States Supreme Court ruled that there were no rights of free speech at shopping malls. Nevertheless, several state supreme courts, including those of California, Oregon, Massachusetts, Colorado, Washington, and New Jersey, have ruled in favor of allowing a certain degree of free-speech activities such as leafleting and canvassing in malls. Mall owners may eventually even support public access, since it has proved neither troublesome nor expensive. In 1991 the giant Hahn Company, which owns and operates thirty malls in California, signed an agreement with the American Civil Liberties Union that permits leafleting and canvassing in designated locations in most of its malls. Robert L. Sorensen, vice-president of Hahn, told *Shopping Centers Today*, an industry monthly: "We haven't seen expenses rise because of it, and I don't think it's costing us shoppers." In other states, although not legally required to do so, mall owners are beginning to make similar provisions. The Rouse Company provided a booth for community activities in its mall in Columbia, Maryland, and encourages its mall managers to make similar accommodations; so did The Edward J. DeBartolo Corporation of Youngstown, Ohio, the nation's largest shopping center developer and manager.

Legal issues aside, it is disingenuous for mall developers to argue that they are merely merchants. They are the new city builders, and as such should be prepared to take the bad – or at least the awkward – with the good. In fact, many malls have been acting more and more like municipal governments, sometimes banning smoking, for example, even in states where they are not legally obliged to do so. This does not mean that shopping malls will become like downtown streets. I think that what attracts people to malls is that they are perceived as public spaces where rules of personal conduct are enforced. In other words, they are more like public streets used to be before police indifference and overzealous protectors of individual rights effectively ensured that any behavior, no matter how antisocial, is tolerated. This is what malls offer: a reasonable (in most eyes) level of public order; the right not to be subjected to outlandish conduct, not to be assaulted and intimidated by boorish adolescents, noisy drunks, and aggressive panhandlers. It does not seem much to ask.

Work and play, shopping and recreation, community service and public protest – more and more of the activities of the traditional downtown have moved to the mall, including that newest of urban industries, tourism. With its skating rink and glass-roofed promenade, Houston's Galleria quickly became a tourist attraction. The builders of the West Edmonton Mall, which includes a resort hotel, also installed a skating rink as well as an aviary, a dolphin pool, an artificial lagoon with a submarine, a floating replica of the *Santa Maria*, an amusement park, and the world's largest indoor water park, complete with artificial beach and rolling surf.

A large portion of the visitors to the Mall of America, the recently opened 4.2-million-square-foot mall in Bloomington, Minnesota, outside Minneapolis, are tourists. The Mall of America is counting on attracting an average of about 100,000 people a day; this was exceeded during the first three months after its opening in August 1992, when nearly a million people a week visited the mall. Indeed, the owners of the Mall of America expect it to outdraw Walt Disney World and the Grand Canyon.

The Mall of America is extremely large – four department stores, about 360 specialty stores to date, more than forty restaurants and food outlets – but the three-level retail area is not particularly remarkable, only bigger than most. What is unusual is the fact that the stores are grouped around a huge (seven-acre) glass-roofed courtyard containing an amusement park complete with twenty-three rides, two theaters, and dozens of smaller attractions. The courtyard design brings to mind another building that combined shopping and recreation – the eighteenth-century incarnation of the Palais Royal in Paris. The instigator of that project was the Duc de Chartres, whose family home in Paris, beside the Louvre, included an extremely large garden. The duke, chronically short of funds, decided to use the garden as the site for a commercial venture. He engaged the architect Victor Louis to design a building to include commercial spaces and rental apartments, and to make the centerpiece of the project a public pleasure park, following the current English fashion. The so-called Palais Royal opened in 1784 to immediate accolades and dominated Parisian social life for fifty years. Like many developers since, the duke did not reap the profits of his brilliant scheme – financially ruined by the heavy investment, he was forced to sell off most of the project.

The Palais Royal consisted of a large landscaped courtyard about one hundred yards by three hundred yards, surrounded on three sides by a five-story building; the fourth side, which was never completed, was temporarily closed by wooden stalls. Facing the garden was a continuous two-story arcade. Within the arcade were glass-fronted shops and a variety of other establishments: cafes, eating places, social clubs, gambling rooms, music rooms, auction houses, a puppet show, a silhouette show, a waxworks, several hotels, a Turkish bath, and a theater (which later became the home of the Comedie Francaise). The upper floors contained apartments and rooms, many of which were rented to the *courtisanes* for whom the Palais became famous. The central pleasure garden contained a roofed amphitheater – the Cirque Royal – used for public performances, concerts, and balls.

In today's language, the eighteenth-century Palais Royal might be described as an upscale mall. Most establishments were luxurious and frequented solely by the rich – or by army officers on a spree – but the arcades and the garden were open to all except the lowest orders (invited in only three days a year), and it was where the aristocrat and the bourgeois mingled. The Palais Royal merged shopping, entertainment, and leisure. "Should an American savage come to the Palais Royal," the Russian novelist Nicolai Karamzin wrote, "in half an hour he would be most beautifully attired and would have a richly furnished house, a carriage, many servants, twenty courses on the table, and, if he wished, a blooming Laïs who each moment would die for love of him."

The Palais Royal still exists, although today it's a sedate place that includes antiquarian stores selling books, prints, and military memorabilia. Where does the average Parisian go for his running shoes or his VCR? He or she drives on the *périphérique* to a giant *hypermarché* out in the suburbs. The Paris of vast shopping marts and high-speed highways is not . . . the Paris of my youthful visits. But it's worth underlining that the *hypermarché* (the model for the warehouse-type shopping mart) is a French invention and not, like Disney World or Macdos (McDonald's hamburgers), an American import. Perhaps our cities are more alike than we imagine.

Kenneth T. Jackson, the author of *Crabgrass Frontier*, a history of the suburbs, argues that shopping malls represent almost the opposite of downtown areas. "They cater exclusively to middle-class tastes," he writes, "and contain no unsavory bars or pornography shops, no threatening-looking characters, no litter, no rain, and no excessive heat or cold." In fact, large malls do appeal to a variety of tastes – they have to. The Mall of America, for example, has The Gap as well as Bloomingdale's, Sam Goody and Brooks Brothers, Radio Shack and Godiva Chocolatier, video arcades and NordicTrack, a maker of expensive exercise machines. The eating establishments also cater to different tastes and budgets: there is an array of fast-food outlets arranged around two food courts; several family restaurants (with occasional live entertainment); an assortment of inexpensive steakhouses; several mainstream Italian eateries; an upscale restaurant serving California cuisine; and a Wolfgang Puck pizza and pasta emporium, the first one outside California. There is also a nightclub area where a sports bar, a comedy club, and a country-and-western supper club are open until one in the morning. It is true that there are no pornography shops in the Mall of America, although the video stores certainly rent soft-core porn, and as for threatening-looking characters, there are plenty of weird-looking (to me, at least) teenagers. Yes, malls (like city streets in Canada and many northern European cities) are clean, but the notion that urbanity is somehow represented by litter is surely a sad comment on the miserable state of American downtowns rather than a serious criticism.

Unquestionably, shopping malls are managed places. They are strictly policed, regularly cleaned, and properly maintained; public washrooms are provided; goods are delivered without disruption; leases are terminated on failing or unprofitable businesses; and as spaces become vacant, new tenants are found to fill them. Mall owners strive to achieve a balanced mix of stores and attract high-profile tenants who will benefit the smaller stores. Special events such as bazaars, concerts, and festivals are organized to attract shoppers; and advertising programs promote the shopping mall as a whole. "In ambiance and retail mix the suburban model of success turned its back on the market-driven chaos of downtown and left little to chance," two MIT professors of urban planning, Bernard Frieden and Lynne Sagalyn, observe. Interestingly, this tactic has proved so popular with the public that it has been emulated by downtown merchants' associations, who realize that the traditional hit-or-miss approach to retailing will no longer do.

And what of the issue of enclosure? Does the fact that shoppers are protected from extremes of heat and cold disqualify malls as urban places? I don't think so. Merchants have been building enclosed shopping spaces for a long time. Glass-roofed arcades, or *passages*, first emerged in the center of Paris in the early 1800s, and were widely imitated across the Continent and in England. Nineteenth-century London and Paris also had shopping bazaars, and Milan, Naples, and other Italian cities had galleries; these were large, often multilevel shopping arcades with independent stalls, covered by impressive roofs constructed of cast iron and glass. One of the few surviving examples of a shopping-bazaar building is the splendid GUM department store in Moscow, completed in 1893. Compared with the high-flown, extravagant interiors of the Victorian shopping bazaars and department stores, which celebrated shopping and consumerism on a scale unrivaled before or since, the architecture of most contemporary shopping malls is downright modest.

Still, Jackson has a point – the lack of extremes of weather does make malls feel artificial. Downtown shopping areas are traditionally made up of open as well as enclosed public spaces, which is one of the appeals of the early shopping centers like Market Square and Country Club Plaza. Part of this atmosphere is undoubtedly produced by the landscaping, the shaded arcades, and the sunny outdoor squares. Contemporary developers, in their rush to build completely enclosed malls, may have missed an opportunity for greater diversity. In Boston, Baltimore, and New York City, The Rouse Company, a major shopping-mall developer, has built so-called festival marketplaces, which are really shopping malls in a waterfront setting. Since the commercial spaces in Faneuil Hall Marketplace, Harborplace, and South Street Seaport are located in rehabilitated dockside buildings, much of the public space is outdoors; in that sense, these malls resemble Lake Forest's Market Square more than Southdale. Indoor/outdoor malls have proved extremely popular as well as commercially successful, and have spawned imitators, like the shopping area in New York's Battery Park City, which has a glass-roofed space that recalls Crystal Palace, and is visually and physically related to the outdoors. Horton Plaza, an urban mall in downtown San Diego, is a large, multilevel shopping complex that dispenses with enclosed traffic areas altogether in favor of arcades and open-air courtyards. Jon Jerde, the architect of Horton Plaza, also designed Citywalk in Universal City, California. Taking a page from Mizner's Palm Beach work, he has created a picturesque pedestrian mall flanked with small-scale buildings. The architecture is not intended to recall Spain, however, but the Los Angeles region, including Sunset Strip, Melrose Avenue, and Venice Beach.

The debate about whether shopping malls could or should replace or augment downtown is academic. In places like Plattsburgh, there is little doubt that the shopping mail *is* the new downtown. The Plattsburgh mall, which was built in the 1970s and greatly enlarged a few years ago, is about two miles from downtown, next to the interstate. It is enclosed, and large enough to qualify as a so-called regional mall – defined as including more than 400,000 square feet of retail space and at least one department store. According to the National Research Bureau, at the end of 1992 there were 38,966 operating shopping centers in the United States, of which 1,835 were regional malls. Although plenty of upscale malls cater to the rich, the mall in Plattsburgh is not an effete oasis of luxury; like most malls it serves Middle America – that is, the broad middle class, which here also includes people from the surrounding farms and small towns.

When I lived in Hemmingford, across the nearby Canadian border, every two or three weeks Shirley and I would drive down the interstate and go to the mall, sometimes to shop, sometimes to go to a movie, sometimes just to stroll. The atmosphere was lively, a marked contrast to the emptiness of downtown's Margaret Street. There were crowds here, excited teenagers swarming to the video arcade, parents trailing children on the way to the movie theater, young couples window-shopping, elderly people walking for exercise or sitting on park benches. On Saturdays, there were usually booths selling the sort of mass-produced crafts that one finds at country fairs: hand-painted ties, varnished wood carvings, junk jewelry. The chamber of commerce occupied a stall and promoted local tourist attractions. There were even Girl Scouts selling cookies.

Families ate lunch in the food court, a sunny space that almost felt like the outdoors thanks to the fairly large trees and the natural light filtering through the stretched fabric roof. The large open area, which was the convivial focus of the mall, was full of tables and seating; on the periphery were counters whose colorful overhead signs proclaimed a variety of take-away foods: TexMex, Chinese, Italian, Middle Eastern. People carried their trays to the tables. Because there was no physical boundary between the eating area and the surrounding mall, the impression was of a giant sidewalk cafe.

I suppose that some people would find this an unsophisticated version of urbanity (although you could get a reasonable espresso here), and some of my academic colleagues would refer darkly to "hyperconsumerism" and artificial reality. But I was more encouraged than depressed by the Plattsburgh mall. I saw people rubbing shoulders and meeting their fellow citizens in a noncombative environment – not behind the wheel of a car, but on foot. As for hyperconsumerism, commercial forces have always formed the center of the American city – the old downtown no less than the new – and it is unclear to me why sitting on a bench in the mall should be considered any more artificial than a bench in the park. Admittedly, I still liked to walk down Margaret Street, but it was a nostalgic urge. When I wanted to be part of a crowd, I went to the mall.

EDWARD SOJA

"Taking Los Angeles Apart: Towards a Postmodern Geography"

from *Postmodern Geographies: The Reassertion of Space in Critical Social Theory* (1989)

Editors' introduction Because of its size, fragmentation, diversity, and dynamism and its role as the epicenter of global image and fantasy, Los Angeles is often held out as the quintessential postmodern city. In this selection University of California geographer and city planning professor Edward Soja "deconstructs" the greater metropolitan Los Angeles region. He takes as his area of enquiry the Sixty-Mile Circle around downtown Los Angeles – a five county area with twelve million people, one hundred thirty-two cities, and a $250 billion economy.

Soja's description of Los Angeles' Sixty-Mile Circle is rich in complexity and paradox. Freeway offramps lead to huge military establishments like Edwards Air Force Base, to an Indian reservation nearly devoid of Indians, and a condor refuge where the last condor has been transplanted to a local zoo to avoid lead poisoning. Within the Sixty-Mile Circle are ski slopes and orange groves; Spanish missions and Latino barrios; leisure towns and prisons; planned communities and unplanned sprawl; all-white stealth communities and ethnic neigborhoods; multiple-center cities and edge cities; science parks, and technopoles and technoburbs employing well-paid, largely white knowledge workers, and industrial zones relying almost exclusively on Hispanic and Asian workers.

Among the many contradictions and paradoxes of the greater Los Angeles metropolitan area Soja notes the area's extraordinary *industrial* dynamism. While William Julius Wilson (p. 112), Mike Savage and Alan Warde (p. 264), and many other writers emphasize the shift of manufacturing jobs away from developed countries to the Third World, Soja characterizes greater Los Angeles as the premier industrial growth pole of the twentieth century. According to Soja Los Angeles may perhaps be accurately described as *post-Fordist*, but hardly *post-industrial*.

Contrary to appearances in this bastion of private capitalism, government investment is important to the economic health of the Los Angeles region. Soja emphasizes the role of the state, particularly the military, in the creation of this "federalized metro-sea of state-rescued capitalism." Defense and aerospace contracts for huge projects like the space shuttle and high-tech weapons systems fuel the growth of the region.

Despite all this growth and change, Soja sees an underlying logic to the region that reflects classic urban geographers' models. There is a still a core central business district in Los Angeles displaying some of the attributes of a concentric zone internal structure as Burgess hypothesized (p. 153). According to Soja there is still a radial logic to the region too. For example Wilshire Boulevard is a twenty-mile "sector" radiating out from downtown Los Angeles in the form real estate economist Homer Hoyt described seventy years ago as underlying city development.

The Los Angeles CBD contains a huge police headquarters building, courts, city hall, the largest women's prison in America, and the central organs of press and pulpit such as the Times/Mirror building complex and St. Vibiana's Cathedral. In the newly emerging "left bank" of Los Angeles, yuppies sip cappuccinos just a short distance from ethnic neighborhoods populated by Vietnamese, Latino, and Chinese immigrant families. Conventioneers in this "dual" city enjoy luxury hotels walled off from panhandlers just outside. Downtown Los Angeles exhibits the cultural contradictions that David Harvey (p. 199) and Sharon Zukin (p.131) describe in postmodern New York City. Los Angeles is a global city, and it expresses the "new geography" that Saskia Sassen describes (p. 208).

Following an approach favored in postmodern critical theory, Soja seeks to "deconstruct" this dazzling array of elements. He seeks to get beneath the symbolism of the area – the semiotic blanket of words and artifacts created by the culture(s) that makes up Los Angeles. In many ways the Los Angeles metropolitan region resembles a theme park. Amidst the whimsy and pastiche, Soja notes that it is tempting to become bemused and detached. But underneath the glitz and glitter, Soja points to what he sees as the hard edge of a capitalist, racist, patriarchial landscape.

This selection is from Edward Soja, *Postmodern Geographies: The Reassertion of Space in Critical Social Theory* (London and New York: Verso, 1989). Other books by Soja include *Thirdspace: Journeys to Los Angeles and Other Real-and-Imagined Places* (Cambridge: Blackwell, 1996) and *The City: Los Angeles and Urban Theory at the End of the Twentieth Century* (Berkeley: University of California Press, 1996), co-edited with Allen J. Scott.

Books on Los Angeles include William Fulton, *The Reluctant Metropolis* (Point Arena, CA: Solano Press, 1998), Norman Smith, *The History of Forgetting: Los Angeles and the Erasure of Memory* (London and New York: Verso, 1997), Dolores Hayden, *The Power of Place: Urban Landscapes and Public History* (Cambridge, MA: MIT Press, 1997), and Kevin Starr's series on California and the American Dream: *Americans and the California Dream, 1850–1915* (New York: Oxford University Press, 1986), *Inventing the Dream: California Through the Progressive Era* (New York: Oxford University Press, 1985), *Material Dreams: California Through the 1920s* (New York: Oxford University Press, 1990), *Endangered Dreams: The Great Depression in California* (New York: Oxford University Press, 1986), and *The Dream Endures: California Enters the 1940s* (New York: Oxford University Press, 1996).

Other books that explore postmodern landscapes in Southern California and elsewhere include Michael Sorkin (ed.), *Variations on a Theme Park: The New American City and the End of Public Space* (New York: Hill and Wang, 1992), Sharon Zukin, *Landscapes of Power: From Detroit to Disney World* (Berkeley: University of California Press, 1993), and Mark Gottdiener, *The Theming of America: Dreams, Visions, and Commercial Spaces* (New York: HarperCollins, 1997).

> 'The Aleph?' I repeated.
> 'Yes, the only place on earth where all places are – seen from every angle, each standing clear, without any confusion or blending' (10–11)
>
> ... Then I saw the Aleph ... And here begins my despair as a writer. All language is a set of symbols whose use among its speakers assumes a shared past. How, then, can I translate into words the limitless Aleph, which my floundering mind can scarcely encompass? (12–13)
>
> (Jorge Luis Borges, 'The Aleph')

Los Angeles, like Borges's Aleph, is exceedingly tough-to-track, peculiarly resistant to conventional description. It is difficult to grasp persuasively in a temporal narrative for it generates too many conflicting images, confounding historicization, always seeming to stretch laterally instead of unfolding sequentially. At the same time, its spatiality challenges orthodox analysis and interpretation, for it too seems limitless and constantly in motion, never still enough to encompass, too filled with 'other spaces' to be informatively described. Looking at Los Angeles from the inside, introspectively, one tends to see only fragments and immediacies, fixed sites of myopic understanding impulsively generalized to represent the whole. To the more far-sighted outsider, the visible aggregate of the whole of Los Angeles churns so confusingly that it induces little more than illusionary stereotypes or self-serving caricatures – if its reality is ever seen at all.

What is this place? Even knowing where to focus, to find a starting point, is not easy, for, perhaps more than any other place, Los Angeles is everywhere. It is global in the fullest sense of the word. Nowhere is this more evident than in its cultural projection and ideological reach, its almost ubiquitous screening of itself as a rectangular dream machine for the world. Los Angeles broadcasts its self-imagery so widely that probably more people have seen this place – or at least fragments of it – than any other on the planet . . .

[. . .]

. . . What follows then is a succession of fragmentary glimpses, a freed association of reflective and interpretive field notes which aim to construct a critical human geography of the Los Angeles urban region. My observations are necessarily and contingently incomplete and ambiguous, but the target I hope will remain clear: to appreciate the specificity and uniqueness of a particularly restless geographical landscape while simultaneously seeking to extract insights at higher levels of abstraction, to explore through Los Angeles glimmers of the fundamental spatiality of social life, the adhesive relations between society and space, history and geography, the splendidly idiographic and the enticingly generalizable features of a postmodern urban geography.

A ROUND AROUND LOS ANGELES

[. . .]

We must have a place to start, to begin reading the context . . . just such a reductionist mapping has popularly presented itself. It is defined by an embracing circle drawn sixty miles (about a hundred kilometres) out from a central point located in the downtown core of the City of Los Angeles.

The Sixty-Mile Circle, so inscribed, covers the thinly sprawling 'built-up' area of five counties, a population of more than 12 million individuals, at least 132 incorporated cities and, it is claimed, the greatest concentration of technocratic expertise and militaristic imagination in the USA. Its workers produce, when last estimated, a gross annual output worth nearly $250 billion, more than the 800 million people of India produce each year. This is certainly Greater Los Angeles, a dizzying world.

CIRCUMSPECTION

Securing the Pacific rim has been the manifest destiny of Los Angeles, a theme which defines its sprawling urbanization perhaps more than any other analytical construct . . . It is not always easy to see the imprint of this imperial history on the material landscape, but an imaginative cruise directly above the contemporary circumference of the Sixty-Mile Circle can be unusually revealing. Figure 1 will help to find the way.

The Circle cuts the south coast at the border between Orange and San Diego Counties, near one of the key checkpoints regularly set up to intercept the northward flow of undocumented migrants, and not far from the San Clemente 'White House' of Richard Nixon and the fitful SONGS of the San Onofre Nuclear Generating Station. The first rampart to watch, however, is Camp Pendleton Marine Corps Base, the largest military base in California in terms of personnel, the freed spouses of whom have helped to build a growing hightechnology complex in northern San Diego County . . .

Another quick hop over Sunnymead, the Box Spring Mountains, and Redlands takes us to Rampart #3, Norton Air Force Base, next to the city of San Bernardino and just south of the almost empty San Manuel Indian Reservation. The guide books tell us that the primary mission of Norton is military airlifts, just in case. To move on we must rise still higher to pass over the skisloped peaks of the San Bernardino Mountains and National Forest, through Cajon

Pass and passing the old Santa Fe Trail, into the picturesque Mojave Desert . . .

The next leg is longer and more serene: over the Antelope Valley and the Los Angeles Aqueduct (tapping the Los Angeles-owned segments of the life-giving but rapidly dying Owens River Valley two hundred miles further away); across Interstate 5 (the main freeway corridor to the north), a long stretch of Los Padres National Forest and the Wild Condor Refuge – the last remaining condors were recently removed to zoos after leadpoisoning threatened their extinction in the 'wild' – to the idyllized town of Ojai (site for the filming of 'Lost Horizon'), and then to the Pacific again at the Mission of San Buenaventura, in Ventura County . . .

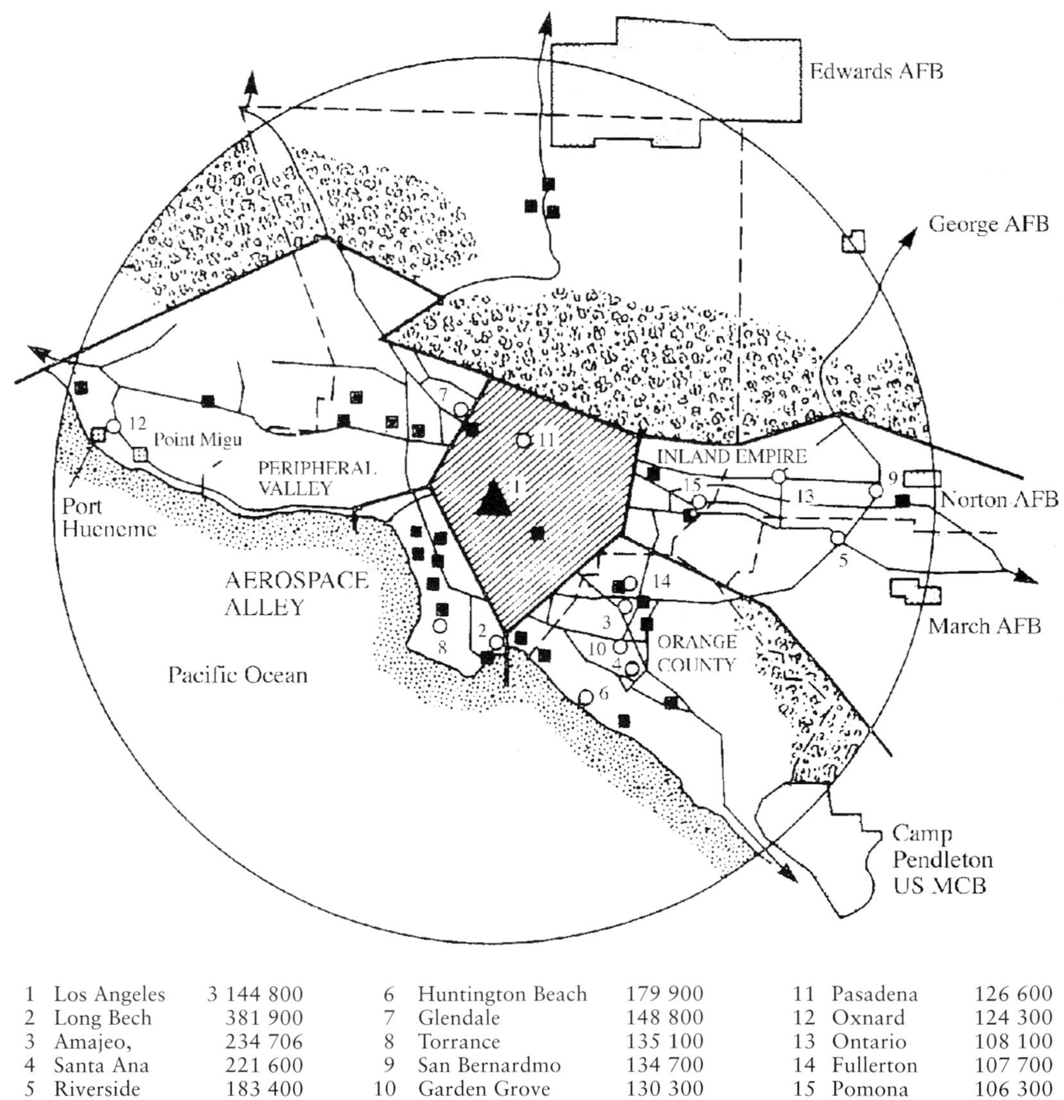

Figure 1 A view of the outer spaces of Los Angeles
The urban core is outlined in the shape of a pentagon, with the Central City denoted by the black triangle. The major military bases on the perimeter of the Sixty-Mile Circle are identified and the black squares are the sites of the largest defence contractors in the region. Also shown are county boundaries, the freeway system outside the central the central pentagon, and the location of all cities with more than 100,000 inhabitants (small open circles)

It is startling how much of the circumference is owned and preserved by the Federal Government in one way or another. Premeditation may be impossible to ascribe, but post-meditation on the circumscriptive federal presence is certainly in order.

ENCLOSURES

[. . .]

If there has emerged a compelling focus to the recent academic literature on Los Angeles, it is the discovery of extraordinary industrial production . . . Yet it is no exaggeration to claim that the Sixty-Mile Circle contains the premier industrial growth pole of the twentieth century, at least within the advanced capitalist countries. Oil, orange groves, films and flying set the scene at the beginning of the century and tend to remain fixed in many contemporary images of industrious, but not industrial, Los Angeles. Since 1930, however, Los Angeles has probably led all other major metropolitan areas in the USA, decade by decade, in the accumulation of new manufacturing employment.

[. . .]

. . . In the past half century, no other area has been so pumped with federal money as Los Angeles, via the Department of Defense to be sure, but also through numerous federal programmes subsidizing suburban consumption (suburbsidizing?) and the development of housing, transportation and water delivery systems. From the last Great Depression to the present, Los Angeles has been the prototypical Keynesian state-city, a federalized metro-sea of state-rescued capitalism enjoying its place in the sunbelt, demonstrating decade by decade its redoubtable ability to go first and multiply the public seed money invested in its promising economic landscape. No wonder it remains so protected. In it are embedded many of the crown jewels of advanced industrial capitalism.

OUTERSPACES

The effulgent Star Wars colony currently blooming around Los Angeles International Airport (LAX) is part of a much larger outer city which has taken shape along the Pacific slope of Los Angeles County. In the context of this landscape, through the story-line of the aerospace industry, can be read the explosive history and geography of the National Security State and what Mike Davis (1984) has called the 'Californianization of Late-Imperial America'.

If there is a single birthplace for this Californianization, it can be found at old Douglas Field in Santa Monica . . . Over half a million people now live in this 'Aerospace Alley', as it has come to be called. During working hours, perhaps 800,000 are present to sustain its global preeminence. Untold millions more lie within its extended orbit.

Attached around the axes of production are the representative locales of the industrialized outer city: the busy international airport; corridors filled with new office buildings, hotels, and global shopping malls; neatly packaged playgrounds and leisure villages; specialized and masterplanned residential communities for the high technocracy; armed and guarded housing estates for top professionals and executives; residual communities of low-pay service workers living in overpriced homes; and the accessible enclaves and ghettoes which provide dependable flows of the cheapest labour power to the bottom bulge of the bimodal local labour market. The LAX – City compage reproduces the segmentation and segregation of the inner city based on race, class, and ethnicity, but manages to break it down still further to fragment residential communities according to specific occupational categories, household composition, and a broad range of individual attributes, affinities, desired lifestyles and moods.

This extraordinary differentiation, fragmentation, and social control over specialized pools of labour is expensive. Housing prices and rental costs in the outer city are easily among the highest in the country and the provision of appropriate housing increasingly absorbs the energy not only of the army of real estate agents but of local corporate and community planners as well, often at the expense of longtime residents fighting to maintain their foothold in 'preferred' locations. From the give and take of this competition have emerged peculiarly intensified urban landscapes. Along the shores of the South Bay, for example, part of what Rayner Banham once called 'Surfurbia', there has developed the largest and most homogenous residential enclave of scientists and engineers in the world. Coincidentally, this beachhead of the high technocracy is also one of the most formidable racial redoubts in the region. Although just a few miles away, across the fortifying boundary of the San Diego freeway, is the edge of the largest and most tightly segregated concentration of Blacks west of Chicago, the sun-belted beach communities stretching south from the airport have remained almost 100 per cent white. The worldview from this highly engineered environment is stereotypically caricatured in Figure 2.

The Sixty-Mile Circle is ringed with a series of these outer cities at varying stages of development, each a laboratory for exploring the contemporaneity of capitalist urbanization . . .

[. . .]

. . . these new territorial complexes seem to be turning the industrial city inside-out, recentering the urban to transform the metropolitan periphery into the core region of advanced industrial production. Decentralization from the inner city has been taking place selectively for at least a century all over the world, but only recently has the peripheral condensation become sufficiently dense to challenge the older urban cores as centres of industrial production, employment nodality, and urbanism. This restructuring process is far from being completed but it is beginning to have some profound repercussions on the way we think about the city, on the words we use to describe urban forms and functions, and on the language of urban theory and analysis.

BACK TO THE CENTRE

[. . .]

To see more of Los Angeles, it is necessary to move away from the riveting periphery and return, literally and figuratively, to the centre of things to the still adhesive core of the urbanized landscape. In Los Angeles as in every city, the nodality of the centre defines and gives substance to the specificity of the urban, its distinctive social and spatial meaning. Urbanization and the spatial divisions of labour associated with it revolve around a socially constructed pattern of nodality and the power of the occupied centres both to cluster and disperse, to centralize and decentralize, to structure spatially all that is social and socially produced. Nodality situates and contextualizes urban society by giving material form to essential social relations. Only with a persistent centrality can there be outer cities and peripheral urbanization. Otherwise, there is no urban at all.

It is easy to overlook the tendential processes of urban structuration that emanate from the centre, especially in the postmodern capitalist landscape. Indeed, in contemporary societies the authoritative and allocative power of the urban centre is purposefully obscured or, alternatively, detached from place, ripped out of context, and given the stretched-out appearance of democratic ubiquity. In addition, as we have seen, the historical development of urbanization over the past century has been marked by a selective dispersal and decentralization, emptying the centre of many of the activities and populations which once aggregated densely around it. For some, this has signalled a negation of nodality, a submergence of the power of central places, perhaps even a Derridean deconstruction of all differences between the 'central' and the 'marginal'.

Yet the centres hold. Even as some things fall apart, dissipate, new nodalities form and old ones are reinforced. The specifying centrifuge is always spinning but the centripetal force of nodality never disappears. And it is the persistent residual of political power which continues to precipitate, specify, and contextualize the urban, making it all stick together.

[. . .]

To maintain adhesiveness, the civic centre has always served as a key surveillant node of the state, supervising locales of production, consumption and exchange. It still continues to do so, even after centuries of urban recomposition and restructuring, after waves of reagglomerative industrialization. It is not production or consumption or exchange in themselves

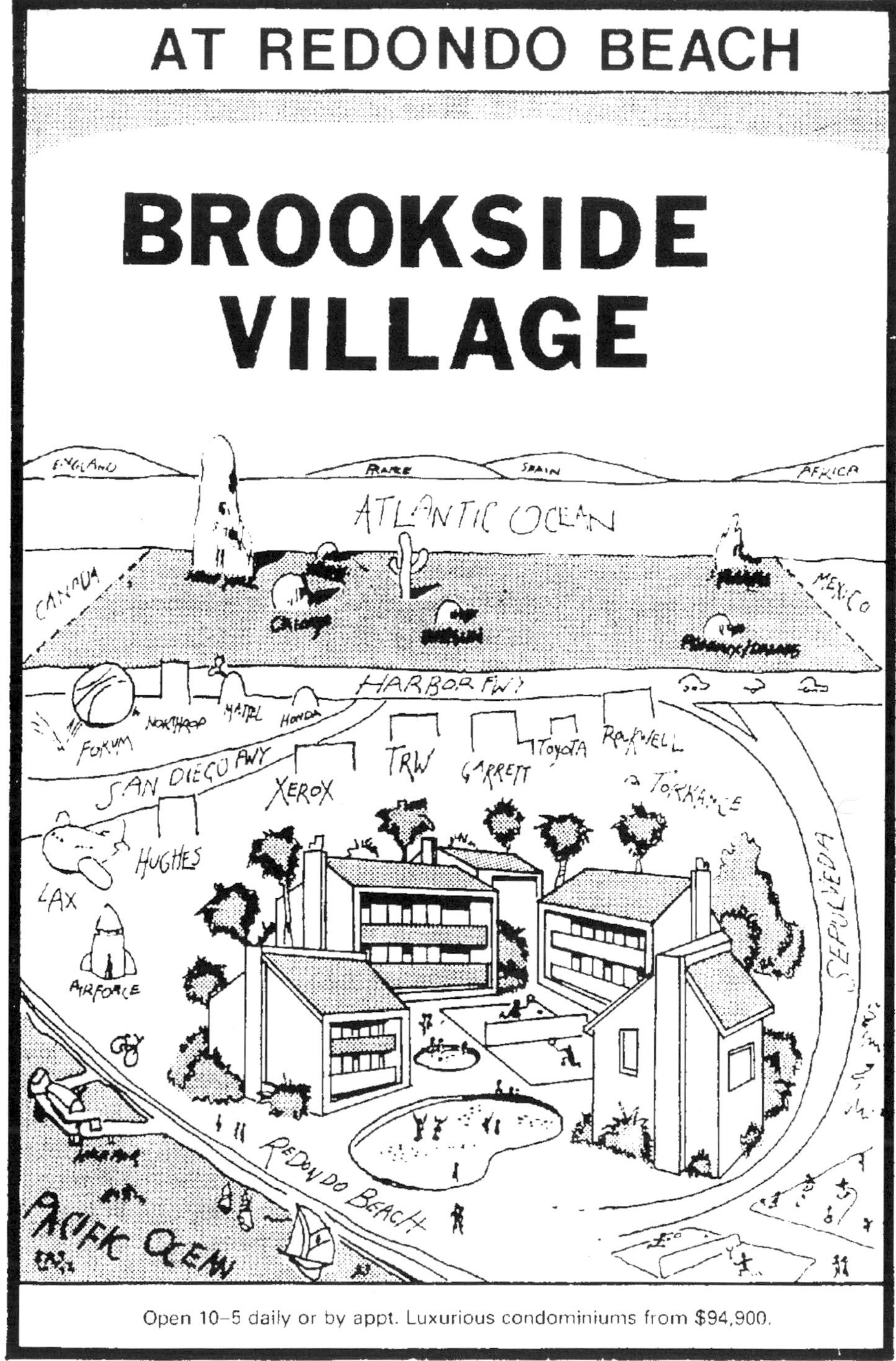

Figure 2 Worldview from Redondo Beach

that specifies the urban, but rather their collective surveillance, supervision and anticipated control within the power-filled context of nodality. In Foucauldian terms, cities are the convergent sites of (social) space, knowledge, and power, the headquarters of societal modes of regulation (from *regula* and *regere,* to rule; the root of our keyword: region).

[. . .]

SIGNIFYING DOWNTOWN

The downtown core of the City of Los Angeles, which the signs call 'Central City' is the agglomerative and symbolic nucleus of the Sixty-Mile Circle, certainly the oldest but also the newest major node in the region. Given what is contained within the Circle, the physical size and appearance of downtown Los Angeles seem almost modest, even today after a period of enormous expansion. As usual, however, appearances can be deceptive.

Perhaps more than ever before, downtown serves in ways no other place can as a strategic vantage point, an urban panopticon counterposed to the encirclement of watchful military ramparts and defensive outer cities. Like the central well in Bentham's eminently utilitarian design for a circular prison, the original panopticon, downtown can be seen (when visibility permits) by each separate individual, from each territorial cell, within its orbit. Only from the advantageous outlook of the centre, however, can the surveillant eye see everyone collectively, disembedded but interconnected. Not surprisingly, from its origin, the central city has been an aggregation of overseers, a primary locale for social control, political administration, cultural codification, ideological surveillance, and the incumbent regionalization of its adherent hinterland.

Looking down and out from City Hall, the site is especially impressive to the observer. Immediately below and around is the largest concentration of government offices and bureaucracy in the country outside the federal capital district. To the east, over a pedestrian skyway, are City Hall East and City Hall South, relatively new civic additions enclosing a shopping mall, some murals, a children's museum, and the Triforium, a splashy sixty-foot fountain of water, light, and music entertaining the lunchtime masses. Just beyond is the imposing police administration building, Parker Center, hallowing the name of a former police chief of note. Looking further outside the central well of downtown but within its eastern salient, one can see an area which houses 25 per cent of California's prison population, at least 12,000 inmates held in four jails designed to hold half that number. Included within this carceral wedge are the largest women's prison in the country (Sybil Brand) and the seventh largest men's prison (Men's Central). More enclosures are being insistently planned by the state to meet the rising demand.

On the south, along First Street, are the State Department of Transportation (CALTRANS) with its electronic wall maps monitoring the arterial freeways of the region, the California State Office Building, and the headquarters of the fourth estate, the monumental Times-Mirror building complex, which many have claimed houses the unofficial governing power of Los Angeles, the source of many stories that mirror the times and spaces of the city. Near the spatial sanctum of the *Los Angeles Times is* also St Vibiana's Cathedral, mother church to one of the largest Catholic archdioceses in the world (nearly four million strong) and controller of another estate of significant proportions. The Pope slept here, across the street from Skid Row missions temporarily closed so that he could not see all his adherents.

Looking westward now, toward the Pacific and the smog-hued sunsets which brilliantly paint the nightfalls of Los Angeles, is first the Criminal Courts Building, then the Hall of Records and Law Library, and next the huge Los Angeles County Courthouse and Hall of Administration, major seats of power for what is by far the country's largest county in total population (now over eight million). Standing across Grand Avenue is the most prominent cultural centre of Los Angeles, described by Unique Media Incorporated in their pictorial booster maps of downtown as 'the cultural crown of Southern California, reigning over orchestral music, vocal performance, opera, theatre and dance'. They add that the Music Center 'tops Bunker Hill like a contemporary Acropolis, one which has dominated civil cultural life since it was inaugurated in 1964. Just beyond this cultural crown is the Department of Water and Power (surrounded by usually waterless fountains) and a multi-level extravaganza of freeway interchanges connecting with every corner of the Sixty-Mile

Circle, a peak point of accessibility within the regional transportation network. On its edge, one of Japan's greatest architects has designed a Gateway Building to punctuate the teeming sea.

Along the northern flank is the Hall of Justice, the US Federal Courthouse, and the Federal Building, completing the ring of local, city, state and federal government authority which comprises the potent civic centre. Sitting more tranquilly just beyond, cut off by a swathe of freeway, is the preserved remains of the old civic centre, now part of El Pueblo de Los Angeles State Historical Park, additional testimony to the lasting power of the central place. Since the origins of Los Angeles the sites described have served as the political citadel, designed with other citadels to command, protect, socialize and dominate the surrounding urban population.

There is still another segment of the citadel–panopticon which cannot be overlooked. Its form and function may be more specific to the contemporary capitalist city but its mercantile roots entwine historically with the citadels of all urbanized societies. Today, it has become the acknowledged symbol of the urbanity of Los Angeles, the visual evidence of the successful 'search for a city' by the surrounding sea of suburbs. This skylined sight contains the bunched castles and cathedrals of corporate power, the gleaming new 'central business district' of the 'central city', pinned next to its ageing predecessor just to the east. Here too the LA-leph's unending eyes are kept open and reflective, reaching out to and mirroring global spheres of influence, localizing the world that is within its reach.

Nearly all the landmarks of the new LA CBD have been built over the past fifteen years and flashily signify the consolidation of Los Angeles as a world city. Now more than half the major properties are in part or wholly foreign owned, although much of this landed presence is shielded from view. The most visible wardens are the banks which light up their logos atop the highest towers: Security Pacific (there again), First Interstate, Bank of America (co-owner of the sleek-black Arco Towers before their recent purchase by the Japanese), Crocker, Union, Wells Fargo, Citicorp (billing itself as 'the newest city in town'). Reading the skyline one sees the usual corporate panorama: large insurance companies (Manulife, Transamerica, Prudential), IBM and major oil companies, the real estate giant Coldwell Banker, the new offices of the Pacific Stock Exchange, all serving as attachment points for silvery webs of financial and commercial transactions extending practically everywhere on earth.

. . . Contrary to popular opinion, Los Angeles is a tightly planned and plotted urban environment, especially with regard to the social and spatial divisions of labour necessary to sustain its pre-eminent industrialization and consumerism. Planning choreographs Los Angeles through the fungible movements of the zoning game and the flexible staging of supportive community participation (when there are communities to be found), a dance filled with honourable intent, dedicated expertise, and selective beneficence. It has excelled, however, as an ambivalent but nonetheless enriching pipeline and place-maker to the domestic and foreign developers of Los Angeles, using its influential reach to prepare the groundwork and facilitate the selling of specialized locations and populations to suit the needs of the most powerful organizers of the urban space economy.

. . . Through a historic act of preservation and renewal, there now exists around downtown a deceptively harmonized showcase of ethni-cities and specialized economic enclaves which play key roles, albeit somewhat noisily at times, in the contemporary redevelopment and internationalization of Los Angeles. Primarily responsible for this packaged and planned production of the inner city is the Community Redevelopment Agency, probably the leading public entrepreneur of the Sixty-Mile Circle.

There is a dazzling array of sites in this compartmentalized corona of the inner city: the Vietnamese shops and Hong Kong housing of a redeveloping Chinatown; the Big Tokyo financed modernization of old Little Tokyo's still resisting remains; the induced pseudo-SoHo of artists' lofts and galleries hovering near the exhibitions of the 'Temporary Contemporary' art warehouse; the protected remains of El Pueblo along Calmexified Olvera Street and in the renewed Old Plaza; the strangely anachronistic wholesale markets for produce and flowers and jewellery growing bigger while other downtowns displace their equivalents; the foetid sweatshops and bustling merchandise marts of the booming garment district; the Latino retail festival along pedestrian-packed Broadway (another preserved zone and inch-for-inch probably the most profitable shopping street in the region); the capital site of urban homelessness in the CRA-gilded skid row district; the enormous muralled barrio stretching eastward to the still unincorporated East Los Angeles; the de-industrializing and virtually resident-less wholesaling City of Vernon to the south filled with chickens and pigs awaiting their slaughter; the Central American and Mexican communities of Pico-Union and Alvarado abutting the high-rises on the west; the obtrusive oil wells and aggressive graffiti in the backyards of predominantly immigrant Temple-Beaudry progressively being eaten away by the spread of Central City West (now being called 'The Left Bank' of downtown); the intentionally yuppifying South Park redevelopment zone hard by the slightly seedy Convention Center; the revenue-milked towers and fortresses of Bunker Hill; the resplendently gentrified pocket of 'Victorian' homes in old Angelino Heights overlooking the citadel; the massive new Koreatown pushing out west and south against the edge of Black Los Angeles; the Filipino pockets to the north-west still uncoalesced into a 'town' of their own; and so much more: a constellation of Foucauldian heterotopias 'capable of juxtaposing in a single real place several spaces, several sites that are in themselves incompatible' but 'function in relation to all the space that remains'.

What stands out from a hard look at the inner city seems almost like an obverse (and perverse) reflection of the outer city, an agglomerative complex of dilapidated and overcrowded housing, low technology workshops, relics and residuals of an older urbanization, a sprinkling of niches for recentred professionals and supervisors, and, above all, the largest concentration of cheap, culturally splintered/occupationally manipulable Third World immigrant labour to be found so tangibly available in any First World urban region. Here in this colonial corona is another of the crown jewels of Los Angeles, carefully watched over, artfully maintained and reproduced to service the continued development of the manufactured region.

The extent and persistence of agglomerated power and ever-watchful eyes in downtown Los Angeles cannot be ignored by either captive participants or outside observers. The industrialization of the urban periphery may be turning the space economy of the region inside-out, but the old centre is more than holding its own as the preeminent political and economic citadel . . .

[. . .]

LATERAL EXTENSIONS

Radiating from the specifying nodality of the central city are the hypothesized pathways of traditional urban theory . . . Formal models of urban morphology have conventionally begun with the assumption of a structuring central place organizing an adherent landscape into discoverable patterns of hinterland development and regionalization . . .

Population densities do mound up around the centres of cities, even in the polycentric archipelago of Los Angeles (where there may be several dozen such mounds, although the most pronounced still falls off from the central city). There is also an accompanying concentric residential rhythm associated with the family life cycle and the relative premiums placed on access to the dense peaks versus the availability of living space in the sparseness of the valleys (at least for those who can afford such freedoms of choice). Land values (when they can be accurately calculated) and some job densities also tend to follow in diminishing peaks outwards from the centre, bringing back to mind those tented webs of the urban geography textbooks.

Adding direction to the decadence of distance reduces the Euclidian elegance of concentric gradations, and many of the most mathematical of urban geometricians have accordingly refused to follow this slightly unsettling path. But direction does indeed induce another fit by pointing out the emanation of fortuitous wedges or sectors from the centre. The sectoral wedges of Los Angeles are especially pronounced once you leave the inner circle around downtown.

The Wilshire Corridor, for example, extends the citadels of the central city almost twenty miles westwards to the Pacific, picking up several other prominent but smaller downtowns en route (the Miracle Mile that initiated this extension, Beverly Hills, Century City,

Westwood, Brentwood, Santa Monica). Watching above it is an even lengthier wedge of the wealthiest residences, running with almost staggering homogeneities to the Pacific Palisades and the privatized beaches of Malibu, sprinkled with announcements of armed responsiveness and signs which say that 'trespassers will be shot'. Here are the hearths of the most vocal homeowners movements, arms raised to slow growth and preserve their putative neighbourhoods in the face of the encroaching, view-blocking, street-clogging, and declassé downtowns.

As if in counterbalance, on the other side of the tracks east of downtown is the salient containing the largest Latino *barrio* in Anglo-America, where many of those who might be shot are carefully barricaded in poverty and there is at least one more prominent wedge, stretching southward from downtown to the twin ports of Los Angeles–Long Beach, still reputed to be one of the largest consistently industrial urban sectors in the world. This is the primary axis of Ruhral Los Angeles.

A third ecological order perturbs the geometrical neatness still further, punching holes into the monocentric gradients and wedges as a result of the territorial segregation of races and ethnicities. Segregation is so noisy that it overloads the conventional statistical methods of urban factorial ecology with scores of tiny but 'significant' eco-components. In Los Angeles, arguably the most segregated city in the country, these components are so numerous that they operate statistically to obscure the spatiality of social class relations deeply embedded in the zones and wedges of the urban landscape, as if they needed to be obscured any further.

These broad social geometries provide an attractive model of the urban geography of Los Angeles, but like most of the inherited overviews of formal urban theory they are seriously diverting and illusory. They mislead not because there is disagreement over their degree of fit – such regular empiricist arguments merely induce a temporary insensibility by forcing debate on to the usually sterile grounds of technical discourse. Instead, they deceive by involuting explanation, by the legerdemain of making the nodality of the urban explain itself through its mere existence, one outcome explaining another. Geographical covariance in the form of empirico-statistical regularity is elevated to causation and frozen into place without a history – and without a human geography which recognizes that the organization of space is a social product filled with politics and ideology, contradiction and struggle, comparable to the making of history. Empirical regularities are there to be found in the surface geometry of any city, including Los Angeles, but they are not explained in the discovery, as is so often assumed.

Different routes and different roots must be explored to achieve a practical understanding and critical reading of urban landscapes. The illusions of empirical opaqueness must be shattered, along with the other disciplining effects of Modern Geography.

DECONSTRUCTION

Back in the centre, shining from its circular turrets of bronzed glass, stands the Bonaventure Hotel, an amazingly storeyed architectural symbol of the splintered labyrinth that stretches sixty miles around it. Like many other Portmanteaus which dot the eyes of urban citadels in New York and San Francisco, Atlanta and Detroit, the Bonaventure has become a concentrated representation of the restructured spatiality of the late capitalist city: fragmented and fragmenting, homogeneous and homogenizing, divertingly packaged yet curiously incomprehensible, seemingly open in presenting itself to view but constantly pressing to enclose, to compartmentalize, to circumscribe, to incarcerate. Everything imaginable appears to be available in this micro-urb, but real places are difficult to find, its spaces confuse an effective cognitive mapping, its pastiche of superficial reflections bewilder co-ordination and encourage submission instead. Entry by land is forbidding to those who carelessly walk but entrance is nevertheless encouraged at many different levels, from the truly pedestrian skyways above to the bunkered inlets below. Once inside, however, it becomes daunting to get out again without bureaucratic assistance. In so many ways, its architecture recapitulates and reflects the sprawling manufactured spaces of Los Angeles.

[. . .]

From the centre to the periphery, in both inner and outer cities, the Sixty-Mile Circle today encloses a shattered metro-sea of fragmented yet homogenized communities, cultures, and economies confusingly arranged to contingently ordered spatial division of labour and power . . .

Over 130 other municipalities and scores of county-administered areas adhere loosely around the irregular City of Los Angeles in a dazzling, sprawling patchwork mosaic. Some have names which are startlingly self-explanatory. Where else can there be a City of Industry and a City of Commerce, so flagrantly commemorating the fractions of capital which guaranteed their incorporation. In other places, names casually try to recapture a romanticized history (as in the many new communities called Rancho something-or-other) or to ensconce the memory of alternative geographies (as in Venice, Naples, Hawaiian Gardens, Ontario, Manhattan Beach, Westminster). In naming, as in so many other contemporary urban processes, time and space, the 'once' and the 'there', are being increasingly played with and packaged to serve the needs of the here and the now, making the lived experience of the urban increasingly vicarious, screened through *simulacra,* those exact copies for which the real originals have been lost.

A recent clipping from the *Los Angeles Times* tells of the 433 signs which bestow identity within the hyperspace of the City of Los Angeles, described as 'A City Divided and Proud of It.' . . . One of the founders of the programme pondered its development: 'At first, in the early 1960s, the Traffic Department took the position that all the communities were part of Los Angeles and we didn't want cities within cities . . . but we finally gave in. Philosophically it made sense. Los Angeles is huge. The city had to recognize that there were communities that needed identification . . .'

For at least fifty years, Los Angeles has been defying conventional categorical description of the urban, of what is city and what is suburb, of what can be identified as community or neighbourhood, of what co-presence means in the elastic urban context. It has in effect been deconstructing the urban into a confusing collage of signs which advertise what are often little more than imaginary communities and outlandish representations of urban locality. I do not mean to say that there are no genuine neighbourhoods to be found in Los Angeles. Indeed, finding them through car-voyages of exploration has become a popular local pastime, especially for those who have become so isolated from propinquitous community in the repetitive sprawl of truly ordinary-looking landscapes that make up most of the region. But again the urban experience becomes increasingly vicarious, adding more layers of opaqueness to *l'espace vécu.*

Underneath this semiotic blanket there remains an economic order, an instrumental nodal structure, an essentially exploitative spatial division of labour, and this spatially organized urban system has for the past half century been more continuously productive than almost any other in the world. But it has also been increasingly obscured from view, imaginatively mystified in an environment more specialized in the production of encompassing mystifications than practically any other you can name. As has so often been the case in the United States, this conservative deconstruction is accompanied by a numbing depoliticization of fundamental class and gender relations and conflicts. When all that is seen is so fragmented and filled with whimsy and pastiche, the hard edges of the capitalist, racist and patriarchal landscape seem to disappear, melt into air.

With exquisite irony, contemporary Los Angeles has come to resemble more than ever before a gigantic agglomeration of theme parks, a lifespace comprised of Disneyworlds. It is a realm divided into showcases of global village cultures and mimetic American landscapes, all-embracing shopping malls and crafty Main Streets, corporation-sponsored magic kingdoms, high-technology-based experimental prototype communities of tomorrow, attractively packaged places for rest and recreation all cleverly hiding the buzzing workstations and labour processes which help to keep it together: like the original 'Happiest Place on Earth', the enclosed spaces are subtly but tightly controlled by invisible overseers despite the open appearance of fantastic freedoms of choice . . .

AFTERWORDS

[. . .]

I have been looking at Los Angeles from many different points of view and each way of seeing assists in sorting out the interjacent medley of the subject landscape. The perspectives explored are purposeful, eclectic, fragmentary, incomplete, and frequently contradictory, but so too is Los Angeles and, indeed, the experienced historical geography of every urban landscape . . . The task of comprehensive, holistic regional description may therefore be impossible, as may be the construction of a compleat historicogeographical materialism.

There is hope nonetheless. The critical and theoretical interpretation of geographical landscapes has recently expanded into realms that functionally had been spatially illiterate for most of the twentieth century. New and avid readers abound as never before, many are directly attuned to the specificity of the urban, and several have significantly turned their eyes to Los Angeles. Moreover, many practised readers of surface geographies have begun to see through the alternatively myopic and hypermetropic distortions of past perspectives to bring new insight to spatial analysis and social theory. Here too Los Angeles has attracted observant readers after a history of neglect and misapprehension, for it insistently presents itself as one of the most informative palimpsests and paradigms of twentieth-century urban-industrial development and popular consciousness.

As I have seen and said in various ways, everything seems to come together in Los Angeles, the totalizing Aleph. Its representations of spatiality and historicity are archetypes of vividness, simultaneity, and interconnection. They beckon inquiry at once into their telling uniqueness and, at the same time, into their assertive but cautionary generalizability. Not all can be understood, appearances as well as essences persistently deceive, and what is real cannot always be captured even in extraordinary language. But this makes the challenge more compelling, especially if once in a while one has the opportunity to take it all apart and reconstruct the context. The reassertion of space in critical social theory – and in critical political praxis – will depend upon a continued deconstruction of a still occlusive historicism and many additional voyages of exploration into the heterotopias of contemporary postmodern geographies.

MIKE DAVIS

"Fortress L.A."

from *City of Quartz: Excavating the Future in Los Angeles* (1990)

Editors' introduction Mike Davis is to contemporary Los Angeles what Friedrich Engels was to mid-nineteenth-century Manchester, England. Engels was a kind of explorer, reporting to an educated, middle-class audience about the horrors of the industrial city and the miserable lives of the new *industrial* proletarian class of modern capitalism. Similarly, Davis, 150 years later, explores the dark side of the postmodern metropolis and reports on the hopelessness and despair of the *postindustrial* "underclass" (largely defined by race, gender, and ethnicity) to an audience largely composed of young, disaffected intellectuals and academics. The parallels are striking. Davis may be the heir to Engels simply because the contemporary metropolis – characterized by wealth and homelessness and divided against itself along the fault lines that separate suburban enclaves from innercity slums – is the heir to the geographical and social-class divisions of the cities of the Industrial Revolution.

Mike Davis, a native Southern Californian who teaches urban theory at the Southern California Institute of Architecture, is an insightful and acerbic social critic who makes his social-class sympathies and anti-establishment bias perfectly clear on every page of *City of Quartz*, which one admirer praised as a "visionary rant."

In *The Condition of the Working Class in England in 1844* (p. 46), Engels noted the boulevards that intersected the city of Manchester and how the facades of those broad thoroughfares served to mask and disguise the hovels of the poor that lay beyond the view of the middle-class commuter. Davis, employing an eclectic "culture studies" methodology, makes a similar but even subtler point about contemporary Los Angeles. The freeways allow middle-class suburbanites to navigate the city as a whole without encountering the lives of the residents of the inner-city neighborhoods. But beyond this, the city itself has become, in the semiotic culture of postmodernism, a vast and continuous system of signs that we read and obey on (mostly) a subconscious level. "Today's upscale, pseudo-public spaces," he writes, ". . . are full of invisible signs warning off the underclass 'Other.' Although architectural critics are usually oblivious to how the built environment contributes to segregation, pariah groups – whether poor Latino families, young Black men, or elderly homeless white females – read the meaning immediately."

Another similarity between Engels and Davis is that both men, as they proceeded to "read the streets" of their respective paradigm cities, seem to reach the limits of language's ability to describe the physical and psychological conditions being reported. Engels adopted the strategy of social science, compiling a mountain of personal observations, journalistic reports, and official survey data to create a virtual catalog of social horror. Davis, on the other hand, relies on a more emotional, even histrionic, strategy. His overheated rhetorical excesses often seem to overwhelm rational discourse. Still, Davis's critique of the contemporary urban social order, however enraged, has a compelling validity quite independent of the received canons of intellectual discourse. Indeed, Davis turned out to be a prophet: the ghettos and barrios of South Central Los Angeles erupted into open rebellion in the Rodney King riots two years after the publication of *City of Quartz*.

Whereas Engels saw massive social dislocations and systematically set about to fashion a theory of revolutionary socialism in response to the observed reality, Davis offers no similarly optimistic solution. In a sense, Davis *personalizes* the horrors of the contemporary city, and his voice is one of desperation and despair offering no obvious way out of the current impasse. He leaves it to others, like David Harvey (p. 199), to propose a theory of social justice that might help resolve some of the conflicts he describes, and to those like Paul Davidoff (p. 423) to propose how city planners might advocate on behalf of the poor and oppressed to make cities better.

Compare Davis's apocalyptic vision with the more balanced description of Los Angeles past and present by Sam Bass Warner (p. 69). Davis's insights regarding the underclass "other" suggest comparisons with such classic analysts of ghettoization as Du Bois (p. 56), and William Julius Wilson (p. 112), but it is difficult to compare Davis with social policy and planning professionals – whether a conservative like Charles Murray (p. 122) or a progressive like Paul Davidoff – whose focus is on influencing the existing system by working within it.

This selection is from Mike Davis, *City of Quartz: Excavating the Future in Los Angeles* (London and New York: Verso, 1990). Mike Davis's most recent book is *The Ecology of Fear: Los Angeles and the Imagination of Disaster* (New York: Metropolitan Books, 1998).

The carefully manicured lawns of Los Angeles' Westside sprout forests of ominous little signs warning: "Armed Response!" Even richer neighborhoods in the canyons and hillsides isolate themselves behind walls guarded by gun-toting private police and state-of-the-art electronic surveillance. Downtown, a publicly subsidized "urban renaissance" has raised the nation's largest corporate citadel, segregated from the poor neighborhoods around it by a monumental architectural glacis. In Hollywood, celebrity architect Frank Gehry, renowned for his "humanism," apotheosizes the siege look in a library designed to resemble a foreign-legion fort. In the Westlake district and the San Fernando Valley the Los Angeles Police barricade streets and seal off poor neighborhoods as part of their "war on drugs." In Watts, developer Alexander Haagen demonstrates his strategy for recolonizing inner-city retail markets: a panopticon shopping mall surrounded by staked metal fences and a substation of the LAPD in a central surveillance tower. Finally, on the horizon of the next millennium, an ex-chief of police crusades for an anti-crime "giant eye" – a geo-synchronous law enforcement satellite – while other cops discreetly tend versions of "Garden Plot," a hoary but still viable 1960s plan for a law-and-order armageddon.

Welcome to post-liberal Los Angeles, where the defense of luxury lifestyles is translated into a proliferation of new repressions in space and movement, undergirded by the ubiquitous "armed response." This obsession with physical security systems, and, collaterally, with the architectural policing of social boundaries, has become a zeitgeist of urban restructuring, a master narrative in the emerging built environment of the 1990s. Yet contemporary urban theory, whether debating the role of electronic technologies in precipitating "postmodern space," or discussing the dispersion of urban functions across poly-centered metropolitan "galaxies," has been strangely silent about the militarization of city life so grimly visible at the street level. Hollywood's pop apocalypses and pulp science fiction have been more realistic, and politically perceptive, in representing the programmed hardening of the urban surface in the wake of the social polarizations of the Reagan era. Images of carceral inner cities (*Escape from New York*, *Running Man*), high-tech police death squads (*Blade Runner*), sentient buildings (*Die Hard*), urban bantustans (*They Live!*), Vietnam-like street wars (*Colors*), and so on, only extrapolate from actually existing trends.

Such dystopian visions grasp the extent to which today's pharaonic scales of residential and commercial security supplant residual hopes for urban reform and social integration. The dire predictions of Richard Nixon's 1969 National Commission on the Causes and Prevention of Violence have been tragically fulfilled: we live in "fortress cities" brutally divided between "fortified cells" of affluent society and "places of terror" where the police battle the criminalized poor. The "Second Civil War" that began in the long hot summers of the 1960s has been institutionalized into the very structure of urban space. The old liberal paradigm of social control, attempting to balance repression with reform, has long been superseded by a rhetoric of social warfare that calculates the interests of the urban poor and the middle classes as a zero-sum game. In cities like Los Angeles, on the bad edge of postmodernity, one observes an unprecedented tendency to merge urban design, architecture and the police apparatus into a single, comprehensive security effort.

This epochal coalescence has far-reaching consequences for the social relations of the built environment. In the first place, the market provision of "security" generates its own paranoid demand. "Security" becomes a positional good defined by income access to private "protective services" and membership in some hardened residential enclave or restricted suburb. As a prestige symbol – and sometimes as the decisive borderline between the merely well-off and the "truly rich" – "security" has less to do with personal safety than with the degree of personal insulation, in residential, work, consumption and travel environments, from "unsavory" groups and individuals, even crowds in general.

Secondly, as William Whyte has observed of social intercourse in New York, "fear proves itself." The social perception of threat becomes a function of the security mobilization itself, not crime rates. Where there is an actual rising arc of street violence, as in Southcentral Los Angeles or Downtown Washington D.C., most of the carnage is self-contained within ethnic or class boundaries. Yet white middle-class imagination, absent from any firsthand knowledge of inner-city conditions, magnifies the perceived threat through a demonological lens. Surveys show that Milwaukee suburbanites are just as worried about violent crime as inner-city Washingtonians, despite a twentyfold difference in relative levels of mayhem. The media, whose function in this arena is to bury and obscure the daily economic violence of the city, ceaselessly throw up spectres of criminal underclasses and psychotic stalkers. Sensationalized accounts of killer youth gangs high on crack and shrilly racist evocations of marauding Willie Hortons foment the moral panics that reinforce and justify urban apartheid.

Moreover, the neo-military syntax of contemporary architecture insinuates violence and conjures imaginary dangers. In many instances the semiotics of so-called "defensible space" are just about as subtle as a swaggering white cop. Today's upscale, pseudo-public spaces – sumptuary malls, office centers, culture acropolises, and so on – are full of invisible signs warning off the underclass "Other." Although architectural critics are usually oblivious to how the built environment contributes to segregation, pariah groups – whether poor Latino families, young Black men, or elderly homeless white females – read the meaning immediately.

THE DESTRUCTION OF PUBLIC SPACE

The universal and ineluctable consequence of this crusade to secure the city is the destruction of accessible public space. The contemporary opprobrium attached to the term "street person" is in itself a harrowing index of the devaluation of public spaces. To reduce contact with untouchables, urban redevelopment has converted once vital pedestrian streets into traffic sewers and transformed public parks into temporary receptacles for the homeless and wretched. The American city, as many critics have recognized, is being systematically turned inside out – or, rather, outside in. The valorized spaces of the new megastructures and super-malls are concentrated in the center, street frontage is denuded, public activity is sorted into strictly functional compartments, and circulation is internalized in corridors under the gaze of private police.

The privatization of the architectural public realm, moreover, is shadowed by parallel restructurings of electronic space, as heavily policed, pay-access "information orders," elite databases and subscription cable services appropriate parts of the invisible agora. Both processes, of course, mirror the deregulation of the economy and the recession of non-market entitlements. The decline of urban liberalism has been accompanied by the death of what might be called the "Olmstedian vision" of public space. Frederick Law Olmsted, it will be recalled, was North America's Haussmann, as well as the Father of Central Park. In the wake of Manhattan's "Commune" of 1863, the great Draft Riot, he conceived public landscapes and parks as social safety-valves, mixing classes and ethnicities in common (bourgeois) recreations and enjoyments. As Manfredo Tafuri has shown in his well-known study of Rockefeller Center, the same principle animated the construction of the canonical urban spaces of the La Guardia–Roosevelt era.

This reformist vision of public space – as the emollient of class struggle, if not the bedrock of the American *polis* – is now as obsolete as Keynesian nostrums of full employment. In regard to the "mixing" of classes, contemporary urban America is more like Victorian England than Walt Whitman's or La Guardia's New York. In Los Angeles, once-upon-a-time a demi-paradise of free beaches, luxurious parks, and "cruising strips," genuinely democratic space is all but extinct. The Oz-like archipelago of Westside pleasure domes – a continuum of tony malls, arts centers and gourmet strips – is reciprocally dependent upon the social imprisonment of the third-world service proletariat who live in increasingly repressive ghettoes and barrios. In a city of several million yearning immigrants, public amenities are radically shrinking, parks are becoming derelict and beaches more segregated, libraries and playgrounds are closing, youth congregations of ordinary kinds are banned, and the streets are becoming more desolate and dangerous.

Unsurprisingly, as in other American cities, municipal policy has taken its lead from the security offensive and the middle-class demand for increased spatial and social insulation. De facto disinvestment in traditional public space and recreation has supported the shift of fiscal resources to corporate-defined redevelopment priorities. A pliant city government – in this case ironically professing to represent a bi-racial coalition of liberal whites and Blacks – has collaborated in the massive privatization of public space and the subsidization of new, racist enclaves (benignly described as "urban villages"). Yet most current, giddy discussions of the "postmodern" scene in Los Angeles neglect entirely these overbearing aspects of counter-urbanization and counter-insurgency. A triumphal gloss – "urban renaissance," "city of the future," and so on – is laid over the brutalization of inner-city neighborhoods and the increasing South Africanization of its spatial relations. Even as the walls have come down in Eastern Europe, they are being erected all over Los Angeles.

The observations that follow take as their thesis the existence of this new class war (sometimes a continuation of the race war of the 1960s) at the level of the built environment. Although this is not a comprehensive account, which would require a thorough analysis of economic and political dynamics, these images and instances are meant to convince the reader that urban form is indeed following a repressive function in the political furrows of the Reagan–Bush era. Los Angeles, in its usual prefigurative mode, offers an especially disquieting catalogue of the emergent liaisons between architecture and the American police state.

THE FORBIDDEN CITY

The first militarist of space in Los Angeles was General Otis of the *Times*. Declaring himself at war with labor, he infused his surroundings with an unrelentingly bellicose air:

> He called his home in Los Angeles the Bivouac. Another house was known as the Outpost. The *Times* was known as the Fortress. The staff of the paper was the Phalanx. The *Times* building itself was more fortress than newspaper plant, there were turrets, battlements, sentry boxes. Inside he stored fifty rifles.

A great, menacing bronze eagle was the *Times*'s crown; a small, functional cannon was installed on the hood of Otis's touring car to intimidate onlookers. Not surprisingly, this overwrought display of aggression produced a response in kind. On 1 October 1910 the heavily fortified *Times* headquarters – citadel of the open shop on the West Coast – was destroyed in a catastrophic explosion blamed on union saboteurs.

Eighty years later, the spirit of General Otis has returned to subtly pervade Los Angeles' new "postmodern" Downtown: the emerging Pacific

Rim financial complex which cascades, in rows of skyscrapers, from Bunker Hill southward along the Figueroa corridor. Redeveloped with public tax increments under the aegis of the powerful and largely unaccountable Community Redevelopment Agency (CRA), the Downtown project is one of the largest postwar urban designs in North America. Site assemblage and clearing on a vast scale, with little mobilized opposition, have resurrected land values, upon which big developers and off-shore capital (increasingly Japanese) have planted a series of billion-dollar, block-square megastructures: Crocker Center, the Bonaventure Hotel and Shopping Mall, the World Trade Center, the Broadway Plaza, Arco Center, CitiCorp Plaza, California Plaza, and so on. With historical landscapes erased, with megastructures and superblocks as primary components, and with an increasingly dense and self-contained circulation system, the new financial district is best conceived as a single, demonically self-referential hyperstructure, a Miesian skyscape raised to dementia.

Like similar megalomaniac complexes, tethered to fragmented and desolated Downtowns (for instance, the Renaissance Center in Detroit, the Peachtree and Omni Centers in Atlanta, and so on), Bunker Hill and the Figueroa corridor have provoked a storm of liberal objections against their abuse of scale and composition, their denigration of street landscape, and their confiscation of so much of the vital life activity of the center, now sequestered within subterranean concourses or privatized malls. Sam Hall Kaplan, the crusty urban critic of the *Times*, has been indefatigable in denouncing the anti-pedestrian bias of the new corporate citadel, with its fascist obliteration of street frontage. In his view the superimposition of "hermetically sealed fortresses" and air-dropped "pieces of suburbia" has "dammed the rivers of life" Downtown.

Yet Kaplan's vigorous defense of pedestrian democracy remains grounded in hackneyed liberal complaints about "bland design" and "elitist planning practices." Like most architectural critics, he rails against the oversights of urban design without recognizing the dimension of foresight, of explicit repressive intention, which has its roots in Los Angeles' ancient history of class and race warfare. Indeed, when Downtown's new "Gold Coast" is viewed en bloc from the standpoint of its interactions with other social areas and landscapes in the central city, the "fortress effect" emerges, not as an inadvertent failure of design, but as deliberate socio-spatial strategy.

The goals of this strategy may be summarized as a double repression: to raze all association with Downtown's past and to prevent any articulation with the non-Anglo urbanity of its future. Everywhere on the perimeter of redevel opment this strategy takes the form of a brutal architectural edge or glacis that defines the new Downtown as a citadel vis-'a-vis the rest of the central city. Los Angeles is unusual amongst major urban renewal centers in preserving, however negligently, most of its circa 1900–30 Beaux Arts commercial core. At immense public cost, the corporate headquarters and financial district was shifted from the old Broadway–Spring corridor six blocks west to the greenfield site created by destroying the Bunker Hill residential neighborhood. To emphasize the "security" of the new Downtown, virtually all the traditional pedestrian links to the old center, including the famous Angels' Flight funicular railroad, were removed.

The logic of this entire operation is revealing. In other cities developers might have attempted to articulate the new skyscape and the old, exploiting the latter's extraordinary inventory of theaters and historic buildings to create a gentrified history – a gaslight district, Faneuil Market or Ghirardelli Square – as a support to middle-class residential colonization. But Los Angeles' redevelopers viewed property values in the old Broadway core as irreversibly eroded by the area's very centrality to public transport, and especially by its heavy use by Black and Mexican poor. In the wake of the Watts rebellion, and the perceived Black threat to crucial nodes of white power (spelled out in lurid detail in the McCone Commission Report), resegregated spatial security became the paramount concern. The Los Angeles Police Department abetted the flight of business from Broadway to the fortified redoubts of Bunker Hill by spreading scare literature typifying Black teenagers as dangerous gang members.

As a result, redevelopment massively reproduced spatial apartheid. The moat of the Harbor Freeway and the regraded palisades of Bunker Hill cut off the new financial core from the poor immigrant neighborhoods that surround it on every side. Along the base of California Plaza, Hill Street became a local Berlin Wall separating the publicly subsidized luxury of Bunker Hill from the lifeworld of Broadway, now reclaimed by Latino immigrants as their primary shopping and entertainment street. Because politically connected speculators are now redeveloping the northern end of the Broadway corridor (sometimes known as "Bunker Hill East"), the CRA is promising to restore pedestrian linkages to the Hill in the 1990s, including the Angels' Flight incline railroad. This, of course, only dramatizes the current bias against accessibility – that is to say, against any spatial interaction between old and new, poor and rich, except in the framework of gentrification or recolonization. Although a few white-collars venture into the Grand Central Market – a popular emporium of tropical produce and fresh foods – Latino shoppers or Saturday strollers never circulate in the Gucci precincts above Hill Street. The occasional appearance of a destitute street nomad in Broadway Plaza or in front of the Museum of Contemporary Art sets off a quiet panic; video cameras turn on their mounts and security guards adjust their belts.

Photographs of the old Downtown in its prime show mixed crowds of Anglo, Black and Latino pedestrians of different ages and classes. The contemporary Downtown "renaissance" is designed to make such heterogeneity virtually impossible. It is intended not just to "kill the street" as Kaplan fears, but to "kill the crowd," to eliminate that democratic admixture on the pavements and in the parks that Olmsted believed was America's antidote to European class polarizations. The Downtown hyperstructure – like some Buckminster Fuller post-Holocaust fantasy – is programmed to ensure a seamless continuum of middle-class work, consumption and recreation, without unwonted exposure to Downtown's working-class street environments. Indeed the totalitarian semiotics of ramparts and battlements, reflective glass and elevated pedways, rebukes any affinity or sympathy between different architectural or human orders. As in Otis's fortress *Times* building, this is the archisemiotics of class war.

DAVID HARVEY

"Social Justice, Postmodernism, and the City"

International Journal of Urban and Regional Research (1992)

Editors' introduction This selection by Oxford geographer David Harvey picks up on Mike Davis's dark vision of "Fortress L.A." with a scene of urban conflict in a New York City park. But Harvey goes beyond Davis's description and critique to wrestle with philosophical issues about how to resolve social and spatial conflicts in the modern city. In this article Harvey nicely bridges two major strands in urban geography – Marxist and postmodern culture studies approaches. The title is a collage of the titles of two of Harvey's books written nearly twenty years apart: *Social Justice and the City* (1973) and *The Condition of Postmodernity* (1990). The article illustrates the intellectual journey Harvey and other geographers have traveled during the last twenty years as the discipline of geography embraced Marxist theory during the 1970s, but by the 1990s increasingly shifted to postmodernist culture studies approaches.

The 1970s were, as Peter Hall points out (p. 362), a period of the "Marxist ascendancy" in urban theory. Racial conflict in US cities and the Vietnam War abroad radicalized Harvey and other urban geographers, political scientists, sociologists, and planners. Theorists in these and other academic disciplines and professional fields rediscovered in Marx a theoretical basis for a critique of late twentieth-century capitalism and the domestic oppression and colonial violence they saw all around them.

Harvey's *Social Justice and the City*, written in the mid-1970s, examines urban conflicts and searches for a coherent philosophical and moral basis to resolve them. Harvey studied the conflicting arguments different groups in Baltimore, Maryland (near Johns Hopkins University where he was then teaching), used to justify their positions. Harvey sought to disentangle their conflicting belief systems and search for over-arching principles of urban social justice.

Harvey found many competing interests clashing around whether to build a new freeway in the Baltimore region and, if so, where to build it. Traffic engineers argued for *efficiency* – moving people from point A to point B as quickly as possible. City officials salivated over the potential *economic growth* better highway access might stimulate in Baltimore's depressed downtown. Advocates for racial minorities and the poor, environmentalists, neighborhood activists, historical preservationists, and others supported or opposed the freeway based on their own redistributive, environmental, communitarian, and preservationist values. Harvey concludes that there is no universal justice; rather, justice varies with time and place and also with individual persons.

In the years since he wrote *Social Justice and the City*, Harvey, like many of his peers, moved away from pure Marxist theory to reliance on postmodern theory. He here uses a postmodernist critical approach to analyze the battle of Tompkins Square Park in New York City. By "deconstructing" the imagery of motorcycle jackets and three piece suits, purple Mohawks and high fashion hairdos, Harvey helps us understand what is going on in this small piece of contested urban turf. Harvey and other postmodernist critics argue that not only do different groups disagree with each other, each group's arguments are based on their own worldview. They are engaged in their own "discourse," and it is a goal of postmodernist critical theory to understand these multiple "discourses."

New York authorities were divided about what posture to take towards the wildly different cultures inhabiting the same urban space: homeless people, skateboarders, basketball players, women with small children, skinheads, Rastafarians, chess players, dog walkers, heavy metal bands, yuppie professionals, crack dealers, and bikers. Ultimately the New York Police Department closed the park and the whole drama came to an end.

Reread Friederich Engels (p. 46) for a Marxist view of class relations in nineteenth-century Manchester. Reconsider the panopticon shopping mall, gated communities, and monumental architectural L.A. hotel glacis Mike Davis describes in "Fortress L.A." (p. 193) as manifestations of oppression calling out for social justice. Does Harvey offer solutions to these problems? See geographer Edward Soja (p. 180) and sociologist Sharon Zukin (p. 131) for other writings adopting a postmodernist culture studies approach to the study of cities.

This article draws upon Harvey's 1970s Marxist approach and his 1990s postmodernist culture studies approach as more fully described in *Social Justice and the City* (Oxford: Basil Blackwell, 1973) and *The Condition of Postmodernity: An Enquiry into the Origins of Cultural Change* (Malden, MA: Blackwell, 1990).

Other books by David Harvey include *Justice, Nature, and the Geography of Difference* (Oxford: Blackwell, 1996), *The Urban Experience* (Oxford: Blackwell, 1989), *The Urbanization of Capital: Studies in the History and Theory of Capitalist Urbanization* (Baltimore: Johns Hopkins University Press, 1985), *Consciousness and the Urban Experience: Studies in the History and Theory of Capitalist Urbanization* (Baltimore: Johns Hopkins University Press, 1985), *The Limits to Capital* (Oxford: Blackwell, 1982), and *Explanation in Geography* (London: Edward Arnold, 1969).

See French Marxist geographer Henri Lefebvre, *The Production of Space* (Oxford: Blackwell, 1991) and *Writings on Cities* (Oxford: Blackwell, 1995) for a Marxist theory of urban geography. John Rawls's *A Theory of Justice* (Cambridge, MA: Harvard University Press, 1971) is the classic philosophical statement on the liberal theory of justice. The five faces of oppression discussed in Harvey's article come from Iris Marion Young's *Justice and the Politics of Difference* (Princeton: Princeton, NJ University Press, 1990).

The title of this essay is a collage of two book titles of mine written nearly 20 years apart, *Social Justice and the City* and *The Condition of Postmodernity*. I here want to consider the relations between them, in part as a way to reflect on the intellectual and political journey many have travelled these last two decades in their attempts to grapple with urban issues, but also to examine how we now might think about urban problems and how by virtue of such thinking we can better position ourselves with respect to solutions. The question of *positionality* is, I shall argue, fundamental to all debates about how to create infrastructures and urban environments for living and working in the twenty-first century.

JUSTICE AND THE POSTMODERN CONDITION

I begin with a report by John Kifner in the *International Herald Tribune* (1 August 1989) concerning the hotly contested spacc of

Tompkins Square Park in New York City – a space which has been repeatedly fought over, often violently, since the 'police riot' of August 1988. The neighbourhood mix around the park was the primary focus of Kifner's attention. Not only were there nearly 300 homeless people, but there were also:

> Skateboarders, basketball players, mothers with small children, radicals looking like 1960s retreads, spikey-haired punk rockers in torn black, skinheads in heavy working boots looking to beat up the radicals and punks, dreadlocked Rastafarians, heavy-metal bands, chess players, dog walkers – all occupy their spaces in the park, along with professionals carrying their drycleaned suits to the renovated 'gentrified' buildings that are changing the character of the neighbourhood.

By night, Kifner notes, the contrasts in the park become even more bizarre:

> The Newcomers Motorcycle Club was having its annual block party at its clubhouse at 12th Street and Avenue B and the street was lined with chromed Harley Davidsons with raised 'apehanger' handlebars and beefy men and hefty women in black leather. A block north a rock concert had spilled out of a 'squat' – an abandoned cityowned building taken over by outlaw renovators, mostly young artists – and the street was filled with young people whose purple hair stood straight up in spikes. At the World Club just off Houston Street near Avenue C, black youths pulled up in the Jeep-type vehicles favored by cash-heavy teen-age crack moguls, high powered speakers blaring. At the corner of Avenue B and Third, considered one of the worst heroin blocks in New York, another concert was going on at an artists' space called The Garage, set in a former gas station walled off by plastic bottles and other found objects. The wall formed an enclosed garden looking up at burned-out, abandoned buildings: there was an eerie resemblance to Beirut. The crowd was white and fashionably dressed, and a police sergeant sent to check on the noise shook his head, bemused: 'It's all yuppies'.

This is, of course, the kind of scene that makes New York such a fascinating place, that makes any great city into a stimulating and exciting maelstrom of cultural conflict and change. It is the kind of scene that many a student of urban subcultures would revel in, even seeing in it, as someone like Iain Chambers (1987) does, the origins of that distinctive perspective we now call 'the postmodern':

> Postmodernism, whatever form its intellectualizing might take, has been fundamentally anticipated in the metropolitan cultures of the last twenty years: among the electronic signifiers of cinema, television and video, in recording studios and record players, in fashion and youth styles, in all those sounds, images and diverse histories that are daily mixed, recycled and 'scratched' together on that giant screen that is the contemporary city.

Armed with that insight, we could take the whole paraphernalia of postmodern argumentation and technique and try to 'deconstruct' the seemingly disparate images on that giant screen which is the city. We could dissect and celebrate the fragmentation, the co-presence of multiple discourses – of music, street and body language, dress and technological accoutrements (such as the Harley Davidsons) – and, perhaps, develop sophisticated empathies with the multiple and contradictory codings with which highly differentiated social beings both present themselves to each other and to the world and live out their daily lives. We could affirm or even celebrate the bifurcations in cultural trajectory, the preservation of pre-existing and the creation of entirely new but distinctive 'othernesses' within an otherwise homogenizing world.

On a good day, we could celebrate the scene within the park as a superb example of urban tolerance for difference . . .

To the degree that the freedom of city life 'leads to group differentiation, to the formation of affinity groups' (Young, 1990: 238) of the sort which Kifner identifies in Tompkins Square, so our conception of social justice 'requires not the melting away of differences, but institutions that promote reproduction of and respect for group differences without oppression' (ibid p. 47) . . .

So what should the urban policy-maker do in the face of these strictures? The best path is to pull out that well-thumbed copy of Jane Jacobs (1961) and insist that we should both respect and provide for 'spontaneous self-diversification among urban populations', in the formulation of our policies and plans. In so doing we can avoid the critical wrath she directs at city designers, who 'seem neither to recognize this force for self-diversification nor to be attracted by the esthetic problems of expressing it' . . . We should not, in short, aim to obliterate differences within the park, homogenize it according to some conception of, say, bourgeois taste or social order. We should engage, rather, with an aesthetics which embraces or stimulates that 'spontaneous self-diversification' of which Jacobs speaks. Yet there is an immediate question mark over that suggestion: in what ways, for example, can homelessness be understood as spontaneous self-diversification, and does this mean that we should respond to that problem with designer-style cardboard boxes to make for more jolly and sightly shelters for the homeless? While Jane Jacobs has a point, and one which many urbanists have absorbed these last few years, there is, evidently, much more to the problem than her arguments encompass.

That difficulty is highlighted on a bad day in the park. So-called forces of law and order battle to evict the homeless, erect barriers between violently clashing factions. The park then becomes a locus of exploitation and oppression, an open wound from which bleed the five faces of oppression which Young defines as exploitation, marginalization, powerlessness, cultural imperialism and violence . . .

[. . .]

On 8 June 1991, the question was resolved by evicting everyone from the park and closing it entirely 'for rehabilitation' under a permanent guard of at least 20 police officers . . .

And what should the policy-maker and planner do in the face of these conditions? Give up planning and join one of those burgeoning cultural studies programmes which revel in chaotic scenes of the Tompkins Square sort while simultaneously disengaging from any commitment to do something about them? Deploy all the critical powers of deconstruction and semiotics to seek new and engaging interpretations of graffiti which say 'Die, Yuppie Scum'? Should we join revolutionary and anarchist groups and fight for the rights of the poor and the culturally marginalized to express their rights and if necessary make a home for themselves in the park? Or should we throw away that dog-eared copy of Jane Jacobs and join with the forces of law and order and help impose some authoritarian solution on the problem?

Decisions of some sort have to be made and actions taken, as about any other facet of urban infrastructure. And while we might all agree that an urban park is a good thing in principle, what are we to make of the fact that the uses turn out to be so conflictual, and that even conceptions as to what the space is for and how it is to be managed diverge radically among competing factions? To hold all the divergent politics of need and desire together within some coherent frame may be a laudable aim, but in practice far too many of the interests are mutually exclusive to allow their mutual accommodation. Even the best shaped compromise (let alone the savagely imposed authoritarian solution) favours one or other factional interest. And that provokes the biggest question of all – what is the *conception* of 'the public' incorporated into the construction of public space?

[. . .]

SOCIAL JUSTICE AND MODERNITY

I now leave this very contemporary situation and its associated conundrums and turn to an older story. It turned up when I unearthed from my files a yellowing manuscript, written sometime in the early 1970s, shortly after I finished *Social Justice and the City.* I there examined the case of a proposal to put a segment of the Interstate Highway System on an east–west trajectory right through the heart of Baltimore – a proposal first set out in the early 1940s and which has still not been fully resolved. I resurrect this case here in part to show that what we would now often depict as a quintessentially modernist problem was even at that time argued about in ways which contained the seeds, if not the essence, of much of what many now view as a distinctively postmodernist form of argumentation.

My interest in the case at that time, having looked at a lot of the discussion, attended hearings and read a lot of documentation, lay initially in the highly differentiated arguments, articulated by all kinds of different groups, concerning the rights and wrongs of the whole project. There were, I found, seven kinds of arguments being put forward:

(1) An *efficiency* argument which concentrated on

the relief of traffic congestion and facilitating the easier flow of goods and people throughout the region as well as within the city;

(2) An *economic growth* argument which looked to a projected increase (or prevention of loss) in investment and employment opportunities in the city consequent upon improvements in the transport system;

(3) An *aesthetic and historical heritage* argument which objected to the way sections of the proposed highway would either destroy or diminish urban environments deemed both attractive and of historical value;

(4) A *social and moral order* argument which held that prioritizing highway investment and subsidizing car owners rather than, for example, investing in housing and health care was quite wrong;

(5) An *environmentalist/ecological* argument which considered the impacts of the proposed highway on air quality, noise pollution and the destruction of certain valued environments (such as a river valley park);

(6) A *distributive justice* argument which dwelt mainly on the benefits to business and predominantly white middle-class suburban commuters to the detriment of low-income and predominantly African-American inner-city residents;

(7) A *neighbourhood and communitarian* argument which considered the way in which close-knit but otherwise fragile and vulnerable communities might be destroyed, divided or disrupted by highway construction.

The arguments were not mutually exclusive, of course, and several of them were merged by proponents of the highway into a common thread – for example, the efficiency of the transport system would stimulate growth and reduce pollution from congestion so as to advantage otherwise disadvantaged inner-city residents. It was also possible to break up each argument into quite distinct parts – the distributive impacts on women with children would be very different from those on male workers.

We would, in these heady postmodern times, be prone to describe these separate arguments as 'discourses', each with its own logic and imperatives. And we would not have to look too closely to see particular 'communities of interest' which articulated a particular discourse as if it was the only one that mattered. The particularistic arguments advanced by such groups proved effective in altering the alignment of the highway but did not stop the highway as a whole. The one group which tried to forge a coalition out of these disparate elements (the *Movement Against Destruction*, otherwise known as *MAD*) and to provide an umbrella for opposition to the highway as a whole turned out to be the least effective in mobilizing people and constituencies even though it was very articulate in its arguments.

The purpose of my own particular enquiry was to see how the arguments (or discourses) for and against the highway worked and if coalitions could be built in principle between seemingly disparate and often highly antagonistic interest groups via the construction of higher order arguments (discourses) which could provide the basis for consensus. The multiplicity of views and forces has to be set against the fact that either the highway is built or it is not, although in Baltimore, with its wonderful way of doing things, we ended up with a portion of the highway that is called a boulevard (to make us understand that this six-lane two-mile segment of a monster cut through, the heart of low-income and predominantly African-American West Baltimore is not what it really is) and another route on a completely different alignment, looping around the city core in such a way as to allay some of the worst political fears of influential communities.

Might there be, then, some higher-order discourse to which everyone could appeal in working out whether or not it made sense to build the highway? A dominant theme in the literature of the 1960s was that it was possible to identify some such higher-order arguments. The phrase that was most frequently used to describe it was *social rationality*. The idea of that did not seem implausible, because each of the seven seemingly distinctive arguments advanced a rational position of some sort and not infrequently appealed to some higher-order rationale to bolster its case. Those arguing on efficiency and growth grounds frequently invoked utilitarian arguments, notions of 'public good' and the greatest benefit to the greatest number, while recognizing (at their best) that individual sacrifices were inevitable and that it was right and proper to offer appropriate compensation for those who would be displaced. Ecologists or communitarians likewise appealed to higher-order arguments – the former to the values inherent in nature and the latter to some higher sense of communitarian values. For all of these reasons, consideration of higher-order arguments over social rationality did not seem unreasonable.

[Harvey discusses "social rationality" theory]

[. . .]

Social justice

Social justice is but one of the seven criteria I worked with and I evidently hoped that careful investigation of it might rescue the argument from the abyss of formless relativism and infinitely variable discourses and interest grouping. But here too the enquiry proved frustrating. It revealed that there are as many competing theories of social justice as there are competing ideals of social rationality. Each ideal has its flaws and strengths . . .

. . . To argue for social justice meant the deployment of some initial criteria to define which theory of social justice was appropriate or more just than another. The infinite regress of higher-order criteria immediately looms, as does, in the other direction, the relative ease of total deconstruction of the notion of justice to the point where it means nothing whatsoever, except whatever people at some particular moment decide they want it to mean. Competing discourses about justice could not be dissassociated from competing discourses about positionality in society.

There seemed two ways to go with that argument. The first was to look at how concepts of justice are embedded in language, and that led me to theories of meaning of the sort which Wittgenstein advanced . . .

From this perspective the concept of justice has to be understood in the way it is embedded in a particular language game. Each language game attaches to the particular social, experiential and perceptual world of the speaker. Justice has no universal meaning, but a whole 'family' of meanings. This finding is completely consistent, of course, with anthropological studies which show that justice among, say, the Nuer, means something completely different from the capitalistic conception of justice. We are back to the point of cultural, linguistic or discourse relativism.

The second path is to admit the relativism of discourses about justice, but to insist that discourses are expressions of social power. In this case the idea of justice has to be the formation of certain hegemonic discourses which derive from the power exercised by any ruling class. This is an idea which goes back to Plato, who in the *Republic* has Thrasymachus argue that:

> Each ruling class makes laws that are in its own interest, a democracy democratic laws, a tyranny tyrannical ones and so on; and in making these laws they define as 'right' for their subjects what is in the interest of themselves, the rulers, and if anyone breaks their laws he is punished as a 'wrong-doer'. That is what I mean when I say that 'right' is the same in all states, namely the interest of the established ruling class . . .
>
> (Plato, 1965)

Consideration of these two paths brought me to accept a position which is most clearly articulated by Engels in the following terms:

> The stick used to measure what is right and what is not is the most abstract expression of right itself, namely *justice* . . . justice is but the ideologized, glorified expression of the existing economic relations, now from their conservative and now from their revolutionary angle. The justice of the Greeks and Romans held slavery to be just; the justice of the bourgeois of 1789 demanded the abolition of feudalism on the ground it was unjust. The conception of eternal justice, therefore, varies not only with time and place, but also with the persons concerned . . .
>
> (Marx and Engels, 1951: 562–4)

. . . Taking capitalistic notions of social rationality or of justice, and treating them as universal values to be deployed under socialism, would merely mean the deeper instanciation of capitalist values by way of the socialist project.

THE TRANSITION FROM MODERNIST TO POSTMODERNIST DISCOURSES

There are two general points I wish to draw out of the argument so far. First, the critique of social rationality and of conceptions such as social justice as policy tools was something that was originated and so ruthlessly pursued by the 'left' (including marxists) in the 1960s that it began to generate radical doubt throughout civil society as to the veracity of all universal claims. From this it was a short, though as I shall shortly argue, unwarranted, step to conclude, as many postmodemists now do, that all forms of metatheory are either misplaced or illegitimate. Both steps in this process were further reinforced by the emergence of the so-called 'new' social movements – the peace and women's movements, the ecologists, the movements against colonization and racism – each of which came to articulate its own definitions of social justice and rationality. There then seemed to be, as Engels had argued, no philosophical, linguistic or logical way to resolve the resulting divergencies in conceptions of rationality and justice, and thereby to find a way to reconcile competing claims or arbitrate between radically different discourses. The effect was to undermine the legitimacy of state policy, attack all conceptions of bureaucratic rationality and at best place social policy formulation in a quandary and at worst render it powerless except to articulate the ideological and value precepts of those in power. Some of those who participated in the revolutionary movements of the 1970s and 1980s considered that rendering transparent the power and class basis of supposedly universal claims was a necessary prelude to mass revolutionary action.

But there is a second and, I think, more subtle point to be made. If Engels is indeed right to insist that the conception of justice 'varies not only with time and place, but also with the persons concerned', then it seems important to look at the ways in which a particular society produces such variation in concepts. In so doing it seems important, following writers as diverse as Wittgenstein and Marx, to look at the material basis for the production of difference, in particular at the production of those radically different experiential worlds out of which divergent language games about social rationality and social justice could arise. This entails the application of historical-geographical materialist methods and principles to understand the production of those power differentials which in turn produce different conceptions of justice and embed them in a struggle over ideological hegemony between classes, races, ethnic and political groupings as well as across the gender divide . . .

From this standpoint we can clearly see that concepts of justice and of rationality have not disappeared from our social and political world these last few years. But their definition and use has changed. The collapse of class compromise in the struggles of the late 1960s and the emergence of the socialist, communist and radical left movements, coinciding as it did with an acute crisis of overaccumulation of capital, posed a serious threat to the stability of the capitalist political-economic system. At the ideological level, the emergence of alternative definitions of both justice and rationality was part of that attack, and it was to this question that my earlier book, *Social Justice and the City*, was addressed. But the recession/depression of 1973–5 signalled not only the savage devaluation of capital stock (through the first wave of deindustrialization visited upon the weaker sectors and regions of a world capitalist economy) but the beginning of an attack upon the power of organized labour via widespread unemployment, austerity programmes, restructuring and, eventually, in some instances (such as Britain) institutional reforms.

. . . [T]he idea that the market is the best way to achieve the most just and the most rational forms of social organization has become a powerful feature of the hegemonic discourses these last 20 years in both the United States and Britain. The collapse of centrally planned economies throughout much of the world has further boosted a market triumphalism which presumes that the rough justice administered through the market in the course of this transition is not only socially just but also deeply rational. The advantage of this solution, of course, is that there is no need for explicit theoretical, political and social argument over what is or is not socially rational just because it can be presumed that, provided the market functions properly, the outcome is nearly always just and rational. Universal claims about rationality and justice have in no way diminished. They are just as frequently asserted in justification of privatization and of market action as they ever were in support of welfare state capitalism.

The dilemmas inherent in reliance on the market are well known and no one holds to it without some qualification. Problems of market breakdown, of externality effects, the provision of public goods and infrastructures, the clear need for some coordination of disparate investment decisions, all of these require some level of government interventionism. Margaret Thatcher may thus have abolished Greater London government, but the business community wants some kind of replacement (though preferably non-elected), because without it city services are disintegrating and London is losing its competitive edge. But there are many voices that go beyond that minimal requirement since free-market capitalism has produced widespread unemployment, radical restructurings and devaluations of capital, slow growth, environmental degradation and a whole host of financial scandals and competitive difficulties, to say nothing of the widening disparities in income distributions in many countries and the social stresses that attach thereto. It is under such conditions that the never quite stilled voice of state regulation, welfare state capitalism, of state management of industrial development, of state planning of environmental quality, land use, transportation systems and physical and social infrastructures, of state incomes and taxation policies which achieve a modicum of redistribution either in kind (via housing, health care, educational services and the like) or through income transfers, is being reasserted. The political questions of social rationality and of social justice over and above that administered through the market are being taken off the back burner and moved to the political agenda in many of the advanced capitalist countries . . .

[. . .]

For my own part, I think Engels had it right. Justice and rationality take on different meanings across space and time and persons, yet the existence of everyday meanings to which people do attach importance and which to them appear unproblematic, gives the terms a political and mobilizing power that can never be neglected. Right and wrong are words that power revolutionary changes and no amount of negative deconstruction of such terms can deny that. So where, then, have the new social movements and the radical left in general got with their own conception, and how does it challenge both market and corporate welfare capitalism?

Young in her *Justice and the Politics of Difference* (1990) provides one of the best recent statements. She redefines the question of justice away from the purely redistributive mode of welfare state capitalism and focuses on what she calls the 'five faces' of oppression, and I think each of them is worth thinking about as we consider the struggle to create liveable cities and workable environments for the twenty-first century.

[Harvey here discusses Young's 'five faces' of oppression from which he draws five propositions which follow in italics:

That just planning and policy practices must confront directly the problem of creating forms of social and political organization and systems of production and consumption which minimize the exploitation of labour power both in the workplace and the living place.

That just planning and policy practices must confront the phenomenon of marginalization in a non-paternalistic mode and find ways to organize and militate within the politics of marginalization in such a way as to liberate captive groups from this distinctive form of oppression.

Just planning and policy practices must empower rather than deprive the oppressed of access to political power and the ability to engage in self-expression.

That just planning and policy practices must be particularly sensitive to issues of cultural imperialism and seek, by a variety of means, to eliminate the imperialist attitude both in the design of urban projects and modes of popular consultation.

A just planning and policy practice must seek out non-exclusionary and non-militarized forms of social control to contain the increasing levels of both personal and institutionalized violence without destroying capacities for empowerment and self-expression.]

Finally, I want to add a sixth principle to those which Young advances. This derives from the fact that all social projects are ecological projects and vice versa. While I resist the view that 'nature has rights' or that nature can be 'oppressed', the justice due to future generations and to other inhabitants of the globe requires intense scrutiny of all social projects for assessment of their ecological consequences. Human beings necessarily appropriate and transform the world around them in the course of making their own history, but they do not have to do so with such reckless abandon as to jeopardize the fate of peoples separated from us in either space or time. The final proposition is, then: *that just planning and policy practices will clearly recognize that the necessary ecological consequences of all social projects have impacts on future generations as well as upon distant peoples and take steps to ensure a reasonable mitigation of negative impacts.*

I do not argue that these six principles can or even should be unified, let alone turned into some convenient and formulaic composite strategy. Indeed, the six dimensions of justice here outlined are frequently in conflict with each other as far as their application to individual persons – the exploited male worker may be a cultural imperialist on matters of race and gender while the thoroughly oppressed person may be the bearer of social injustice as violence. On the other hand, I do not believe the principles can be applied in isolation from each other either. Simply to leave matters at the level of a "non-consensual' conception of justice, as someone like Lyotard (1984) would do, is not to confront some central issues of the social processes which produce such a differentiated conception of justice in the first place. This then suggests that social policy and planning has to work at two levels. The different faces of oppression have to be confronted for what they are and as they are manifest in daily life, but in the longer term and at the same time the underlying sources of the different forms of oppression in the heart of the political economy of capitalism must also be confronted, not as the fount of all evil but in terms of capitalism's revolutionary dynamic which transforms, disrupts, deconstructs and reconstructs ways of living, working, relating to each other and to the environment. From such a standpoint the issue is never about whether or not there shall be change, but what sort of change we can anticipate, plan for, and proactively shape in the years to come.

I would hope that consideration of the varieties of justice as well as of this deeper problematic might set the tone for present deliberations. By appeal to them, we might see ways to break with the political, imaginative and institutional constraints which have for too long inhibited the advanced capitalist societies in their developmental path. The critique of universal notions of justice and rationality, no matter whether embedded in the market or in state welfare capitalism, still stands. But it is both valuable and potentially liberating to look at alternative conceptions of both justice and rationality as these have emerged within the new social movements these last two decades. And while it will in the end ever be true, as Marx and Plato observed, that 'between equal rights force decides', the authoritarian imposition of solutions to many of our urban ills these past few years and the inability to listen to alternative conceptions of both justice and rationality is very much a part of the problem. The conceptions I have outlined speak to many of the marginalized, the oppressed and the exploited in this time and place. For many of us, and for many of them, the formulations may well appear obvious, unproblematic and just plain common sense. And it is precisely because of such widely held conceptions that so much welfare-state paternalism and market rhetoric fails. It is, by the same token, precisely out of such conceptions that a genuinely liberatory and transformative politics can be made. 'Seize the time and the place', they would say around Tompkins Square park, and this does indeed appear an appropriate time and place to do so. If some of the walls are coming down all over eastern Europe, then surely we can set about bringing them down in our own cities as well.

REFERENCES

Chambers, I. (1987) "Maps for the metropolis: a possible guide to the present". *Cultural Studies* 1, 1–22.

Engels, F. (1951) *Selected works, vol.* 1. Progress Publishers, Moscow.

Harvey, D. (1973) *Social Justice and the City.* Edward Arnold, London.

——— (1989) *The Condition Of Postmodernity.* Blackwell, Oxford.

Jacobs, J. (1961) *The Death And Life Of Great American Cities.* Vintage, New York.

Lyotard, J. (1984) *The Postmodern Condition.* Manchester University Press, Manchester.

Marx, K. and Engels, F. (1967) *Capital, vol.* 1. International Publishers, New York. (1951)

Plato (1965) *The Republic.* Penguin Books, Harmondsworth, Middlesex.

Whey, J. (1989) "No miracles in the park: homeless New Yorkers amid drug lords and slumlords". *International Herald Tribune,* 1 August 1989, p. 6.

Young, I. M. (1990) *Justice and the Politics of Difference.* Princeton University Press, Princeton, NJ.

SASKIA SASSEN

"A New Geography of Centers and Margins: Summary and Implications"

from *Cities in a World Economy* (1994)

Editors' introduction Contemporary Los Angeles is just one world city that reflects a massive restructuring of the world economy during the last twenty years. Banking, finance, corporate planning and management are increasingly concentrated in powerful global cities. Manufacturing is dispersing around the globe, although, as Edward Soja reminds us, not all manufacturing has fled supposedly post-industrial cities such as Los Angeles (p. 180). The global distribution of wealth and power is in flux. This has enormous implications for the structure of the global system of cities, the functions cities perform, and the nature of social life within them. These issues are explored in greater depth in Part 4, Urban Politics, Governance, and Economics, and Part 8, The Future of the City.

Saskia Sassen, a professor of urban and regional planning at Columbia University, provides insight into the way in which the rapid and profound changes in the world economy have affected the evolution of cities today. Sassen has carefully examined data on the economies and workforce characteristics of the largest global cities and the way in which they connect to other cities in the world economy. In the following selection Sassen describes the nature of the new global system of cities.

Sassen argues that global cities – where banks, corporate headquarters, and other command functions and high-level producer-service firms such as law firms and advertising agencies oriented to world markets are concentrated – have emerged as strategic sites in the world economy. Decisions made in London, New York, Tokyo, or Sydney affect jobs, wages, and the economic health of locations as remote as Kuala Lumpur, Malaysia, or Santiago, Chile.

Beginning with Melvin Webber (p. 535), some writers argued that instant global telecommunications and an interconnected world economy may make place unimportant and portend the end of cities. But according to Sassen that is not what has happened so far. Her research indicates that global cities such as New York, London, Los Angeles, and Tokyo have become more, not less, dense recently. And their wealth and power is growing, not declining. On the other hand, many cities that have historically served as manufacturing centers in Europe, North America, and Australia are in economic decline as manufacturing shifts to Asia, South and Central America and elsewhere in the Third World.

One of Sassen's most important theoretical contributions to the study of cities is her sharp questioning of the whole notion of "rich" countries and "rich" cities; places central to the world economy, and those that exist at the margin. Sassen argues that cities at the center of the world economy are increasingly both rich and poor and that many Third World cities – while economically subordinate to global command centers – are also stratified by income.

Sassen's work is central to debates about the effects of global economic restructuring on cities. Some writers on information technology stress the decentralizing effects communications technology has had and will likely have in the future. They disagree with Sassen's assessment that command functions will be increasingly centralized. Marxist and neo-Marxist writers such as David Harvey, Doreen Massey, Richard Walker, and Michael Storper see the changes in the global system of cities as the inevitable result of current capitalist development, though they do not agree among themselves on how capitalist processes are unfolding. Their views are explored further in Mike Savage and Alan Warde's discussion of uneven development (p. 264).

This selection is from Saskia Sassen, *Cities in A Global Economy* (Thousand Oaks, CA: Pine Forge Press, 1994). Other books by Saskia Sassen include *Globalization and its Discontents* (New York: New Press, 1998), with Anthony Appiah, *Losing Control?: Sovereignty in an Age of Globalization* (New York: Columbia University Press, 1996), *The Mobility of Labor and Capital: A Study in International Investment and Labor Flows* (New York: Cambridge University Press, 1988) and *The Global City* (Princeton, NJ: Princeton University Press, 1993).

For additional information on global cities see David Clark (p. 579). Neil Peirce's *Citistates* (Washington, DC: Seven Locks Press, 1993) describes how US regions are managing their transformations in relation to the world economy.

Three important developments over the last 20 years laid the foundation for [the following analysis] of cities in the world economy.

1 *The territorial dispersal of economic activities, of which globalization is one form, contributes to the growth of centralized functions and operations.* We find here a new logic for agglomeration and key conditions for the renewed centrality of cities in advanced economies. Information technologies, often thought of as neutralizing geography, actually contribute to spatial concentration. They make possible the geographic dispersal and simultaneous integration of many activities. But the particular conditions under which such facilities are available have promoted centralization of the most advanced users in the most advanced telecommunications centers. We see parallel developments in cities that function as regional nodes – that is, at smaller geographic scales and lower levels of complexity than global cities.

2 *Centralized control and management over a geographically dispersed array of economic operations does not come about inevitably as part of a "world system."* It requires the production of a vast range of highly specialized services, telecommunications infrastructure, and industrial services. Major cities are centers for the servicing and financing of international trade, investment, and headquarters operations. And in this sense they are strategic production sites for today's leading economic sectors. This function is reflected in the ascendance of these activities in their economies. Again, cities that serve as regional centers exhibit similar developments. This is the way in which the spatial effects of the growing service intensity in the organization of all industries materialize in cities.

3 *Economic globalization has contributed to a new geography of centrality and marginality.* This new geography assumes many forms and operates in many terrains, from the distribution of telecommunications facilities to the structure of the economy and of employment. Global cities become the sites of immense concentrations of economic power, while cities that were once major manufacturing centers suffer inordinate declines; highly educated workers see their incomes rise to unusually high levels, while low- or medium-skilled workers see theirs sink. Financial services produce superprofits while industrial services barely survive.

Let us look more closely now at this last and most encompassing of the propositions.

THE LOCUS OF THE PERIPHERAL

The sharpening distance between the extremes evident in all major cities of developed countries raises questions about the notion of "rich" countries and "rich" cities. It suggests that the geography of centrality and marginality, which in the past was seen in terms of the duality of highly developed and less developed countries, is now also evident within developed countries and especially within their major cities.

One line of theorization posits that the intensified inequalities . . . represent a transformation in the geography of center and periphery. They signal that peripheralization processes are occurring inside areas that were once conceived of as "core" areas – whether at the global, regional, or urban level – and that alongside the sharpening of peripheralization processes, centrality has also become sharper at all three levels.

The condition of being peripheral is installed in different geographic terrains depending on the prevailing economic dynamic. We see new forms of peripheralization at the center of major cities in developed countries not far from some of the most expensive commercial land in the world: "inner cities" are evident not only in the United States and large European cities, but also now in Tokyo. Furthermore, we can see peripheralization operating at the center in organizational terms as well. We have long known about segmented labor markets, but the manufacturing decline and the kind of devaluing of nonprofessional workers in leading industries that we see today in these cities go beyond segmentation and in fact represent an instance of peripheralization.

Furthermore, the new forms of growth evident at the urban perimeter also mean crisis: violence in the immigrant ghetto of the *banlieus* (the French term for *suburbs*), exurbanites clamoring for control over growth to protect their environment, new forms of urban governance. The regional mode of regulation in many of these cities is based on the old center/suburb model and may hence become increasingly inadequate to deal with intraperipheral conflicts – conflicts among different types of constituencies at the urban perimeter or urban region. Frankfurt, for example, is a city that cannot function without its region's towns; yet this particular urban region would not have emerged without the specific forms of growth in Frankfurt's center. Keil and Ronneberger (1993) note the ideological motivation in the call by politicians to officially recognize the region so as to strengthen Frankfurt's position in the global interurban competition. This call also provides a rationale for coherence and the idea of common interests among the many objectively disparate interests in the region: it displaces the conflicts among unequally advantaged sectors onto a project of regional competition with other regions. Regionalism then emerges as the concept for bridging the global orientation of leading sectors with the various local agendas of various constituencies in the region.

In contrast, the city discourse rather than the ideology of regionalism dominates in cities such as New York or São Paulo. The challenge is how to bridge the inner city, or the squatters at the urban perimeter, with the center. In multiracial cities, multiculturalism has emerged as one form of this bridging. A "regional" discourse is perhaps beginning to emerge, but it has until now been totally submerged under the suburbanization banner, a concept that suggests both escape from and dependence on the city. The notion of conflict within the urban periphery among diverse interests and constituencies has not really been much of a factor in the United States. The delicate point at the level of the region has rather been the articulation between the residential suburbs and the city.

CONTESTED SPACE

Large cities have emerged as strategic territories for these developments. *First, cities are the sites for concrete operations of the economy*. For our purposes we can distinguish two forms of such concrete operations: (1) in terms of economic globalization and place, cities are strategic places that concentrate command functions, global markets, and . . . production sites for the advanced corporate service industries. (2) In terms of day-to-day work in the leading industrial complex, finance, and specialized services . . . a large share of the jobs involved are low paid and manual, and many are held by women and immigrants. Although these types of workers and jobs are never represented as part of the global economy, they are in fact as much a part of globalization as international finance is. We see at work here a dynamic of valorization that has sharply increased the distance between the devalorized and the valorized – indeed overvalorized – sectors of the economy. These joint presences have made cities a contested terrain.

The structure of economic activity has brought about changes in the organization of work that are reflected in a pronounced shift in the job supply, with strong polarization occurring in the income distribution and occupational distribution of workers. Major growth industries show a greater incidence of jobs at the high- and low-paying ends of the scale than do the older industries now in decline. Almost half the jobs in the producer services are lower-income jobs, and the other half are in the two highest earnings classes. On the other hand, a large share of manufacturing workers were in middle-earning jobs during the postwar period of high growth in these industries in the United States and the United Kingdom.

One particular concern here was to understand how new forms of inequality actually are constituted into new social forms, such as gentrified neighborhoods, informal economies, or downgraded manufacturing sectors. To what extent these developments are connected to the consolidation of an economic complex oriented to the global market is difficult to say. Precise empirical documentation of the linkages or impacts is impossible; the effort here is focused, then, on a more general attempt to understand the consequences of both the ascendance of such an international economic complex and the general move to a service economy.

Second, the city concentrates diversity. Its spaces are inscribed with the dominant corporate culture but also with a multiplicity of other cultures and identities, notably through immigration. The slippage is evident: the dominant culture can encompass only part of the city. And while corporate power inscribes noncorporate cultures and identities with "otherness," thereby devaluing them, they are present everywhere. The immigrant communities and informal economy . . . are only two instances. Diverse cultures and ethnicities are especially strong in major cities in the United States and western Europe; these also have the largest concentrations of corporate power.

We see here an interesting correspondence between great concentrations of corporate power and large concentrations of "others." It invites us to see that globalization is not only constituted in terms of capital and the new international corporate culture (international finance, telecommunications, information flows) but also in terms of people and noncorporate cultures. There is a whole infrastructure of low-wage, non-professional jobs and activities that constitutes a crucial part of the so-called corporate economy.

A focus on the work behind command functions, on production in the finance and services complex, and on marketplaces has the effect of incorporating the material facilities underlying globalization and the whole infrastructure of jobs and workers typically not seen as belonging to the corporate sector of the economy: secretaries and cleaners, the truckers who deliver the software, the variety of technicians and repair workers, and all the jobs having to do with the maintenance, painting, and renovation of the buildings where it is all housed.

This expanded focus can lead to the recognition that a multiplicity of economies is involved in constituting the so-called global information economy. It recognizes the types of activities, workers, and firms that have the "center" of the economy or that have been evicted from that center in the restructuring of the 1980s and have therefore been devalued in a system that puts too much weight on a narrow conception of the center of the economy. Globalization can, then, be seen as a process that involves multiple economies and work cultures.

. . . [C]ities are of great importance to the dominant economic sectors. Large cities in the highly developed world are the places where globalization processes assume concrete localized forms. These localized forms are, in good part, what globalization is about. We can then think of cities also as the place where the contradictions of the internationalization of capital either come to rest or conflict. If we consider, further, that large cities also concentrate a growing share of disadvantaged populations – immigrants in both Europe and the United States; African Americans and Latinos in the United States – then we can see that cities have become a strategic terrain for a whole series of conflicts and contradictions.

On one hand, they concentrate a disproportionate share of corporate power and are one of the key sites for the overvalorization of the corporate economy; on the other, they concentrate a disproportionate share of the disadvantaged and are one of the key sites for their devalorization. This joint presence happens in a context where (1) the internationalization of the economy has grown sharply and cities have become increasingly strategic for global capital; and (2) marginalized people have come into representation and are making claims on the city as well. This joint presence is further brought into focus by the sharpening of the distance between the two. The center now concentrates immense power, a power that rests on the capability for global control and the capability to produce superprofits. And marginality, notwithstanding weak economic and political power, has become an increasingly strong presence through the new politics of culture and identity.

If cities were irrelevant to the globalization of economic activity, the center could simply abandon them and not be bothered by all of this. Indeed, this is precisely what some politicians argue – that cities have become hopeless reservoirs for all kinds of social despair. It is interesting to note again how the dominant economic narrative argues that place no longer matters, that firms can be located anywhere thanks to telematics, that major industries now are information-based and hence not place-bound. This line of argument devalues cities at a time when they are major sites for the new cultural politics. It also allows the corporate economy to extract major concessions from city governments under the notion that firms can simply leave and relocate elsewhere, which is not quite the case for a whole complex of firms . . .

In seeking to show that (1) cities are strategic to economic globalization because they are command points, global marketplaces, and production sites for the information economy; and (2) many of the devalued sectors of the urban economy actually fulfill crucial functions for the center, this book attempts to recover the importance of cities specifically in a globalized economic system and the importance of those overlooked sectors that rest largely on the labor of women, immigrants, and, in the case of large U.S. cities, African Americans and Latinos. In fact it is the intermediary sectors of the economy (such as routine office work, headquarters that are not geared to the world markets, the variety of services demanded by the largely suburbanized middle class) and of the urban population (the middle class) that can and have left cities. The two sectors that have stayed, the center and the "other," find in the city the strategic terrain for their operations.

REFERENCE

Keil, Roger and Ronneberger, Klaus (1993) "City Turned Inside Out: Spatial Strategies and Local Politics" in H. Hitz, R. Keil, V. Lehrer, K. Ronneberger, C. Schmid, and R. Wolff (eds.), *Financial Metropoles in Restructuring: Zurich and Frankfurt en Route to Post-Fordism*. Zurich: Rotpunkt Publishers.

PART 4

Urban Politics, Governance, and Economics

PART FOUR

INTRODUCTION

The material on urban space in Part 3 and on urban society and culture in Part 2 raises important issues about the economy of cities and how cities should be governed. The conflicts related to a new freeway for Baltimore and appropriate use of Tompkins Square Park which geographer David Harvey discusses (p. 199) pose political questions: what *should* government, particularly local government, do when different groups want to use urban space in different ways? The debate between sociologist William Julius Wilson (p. 112) and conservative policy analyst Charles Murray (p. 122) described in Part 2 illustrates how sociological issues of race and class are intimately related to both economic questions and questions of urban *politics and governance*. It is to these questions of urban politics, governance, and economics that we now turn.

Urban politics is a distinct subfield within the social science discipline of political science and political sociology is a subfield of sociology. In the 1950s and 1960s seminal work by sociologist Floyd Hunter and political scientist Robert Dahl began a major debate between two schools of thought. In his book *Community Power Structure* (1953), based on research in Atlanta, Georgia, Hunter concluded that a small, interlocking elite consisting of key businessmen and members of established and socially prominent families sat on the boards of each other's corporations, chatted at the same social clubs, and made all the really important decisions about Atlanta, including the governmental decisions. In *Who Governs?* (1961) Dahl reached almost diametrically opposite conclusions. He concluded that local political power in New Haven, Connecticut – and by implication other cities – was fragmented. Dahl reported that many different people from a variety of walks of life were involved in decision-making by the New Haven city government and influenced the outcome of different political decisions. In sum, Dahl advanced a *pluralist model* of urban community power. Because Hunter concluded that a small elite ruled Atlanta, his model of urban community power is referred to as the *elitist* model of community power and he is referred to as a *structuralist*. During the next three decades, Dahl's and Hunter's competing models stimulated debate between elitists and pluralists and the further elaboration of structuralist theories of urban politics.

John Mollenkopf, a political scientist at the Graduate School of the City University of New York, reviews the current status of the debate between pluralists and structuralists in the first selection in this section: "How to Study Urban Political Power." He notes that pluralist explanations were widely accepted in the 1960s and early 1970s. Marxist and other structuralist explanations of urban political power gained wider acceptance during the 1970s and 1980s. Extreme structuralists even questioned whether the study of urban politics itself makes sense. They felt that urban outcomes are so largely determined by economic and social structures that urban politics is irrelevant. Many political scientists today draw on both structuralist and pluralist theory. While they see urban politics as influenced by economic and societal forces that they cannot control, these theorists argue that local elected officials and political interest groups can shape their cities' futures.

Mollenkopf rejects the extreme structuralist interpretation that urban politics is irrelevant. His own answer to the question of how to study urban political power is synthetic and subtle. It acknowledges the importance of structural constraints, but notes that the real world of urban politics involves many conflicting forces and leaves plenty of room for politics. Mollenkopf stresses the connection between formal political structures and processes, corporate decision-making, and the exercise of power by many other institutions such as churches, unions, professional associations, and neighborhood groups which may form alliances with elected officials in order to get things done. Like earlier pluralists, Mollenkopf marshals evidence that there is broad participation in local government, and he believes that many participants do influence decisions. Mollenkopf discusses and borrows heavily from the most important new paradigm in urban politics and community power studies: regime theory.

Some theorists argue that the world economy is moving from a "Fordist" mode of production, dominated by large corporations and supported by the nation state, towards a new "post-Fordist" economy with smaller, more flexible firms existing in a world of permeable sovereignty where the nation state has less influence. Using post-Fordist analysis, Margit Mayer, a political scientist from the Free University of Berlin, describes the way in which urban politics in Western European countries and the United States is changing. Mayer argues that as the national governments of Germany, the UK, France, other European countries and the United States lose the ability to manage the increasingly complex global system, local politics is becoming more, rather than less important. This may appear paradoxical as the budgets of local governments are shrinking and many local governments are downsizing or shedding historic social consumption and welfare functions like the provision of libraries, garbage collection, welfare and housing subsidies. Mayer's answer to this apparent paradox is closely related to the discussion of urban regimes. In her view, as they become leaner, local governments are assuming new command functions over activities carried out by a combination of local government itself, private sector corporations and nonprofit organizations. Local governments are now setting agendas that are implemented by public–private sector partnerships and nongovernmental organizations. Local governments are increasingly bargaining with private corporations and neighborhood groups about economic development, job training, and other governmental programs. Mayer notes that the new bargaining structures in different cities vary in their inclusiveness and responsiveness. Neighborhood groups may not agree with real estate developers about a proposed project. There may also be competing agendas within a given community. For example, local merchants may disagree with public housing residents about whether community development funds are best spent to spruce up street lighting on the main shopping street or to build a recreation area for children. Government may be called upon to bargain with multiple parties using the strategies John Forester (p. 410) describes.

Mayer argues that local post-Fordist political regimes will be more or less responsive depending on how actors at the local level seize and struggle over the opportunities post-Fordism provides. Her selection provides a good context for Sherry Arnstein's selection, "A Ladder of Citizen Participation," (p. 240) which describes how local citizens can participate in urban politics in ways that will really empower them and affect urban policy rather than being duped or manipulated.

During the 1960s Sherry Arnstein was the Chief Advisor on Citizen Participation in the US

Department of Housing and Urban Development's Model Cities Program. Arnstein begins her article by noting that "The idea of citizen participation is a little like eating spinach: no one is against it in principle because it is good for you." But whether or not city dwellers participate effectively in programs that affect them and their neighborhoods has varied greatly. A little background will help clarify Arnstein's selection. Many urban renewal programs in the United States during the 1950s and 1960s were intended to and did remove low-income and minority residents and replace their homes with office buildings, luxury housing, garages, and other developments totally unrelated to them. Invitations to neighborhood residents to help decide what such urban renewal projects should be like were a sham. In sharp contrast, President Johnson's "war on poverty" in the late 1960s emphasized "maximum feasible participation of the poor." Many decisions about how to use federal anti-poverty funds were made by residents themselves. While Daniel Patrick Moynihan's judgment that the "War on poverty" led to "maximum feasible misunderstanding" may be too harsh, many poverty programs lacked the capacity to manage programs or dissolved into internal feuding. Cutting local elected officials and established agencies out of the loop to design programs and manage anti-poverty funds certainly did create a major political backlash. The Model Cities program sought to strike a balance between the top-down urban renewal approach in which real citizen participation in decisions rarely occurred and the war on poverty model in which decentralization often led to confusion and ineffectiveness and provoked a political backlash.

Arnstein suggests that the reader imagine a "ladder" of different degrees of citizen participation. The rungs of the ladder range from non-participation (*manipulation; therapy*) at the bottom to citizen control at the top. While Arnstein herself favors citizen control as the ultimate objective of urban programs, she acknowledges all but the bottom two rungs of her ladder as useful to varying degrees. In the thirty plus years since this article was written there has been a succession of urban development programs in the United States and Western European countries, most recently the UK's Enterprize Zone program and the US Empowerment Zones/ Enterprise Communities program. Each of these programs calls for some degree of citizen participation.

Arnstein's ladder is useful in understanding how to create meaningful citizen participation in these government programs that are explicitly designed for urban development. It is also helpful in understanding the theoretical concept of pluralism as Mollenkopf (p. 219) discusses it at the grassroots level and in understanding what working within local urban political regimes and post-Fordist bargaining situations as conceptualized by Mayer (p. 229) could be like.

One of the most important – and certainly among the most sensitive – local government functions anywhere in the world is police work. In "Broken Windows" (p. 253) James Q. Wilson and George L. Kelling propose a theory about *how* crime comes to dominate declining neighborhoods and what to do about it. Wilson and Kelling emphasize the importance of citizen perceptions of crime. They argue that whether or not crime has actually increased in a neighborhood, if residents think it has, they become more reclusive and less involved in the community. And that in turn will open the door to real crime. A single broken window may be a trivial problem, Wilson and Kelling argue, but if it is not fixed the signal that no one cares enough to fix it will lead to more broken windows and then to drug dealing and violent crime. The remedy Wilson and Kelling suggest – which has been widely adopted – is a "community policing" model where police regularly patrol marginal neighborhoods on foot. They are highly visible and get to know the neighborhood and its residents very well. The authors go on to advance the controversial view that these community police should informally enforce community norms of appropriate civil behavior as the neighborhood itself defines them even if that calls for extralegal or perhaps illegal controls. Wilson and Kelling note that a transitional neighborhood may be inhabited both by "the regulars" and by strangers. The regulars in turn consist of what Wilson and Kelling term "decent folk" and derelicts and drunks who are not so decent but "know their place." So long as questionable street behavior stays within neighborhood-defined norms, community police will look the other way. But if rowdy teenagers, prostitutes, dope dealers, or other strangers violate community norms (regardless of whether they violate the law) community police will intervene, perhaps by ordering the strangers to leave even if there is no legal basis for such an order.

The kinds of neighborhoods in transition that Wilson and Kelling discuss and the residents they contain are products of a changing world economy. The world economy has become global and the fate of individual cities is now linked to world trends as never before. In this new global economy there has been a massive restructuring. Many cities are performing economic roles quite different from what they were just a short time ago. Two of the most significant trends in the emerging global economy are a shift of manufacturing jobs from the developed world to Third World countries and the growth of high-tech, information management-oriented advanced service jobs in many cities, within the developed world. These trends create problems of job loss and unemployment among some sectors in some cities and boomtowns elsewhere. Both types of cities imply dramatic changes in the wealth and class structure of cities. The selection in this section by Mike Savage and Alan Warde (p. 264) reviews the literature on "uneven economic development." Savage and Warde draw upon works by geographers, sociologists, economists, and many researchers who blend all of these disciplines. Capitalist and Marxist critics agree that economic power exerts a tremendous influence over political decision-making. But whereas the Marxists see the economic forces as ever-dominant – reducing urban politics to an ongoing exercise in managing the class struggle – the capitalist analysts argue that free market systems afford everyone some degree of economic opportunity.

Savage and Warde raise issues of uneven development and urban inequality at a theoretical level and on a global scale. They blame the capitalist system itself for creating inequality. The final selection in this section, by Harvard business professor Michael Porter (p. 278), approaches the issue of urban inequality and what to do about it from a quite different perspective. Porter believes that the economic and social health of inner cities depends upon economic development by the private sector. To succeed, he argues, economic development must be based on the economic self-interest of private firms rather than phony businesses propped up by government subsidies and preference programs. The key to success, according to Porter, lies in capitalizing on the competitive advantage which inner city neighborhoods possess: strategic location, local market demand, their capacity to integrate with regional business clusters, and human resources. Porter urges an economic, not a social, model for development. He counsels corporations to shift their philanthropic priorities away from providing social services, such as daycare or food for the homeless, to providing managerial expertise to develop neighborhood economies.

The varied selections in this section demonstrate that both economics and politics matter a great deal to cities. Cities are affected by the forces of global economic restructuring which they often barely comprehend and cannot control. For the developed world the Fordist age is over and a post-Fordist age has begun. The role of local government in the new world order is changing. The writers in this section point towards a new global economy of permeable sovereignty where understanding regimes, coalitions, and bargaining is critical.

JOHN MOLLENKOPF

"How to Study Urban Political Power"

from *A Phoenix in the Ashes: The Rise and Fall of the Koch Coalition in New York City Politics* (1992)

Editors' introduction "Is urban politics worth studying at all, or is the urban political realm so subordinate to, dependent on, and constrained by its economic and social context that factors from this domain have little independent explanatory power?" In posing this question, John Mollenkopf raises the fundamental issue that has dominated urban political thought since at least the period of the Industrial Revolution. Indeed, the question of whether urban residents are actually members of a self-governing political community – citizens of a modern polis – or merely the helpless pawns of larger, faceless forces is an issue as old as Aristotle's definition of man as a *zoon politikon*.

John Mollenkopf is a professor of political science and director of the Public Policy Program at the Graduate School of the City University of New York. He previously taught at Stanford University, where he was one of the founders of the Program on Urban Studies. In 1983, he published *The Contested City*, a pioneering study of postwar American urban politics. Based on the experience of Boston and San Francisco, Mollenkopf outlined the process by which federal development programs were employed to help forge powerful pro-growth political coalitions at the local level. In *A Phoenix in the Ashes*, he applies a similar analysis to the politics of the city of New York during the mayoralty of Edward Koch.

"How to Study Urban Political Power" is a kind of prolegomena to Mollenkopf's analysis of New York politics and to the study of urban politics generally. Analysts of city politics, he explains, fall into two camps – pluralists and structuralists – and the recent history of urban political theory has been a back-and-forth struggle between the contending conceptual frameworks.

To begin with, writes Mollenkopf, an extensive body of theory by both sociologists and political scientists argued that local decision-making is dominated by entrenched elites. In opposition to this prevailing orthodoxy, a number of pluralist scholars went into the field where no elitist model of governance could easily explain what they saw. Noting that almost one-quarter of the gross national product of the United States "passes through the public sector . . . much of it through urban governments," pluralist political analysts tended to see urban politics "as an autonomous realm that possessed real authority and commanded important resources."

But in an intellectual counterattack, a new body of structuralist theory arose out of the social upheavals of the 1960s and 1970s that argued that the pluralist dispersion of power was mostly illusion, especially to inner-city minority communities, and that the imperatives of capitalism, in both the economic and social structural realms, repeatedly and inevitably established the basic parameters of local development policy.

After exploring the two poles of urban political theory, and providing a catalog of their principal exponents, Mollenkopf proposes a synthesis that simultaneously avoids the pitfalls of one-dimensionality and recognizes the legitimate claims of each school. "How can we develop a vocabulary for analyzing politics and state action," he asks, "that reconciles the political system's independent impact on social outcomes with its observed systemic bias in favor of capital?" And he returns to his initial question – whether urban politics is worth studying at all – and answers in the affirmative. Although the structuralists are correct in stressing the importance of underlying economic forces, they typically lack "a well developed theory of the state," and thus Mollenkopf concludes that "economy-centered theorizing" must always be tempered by "polity-centered" thinking.

This selection is from *A Phoenix in the Ashes: The Rise and Fall of the Koch Coalition in New York City Politics* (Princeton, NJ: Princeton University Press, 1992). Other books by John Mollenkopf include *New York City in the 1980s: A Social, Economic, and Political Atlas* (New York: Simon & Schuster, 1993), *Power, Culture, and Place: Essays on New York City* (New York: Russell Sage, 1988), *The Contested City* (Princeton, NJ: Princeton University Press, 1983), and *Dual City: Restructuring New York* (New York: Russell Sage, 1991) co-edited with Manuel Castells.

The classic elitist/structuralist study of urban community power is Floyd Hunter's, *Community Power Structure* (New York: Anchor, 1953). The classic pluralist study of urban community power is Robert Dahl's, *Who Governs?* (New Haven, CT: Yale University Press, 1961).

Other studies of urban politics include Douglas Yates, *The Ungovernable City* (Cambridge, MA: MIT Press, 1977), Paul E. Peterson, *City Limits* (Chicago: University of Chicago Press, 1981), Barbara Ferman, *Governing the Ungovernable City* (Philadelphia: Temple University Press, 1985), John R. Logan and Harvey L. Molotch, *Urban Fortunes: The Political Economy of Place* (Berkeley and Los Angeles: University of California Press, 1987), Timothy Barnekov, Robin Boyle, and Daniel Rich, *Privatism and Urban Policy in Britain and the United States* (Oxford: Oxford University Press, 1989), and John R. Logan and Todd Swandstrom, *Beyond the City Limits: Urban Policy and Economic Restructuring in Comparative Perspective* (Philadelphia: Temple University Press, 1990).

What is the appropriate way to conceptualize the organization of political power in New York City during the Koch era? The dialogue between the pluralist interpreters of urban power and their structuralist critics has produced a rich variety of answers to this question. In the early 1960s, pluralist political scientists launched an attack on the previously accepted view, established by sociologists, that socioeconomic elites dominated urban politics. The success of this assault enabled pluralists to establish their view as the norm in political science.

From the mid-1970s onward, however, a new generation of structurally oriented critics challenged the pluralist point of view. While they were able to undermine the prevailing wisdom, they did not manage to supplant it with a new one, in part because of defects in their arguments that pluralists were quick to point out. More recently, students of urban politics have attempted to synthesize the strengths of both approaches. With respect to framing the study of how the Koch administration amassed and exercised political power, the debate between pluralists and their critics focuses our attention on four interrelated questions:

1 Is urban politics worth studying at all, or is the urban political realm so subordinate to, dependent on, and constrained by its economic and social context that factors from this domain have little independent explanatory power?

2 If urban politics does have an independent impact, how should we conceptualize power relations among interests or actors?

3 In particular, what factors govern the construction of a dominant political coalition within a given set of structural constraints and opportunities?

4 In constructing such a coalition, how important is promoting private investment compared to

other strategies, such as increasing social spending to incorporate potentially insurgent groups?

THE PLURALIST CONCEPTION OF THE URBAN POLITICAL ORDER

The classic pluralist studies of a generation ago, like Banfield's *Political Influence*, Dahl's *Who Governs?*, or Sayre and Kaufman's *Governing New York City*, made important theoretical and methodological advances over the so-called elitists they attacked. They did not deduce power relations from the interlocks between economic and political elites. Instead, they went into the field to examine the tangled complexity of interest alignments around actual policy decisions and disputes. Pluralist scholars showed that no model of direct control by a unified economic or status elite could easily explain what they saw.

While most pluralists did not dwell theoretically on the larger relationship between the state and the economy, they implicitly rejected the notion that some underlying structural logic subordinated local politics to the private economy. They saw politics as an autonomous realm that possessed real authority and commanded important resources. They explicitly rejected the notion that economic or social notables controlled the state in any instrumental sense. Since they argued that every "legitimate" group commanded some important resource (if only the capacity to resist) and no one group commanded sufficient resources to control all others, pluralists argued that the bargaining among a multiplicity of groups defined the urban power structure.

In this view, coalition building was central to the definition of power. Political leaders and private interests built coalitions around specific issues, the coalitions varied from issue to issue, and they tended to be short lived. By selecting a range of different policy decisions as case studies for research, pluralists seemed to imply that urban development and social service issues had an equal importance in organizing political competition.

In the face of examples where entrenched interest groups dominated their own particular, fragmented policy areas over time to the exclusion of the public interest, the pluralist approach developed a clearly critical strand of analysis. But these scholars simply saw the dark side of the pluralist worldview without fundamentally challenging its basic assumptions or deflating the optimistic claims about system openness or responsiveness prevailing among other pluralists.

[. . .]

While the pluralist studies may have been convincing and accurate portraits of urban politics in the 1950s and early 1960s, the eruption of turmoil and political mobilization in the 1960s and the fiscal crisis of the 1970s soon revealed basic flaws in the pluralist analysis. Except for Robert Dahl's work, *Who Governs?*, these studies lacked a context in economic and political development. Despite obligatory opening chapters covering economic, social, and political trends, pluralist studies such as Sayre and Kaufman's did not treat the changing structure of urban economies or racial succession as problematic for the urban political order. It would, they thought, simply absorb and adapt to these changes. While Dahl provided a fine treatment of the transition from patrician dominance to what he argued were the dispersed inequalities of pluralist democracy in New Haven, he also failed to see that blacks might be led to challenge the system, not just participate in it as a minor interest. Neither Dahl nor his colleagues foresaw how economic transormation and racial succession might fundamentally challenge the previously observed "normal" patterns.

[. . .]

Dahl explicitly denied that economic and social inequalities would overlap and reinforce each other in the political arena. Other pluralist scholars also did not recognize the possibility that nonelite elements of the urban population would feel systematically excluded from power and would react by pressing for greater representation and more vigorously redistributive policies. As a result, the urban battles that erupted in the latter 1960s in New York City and elsewhere made their relatively tranquil picture of urban politics as a kind of market equilibrium-reaching mechanism seem anachronistic.

STRUCTURALIST CRITIQUES

As the pluralist political equilibrium unraveled on the ground, it came under increasing challenge from structuralist critics. The broad outlines of their progress may be traced from Peter Bachrach and Morton Baratz's classic essay on the "two faces of power" to Clarence Stone's work on "systemic power" to John Manley's "class analysis of pluralism." Bachrach and Baratz attacked pluralists for focusing on the "first face" of power, namely its exercise, while ignoring the second, namely the way that the relationship between the state and the underlying socioeconomic system shapes the political agenda. "Power may be, and often is," they said, "exercised by confining the scope of decision-making to relatively 'safe' issues." But while making a case for analyzing how the values embedded in institutional practices bias the rules of the game, they do not specify the mechanisms that promote some interests and issues while dampening others.

Stone advanced this line of thought by shifting the locus of analysis from decisions ("market exchange") toward the mechanisms that create systemic or strategic advantages for some interests over others ("production"). The un equal distribution of private resources, he argues, creates a differential capacity among political actors to shape the flow of benefits from the basic rules of the game, the construction of particular agendas, and the making of specific decisions. Business, in particular, derives systemic power not only from its juridical status and economic resources but from its attractiveness as an ally for those who advance any policy change and from the shared subculture from which private and public officials both emerged.

Despite the structuralist leaning in his concept of "systemic power," Stone did not break decisively with the pluralist interplay of interests around decisions. Manley's Marxist critique does make this break. He embraced the argument that the legal and structural primacy enjoyed by private ownership of capital requires the state to reinforce the systemic inequalities that result from the drive for private profit. He attacked pluralists, even the later work of Dahl and Lindblom that concedes that business enjoys a privileged position in pluralist competition, for lacking a theory of exploitation and, hence, an objective standard of a just or equal distribution. In Manley's view, the juridical protection of private property inevitably commits the state to control workers and promote capital.

While neo-Marxist work similar to Manley's stressed the systematic subordination of the state and politics to capital accumulation and the private market, a parallel and quite nonradical strand of public choice analysis reached quite similar conclusions. Focusing on the notion that cities compete to attract well-off residents and private investment, this line of analysis . . . is logically quite similar to some neo-Marxist critiques of pluralism.

NEO-MARXIST CRITIQUES

Structuralists have decisively transcended the pluralist vocabulary. They provided the social and economic context missing from pluralism and highlighted the ways that private property, market competition, wealth and income inequality, the corporate system, and the stage of capitalist development pervasively shape the terrain on which political competition occurs. They underscored the need to analyze how basic patterns of the economic, political, or cultural rules of the game bias the capacity of different interests to realize their ends through politics and the state.

Most importantly, neo-Marxist structuralists were able to empirically investigate these mechanisms, refuting the pluralist retort that "nondecisions" either must be studied just like decisions or else are unobservable ideological constructs. They have shown cases in which the systemic and cumulative inequality of political capacity undergirded, and indeed was ideologically reinforced by, a superficial pluralism. Structuralist studies may be flawed by economic determinism, but they are factually on target in observing and describing mechanisms that generate systemic, cumulative, political inequality, which has a more profound impact on outcomes than the coalition patterns studied by pluralists. Such critiques won relatively broad support among the younger generation of scholars, if not their elders. They may be subclassified into theories that stress the political logic of capital accumulation, social control, or the interplay of accumulation and legitimation. Each offers a different perspective on the central mechanisms that generate cumulative political inequality.

Theorists influenced by Marx's economic works have tended to argue that the mode of production stamps its pattern more or less directly on the organization of the state and on the dynamics of political competition. Marxists as different as David Harvey and David Gordon have both argued that the stage of capitalist development and the circuits of capital have determined urban spatial patterns, the bureaucratic state, and for Harvey even urban consciousness. While this strand of Marxist thinking made a breakthrough in orienting analysts to the importance of the process of capital accumulation, it has generally lacked a well-developed theory of the state that either identifies the instrumental mechanisms that link state actions to the power of capital or grants the relative autonomy to the state.

This literature does stress one mechanism, however: the state's dependence on private investment for public revenues. If the mobility of capital can discipline the state and constrain political competition, then competition among polities (whether cities or nations) to attract investment leads them to grant systematic benefits for capital, a dynamic that Alford and Friedland have called "power without participation." As Harvey wrote,

> The successful urban region is one that evolves the right mix of life-styles and cultural, social, and political forms to fit with the dynamics of capital accumulation ... Urban regions racked by class struggle or ruled by class alliances that take paths antagonistic to accumulation ... at some point have to face the realities of competition for jobs, trade, money, investments, services, and so forth.

Sooner or later, the state and political competition will be subordinated to the needs of capital.

Several analysts, including Friedland and Palmer as well as Molotch and Logan, abstracted this mechanism from the larger Marxian vocabulary and made it central to their analysis of urban power. Friedland and Palmer argued that, while businesses do directly influence policy-making, such intervention is logically secondary. "The growth of locales depends on the fortunes of their firms," according to Friedland and Palmer, thus "dominant and mobile [corporate] actors set the boundaries within which debate over public policy takes place." As capital has become more mobile and less tied to specific locations, the need for business to intervene directly in politics has waned, while the structural subordination of local government to the general interests of business has waxed.

Molotch and Logan took a different tack on the same course. While conceding that the mobility of capital gives local government a powerful incentive to defer to capitalists, they argued that certain classes of business are not mobile: real estate developers, utilities, newspapers, and others with a fixed relationship to a place. Large sunk costs give these interests a powerful incentive to intervene in and dominate local politics in order to get local government to promote new investment. They saw this "growth machine" as a ubiquitous, inevitable, and at best weakly challenged feature of American cities ...

[. . .]

While this strand of thinking argued that the multiplicity of competing local governments forces the state to reproduce and protect basic features of the advanced capitalist economy, a second, equally important school of neo-Marxist thinking stressed the way urban politics serves to dampen and regulate the conflicts inevitably generated by capitalist urbanization. Castells's work on "collective consumption" and urban social movements, Piven and Cloward's studies of urban protest, and Katznelson's studies of the absorptive capacity of local bureaucracies and the bias against class issues in urban politics represent the best of this work.

While these analysts differed over how the state coopts movements that challenge urban governments, they share the idea that this pro cess is a central feature of urban politics in advanced capitalist societies. Not everyone, even on the left, has agreed with these contentions. Theret, Mingione, and Gottdiener have criticized the explanatory power of the notion of collective consumption, while Ceccarelli has argued that urban social movements did not turn out to be the force in west European urban politics that Castells portrayed them to be. Whatever the situation in

Europe, the civil rights movement, urban unrest, and community organization clearly had a profound impact on urban politics in the United States after the 1960s, particularly in the rise of programs designed to absorb and deflect these forces.

[. . .]

PUBLIC CHOICE CRITIQUES

Neo-Marxist thinking is not the only source of structural criticism of the pluralist paradigm, however. Microeconomics, in the form of public choice theory, has contributed its own critique. Tiebout's seminal work led to Forrester's simulation of urban systems and ultimately to Paul Peterson's sophisticated "unitary" theory of urban politics. This tradition, born of the economists' distrust of state allocation of resources, has sought a functional equivalent to the marketplace in the multiplicity of local governments. They would compete, Tiebout argued, for residents of different means and desires by providing different service packages at various tax costs. An equilibrium would thus be reached in the sorting of populations across urban and suburban jurisdictions within the metropolis. This equilibrium would represent an efficient production of public goods, matching the marginal prospective resident with the jurisdiction's need to add (or subtract) residents on its own margin to provide services at the most efficient scale.

Such thinking has undergirded much of the orthodox literature on urban economics and local public finance. Urban housing, for example, has been analyzed as a function of how consumers trade off housing and commuting costs, given various levels of residential amenities. Forrester built the underlying assumptions into a model, influential for a time, that implied that whatever cities do to provide housing or social services for the poor will attract more of them, drive out the better off, and erode the tax base.

[. . .]

STRUCTURALISM RECONSIDERED

By providing the missing economic and social-structural context, these structuralist critiques achieved a considerable advance over pluralist analysis. Cities can no longer be taken as independent entities isolated from the larger economic and social forces that operate on them. Analysts can no longer ignore the impact of global and national economic restructuring on large central cities. Since cities cannot retard these global economic trends (though New York and others may propagate them), nor remake their populations at will, they clearly navigate in a sea of externally generated constraints and imperatives.

The structuralist critiques also make it clear that urban politics can no longer be considered to be unrelated to the cumulative pattern of inequality in the economy and society. They have focused attention on how the state's dependence on private investment fosters political outcomes that systematically favor business interests. Structuralists have explored specific mechanisms that produce this result, such as the invidious competition among fragmented, autonomous urban governments for investment, the segregation of local government functions into quasi-private agencies that promote investment and politically exposed agencies that absorb and deflect protest, and the organization of the channels of political representation so as to articulate interests in some ways but not others. By stressing that advanced capitalism characteristically generates urban social movements and political conflicts, some structuralists have also implied that political action can alter some of the constraints capitalism imposes on democracy.

Despite these strengths, however, structuralist perspectives also have grave flaws. The assertion that the state "must" undertake activities that favor capital tends to be functionalist. Such a standpoint begs the question of how these "imperatives" are put in place and reproduced over time, which inevitably must be through the medium of politics. As a result, structuralists may not see that political actors can fail to fulfill or to maximize their supposed imperatives. Dominant urban political coalitions have certainly done things that cost them elections and the ability to exercise power; they have persisted in increasing the tax burden on private capital and imposing exactions on private developers even after the point that they diminish further investment. Others have chosen to increase budget deficits and risk their bond ratings.

Given the right conditions, nothing is inevitable about an administration's pursuit of electoral success, private investment, well-managed social tensions, or even good bond ratings. Nothing guarantees that city government will be willing or able to fulfill the functions structuralists have assigned to it. As Piven and Friedland observed in rejecting a "smoothly functioning determinism," a structural analysis cannot be adequate until it specifies "the political processes through which . . . systematic imperatives are translated into government policies."

Structural critiques also tend not to be disconfirmable. For example, if structuralists argue that the use of legal injunctions by the conservative Republican administrations before the New Deal illustrates how the state supports capitalism, while the New Deal's recognition and promotion of trade unions also illustrate state support for capitalism, then they are explaining everything and nothing. Put another way, structural theories tend to have a hard time explaining the real and important variation over time and across places. The basic features of capitalism are common across nations and evolve slowly, while the political outcomes that capitalism is supposed to drive are highly varied and change more quickly.

Finally, when structuralists appeal to an ultimate economic determinism, they aggravate these problems. Agency fades out of the analytic picture. To be sure, the most attractive variants of neo-Marxism sought to avoid this trap by using the concepts of the "relative autonomy of the state" and "conflicting imperatives of accumulation and legitimation" to introduce a political dimension into an otherwise inadequate economic determinism. Yet in their discomfort with granting politics a co-equal causal role, even these variants ultimately retreat to the view that politics is subordinate to economics: autonomy is after all only relative. From Gottdiener at one end of the spectrum to Peterson on the other, structuralist analysts have made no bones about calling politics analytically irrelevant in the face of the economy's ability to constrain and impel.

To summarize, for all their strengths, structuralists conceptualized the political system as ultimately subordinate to economic structure. They tended to reduce urban politics to the fulfillment of economic imperatives; even social control achieved through political means serves capitalist ends. The most promising threads of structuralist thinking examined how systemic imperatives might conflict with each other or generate system-threatening conflict, thus opening the way for political indeterminacy. Here, however, they risked moving outside and beyond a structuralist paradigm. Indeed, orthodox Marxism (or for that matter orthodox neoclassical economics) simply does not provide a good basis for building a theory of politics. To the extent that structuralist theorists held true to the logic of their argument, they underplayed the importance of politics. They did not appreciate that policies that promote private investment must be constructed in a political environment that may favor but by no means guarantees this outcome. Indeed, popular, social, and communal forces pressure the state and the political process just as strongly in different, and often opposed, directions.

This tendency to trivialize politics removes a way to explain why outcomes vary even though capitalism is constant. States may be constrained, but they are also sovereign. They exercise a monopoly on the legitimate use of force, establish the juridical basis for private property, and shape economic development in myriad ways. Economies are delicate. They depend on political order and have been deformed or smashed by political disorder. State actions may be conditioned by economic structure, but they cannot be reduced to it. Many substantially different capitalisms are possible, and politics determine which ones evolve. Just as the state is dependent on the economy, economic institutions depend on and are vulnerable to the state and its changing political circumstances.

[. . .]

"Polity-centered" thinking must thus augment the "economy-centered" theorizing of the structuralist critiques. This does not require an equally one-sided political determinism. Rather, it requires us to extend the lines of structuralist thinking that stress conflict among imperatives or developmental tendencies until we go beyond the limits of economic determinism. We must recognize that "state power is *sui generis*, not reducible to class power," as Block put it. Or as Manuel Castells recently reflected, "experience was right and Marxist theory was wrong" about the central theoretical importance of urban social movements and the impossibility of reducing them to a class basis.

But if we give politics an analytic weight equal to that of economic structure, how can we avoid returning to a voluntaristic pluralism? How can we develop a vocabulary for analyzing politics and state action that reconciles the political system's independent impact on social outcomes with its observed systemic bias in favor of capital? A satisfactory approach must operate at three interrelated levels: (1) how the local state's relationship to the economy and society conditions its capacity to act; (2) how the "rules of the game" of local politics shape the competition among interests and actors to construct a dominant political coalition able to exercise that capacity to act; and (3) how economic and social change and the organization of political competition shape the mobilization of these interests.

TOWARD A THEORETICAL SYNTHESIS

We can begin to build such an approach by recognizing that city government and its political leaders interact with the resident population and constituency interests in its political and electoral operating environment and with market forces and business interests in its economic operating environments. This approach emphasizes two primary interactions: first, between the leaders of city government and their political/electoral base; and second, between the leaders of city government and their economic environment. It also suggests that political entrepreneurs who seek to direct the actions of city government must contend with three distinct sets of interests: (1) public sector producer interests inside local government; (2) popular or constituency interests (which are also public sector consumer interests), especially as they are organized in the electoral system; and (3) private market interests, particularly corporations with discretion over capital investment, as they are organized in the local economy.

To be sure, these interests are highly complex in a city like New York and cannot be captured by simple dichotomies like black versus white or capitalist versus worker . . . The city's residential communities are highly heterogeneous. Terms like "minority" hide far more than they reveal; even "black" or "Latino" blur important distinctions regarding nativity and ethnicity. Business interests come in many sizes, industries, and competitive situations; even corporate elites vary greatly. Still, a focus on the relationships among state, citizenry, and marketplace provides an entry point for analyzing what deter mines the shape of the urban political arena.

The concept of a "dominant political coalition" gives us a focal point for this analysis. A dominant political coalition is a working alliance among different interests that can win elections for executive office and secure the cooperation it needs from other public and private power centers in order to govern. To have an opportunity to become dominant, it must first win election to the chief executive office. To remain dominant, it must use the powers of government to consolidate its electoral base, win subsequent elections, and gain support from those other wielders of public authority and private resources whose cooperation is necessary for state action to go forward. Put another way, a dominant coalition must organize working control over both its political and its private market operating environments.

This formulation improves on the pluralist approach by directing our attention toward how the relationship between politics and markets biases outcomes in favor of private market interests, as structuralist approaches have pointed out. The notion of a dominant political coalition would not sit well with pluralists, who have argued that coalitions are unstable, form or re-form according to the issue, and may be stymied by the capacity of any sizable group to resist. We posit instead that coalitions can be stable, operate across issues, and create persistent winners and losers. Challenging and supplanting such coalitions have generally been difficult, particularly for constituencies that lack resources or are particularly vulnerable to sanction. Effective challenges generally arise only at moments of crisis in periods of rapid social and economic change.

This formulation also improves on the structuralist approach by according the political/electoral arena an influence equal to that of economic forces. It also points us toward how strategies to control the direction of city government are shaped by (and in turn shape) the political environment and by the public sector producer interests that have a permanent stake in its operation. It posits a scope for political choice and innovation that is lacking in the structuralist perspective.

This approach points us toward the following central questions: how do political entrepreneurs seek to organize such coalitions, what enables them to succeed in the first instance, and how do they sustain success over time? In what ways can such coalitions be bound together? What interests do dominant coalitions include and exclude and why? How do the economic and political contexts affect these binding relationships? And what tensions or conflicts undermine dominant coalitions, opening the way for power realignment?

As structuralists have shown, one part of the answer to these questions lies in the relationship of politics to the structure of economic interests. Efforts to explore this relationship may be found in Stone's studies of Atlanta, Shefter's study of New York, and my own work on pro-growth coalitions in Boston and San Francisco. Stone distinguished three levels at which to analyze the relationship between a dominant coalition and various urban interests. The least interesting is the pluralist domain of individual decisions or "command power" in which one actor induces or coerces others to follow his or her bidding. The two others are more relevant to this analysis.

Political actors wield "coalition power" when they join together to exercise the policy powers of the state to produce a steady flow of benefits to their allies, without the need for coercing or inducing specific actions. Stone showed how the Atlanta regime used public and private subcontracts to minority business enterprises to cement its political support, but he gave relatively little attention to other aspects of how the coalition tried to dominate its political operating environment.

Instead, he emphasized the "preemptive" or "systemic power" enjoyed by private interests whose command over private resources is so great as to make their support crucial to the dominant coalition. Among the mechanisms of preemptive power in Atlanta, Stone identified the unity of a well-organized downtown business community, newspaper and television support for policies that favored downtown development, the reliance of politicians on campaign contributions from developers, business control over the equity and credit that government needed to carry out its plans, and the business community's ability to provide or deny access to upward mobility for the black middle class. These systemic powers made corporate interests ideal allies for politicians seeking to achieve and sustain political dominance.

While the arrival of a black majority in the electorate, militant new black leadership, and neighborhood mobilization eventually destabilized and modified the tradition of white dominance that prevailed in Atlanta until Maynard Jackson was elected mayor in 1973, Stone argued that they did not overturn the preemptive power of corporate interests. Jackson's successor, Andrew Young, chose to abandon the fragmented and undisciplined neighborhood movement in favor of pro-growth politics with a new face, consolidated by white business support for set-asides to minority entrepreneurs.

[. . .]

My own study of how political entrepreneurs constructed pro-growth coalitions in Boston, San Francisco, and other large cities in the late 1950s and 1960s also advanced reasons why politicians would want to forge alliances with private sector elites. Promoting private development was an obvious way to bring together such otherwise disparate elements as a Republican corporate elite, regular Democratic party organizations, and reform-oriented rising public sector and nonprofit professionals.

These perspectives on how dominant political coalitions shape development politics and budget policy to secure business support, while convincing, remain incomplete. The "preemptive power" of business interests only explains part of how political actors construct a coalition to direct city government in the exercise of its powers. As Lincoln Steffens long ago observed, dominant coalitions needed to develop a grassroots base of legitimacy as well as support from elite interests. However much they may need corporate support, dominant coalitions must also have support from popular constituencies organized by such organizations as political parties, labor unions, and community organizations.

Mayors can lead dominant political coalitions only when they win electoral majorities and keep potential sources of electoral challenge fragmented or demobilized. The mobilization of blacks and Latinos and the neighborhood organization that began in the mid-1960s and continue today have prompted many currently dominant political coalitions to adopt policies that do not follow from a devotion to private market interests or public sector producer interests. For example, dominant coalitions must respond to mobilizations against the negative impacts of downtown growth and inadequate public services, or demands for government programs that provide upward mobility for excluded groups.

Rufus Browning, Dale Marshall, and David Tabb have argued, for example, that when insurgent liberal biracial coalitions came to power in a few of the northern California cities they studied, the coalitions shifted policies in favor of the formerly underrepresented groups that helped to elect them. Albert Karnig and Susan Welch's and Peter Eisinger's studies of the impact of black mayors reach a similar conclusion. My work on Boston and San Francisco showed that, while neighborhood protest and greater mayoral sensitivity toward neighborhood concerns did not halt the transformation of these two cities, they did produce numerous specific policy changes. In a comparative context, studies by Michael Aiken and Guido Martinotti and Edmond Preteceille suggest that left-wing local governments in Europe produced more progressive policy outputs. Castells concludes that "grassroots mobilization has been a crucial factor in the shaping of the city, as well as the decisive element in urban innovation against prevailing social interests."

[. . .]

In sum, the analysis of how Edward Koch and his allies constructed a new dominant political coalition in New York City must be framed in terms of three broad sets of factors. Building on the structuralists, it must understand how the local political system's interaction with private interests creates constraints and imperatives for the local state but also opportunities that astute political entrepreneurs can seize. Second, it must go beyond the structuralists by recognizing that how popular constituencies are organized in the city's political and electoral arena has an equally strong impact on the strategies pursued by coalition builders. Finally, a sound theory must be sensitive to how the organization of interests within the public sector, embodied in political practices as well as formal authority, also influenced their choices and actions . . .

MARGIT MAYER

"Post-Fordist City Politics"

from Ash Amin (ed.), *Post-Fordism: A Reader* (1994)

Editors' introduction Some political economists describe the new world order today as "post-Fordist." The following selection by German political scientist Margit Mayer introduces the notion of post-Fordism and describes what post-Fordist city politics are like.

According to Mayer and other post-Fordist theorists, auto magnate Henry Ford's revolution of the auto industry defined the "mode of production" that began shortly before World War I and continued through the 1960s. Work was mechanized. Machines replaced humans wherever possible. Workers performing the same task over and over again quickly and efficiently assembled auto components. One worker's sole job might be to attach left rear hubcaps while another was only responsible for inserting windshield wiper blades.

According to post-Fordist analysis, world capitalism is inherently unstable and prone to crises, both in production and in the role of the welfare state. By the 1970s the Fordist system appeared to be breaking down and capitalism entered a period of crisis which has led to global restructuring to produce a new paradigm of economic, social, and political order.

What is the role of local government in the new post-Fordist system? What should progressive politics be like in the new post-Fordist system? These are the questions Free University of Berlin politics professor Margit Mayer poses in the following selection. Based on her review of city politics in Western Europe and the United States, Mayer argues that local governments are assuming entirely new roles and ways of governing in response to post-Fordism. The new system poses both threats and opportunities for progressive politics.

Throughout Western Europe and the United States, cities are de-emphasizing redistributive policies. They are spending less on public services like bus systems and garbage collection. Some are privatizing such functions entirely. Most are also cutting welfare benefits and other social consumption expenditures, which are generally funded by the nation state, but delivered by city government.

At the same time as these cutbacks in the arenas of public services and social consumption expenditures, Mayer notes that local politics has become more important with respect to proactive economic development strategies. There has been a shift towards the "entrepreneurial city."

Today city governments are forming partnerships with private corporations aimed at economic development and partnering with nonprofit corporations to carry out some of the public service and social consumption functions they are divesting themselves of and to support the economic development projects they are promoting. Garbage collection, for example, has become a purely private function in many cities. Just as many cities are turning to private enterprise to carry out what were local government service functions, many cities are turning to nonprofits to build and operate assisted housing, run homeless shelters, and even to distribute welfare checks.

Emerging post-Fordist local government systems are ambiguous and in flux. Exactly what functions will be public and what will be private in the future is uncertain. How much funding nonprofits will get for economic development and job training programs is uncertain. What development will occur and on what terms is contested. In this context Mayer emphasizes the importance of local groups bargaining effectively. The emerging post-Fordist system offers both opportunity and danger.

Consider Michael Porter's views on the competitive advantage of the inner city (p. 278) in light of Mayer's description of post-Fordist city politics. How might nonprofits, citizens groups, and low-income and minority groups engaged in the kind of bargaining with local government and the private sector Mayer describes benefit from Porter's analysis? Read the selection by Sherry Arnstein (p. 240) on the "ladder" of citizen participation for insight into how citizen groups might better participate in post-Fordist city politics.

This selection is from Ash Amin (ed.), *Post-Fordism: A Reader* (Oxford and Cambridge, MA: Blackwell, 1995). Other books on post-Fordism and city government include Alain Lipietz, *Towards a New Economic Order: Post-Fordism, Ecology and Democracy* (Oxford: Oxford University Press, 1992), Alain Liepitz, *Mirages and Miracles: The Crises of Global Fordism* (London: Verso, 1987), Michael Storper and A. J. Scott (eds.), *Pathways to Industrialization and Regional Development* (London: Routledge, 1992), Bob Jessop *et al.*, *The Politics of Flexibility: Restructuring State and Industry in Britain, Germany and Scandinavia* (London: Routledge, 1992), and Ash Amin and Nigel Thrift (eds.), *Globalization, Institutions and Regional Development in Europe* (Oxford: Oxford University Press, 1994).

This chapter looks at local (urban) institutions and politics under post-Fordism. It first identifies the new practices and forms of urban governance observable in most Western European nations as well as in the United States over the past two decades. It assumes that we are in a transitional period of experimenting with ways that might resolve the current economic and welfare state crisis, to screen the changes that have occurred in the context of urban governance in order to examine whether they contribute to resolving the current dilemmas of urban politics in consistency with the logic of a new 'growth model' . . . For example, in terms of the organization of capital–labour relations, some countries have developed more negotiated involvement, while others have adopted less consensual flexibility strategies. Applying such a regulationist analysis to urban politics therefore points to strategic implications (spelled out in the final section of this chapter) for the social movements and actors working to develop democratic concepts for local politics and management.

CHANGES IN URBAN GOVERNMENT

Many changes have affected local politics over the past two decades, some of which have congealed into patterns common across national and regional particularities. At least three parallel trends have been identified in the recent literature on urban politics.

First, in all advanced Western nations local politics have gained in importance as a focus for proactive economic development strategies. The background for these developments is changes in capital mobility and shifts in the technological and social organization of production. One of the effects relevant for the local level has been that the changes have made it increasingly impossible for particular (re)production conditions to be organized or coordinated by the central state. While under Fordism local modes of regulation played a minor and subordinate role in assuring the coherence of the overall regime (the central state and other larger-scale modes of regulation played the crucial roles), efforts to respond to the crises of Fordism have involved a shift in this 'division of labour'. The specific local conditions of production and reproduction required by globally mobile capital cannot be orchestrated by the central state. Hence local political organizations, their skills in negotiating with supraregional and multinational capital, and the effectiveness with which they tailor the particular set of local conditions of production have become decisive factors in shaping a city's profile as well as its place in the international urban hierarchy.

Second, there has been an increasing mobilization of local politics in support of economic development and a concomitant subordination of social policies to economic and labour market policies. This shift in emphasis between different policy fields has often been labelled as a shift towards the 'entrepreneurial' city, and it goes hand in hand with a restructuring of the provision of social services. Both in the local economic interventionism and in the reorganization of public services the local state now involves other, non-governmental, actors in key roles.

This constitutes the third novel trend in urban governance, namely the expansion of the sphere of local political action to involve not only the local authority but a range of private and semipublic actors. To coordinate these various policy fields and functional interests, new bargaining systems have emerged, and new forms of public–private collaboration, in which the role of the local authority in respect of business and real estate interests, and the voluntary sector and community groups, is becoming redefined.

The first trend, the development of a 'perforated sovereignty' whereby nations become more open to trans-sovereign contacts by subnational governments, and regional/local forces become more active in advancing their own locational policy strategies oriented directly to the world market, is seen by many observers to contribute to a greater salience of the local state (as well as other local institutions of governance and economic relations . . .). 'Greater salience' does not mean greater strength, autonomy or a shift in the balance of central–local relations; in fact, local authorities have extended their strategic and active intervention at a time when they have been under increasing political pressure – in the UK there is even a question mark over their very survival. Despite or because of this, there is a resurgence of local politics, which provides the basis for the other two changes in urban governance, which I will present in some more detail.

SHIFTS IN EMPHASIS BETWEEN DIFFERENT POLICY FIELDS

Increased engagement of the local authority in economic development

With central government grants decreasing since the mid-1970s, local authorities have sought to respond to whatever restructuring problems were manifest in their region. In the declining old industrial areas, anti-unemployment programmes and local labour market policies were put into place: diverse strategies were explored to foster a more favourable business climate; many cities increased spending on culture and leisure facilities, or implemented strategies to upgrade the 'image' or the ambiance of a town. Some local governments seem to be aware of the increasingly polarized occupational and class structure of their cities and seek to counteract the attendant social disintegration with consciously chosen strategies to stimulate growth. From case studies we can gather that the urban leaders engaged in diverse local economic development activities were often far from certain as to how precisely an improvement in the course of urban development might be brought about, except in agreeing that 'industry and employment matters should be important'. Gradually, these activities have consolidated into a more systematic economic development policy strategy oriented explicitly towards nurturing 'growth' and, supposedly, employment.

This increased local economic interventionism is expressed not merely in the quantitative growth of local government spending for economic development, but, more importantly, in qualitatively different approaches to economic intervention, which seek to make use of indigenous skills and entrepreneurship, which emphasize innovation and new technologies, and which involve non-state actors in the organization of conditions for local economic development. While traditionally the economic development measures of local authorities would focus on attracting mobile capital (with conventional location inducements such as financial and tax incentives, infrastructure improvement or assistance with site selection), a shift in the approach of local economic development offices is now obvious. Subsidies are now targeted to industries promising innovation and growth; more public resources are focused on stimulating research, consulting and technology transfer, as well as on building alliances embracing universities, polytechnics, chambers of commerce and unions; land is no longer a cheap resource to be offered generously, but a precious one to be developed strategically. Instead of seeking to attract capital from elsewhere, strategies focus on new business formation and small business expansion; thus, instead of competing with other jurisdictions for the same investment, cities make efforts to strengthen existing and potential indigenous resources. Going beyond traditional booster campaigns used by development officials to publicize the virtues of their respective business climate, cities increasingly 'market' themselves in the global economy. Finally, the new development strategies frequently include employment strategies involving the so-called 'third' or 'alternative' sector.

These diverse efforts to mobilize and coordinate local potential for economic growth together have produced the effect of gradually undermining the traditional sharp distinctions between different policy areas. This is particularly true in the case of labour market and social policy domains, but equally, educational, environmental and cultural policies have become more integrated with, and are often part and parcel of, economic development measures. In addition, the new efforts have introduced institutional changes: new departments and interagency networks have been created within the administration, and new institutions which contribute in significant ways to the shaping of local politics have been established and/or supported outside of the local authority (e.g. urban development corporations, training and enterprise councils, technology centres, growth alliances, local 'round tables').

Restructuring and subordination of social consumption

In addition to the mobilization of local politics for economic development, whereby the local state seeks to organize private capital accumulation by including relevant private actors, the local state has also been significantly restructured in its public services and welfare functions (social consumption). The pressures exerted by economic restructuring and mass unemployment on the one hand and by shrinking subsidies from central government on the other and the willingness to accord priority to economic development policies have pushed into the background one of the formerly central functions of local state politics, namely the provision of social consumption goods and welfare services. Not only has local government spending for social consumption declined as a proportion of overall expenditure, but a qualitative restructuring has taken place involving an increase in the importance of non-state (private and voluntary sector) organizations or of public agencies directed by market criteria (quasi-governmental agencies) in the provision of public services. In various policy fields where the local state used to be the exclusive provider of a service, non-governmental agencies have been upgraded or private markets have emerged (e.g. in waste disposal). In urban renewal, environmental and social policies local authorities cooperate more and more frequently with neighbourhood initiatives, self-help or other social movement organizations.

As in the sphere of economic development, in the sphere of social reproduction once public sector led forms of service provision and management have been scaled down and complemented or replaced by a variety of private, voluntary and semi-public agencies and initiatives, parallel coordinating structures have begun to emerge. What is more, the traditional redistributive policies of the welfare state have been supplemented by employment and labour market policies designed to promote labour force flexibility. For example, in many cities attempts are being made to switch from unemployment compensation to job creation and retraining programmes, and to generate employment opportunities for specific social groups (which directly supplement or replace traditional welfare policy). A plethora of municipally funded programmes have been established in social, environmental and urban renewal policy domains, which tend to be hybrid programmes emphasizing workfare and job creation while burdening non-profit (third sector) organizations with the delivery and implementation of urban repair or social service functions. Though they are quantitatively rather insignificant, municipal employment and training programmes have served to mobilize and integrate the job-creating potentials from different policy areas. Active labour market policy measures of this kind therefore imply a blurring of the traditional distinction between economic and social policies, as they create a real link between the local economy and the local operation of the welfare state: welfare becomes increasingly redefined in the direction of the economic success of a local area.

This means that social welfare measures which used to be relatively universal and guaranteed by the national welfare state (but delivered by the local state) are now an arena of struggle, and are implemented in a fragmented fashion. This shift away from service provision through unitary and elected authorities towards more fragmented structures with increased involvement of local business, as well as of other private and voluntary sector agencies, has turned local government into merely one part – though perhaps the 'enabling' part – of broader 'growth coalitions'. Further, the new mix of unpaid self-service labour and private and public sector paid labour contributes to the development of a new consumption norm which supports the commodification and/or the self-servicing of welfare functions.

Thus, the new public–private forms of co-operation in the area of social consumption are also part of structural changes in the repertoire of municipal action. Whether the local struggles and bargaining processes will result in more egalitarian and accountable models responsive to broad local needs, or in divisive models enforcing processes of polarization and marginalization, one of the certain new characteristics of the emerging local 'welfare state' that distinguishes it from the past is its role in enabling negotiation with, and initiating activities by, 'outside' actors.

EXPANSION OF THE SPHERE OF LOCAL POLITICAL ACTION: NEW BARGAINING SYSTEMS AND PUBLIC–PRIVATE PARTNERSHIPS

The strategies developed to mobilize local potential for economic growth involve actors way beyond those of classical municipal politics. Labour market policy, for example, now involves not only the local authority, but also federal or national employment offices, individual state programmes (and their local participants), social welfare associations, churches, unions and in many cases individual companies and newly created consultancies. Urban development policy now involves private actors as early as the planning stage, while the local authority also has a say in implementation processes. And urban social programmes, emphasizing self-organized and community-based forms of social service provision, and relying on funds from diverse state and other sources, require novel types of cooperation between different municipal actors as well as between municipal and private agencies.

In these novel cooperation processes, spanning different policy fields and bringing together actors from very different backgrounds, bargaining systems have emerged which exhibit round-table structures and are characterized by a cooperative style of policy-making where, instead of giving orders, the local authority moderates or initiates cooperation. Such a non-hierarchical style seems to have been recognized as essential for identifying and acting on the intersecting areas of interest of the different actors. The novelty consists in the fact that bargaining and decision-making processes increasingly take place outside of traditional local government structures, and that urban governance becomes based on the explicit representation and coordination of functional interests active at the local level.

The actors participating in the definition and implementation of economic development and technological modernization programmes tend to be business associations, chambers of commerce, local companies, banks, research institutes, universities and unions. The restructuring of the local welfare state, on the other hand, has expanded the sphere of local political action to include an additional set of actors: welfare associations, churches and frequently grassroots initiatives and community organizations. Given the new employment structures, the growth of precarious and casualized job relations and structural long-term unemployment, the traditional distinction between these 'soft' and 'hard' policy spheres, however, has been eroded as municipal programmes seek to address 'social' problems in the context of economic development and labour market policies.

Alongside the new forms of public–private collaboration in economic development and in social service provision, explicit public–private partnerships have also emerged in urban renewal and urban physical development programmes. Faced with both tight budgets and increasing redevelopment tasks many city governments have explored new ways of planning and financing urban redevelopment. In order to upgrade their central business districts, to refashion old industrial sites and to develop attractive new projects, they have entered into partnerships with large investors, developers and consortia of private firms.

There is no 'typical' public–private partnership, but more or less intensive forms of co-operation and more or less traditional forms of partnership. The new partnership embraces a range of forms of collaboration, from mere transfer of subsidy from the local authority to particular firms, in which local government plays the role of a 'junior partner', to joint ventures where state and firms share risks and equity interests on a relatively equal footing. Partnership projects most frequently focus on the physical upgrading of a large area near the central business district, but increasingly they involve development planning and implementation in more neglected neighbourhoods, which include community development corporations and other neighbourhood-based groups.

In any event, the partnership rests on a 'deal' between the public and the private participants: in exchange for the local authority's subsidy, use of governmental powers (planning, assembling of properties, tax concessions), interpretation of government regulations (zoning, land usage) etc., the private partner is expected to meet certain project goals and to take on later management tasks. The private partner also has to share project returns with the local authority. This may occur through later lease or tax payments, through the provision of public infrastructure (e.g. subway stations), or through the hiring of local (often minority) workforces in project construction or maintenance.

Private investors gain from such a deal because the local authority's resources offer them attractive ways to expand their activities. In areas with intense physical development pressures, urban redevelopment provides highly profitable opportunities for private developers, who need access to promising real estate as well as land titles. Large investors, such as banks, insurance companies and construction contractors have recognized the potential of this municipal market for some time.

City governments gain from this deal because it allows them to attract more financial resources into urban development and to increase their effectiveness in achieving development goals. By combining public powers with entrepreneurial flexibility, organizational capacity and additional private (venture) capital, complex urban development tasks can be carried out more quickly and efficiently. Further, city governments can decrease their dependency on the national government and are able to tailor development more directly to particular local needs. Pressure on limited municipal administrative capacities is relieved and partnerships often work to increase the qualification and flexibilization of public administrations. In contrast to the total privatization of public tasks, the city retains, despite limited finances, some control and influence. In fact, over the years, public negotiators have become more skilled in obtaining concessions from developers and in holding private partners responsible for meeting performance obligations.

Nevertheless, this 'deal' between the public and private sectors contains a high level of ambiguity, as partnership schemes remain sites of continuing political and economic renegotiation: 'In effect, what is going on in partnerships is a version of the broader conflict over the future organization and scope of the public sector.' Precisely this ambiguous character, however, leaves space for a strategic role for local government and other 'public interest' organizations.

Both community-oriented partnerships and redevelopment partnerships in growth-promising central areas vary greatly in terms of their openness and responsiveness to affected interests, depending on local political traditions and prevailing balances of power. The more horizontal style of the new bargaining systems and project-specific partnerships does not necessarily imply greater openness to democratic influence or accountability to local social or environmental needs. On the contrary, the participants may form an exclusive group representing only selected interests. While there remain significant differences in the relative power of business, unions and community groups, as well as between 'established' community groups and more marginalized, unorganized interests, and while new bargaining systems and partnerships continue to vary in their inclusiveness, the new institutional relations and arenas of urban management have altered the political terrain and opportunities for all local political actors. Politics in the sense of arriving at and implementing binding decisions occurs more and more via negotiation and renegotiation between different public and private actors, both of whom are affected by the process of bargaining, as the partners try to 'move the objectives and culture of the other towards their own ideas'.

In these partnerships, the distinction between urban (re)development projects and economic development strategies, as described earlier, is increasingly blurry, especially in the case of community development projects or corporations, which now are typically as concerned with industrial and commercial development objectives as with housing and physical renewal. Such partnerships, which usually include some form of community representation, may offer services and technical assistance to local (small) businesses, run job placement services or help with developing export programmes for local businesses. They seek to tap whatever local economic development potential exists, thus contributing to the municipal strategy of mobilizing indigenous potential for economic growth and regeneration.

On the other hand, the expansion of development corporations concerned with improving housing and social conditions and the quality of life in neglected neighbourhoods may also be considered as part of the restructuring of the local welfare state along the lines described earlier. In the past, municipalities have used non-profit organizations to different degrees in different nations, primarily for the delivery of services. But since public funds for community development have dried up everywhere, broader partnerships have been forged, involving banks, investors and voluntary association – with community development corporations (CDCs) as catalytic actors within them. Now, they are involved in the planning as well as implementation of (social and physical) renewal of urban communities; their intermediary organizations and renewal agents combine social, environmental and revitalization work while also performing lobbying and political functions.

In addition to private (market) actors and public (state) actors, these partnerships importantly involve the so-called voluntary or third sector. A boom of 'third sector' literature reflects the 'discovery' of this sector at a time when politicians began to reconsider the division of labour between public and private sectors, and to examine ways of reducing state responsibility. However, while research identifies an explosive growth in the non-market, non-government organizations and activities lumped together under this label, less attention has been paid to the parallel penetration of this sector by the logic of the state and/or the market. Simultaneously, while traditional third sector organizations (previously dominated by the Fordist welfare state) tend increasingly to be run for profit like capitalist enterprises, newer organizations also function as elements in an alternative economy, which, in turn, is tied increasingly to municipal programmes. Both cases, while serving to make the welfare state more flexible through less rigid bureaucratic forms and more competition, also enlarge and restructure the sphere of local political action. In this expanded system of local politics the public sector reduces its functions, yet plays a more activist role in its interaction with the non-state sectors. No longer the centre of decision-making, as bargaining and decision-making processes occur outside traditional, local government structures, local government takes on the role of a moderator, managing the intersecting areas of interest and – in successful cases – exerting more leadership and control as it provides its resources on a conditional basis.

While, in the more traditional collaboration between the public and the private sector, cities would seek to attract investors with cheap land, low taxes and capital subsidies (without expecting to influence the firms' future behaviour and decisions), the recent urban economic development programmes focus public resources on firms and industries that promise growth, and they hold the private partner responsible for meeting contractual and other obligations. While in traditional urban development the redevelopment process was subject to approval by federal bureaucrats, now things are 'entirely up to the locality, where communities are mobilized or have gained access to City Hall, they have the potential to influence programs'.

Thus, the role of the municipality has changed from being the (more or less redistributive) local 'arm' of the welfare state to acting as the catalyst of processes of innovation and cooperation, which it seeks to steer in the direction of improving the city's (or community's) economic and social situation. These forms of cooperation are increasingly replacing state-provided functions to ensure social reproduction. In order to win the resources and competences of various private actors, the local authority has to respect to some degree the peculiar character and particular functional conditions of these nonstate organizations.

REGULATIONIST ANALYSIS: IDENTIFYING THE CONSTRAINTS AND OPTIONS FOR LOCAL POLITICS

A regulationist analysis helps in disentangling the implications of the identified changes in the forms and institutions of urban governance for political action. However, there has been a lot of confusion over 'post-Fordism' because the theoretical language of the regulationist framework, as originally developed by French political economists during the 1970s, has been adopted and redefined by many other writers. The British debate, in particular, has been massively influenced by a version put forth by the journal *Marxism Today*, which sees the breakdown of the monolithic methods of production under Fordism leading inevitably to the success of 'post-Fordist' political and social aims. This variant seeks to replace the old debate between the Left and Right by a new opposition between past and future (hence its use of the term New *Times*): the reorganization of production around new methods of flexible specialization is supposed to bear greater individual freedom and the end of centralized bureaucracies; post-Fordism is seen as a preordained successor to Fordism.

Against such a 'mistranslation' of the regulationist approach, which has influenced the debate concerning local state restructuring in Britain, this chapter draws on the original French analysis of post-Fordism, which offers a framework for assessing the possibility of *different* compromises under the conditions thrown up by a new accumulation regime and new social modes of regulation. While there are different theoretical *explanations within* the regulation approach, it is generally assumed that the Fordist regime of accumulation has been in crisis since the mid-1970s and that – without major restructuring and new modes of regulation – the crisis cannot be transformed into a new constellation of prosperity. By focusing on the correspondence between the system of accumulation and modes of social regulation, and by seeing the latter playing a crucial role in securing (temporary) stability and coherence in the capitalist system which is highly dynamic and in principle unstable, the regulation approach provides the opportunity to explore whether emerging elements of regulation are helping to resolve crisis tendencies and address the limits of the Fordist models and whether they contribute to securing the conditions for a post-Fordist 'virtuous circle' to operate. Further, by identifying the compatibility requirements of a new mode of regulation, the regulation approach allows us to explore the variety of options and scenarios theoretically possible within this mode and to recognize the conditions under which more progressive/democratic or more conservative/exclusionary models would emerge. The issue of compatibility and the issue of versions of post-Fordist models of regulation are discussed in turn below.

Compatibility

Can the new entrepreneurial local state outlined in the previous section be described as post-Fordist? If it can be shown that both the new forms of state intervention and the new institutional relations at the local level described in the first half of this chapter address the limits and solve the crises of the traditional model, and contribute to securing the conditions for a new growth model, then they may indeed be said to prefigure forms of urban governance capable of delivering a new coherent framework for urban management rather than being mere transitional forms of crisis management.

As we have seen, the new forms of economic intervention, which focus on competitiveness, seek to promote primarily technological innovation, *new* sectors, or new processes in established or restructured sectors, and thereby do address the problem of insufficient productivity; they move away from the traditional (Keynesian, central government led) interventionism designed to maintain levels of aggregate demand compatible with full employment (seeking, even, to maintain employment in declining sectors), which contributed to the stagflation of the 1970s and to disrupting the Fordist growth dynamic. Further, with its restructuring of social welfare in the direction of subordinating welfare policy to the demands of flexible labour markets and structural competitiveness, and of promoting more flexible and innovative provision of collective consumption, the entrepreneurial local state not only reduces social consumption expenditure (which had triggered the fiscal crisis of the Keynesian welfare state), but also reorients social policy away from generalizing the norms of mass consumption and the forms of collective consumption that supported the Fordist growth dynamic. Instead, a fragmented and potentially highly uneven provision of social consumption – tied to economic performance – is established, depending on the skills, political priorities and mobilization of local political actors.

The new *institutional relations* also contribute to resolving crisis tendencies of the traditional local state by replacing the overbearing, hierarchical state with a more pluralistic and, in some ways, more egalitarian version. This reorganization of the local political system reflects the new requirement to make connections between different policy areas, in particular between economic and technology policies, and policies on education, manpower training, infrastructural provision and so on. The sphere of local political action has been expanded: local unions, chambers of commerce, investors, education bodies and research centres have entered into partnership arrangements of different kinds with the local state to regenerate the local economy, and new bargaining systems based on negotiation have evolved. These local networks and bargaining systems address the limits of the centralized, hierarchical, bureaucratic-corporative structures that were characteristic of the Fordist state and that ended up producing huge costs, inefficiency and waste, as well as protest by new social movements. Further, the distribution of territorial management activities among a *range* of private and semi-public agencies as well as local government might prove more capable of contributing to stable reproduction under the new conditions of sharpened interregional and intercity competition.

Identifying these features of compatibility in the local mode of regulation implies that we can equate the requirements for 'local economic integrity' or for 'success' more widely than with mere institutional capacity (i.e. the presence of many institutions of different kinds, with high levels of interaction and an awareness of a common enterprise). 'Institutional thickness' would be the condition for success, but specific institutions compatible with and oriented towards supporting the emerging regime of accumulation must be present. The examples of the North of England or the Ruhr Valley illustrate that the presence of countless regeneration-oriented institutions (*and* national government subsidies) do not bring about successful regional restructuring. It appears that the persistence of 'old-fashioned' unions, strong Keynesian welfare institutions and long-entrenched social-democratic labour coalitions is blocking rather than aiding the generation of a local institutional framework conducive to successful restructuring. Thus, the 'new mechanisms for attaining some form of local economic integrity' may be captured more adequately than in terms of institutional thickness by focusing on, as regulation theory does, the necessary correspondence between the mode of social regulation and the structures of an emerging post-Fordist accumulation regime.

However, even though this new pattern of urban entrepreneurialism and partnership is on the agenda of all post-Fordist scenarios (liberal, progressive, conservative), it is quite another question whether social and political conflict will allow the actual establishment of these new arrangements as elements of a *dominant* mode of social regulation. The entrenched habits of those in power, routinized forms of party political competition, occasional powerful political support for declining sectors (where these are strong, the need for new products and processes does not get fully articulated) and institutional inertia are renowned as stumbling blocks to the actual implementation of strategies that are meanwhile widely applauded in political discourse. But institutions and policy interventions which do not take into account the constraints of the emerging accumulation regime and the elements of the new mode of regulation face the likelihood of failure and the huge costs associated with such failure.

Possible versions of post-Fordist modes of regulation

As indicated above, a variety of political platforms pursue this post-Fordist scenario: whether dominated by the Left or by the Right, city governments now commonly give priority to economic development policies (via the entrepreneurial mobilization of indigenous potential), thereby pushing one of the formerly central functions of local state politics, namely the provision of collective consumption goods and welfare services, into the background. This devolution or privatization of the local (welfare) state and its increased engagement in the arena of economic development tends to occur via new forms of negotiation and implementation privileging non-governmental (intermediary) organizations.

We find this basic model experimented everywhere, long-held political traditions notwithstanding. Subnational state intervention to encourage growth and employment is pursued even in the most liberal, so-called non-interventionist, environments, and the post-Fordist welfare-workfare state is present on the most diverse political agendas: the Right finds it attractive because it involves voluntary action and workfare, allowing state shrinkage; the Left because it is 'enabling' people to exercise power for themselves; and the liberals because it emphasizes local community action. Furthermore, as we have seen, new bargaining structures have become a reality in many different cities even if they contrast starkly in terms of their inclusiveness and responsiveness with regard to interests outside those in the central business district, real estate and the large investor sector. In addition, cleavages have become apparent not just between neighbourhoods and large developers or large firms, but also between newly included community interests and groups peripheral to the new arrangements. In any event, city governments can play a more initiating and more active role than in the past, and local state activity is bound to reflect the power struggles and political conflicts within a locality.

In other words, more or less democratic versions of this basic model are possible – without seriously restraining the transition to post-Fordism. Indeed, it is not a requirement that the new institutions, in order to contribute to a new temporary stability, must prefigure political empowerment within localities, nor is it the case that the new bargaining structures per se are more biased towards private business than the old form of urban governance, which emphasized the separation between public benefit and private profit. Concrete developments and the degree of responsiveness and openness of versions will depend on how actors at the local level seize and struggle over the opportunities and forms provided within this basic model.

So what should the (environmental and democratic) movements be arguing for as key elements and practices of urban management? Different proposals have been put on the table: Lipietz . . . for instance, argues for the creation of a new sector dedicated to socially useful tasks of the kind which are provided expensively by the welfare state, by unpaid female work or not at all. Others argue for a strengthening of national redistributive policies and for challenging the political ideology 'that eschews state ownership of housing and industry'.

Our analysis, however, shows that the situation has already become more complicated. The 'new alternative sector' envisioned by Lipietz to be the way forward is already a widespread practice within many of the municipal programmes which are tying third sector groups and their polyvalent work to state employment policies. It has already differentiated into a multilayered, conflictual set of arrangements, subject to the pressures of both market and public sector demands, and characterized by internal tensions and cleavages. The task for movements, then, is not to create such an 'alternative sector' but to make it accessible to and resourceful for marginalized groups threatened by the powerful polarization processes of post-Fordism. Social movements need to use the new channels and forums provided by the new bargaining systems to challenge the powerful post-Fordist trend towards inequality and to attack its social divisions and its political forms of exclusion in order to strengthen the democratic potential of the new forms of urban governance.

On the other hand, while one may conceive of national strategies that redistribute resources from wealthy to poor areas or groups as a prop for democratic movements, demands for such national projects are improbable today, given the national welfare state's overbearing form (which contributed to the crisis of Fordism) and the disappearance of the preconditions for a Fordist 'deal' embracing the big social blocks (unions, employers, the state). In any case, such demands have to confront the erosion ('hollowing out') of the nation state form, 'especially in its Keynesian welfare state guise'. These trends have to be taken into account, so a more appropriate strategy might be to make use of the forms and structures that have become available at the subnational level.

Instead of hanging on to 'old-fashioned' large-scale, nationally oriented strategies, instead of demanding unspecified third sector or community representation, social movements will need to use their own card within the structure of the new bargaining systems. Since urban governance has become based on the representation of functional interests active at the local level, and since the local authority has to respect to some degree the particular functional characteristics of the other actors involved in the new 'partnerships', and since all the involved participants control resources that are necessary for the policies to be effective, even social movement groups have a real basis for negotiation. But 'negotiation' may be a mild term for the struggle at hand. The emerging post-Fordist regime, with the new social modes of regulation including the new forms of urban governance described in this chapter, may function with some temporary stability, but it poses enormous long-term problems of social disintegration. The emphasis on economic innovation and competition, and the subordination of all social programmes to these economic priorities, will tend to produce deep divisions in society and threaten the decay of civil society (which, of course, in the long run causes difficulty for economic stability). Given emerging increasingly polarized class relations and the fragmented local situations, social movements need to mobilize to create pressure on the local authority, first, to develop strategic plans that make every effort to avoid social segregation and marginalization, and, second, to use the resources of large private investors to meet local social and environmental needs. If they manage to seize the opportunities and spaces provided by the new, fragmented political arrangements, they may yet influence the concrete shape of the post-Fordist development path.

SHERRY ARNSTEIN

"A Ladder of Citizen Participation"

***Journal of the American Institute of Planners* (1969)**

Editors' introduction The above selections by Mollenkopf (p. 219), and Mayer (p. 229) argue that local government is important and that plural actors can influence the outcome of policies and programs that affect their lives. They depict local government as an increasingly important part of a new global order where public, private, and nonprofit sectors work together in complex regimes. This new order raises local citizens' stakes in having their interests taken into account by local decision-makers. Guidance as to how this might best be done comes from a classic article by Sherry Arnstein titled "A Ladder of Citizen Participation."

Arnstein had firsthand experience with neighborhood organizations and the tumultuous urban politics of the 1960s. She uses the metaphor of a ladder to describe gradations of "citizen participation" in urban programs that affect their lives. She makes clear her own personal commitment to a redistribution of power from haves to have-nots by empowering the poor and powerless.

At the lowest level of Arnstein's ladder are two forms of nonparticipation, which she terms *manipulation* and *therapy*. According to Arnstein, some governmental organizations have contrived phony forms of participation, which are really aimed at educating citizens to accept a predetermined course of action. Tied for bottom of the ladder is another form of nonparticipation, which Arnstein identifies as therapy. Arnstein brands this form of nonparticipation both dishonest and arrogant. Here the intent is to "cure" participants of attitudes and behaviors an agency itself does not like under the guise of seeking their advice.

Legitimate, but low, rungs of the ladder are *informing* and *consultation*. Informing citizens of the facts about a government program and their rights, responsibilities and options is a good first step, particularly if it is designed to go beyond a one-way flow of information. Consultation – getting citizens' opinions – is also helpful if it is genuine. Surveys, for example, may provide real input from citizens to decision-makers, but if that is the only form of participation it would not go far in assuring that citizen views carry weight. *Placation* – in which government gives in to some citizen demands – goes a step further. But a model in which weak citizens only get part of what they want to placate them is not satisfactory.

The highest rungs on Arnstein's ladder are *partnership*, two rungs from the top, *delegated power*, one rung below the top, and *citizen control* at the very top. During the "war on poverty" in the 1960s some citizen groups were given control over programs, or were delegated power to carry out programs. Citizen control and delegated power have been rare since that time. Opponents of citizen control advance many of the arguments that Arnstein identified – that citizen control arguably balkanizes public services, may be costly and inefficient, can enable citizen hustlers to be opportunistic, and may be symbolic politics. Today public/private/nonprofit partnerships are popular. Arnstein places true partnerships relatively high on her eight-rung ladder because they represent a redistribution of power arrived at through negotiation.

Where the odd bedfellows of local government, private corporations, and neighborhood nonprofit community-based organizations form joint planning and decision-making structures, citizen views have real weight.

Both Sherry Arnstein and Paul Davidoff were engaged liberals who wrote their classic statements about citizen participation and advocacy planning in the late 1960s. Compare the approach of Davidoff, the lawyer, who argues in favor of skilled professionals, advocating on behalf of powerless clients (p. 423) with Arnstein, the social work professional who favors empowering individuals and communities by involving them directly in planning and decision-making.

Other books on citizen participation in urban planning and programs include Milan J. Dluhy, *Building Coalitions in the Human Services* (Thousand Oaks, CA: Sage, 1990), Michael Fagence, *Citizen Participation in Planning* (Oxford and New York: Pergamon Press, 1977), Ronald L. Thomas, Mary C. Means, Margaret A. Grieve, and Neil R. Peirce, *Taking Charge: How Communities are Planning their Futures: A Special Report on Long Range/Strategic Planning Trends and Innovations* (Washington, DC: International City/County Management Association, 1988).

Peter Marris and Martin Rein's *Dilemmas of Social Reform*, 2nd edn. (Chicago: University of Chicago Press, 1982) describes community-based urban programs and articulates a philosophy of social change that influenced US urban policy in the 1960s. Two views on the US "war on poverty" are Sar Levitan, *The Great Society's Poor Law* (Baltimore: Johns Hopkins, 1969), and Daniel Patrick Moynihan, *Maximum Feasible Misunderstanding* (New York: Free Press, 1969). The US Model Cities program, its antecedents and the initial phase of the successor Community Development Block Grant program are discussed in Bernard J. Frieden and Marshal Kaplan, *The Politics of Neglect: Urban Aid from Model Cities to Revenue Sharing* (Cambridge, MA: MIT Press, 1975).

Books on public participation in urban planning and programs in Europe include James Barlow, *Public Participation in Urban Development: The European Experience* (Washington, DC: Brookings, 1995), Her Majesty's Stationery Office, *Community Involvement in Planning and Development Processes* (London: HMSO, 1995), and Albert Mabileau, *Local Politics and Participation in Britain and France* (Cambridge: Cambridge University Press, 1990).

The idea of citizen participation is a little like eating spinach: no one is against it in principle because it is good for you. Participation of the governed in their government is, in theory, the cornerstone of democracy – a revered idea that is vigorously applauded by virtually everyone. The applause is reduced to polite handclaps, however, when this principle is advocated by the have-not blacks, Mexican Americans, Puerto Ricans, Indians, Eskimos, and whites. And when the have-nots define participation as redistribution of power, the American consensus on the fundamental principle explodes into many shades of outright racial, ethnic, ideological, and political opposition.

There have been many recent speeches, articles, and books which explore in detail *who* are the have-nots of our time. There has been much recent documentation of *why* the have-nots have become so offended and embittered by their powerlessness to deal with the profound inequities and injustices pervading their daily lives. But there has been very little analysis of the content of the current controversial slogan: "citizen participation" or "maximum feasible participation." In short: *What* is citizen participation and what is its relationship to the social imperatives of our time?

Citizen participation is citizen power

Because the question has been a bone of political contention, most of the answers have been purposely buried in innocuous euphemisms like "self-help" or "citizen involvement." Still others have been embellished with misleading rhetoric like "absolute control" which is something no one – including the President of the United States – has or can have. Between understated euphemisms and exacerbated rhetoric, even scholars have found it difficult to follow the controversy. To the headline reading public, it is simply bewildering.

My answer to the critical *what* question is simply that citizen participation is a categorical term for citizen power. It is the redistribution of power that enables the have-not citizens, presently excluded from the political and economic processes, to be deliberately included in the future. It is the strategy by which the have-nots join in determining how information is shared, goals and policies are set, tax resources are allocated, programs are operated, and benefits like contracts and patronage are parceled out. In short, it is the means by which they can induce significant social reform which enables them to share in the benefits of the affluent society.

Figure 1 French student poster. In English, "I participate, you participate, he participates, we participate, you participate . . . they profit."

EMPTY REFUSAL VERSUS BENEFIT

There is a critical difference between going through the empty ritual of participation and having the real power needed to affect the outcome of the process. This difference is brilliantly capsulized in a poster painted last spring [1968] by the French students to explain the student-worker rebellion. (See Figure 1.) The poster highlights the fundamental point that participation without redistribution of power is an empty and frustrating process for the powerless. It allows the powerholders to claim that all sides were considered, but makes it possible for only some of those sides to benefit. It maintains the status quo. Essentially, it is what has been happening in most of the 1,000 Community Action Programs, and what promises to be repeated in the vast majority of the 150 Model Cities programs.

Types of participation and "nonparticipation"

A typology of eight *levels* of participation may help in analysis of this confused issue. For illustrative purposes the eight types are arranged in a ladder pattern with each rung corresponding to the extent of citizens' power in determining the end product. (See Figure 2.)

The bottom rungs of the ladder are (1) *Manipulation* and (2) *Therapy.* These two rungs describe levels of "non-participation" that have been contrived by some to substitute for genuine participation. Their real objective is not to enable people to participate in planning or conducting programs, but to enable powerholders to "educate" or "cure" the participants. Rungs 3 and 4 progress to levels of "tokenism" that allow the have-nots to hear and to have a voice: (3) *Informing* and (4) *Consultation.* When they are proffered by powerholders as the total extent of participation, citizens may indeed hear and be heard. But under these conditions they lack the power to insure that their views will be *heeded* by the powerful. When participation is restricted to these levels, there is no follow-through, no "muscle," hence no assurance of changing the status quo. Rung (5) *Placation* is

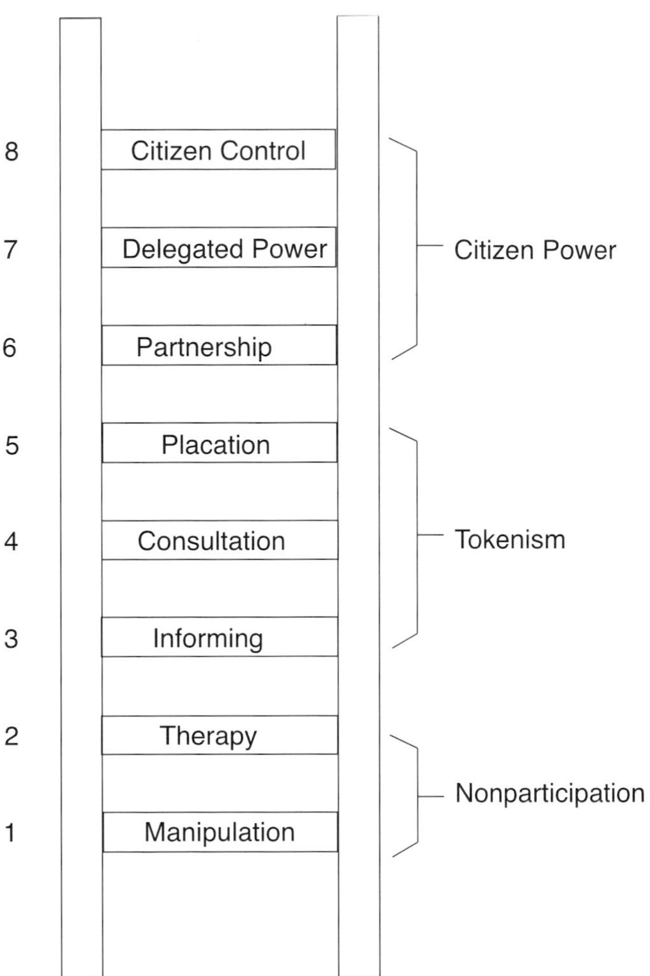

Figure 2 Eight rungs on the ladder of citizen participation

simply a higher level tokenism because the groundrules allow have-nots to advise, but retain for the powerholders the continued right to decide.

Further up the ladder are levels of citizen power with increasing degrees of decision-making clout. Citizens can enter into a (6) *Partnership* that enables them to negotiate and engage in trade-offs with traditional power holders. At the topmost rungs, (7) *Delegated Power* and (8) *Citizen Control*, have-not citizens obtain the majority of decision-making seats, or full managerial power.

Obviously, the eight-rung ladder is a simplification, but it helps to illustrate the point that so many have missed – that there are significant gradations of citizen participation. Knowing these gradations makes it possible to cut through the hyperbole to understand the increasingly strident demands for participation from the have-nots as well as the gamut of confusing responses from the powerholders.

Though the typology uses examples from federal programs such as urban renewal, anti-poverty, and Model Cities, it could just as easily be illustrated in the church, currently facing demands for power from priests and laymen who seek to change its mission; colleges and universities which in some cases have become literal battlegrounds over the issue of student power; or public schools, city halls, and police departments (or big business which is likely to be next on the expanding list of targets). The underlying issues are essentially the same – "nobodies" in several arenas are trying to become "somebodies" with enough power to make the target institutions responsive to their views, aspirations, and needs.

LIMITATIONS OF THE TYPOLOGY

The ladder juxtaposes powerless citizens with the powerful in order to highlight the fundamental divisions between them. In actuality, neither the have-nots nor the powerholders are homogeneous blocs. Each group encompasses a host of divergent points of view, significant cleavages, competing vested interests, and splintered subgroups. The justification for using such simplistic abstractions is that in most cases the have-nots really do perceive the powerful as a monolithic "system," and powerholders actually do view the have-nots as a sea of "those people," with little comprehension of the class and caste differences among them.

It should be noted that the typology does not include an analysis of the most significant roadblocks to achieving genuine levels of participation. These roadblocks lie on both sides of the simplistic fence. On the powerholders' side, they include racism, paternalism, and resistance to power redistribution. On the have-nots' side, they include inadequacies of the poor community's political socioeconomic infrastructure and knowledge-base, plus difficulties of organizing a representative and accountable citizens' group in the face of futility, alienation, and distrust.

Another caution about the eight separate rungs on the ladder: In the real world of people and programs, there might be 150 rungs with less sharp and "pure" distinctions among them. Furthermore, some of the characteristics used to illustrate each of the eight types might be applicable to other rungs. For example, employment of the have-nots in a program or on a planning staff could occur at any of the eight rungs and could represent either a legitimate or illegitimate characteristic of citizen participation. Depending on their motives, powerholders can hire poor people to coopt them, to placate them, or to utilize the have-nots' special skills and insights. Some mayors, in private, actually boast of their strategy in hiring militant black leaders to muzzle them while destroying their credibility in the black community.

Characteristics and illustrations

It is in this context of power and powerlessness that the characteristics of the eight rungs are illustrated by examples from current federal social programs.

1. MANIPULATION

In the name of citizen participation, people are placed on rubberstamp advisory committees or advisory boards for the express purpose of "educating" them or engineering their support. Instead of genuine citizen participation, the bottom rung of the ladder signifies the distortion of participation into a public relations vehicle by powerholders.

This illusory form of "participation" initially came into vogue with urban renewal when the socially elite were invited by city housing officials to serve on Citizen Advisory Committees (CACs). Another target of manipulation were the CAC subcommittees on minority groups, which in theory were to protect the rights of Negroes in the renewal program. In practice, these subcommittees, like their parent CACs, functioned mostly as letterheads, trotted forward at appropriate times to promote urban renewal plans (in recent years known as Negro removal plans).

At meetings of the Citizen *Advisory* Committees, it was the officials who educated, persuaded, and advised the citizens, not the reverse. Federal guidelines for the renewal programs legitimized the manipulative agenda by emphasizing the terms "information-gathering," public relations," and "support" as the explicit functions of the committees.

This style of nonparticipation has since been applied to other programs encompassing the poor. Examples of this are seen in Community Action Agencies (CAAs) which have created structures called "neighborhood councils" or "neighborhood advisory groups." These bodies frequently have no legitimate function or power. The CAAs use them to "prove" that "grassroots people" are involved in the program. But the program may not have been discussed with "the people." Or it may have been described at a meeting in the most general terms; "We need your signatures on this proposal for a multiservice center which will house, under one roof, doctors from the health department, workers from the welfare department, and specialists from the employment service."

The signatories are not informed that the $2 million-per-year center will only refer residents to the same old waiting lines at the same old agencies across town. No one is asked if such a referral center is really needed in his neighborhood. No one realizes that the contractor for the building is the mayor's brother-in-law, or that the new director of the center will be the same old community organization specialist from the urban renewal agency.

After signing their names, the proud grassrooters dutifully spread the word that they have "participated" in bringing a new and wonderful center to the neighborhood to provide people with drastically needed jobs and health and welfare services. Only after the ribbon-cutting ceremony do the members of the neighborhood council realize that they didn't ask the important questions, and that they had no technical advisors of their own to help them grasp the fine legal print. The new center, which is open 9 to 5 on weekdays only, actually adds to their problems. Now the old agencies across town won't talk with them unless they have a pink paper slip to prove that they have been referred by "their" shiny new neighborhood center.

Unfortunately, this chicanery is not a unique example. Instead it is almost typical of what has been perpetrated in the name of high-sounding rhetoric like "grassroots participation." This sham lies at the heart of the deep-seated exasperation and hostility of the have-nots toward the powerholders.

One hopeful note is that, having been so grossly affronted, some citizens have learned the Mickey Mouse game, and now they too know how to play. As a result of this knowledge, they are demanding genuine levels of participation to assure them that public programs are relevant to their needs and responsive to their priorities.

2. THERAPY

In some respects group therapy, masked as citizen participation, should be on the lowest rung of the ladder because it is both dishonest and arrogant. Its administrators – mental health experts from social workers to psychiatrists – assume that powerlessness is synonymous with mental illness. On this assumption, under a masquerade of involving citizens in planning, the experts subject the citizens to clinical group therapy. What makes this form of "participation" so invidious is that citizens are engaged in extensive activity, but the focus of it is on curing them of their "pathology" rather than changing the racism and victimization that create their "pathologies."

Consider an incident that occurred in Pennsylvania less than one year ago. When a father took his seriously ill baby to the emergency clinic of a local hospital, a young resident physician on duty instructed him to take the baby home and feed it sugar water. The baby died that afternoon of pneumonia and dehydration. The overwrought father complained to the board of the local Community Action Agency. Instead of launching an investigation of the hospital to determine what changes would prevent similar deaths or other forms of malpractice, the board invited the father to attend the CAA's (therapy) child-care sessions for parents, and promised him that someone would "telephone the hospital director to see that it never happens again."

Less dramatic, but more common examples of therapy, masquerading as citizen participation, may be seen in public housing programs where tenant groups are used as vehicles for promoting control-your-child or cleanup campaigns. The tenants are brought together to help them "adjust their values and attitudes to those of the larger society." Under these ground rules, they are diverted from dealing with such important matters as: arbitrary evictions; segregation of the housing project; or why is there a three-month time lapse to get a broken window replaced in winter.

The complexity of the concept of mental illness in our time can be seen in the experiences of student/civil rights workers facing guns, whips, and other forms of terror in the South. They needed the help of socially attuned psychiatrists to deal with their fears and to avoid paranoia.

3. INFORMING

Informing citizens of their rights, responsibilities, and options can be the most important first step toward legitimate citizen participation. However, too frequently the emphasis is placed on a one-way flow of information – from officials to

citizens – with no channel provided for feedback and no power for negotiation. Under these conditions, particularly when information is provided at a late stage in planning, people have little opportunity to influence the program designed "for their benefit." The most frequent tools used for such one-way communication are the news media, pamphlets, posters, and responses to inquiries.

Meetings can also be turned into vehicles for one-way communication by the simple device of providing superficial information, discouraging questions, or giving irrelevant answers. At a recent Model Cities citizen planning meeting in Providence, Rhode Island, the topic was "tot-lots." A group of elected citizen representatives, almost all of whom were attending three to five meetings a week, devoted an hour to a discussion of the placement of six tot-lots. The neighborhood is half black, half white. Several of the black representatives noted that four tot-lots were proposed for the white district and only two for the black. The city official responded with a lengthy, highly technical explanation about costs per square foot and available property. It was clear that most of the residents did not understand his explanation. And it was clear to observers from the Office of Economic Opportunity that other options did exist which, considering available funds, would have brought about a more equitable distribution of facilities. Intimidated by futility, legalistic jargon, and prestige of the official, the citizens accepted the "information" and endorsed the agency's proposal to place four lots in the white neighborhood.

4. CONSULTATION

Inviting citizens' opinions, like informing them, can be a legitimate step toward their full participation. But if consulting them is not combined with other modes of participation, this rung of the ladder is still a sham since it offers no assurance that citizen concerns and ideas will be taken into account. The most frequent methods used for consulting people are attitude surveys, neighborhood meetings, and public hearings.

When powerholders restrict the input of citizens' ideas solely to this level, participation remains just a window-dressing ritual. People are primarily perceived as statistical abstractions, and participation is measured by how many come to meetings, take brochures home, or answer a questionnaire. What citizens achieve in all this activity is that they have "participated in participation." And what powerholders achieve is the evidence that they have gone through the required motions of involving "those people."

Attitude surveys have become a particular bone of contention in ghetto neighborhoods. Residents are increasingly unhappy about the number of times per week they are surveyed about their problems and hopes. As one woman put it: "Nothing ever happens with those damned questions, except the surveyor gets $3 an hour, and my washing doesn't get done that day." In some communities, residents are so annoyed that they are demanding a fee for research interviews.

Attitude surveys are not very valid indicators of community opinion when used without other input from citizens. Survey after survey (paid for out of anti-poverty funds) has "documented" that poor housewives most want tot-lots in their neighborhood where young children can play safely. But most of the women answered these questionnaires without knowing what their options were. They assumed that if they asked for something small, they might just get something useful in the neighborhood. Had the mothers known that a free prepaid health insurance plan was a possible option, they might not have put tot-lots so high on their wish lists.

A classic misuse of the consultation rung occurred at a New Haven, Connecticut, community meeting held to consult citizens on a proposed Model Cities grant. James V. Cunningham, in an unpublished report to the Ford Foundation, described the crowd as large and mostly hostile:

> Members of The Hill Parents Association demanded to know why residents had not participated in drawing up the proposal. CAA director Spitz explained that it was merely a proposal for seeking Federal planning funds – that once funds were obtained, residents would be deeply involved in the planning. An outside observer who sat in the audience described the meeting this way:
>
> "Spitz and Mel Adams ran the meeting on their own. No representatives of a Hill group moderated or even sat on the stage. Spitz told the 300 residents that this huge meeting was an

example of 'participation in planning.' To prove this, since there was a lot of dissatisfaction in the audience, he called for a 'vote' on each component of the proposal. The vote took this form: 'Can I see the hands of all those in favor of a health clinic? All those opposed?' It was a little like asking who favors motherhood."

It was a combination of the deep suspicion aroused at this meeting and a long history of similar forms of "window-dressing participation" that led New Haven residents to demand control of the program.

By way of contrast, it is useful to look at Denver where technicians learned that even the best intentioned among them are often unfamiliar with, and even insensitive to, the problems and aspirations of the poor. The technical director of the Model Cities program has described the way professional planners assumed that the residents, victimized by high-priced local storekeepers, "badly needed consumer education." The residents, on the other hand, pointed out that the local storekeepers performed a valuable function. Although they overcharged, they also gave credit, offered advice, and frequently were the only neighborhood place to cash welfare or salary checks. As a result of this consultation, technicians and residents agreed to substitute the creation of needed credit institutions in the neighborhood for a consumer education program.

5. PLACATION

It is at this level that citizens begin to have some degree of influence though tokenism is still apparent. An example of placation strategy is to place a few hand-picked "worthy" poor on boards of Community Action Agencies or on public bodies like the board of education, police commission, or housing authority. If they are not accountable to a constituency in the community and if the traditional power elite hold the majority of seats, the have-nots can be easily outvoted and outfoxed. Another example is the Model Cities advisory and planning committees. They allow citizens to advise or plan ad infinitum but retain for powerholders the right to judge the legitimacy or feasibility of the advice. The degree to which citizens are actually placated, of course, depends largely on two factors: the quality of technical assistance they have in articulating their priorities; and the extent to which the community has been organized to press for those priorities.

It is not surprising that the level of citizen participation in the vast majority of Model Cities programs is at the placation rung of the ladder or below. Policy-makers at the Department of Housing and Urban Development (HUD) were determined to return the genie of citizen power to the bottle from which it had escaped (in a few cities) as a result of the provision stipulating "maximum feasible participation" in poverty programs. Therefore, HUD channeled its physical-social-economic rejuvenation approach for blighted neighborhoods through city hall. It drafted legislation requiring that all Model Cities' money flow to a local City Demonstration Agency (CDA) through the elected city council. As enacted by Congress, this gave local city councils final veto power over planning and programming and ruled out any direct funding relationship between community groups and HUD.

HUD required the CDAs to create coalition, policy-making boards that would include necessary local powerholders to create a comprehensive physical-social plan during the first year. The plan was to be carried out in a subsequent five-year action phase. HUD, unlike OEO, did not require that have-not citizens be included on the CDA decision-making boards. HUD's Performance Standards for Citizen Participation only demanded that "citizens have clear and direct access to the decision-making process."

Accordingly, the CDAs structured their policy-making boards to include some combination of elected officials; school representatives; housing, health, and welfare officials; employment and police department representatives; and various civic, labor, and business leaders. Some CDAs included citizens from the neighborhood. Many mayors correctly interpreted the HUD provision for "access to the decision-making process" as the escape hatch they sought to relegate citizens to the traditional advisory role.

Most CDAs created residents' advisory committees. An alarmingly significant number created citizens' policy boards and citizens' policy committees which are totally misnamed as they have either no policy-making function or only a very limited authority. Almost every CDA created about a dozen planning committees or task forces on functional lines: health, welfare, education, housing, and unemployment. In most cases, have-not citizens were invited to serve on these committees along with technicians from relevant public agencies. Some CDAs, on the other hand, structured planning committees of technicians and parallel committees of citizens.

In most Model Cities programs, endless time has been spent fashioning complicated board, committee, and task force structures for the planning year. But the rights and responsibilities of the various elements of those structures are not defined and are ambiguous. Such ambiguity is likely to cause considerable conflict at the end of the one-year planning process. For at this point, citizens may realize that they have once again extensively "participated" but have not profited beyond the extent the powerholders decide to placate them.

Results of a staff study (conducted in the summer of 1968 before the second round of seventy-five planning grants were awarded) were released in a December 1968 HUD bulletin. Though this public document uses much more delicate and diplomatic language, it attests to the already cited criticisms of non-policy-making policy boards and ambiguous complicated structures, in addition to the following findings:

1. Most CDAs did not negotiate citizen participation requirements with residents.

2. Citizens, drawing on past negative experiences with local powerholders, were extremely suspicious of this new panacea program. They were legitimately distrustful of city hall's motives.

3. Most CDAs were not working with citizens' groups that were genuinely representative of model neighborhoods and accountable to neighborhood constituencies. As in so many of the poverty programs, those who were involved were more representative of the upwardly mobile working-class. Thus their acquiescence to plans prepared by city agencies was not likely to reflect the views of the unemployed, the young, the more militant residents, and the hard-core poor.

4. Residents who were participating in as many as three to five meetings per week were unaware of their minimum rights, responsibilities, and the options available to them under the program. For example, they did not realize that they were not required to accept technical help from city technicians they distrusted.

5. Most of the technical assistance provided by CDAs and city agencies was of third-rate quality, paternalistic, and condescending. Agency technicians did not suggest innovative options. They reacted bureaucratically when the residents pressed for innovative approaches. The vested interests of the old-line city agencies were a major – albeit hidden – agenda.

6. Most CDAs were not engaged in planning that was comprehensive enough to expose and deal with the roots of urban decay. They engaged in "meetingitis" and were supporting strategies that resulted in "projectitis," the outcome of which was a "laundry list" of traditional programs to be conducted by traditional agencies in the traditional manner under which slums emerged in the first place.

7. Residents were not getting enough information from CDAs to enable them to review CDA developed plans or to initiate plans of their own as required by HUD. At best, they were getting superficial information. At worst, they were not even getting copies of official HUD materials.

8. Most residents were unaware of their rights to be reimbursed for expenses incurred because of participation – babysitting, transportation costs, and so on. The training of residents, which would enable them to understand the labyrinth of the federal-state-city systems and networks of subsystems, was an item that most CDAs did not even consider.

These findings led to a new public interpretation of HUD's approach to citizen participation. Though the requirements for the seventy-five "second-round" Model City grantees were not changed, HUD's twenty-seven page technical bulletin on citizen participation repeatedly advocated that cities share power with residents. It also urged CDAs to experiment with subcontracts under which the residents' groups could hire their own trusted technicians.

A more recent evaluation was circulated in February 1969 by OSTI, a private firm that entered into a contract with OEO to provide technical assistance and training to citizens involved in Model Cities programs in the northeast region of the country. OSTI's report to OEO corroborates the earlier study. In addition it states:

> In practically no Model Cities structure does citizen participation mean truly shared decision-making, such that citizens might view themselves as "the partners in this program. . . ."
>
> In general, citizens are finding it impossible to have a significant impact on the comprehensive planning which is going on. In most cases the staff planners of the CDA and the planners of existing agencies are carrying out the actual planning with citizens having a peripheral role of watchdog and, ultimately, the "rubber stamp" of the plan generated. In cases where citizens have the direct responsibility for generating program plans, the time period allowed and the independent technical resources being made available to them are not adequate to allow them to do anything more than generate very traditional approaches to the problems they are attempting to solve.
>
> In general, little or no thought has been given to the means of insuring continued citizen participation during the stage of implementation. In most cases, traditional agencies are envisaged as the implementors of Model Cities programs and few mechanisms have been developed for encouraging organizational change or change in the method of program delivery within these agencies or for insuring that citizens will have some influence over these agencies as they implement Model Cities programs . . . By and large, people are once again being planned *for*. In most situations the major planning decisions are being made by CDA staff and approved in a formalistic way by policy boards.

6. PARTNERSHIP

At this rung of the ladder, power is in fact redistributed through negotiation between citizens and powerholders. They agree to share planning and decision-making responsibilities through such structures as joint policy boards, planning committees and mechanisms for resolving impasses. After the groundrules have been established through some form of give-and-take, they are not subject to unilateral change.

Partnership can work most effectively when there is an organized power-base in the community to which the citizen leaders are accountable; when the citizens group has the financial resources to pay its leaders reasonable honoraria for their time-consuming efforts; and when the group has the resources to hire (and fire) its own technicians, lawyers, and community organizers. With these ingredients, citizens have some genuine bargaining influence over the outcome of the plan (as long as both parties find it useful to maintain the partnership). One community leader described it "like coming to city hall with hat on head instead of in hand."

In the Model Cities program only about fifteen of the so-called first generation of seventy-five cities have reached some significant degree of power-sharing with residents. In all but one of those cities, it was angry citizen demands, rather than city initiative, that led to the negotiated sharing of power. The negotiations were triggered by citizens who had been enraged by previous forms of alleged participation. They were both angry and sophisticated enough to refuse to be "conned" again. They threatened to oppose the awarding of a planning grant to the city. They sent delegations to HUD in Washington. They used abrasive language. Negotiation took place under a cloud of suspicion and rancor.

In most cases where power has come to be shared it was *taken by the citizens,* not given by the city. There is nothing new about that process. Since those who have power normally want to hang onto it, historically it has had to be wrested by the powerless rather than proffered by the powerful.

Such a working partnership was negotiated by the residents in the Philadelphia model neighborhood. Like most applicants for a Model Cities grant, Philadelphia wrote its more than 400 page application and waved it at a hastily called meeting of community leaders. When those present were asked for an endorsement, they angrily protested the city's failure to consult them on preparation of the extensive application. A community spokesman threatened to mobilize a neighborhood protest *against* the application unless the city agreed to give the citizens a couple of weeks to review the application and recommend changes. The officials agreed.

At their next meeting, citizens handed the city officials a substitute citizen participation section that changed the groundrules from a weak citizens' advisory role to a strong shared power agreement. Philadelphia's application to HUD included the citizens' substitution word for word. (It also included a new citizen prepared introductory chapter that changed the city's description of the model neighborhood from a paternalistic description of problems to a realistic analysis of its strengths, weaknesses, and potentials.) Consequently, the proposed policy-making committee of the Philadelphia CDA was revamped to give five out of eleven seats to the residents' organization, which is called the Area Wide Council (AWC). The AWC obtained a subcontract from the CDA for more than $20,000 per month, which it used to maintain the neighborhood organization, to pay citizen leaders $7 per meeting for their planning services, and to pay the salaries of a staff of community organizers, planners, and other technicians. AWC has the power to initiate plans of its own, to engage in joint planning with CDA committees, and to review plans initiated by city agencies. It has a veto power in that no plans may be submitted by the CDA to the city council until they have been reviewed, and any differences of opinion have been successfully negotiated with the AWC. Representatives of the AWC (which is a federation of neighborhood organizations grouped into sixteen neighborhood "hubs") may attend all meetings of CDA task forces, planning committees, or subcommittees.

Though the city council has final veto power over the plan (by federal law), the AWC believes it has a neighborhood constituency that is strong enough to negotiate any eleventh-hour objections the city council might raise when it considers such AWC proposed innovations as an AWC Land Bank, an AWC Economic Development Corporation, and an experimental income maintenance program for 900 poor families.

7. DELEGATED POWER

Negotiations between citizens and public officials can also result in citizens achieving dominant decision-making authority over a particular plan or program. Model City policy boards or CAA delegate agencies on which citizens have a clear majority of seats and genuine specified powers are typical examples. At this level, the ladder has been scaled to the point where citizens hold the significant cards to assure accountability of the program to them. To resolve differences, powerholders need to start the bargaining process rather than respond to pressure from the other end.

Such a dominant decision-making role has been attained by residents in a handful of Model Cities including Cambridge, Massachusetts; Dayton, and Columbus, Ohio; Minneapolis, Minnesota; St. Louis, Missouri; Hartford and New Haven, Connecticut; and Oakland, California.

In New Haven, residents of the Hill neighborhood have created a corporation that has been delegated the power to prepare the entire Model Cities plan. The city, which received a $117,000 planning grant from HUD, has subcontracted $110,000 of it to the neighborhood corporation to hire its own planning staff and consultants. The Hill Neighborhood Corporation has eleven representatives on the twenty-one-member CDA board which assures it a majority voice when its proposed plan is reviewed by the CDA.

Another model of delegated power is separate and parallel groups of citizens and powerholders, with provision for citizen veto if differences of opinion cannot be resolved through negotiation. This is a particularly interesting coexistence model for hostile citizen groups too embittered toward city hall – as a result of past "collaborative efforts" – to engage in joint planning.

Since all Model Cities programs require approval by the city council before HUD will fund them, city councils have final veto powers even when citizens have the majority of seats on the CDA Board. In Richmond, California, the city council agreed to a citizens' counter-veto, but the details of that agreement are ambiguous and have not been tested.

Various delegated power arrangements are also emerging in the Community Action Program as a result of demands from the neighborhoods and OEO's most recent instruction guidelines which urged CAAs "to exceed (the) basic requirements" for resident participation. In some cities, CAAs have issued subcontracts to resident dominated groups to plan and/or operate one or more decentralized neighborhood program components like a multipurpose service center or a Headstart program. These contracts usually include an agreed upon line-by-line budget and program specifications. They also usually include a specific statement of the significant powers that have been delegated, for example: policy-making; hiring and firing; issuing subcontracts for building, buying, or leasing. (Some of the subcontracts are so broad that they verge on models for citizen control.)

8. CITIZEN CONTROL

Demands for community controlled schools, black control, and neighborhood control are on the increase. Though no one in the nation has absolute control, it is very important that the rhetoric not be confused with intent. People are simply demanding that degree of power (or control) which guarantees that participants or residents can govern a program or an institution, be in full charge of policy and managerial aspects, and be able to negotiate the conditions under which "outsiders" may change them.

A neighborhood corporation with no intermediaries between it and the source of funds is the model most frequently advocated. A small number of such experimental corporations are already producing goods and/or social services. Several others are reportedly in the development stage, and new models for control will undoubtedly emerge as the have-nots continue to press for greater degrees of power over their lives.

Though the bitter struggle for community control of the Ocean Hill-Brownsville schools in New York City has aroused great fears in the headline reading public, less publicized experiments are demonstrating that the have-nots can indeed improve their lot by handling the entire job of planning, policy-making, and managing a program. Some are even demonstrating that they can do all this with just one arm because they are forced to use their other one to deal with a continuing barrage of local opposition triggered by the announcement that a federal grant has been given to a community group or an all black group.

Most of these experimental programs have been capitalized with research and demonstration funds from the Office of Economic Opportunity in cooperation with other federal agencies. Examples include:

1. A $1.8 million grant was awarded to the Hough Area Development Corporation in Cleveland to plan economic development programs in the ghetto and to develop a series of economic enterprises ranging from a novel combination shopping-center-public-housing project to a loan guarantee program for local building contractors. The membership and board of the nonprofit corporation is composed of leaders of major community organizations in the black neighborhood.

2. Approximately $1 million ($595,751 for the second year) was awarded to the Southwest Alabama Farmers' Cooperative Association (SWAFCA) in Selma, Alabama, for a ten-county marketing cooperative for food and livestock. Despite local attempts to intimidate the coop (which included the use of force to stop trucks on the way to market) first year membership grew to 1,150 farmers who earned $52,000 on the sale of their new crops. The elected coop board is composed of two poor black farmers from each of the ten economically depressed counties.

3. Approximately $600,000 ($300,000 in a supplemental grant) was granted to the Albina Corporation and the Albina Investment Trust to create a black-operated, black-owned manufacturing concern using inexperienced management and unskilled minority group personnel from the Albina district. The profitmaking wool and metal fabrication plant will be owned by its employees through a deferred compensation trust plan.

4. Approximately $800,000 ($400,000 for the second year) was awarded to the Harlem Commonwealth Council to demonstrate that a community-based development corporation can catalyze and implement an economic development program with broad community support and participation. After only eighteen months of program development and negotiation, the council will soon launch several large-scale ventures including operation of two supermarkets, an auto service and repair center (with built-in manpower training program), a finance company for families earning less than $4,000 per year, and a data processing company. The all black Harlem-based board is already managing a metal castings foundry.

Though several citizen groups (and their mayors) use the rhetoric of citizen control, no Model City can meet the criteria of citizen control since final approval power and accountability rest with the city council.

Daniel P. Moynihan argues that city councils are representative of the community, but Adam Walinsky illustrates the nonrepresentativeness of this kind of representation:

> Who ... exercises "control" through the representative process? In the Bedford-Stuyvesant ghetto of New York there are 450,000 people – as many as in the entire city of Cincinnati, more than in the entire state of Vermont. Yet the area has only one high school, and 80 per cent of its teenagers are dropouts; the infant mortality rate is twice the national average; there are over 8000 buildings abandoned by everyone but the rats, yet the area received not one dollar of urban renewal funds during the entire first 15 years of that program's operation; the unemployment rate is known only to God.
>
> Clearly, Bedford-Stuyvesant has some special needs; yet it has always been lost in the midst of the city's eight million. In fact, it took a lawsuit to win for this vast area, in the year 1968, its first Congressman. In what sense can the representative system be said to have "spoken for" this community, during the long years of neglect and decay?

Walinsky's point on Bedford-Stuyvesant has general applicability to the ghettos from coast to coast. It is therefore likely that in those ghettos where residents have achieved a significant degree of power in the Model Cities planning process, the first-year action plans will call for the creation of some new community institutions entirely governed by residents with a specified sum of money contracted to them. If the groundrules for these programs are clear and if citizens understand that achieving a genuine place in the pluralistic scene subjects them to its legitimate forms of give-and-take, then these kinds of programs might begin to demonstrate how to counteract the various corrosive political and socioeconomic forces that plague the poor.

In cities likely to become predominantly black through population growth, it is unlikely that strident citizens' groups like AWC of Philadelphia will eventually demand legal power for neighborhood self-government. Their grand design is more likely to call for a black city achieved by the elective process. In cities destined to remain predominantly white for the foreseeable future, it is quite likely that counterpart groups to AWC will press for separatist forms of neighborhood government that can create and control decentralized public services such as police protection, education systems, and health facilities. Much may depend on the willingness of city governments to entertain demands for resource allocation weighted in favor of the poor, reversing gross imbalances of the past.

Among the arguments against community control are: it supports separatism; it creates balkanization of public services; it is more costly and less efficient; it enables minority group "hustlers" to be just as opportunistic and disdainful of the have-nots as their white predecessors; it is incompatible with merit systems and professionalism; and ironically enough, it can turn out to be a new Mickey Mouse game for the have-nots by allowing them to gain control but not allowing them sufficient dollar resources to succeed. These arguments are not to be taken lightly. But neither can we take lightly the arguments of embittered advocates of community control – that every other means of trying to end their victimization has failed!

JAMES Q. WILSON AND GEORGE L. KELLING

"Broken Windows"

Atlantic Monthly (1982)

Editors' introduction Why is urban crime a problem in inner-city neighborhoods? What can and should government do about it? That is the subject of the following selection James Q. Wilson and George L. Kelling.

Wilson and Kelling studied police behavior in troubled neighborhoods – particularly in Newark, New Jersey. One of the authors (Kelling) spent many months walking Newark neighborhoods with local police observing what was going on in the neighborhoods and how the police handled neighborhood problems. Wilson and Kelling were particularly interested in police discretion and how the police dealt with troublesome behavior at the borderline between acceptable and unacceptable, legal and illegal. Wilson and Kelling are interested as much in neighborhood residents' perceptions of crime as in crime itself. A particular focus of their research and theory building is to develop a model of policing which neighborhood residents (at least those residents the authors term "decent folk") will support. Based on this research, they advanced a controversial theory of neighborhood transition and an influential model for community policing which has been implemented in many communities.

The central metaphor of this selection is a single broken window. Imagine a diverse city neighborhood like the area around Tomkins Square Park as David Harvey describes it (p. 199) a diverse, viable, exciting urban neighborhood with problems. Most of the people in the neighborhood are regulars – people who live or work there or are frequently in the area. Some are strangers. Most of the people in the Tompkins Square neighborhood are not criminals and do not engage in behavior that harms other people, though the behavior of some of the subcultures – punks, motorcycle club members, skateboarders – might scare conventional New Yorkers. Some of the people in the area, however, are dangerous and do engage in criminal behavior: drug dealing, prostitution, theft, assault and other serious crimes. Civil order exists, but it is fragile. Imagine that someone breaks a single window in the Tompkins Square Park area. According to Wilson and Kelling, how the police respond to that trivial, but unacceptable, act is fraught with consequences. One response is to do nothing. This, Wilson and Kelling argue, will signal that no one cares about unacceptable conduct. It will be an invitation for people to break more windows. Neighborhood residents will start avoiding one another, stop participating in neighborhood block parties and cower in their own homes.

Wilson and Kelling argue that historically the success of police activity was judged by whether it succeeded in maintaining order. Governments often tolerated (or encouraged) police to keep order through *informal means* without much regard to legal niceties. A drunk might have a *legal* right to sit on a neighborhood park bench. But if his presence sufficiently disturbed "decent" neighborhood residents, the local cop was encouraged to make him move along elsewhere. Gang members in Chicago's crime-ridden Robert Taylor homes might have a legal right to loiter by a playground, but the authors argue that project residents would want the local police to get rid of them ("kick ass" is the term one police officer uses). Wilson and Kelling take the controversial position that broad police discretion to enforce neighborhood standards is acceptable, even desirable.

Compare Wilson and Kelling's approach to Mike Davis's description of how he feels the Los Angeles police oppress low-income minorities (p. 193). How would you draw the line between "decent" and not-so-decent folk in the Tompkins Square Park neighborhood as David Harvey describes it (p. 199)? In a city as culturally heterogeneous as Zukin (p. 131) describes New York to be, whose culture should define neighborhood norms?

The argument in this selection is further developed in George L. Kelling and Catherine M. Coles, *Fixing Broken Windows* (New York: Martin Kessler, 1996). Other books by James Q. Wilson related to crime prevention include, co-edited with Joan Petersilia, *Crime* (San Francisco: ICS Press, 1995), *Crime and Public Policy* (San Francisco: ICS Press, 1983), co-edited with Michael Tonry, *Drugs and Crime* (Chicago: University of Chicago Press, 1990), co-edited with Glenn C. Loury, *Families, Schools, and Delinquency Prevention* (New York: Springer-Verlag, 1987), and, co-edited with David P. Farrington and Lloyd E. Ohlin, *Understanding and Controlling Crime* (New York: Springer-Verlag, 1986).

Books on community policing include Elizabeth M. Watson, Alfred R. Stone, and Stuart M. DeLuca, *Strategies for Community Policing* (Upper Saddle River, NJ: Prentice-Hall, 1998), Wesley G. Skogan and Susan M. Hartnett, *Community Policing, Chicago Style* (New York: Oxford University Press, 1997), Kenneth J. Peak and Ronald W. Glensor, *Community Policing and Problem Solving: Strategies and Practices* (Upper Saddle River, NJ: Prentice-Hall, 1996), Nigel Fielding, *Community Policing* (Oxford and New York: Oxford University Press, 1996).

In the mid-1970s, the state of New Jersey announced a "Safe and Clean Neighborhoods Program," designed to improve the quality of community life in twenty-eight cities. As part of that program, the state provided money to help cities take police officers out of their patrol cars and assign them to walking beats. The governor and other state officials were enthusiastic about using foot patrol as a way of cutting crime, but many police chiefs were skeptical. Foot patrol, in their eyes, had been pretty much discredited. It reduced the mobility of the police, who thus had difficulty responding to citizen calls for service, and it weakened headquarters control over patrol officers.

Many police officers also disliked foot patrol, but for different reasons: it was hard work, it kept them outside on cold, rainy nights, and it reduced their chances for making a "good pinch." In some departments, assigning officers to foot patrol had been used as a form of punishment. And academic experts on policing doubted that foot patrol would have any impact on crime rates; it was, in the opinion of most, little more than a sop to public opinion. But since the state was paying for it, the local authorities were willing to go along.

Five years after the program started, the Police Foundation, in Washington, DC, published an evaluation of the foot-patrol project. Based on its analysis of a carefully controlled experiment carried out chiefly in Newark, the foundation concluded, to the surprise of hardly anyone, that foot patrol had not reduced crime rates. But residents of the foot-patrolled neighborhoods seemed to feel more secure than persons in other areas, tended to believe that crime had been reduced, and seemed to take fewer steps to protect themselves from crime (staying at home with the doors locked, for example). Moreover, citizens in the foot-patrol areas had a more favorable opinion of the police than did those living elsewhere. And officers walking beats had higher morale, greater job satisfaction, and a more favorable attitude toward citizens in their neighborhoods than did officers assigned to patrol cars.

These findings may be taken as evidence that

the skeptics were right – foot patrol has no effect on crime; it merely fools the citizens into thinking that they are safer. But in our view, and in the view of the authors of the Police Foundation study (of whom Kelling was one), the citizens of Newark were not fooled at all. They knew what the foot-patrol officers were doing, they knew it was different from what motorized officers do, and they knew that having officers walk beats did in fact make their neighborhoods safer.

But how can a neighborhood be "safer" when the crime rate has not gone down – in fact, may have gone up? Finding the answer requires first that we understand what most often frightens people in public places. Many citizens, of course, are primarily frightened by crime, especially crime involving a sudden, violent attack by a stranger. This risk is very real, in Newark as in many large cities. But we tend to overlook or forget another source of fear – the fear of being bothered by disorderly people. Not violent people, nor, necessarily, criminals, but disreputable or obstreperous or unpredictable people: panhandlers, drunks, addicts, rowdy teenagers, prostitutes, loiterers, the mentally disturbed.

What foot-patrol officers did was to elevate, to the extent they could, the level of public order in these neighborhoods. Though the neighborhoods were predominantly black and the foot patrolmen were mostly white, this "order-maintenance" function of the police was performed to the general satisfaction of both parties.

One of us (Kelling) spent many hours walking with Newark foot-patrol officers to see how they defined "order" and what they did to maintain it. One beat was typical: a busy but dilapidated area in the heart of Newark, with many abandoned buildings, marginal shops (several of which prominently displayed knives and straight-edged razors in their windows), one large department store, and, most important, a train station and several major bus-stops. Though the area was run-down, its streets were filled with people, because it was a major transportation center. The good order of this area was important not only to those who lived and worked there but also to many others, who had to move through it on their way home, to supermarkets, or to factories.

The people on the street were primarily black; the officer who walked the street was white. The people were made up of "regulars" and "strangers." Regulars included both "decent folk" and some drunks and derelicts who were always there but who "knew their place." Strangers were, well, strangers, and viewed suspiciously, sometimes apprehensively. The officer – call him Kelly – knew who the regulars were, and they knew him. As he saw his job, he was to keep an eye on strangers, and make certain that the disreputable regulars observed some informal but widely understood rules. Drunks and addicts could sit on the stoops, but could not lie down. People could drink on side streets, but not at the main intersection. Bottles had to be in paper bags. Talking to, bothering, or begging from people waiting at the bus stop was strictly forbidden. If a dispute erupted between a businessman and a customer, the businessman was assumed to be right, especially if the customer was a stranger. If a stranger loitered, Kelly would ask him if he had any means of support and what his business was; if he gave unsatisfactory answers, he was sent on his way. Persons who broke the informal rules, especially those who bothered people waiting at bus stops, were arrested for vagrancy. Noisy teenagers were told to keep quiet.

These rules were defined and enforced in collaboration with the "regulars" on the street. Another neighborhood might have different rules, but these, everybody understood, were the rules for this neighborhood. If someone violated them, the regulars not only turned to Kelly for help but also ridiculed the violator. Sometimes what Kelly did could be described as "enforcing the law," but just as often it involved taking informal or extralegal steps to help protect what the neighborhood had decided was the appropriate level of public order. Some of the things he did probably would not withstand a legal challenge.

A determined skeptic might acknowledge that a skilled foot-patrol officer can maintain order but still insist that this sort of "order" has little to do with the real sources of community fear – that is, with violent crime. To a degree, that is true. But two things must be borne in mind. First, outside observers should not assume that they know how much of the anxiety now endemic in many big-city neighborhoods stems from a fear of "real" crime and how much from a sense that the street is disorderly, a source of distasteful, worrisome encounters. The people of Newark, to judge from their behavior and their remarks to interviewers, apparently assign a high value to public order, and feel relieved and reassured when the police help them maintain that order.

Second, at the community level, disorder and crime are usually inextricably linked, in a kind of developmental sequence. Social psychologists and police officers tend to agree that if a window in a building is broken *and is left unrepaired*, all the rest of the windows will soon be broken. This is as true in nice neighborhoods as in run-down ones. Window-breaking does not necessarily occur on a large scale because some areas are inhabited by determined window-breakers whereas others are populated by window-lovers; rather, one unrepaired broken window is a signal that no one cares, and so breaking more windows costs nothing. (It has always been fun.)

Philip Zimbardo, a Stanford psychologist, reported in 1969 on some experiments testing the broken-window theory. He arranged to have an automobile without license plates parked with its hood up on a street in the Bronx and a comparable automobile on a street in Palo Alto, California. The car in the Bronx was attacked by "vandals" within ten minutes of its "abandonment." The first to arrive were a family – father, mother, and young son – who removed the radiator and battery. Within twenty-four hours, virtually everything of value had been removed. Then random destruction began – windows were smashed, parts torn off, upholstery ripped. Children began to use the car as a playground. Most of the adult "vandals" were well-dressed, apparently clean-cut whites. The car in Palo Alto sat untouched for more than a week. Then Zimbardo smashed part of it with a sledgehammer. Soon, passersby were joining in. Within a few hours, the car had been turned upside down and utterly destroyed. Again, the "vandals" appeared to be primarily respectable whites.

Untended property becomes fair game for people out for fun or plunder, and even for people who ordinarily would not dream of doing such things and who probably consider themselves law-abiding. Because of the nature of community life in the Bronx – its anonymity, the frequency with which cars are abandoned and things are stolen or broken, the past experience of "no one caring" – vandalism begins much more quickly than it does in staid Palo Alto, where people have come to believe that private possessions are cared for, and that mischievous behavior is costly. But vandalism can occur anywhere once communal barriers – the sense of mutual regard and the obligations of civility – are lowered by actions that seem to signal that "no one cares."

We suggest that "untended" behavior also leads to the breakdown of community controls. A stable neighborhood of families who care for their homes, mind each other's children, and confidently frown on unwanted intruders can change, in a few years or even a few months, to an inhospitable and frightening jungle. A piece of property is abandoned, weeds grow up, a window is smashed. Adults stop scolding rowdy children; the children, emboldened, become more rowdy. Families move out, unattached adults move in. Teenagers gather in front of the corner store. The merchant asks them to move; they refuse. Fights occur. Litter accumulates. People start drinking in front of the grocery; in time, an inebriate slumps to the sidewalk and is allowed to sleep it off. Pedestrians are approached by panhandlers.

At this point it is not inevitable that serious crime will flourish or violent attacks on strangers will occur. But many residents will think that crime, especially violent crime, is on the rise, and they will modify their behavior accordingly. They will use the streets less often, and when on the streets will stay apart from their fellows, moving with averted eyes, silent lips, and hurried steps. "Don't get involved." For some residents, this growing atomization will matter little, because the neighborhood is not their "home" but "the place where they live." Their interests are elsewhere; they are cosmopolitans. But it will matter greatly to other people, whose lives derive meaning and satisfaction from local attachments rather than worldly involvement; for them, the neighborhood will cease to exist except for a few reliable friends whom they arrange to meet.

Such an area is vulnerable to criminal invasion. Though it is not inevitable, it is more likely that here, rather than in places where people are confident they can regulate public behavior by informal controls, drugs will change hands, prostitutes will solicit, and cars will be stripped. That the drunks will be robbed by boys who do it as a lark, and the prostitutes' customers will be robbed by men who do it purposefully and perhaps violently. That muggings will occur.

Among those who often find it difficult to move away from this are the elderly. Surveys of citizens suggest that the elderly are much less likely to be the victims of crime than younger persons, and some have inferred from this that the well-known fear of crime voiced by the elderly is an exaggeration: perhaps we ought not to design special programs to protect older persons; perhaps we should even try to talk them out of their mistaken fears. This argument misses the point. The prospect of a confrontation with an obstreperous teenager or a drunken panhandler can be as fear-inducing for defenseless persons as the prospect of meeting an actual robber; indeed, to a defenseless person, the two kinds of confrontation are often indistinguishable. Moreover, the lower rate at which the elderly are victimized is a measure of the steps they have already taken – chiefly, staying behind locked doors – to minimize the risks they face. Young men are more frequently attacked than older women, not because they are easier or more lucrative targets but because they are on the streets more.

Nor is the connection between disorderliness and fear made only by the elderly. Susan Estrich, of the Harvard Law School, has recently gathered together a number of surveys on the sources of public fear. One, done in Portland, Oregon, indicated that three-fourths of the adults interviewed cross to the other side of a street when they see a gang of teenagers; another survey, in Baltimore, discovered that nearly half would cross the street to avoid even a single strange youth. When an interviewer asked people in a housing project where the most dangerous spot was, they mentioned a place where young persons gathered to drink and play music, despite the fact that not a single crime had occurred there. In Boston public housing projects, the greatest fear was expressed by persons living in the buildings where disorderliness and incivility, not crime, were the greatest. Knowing this helps one understand the significance of such otherwise harmless displays as subway graffiti. As Nathan Glazer has written, the proliferation of graffiti, even when not obscene, confronts the subway rider with the "inescapable knowledge that the environment he must endure for an hour or more a day is uncontrolled and uncontrollable, and that anyone can invade it to do whatever damage and mischief the mind suggests."

In response to fear, people avoid one another, weakening controls. Sometimes they call the police. Patrol cars arrive, an occasional arrest occurs, but crime continues and disorder is not abated. Citizens complain to the police chief, but he explains that his department is low on personnel and that the courts do not punish petty or first-time offenders. To the residents, the police who arrive in squad cars are either ineffective or uncaring; to the police, the residents are animals who deserve each other. The citizens may soon stop calling the police, because "they can't do anything."

The process we call urban decay has occurred for centuries in every city. But what is happening today is different in at least two important respects. First, in the period before, say, World War II, city dwellers – because of money costs, transportation difficulties, familial and church connections – could rarely move away from neighborhood problems. When movement did occur, it tended to be along public-transit routes. Now mobility has become exceptionally easy for all but the poorest or those who are blocked by racial prejudice. Earlier crime waves had a kind of built-in self-correcting mechanism: the determination of a neighborhood or community to reassert control over its turf. Areas in Chicago, New York, and Boston would experience crime and gang wars, and then normalcy would return, as the families for whom no alternative residences were possible reclaimed their authority over the streets.

Second, the police in this earlier period assisted in that reassertion of authority by acting, sometimes violently, on behalf of the community. Young toughs were roughed up, people were arrested "on suspicion" or for vagrancy, and prostitutes and petty thieves were routed. "Rights" were something enjoyed by decent folk, and perhaps also by the serious professional criminal, who avoided violence and could afford a lawyer.

This pattern of policing was not an aberration or the result of occasional excess. From the earliest days of the nation, the police function was seen primarily as that of a night watchman: to maintain order against the chief threats to order – fire, wild animals, and disreputable behavior. Solving crimes was viewed not as a police responsibility but as a private one. In the March, 1969, Atlantic, one of us (Wilson) wrote a brief account of how the police role had slowly changed from maintaining order to fighting crimes. The change began with the creation of private detectives (often ex-criminals), who worked on a contingency-fee basis for individuals who had suffered losses. In time, the detectives were absorbed into municipal police agencies and paid a regular salary; simultaneously, the responsibility for prosecuting thieves was shifted from the aggrieved private citizen to the professional prosecutor. This process was not complete in most places until the twentieth century.

In the 1960s, when urban riots were a major problem, social scientists began to explore carefully the order-maintenance function of the police, and to suggest ways of improving it – not to make streets safer (its original function) but to reduce the incidence of mass violence. Order-maintenance became, to a degree, coterminous with "community relations." But, as the crime wave that began in the early 1960s continued without abatement throughout the decade and into the 1970s, attention shifted to the role of the police as crime-fighters. Studies of police behavior ceased, by and large, to be accounts of the order-maintenance function and became, instead, efforts to propose and test ways whereby the police could solve more crimes, make more arrests, and gather better evidence. If these things could be done, social scientists assumed, citizens would be less fearful.

A great deal was accomplished during this transition, as both police chiefs and outside experts emphasized the crime-fighting function in their plans, in the allocation of resources, and in deployment of personnel. The police may well have become better crime-fighters as a result. And doubtless they remained aware of their responsibility for order. But the link between order-maintenance and crime-prevention, so obvious to earlier generations, was forgotten.

That link is similar to the process whereby one broken window becomes many. The citizen who fears the ill-smelling drunk, the rowdy teenager, or the importuning beggar is not merely expressing his distaste for unseemly behavior; he is also giving voice to a bit of folk wisdom that happens to be a correct generalization – namely, that serious street crime flourishes in areas in which disorderly behavior goes unchecked. The unchecked panhandler is, in effect, the first broken window. Muggers and robbers, whether opportunistic or professional, believe they reduce their chances of being caught or even identified if they operate on streets where potential victims are already intimidated by prevailing conditions. If the neighborhood cannot keep a bothersome panhandler from annoying passersby, the thief may reason, it is even less likely to call the police to identify a potential mugger or to interfere if the mugging actually takes place.

Some police administrators concede that this process occurs, but argue that motorized-patrol officers can deal with it as effectively as foot-patrol officers. We are not so sure. In theory, an officer in a squad car can observe as much as an officer on foot; in theory, the former can talk to as many people as the latter. But the reality of police–citizen encounters is powerfully altered by the automobile. An officer on foot cannot separate himself from the street people; if he is approached, only his uniform and his personality can help him manage whatever is about to happen. And he can never be certain what that will be – a request for directions, a plea for help, an angry denunciation, a teasing remark, a confused babble, a threatening gesture.

In a car, an officer is more likely to deal with street people by rolling down the window and looking at them. The door and the window exclude the approaching citizen; they are a barrier. Some officers take advantage of this barrier, perhaps unconsciously, by acting differently if in the car than they would on foot. We have seen this countless times. The police car pulls up to a corner where teenagers are gathered. The window is rolled down. The officer stares at the youths. They stare back. The officer says to one, "C'mere." He saunters over, conveying to his friends by his elaborately casual style the idea that he is not intimidated by authority "What's your name?" "Chuck." "Chuck who?" "Chuck Jones." "What'ya doing, Chuck?" "Nothin'." "Got a P.O. [parole officer]?" "Nah." "Sure?" "Yeah." "Stay out of trouble, Chuckie." Meanwhile, the other boys laugh and exchange comments among themselves, probably at the officer's expense. The officer stares harder. He cannot be certain what is being said, nor can he join in and, by displaying his own skill at street banter, prove that he cannot be "put down." In the process, the officer has learned almost nothing, and the boys have decided the officer is an alien force who can safely be disregarded, even mocked.

Our experience is that most citizens like to talk to a police officer. Such exchanges give them a sense of importance, provide them with the basis for gossip, and allow them to explain to the authorities what is worrying them (whereby they gain a modest but significant sense of having "done something" about the problem). You approach a person on foot more easily, and talk to him more readily, than you do a person in a car. Moreover, you can more easily retain some anonymity if you draw an officer aside for a private chat. Suppose you want to pass on a tip about who is stealing handbags, or who offered to sell you a stolen TV. In the inner city, the culprit, in all likelihood, lives nearby. To walk up to a marked patrol car and lean in the window is to convey a visible signal that you are a "fink."

The essence of the police role in maintaining order is to reinforce the informal control mechanisms of the community itself. The police cannot, without committing extraordinary resources, provide a substitute for that informal control. On the other hand, to reinforce those natural forces the police must accommodate them. And therein lies the problem.

Should police activity on the street be shaped, in important ways, by the standards of the neighborhood rather than by the rules of the state? Over the past two decades, the shift of police from order-maintenance to law-enforcement has brought them increasingly under the influence of legal restrictions, provoked by media complaints and enforced by court decisions and departmental orders. As a consequence, the order-maintenance functions of the police are now governed by rules developed to control police relations with suspected criminals. This is, we think, an entirely new development. For centuries, the role of the police as watchmen was judged primarily not in terms of its compliance with appropriate procedures but rather in terms of its attaining a desired objective. The objective was order, an inherently ambiguous term but a condition that people in a given community recognized when they saw it. The means were the same as those the community itself would employ, if its members were sufficiently determined, courageous, and authoritative. Detecting and apprehending criminals, by contrast, was a means to an end, not an end in itself; a judicial determination of guilt or innocence was the hoped-for result of the law-enforcement mode. From the first, the police were expected to follow rules defining that process, though states differed in how stringent the rules should be. The criminal-apprehension process was always understood to involve individual rights, the violation of which was unacceptable because it meant that the violating officer would be acting as a judge and jury – and that was not his job. Guilt or innocence was to be determined by universal standards under special procedures.

Ordinarily, no judge or jury ever sees the persons caught up in a dispute over the appropriate level of neighborhood order. That is true not only because most cases are handled informally on the street but also because no universal standards are available to settle arguments over disorder, and thus a judge may not be any wiser or more effective than a police officer. Until quite recently in many states, and even today in some places, the police make arrests on such charges as "suspicious person" or "vagrancy" or "public drunkenness" – charges with scarcely any legal meaning. These charges exist not because society wants judges to punish vagrants or drunks but because it wants an officer to have the legal tools to remove undesirable persons from a neighborhood when informal efforts to preserve order in the streets have failed.

Once we begin to think of all aspects of police work as involving the application of universal rules under special procedures, we inevitably ask what constitutes an "undesirable person" and why we should "criminalize" vagrancy or drunkenness. A strong and commendable desire to see that people are treated fairly makes us worry about allowing the police to rout persons who are undesirable by some vague or parochial standard. A growing and not-so-commendable utilitarianism leads us to doubt that any behavior that does not "hurt" another person should be made illegal. And thus many of us who watch over the police are reluctant to allow them to perform, in the only way they can, a function that every neighborhood desperately wants them to perform.

This wish to "decriminalize" disreputable behavior that "harms no one" – and thus remove the ultimate sanction the police can employ to maintain neighborhood order – is, we think, a mistake. Arresting a single drunk or a single vagrant who has harmed no identifiable person seems unjust, and in a sense it is. But failing to do anything about a score of drunks or a hundred vagrants may destroy an entire community. A particular rule that seems to make sense in the individual case makes no sense when it is made a universal rule and applied to all cases. It makes no sense because it fails to take into account the connection between one broken window left untended and a thousand broken windows. Of course, agencies other than the police could attend to the problems posed by drunks or the mentally ill, but in most communities – especially where the "deinstitutionalization" movement has been strong – they do not.

The concern about equity is more serious. We might agree that certain behavior makes one person more undesirable than another, but how do we ensure that age or skin color or national origin or harmless mannerisms will not also become the basis for distinguishing the undesirable from the desirable? How do we ensure, in short, that the police do not become the agents of neighborhood bigotry?

We can offer no wholly satisfactory answer to this important question. We are not confident that there is a satisfactory answer, except to hope that by their selection, training, and supervision, the police will be inculcated with a clear sense of the outer limit of their discretionary authority. That limit, roughly, is this – the police exist to help regulate behavior, not to maintain the racial or ethnic purity of a neighborhood.

Consider the case of the Robert Taylor Homes in Chicago, one of the largest public-housing projects in the country. It is home for nearly 20,000 people, all black, and extends over ninety-two acres along South State Street. It was named after a distinguished black who had been, during the 1940s, chairman of the Chicago Housing Authority. Not long after it opened, in 1962, relations between project residents and the police deteriorated badly. The citizens felt that the police were insensitive or brutal; the police, in turn, complained of unprovoked attacks on them. Some Chicago officers tell of times when they were afraid to enter the Homes. Crime rates soared.

Today, the atmosphere has changed. Police–citizen relations have improved – apparently, both sides learned something from the earlier experience. Recently, a boy stole a purse and ran off. Several young persons who saw the theft voluntarily passed along to the police information on the identity and residence of the thief, and they did this publicly, with friends and neighbors looking on. But problems persist, chief among them the presence of youth gangs that terrorize residents and recruit members in the project. The people expect the police to "do something" about this, and the police are determined to do just that.

But do what? Though the police can obviously make arrests whenever a gang member breaks the law, a gang can form, recruit, and congregate without breaking the law. And only a tiny fraction of gang-related crimes can be solved by an arrest; thus, if an arrest is the only recourse for the police, the residents' fears will go unassuaged. The police will soon feel helpless, and the residents will again believe that the police "do nothing." What the police in fact do is to chase known gang members out of the project. In the words of one officer, "We kick ass."

Project residents both know and approve of this. The tacit police–citizen alliance in the project is reinforced by the police view that the cops and the gangs are the two rival sources of power in the area, and that the gangs are not going to win.

None of this is easily reconciled with any conception of due process or fair treatment. Since both residents and gang members are black, race is not a factor. But it could be. Suppose a white project confronted a black gang, or vice versa. We would be apprehensive about the police taking sides. But the substantive problem remains the same: how can the police strengthen the informal social-control mechanisms of natural communities in order to minimize fear in public places? Law enforcement, per se, is no answer. A gang can weaken or destroy a community by standing about in a menacing fashion and speaking rudely to passersby without breaking the law.

We have difficulty thinking about such matters, not simply because the ethical and legal issues are so complex but because we have become accustomed to thinking of the law in essentially individualistic terms. The law defines *my* rights, punishes *his* behavior, and is applied by *that* officer because of *this* harm. We assume, in thinking this way, that what is good for the individual will be good for the community, and what doesn't matter when it happens to one person won't matter if it happens to many. Ordinarily, those are plausible assumptions. But in cases where behavior that is tolerable to one person is intolerable to many others, the reactions of the others – fear, withdrawal, flight – may ultimately make matters worse for everyone, including the individual who first professed his indifference.

It may be their greater sensitivity to communal as opposed to individual needs that helps explain why the residents of small communities are more satisfied with their police than are the residents of similar neighborhoods in big cities. Elinor Ostrom and her co-workers at Indiana University compared the perception of police services in two poor, all-black Illinois towns – Phoenix and East Chicago Heights – with those of three comparable all-black neighborhoods in Chicago. The level of criminal victimization and the quality of police–community relations appeared to be about the same in the towns and the Chicago neighborhoods, but the citizens living in their own villages were much more likely than those living in the Chicago neighborhoods to say that they do not stay at home for fear of crime, to agree that the local police have "the right to take any action necessary" to deal with problems, and to agree that the police "look out for the needs of the average citizen." It is possible that the residents and the police of the small towns saw themselves as engaged in a collaborative effort to maintain a certain standard of communal life, whereas those of the big city felt themselves to be simply requesting and supplying particular services on an individual basis.

If this is true, how should a wise police chief deploy his meager forces? The first answer is that nobody knows for certain, and the most prudent course of action would be to try further variations on the Newark experiment, to see more precisely what works in what kinds of neighborhoods. The second answer is also a hedge – many aspects of order-maintenance in neighborhoods can probably best be handled in ways that involve the police minimally, if at all. A busy, bustling shopping center and a quiet, well-tended suburb may need almost no visible police presence. In both cases, the ratio of respectable to disreputable people is ordinarily so high as to make informal social control effective.

Even in areas that are in jeopardy from disorderly elements, citizen action without substantial police involvement may be sufficient. Meetings between teenagers who like to hang out on a particular corner and adults who want to use that corner might well lead to an amicable agreement on a set of rules about how many people can be allowed to congregate, where, and when.

Where no understanding is possible – or if possible, not observed – citizen patrols may be a sufficient response. There are two traditions of communal involvement in maintaining order. One, that of the "community watchmen," is as old as the first settlement of the New World. Until well into the nineteenth century, volunteer watchmen, not policemen, patrolled their communities to keep order. They did so, by and large, without taking the law into their own hands – without, that is, punishing persons or using force. Their presence deterred disorder or alerted the community to disorder that could not be deterred. There are hundreds of such efforts today in communities all across the nation. Perhaps the best known is that of the Guardian Angels, a group of unarmed young persons in distinctive berets and T-shirts, who first came to public attention when they began patrolling the New York City subways but who claim now to have chapters in more than thirty American cities. Unfortunately, we have little information about the effect of these groups on crime. It is possible, however, that whatever their effect on crime, citizens find their presence reassuring, and that they thus contribute to maintaining a sense of order and civility.

The second tradition is that of the "vigilante." Rarely a feature of the settled communities of the East, it was primarily to be found in those frontier towns that grew up in advance of the reach of government. More than 350 vigilante groups are known to have existed; their distinctive feature was that their members did take the law into their own hands, by acting as judge, jury, and often executioner as well as policeman. Today, the vigilante movement is conspicuous by its rarity, despite the great fear expressed by citizens that the older cities are becoming "urban frontiers." But some community-watchmen groups have skirted the line, and others may cross it in the future. An ambiguous case, reported in *The Wall Street Journal,* involved a citizens' patrol in the Silver Lake area of Belleville, New Jersey. A leader told the reporter, "We look for outsiders." If a few teenagers from outside the neighborhood enter it, "we ask them their business," he said. "if they say they're going down the street to see Mrs. Jones, fine, we let them pass. But then we follow them down the block to make sure they're really going to see Mrs. Jones."

Though citizens can do a great deal, the police are plainly the key to order-maintenance. For one thing, many communities, such as the Robert Taylor Homes, cannot do the job by themselves. For another, no citizen in a neighborhood, even an organized one, is likely to feel the sense of responsibility that wearing a badge confers. Psychologists have done many studies on why people fail to go to the aid of persons being attacked or seeking help, and they have learned that the cause is not "apathy" or "selfishness" but the absence of some plausible grounds for feeling that one must personally accept responsibility. Ironically, avoiding responsibility is easier when a lot of people are standing about. On streets and in public places, where order is so important, many people are likely to be "around," a fact that reduces the chance of any one person acting as the agent of the community. The police officer's uniform singles him out as a person who must accept responsibility if asked. In addition, officers, more easily than their fellow citizens, can be expected to distinguish between what is necessary to protect the safety of the street and what merely protects its ethnic purity.

But the police forces of America are losing, not gaining, members. Some cities have suffered substantial cuts in the number of officers available for duty. These cuts are not likely to be reversed in the near future. Therefore, each department must assign its existing officers with great care. Some neighborhoods are so demoralized and crime-ridden as to make foot patrol useless; the best the police can do with limited resources is respond to the enormous number of calls for service. Other neighborhoods are so stable and serene as to make foot patrol unnecessary. The key is to identify neighborhoods at the tipping point where the public order is deteriorating but not unreclaimable, where the streets are used frequently but by apprehensive people, where a window is likely to be broken at any time, and must quickly be fixed if all are not to be shattered.

Most police departments do not have ways of systematically identifying such areas and assigning officers to them. Officers are assigned on the basis of crime rates (meaning that marginally threatened areas are often stripped so that police can investigate crimes in areas where the situation is hopeless) or on the basis of calls for service (despite the fact that most citizens do not call the police when they are merely frightened or annoyed). To allocate patrols wisely, the department must look at the neighborhoods and decide, from first-hand evidence, where an additional officer will make the greatest difference in promoting a sense of safety.

One way to stretch limited police resources is being tried in some public-housing projects. Tenant organizations hire off-duty police officers for patrol work in their buildings. The costs are not high (at least not per resident), the officer likes the additional income, and the residents feel safer. Such arrangements are probably more successful than hiring private watchmen, and the Newark experiment helps us understand why. A private security guard may deter crime or misconduct by his presence, and he may go to the aid of persons needing help, but he may well not intervene – that is, control or drive away someone challenging community standards. Being a sworn officer – a "real cop" – seems to give one the confidence, the sense of duty, and the aura of authority necessary to perform this difficult task.

Patrol officers might be encouraged to go to and from duty stations on public transportation and, while on the bus or subway car, enforce rules about smoking, drinking, disorderly conduct, and the like. The enforcement need involve nothing more than ejecting the offender (the offense, after all, is not one with which a booking officer or a judge wishes to be bothered). Perhaps the random but relentless maintenance of standards on buses would lead to conditions on buses that approximate the level of civility we now take for granted on airplanes.

But the most important requirement is to think that to maintain order in precarious situations is a vital job. The police know this is one of their functions, and they also believe, correctly, that it cannot be done to the exclusion of criminal investigation and responding to calls. We may have encouraged them to suppose, however, on the basis of our oft-repeated concerns about serious, violent crime, that they will be judged exclusively on their capacity as crime-fighters. To the extent that this is the case, police administrators will continue to concentrate police personnel in the highest-crime areas (though not necessarily in the areas most vulnerable to criminal invasion), emphasize their training in the law and criminal apprehension (and not their training in managing street life), and join too quickly in campaigns to decriminalize "harmless" behavior (though public drunkenness, street prostitution, and pornographic displays can destroy a community more quickly than any team of professional burglars).

Above all, we must return to our long-abandoned view that the police ought to protect communities as well as individuals. Our crime statistics and victimization surveys measure individual losses, but they do not measure communal losses. Just as physicians now recognize the importance of fostering health rather than simply treating illness, so the police – and the rest of us – ought to recognize the importance of maintaining, intact, communities without broken windows.

MIKE SAVAGE AND ALAN WARDE

"Cities and Uneven Economic Development"

from *Urban Sociology, Capitalism, and Modernity* (1993)

Editors' introduction As the global economy becomes increasingly interconnected, the economic functions of different cities in the world become ever more dramatically different from each other. A very few Third World cities like Timbuktu have economies still based on trade or primitive manufacturing. The economies of these cities continue to operate as they have for centuries. But many other Third World cities like Kuala Lumpur, Malaysia, and Bangkok, Thailand, have economies increasingly based on manufacturing computers, phone systems, televisions and fax machines destined for export in the global economy. Some cities in the developed world such as New York, London, and Tokyo have emerged as economically potent centers for manipulation of information, technological innovation, and global financial command functions. Once prosperous cities in the English Midlands and the US Midwest are in steep economic decline.

What explains this "uneven" economic development of cities? This question has intrigued many economists, geographers, and sociologists. Can world cities be classified into clearly distinct *types* based on their economic function? Is there a regular progression from pre-industrial to post-industrial status which cities move through over time? Are there structural aspects of advanced capitalism that explain what is occurring? Have we moved from the "Fordist" eras to a "post-Fordist" era as Mayer describes (p. 229)? In the following section sociologists Mike Savage and Alan Warde review the literature on uneven economic development.

There have been many attempts to classify cities into categories based on their economic function. The authors of this selection identify five types – Third World cities, global cities, older industrial cities, new industrial districts, and cities in socialist countries. While such broad categories may be helpful, the authors are quick to note that such a typology is not exhaustive, and its categories are neither mutually exclusive nor static. Cities in China are developing mixed economies or their own brand of capitalism. The old industrial city of Glasgow, Scotland, has developed "Silicon Glen" with billions of dollars of high-tech industry. Saskia Sassen's research (p. 208) suggests that parts of New York and London function like Third World cities. And Edward Soja (p. 180) reminds us that there are large industrial districts in postmodern Los Angeles. Among urbanists who seek to classify cities based on their economic function, there are continuing debates about what the categories should be. Should Glasgow be considered an older industrial city, a new industrial district, or what? Because cities' economies are so diverse, some urban sociologists and urban economists do not think that meaningful systems to classify cities based on their economic functions can be developed at all.

Another approach is to look at the evolution of cities' economies through time. We know from V. Gordon Childe (p. 22) that Ur and other Mesopotamian cities had revolutionized the economies of their regions by 3500 BCE, from Pirenne (p. 37) that by the eleventh century some medieval European cities had evolved elaborate economics based on the revival of trade, from Engels (p. 46) that mid-nineteenth-century Manchester had become an industrial city with enormous disparities of wealth, and from Saskia Sassen (p. 208) that some global cities now base their economies on managing information and global financial services. Savage and Warde discuss theorists who have looked carefully at the stages in the evolution (and devolution) of industrial cities. They propose an 11,000 year stage of "primordial urbanization" before 4000 BCE, followed by an epoch of pre-industrial "definitive urbanization" which lasted until CE 1700, and a second epoch of industrial "definitive urbanization" after 1700. Their evolutionary typology is quite different in terminology and periods than those developed by V. Gordon Childe (p. 22).

A third approach to the question of "uneven economic development" stresses the international division of labor, particularly the movement of manufacturing from developed countries to the Third World. Good descriptive work has been done on changes at the local level in both cities in developed countries which are exporting their manufacturing bases and on Third World countries where the manufacturing is going.

One of the most important wellsprings of theory about the changing economy of cities is Marxism. As Peter Hall describes in "The City of Theory" (p. 362), Marxist and neo-Marxist urban theory experienced a resurgence during the 1970s and 1980s, and British geographer David Harvey applied Marxist theory to help understand approaches to social justice in cities (p. 199). As Savage and Warde describe, other theorists in the UK, the US, continential Europe and elsewhere are employing Marxist insights to uneven development today.

Other writings on uneven economic development include Folker Froebel, Jurgen Heinrichs and Otto Kreye, *The New International Division of Labor: Structural Unemployment in Industrialized Countries and Industrialization in Developing Countries* (Cambridge: Cambridge University Press, 1980), Doreen Massey, *Spatial Divisions of Labour* (London: Macmillan, 1984), Michael Smith and Joe Feagin, *The Capitalist City: Global Restructuring and Territorial Development* (London: Sage, 1987) and a number of David Harvey's books cited in our introduction to "Social Justice, Postmodernism, and the City" (p. 199).

EVOLUTIONARY THEORIES OF CITIES

In the work of . . . the Chicago School, cities represented the new and the modern, epitomes of the emergent economic and social order produced by industrial capitalism. Implicitly they drew upon an evolutionary model of economic change. The city of Chicago, in particular, was taken as representative of the modern industrial city, and attempts to apply the concentric ring model (developed by Burgess and modified by others) to other industrial cities were legion. Within this frame of thought the city was seen as the product of the elaborate division of labour characteristic of modern industrial society. Cities owed their economic role to their pivotal place in this new industrial order as centres of commerce, sites of production, and bases for the most specialized economic activities. In this line of reasoning the city was the most advanced manifestation of an evolutionary process of economic change, 'the workshops of civilization' in Park's words (see Harvey 1973, p. 195).

Evolutionary approaches to urban development argued that the industrial city was the culmination of a long evolutionary process, stretching back to the earliest historical periods. Lampard (1965) distinguished two urban epochs in human history. These were, first, 'primordial urbanization' where settlements first emerged in the years between 15000 BC and 4000 BC, as a collective form of organization additional to the usual migratory agricultural activities. The importance of the second period of 'definitive urbanization', which began in Mesopotomia after 4000 BC, was that cities developed as fixed sites, in which 'by means of its capacity to generate, store, and utilise social saving, the definitive city artefact is capable of transplanting itself out of its native uterine environments' (Lampard 1965, p. 523). This period of 'definitive urbanization' is itself split into two epochs, before and after 1700 AD. In the first of these, cities were centres for a hinterland and existed in a stable hierarchy, in which hamlets formed a hinterland for villages, villages for towns, towns for cities, and cities for capital cities. Urban expansion was limited since cities were essentially parasitic on a limited agricultural economy. After 1700, the industrial city emerged as a dynamic force, able to increase in size because of the ability of economic production based in cities to sever their dependency on agriculture.

The industrial city was hence seen as the locus of the new industrial society and as ushering in a new period in history when urban growth could continue at a vastly expanded level. Yet since the 1930s the industrial societies which cities were seen to embody have themselves been transformed by deindustrialization – manufacturing industries in many urban heartlands have collapsed; service industries have arisen and industrial production has developed in new, rural, areas, appearing to cut the apparently close connection between cities and industry on which the evolutionary ideas were based.

Attempts to apply evolutionary lines of thinking have persisted into the present day and have taken a new turn as industrial economies have changed. A good example is the work of Peter Hall who has developed the evolutionary model of the city to encompass deindustrialization as well as industrialization (see Hall and Hay, 1980; Hall, 1988). Hall begins by arguing that the urban system has been massively transformed in recent decades. Drawing upon American evidence he argues that four linked processes have undermined the centrality of the large, industrial urban conurbation which characterized earlier periods of industrial capitalism. These are:

1 suburbanization, where urban growth takes place in suburban rather than central urban areas;
2 deurbanization, where the urban population reduces relative to the population of rural and nonurban areas;
3 the contraction of the largest cities; and
4 the rise of new regions and the decline of old.

Hall explains this transformation by distinguishing six evolutionary stages through which cities go as industrial economies change and decline. His emphasis is upon the way in which, as regions industrialize, cities develop in size and concentration. After a period of time, however, any industrial area begins to stagnate as innovation occurs elsewhere. Hence cities begin to decline. Because this process of industrial growth and fall is inevitable, all cities pass through the same six-stage cycle.

The six stages Hall specifies are divided into two groups. The first three stages occur during industrialization, the last three when deindustrialization begins to take effect in any given region:

1 The stage of 'centralization during loss' happens during early industrialization. People migrate from the country to the city, leading to a growing urban population, but the overall population in any region is in net decline as more people leave the region overall.
2 As industrialization continues, the overall proportion of people living in cities within regions increases.
3 'Relative centralization' occurs when the city stretches over its boundaries and begins to develop suburbs. Nevertheless the proportion of urban dwellers continues to grow. This is the type of city which was the focus of the Chicago School studies, where there were large and dense urban populations and suburbs had begun to emerge.

Hall's argument, however, is that urban evolution has now continued beyond this, and a process of urban decline marks a new stage from that studied by the Chicago writers:

4 Suburbs begin to grow faster than the urban core, so that 'relative decentralization' occurs as people move to the outer reaches of cities.
5 Starting about 1900 in the largest European cities (but generally much more recently) 'absolute decentralization' occurs as people begin to move out of the inner city as it becomes increasingly specialized around office and commercial functions.
6 The entire city begins to decline as people begin to move out to the rural areas as deindustrialization proceeds.

This process of 'counterurbanization' has been much debated since the 1960s (Fielding, 1982). The period of industrial urban expansion, which earlier writers had expected to continue unabated, gives way, in Hall's view, to a situation of urban decline.

Hall is wary about applying his evolutionary model. It is derived from research in the USA chronicling the decline of large cities from the 1960s. In Western Europe there are different patterns, and 'the different countries' urban systems . . . display marked differences from one another' (Hall, 1988, p. 116). In Britain and Germany the largest cities were declining in population by the 1970s as Hall would have expected. However in France, Italy and the Benelux countries they were not.

Although many cities are seeing significant population loss, there are a number of difficulties with an evolutionary model such as Hall's. First, there is a problem with the way that Hall, in common with other writers referring to the phenomenon of 'counterurbanization', characterizes the decline of cities in the current period. There is no doubt that in many parts of the developed world population and employment is moving from central urban locations, but whether this should be seen as testifying to the decline of cities rather than their further expansion into new areas is a moot point . . .

Second, there is a problem about generalizing from Hall's study of urban trends in twentieth-century Western Europe. It could be argued that contemporary cities are becoming increasingly differentiated according to their role in the world economy, which makes it unhelpful to generalize about a single evolutionary path for all. Five prominent urban types stand out – Third World cities, global cities, declining industrial cities, new industrial districts, and socialist cities – all of which have a different character.

Third World cities are themselves heterogeneous, but tend to possess a number of distinctive features. They are 'over-urbanized' (Timberlake, 1987). This means that they tend to be extremely large relative to the population of the particular country – a result of the fact that inward capitalist investment often focuses upon these capital city sites, a phenomenon described as 'urban bias'. They also tend to be 'dualistic', with major divisions between the formal and informal economy, between city and country, and between social groups. This dualistic format is related, in many cases, to the colonial legacy of 'urban apartheid' (Abu-Lughod, 1980; King, 1990), where colonial rulers lived in separate parts of the city and were subject to a different jurisdiction from that applying to native dwellers.

'*Global cities*' (or world cities) are ones which increasingly depend on multinational financial services and are linked to the circulation and realization of wealth. They are frequently the location of corporate headquarters of major multinational enterprises and are the sites of what Massey (1988) refers to as 'control functions', whence the control and management of corporate enterprise is directed. London, New York, Frankfurt and Tokyo are examples of this type of city. They tend to be large, centralized (with a distinct urban core specializing in international financial services), and contain both an elite group of workers and lower-paid servicing workers (Kasarda, 1988).

Older industrial cities, now in precipitate decline following the collapse of urban manufacturing, constitute the third type. Britain has many of the most dramatic examples – Glasgow, Liverpool (a trading rather than industrial city) and Bradford being especially prominent. Other noted examples have been found in the North-east and Mid-West of America (Detroit, Buffalo, Cleveland), and in Germany (Essen, Duisberg). These cities are characterized by decay and dereliction, high levels of unemployment, poor housing conditions and so forth.

'*New industrial districts*' have recently been given a great deal of attention. These are distinctively new urban developments (colonial cities, global cities, and older industrial cities being adaptations of older urban forms), which are not organized around an urban core with a suburban hinterland, but are more decentralized and cover a larger area. Here much development takes place round neighbourhood centres, and around the major transport networks, in the form of out-of-town shopping malls, employment centres and suburban housing. Examples include the Los Angeles area of the USA, and parts of the Home Counties in South-East England.

Cities in socialist countries have also experienced dynamics very different from those in the capitalist world. They have tended to grow more slowly than their capitalist counterparts. Many socialist regimes have been explicitly anti-urban (Forbes and Thrift, 1987). The immediate post-revolutionary period tended to freeze, and in some cases reduce, urban population growth. These cities have been subject to greater planning and zoning.

The foregoing typology is not exhaustive. Many urban centres fall into several of these categories. The point is that it is impossible to see one form of city as archetypal of the current economic and social order in the way in which Chicago was taken as an exemplar of industrial capitalism in the early twentieth century. It is not true that all cities experience the same logic of development, but rather that some cities obtain distinct roles in the world economy, and once established they become differentiated from other cities occupying different roles within the same environment.

At the heart of the analysis is the fact that cities exist within a wider world system. The dynamics of this world system affect the way that cities develop and decline. A recognition of this belies a linear historical view of urban differentiation – where different urban forms are reflections of the specific period which any given city has reached in an evolutionary urban cycle – implying instead that spatial dynamics of the world system profoundly shape urban form. It is to a greater consideration of these processes that we now turn.

COMPETING THEORIES OF URBAN DEVELOPMENT

We have shown that evolutionary views fail to recognize the specificity of cities and the distinct roles they perform in a wider world economy. Let us consider in greater detail how these differences are sustained by spatial processes of uneven development. Various theories address this issue. Many are of Marxist provenance, emerging from the revived intellectual reputation of Marxist analysis in the social sciences in the 1970s. The effect was to focus attention on the specifically capitalist mechanisms operating to create the geography of economic life. Thus, rather than beginning from the nature of industrialism, as did much orthodox economic sociology and geography in the post-war period, the central concerns were ones of capitalist accumulation, competition, exploitation and restructuring. When applied to the area of urban studies this constituted a more rigorous and detailed approach to the economic bases of urban systems.

[. . .]

The new international division of labour thesis

One of the earliest and most original of the new accounts of the contemporary spatial division of labour was presented by Frobel, Heinrichs and Kreve in their *New International Division of Labour* (NIDL), first published in German in 1977. Their concern was with the growing internationalization of production since 1945 and its effects on the world economic system. Their main point was that manufacturing production processes which had once been undertaken in core countries in Western Europe were increasingly located in the Third World, which as peripheral countries within the world economy had previously concentrated on agricultural produce and raw materials for export to the advanced countries. Whereas in the 1950s Western Europe imported scarcely any manufactured goods, by 1975 much of the production in certain industrial sectors, like textiles and electrical goods, was carried out overseas, financed and controlled by metropolitan companies.

> The development of the world economy has increasingly created conditions in which the survival of more and more companies can only be assured through the relocation of production to new industrial sites, where labour-power is cheap to buy, abundant and well-disciplined; in short, through the transnational reorganization of production.
>
> (Frobel *et al.*, 1980, p. 15)

This process seemed to mark a new phase in the relationships between core and periphery which Wallerstein (1974) had observed. The prime reason for the emergence of the NIDL according to Frobel *et al.*, was the change in the labour process as levels of skill involved in manufacturing production were reduced sharply. In such circumstances, a vast pool of unemployed or underemployed unskilled labour could be exploited on a world scale. The terms of employment of unskilled labour in Third World countries were especially favourable to capital: wages are much lower, working conditions poorer, trade unions weaker, labour forces easier to discipline, etc., than in the West. The improvement in methods of communication and transport made it possible to exploit these new reserves of labour. Other factors such as tax concessions to multinationals, absence of pollution control, and the absence of health and safety legislation enhance the attractiveness of these locations. Also, certain other conditions have to be fulfilled to make overseas sites acceptable: transport costs which depend on the size and weight of the product; the political ability of overseas political regimes: property law; the corruptibility of officials, etc. (see Frobel *et al.*, 1980, pp. 145–47 for a list). But, where such conditions are met, it becomes profitable to transfer machinery to sites outside Europe to take advantage of favourable labour conditions.

The ramifications of this new international division of labour were thought very considerable, both for metropolitan and peripheral countries. One was the changing industrial structure in metropolitan countries especially the decline of manufacturing employment. Deindustrialization, as the process was first known, had implications also for levels and types of occupational opportunity, with fewer skilled and unskilled manufacturing jobs available at home.

The NIDL thesis was intellectually of enormous importance. It brought to scholarly attention a new form of internationalization of the capitalist economy, explained recent changes in patterns of employment and indicated how multinational and transnational corporations could exploit spatial differences in labour markets in conjunction with a new technical division of labour within particular industrial sectors. It offered a relatively simple explanation of the phenomenon of deindustrialization. Derived in part from the neo-Marxist world-systems theory of Wallerstein it did not depend on any particularly sophisticated economic theory. As Froebel *et al.* express their premises:

> The determining force, the prime mover, behind capitalist development is therefore the valorization and accumulation process of capital, and not for example, any alleged tendency towards the extension and deepening of the wage labour/capital relation or the 'unfolding' of the productive forces.
>
> (Frobel *et al.*, 1980, p. 25)

From that point of view the dynamic was mostly one of profitseeking and minimizing labour costs in deskilled production processes.

What is of particular concern to us is the implication of the NIDL for urban systems. The NIDL thesis can be used, in some ways, to explain the differentiation of cities in different parts of the world. Rather than see cities inevitably decline as an evolutionary concomitant of deindustrialization, as Hall suggests, the NIDL thesis is able to explain the differential fate of cities in various parts of the globe. At one level, the prime position of Western capital cities could be explained by their coordinating role in the new international division of labour. At another level the growth of large cities, such as Mexico City, in the periphery could also be explained by the role they played as sites for the new decentralized production.

Other important work has focused upon the impacts of NIDL on Western cities and urban systems. The collection of essays by Smith and Feagin *The Capitalist City* (1987), perhaps offers the best access to work in this vein. In some old manufacturing towns, such as Buffalo, New York, USA, the removal of production to peripheral locations has led to the collapse of employment (Perry, 1987). In Buffalo employment in manufacturing fell from 200,000 in the 1950s to 100,000 by the early 1980s. In other, often neighbouring, cities, economic expansion – particularly in the service sector – has followed the consolidation of new controlling activities in the NIDL.

It is perhaps because the NIDL thesis is so closely specified in terms of *manufacturing* that it has been applied most often by US and British scholars for it is in the USA and UK that deindustrialization has been most severe as old manufacturing towns have been affected detrimentally. There the very force of a description of a changing local economic situation and its obvious impact upon employment opportunities, living standards and social relations is sufficient to demonstrate the effects of corporate restructuring in a global economy.

However, since many countries have not deindustrialized, and since most formal economic activity is in the service sector in all Western societies, the overall impact of economic transformations on cities is not fully grasped.

[. . .]

David Harvey, the second circuit of capital and urbanization

In the 1970s David Harvey attempted an ambitious theoretical approach to the analysis of uneven development, derived from a new appreciation of Marx's economic theory and its implications for urban growth. In many ways it offered a powerful contrast to the NIDL thesis, since it tried to build a theory which is historically sensitive, aware of urban specificity, and deliberately emphasizing the importance of social conflict for urban development.

Harvey's starting-point was to develop Marx's own analysis of capital accumulation and draw our the implications for the urban structure. This primarily involved an examination of landed property and its role in capital accumulation, a subject about which Marx said relatively little. In Harvey's early work (1973) he specified the distinctive nature of land as a commodity in capitalist society: while it is something which can be bought and sold – like any other commodity – it has a number of peculiarities. The most important of these are that it is spatially fixed, since land cannot be transported: it is necessary to human life, since we all need to live somewhere: it allows assets and improvements to be stored; and it is relatively permanent, since improvements to land (e.g. buildings) tend to survive considerable periods of time, longer than the time it takes for clothes to wear out or food to be eaten, for instance!

Much of Harvey's work can be seen as an exploration of the implications of the specific character of capital investment in land rather than in other areas. He emphasized that such investment is both highly significant for the functioning of the capitalist economy – since a great deal of capital is usually tied up in the built environment – and also that such investment leaves a relatively enduring physical legacy. The resulting built form can help to aid capital accumulation, if it is a profitable avenue for investment, but can also be a barrier to it, when its enduring qualities render it outdated and anachronistic in a relatively short period of time. Much of Harvey's work can be seen as an elaboration of this idea of the double-edged nature of property for capital accumulation.

Harvey, in later work (1977, 1982), developed his analysis of the precise role of land for capital accumulation by examining the three circuits of capital. The primary circuit – the production of commodities within manufacturing – is the one to which Marx gives greatest attention. Harvey emphasized how the accumulation of profit by the exploitation of labour within capitalist enterprises runs into severe contradictions, most notably when goods are overproduced without adequate money in the economy to purchase them. As a result of this, profits may fall and capital lie idle. It is this crisis of overaccumulation that causes capital to be switched into the "second circuit" – where capital is fixed in the built environment money is moved from the primary circuit to the secondary – so long as a supportive framework for this transition exists, as when a state encourages such investment. The tertiary circuit of capital involves scientific knowledge and expenditures to reproduce labour power. Expenditure in this circuit is often the result of social struggle rather than being a direct opportunity for capital to find new avenues for accumulation.

Harvey's analysis illuminated urban processes in two ways. First, it conceptualized the significance of investment in the built environment in relation to other economic processes, suggesting links between urban restructuring and economic restructuring. Harvey's principal example attributed the growth of suburbs in America after the Second World War to the switching of capital out of the primary circuit, where cases of overaccumulation were emerging. The changing structure of the capitalist city was thus related to broader trends in the capitalist economy. The property boom of the early 1970s in the USA and Britain, which saw the development of office blocks in many urban centres, owed much to similar pressures.

The built environment, however, is not simply a means of resolving crises in capital accumulation: it can, in turn, cause further crises. As capital is invested in the built environment and hence the economy is more generally 'cooled down', new opportunities for capital accumulation in the primary circuit open up again: capital moves back into this circuit, capital of the secondary circuit is devalued and it becomes a less attractive avenue for investment. Once constructed, the existing built environment is no longer as efficient as new building and may prove a barrier to effective capital accumulation, so causing capital investment to move to newer and more advanced sites. One result is that the built environment concerned is abandoned or downgraded such that capital moves elsewhere to restore profitability.

Harvey's model of the urban process under capitalism is hence the very opposite of the evolutionary view we discussed above. For Harvey investment in the urban form offers a temporary solution to crises in capitalism, but then in turn it becomes a problem which needs to be addressed by switching capital investment elsewhere. Cities – and other spatial units – hence grow and decline in an almost cyclical way. Yet Harvey is also attuned to the social and political struggles that can attempt to 'fix' the role of a particular city, against particular economic forces. Struggles by social groups threatened by the removal of capital can prevent capital flight and ensure the survival of an urban infrastructure. The miners' strike in Britain in 1984/85 is an example of a failed attempt to fix investment to particular traditional coal-mining areas. In other cases 'growth coalitions' may succeed in attracting investment. Ultimately, it is a matter of political struggle as to the way that the tendencies within capitalism to make *and* break places work out in practice . . .

Harvey also helped to draw attention to the social and political role of the bourgeoisie – landlords – who had a particular stake within any one place. Capitalists owning land are committed to keeping their investment in a specific place. They often play a crucial role in defending local economies and engage in civic 'boosterism' to encourage the economic prosperity of their place, which will enhance property prices and the value of their land. This theme has been developed by American writers such as Gottdiener (1985) and Logan and Molotch (1987) who identify the central role of landed interests in affecting urban fortunes.

The strengths of Harvey's account are several. First, it is possible to use his ideas to explore the *variety* of urban processes in the contemporary world. Whilst his discussion of the tendencies of capital to move between circuits very usefully explores the bases of switches of investment in the built environment, he is also cognizant of the role of political struggle. Thus, he is able to show how social and political forces in a particular city may act to modify or even thwart attempts by capitalists to disinvest. His stress on the way in which the built environment is at different times a help and a hindrance for capital accumulation, and thus how dramatic changes can occur to the same city within relatively short time-spans, makes sense of dramatic episodes of contemporary urban change. His theory of uneven development allows historical specificity and recognizes the role of human agency.

Harvey's analysis is not without problems, however. The major one is that his work is empirically largely unsubstantiated, for little research has actually used Harvey's insights to shed light on processes of urban change. The main exception to this concerns studies of suburbanization and gentrification, which we examine in the next chapter. Harvey's own case studies, such as that of Paris in the nineteenth century, seem to lapse all too quickly into detailed historical accounts . . .

Industrial restructuring and class struggle

The relationship between social conflict and capitalist restructuring lies at the heart of a third account of uneven development, pioneered by Doreen Massey. Sometimes called the 'restructuring' approach (see Bagguley *et al.*, 1990), it led to a large amount of empirical research, particularly in the UK, concerning the relationship between economic restructuring, urban and regional change, and political conflict.

Massey's approach differs from those discussed above in being concerned less with the abstract logic of capital accumulation, and more with how the strategies adopted by enterprises to survive and prosper in the world capitalist economy affect patterns of spatial inequality. She examines the ways in which organizations restructure in response to changes in their economic environment and the spatial consequences. Whilst the other theories operate at a macro-level, Massey's work occupies a middle ground, providing conceptual guidance as to how specific places are affected by differing types of restructuring.

In her earlier work with Richard Meegan (Massey and Meegan, 1979, 1982), it was argued that firms in different sectors of the economy responded to international economic pressures by adopting different strategies. The most important of these were rationalization (the closure of specific units of production and centralization of production in other sites), intensification (making employees work harder), and investment and technical change (involving capital investment and better productivity). These strategies make for uneven development, for some areas lose employment as production is rationalized away from them, whilst others gain employment because they are subject to fresh investment. Spatial differentiation is also linked to the way in which firms deal with resistance to their restructuring strategies. One repeatedly used strategy is to shed skilled workers in one location and replace them, when necessary, with unskilled people somewhere else. Thus, work-places in the inner cities, often employing union-organized skilled workers might be closed down and the production process, with perhaps new technology, shifted to, or expanded in, other areas where new, unskilled, often inexperienced, and often female labour will be engaged. There are plenty of examples of this in Britain: rural regions like East Anglia and North Wales have been fastest growing in terms of manufacturing employment in the past twenty years. Again, car production in the USA has been moved out of Detroit and Chicago to sites further south where labour is more docile. For Massey, labour becomes locally (or perhaps more correctly, regionally) specialized as workers with specific skills congregate together.

In her best-known work, *Spatial Divisions of Labour* (1984), Massey developed and systematized this argument by showing how, as firms restructured, they tended to specialize activities in those areas where the cheapest and most pliable labour force could be found. Research and development work, along with the administrative functions of Head Office, was located in those areas where professional and managerial workers were plentiful and near the corridors of power. As a result, she argued, Britain could increasingly be seen as a country divided between a prosperous South-East, where the 'control functions' of large organizations were concentrated, and the depressed peripheral regions, where employment tended to be concentrated in branch plants and largely involved unskilled workers. This polarization marked a new spatial division of labour and was a major change from the older patterns where differing parts of Britain had semi-autonomous regional economies, typically based on a specific product (textiles in north-west England, shipbuilding in north-east England, and so on), and in which skilled, unskilled, and managerial workers were employed in smaller, less spatially desegregated firms.

The logic of Massey's account is that capital has come to use spatial differentiation in the competitive search for profit, as it invests in those areas where it can draw upon a suitable labour force. Spatial advantage is most readily obtained by discriminating among available labour forces. This acknowledges that capital is nowadays highly mobile, and certainly more mobile than labour, thus implying that many constraints on industrial location which pertained in earlier epochs have been overcome.

Unlike the NIDL theorists and David Harvey, Massey avoids a purely economistic account, and finds a way of explaining how the social character of specific places impacts on processes of restructuring. The social qualities of labour are significant in repelling or attracting capital and hence, Massey argues, it is important to consider how local work cultures are formed and how they facilitate types of militancy or passivity. In the UK, for instance, industrial employment in the Home Counties expanded in the 1980s partly because firms chose to locate to areas without trade-union traditions where workers might be more compliant. Trade-union membership has become much more dispersed recently, indicating the demise of densely unionized towns and regions. As their population have declined, some of the larger industrial cities have lost some bases for labour militancy.

[. . .]

Regulation Marxism and the California School

The final approach to uneven development and urban differentiation which we consider is associated with the Regulation School, in particular through its impact on neo-Marxist geographers such as Allen Scott, Michael Storper, Richard Walker, and the more recent work of David Harvey. As with work inspired by Massey, one of the main attractions of this theoretical current is its ability to support a wide-ranging research programme. Also in common with Massey this approach is historically sensitive and attuned to urban specificity.

Regulation School theory is descended from French structural Marxism of the 1970s (see Jessop, 1990, for an overview). Its principal figures, Aglietta, Lipietz, Boyer, and others, have employed a distinctive set of theoretically generated concepts – regime of accumulation, mode of regulation, Fordism – to explore relationships between capital, labour and the state. The main starting-point for these writers is the argument that nation-states play a crucial role in regulating capital accumulation, and they see the differing ways in which capitalism is regulated as historically specific 'regimes of accumulation'. Much of their work is thus an historically grounded attempt to consider the implications of the contemporary shift from one 'regime of accumulation' – Fordism, to another 'regime of accumulation' – neo-Fordism, or post-Fordism.

The Italian Marxist Gramsci apparently coined the phrase 'Fordism' to characterize the mass-production methods pioneered by Henry Ford in the inter-war years of the twentieth century, and some of their effects on social and family life in Italy. The concept re-entered contemporary social and economic thought through the writings of the Regulation School who referred to a complete era in capitalist development as Fordist. Their argument is that Fordism was the dominant mode of industrial organization in the mid-twentieth century and that it constituted a distinctive 'regime of accumulation'. The regime of accumulation is based on a specific 'mode of regulation' (whence the name of the School), where regulation refers to things like the forms of the state, the nature of intervention, welfare arrangements, legal forms, and so forth. In addition, for the Regulation School, phases of capitalist development are determined both by the mode of production and consumption. The Fordist era was characterized by mass production and mass consumption. However, they argued that in the 1980s this regime was gradually giving way to a neo- or post-Fordist one with less demand for mass-produced goods and in which competitive pressures required much more flexible methods of production.

The concept of post-Fordism, like many other concepts is primarily constructed as a negative ideal-type, identifying characteristics that were not present in a preceding, and better understood institutional setting. The model of Fordism is relatively well-established, and many commentators would think of Fordist arrangements as characterizing the leading manufacturing firms from the 1930s through to the 1970s. The Fordist firm is one characterized by scientific management, economies of scale, mass production, and technical control. Post-Fordist production arrangements are associated with the declining size of production units, small batch production, customized products, flexible working practices, greater worker discretion and more responsible autonomy.

Critics of regulation theory see it as bearing many of the alleged defects of its structuralist predecessors: functionalist, economist, reductionist, excessively abstract, ignoring individual action and underemphasizing social struggles. Nevertheless, the technical vocabulary of the Regulation School is frequently slipped into discussions of new flexible forms of production, though often in a highly eclectic way (e.g. *Society and Space*, 1988). Quite often the concepts are invoked without regard to the theoretical scheme from which they were derived. Nevertheless the notions of Fordism, post-Fordism and flexibility have been widely taken up to analyse new patterns of spatial inequality. David Harvey's book *The Condition of Postmodernity* (1989) is an important example.

[. . .]

Harvey does not, in this book, consider in detail the urban transformations brought about by the new form of flexible accumulation. Whilst Harvey might have strengthened his analysis of the contemporary transformation of capitalism, he has not applied his framework to uneven development and urban change in any detail.

It is in this field that the geographers of the 'California School' have made greater strides. They might be seen as attempting to prove a theoretical account of the dramatic development of the California urban conglomerations of Los Angeles and (to a lesser extent) San Francisco. Los Angeles is perhaps the most discussed city, of the late twentieth-century world (Davis, 1990; Soja, 1989; Jameson, 1984, etc.): for the School, Los Angeles is to the 1980s and 1990s what Chicago was to the early twentieth century, a particularly stark example of the urbanizing processes which are to be found throughout much of the world economy. They see the rise of the California economy as tied to the decline of the old industrial regions of the north-east of the USA (the 'Rust Belt') and the fact that new industries, such as electronics and defence, are located in California, while contracting ones, like shipbuilding, are in the Rust Belt. They concentrate on the experience of recently grown industrial sectors and argue that establishments in these sectors are tending to cluster in 'new industrial districts'. They provide evidence for a variety of sectors – for example, motion pictures (Christopherson and Storper, 1986) animated pictures (Scott, 1988b) printed-circuit fabrication (Scott, 1988b) – where factories tend to cluster together in the same district of a large metropolitan area. The reason for this is to obtain economies of scope rather than the economies of scale that were the objective of Fordist mass production.

Although the Californians deploy Marxist concepts (particularly of regulation theory), the core of their current position is a theory of the firm associated with the economist Oliver Williamson (for a summary see Williamson, 1990), who has developed the theory of 'transaction cost analysis'. Very simply their theory distinguishes those situations under which firms find it best to internalize contributory activities (such as marketing, or research, or various production functions), and those where it is best to externalize them, by using sub-contractors or buying services on the market. Scott (1988b) pursues the spatial implications of this contrast, observing that when firms externalize their activities they tend to congregate close to the other firms involved in their production network, leading to agglomeration economies and the emergence of New Industrial Districts. Alternatively, if activities are internalized, firms may be able to separate functions spatially onto different sites. Massey's account of the spatial separation of production functions will apply only to such cases.

[. . .]

Much of the work of the California School is directed purely towards explaining industrial location and tends to avoid any wider discussion of its effects on urban development. An exception to this is Scott's (1988a) *Metropolis: From the Division of Labor to Urban Form*. Having outlined the process whereby firms reorganize, compelled by the benefits of vertical disintegration he makes a series of claims about the way in which the concentration of workers' residences near to the 'neo-Marshallian' industrial districts in which they work has effects on social segregation, ethnic differentiation and community formation.

Beginning from the premise of the spatial separation of home and workplace in capitalist economies, Scott (1988a, pp. 217–30) argues that the employment relation is a key determinant of residential location. He uses data to show that although there are other cross-cutting bases of residential segregation, occupation is primary, universal and constant in large cities of the advanced capitalist societies.

[. . .]

Scott's account is not entirely convincing simply because firm reorganization and labour market are insufficient as basic mechanisms to generate the complete range of social effects. The Californians, unlike Harvey and certainly Massey, say very little about social conflict and its impact on economic restructuring and social change. Their analysis is conducted at the level of economic theory and even at that level it is probably too narrow. The Californians take little notice of trends in the service industries and their role in employment, since they see services as largely dependent on manufacturing production (see Sayer and Walker, 1992). If urbanization is connected only to industrialization, as it is by Scott, then we have a limited grasp on the impact of most economic activity.

One final problem with the Californian

account, and another sense in which it may be seen as unduly economistic, is that it almost entirely ignores the state. Here again this may be the result of focusing attention on one country, the USA, where the federal government in particular is relatively non-interventionist in regional planning. Nevertheless the state does many things short of direct intervention that set the framework for business activity. This criticism is developed by Feagin and Smith (1987) and Gottdiener (1989).

CONCLUSION

In the past decade, theories of uneven development have become increasingly sophisticated and aware of the problems raised at the beginning of [this extract]. Accounts have grown more sensitive historically and have identified explicitly how different places may be affected in diverse ways by uneven development. It has been demonstrated that urban development is not some evolutionary process through which all cities pass. Rather these new theories have demonstrated the instability of urban fortunes and the reasons why cities rise and fall, fall and rise. Causes include the dynamics of the world capitalist economy which allow the relocation or industry across the globe; the cycles of investment and disinvestment in the built environment; forms of corporate restructuring; and the dynamics of product innovation. As a result, particular cities cannot be deemed emblematic of a form of social organization, in the way that the city of Chicago stood for industrial capitalism. Instead we should recognize the inherent impermanence of the economic foundations of cities and the multiple roles of cities in a world capitalist economy.

Jointly, these theories succeeded in analysing the economic foundations of urban change and identifying a series of forces which derive specifically from mechanisms of the capitalist organization of production. As such they have proved an important corrective to the previous neglect by urban sociology of such matters. Individually, each seems to have identified some characteristic recent strategies and processes of the global economy. Their disagreements stem partly from concentrating on different nations and different industrial sectors, though there are more fundamental theoretical sources of dispute too.

Theories of uneven development have been far less successful at explaining the sources of intraurban change and social change within cities. Once they move away from delineating the economic position of particular places, and begin to refer to the impact of uneven development on their urban structure, social order and cultural patterns, they begin to falter. Although Harvey sought to capture the importance of social conflict for urban development, Massey sought to show how economic restructuring is related to local social and political change and Scott sought to try to demonstrate how neighbourhoods are produced by industrial location, their solutions are at best partial.

Theories of uneven development need to be supplemented by a much fuller analysis of the social, cultural and political processes which shape and are themselves shaped by cities. Much might be gained by uniting some aspects of classical urban sociology with the enhanced understanding of capitalist spatial development. Subsequent chapters examine material inequality, sociation, the cultural specificity of place and the nature of political conflict in the contemporary city, all themes that have featured prominently in urban sociology. Typically though they were explored through analysis of the nature of modernity, rather than of capitalism. What is required is better specification of the relationship between capitalist dynamics and the social conditions of modernity. A principal connection is through the analysis of the inequalities constantly generated by the mechanisms of accumulation which are reproduced, modulated or transformed in the course of the mundane practices of daily life captured by analyses of the experience of modernity.

REFERENCES

Abu-Lughod, J. (1980) *Urban Apartheid: A Study of Rabat*. Cambridge, Mass.: MIT Press.

Bagguley, P., Mark-Lawson, J., Shapiro, D., Urry, J., Walby, S. and Warde, A. (1990) *Restructuring: Place, Class and Gender*. London: Sage.

Castells, M. (1983) *The City and the Grassroots*. London: Edward Arnold.

Christopher, S. and Storper, M. (1986) 'The City as

Studio: The World as Back Lot: The Impact of Vertical Disintegration on the Location of the Modern Picture Industry', *Environment and Planning D: Society and Space* 4(3): 305–20.

Cooke, P. (ed.) (1986) *Global Restructuring, Local Response*. London: ESRC.

Cooke, P. (1989a) "Locality, Economic Restructuring and World Development" in P. Cooke (ed.) *Localities*. London: Unwin Hyman, 1–44.

Cooke, P. (ed.) (1989b) *Localities*. London: Unwin Hyman.

Davis, M. (1990) *City of Quartz: Excavating the Future in Los Angeles*. London: Verso.

Fainstein, S. (1987) "Local Mobilisation and Economic Discontent" in M. P. Smith and J. R. Feagin (eds) *The Capitalist City*. Oxford: Basil Blackwell, 323–42.

Feagin, J. R. and Smith, M. P. (1987) "Cities and the New International Division of Labour: An Overview" in M. P. Smith and J. R. Feagin (eds) *The Capitalist City*. Oxford: Basil Blackwell, 3–36.

Fielding, A. J. (1982) *Counter Urbanisation*. London: Methuen.

Forbes, D. and Thrift, N. (1987) *The Socialist Third World*. Oxford: Basil Blackwell.

Frobel, F., Heinrichs, J. and Kreye, K. (1980) *The New International Division of Labour: Structural Unemployment in Industrial Countries and Industrialisation in Developing Countries*. Cambridge: Cambridge University Press.

Gottdiener, M. (1985) *The Social Production of Urban Space*. Austin: University of Texas Press.

Hall, P. (1988) "Urban Growth in Western Europe" in M. Dogan and J. D. Kasarda (eds) *The Metropolis Era*, vol. 1. New York: Sage, 111–27.

Hall, P. and Hay, P. (1980) *Growth Centers in European Urban Systems*. Berkeley: University of California Press.

Harvey, D. (1973) *Social Justice and the City*. London: Edward Arnold.

Harvey, D. (1977) "Labour, Capital and Class Struggle around the Built Environment in Advanced Capitalist Societies", *Politics and Society* 6: 265–95.

Harvey, D. (1982) *The Limits to Capital*. Oxford: Basil Blackwell.

Harvey, D. (1985a) *The Urbanisation of Capital*. Oxford: Basil Blackwell.

Harvey, D. (1985b) *Consciousness and the Urban Experience*. Oxford: Basil Blackwell.

Harvey, D. (1989) *The Condition of Postmodernity*. Oxford: Basil Blackwell.

Jameson, F. (1984) "Postmodernism, or the Cultural Logic of Late Capitalism", *New Left Review* 146: 53–92.

Jessop, B. (1990) "Regulation Theories in Retrospect and Prospect", *Economy and Society* 19(2): 153–216.

Kasarda, J. (1988) "Economic Restructuring and the American Urban Dilemma", in M. Dogan and J. Kasarda (eds) *The Metropolis Era*. Newbury Park, Calif.: Sage, 56–84.

King, A. (1990) *World Cities*. London: Routledge.

Lampard, E. (1965) "Historical Aspects of Urbanisation" in P. M. Hauser and L. F. Schnore (eds) *The Study of Urbanization*. New York: Wiley, 519–54.

Logan, J. and Molotch, H. (1987) *Urban Fortunes: The Political Economy of Place*. Berkeley: University of California Press.

Marshall, G., Rose, G., Newby, H. and Vogler, C. (1988) *Social Class in Britain*. London: Unwin Hyman.

Massey, D. (1984) *Spatial Divisions of Labour*. London: Macmillan.

Massey, D. (1988) "Uneven Redevelopment: Social Change and Spatial Divisions of Labour" in D. Massey and J. Allen (eds) *Uneven Redevelopment*. London: Hodder & Stoughton.

Massey, D. and Meegan, R. (1979) "The Geography of Industrial Reorganisation", *Progress in Planning* 10: 159–237.

Massey, D. and Meegan, R. (1982) *The Anatomy of Job Loss: The How, Why and Where of Employment Decline*. London: Macmillan.

Murgatroyd, L. and Urry, J. (1985) "The Class and Gender Restructuring of Lancaster" in L. Murgatroyd, M. Savage, D. Shapiro, J. Urry, S. Walby and A. Warde, *Localities, Class and Gender*. London: Pion, 30–53.

Perry, D. C. (1987) "The Politics of Dependency in Deindustrialising America: The Case of Buffalo, New York" in M. P. Smith and J. R. Feagin (eds), *The Capitalist City*. Oxford: Basil Blackwell.

Pinch, S. (1989) "The Restructuring Thesis and the Study of Public Services", *Environment and Planning A* 21(7): 905–26.

Piore, M. and Sabel, C. (1984) *The Second Industrial Divide: Possibilities for Prosperity*. New York: Basic Books.

Saunders, P. (1986) *Social Theory and the Urban Question*, 2nd edn. London: Hutchinson.

Sayer, A. and Walker, R. (1992) *The New Social Economy: Reworking the Division of Labour*. Oxford: Basil Blackwell.

Scott, A. J. (1988a) *Metropolis: From the Division of Labor to Urban Form*. Berkeley: University of California Press.

Scott, A. J. (1988b) *New Industrial Spaces: Flexible Production, Organization and Regional Development in North America and Western Europe*. London: Pion.

Smith, M. P. and Feagin, J. (eds) (1987) *The Capitalist City: Global Restructuring and Community Politics*. Oxford: Basil Blackwell.

Smith, M. P. and Tardanico, R. (1987) "Urban Theory Reconsidered: Production, Reproduction and Collective Action" in M. P. Smith and J. R. Feagin (eds) *The Capitalist City*. Oxford: Basil Blackwell, 87–112.

Soja, E. (1989) *Postmodern Geographies*. London: Verso.

Storper, M. and Walker, R. (1989) *The Capitalist*

Imperative: Territory, Technology and Industrial Growth. Oxford: Basil Blackwell.

Timberlake, M. (1987) "World Systems Theory and Comparative Urbanisation" in M. P. Smith and J. Feagin (eds) *The Capitalist City*. Oxford: Basil Blackwell, 37–65.

Wallerstein, I. (1974) *The Modern World System*, vol. 1. New York: Academic Press.

Williamson, O. E. (1990) "The Firm as a Nexus of Treaties: An Introduction" in M. Aoki, B. Gustafsson and O. Williamson (eds) *The Firm as a Nexus of Treaties*. London: Sage, 1–25.

MICHAEL E. PORTER

"The Competitive Advantage of the Inner City"

Harvard Business Review (1995)

Editors' introduction Harvard Business School professor Michael Porter declares that the economic distress of inner-city neighborhoods may be the most pressing issue facing the United States. Like William Julius Wilson (p. 112) Porter sees lack of jobs as the root cause of drug abuse, crime, and other social problems. And like Wilson, Porter feels that government response to the problem of inner-city decline and lack of jobs has been ineffective.

Porter characterizes the current approach as based on a *social model* aimed at the individual. Governments have historically provided "relief" for the effects economic problems have had on individuals in the form of welfare payments, subsidized housing, food stamps, or free medical assistance. Corporate philanthropy has often done the same.

Porter feels that government programs to create jobs in inner city neighborhoods and train local residents for them have been fragmented and inefficient. They have consisted of subsidies, preference programs, and expensive efforts to stimulate economic activity. Governments have dumped money on small marginal inner-city businesses which could not turn a profit without government help. They have hired incompetent workers or required private sector contractors to hire them. Or they have spent large sums to get blighted redevelopment areas and contaminated brownfields back into useable condition. Often the subsidized firms fail, the workers hired through preference programs are fired, and the brownfields and redevelopment project areas sit vacant.

Porter argues for a new *economic* model of inner-city revitalization. He favors private, for-profit initiatives based on economic self-interest, not artificial inducements, charity, and government mandates. Such an approach will only work, he says, if it takes advantage of the true competitive advantages of the inner city.

Porter punctures the myth that inner-city real estate or labor costs are sufficiently lower to make a compelling reason for firms to locate in inner-city neighborhoods rather than in suburban or exurban locations. Rather, he argues, there are four true advantages to inner city locations: (a) their strategic location, (b) local market demand the areas themselves possess, (c) possibilities of integration with regional job clusters, and (d) an industrious labor force which is eager to work. Firms that can take advantage of these inner-city attributes may find inner-city locations the best place to do business. Unlike Wilson, who emphasizes the loss of jobs that people with limited education can perform, Porter argues that there are still firms that can use unskilled labor for warehouse and production line workers, as truck drivers and in related unskilled jobs.

Porter places a large part of the blame for high costs and the difficulties of doing business in inner cities on government regulation and anti-business attitudes. He argues that local governments can improve the economic climate and make inner-city neighborhoods more attractive for private investment by reducing regulation. This has been a theme in British Enterprise Zone and US

Empowerment Zone/Enterprise Community Programs. In these programs government works to reduce government regulations in distressed neighborhoods in order to lure in business. Government can also enter into the kind of public–private partnerships Mayer (p. 229) describes based on market realities and economic self-interest.

What would his Harvard colleague William Julius Wilson (p. 112) think of Porter's characterization of the social model and preference programs. Given historic racism, bad schools, and the lack of skills many inner-city Black ghetto residents possess, aren't preference programs necesssary?

Michael Porter is the C. Roland Christensen Professor of Business Administration at the Harvard Business School. He is the author of *Competitive Advantage: Creating and Sustaining Superior Performance* (New York: Free Press, 1998), *Competitive Strategy: Techniques for Analyzing Industries and Competitors* (New York: Free Press, 1998), *On Competition* (Boston: Harvard Business School Press, 1998), *The Competitive Advantage of Nations* (New York: Free Press, 1998), an edited anthology, *Strategy: Seeking and Securing Competitive Advantage* (Boston: Harvard Business School Press, 1991) and many other books.

Other books on community-based economic development include Michael H. Shuman, *Going Local: Creating Self-Reliant Communities in a Global Age* (New York: Free Press, 1998), Douglas Henton, John Melville, and Kimberly Walesh, *Grassroots Leaders for a New Economy* (San Francisco: Jossey-Bass, 1997), Richard P. Taub, *Community Capitalism* (Boston: Harvard Business School Press, 1994), Gregory D. Squires (ed.), *From Redlining to Reinvestment* (Philadelphia: Temple University Press, 1992), Edward J. Blakeley, *Planning Local Economic Development* (Thousand Oaks, CA: Sage, 1994), and Julia Ann Parzen, Michael Hall Kieschnick (contributor), *Credit Where It's Due* (Philadelphia: Temple University Press, 1993).

The economic distress of America's inner cities may be the most pressing issue facing the nation. The lack of businesses and jobs in disadvantaged urban areas fuels not only a crushing cycle of poverty but also crippling social problems, such as drug abuse and crime. And, as the inner cities continue to deteriorate, the debate on how to aid them grows increasingly divisive.

The sad reality is that the efforts of the past few decades to revitalize the inner cities have failed. The establishment of a sustainable economic base and with it employment opportunities, wealth creation, role models, and improved local infrastructure – still eludes us despite the investment of substantial resources.

Past efforts have been guided by a social model built around meeting the needs of individuals. Aid to inner cities, then, has largely taken the form of relief programs such as income assistance, housing subsidies, and food stamps, all of which address highly visible – and real – social needs.

Programs aimed more directly at economic development have been fragmented and ineffective. These piecemeal approaches have usually taken the form of subsidies, preference programs, or expensive efforts to stimulate economic activity in tangential fields such as housing, real estate, and neighborhood development. Lacking an overall strategy, such programs have treated

 (Portions of the original article which have been omitted from this edited selection are indicated by ellipses in the form of three dots for material within a single paragraph, or three dots enclosed in brackets for one or more entire paragraphs.)

the inner city as an island isolated from the surrounding economy and subject to its own unique laws of competition. They have encouraged and supported small, subscale businesses designed to serve the local community but ill equipped to attract the community's own spending power, much less export outside it. In short, the social model has inadvertently undermined the creation of economically viable companies. Without such companies and the jobs they create, the social problems will only worsen.

The time has come to recognize that revitalizing the inner city will require a radically different approach. While social programs will continue to play a critical role in meeting human needs and improving education, they must support – and not undermine – a coherent economic strategy. The question we should be asking is how inner-city-based businesses and nearby employment opportunities for inner city residents can proliferate and grow. A sustainable economic base *can* be created in the inner city, but only as it has been created elsewhere: through private, for-profit initiatives and investment based on economic self-interest and genuine competitive advantage – not through artificial inducements, charity, or government mandates.

We must stop trying to cure the inner city's problems by perpetually increasing social investment and hoping for economic activity to follow. Instead, an economic model must begin with the premise that inner city businesses should be profitable and should be positioned to compete on a regional, national, and even international scale. These businesses should be capable not only of serving the local community but also of exporting goods and services to the surrounding economy. The cornerstone of such a model is to identify and exploit the competitive advantages of inner cities that will translate into truly profitable businesses.

Our policies and programs have fallen into the trap of redistributing wealth. The real need – and the real opportunity – is to create wealth.

TOWARDS A NEW MODEL: LOCATION AND BUSINESS DEVELOPMENT

Economic activity in and around inner cities will take root if it enjoys a competitive advantage and occupies a niche that is hard to replicate elsewhere. If companies are to prosper, they must find a compelling competitive reason for locating in the inner city. A coherent strategy for development starts with that fundamental economic principle, as the contrasting experiences of the following companies illustrate.

Alpha Electronics (the company's name has been disguised), a 28-person company that designed and manufactured multimedia computer peripherals, was initially based in lower Manhattan. In 1987, the New York City Office of Economic Development set out to orchestrate an economic "renaissance" in the South Bronx by inducing companies to relocate there. Alpha, a small but growing company, was sincerely interested in contributing to the community and eager to take advantage of the city's willingness to subsidize its operations. The city, in turn, was happy that a high-tech company would begin to stabilize a distressed neighborhood and create jobs. In exchange for relocating, the city provided Alpha with numerous incentives that would lower costs and boost profits. It appeared to be an ideal strategy.

By 1994, however, the relocation effort had proved a failure for all concerned. Despite the rapid growth of its industry, Alpha was left with only 8 of its original 28 employees. Unable to attract high quality employees to the South Bronx or to train local residents, the company was forced to outsource its manufacturing and some of its design work. Potential suppliers and customers refused to visit Alpha's offices. Without the city's attention to security, the company was plagued by theft.

What went wrong? Good intentions notwithstanding, the arrangement failed the test of business logic. Before undertaking the move, Alpha and the city would have been wise to ask themselves why none of the South Bronx's thriving businesses was in electronics. The South Bronx as a location offered no specific advantages to support Alpha's business, and it had several disadvantages that would prove fatal. Isolated from the lower Manhattan hub of computer-design and software companies, Alpha was cut off from vital connections with customers, suppliers, and electronic designers.

In contrast, Matrix Exhibits, a $2.2 million supplier of trade-show exhibits that has 30 employees, is thriving in Atlanta's inner city. When Tennessee based Matrix decided to enter the Atlanta market in 1985, it could have chosen a variety of locations. All the other companies that create and rent trade show exhibits are based in Atlanta's suburbs. But the Atlanta World Congress Center, the city's major exhibition space, is just a six-minute drive from the inner city, and Matrix chose the location because it provided a real competitive advantage. Today Matrix offers customers superior response time, delivering trade-show exhibits faster than its suburban competitors. Matrix benefits from low rental rates for warehouse space – about half the rate its competitors pay for similar space in the suburbs – and draws half its employees from the local community. The commitment of local police has helped the company avoid any serious security problems. Today Matrix is one of the top five exhibition houses in Georgia.

Alpha and Matrix demonstrate how location can be critical to the success or failure of a business. Every location – whether it be a nation, a region, or a city – has a set of unique local conditions that underpin the ability of companies based there to compete in a particular field. The competitive advantage of a location does not usually arise in isolated companies but in clusters of companies – in other words, in companies that are in the same industry or otherwise linked together through customer, supplier, or similar relationships. Clusters represent critical masses of skill, information, relationships, and infrastructure in a given field. Unusual or sophisticated local demand gives companies insight into customers' needs. Take Massachusetts's highly competitive cluster of information-technology industries: it includes companies specializing in semiconductors, workstations, supercomputers, software, networking equipment, databases, market research, and computer magazines.

Clusters arise in a particular location for specific historical or geographic reasons – reasons that may cease to matter over time as the cluster itself becomes powerful and competitively self-sustaining. In successful clusters such as Hollywood, Silicon Valley, Wall Street, and Detroit, several competitors often push one another to improve products and processes. The presence of a group of competing companies contributes to the formation of new suppliers, the growth of companies in related fields, the formation of specialized training programs, and the emergence of technological centers of excellence in colleges and universities. The clusters also provide newcomers with access to expertise, connections, and infrastructure that they in turn can learn and exploit to their own economic advantage.

If locations (and the events of history) give rise to clusters, it is clusters that drive economic development. They create new capabilities, new companies, and new industries. I initially described this theory of location in *The Competitive Advantage of Nations* (Free Press, 1990), applying it to the relatively large geographic areas of nations and states. But it is just as relevant to smaller areas such as the inner city. To bring the theory to bear on the inner city, we must first identify the inner city's competitive advantages and the ways inner city businesses can forge connections with the surrounding urban and regional economies.

THE TRUE ADVANTAGES OF THE INNER CITY

The first step toward developing an economic model is identifying the inner city's true competitive advantages. There is a common misperception that the inner city enjoys two main advantages: low-cost real estate and labor. These so-called advantages are more illusory than real. Real estate and labor costs are often higher in the inner city than in suburban and rural areas. And even if inner cities were able to offer lower-cost labor and real estate compared with other locations in the United States, basic input costs can no longer give companies from relatively prosperous nations a competitive edge in the global economy. Inner cities would inevitably lose jobs to countries like Mexico or China, where labor and real estate are far cheaper.

Only attributes that are unique to inner cities will support viable businesses. My ongoing research of urban areas across the United States identifies four main advantages of the inner city: strategic location, local market demand, integration with regional clusters, and human resources. Various companies and programs have identified and exploited each of those advantages from time to time. To date, however, no systematic effort has been mounted to harness them.

Strategic location

Inner cities are located in what *should* be economically valuable areas. They sit near congested high-rent areas, major business centers, and transportation and communications nodes. As a result, inner cities can offer a competitive edge to companies that benefit from proximity to downtown business districts, logistical infrastructure, entertainment or tourist centers, and concentrations of companies.

Local market demand

The inner city market itself represents the most immediate opportunity for inner-city-based entrepreneurs and businesses. At a time when most other markets are saturated, inner city markets remain poorly served – especially in retailing, financial services, and personal services. In Los Angeles, for example, retail penetration per resident in the inner city compared with the rest of the city is 35% in supermarkets, 40% in department stores, and 50% in hobby, toy, and game stores.

The first notable quality of the inner city market is its size. Even though average inner city incomes are relatively low, high population density translates into an immense market with substantial purchasing power. Boston's inner city, for example, has an estimated total family income of $3.4 billion.

Spending power per acre is comparable with the rest of the city despite a 21% lower average household income level than in the rest of Boston, and, more significantly, higher than in the surrounding suburbs. In addition, the market is young and growing rapidly, owing in part to immigration and relatively high birth rates.

Integration with regional clusters

The most exciting prospects for the future of inner city economic development lie in capitalizing on nearby regional clusters: those unique-to-a-region collections of related companies that are competitive nationally and even globally. For example, Boston's inner city is next door to world-class financial-services and health-care clusters. South Central Los Angeles is close to an enormous entertainment cluster and a large logistical-services and wholesaling complex.

The ability to access competitive clusters is a very different attribute – and one much more far reaching in economic implication – than the more generic advantage of proximity to a large downtown area with concentrated activity. Competitive clusters create two types of potential advantages. The first is for business formation. Companies providing supplies, components, and support services could be created to take advantage of the inner city's proximity to multiple nearby customers in the cluster . . .

The second advantage of these clusters is the potential they offer inner city companies to compete in downstream products and services. For example, an inner city company could draw on Boston's strength in financial services to provide services tailored to inner city needs – such as secured credit cards, factoring, and mutual funds – both within and outside the inner city in Boston and elsewhere in the country.

[. . .]

Human resources

The inner city's fourth advantage takes on a number of deeply entrenched myths about the nature of its residents. The first myth is that inner city residents do not want to work and opt for welfare over gainful employment. Although there is a pressing need to deal with inner city residents who are unprepared for work, most inner city residents are industrious and eager to work. For moderate-wage jobs ($6 to $10 per hour) that require little formal education (for instance, warehouse workers, production-line workers, and truck drivers), employers report that they find hardworking, dedicated employees in the inner city. For example, a company in Boston's inner city neighborhood of Dorchester bakes and decorates cakes sold to supermarkets throughout the region. It attracts and retains area residents at $7 to $8 per hour (plus contributions to pensions and health insurance) and has almost 100 local employees. The loyalty of its labor pool is one of the factors that has allowed the bakery to thrive.

Admittedly, many of the jobs currently

available to inner city residents provide limited opportunities for advancement. But the fact is that they are jobs; and the inner city and its residents need many more of them close to home. Proposals that workers commute to jobs in distant suburbs – or move to be near those jobs – underestimate the barriers that travel time and relative skill level represent for inner city residents. Moreover, in deciding what types of businesses are appropriate to locate in the inner city, it is critical to be realistic about the pool of potential employees. Attracting high-tech companies might make for better press, but it is of little benefit to inner city residents. Recall the contrasting experiences of Alpha Electronics and Matrix Exhibits. In the case of Alpha, there was a complete mismatch between the company's need for highly skilled professionals and the available labor pool in the local community. In contrast, Matrix carefully considered the available workforce when it established its Atlanta office. Unlike the Tennessee headquarters, which custom-designs and creates exhibits for each client, the Atlanta office specializes in rentals made from prefabricated components – work requiring less-skilled labor, which can be drawn from the inner city. Given the workforce, low-skill jobs are realistic and economically viable: they represent the first rung on the economic ladder for many individuals who otherwise would be unemployed. Over time, successful job creation will trigger a self-reinforcing process that raises skill and wage levels.

The second myth is that the inner city's only entrepreneurs are drug dealers. In fact, there is a real capacity for legitimate entrepreneurship among inner city residents, most of which has been channeled into the provision of social services. For instance, Boston's inner city has numerous social service providers as well as social, fraternal, and religious organizations. Behind the creation and building of those organizations is a whole cadre of local entrepreneurs who have responded to intense local demand for social services and to funding opportunities provided by government, foundations, and private sector sponsors. The challenge is to redirect some of that talent and energy toward building for-profit businesses and creating wealth.

The third myth is that skilled minorities, many of whom grew up in or near inner cities, have abandoned their roots. Today's large and growing pool of talented minority managers represents a new generation of potential inner city entrepreneurs. Many have been trained at the nation's leading business schools and have gained experience in the nation's leading companies. Approximately 2,800 African Americans and 1,400 Hispanics graduate from M.B.A. programs every year compared with only a handful 20 years ago. Thousands of highly trained minorities are working at leading companies such as Morgan Stanley, Citibank, Ford, HewlettPackard, and McKinsey & Company. Many of these managers have developed the skills, net-work, capital base, and confidence to begin thinking about joining or starting entrepreneurial companies in the inner city . . .

THE REAL DISADVANTAGES OF THE INNER CITY

The second step toward creating a coherent economic strategy is addressing the very real disadvantages of locating businesses in the inner city. The inescapable fact is that businesses operating in the inner city face greater obstacles than those based elsewhere. Many of those obstacles are needlessly inflicted by government. Unless the disadvantages are addressed directly, instead of indirectly through subsidies or mandates, the inner city's competitive advantages will continue to erode.

Land

Although vacant property is abundant in inner cities, much of it is not economically usable. Assembling small parcels into meaningful sites can be prohibitively expensive and is further complicated by the fact that a number of city, state, and federal agencies each control land and fight over turf . . .

Building costs

The cost of building in the inner city is significantly higher than in the suburbs because of the costs and delays associated with logistics, negotiations with *community* groups, and strict urban regulations: restrictive zoning, architectural codes, permits, inspections, and government-required union contracts and minority setasides.

New Model	Old Model
Economic: create wealth	Social: redistribute wealth
Private sector	Government and social service organizations
Profitable businesses	Subsidized businesses
Integration with the regional economy	Isolation from the larger economy
Companies that are export oriented	Companies that serve the local community
Skilled and experienced minorities engaged in building businesses	Skilled and experienced minorities engaged in the social service sector
Mainstream, private sector institutions enlisted	Special institutions created
Inner city disadvantages addressed directly	Inner city disadvantages counterbalanced
Government focuses on improving the environment for business	Government directly involved with providing services or funding

Figure 1 Inner city economic development

Ironically, despite the desperate need for new projects, construction in inner cities is far more regulated than it is in the suburbs – a legacy of big city politics and entrenched bureaucracies.

[. . .]

Other costs

Compared with the suburbs, inner cities have high costs for water, other utilities, workers' compensation, health care, insurance, permitting and other fees, real estate and other taxes, OSHA compliance, and neighborhood hiring requirements. For example, Russer Foods, a manufacturing company located in Boston's inner city, operates a comparable plant in upstate New York. The Boston plant's expenses are 55% higher for workers' compensation, 50% higher for family medical insurance, 166% higher for unemployment insurance, 340% higher for water, and 67% higher for electricity. High costs like these drive away companies and hold down wages. Some costs, such as those for workers' compensation, apply to the state or region as a whole. Others, such as real estate taxes, apply citywide. Still others, such as property insurance, are specific to the inner city. All are devastating to maintaining fragile inner city companies and to attracting new businesses.

It is an unfortunate reality that many cities – because they have a greater proportion of residents dependent on welfare, Medicaid, and other social programs – require higher government spending and, as a result, higher corporate taxes. The resulting tax burden feeds a vicious cycle – driving out more companies while requiring even higher taxes from those that remain. Cities have been reluctant to challenge entrenched bureaucracies and unions, as well as inefficient and outdated government departments, all of which unduly raise city costs.

Finally, excessive regulation not only drives up building and other costs but also hampers almost all facets of business life in the inner city, from putting up an awning over a shop window to operating a pushcart to making site improvements. Regulation also stunts inner city entrepreneurship, serving as a formidable barrier to small and start-up companies. Restrictive licensing and permitting, high licensing fees, and archaic safety and health regulations create barriers to entry into the very types of businesses that are logical and appropriate for creating jobs and wealth in the inner city.

Security

Both the reality and the perception of crime represent profound impediments to urban economic development. First, crime against property raises costs. For example, the Shops at Church Square, an inner city strip shopping center in Cleveland, Ohio, spends $2 per square foot more than a comparable suburban center for a full-time security guard, increased lighting, and continuous cleaning – raising overall costs by more than 20%. Second, crime against employees and customers creates an unwillingness to work in and patronize inner city establishments and restricts companies' hours of operation. Fear of crime ranks among the most important reasons why companies opening new facilities failed to consider inner city locations and why companies already located in the inner city left. Currently, police devote most of their resources to the security of residential areas, largely overlooking commercial and industrial sites.

Infrastructure

Transportation infrastructure planning, which today focuses primarily on the mobility of residents for shopping and commuting, should consider equally the mobility of goods and the ease of commercial transactions. The most critical aspects of the new economic model – the importance of the location of the inner city, the connections between inner city businesses and regional clusters, and the development of export-oriented businesses – require the presence of strong logistical links between inner city business sites and the surrounding economy. Unfortunately, the business infrastructure of the inner city has fallen into disrepair. The capacity of roads, the frequency and location of highway on-ramps and off-ramps, the links to downtown, and the access to railways, airports, and regional logistical networks are inadequate.

Employee skills

Because their average education levels are low, many inner city residents lack the skills to work in any but the most unskilled occupations. To make matters worse, employment opportunities for less educated workers have fallen markedly. In Boston between 1970 and 1990, for example, the percentage of jobs held by people without high school diplomas dropped from 29% to 7%, while those held by college graduates climbed from 18% to 44%. And the unemployment rate for African-American men aged 16 to 64 with less than a high school education in major northeastern cities rose from 19% in 1970 to 57% in 1990.

Management skills

The managers of most inner city companies lack formal business training. That problem, however, is not unique to the inner city; it is a characteristic of small businesses in general. Many individuals with extensive work histories but little or no formal managerial training start businesses. Inner city companies without well trained managers experience a series of predictable problems that are similar to those that affect many small businesses: weaknesses in strategy development, market segmentation, customer-needs evaluation, introduction of information technology, process design, cost control, securing or restructuring financing, interaction with lenders and government regulatory agencies, crafting business plans, and employee training. Local community colleges often offer management courses, but their quality is uneven, and entrepreneurs are hard-pressed for time to attend them.

Capital

Access to debt and equity capital represents a formidable barrier to entrepreneurship and company growth in inner city areas.

First, most inner city businesses still suffer from poor access to debt funding because of the limited attention that mainstream banks paid them historically. Even in the best of circumstances, small business lending is only marginally profitable to banks because transaction costs are high relative to loan amounts. Many banks remain in small-business lending only to attract deposits and to help sell other more profitable products.

The federal government has made several efforts to address the inner city's problem of debt capital. As a result of legislation like the Community Reinvestment Act, passed in order to overcome bias in lending, banks have begun to pay much more attention to inner city areas. In Boston, for example, leading banks are competing fiercely to lend in the inner city – and some claim to be doing so profitably. Direct financing efforts by government, however, have proved ineffective. The proliferation of government loan pools and quasi-public lending organizations has produced fragmentation, market confusion, and duplication of overhead. Business loans that would provide scale to private sector lenders are siphoned off by these organizations, many of which are high-cost, bureaucratic, and risk-averse. In the end, the development of high quality private sector expertise in inner city business financing has been undermined.

Second, equity capital has been all but absent. Inner city entrepreneurs often lack personal or family savings and networks of individuals to draw on for capital. Institutional sources of equity capital are scarce for minority-owned companies and have virtually ignored inner city business opportunities.

Attitudes

A final obstacle to companies in the inner city is antibusiness attitudes. Some workers perceive businesses as exploitative, a view that guarantees poor relations between labor and management. Equally debilitating are the antibusiness attitudes held by community leaders and social activists. These attitudes are the legacy of a regrettable history of poor treatment of workers, departures of companies, and damage to the environment. But holding on to these views today is counterproductive. Too often, community leaders mistakenly view businesses as a means of directly meeting social needs; as a result, they have unrealistic expectations for corporate involvement in the community . . .

Demanding linkage payments and contributions and stirring up antibusiness sentiment are political tools that brought questionable results in the past when owners had less discretion about where they chose to locate their companies. In today's increasingly competitive business environment, such tactics will serve only to stunt economic growth.

Overcoming the business disadvantages of the inner city as well as building on its inherent advantages will require the commitment and involvement of business, government, and the nonprofit sector. Each will have to abandon deeply held beliefs and past approaches. Each must be willing to accept a new model for the inner city based on an economic rather than social perspective. The private sector, nongovernment or social service organizations, must be the focus of the new model.

The new role of the private sector

The economic model challenges the private sector to assume the leading role. First, however, it must adopt new attitudes toward the inner city. Most private sector initiatives today are driven by preference programs or charity. Such activities would never stand on their own merits in the marketplace. It is inevitable, then, that they contribute to growing cynicism. The private sector will be most effective if it focuses on what it does best: creating and supporting economically viable businesses built on true competitive advantage. It should pursue four immediate opportunities as it assumes its new role.

1. *Create and expand business activity in the inner city.* The most important contribution companies can make to inner cities is simply to do business there. Inner cities hold untapped potential for profitable businesses. Companies and entrepreneurs must seek out and seize those opportunities that build on the true advantages of the inner city. In particular, retailers, franchisers, and financial services companies have immediate opportunities. Franchises represent an especially attractive model for inner city entrepreneurship because they provide not only a business concept but also training and support.

Businesses can learn from the mistakes that many outside companies have made in the inner city. One error is the failure of retail and service businesses to tailor their goods and services to the local market . . .

Another common mistake is the failure to build relationships within the community and to hire locally. Hiring local residents builds loyalty from neighborhood customers, and local employees of retail and service businesses can help stores customize their products. Evidence suggests that companies that were perceived to be in touch with the community had far fewer security problems, whether or not the owners lived in the community.

[. . .]

2. *Establish business relationships with inner city companies.* By entering into joint ventures or customer–supplier relationships, outside companies will help inner city companies by encouraging them to export and by forcing them to be competitive. In the long run, both sides will benefit . . . Such relationships, based not on charity but on mutual self-interest, are sustainable ones; every major company should develop them.

3. *Redirect corporate philanthropy from social services to business-to-business efforts.* Countless companies give many millions of dollars each year to worthy inner-city social-service agencies. But philanthropic efforts will be more effective if they also focus on building business-to-business relationships that, in the long run, will reduce the need for social services.

First, corporations could have a tremendous impact on training. The existing system for job training in the United States is ineffective. Training programs are fragmented, overhead intensive, and disconnected from the needs of industry. Many programs train people for nonexistent jobs in industries with no projected growth. Although reforming training will require the help of government, the private sector must determine how and where resources should be allocated to ensure that the specific employment needs of local and regional businesses are met. Ultimately, employers, not government, should certify all training programs based on relevant criteria and likely job availability.

Training programs led by the private sector could be built around industry clusters located in both the inner city (for example, restaurants, food service, and food processing in Boston) and the nearby regional economy (for example, financial services and health care in Boston). Industry associations and trade groups, supported by government incentives, could sponsor their own training programs in collaboration with local training institutions.

[. . .]

Second, the private sector could make an equally substantial impact by providing management assistance to inner city companies. As with training, current programs financed or operated by the government are inadequate. Outside companies have much to offer companies in the inner city: talent, know-how, and contacts. One approach to upgrading management skills is to emphasize networking with companies in the regional economy that either are part of the same cluster (customers, suppliers, and related businesses) or have expertise in needed areas. An inner city company could team up with a partner in the region who provides management assistance; or a consortium of companies with a required expertise, such as information technology, could provide assistance to inner city businesses in need of upgrading their systems.

[. . .]

4. *Adopt the right model for equity capital investments.* The investment community – especially venture capitalists – must be convinced of the viability of investing in the inner city. There is a small but growing number of minority-oriented equity providers (although none specifically focus on inner cities). A successful model for inner city investing will probably not look like the familiar venture-capital model created primarily for technology companies. Instead, it may resemble the equity funds operating in the emerging economies of Russia or Hungary – investing in such mundane but potentially profitable projects as supermarkets and laundries. Ultimately, inner-city-based businesses that follow the principles of competitive advantage will generate appropriate returns to investors – particularly if aided by appropriate incentives, such as tax exclusions for capital gains and dividends for qualifying inner city businesses.

The new role of government

To date, government has assumed primary responsibility for bringing about the economic revitalization of the inner city. Existing programs at the federal, state, and local levels designed to create jobs and attract businesses have been piecemeal and fragmented at best. Still worse, these programs have been based on subsidies and mandates rather than on marketplace realities. Unless we find new approaches, the inner city will continue to drain our rapidly shrinking public coffers.

Undeniably, inner cities suffer from a long history of discrimination. However, the way for government to move forward is not by looking behind. Government can assume a more effective role by supporting the private sector in new economic initiatives. It must shift its focus from direct involvement and intervention to creating a favorable environment for business. This is not to say that public funds will not be necessary. But subsidies must be spent in ways that do not distort business incentives, focusing instead on providing the infrastructure to support genuinely profitable businesses. Government at all levels should focus on four goals as it takes on its new role.

1. Direct resources to the areas of greatest economic need. The crisis in our inner cities demands that they be first in line for government assistance. This may seem an obvious assertion. But the fact is that many programs in areas such as infrastructure, crime prevention, environmental cleanup, land development, and purchasing preference spread funds across constituencies for political reasons. For example, most transportation infrastructure spending goes to creating still more attractive suburban areas. In addition, A majority of preference-program assistance does not go to companies located in low income neighborhoods.

[. . .]

Unfortunately, the qualifying criteria for current government assistance programs are not properly designed to channel resources where they are most needed. Preference programs support business based on the race, ethnicity, or gender of their owners rather than on economic need. In addition to directing resources away from the inner city, such race-based or gender-based distinctions reinforce inappropriate stereotypes and attitudes, breed resentment, and increase the risk that programs will be manipulated to serve unintended populations. Location in an economically distressed area and employment of a significant percentage of its residents should be the qualification for government assistance and preference programs. Shifting the focus to economic distress in this way will help enlist all segments of the private sector in the solutions to the inner city's problems.

2. Increase the economic value of the inner city as a business location. In order to stimulate economic development, government must recognize that it is a part of the problem. Today its priorities often run counter to business needs. Artificial and outdated government-induced costs must be stripped away in the effort to make the inner city a profitable location for business. Doing so will require rethinking policies and programs in a wide range of areas . . .

Indeed, there are numerous possibilities for reform. Imagine, for example, policy aimed at eliminating the substantial land and building cost penalties that businesses face in the inner city. Ongoing rent subsidies run the risk of attracting companies for which an inner city location offers no other economic value. Instead, the goal should be to provide building-ready sites at market prices. A single government entity could be charged with assembling parcels of land and with subsidizing demolition, environmental cleanup, and other costs. The same entity could also streamline all aspects of building – including zoning, permitting, inspections, and other approvals.

That kind of policy would require further progress on the environmental front. A growing number of cities – including Detroit, Chicago, Indianapolis, Minneapolis, and Wichita, Kansas – have successfully developed so-called brownfield urban areas by making environmental cleanup standards more flexible depending on land use, indemnifying land owners against additional costs if contamination is found on a site after a cleanup, and using tax-increment financing to help fund cleanup and redevelopment costs.

Government entities could also develop a more strategic approach to developing transportation and communications infrastructures, which would facilitate the fluid movement of goods, employees, customers, and suppliers within and beyond the inner city. Two projects in Boston are prime examples: first, a new exit ramp connecting the inner city to the nearby Massachusetts Turnpike, which in turn connects to the surrounding region and beyond; and a direct access road to the harbor tunnel, which connects to Logan International Airport. Though inexpensive, both projects are stalled because the city does not have a clear vision of their economic importance.

3. *Deliver economic development programs and services through mainstream, private sector institutions.* There has been a tendency to rely on small community-based nonprofits, quasi-governmental organizations, and special-purpose entities, such as community development banks and specialized small-business investment corporations, to provide capital and business-related services. Social service institutions have a role, but it is not this. With few exceptions nonprofit and government organizations cannot provide the quality of training, advice, and support to substantial companies that mainstream, private sector organizations can. Compared with private sector entities such as commercial banks and venture capital companies, special-purpose institutions and nonprofits are plagued by high overhead costs; they have difficulty attracting and retaining high-quality personnel, providing competitive compensation, or offering a breadth of experience in dealing with companies of scale.

[. . .]

The most important way to bring debt and equity investment to the inner city is by engaging the private sector. Resources currently going to government or quasi-public financing would be better channeled through other private financial institutions or directed at recapitalizing minority-owned banks focusing on the inner city, provided that there were matching private sector investors. Minority-owned banks that have superior knowledge of the inner city market could gain a competitive advantage by developing business-lending expertise in inner city areas.

As in lending, the best approach to increase the supply of equity capital to the inner cities is to provide private sector incentives consistent with building economically sustainable businesses. One approach would be for both federal and state governments to eliminate the tax on capital gains and dividends from long-term equity investments in inner-city-based businesses or subsidiaries that employ a minimum percentage of inner city residents. Such tax incentives, which are based on the premise of profit, can play a vital role in speeding up private sector investment. Private sector sources of equity will be attracted to inner city investment only when the creation of genuinely profitable businesses is encouraged.

4. *Align incentives built into government programs with true economic performance.* Aligning incentives with business principles should be the goal of every government program. Most programs today would fail such a test. For example, preference programs in effect guarantee companies a market. Like other forms of protectionism, they dull motivation and retard cost and quality improvement. A 1988 General Accounting Office report found that within six months of graduating from the Small Business Association's purchasing preference program, 30% of the companies had gone out of business. An additional 58% of the remaining companies claimed that the withdrawal of the SBA's support had had a devastating impact on business. To align incentives with economic performance, preference programs should be rewritten to require an increasing amount of non-set-aside business over time.

Direct subsidies to businesses do not work. Instead, government funds should be used for site assembly, extra security, environmental cleanup, and other investments designed to improve the business environment. Companies then will be left to make decisions based on true profit.

The new role of community-based organizations

Recently, there has been renewed activity among community-based organizations (CBOs) to become directly involved in business development. CBOs can, and must, play an important supporting role in the process. But choosing the proper strategy is critical, and many CBOs will have to change fundamentally the way they operate. While it is difficult to make a general set of recommendations to such a diverse group of organizations, four principles should guide community-based organizations in developing their new role.

1. *Identify and build on strengths.* Like every

other player, CBOs must identify their unique competitive advantages and participate in economic development based on a realistic assessment of their capabilities, resources, and limitations. Community-based organizations have played a much-needed role in developing low-income housing, social programs, and civic infrastructure. However, while there have been a few notable successes, the vast majority of businesses owned or managed by CBOs have been failures. Most CBOs lack the skills, attitudes, and incentives to advise, lend to, or operate substantial businesses. They were able to master low-income housing development, in which there were major public subsidies and a vacuum of institutional capabilities. But, when it comes to financing and assisting for-profit business development, CBOs simply can't compete with existing private sector institutions.

Moreover, CBOs naturally tend to focus on community entrepreneurship: small retail and service businesses that are often owned by neighborhood residents. The relatively limited resources of CBOs, as well as their focus on relatively small neighborhoods, is not well-suited to developing the more substantial companies that are necessary for economic vitality.

Finally, the competitive imperatives of for-profit business activity will raise inevitable conflicts for CBOs whose mission rests with the community. Turning down local residents in favor of better qualified outside entrepreneurs, supporting necessary layoffs or the dismissal of poorly performing workers, assigning prime sites for business instead of social uses, and approving large salaries to successful entrepreneurs and managers are only a handful of the necessary choices. Given these organizations' roots in meeting the social needs of neighborhoods, it will be difficult for them to put profit ahead of their traditional mission.

2. *Work to change workforce and community attitudes.* Community-based organizations have a unique advantage in their intimate knowledge of and influence within inner city communities, and they can use that advantage to help promote business development. CBOs can help create a hospitable environment for business by working to change community and workforce attitudes and acting as a liaison with residents to quell unfounded opposition to new businesses . . .

3. *Create work-readiness and job referral systems.* Community-based organizations can play an active role in preparing, screening, and referring employees to local businesses. A pressing need among many inner city residents is work-readiness training, which includes communication, self-development, and workplace practices. CBOs, with their intimate knowledge of the local community, are well equipped to provide this service in close collaboration with industry . . .

CBOs can also help inner city residents by actively developing screening and referral systems. Admittedly, some inner-city-based businesses do not hire many local residents. The reasons are varied and complex but seem to revolve around a few bad experiences that owners have had with individual employees and their work attitudes, absenteeism, false injury claims, or drug use . . .

4. *Facilitate commercial site improvement and development.* Community-based organizations (especially community development corporations) can also leverage their expertise in real estate and act as a catalyst to facilitate environmental cleanup and the development of commercial and industrial property . . .

OVERCOMING IMPEDIMENTS TO PROGRESS

This economic model provides a new and comprehensive approach to reviving our nation's distressed urban communities. However, agreeing on and implementing it will not be without its challenges. The private sector, government, inner city residents, and the public at large all hold entrenched attitudes and prejudices about the inner city and its problems. These will be slow to change. Rethinking the inner city in economic rather than social terms will be uncomfortable for many who have devoted years to social causes and who view profit and business in general with suspicion. Activists accustomed to lobbying for more government resources will find it difficult to embrace a strategy for fostering wealth creation. Elected officials used to framing urban problems in social terms will be resistant to changing legislation, redirecting resources, and taking on recalcitrant bureaucracies. Government entities may find it hard to cede power and control accumulated through past programs. Local leaders who have built social service organizations and merchants who have run mom-and-pop stores could feel threatened by the creation of new initiatives and centers of power. Local politicians schooled in old-style community organizing and confrontational politics will have to tread unfamiliar ground in facilitating cooperation between business and residents.

These changes will be difficult ones for both individuals and institutions. Nonetheless, they must be made. The private sector, government, and community-based organizations all have vital new parts to play in revitalizing the economy of the inner city. Businesspeople, entrepreneurs, and investors must assume a lead role; and community activists, social service providers, and government bureaucrats must support them. The time has come to embrace a rational economic strategy and to stem the intolerable costs of outdated approaches.

THE CRISIS,

OR THE CHANGE FROM ERROR AND MISERY, TO TRUTH AND HAPPINESS

1832.

IF WE CANNOT YET RECONCILE ALL OPINIONS,

LET US ENDEAVOUR TO UNITE ALL HEARTS

IT IS OF ALL TRUTHS THE MOST IMPORTANT, THAT THE CHARACTER OF MAN IS FORMED FOR — NOT BY HIMSELF.

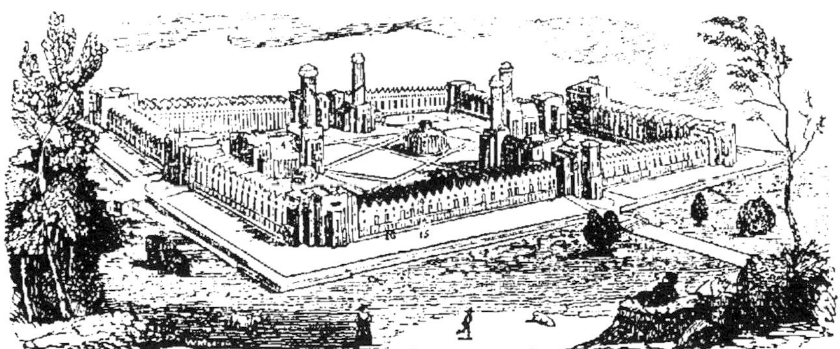

Design of a Community of 2,000 Persons, founded upon a principle, commended by Plato, Lord Bacon, Sir T. More, & R. Owen

EDITED BY

ROBERT OWEN AND ROBERT DALE OWEN.

London:

PRINTED AND PUBLISHED BY J. EAMONSON, 15, CHICHESTER PLACE GRAY'S INN ROAD.

STRANGE, PATERNOSTER ROW, PURKISS, OLD COMPTON STREET. AND MAY BE HAD OF ALL BOOKSELLERS.

PART 5

Urban Planning History and Visions

PART FIVE

INTRODUCTION

The effects of urban planning are perhaps the greatest – and, at the same time, the most invisible – influences on human life and culture. In the words of Paul and Percival Goodman, the co-authors of *Communitas: Means of Livelihood and Ways of Life* (1947), we hardly realize as we go about the daily round of our lives "that somebody once drew some lines on a piece of paper who might have drawn otherwise" and that "now, as engineer and architect once drew, people have to walk and live."

When the Sumerian kings built the walls of Eridu and Uruk, they engaged in acts of urban planning and thus determined how their people would "walk and live." The walls provided safety and protection for the people of the city and also defined the new political unity of the city-state. The associated roads, bridges, irrigation systems, and centers for market and ceremonial functions all served a dual function in that they met the practical social needs of the urban population in general and fulfilled the power aspirations of the urban elites in particular. The ancient citadels were centers of religious meaning, as well as economic and political power, and thus a third component of urban planning – an idealized, often spiritual vision of what constitutes the best possible state of human existence – was present at the very beginning of city building.

The origins of modern urban planning are complex. On one level, modern planning is a direct extension of the ancient and pre-modern models: imposing order on nature for the health, safety, and amenity of the urban masses, for the political benefit of the urban elites, and as a way of expressing each culture's highest spiritual ideals. On another level, however, modern planning is far more complex than anything that had ever gone on before. Modern planning operates, by and large, in a politically and economically pluralistic environment, making every alteration of the physical arrangements of the city a complex negotiation between competing interests. And the practice of modern urban planning also takes place at a stage of human development when the planner's defining goal is no longer merely to impose human order on nature, but continuously to impose order on the city itself.

All the goals and functions of planning – both the ancient hold-overs and the modern elaborations – are present in the first planning responses to the urban conditions associated with the Industrial Revolution. As Friedrich Engels (p. 46) and other contemporary observers described, the cities of the new industrialism were characterized by horrendous overcrowding, ubiquitous misery and despair. There were daily threats to the public health and safety, not just for the impoverished working class but for the capitalist middle class as well. These conditions gave rise to movements for housing reform, to great advances in the technologies of water supply and sewage disposal and to the emergence of middle-class suburbs. They also led to the construction of model "company towns" by various industrial firms in both Europe and America, and to the development of a modern urban planning profession.

In "Modernism and Early Urban Planning, 1870–1940," Richard LeGates and Frederic Stout (the co-editors of this volume) review the history of modern urban planning from its origins in the public health, utopian reform, and aesthetic movements of the nineteenth century to the professionalization of planning as an essential state activity during the early twentieth century. Even in its earliest stages, the practice of city planning exhibited certain characteristic concerns and methodologies that continue to be the focus of the work of planning in the contemporary world. Hence, "the classic texts of early urban planning history often seem surprisingly modern."

An example of the surprising modernity of early urban planning is the nineteenth-century parks movement, especially the work of Frederick Law Olmsted, which gave rise to something very like comprehensive urban planning practice. Projects like Central Park in New York (Plate 23) represented a transplantation and democratization of European landscape gardening traditions, to be sure, but Olmsted's goal was not merely to bring nature into the city. Rather, Olmsted repeatedly appealed to the political and economic leadership of American cities to create parks that would achieve a whole range of public benefits: they would contribute to the public health by serving as the "lungs" of the city; they would be practical and necessary additions to the physical infrastructure of the metropolis, providing a general recreation ground; their ponds and reservoirs would serve as adjuncts to municipal water-supply systems; and they would soften and tame human nature, by providing wholesome alternatives to the vulgar street amusements that daily tempted poor and working-class youth.

Olmsted was a visionary and a reformer, but he was also a successful businessman and a canny political operative capable of offering his clients useful strategic advice on how to fund and build constituencies in favor of large municipal projects. Somewhat less practical, but even more visionary, were a group of architects, planners, and activists who may be termed, collectively, the utopian modernists. Three of these – Ebenezer Howard, Le Corbusier, and Frank Lloyd Wright – define the mainstream of that utopian tradition. Not one of them had his utopian vision realized in its entirety, but each had an enormous influence on the way contemporary cities, and city life, developed in the twentieth century. A fourth, the Spanish engineer/planner Soria y Mata, is influential for his vision of the relationship between transportation systems and land use. Plate 24 illustrates Soria y Mata's vision of a "linear city" developed along a central spine containing an electric streetcar line and other utilities.

Ebenezer Howard prided himself on being "the inventor of the Garden City idea," and his tireless devotion to the project of decongesting the modern metropolis by building small, self-contained, green-belted cities in the rural countryside is one of the marvels of modern urban planning history. Plate 25 from *Garden Cities of To-morrow* illustrates Howard's vision of "a group of slumless, smokeless cities." Howard originally wanted his garden cities to be cooperatively owned. He wanted the surrounding greenbelt to be much larger than the built-up part of the city itself. And he wanted his cities to be economically independent, not commuter suburbs. In the process of actually building Letchworth and Welwyn – the two garden cities constructed before his death in 1924 – Howard had to compromise many of his original goals. Building lots and businesses were privately owned; the greenbelt became more of a park than an extensive rural buffer zone; and neither of the original garden cities ever became a fully independent economic entity. Nonetheless as Plate 26, the plan for Welwyn illustrates, these were fully planned communities that embodied many of Howard's ideals. Moreover, the Garden City experiment gave rise to a larger movement of town planning, and disciples of Howard spread his ideas and his example worldwide.

Among the disciples and popularizers of Howard's ideas was an indefatigable prophet of regionalism, Patrick Geddes. In the process of promoting the Garden City vision, Geddes helped to establish some of the basic principles of modern planning practice – such as conducting a comprehensive survey and soliciting informed public input before drawing up the plan. His writings, including *City Development* (1904) and *Cities in Evolution* (1915), are among the classics of urban planning history still taught in planning schools today.

Charles Edouard Jeanneret, better known as Le Corbusier, was another utopian visionary who never saw his ideal plans fully developed but who was enormously influential nonetheless. Le Corbusier wanted his "Contemporary City of Three Million," illustrated in Plate 27, to be a series of exquisite towers, geometrically arranged in a surrounding park, and he spent years looking for governmental and industrial sponsors for his plan. Many "Corbusian" high-rise urban developments have been built throughout the world, of course. Indeed, the "International Style" of modern architecture and the principles of the International Congress of Modern Architecture (CIAM) which he pioneered have become global standards of urban development. But in almost every case, the surrounding park has been compromised away in the process of realization. In case after case, the tower in the park has become the tower without the park or, even worse, the tower in the parking lot!

While Le Corbusier was issuing his manifestos and shocking the architectural and planning establishment with his modernist plans, American planners Clarence Stein and Henry Wright were also wrestling with the problem of how to adapt urban form to the automobile. In their influential plan for Radburn, New Jersey, illustrated in Plate 28, Stein and Wright invented and implemented a series of planning concepts including superblocks and the separation of pedestrian and vehicular traffic.

Frank Lloyd Wright, the originator of the visionary "Broadacre City" plan, was also responding to the automobile. Wright called for a city composed of family homesteads – one full acre per person – and the withering away of the dense and crowded traditional cities. Wright's 1935 plan for Broadacre City is shown in Plate 29. The private automobile, Wright thought, would virtually abolish distance and allow for a new kind of community based on individualism and self-reliance. What actually became of Wright's Broadacre was sprawl suburbia. One acre per person became one-eighth acre per family or less; the core cities refused to wither away; the transportation monoculture of the automobile became a new form of dependency, rather than a technology of liberation; and the family-oriented suburban community became problematic at best, an object of ridicule at worst.

The utopian visionaries were more that just planners, if they can be said to be planners at all. Even Ebenezer Howard, the most moderate of the group, was a dreamer. Together, the utopian modernists concerned themselves with great philosophical issues such as the connection between Mankind and Nature, the relationship of city plan to moral reform, and role of urban design and technologies to the evolutionary transformation of society. It would be left to more practical men and women – the actual members of the urban planning profession as it developed in the twentieth century – to address the real-world problems of ever-changing cities and metropolitan regions. If the utopian modernists established the lofty goals, the professional planners attended to the details.

Still, the role of visionary projections of better lives through better urban planning persists as an important motivating force in contemporary urban planning. Establishing a good planning vision and sticking to it can have profound positive impacts. Plate 30 illustrates the famous "Paseo del Rio" of San Antonio, Texas. Like hundreds of other cities San Antonio had a blighted river-turned-drainage-ditch disfiguring the downtown. But unlike other cities San Antonio developed a vision of turning the problem area into a magnificent area of riverfront provides a pleasant place to sit, stroll, paddle, and shop. Boston, Massachusetts worked the same kind of urban planning magic by collaborating with developer James Rouse to turn a seedy and obsolete marketplace around Quincy Market into a magnificent center for strolling, shopping, dining, and cultural events. Quincy Market today is illustrated in Plate 31.

The tradition of urban planning visions is alive and well today. In addition to visions of sustainable urban development discussed throughout this anthology (p. 434, p. 519, and p. 540), the "new urbanism" is both a profit-making process of real-estate development that emphasizes small-town scale as the basis of new communities from Florida to California and a practical application of visionary ideas. Architect/planner Peter Calthorpe advocates a vision of transit-oriented development (TOD) in which land use and transportation systems are designed together in harmony with the natural environment to produce livable communities. One of Calthorpe's visions, which is being implemented in a number of projects throughout the United States, is for "Pedestrian Pockets." Pedestrian pockets such as that illustrated in Plate 32 are being built not as a further extension of traditional suburbia but in and for what he has variously called the "Next American Metropolis" and the "Post-Suburban Metropolis." In one sense, Calthorpe's pedestrian pockets and transit-oriented developments look backward to the garden cities of Ebenezer Howard – especially in their use of light-rail mass-transit options – but, in another sense, they look forward to an entirely new relationship between city and region, between individual and community. In short, Calthorpe's vision is a response to a future already in the process of becoming, characterized by "a dramatic shift in the nature and location of our work place and a fundamental deviation in the character of our increasingly diverse households."

RICHARD T. LEGATES and FREDERIC STOUT

"Modernism and Early Urban Planning, 1870–1940"

adapted from "Editors' Introduction" to Richard T. LeGates and Frederic Stout (eds.), *Early Urban Planning, 1870–1940* (1998)

Editors' introduction Although urban planning is as old as cities themselves, it was the modern period and the modernist movement that gave rise to the professional practice of planning that we recognize today as an essential component of social organization and political governance. First with the rise of rationalism and science during the Renaissance and the Enlightenment, and later with the emergence of the chaotic conditions of the industrial city, urban planning developed as a way of organizing and stabilizing urban society through the rational design of space and the systematic ordering of human activity.

In "Modernism and Early Urban Planning," Richard T. LeGates and Frederic Stout (the co-editors of this volume) outline the development of planning from the early urban reform movements of the nineteenth century to the elaboration and professionalization of planning practice in the first half of the twentieth century. They describe various responses to the horrors of the nineteenth century industrial city. Most notable of these movements were the parks movement led by Frederick Law Olmsted, public hygiene proposals, and agitation for decongestion and housing reform by social reformers on both sides of the Atlantic. These early efforts at the efficient planning of cities combined with aesthetic movements such as the proponents of "city beautiful" planning and "civic art." Social utopianists such as Arturo Soria y Mata and Ebenezer Howard added their linear and garden city proposals. From all of these strands, professional, comprehensive planning practice that we are familiar with today began to arise first in Germany, England and the United States; later everywhere in the world. Great implementers like Raymond Unwin and Patrick Abercrombie in England and Clarence Stein and Henry Wright in the United States designed and built communities that incorporated many of the planning visions.

Although extraordinary advances have been made in planning theory and practice in the years since World War II – see Peter Hall's "The City of Theory" (p. 362) and the other selections in Part 6 of this volume – many of the basic concerns of modern urban planning – even advocacy, citizen participation, and systems thinking – were present from the beginning. Planners today are turning back to the ecologically oriented planning advocated by Patrick Geddes, John Nolan's proto-"new urbanism," and especially Lewis Mumford's forward looking ideas on regional planning.

Richard T. LeGates is a professor of political science and urban studies at San Francisco State University and has published widely in the fields of urban housing, planning, and law. Frederic Stout is a lecturer in urban studies at Stanford University, where he also works with the Center for Teaching and Learning. A longer version of "Modernism and Early Urban Planning" appears as the Editors' Introduction to *Early Urban Planning, 1870–1940*, a nine-volume compilation of classic texts in urban planning history (London: Routledge/Thoemmes Press, 1998).

Other books on urban planning history and visions include Peter Hall, *Cities of Tomorrow*, 2nd edn, (Oxford and Cambridge, MA: Blackwell, 1998), Donald A. Krueckeberg, *The American Planner* (New York: Methuen, 1994), Christine Boyer, *Dreaming the Rational City* (Cambridge, MA: MIT Press, 1986), and John Reps, *The Making of Urban America* (Princeton, NJ: Princeton University Press, 1965).

For another analysis of the relationship between modernism and urbanism, see Stephen Toulmin, *Cosmopolis: The Hidden Agenda of Modernity* (New York: Free Press, 1990).

INTRODUCTION

In the years since World War II, urban planning has made enormous advances but has, at century's end, reached something of a conceptual plateau and now faces an uncertain future. Increasingly, the extraordinary accomplishments of recent urban planning have been paralleled by equally notable failures: the destructive, unintended consequences of many massive urban renewal and redevelopment schemes; a perceived lack of democratic participation on the part of affected populations; and a persistent inability to solve the seemingly intractable problems of inner-city poverty, crime, and social alienation. Confronted with this apparent impasse, a review of urban planning literature from the mid-nineteenth century up to World War II will permit us to evaluate the history of urban planning as one of the great, indeed paradigm projects of the modern era and may lay the groundwork for reasonable speculation about the future of urban planning in a rapidly emerging new world that will be characterized by postmodernist social and cultural norms.

Urban planning and modernism

In *Communitas* (1947), Percival and Paul Goodman wrote that "the works of engineering and architecture and town plan are the heaviest and biggest part of what we experience. They lie underneath, they loom around, as the prepared place of our activity." "[S]omebody once drew some lines on a piece of paper," they observed, and "now, as engineer and architect once drew, people have to walk and live." This simple statement eloquently captures both the essence and the scope of the modern urban planning project: to shape and control the human physical and social world through rational design and policy.

City building has preoccupied kings and cardinals, mayors and burghers, for thousands of years. But it was only in the modern period that urban planning became an accepted profession and a well-defined field of study. Although the origins of modernism lie in the Renaissance re-discovery of classical learning and the Enlightenment attempt to impose a rational order on both external nature and the social nature of mankind, what we call modern urban planning – no less than modern art, or modern politics, or modern family life – was born of the fundamental transformations of economic, social, and political life that propelled and grew out of the Industrial Revolution.

Among the social and cultural dislocations of industrial modernism were internal and transnational immigration; the emergence of the industrial working class and a persistent urban underclass; an unequal global division of resources between the industrially advanced nations of Europe and North America and the less developed countries of Latin America, Asia, and Africa; and the development of new political and economic systems – communism, democratic socialism, welfare-state liberal capitalism – aimed at planning and regulating the potentially explosive social reality associated with the modern bureaucratic state. Insofar as urban planning has, in one form or another, addressed all of these issues and concerns from the very beginning, the classic texts of early urban planning history often seem surprisingly modern. Responding to the squalor and disorder of the industrial city, European and American professionals of every stripe – landscape architects, settlement house workers, public health officials, land use legal experts, philanthropists, local elected officials, and business leaders – asked what could be done to make cities more healthy, beautiful, efficient, governable, just, and humane. How, they wondered, might cities be related to their natural environments and to other cities in their regions? And how might the terrible gulf between capitalists and proletarians, between haves and have-nots, be narrowed and the social wound healed so as to create a better society and a higher sense of community for all?

The view ahead

The end of a century naturally invites exercises in retrospect and prospect. Today, a new *post*modernist era is dawning. Far more than merely an architectural style or that congeries of novel analytical theories and methodologies that so excites contemporary academia, post-modernism, as used in this essay, refers to the compelling and widely felt sense of profound cultural change that currently grips the imaginations of people worldwide as we experience the transition from the industrial to a post-industrial period of history.

The collapse of the Soviet Union and the end of the Cold War have swept away an old "world order" without clearly defining the terms of a new one. An invigorated global marketplace calls received notions of nationality and polity into grave question. And the rapidly developing, ever newer technologies of computing and telecommunications herald the coming of an information-based economic system that is rushing toward us, ready or not. In this transformational context, the theory and practice of modern urban planning are called into question as never before. Today, historians of urban planning are faced with a series of inescapable questions: In a postmodern, post-industrial world, what will urban planning look like? How will its claims to authority be revised in the light of experience? What will be its new challenges? And what will be its broad cultural purposes as the activity that shapes and defines "the prepared place" of human activity?

NINETEENTH-CENTURY URBAN CRISIS AND REFORM

Modernism and the Industrial Revolution

Of all the contradictions that the burgeoning European and American cities of the late nineteenth and early twentieth century presented, none was more striking than the contrast between the astonishing advances in technology and industry and the equally astonishing disorder and distress in the cities that were the nesting grounds of capitalist industrialization.

Manchester, in the English Midlands, has long been considered the prototype city of the rapid urbanization process that accompanied the Industrial Revolution. Industrial Manchester was observed in minute and critical detail by Friedrich Engels (1820–1895) in *The Condition of the Working Class in England in 1844*. Engels presents a devastating critique of unrestrained capitalist social aggression and a call to arms in support of the oppressed workers. He also describes in sickening detail the environmental pollution caused by unregulated industrial enterprises and grossly unsanitary conditions caused by a total lack of municipal sewer and water-supply systems; the overcrowding that leads to various forms of ill-health and moral depravity; the sense of personal alienation and social anomie that results from the hub-bub of modern urban life; and how specific planning interventions – such as the uniformity of facade treatments along the long boulevards that run from the central city to the outlying suburbs – psychologically separate the classes by quarantining the hovels of the poor from the sensibilities of the privileged. In all of this, *The Condition of the Working Class* – the ur-document of modern urban studies – addresses issues that remain relevant a century and a half later.

The parks movement

One of the first responses to the horrors and social dislocations of industrial urbanism was romantic utopianism. Indeed, utopian visions have played a continuing role in the history of urban social and environmental reform. Strongly influenced by romantic notions of social perfectionism, the parks movement was one of the earliest responses to the social and environmental issues raised by industrial urbanism. Landscape gardening had existed as a sub-category of architecture long before urban planning came into being as a named and defined profession, and many landscape architects saw beyond the design of gardens to a larger canvas on which to paint an environmental context for human social activity. Beginning in the Renaissance, monarchs and aristocrats had begun to open up and embellish dense medieval cities with formal squares and gardens.

Eventually the practice of park building evolved in more democratic directions. In 1844, the city of Liverpool engaged the gardener Joseph Paxton to lay out Birkenhead Park as the first urban garden, complete with recreation areas for sports, open to the general public. In London, Victoria Park was opened in 1845, and in 1872 it was enlarged to encompass more than 200 acres of gardens, walkways, ponds, meadows and woods. And in Paris, the extensive demolition and redesign project of Baron Haussmann also involved extensive landscape design elements. Although an ambitious 1859 project to encircle the entire city with a greenbelt was defeated, Haussmann's broad boulevards and tree-lined avenues brought well-tended greenery into the city and connected both existing suburban parklands such as the Bois de Boulogne and the Bois de Vincennes and new parks such as Montsoris and Buttes-Chaumont to within the reach of the city's burgeoning population.

In America, the man who transformed landscape gardening from a skill in service to the landed aristocracy to a vehicle of democratic social reform was Frederick Law Olmsted (1822–1903). In 1811, the City of New York published a plan for the eventual development of the whole of Manhattan Island. Although the inhabited city only occupied the tip of Manhattan at the time – with the rest of the island given over to farms, grazing lands, and a few isolated villages – the Commissioners' Plan proposed an unrelieved gridiron platting that extended from Wall Street to the northernmost tip of the island and from the East River to the Hudson. Almost immediately, the Commissioners' Plan was opposed by public-spirited community leaders, and over a period of three decades the opposition coalesced into an organized popular movement to build a great Central Park for the citizens of New York. Eventually Olmsted and his partner, the young British architect Calvert Vaux, won the design competition and together began work in 1857.

As it neared completion in 1863, Central Park was recognized as a masterpiece, and it remains today one of the most successful examples of the enhancement of urban space by the intervention of artfully designed nature. Composed primarily in the naturalistic English landscape tradition, Central Park also contains formal gardens and elegant esplanades. But Olmsted was much more than just a landscape naturalist. For Olmsted, park building became the genesis of a much larger conception of urban design. The Central Park design mixes spaces meant for picnics and baseball with spaces meant for solitary walks and quiet contemplation. It pioneered the use of a multi-level transportation network that separated pedestrian traffic from carriages and that permitted cross-town traffic to transverse the park unobtrusively. In addition, it functioned as an integral part of the great Croton Reservoir system that provided fresh water to the whole of Manhattan.

In an address published as "Public Parks and the Enlargement of Towns," Olmsted demonstrated the extraordinary breadth and richness of his understanding of urban life in all its dimensions – the physical, the social, the political, and the cultural – and of the inter-connectedness of those aspects of the urban whole. Olmsted was fascinated by the possibilities inherent in the latest technologies such as forced air heating systems, the telegraph, and networks of pneumatic tubes for the dissemination of letters and small packages. And for Olmsted, parks and good urban design also meant the nurturing and preservation of social morality. The park was to be a vehicle to control vice and provide healthy outlets for the city's poor and working-class populations. In his view the park would exercise "a distinctly harmonizing and refining influence upon the most unfortunate and most lawless classes of the city." Offering both active recreational opportunities (baseball) and passive entertainment (concerts), the urban park would be an alternative to the grog shop.

Planning for public health

The parks movement sought to provide the congested city with "lungs," and this concern tied Olmsted and other landscape designers to an allied approach to planning, the public hygiene movement. Cities have always had higher mortality rates than rural areas, largely because of epidemics and disease caused by the crowding together of large numbers of people in unhealthy conditions. But the enormous increase of urban populations in the nineteenth century and the misery entailed by the Industrial Revolution greatly compounded urban health problems. Before Haussmann's transformation of Paris in the 1850s and 1860s, cemeteries within the city limits leached into the aquifer, tens of thousands of households were crowded into cellar tenements without running water, and city sewers emptied into the Seine. Conditions in London, Manchester, and the other industrial centers of England were worse. In Europe and America, cholera and yellow fever epidemics periodically killed large numbers of people, leaping across class boundaries to spread terror to middle- and upper-class sections of cities as well as the slums. The new generation of city dwellers brought up breathing air heavy with coal dust, drinking polluted water, living in lightless, airless tenements, and subjected to long hours of physical labor without concern for occupational health and safety were increasingly found to be unfit for military service or any form of work.

One of the most interesting statements on planning for improved municipal public health was *Hygeia, City of Health* (1876), by the physician Benjamin Ward Richardson (1828–1896). Richardson saw disease and ill-health as the principal curse of modern industrialism and sought to discover "the conditions which lead to the pain and penalty of disease." In *Hygeia*, he designed a city that would achieve "the co-existence of the lowest possible general mortality with the highest possible individual longevity." Hygeia would be a new city of 100,000 people on 4,000 acres in which there would be a total prohibition of alcohol and tobacco, elaborate technologies to eliminate air pollution from chimneys, "ozone generators" to purify the water supply, public laundries and slaughter houses, and a public health system overseen by a duly qualified medical man as principal sanitary officer to "watch over the sanitary welfare of the place." Each house would be built on arches of solid brickwork with subways through which the air would flow freely, and all currents of water would be carried away. As a result of these hygienic designs, Richardson was convinced that the moral health of the community, especially its children, would be improved along with its physical health. "Gutter children are an impossibility," he wrote, "in a place where there are no gutters . . . instead of the gutter, the poorest child has the garden."

Ebenezer Howard and the Garden City ideal

Marx and Engels included the "gradual abolition of the distinction between town and country" as one of their basic goals in *The Communist Manifesto* of 1848. For the Spanish engineer Arturo Soria y Mata (ca. 1890), the key to improved public health was an urban plan that eliminated congestion and kept the open countryside close at hand, and a very similar solution was proposed by the Kansas eccentric Henry Olerich in *A Cityless and Countryless World* (1893). But of all of the nineteenth century urban utopias that sought to reintegrate the urban and the rural, the Garden City plan of Ebenezer Howard (1850–1928) is far and away the most important, both as a unified vision addressing the full range of urban development issues and as an initiator and formulator of modern urban planning as a profession and a body of theory.

Howard was a modest man who worked tirelessly and selflessly for three decades to make his dream a reality and lived to see two of his garden cities, Letchworth and Welwyn, actually built. A parliamentary stenographer by trade, Howard was caught up in the heady spirit of radical reform and borrowed freely from other visionaries and activists. Referring to himself only as "the inventor of the Garden City idea," he never claimed originality for what he called his "unique combination of proposals."

Howard's plan was first published in 1898 as *To-morrow: A Peaceful Path to Real Reform* (republished, in 1902 and subsequently, as *Garden Cities of To-morrow*). The book begins with the observation that the one question "on which all persons, no matter of what political party, or of what shade of sociological opinion, would be found to be fully and entirely agreed" is "that it is deeply to be deplored that the people should continue to stream into the already over-crowded cities, and should thus further deplete the country districts." Howard explained this movement from the country to the city – and the basic premise of his proposed solution – by reference to his now-famous metaphor of "the three magnets." The attractions of the town magnet are high wages, social opportunity, and places of amusement, along with high rents, foul air, and social isolation. The attractions of the country magnet are natural beauty, low rents, and fresh air, combined with long hours, no amusements, and a lack of society. Only the town–country magnet as found in the Garden City will combine the best of both urban and rural–low rents, high wages, beauty of nature, social opportunity – and attract the people to a new, more healthy, more self-fulfilling way of life.

In Howard's original vision, the Garden City would consist of 6,000 acres – a town of 1,000 acres surrounded by a permanent greenbelt of 5,000 acres – supporting a population of 32,000. All land would be collectively owned, with start-up loans retired over time from yearly municipal revenues. Eventually, Howard argued, the municipality should capture the increment in land values achieved from buying land at its agricultural value and creating value by successfully building the Garden City. The city itself would feature a complete array of municipal services and amenities: parks, public gardens, tree-lined boulevards, hospitals and asylums, and an enclosed, centrally-located Crystal Palace-style emporium. And although the Garden City would be connected to a larger system of "social cities" by rail lines and canals, it would be economically self-sufficient, with its own factories and workshops, not a bedroom suburb for commuters or a satellite to an existing urban center.

The Garden City movement begins

Ebenezer Howard attracted a cohort of dedicated followers, including Raymond Unwin and Patrick Geddes in Britain and Lewis Mumford, Henry Wright, and Clarence Stein in America. And within a few years of the publication of *To-morrow*, he had attracted a number of financial backers as well – the influential lawyer Ralph Neville and two prominent industrialists who had themselves built model company towns, chocolate and soap magnates George Cadbury and W. H. Lever.

Working together, Howard's backers bought up some 3,800 acres in Hertfordshire, not far from London, and began to build Letchworth, the world's first garden city. Success, however, came at a price. In order to make Letchworth a sound investment opportunity, Howard was forced to abandon some of the more radical elements of his original plan including publicly owned land rented with 1,000 year leases and housing provided by cooperatives. The greenbelt was kept as a planning element, but greatly reduced in size and function.

Raymond Unwin (1863–1940) – the author of *Town Planning in Practice* (1909) and the influential pamphlet *Nothing Gained by Overcrowding* (1912) – along with his cousin Barry Parker, two young architects who had been influenced by the utopianism and arts-and-crafts aesthetics of William Morris, were chosen to design Letchworth. Parker and Unwin created an intelligent and sensitive interpretation of the Howard model. Housing styles suggested a medieval English village, and factories and workshops were placed in a separate zone near the railway. By 1904, the first residents had moved in.

Howard hoped that the Town Planning Law, passed by Parliament in 1909, would spur the construction of dozens of new garden cities, but it was not until after World War I, in 1919, that land for Britain's second garden city, Welwyn, was purchased and construction begun on a bland neo-Georgian plan by the architect Louis de Soissons. But if the movement began slowly, it had nonetheless begun.

THE PROFESSIONALIZATION OF AMERICAN PLANNING

While garden city planning advanced in Britain, a growing number of professional city planners, regular annual city planning conferences, legal recognition of the legitimacy of city planning institutions and plans, and a proliferating body of academic literature – all helped to advance city planning as a profession in both Europe and America. At the turn of the twentieth century there were no standards concerning what a city planner was or what he or she should do. By World War II, planning had become a recognized profession.

Urban aestheticism and the City Beautiful movement

In the 1890s, concern with the "adornment" of

cities, with "civic design", "municipal art," and "the city beautiful" supplanted parks and public health as the dominant concern in city planning. This trend was strongly influenced by L'Enfant's plan for Washington, DC, and the work of Haussmann in Paris, particularly his grand public buildings and boulevards lined with neo-classical and neo-Baroque apartment buildings. But another branch of the aestheticist movement was equally strong: the neo-medievalism that was exemplified in England by the art history of John Ruskin, the romantic utopian fantasies of William Morris, and the Arts and Crafts movement. On the continent, this tendency reached its peak with the brilliant writings and designs of Camillo Sitte. In 1889, Sitte published *Der Stadte-Bau nach seinen kunstlerischen Grundsatzen* ("City Planning According to Artistic Principles") in which he carried out a systematic spatial analysis of existing historic cities. Sitte paid special attention to buildings as parts of a larger compositional arrangement and to the way streets flowed into squares and plazas to form a pleasing, inter-connected whole.

Daniel Burnham of Chicago

Late in his career, Frederick Law Olmsted accepted a commission to lay out the grounds of a new World's Fair to be held in Chicago in 1893. Called the Columbian Exposition in commemoration of the 400th anniversary of the European discovery of America, the Chicago Fair became an immediate and compelling symbol of the United States as it was coming of age as a world power. The design of the fairgrounds and pavilions, extravagant with neo-classical splendor and Beaux-Arts pomposity, attracted some of the most talented designers, architects, and planners in the United States: in addition to Olmsted, men like Louis Sullivan, Dankmar Adler, Charles McKim, and, most especially, Daniel Burnham (1846–1912).

The "White City" on the city's lakefront was both an outstanding achievement in integrated design and a clear expression of the new capitalist order's sense of self. In a city where Jane Addams of Hull House ministered to the poor and where Upton Sinclair was to find the inspiration for *The Jungle* (1906), a gleaming temporary urban fantasy was constructed that attracted worldwide interest. Here America displayed its power and its technological achievements: telephones, electric lights, horseless carriages, and even a prominently placed "Hygeia Fountain." Millions of visitors from North American and abroad marveled at this image of how beautiful a city might be.

The principal elements of City Beautiful design, and its allied Civic Art movement, were strong axial arrangements, magnificent boulevards, and impressive public buildings. The culmination of Burnham's career, his crowning achievement as a designer of cities, was the Chicago Plan of 1909. The plan featured an elaborate system of public parks and lagoons, an imposing civic center, harbor improvements, diagonal streets, a stately yacht harbor, and transit and open space connections throughout the metropolitan area. Lewis Mumford called Burnham's Chicago plan "magnificent in its outlines, narrow in it social purposes," and this has remained a familiar judgment on much of the work of the City Beautiful movement. In part, the judgment is correct. Burnham's work places too much emphasis on the pomp and parade of elite public space and too clearly serves the interests of the rich and powerful. But other aspects of city beautiful planning deserve more respect, especially the emphasis on city planning as a comprehensive and unified process. Burnham is famous for the motto "make no little plans for they have no magic to stir men's blood." But it was not just bigness that characterized Burnham's grand conceptions, but a truly visionary sense of the city as a metropolitan whole. And while housing and poverty did not feature prominently in Burnham's thinking or in the Chicago Plan, the plan does advocate imposing restrictions on overcrowding, enforcing sanitary regulations, limiting the amount of a lot tenements could cover, and advocates municipal housing for slum dwellers as housing of last resort – an idea that did not gain real currency in America until Catherine Bauer's *Modern Housing* twenty-five years later.

Planning comes of age

One radical reformer who played a brief, but influential role in the creation of modern city planning was Benjamin Clark Marsh, author of an extraordinarily superficial and opinionated book titled *An Introduction to City Planning*

(1909). The moralistic son of missionaries, Marsh saw population congestion as a prime national evil. He helped to organize the New York Committee on Congestion of the Population in 1907. Marsh's concern for what is today termed equity planning is best expressed in "City Planning in Justice to the Working Population" (1908). Noting that city beautiful projects have little effect on the daily lives of working-class people, Marsh argued that all public improvements should be scrutinized with a view to the benefits they will confer upon those most in need. While many of Marsh's ideas, such as zoning districts for factories, residential height limits, increased public transportation, and more parks were quite conventional, his advocacy of Henry George's "single tax" ideas was too radical for most of his peers.

Marsh's most important contribution to American city planning was organizing the First (US) National Conference on City Planning which took place in Washington, DC in May 1909. Unfortunately, Marsh's confrontational personal style helped splinter the participants in this important first national meeting. The professional planners did not care to be associated with Marsh's amateurish book, brash style, and single tax radicalism. When they organized a second and subsequent annual conferences they dropped all mention of the problems of housing and population congestion from the conference title and the conference agenda.

Progressivism and the city efficient

During the years before and after World War I, United States politics witnessed the rise of a new, activist philosophy of government. In part, Progressivism, as the movement was called, grew out of "good government" reformers who battled the corruption and managerial inefficiency of big city, immigrant-based political machines. Often frankly representing middle-class social and economic interests, the Progressives sought to apply the best scientific thinking – including new social scientific theories of education, welfare, and social work – to the management of America's cities. With the rise of the new social and political ethos, city beautiful concerns about urban aesthetics during the early urban planning period gave way to an emphasis on making the modern city function efficiently. George B. Ford's address to the Fifth National City Planning Conference in 1913, titled "The City Scientific," and Nelson P. Lewis's *Planning the Modern City* (1916) are exemplary of "city scientific" thinking. The City Scientific movement shared the boundless, modernist faith in the power of rational, scientific decision making to determine the one best solution to any urban problem and the applicability of uniform standards across cities.

Other writers concerned with modern scientific efficiency focused attention on the theory of planning as a social activity. Among the most important is Frederick Law Olmsted, Jr., the son of the great park planner. In an address titled "A City Planning Program" to the Fifth National Conference on Planning in Chicago in 1913, Olmsted laid out a remarkably subtle and visionary city planning program looking ahead to what city planning might become. Imagining city planning as it might exist in fifty years, he saw the city plan as "a live thing – a growing and gradually changing aggregation of accepted ideas or projects for physical changes in the city, all consistent with each other and each surviving by virtue of its own inherent merit and by virtue of its harmonizing with the rest." Olmsted distinguished between *real* plans which actually express the collective will of the community and nominal paper plans. He envisaged a "city plan office" fulfilling three main functions: a custodian of ideas, an interpreter to make the plan consistent, and an amender of the plan.

Edward Bassett and the master plan

As American city planning matured, city planning departments began to develop general, comprehensive plans for many cities. But the rationale for these plans and what they should contain was not well articulated until Edward Bassett sought to define what a general plan should contain and its relationship to the processes of city government.

Bassett's answer about what the master plan should be, articulated in *The Master Plan* (1935), was a general, flexible document, adopted by the local planning commission, but deliberately not adopted by the local legislative body. The plan would consist of both map and text. The text would be organized in relation to a small number of plan elements, and the plan need not be consistent with existing zoning. Bassett felt that the plan should have a certain visionary quality but that it should emphasize physical land use planning, not social or economic planning or strategic or program planning. He distinguished between a plan, which he felt should be easy to change, and an official map of streets, highways, plazas, and parks, which would be much more permanent.

At the time that Bassett wrote, the idea that a plan should be divided into "elements" was not new. But how many elements there should be and what they should consist of were far from standard. Bassett proposed seven elements and gave them functional definitions. He emphasized streets and other public infrastructure and the "fixing of boundaries," not public control of private property as the defining feature of planning. Bassett made zoning a plan element, not a plan implementation device. He angered housing reformers by advocating that there should be no subsidized housing at all, that all housing should be "economic housing." But despite his conservative stance, Bassett's concept of the master plan as the core document of city planning agencies was both the culmination of the process that led to the professionalization of planning and the starting point of all future planning theory and practice.

NEW TOWNS AND REGIONALISM

While much early city planning thought focused on a single city as the unit of analysis, some of the new urban theorists and professional practitioners of city planning were concerned with entire regions. In this framework new towns became building blocks for a broader restructuring of modern urban society.

The contribution of Patrick Geddes

The construction of Letchworth and Welwyn, along with the important advocacy and implementation work of Raymond Unwin and others, established garden city planning in the mainstream of British urban planning practice. But other important tributaries of planning thought were soon to join with the Garden City movement to create a truly balanced and comprehensive approach to urban planning theory and practice. Much of this new thought came from a brilliant, eccentric Scot named Patrick Geddes (1854–1932).

Geddes outlined his theories of urban development and planning in *City Development: A Study of Parks, Gardens, and Culture Institutes* (1904) and *Cities in Evolution: An Introduction to the Town Planning Movement and to the Study of Civics* (1915). The subtitle of this second book, with its explicit conflating of physical planning and social civics, is particularly revealing of his approach to urban issues. Geddes fully developed the regional vision that was implicit in Howard's system of "social cities" and brought the abstraction down to earth. Before any changes could be made to a city or its neighborhoods, a survey would place the city within the environmental context of its region's surrounding eco-systems. Water supply and climate, for example, would be analyzed in terms of watersheds and recurring weather patterns. The survey would also encompass the human history of any targeted urban place – the city's growth over centuries, the unique social characteristics of its people – and all this information, both natural and social, would be combined into graphic public displays which would constitute the City Exhibition. That these innovations seem second nature to planners today simply confirms how important and long lasting the contribution of Patrick Geddes has been.

New towns for America

Of the many disciples that Geddes attracted, perhaps none was more brilliant and influential than Lewis Mumford (1895–1990). Mumford first read Geddes's work when he was only nineteen. Almost immediately, he wrote to Edinburgh seeking information on how to enroll in courses at the Outlook Tower. What followed was a correspondence and friendship that spanned eight decades.

Mumford carried the ideas of both Howard and Geddes into his own philosophy of urban development and helped to popularize those ideas in America. He saw that the power of transportation and communication technologies did not inevitably lead to urban sprawl, but could actually permit decentralization of population and industry throughout regions. At the center of a small, but extremely influential group of intellectuals called the Regional Plan Association of America (RPAA) – a group that included Clarence Stein, Henry Wright, and Benton MacKaye – Mumford developed a powerful vision of regional planning that would turn existing urban development away from megalopolitan sprawl toward a clearer, more humane pattern of small cities that would fit harmoniously within the greater New York region. The RPAA's first important project was Sunnyside Gardens in Queens, New York, where Mumford and his family lived for six years, begun in 1924. This was followed by the even more ambitious Radburn, New Jersey, project begun in 1928.

In the Radburn Plan, Stein and Wright proposed a city using "superblocks" in place of the characteristic narrow rectangular blocks, roads for different uses (service lanes, secondary collector roads, main roads, and parkways), a complete separation of pedestrian and automobile traffic, and houses turned away from the street to face a series of parks forming the backbone of the community. The Radburn Plan and other Stein/Wright ideas and achievements are discussed in Stein's *Towards New Towns for America* (1950).

Architect Clarence Perry (1872–1944) took the ideas of human scale development further and thought deeply about how to design neighborhoods that would function well in the automobile age. His thoughts are summarized in "The Neighborhood Unit" (1931). Perry envisioned the school as the centerpiece of the neighborhood, performing a role in the community well beyond educating primary school children, and argued that the neighborhood should have sufficient population to support one elementary school. Perry gave a good deal of attention to the relationship between the neighborhood and streets. He suggested that neighborhoods should be bounded on all sides by arterial streets for through traffic, but internal street systems should be almost exclusively for use by the residents. The use of culs-de-sac and careful separation of streets from pedestrian ways would harmonize transportation with living space. Perry saw the residential environment's character as an extension of a person's personality. Local shops, neighborhood-serving parks and playgrounds, and skillful landscaping would make a most humane living environment. Many of the best new towns of this period are described in John Nolen's *New Towns for Old* (1927).

The plan for New York and environs

Concerned with the nature of development in the New York region, the Russell Sage Foundation invested in a number of city planning and housing programs and in 1922 funded a monumental nine-year study of the New York region. As chair of the Russell Sage Foundation's Committee on the Regional Plan of New York and Environs, Charles Dyer Norton advocated a monumental planning effort and convinced other members of the Foundation board to fund this effort on a massive scale. Norton had been active in the Chicago Commercial Club and became deeply involved in Burnham's 1909 Plan for the City of Chicago. The person he chose to head this regional planning effort is one of the most influential of twentieth-century planners: Thomas Adams (1871–1941).

As a young man, Adams had worked with Ebenezer Howard and Raymond Unwin on Letchworth and other garden city projects. He served as the first chair of the Garden City Association and became the first president of the British Town Planning Institute in 1904. From 1913 to 1921, he was the town planning advisor to the Governor of Canada and in this capacity developed the first real regional plan in North America for the area around Niagara Falls. Adams assumed the position of general director of plans and surveys for the New York regional planning effort in 1923.

In one of the most celebrated conflicts in American planning history, Lewis Mumford and other members of the RPAA attacked the completed regional plan. After poring over each sentence in the massive *Plan for the New York Region*, Mumford declared that he found little of value in the whole exercise. He dismissed the plan as an essentially conservative document which dodged hard choices, accepted continuation of the status quo as inevitable, and failed in its goal of providing a real vision of regional development. Although the plan talked about garden cities, Mumford argued, it really was a prescription for more congestion. Although it contained Clarence Perry's splendid background study of the neighborhood unit, it really called for more chaotic land subdivision. And although it proposed standards for more light and air, it actually would permit more, not less, over-crowded land development.

PROPHETS OF HIGH MODERNISM

Utopian modernism

As urban planning became professionalized and regularized both in Europe and America, a new urban utopianism emerged to reinvigorate the movement at the level of theory. While the professionals planned for today, new dreamers planned for the city of tomorrow.

Le Corbusier and the International Style

Charles Edouard Jeanneret, who reinvented himself as Le Corbusier, was the prophet of a higher, later stage of modernism: the city as the administrative center of the bureaucratic, technocratic state. Born in provincial La Chaux-de-Fonds, Le Corbusier took his famous pseudonym as a young man in Paris engaged in painting, criticism, and cultural revolution in the name of a triumphant machine-age modernism that would sweep before it all that had come before. In 1920, he began publishing *L'espirit nouveau*, announcing that "A GREAT EPOCH HAS BEGUN!" and in 1922, he proposed "A Contemporary City for Three Million People." This was a breath-taking, totally modern vision of spare, undecorated skyscrapers, evenly spaced in a park, that astonished the people of Paris and that still seems futuristic today.

Le Corbusier was an accomplished provocateur and brilliant publicist. By announcing that his city would house three million people – about 100 times the population of Letchworth – he consciously flew directly in the face of the Garden City advocates while, at the same time, advocating many of their own ideals: simultaneously decongesting cities while maintaining their density. By proposing the use of skyscrapers, he incorporated a new element associated with the crass business culture of America as a solution for European urbanism. And in 1925, he boldly announced a new version of the Contemporary City plan, the Plan Voisin, that was to be built on a site in the middle of Paris, previously cleared and leveled by bulldozers! The popular response was, of course, outrage, but Le Corbusier became instantly famous, a spokesman for a new, uncompromising modernism.

Throughout the 1920s and 1930s, Le Corbusier sought backers for his bold new vision in what became a long, mostly unsuccessful series of supplications before the various thrones of twentieth-century power – capitalists, fascists, and communists. It was only during World War II (when he secured a few commissions from the Vichy government of occupied France) and after the war, when Corbusian principles were adopted by governments worldwide as a quick and easy response to the demands of reconstruction, that Le Corbusier became accepted as a true prophet of modern urbanism. Today, the skyscraper in the park (as often as not reinterpreted as the skyscraper in the parking lot!) is one of the standard and ubiquitous realities of modern cities everywhere. In Marseilles, Brasilia, Teheran, Moscow, Istanbul, London, and New York, the Corbusian office-and-residential tower, similar to the "unité d'habitation" he constructed in Marseilles in 1945, is a modernist commonplace. A technocrat and a syndicalist, Le Corbusier believed wholeheartedly in the eventual triumph of rationality and order. He even dedicated one of his major plans "To Authority." Perhaps simply because Le Corbusier embraced bureaucracy and the command-and-control functions of political and economic elites everywhere, his style has truly become the International Style of our time.

Frank Lloyd Wright's alternative vision

While Le Corbusier promoted a vision of the city of tomorrow that embraced skyscraper development and command-and-control bureaucratic states of all political persuasions, Frank Lloyd Wright was the prophet of middle-class urban flight and automobile-based sprawl suburbia. An older man than Le Corbusier, Wright had already begun a distinguished career as an architect when Louis Sullivan was creating the White City for the 1893 Chicago World's Fair. By the 1920s, he had achieved a celebrity unequaled by almost any architect before or since. Assiduously cultivating his dramatic personal eccentricities and his reputation as a universal genius, Wright advocated a naturalistic architectural style, as well as a vision of urbanism, that were totally at odds with Le Corbusier's hard-edge cubist conceptions.

Announced as early as 1932 in *The Disappearing City*, Wright's Broadacre City allocated a minimum of one acre per person, with no large urban concentrations whatsoever. Wright the individualist proposed a "new community" that would not be dominated by the urban "mobocracy" but which would give free rein to personal self-creation. Broadacre City would be family based, and Wright designed an extraordinary small house with an attached carport – the Usonian house – that subsequently became the model for millions of suburban houses in the decades following World War II. Broadacre City was decried as an anti-city, as no city at all, and Wright was happy to proclaim his distaste for cities in general and modern cities in particular. But the Broadacre model, with its emphasis on the automobile and the telephone as annihilators of space and time, was prophetic of a new urban/suburban reality that would dominate the planning of the future.

Clearly, neither Frank Lloyd Wright nor Le Corbusier would be pleased by what world urban centers and middle-class suburbs actually became in the last half of the twentieth century. Residential lots were too small, the lives of the residents too conformist, to match Wright's Jeffersonian-Emersonian standards. And Corbusian reality never really achieved the purity and sublimity of the Corbusian dream. But the regionalism and decentralization proposed by the garden city and new town advocates now faced two rival approaches that would help to define the urbanism of the twentieth century and the traditions of modern urban planning.

PLANNING AND THE GREAT DEPRESSION

Cities and the crisis of capitalism

Throughout the 1920s and 1930s, the example of the Soviet Union was a powerful force in planning theory. There, few real innovations were accomplished in the area of urban planning, but planning that directed the entire society and economy, including the provision of great public works and new community development, was incorporated into a series of sweeping Five Year Plans. In Hitler's Germany and Mussolini's Italy, fascist regimes put into motion enormous public works programs that helped to glorify the power of the state and the ruling regimes. Mussolini's historic preservation projects in Rome, Rhodes, and elsewhere sought to rekindle imperialist sentiment in the Italian people.

In all the capitalist democracies, the Great Depression of the 1930s called for a fundamental re-evaluation of the relationship between government and the existing social order. Faced with near-total economic collapse – and properly alarmed by the rising tide of totalitarianism elsewhere – democratic governments in Europe and America sought new ways to stabilize themselves, to protect the lower strata of their populations from utter destitution, and to invest in massive new programs of social reform and infrastructure development. In the United States, Franklin Roosevelt's New Deal included a Public Works Administration that constructed thousands of post offices, courthouses, and hydroelectric dams throughout the nation in an impressive and uniform federal style. One of the New Deal's leading theorists, Rexford Guy Tugwell, even proposed that the US Constitution should be amended to include an all-powerful, directive "fourth estate" to carry out the planning function of government. This was a climate in which urban planning, indeed all forms of planning, made major strides.

Modern housing for the depression poor

One of the most important figures in New Deal urban planning was Catherine Bauer (1905–1964). Upper-class and Vassar-educated, Bauer profoundly affected US housing and urban development policy throughout the 1930s and 1940s. After studying post-World War I housing in Europe, Bauer felt that the United States lagged far behind – four million "modern" postwar housing units built in Europe between the end of World War I and the beginning of the Great Depression compared to the paltry parallel US record of no more than 10,000 comparable units completed during the same time. While not as flamboyant a modernist as Le Corbusier,

Bauer was a true modernist with a faith in large scale, rationalized housing using the most advanced building methods and materials – cement slabs, glass, and iron. Central to her thinking was the importance of scale. "The complete neighborhood" she wrote, "not the individual home or apartment building, must be the unit of planning, of finance, of construction, and administration." She saw housing units as intimately related to schools, shops, laundries, public open space for recreation, and gardens. She even included a cafe as a required minimum for inclusion in a neighborhood unit!

Bauer's *Modern Housing* (1934) was timely and drew her immediately into an important role in US housing policy. At her urging, the National Association of Housing Officials brought Raymond Unwin and other European housing planners to the United States in 1934, and Bauer played a lead role in the National Conference that worked with the Europeans to formulate a housing program for the United States. Later, she helped mobilize national support for housing legislation and played a leading role in formulating and securing passage of the critical US housing acts of 1937 and 1949 which created the US Public Housing and Urban Renewal programs.

Patrick Abercrombie and the Barlow Report

In Britain, Patrick Abercrombie (1879–1957) was drawn from architecture into town planning as the recipient of a research fellowship established at the University of Liverpool by soap magnate William H. Lever shortly after the first courses in urban planning were established there in 1909. A brilliant student, Abercrombie was appointed a professor of civic design at the University of Liverpool in 1914. In that position and as editor of *Town Planning Review* he established a reputation as Britain's leading academic planning theorist. Abercrombie was also a practitioner who developed many town and country planning schemes culminating in the monumental 1944 Greater London Plan.

Abercrombie's *Town and Country Planning* (1933) is a masterful synthesis of the best theoretical and applied material on planning through that time. It foreshadows the regional approach Abercrombie would successfully advocate for planning postwar London and its system of satellite new towns. Abercrombie developed a philosophy of planning responsive to democratic impulses, but which would give the state power to implement plans, which had been democratically arrived at. He argued that in a democracy, "Town and country planning seeks to proffer a guiding hand to the trend of natural evolution, as a result of careful study of the place itself and its external relationships. The result is to be more than a piece of skillful engineering, or satisfactory hygiene or successful economics: it should be a social organism and a work of art."

While Tugwell's "fourth estate" never gained real currency in the United States, in Britain national level urban planning was accepted in national discourse during the 1930s and national level planning achieved almost the status of a "fourth estate" after World War II. Prime Minister Neville Chamberlain had long supported regional planning and garden cities. In 1937, he appointed a Royal Commission on the Distribution of the Industrial Population, popularly called the Barlow Commission, which undertook a monumental review of the location of industry and housing throughout Britain. While the Commission's concerns were to develop fundamental policy for industrial location that went well beyond immediate strategic concerns it was the danger of industrial concentration at the outbreak of the war and the perceived need for strong, centralized planning for postwar reconstruction that made national-level city and regional planning possible in Britain.

POSTSCRIPT

Planning since World War II

The Barlow Report was not issued until January 1940, and wartime priorities made any further work in the area of urban planning virtually impossible. In 1944, however, as the war in Europe was coming to a successful conclusion, the need for concerted policy to rebuild the war-shattered nation was of great urgency. The great premises and programs of the Barlow Report were reiterated in Patrick Abercrombie's historic Plan for Greater London. The Abercrombie Plan called for the creation of new towns outside of a decongested, greenbelted

London. It became the centerpiece of the new Labour Party government's social policy. Although other aspects of that social policy – the nationalization of key industries and the expansion of the welfare state – were much debated and modified in subsequent Conservative and Labour regimes, the urban development policy remained largely intact until the 1980s. So it was in Britain, not the United States, that something like Tugwell's "fourth estate" in planning at the national level developed.

In Britain and elsewhere in Europe, planners saw regionalism and new towns policies, along with parallel increases in welfarism, that helped in the rebuilding process that was the inevitable work of postwar reconstruction. And in the United States, the 1949 housing act, strongly influenced by Bauer, called for an expansion of public housing and instituted urban renewal. As if continuing the wartime total mobilization, massive new efforts at slum clearance and inner-city redevelopment were undertaken under this important legislation. Large scale inner-city reconstruction projects borrowed heavily from Le Corbusier's ideas. And postwar prosperity also brought an extraordinary expansion of suburban tract-home communities, borrowing the energy and focus of wartime mobilization and applying them to domestic needs. Broadacre City became Levittown.

Beginning in the 1960s, the conceptual and political underpinnings of democratic socialism and welfare-state liberalism began to erode significantly. In the United States, major new urban programs were initiated as a part of the War on Poverty and the Model Cities program of Lyndon Johnson's Great Society, but increasingly the politics of resistance and opposition to urban planning initiatives imposed from above supplanted an earlier faith in the glories of a scientifically planned future. Later, as the Soviet Union stumbled toward eventual collapse, and as new communications technologies created the preconditions for a new world order and truly global economy, conservative regimes such as the administrations of Margaret Thatcher in Britain and Ronald Reagan in the United States retreated from social welfarism, and many of the accepted truisms of modernism were called into question.

Failures, limitations, and accomplishments

A review of early urban planning contains both surprises and confirmations. If the concerns of the early planners, and even some of their solutions, strike us as remarkably modern, urban planning proceeds through several distinct stages of development – from the first attempts to control and beautify the industrial city, through the gradual evolution of a vision of comprehensive physical planning, to the eventual merging of the goals of physical design and social control. As with other forms of evolutionary development, many of the characteristics of full maturity were present even at birth.

One reason to study urban planning history is in order not to repeat the errors of the past, and the failures of recent urban planning have in some cases been nothing less than spectacular. The dynamiting of the infamous Pruitt-Igoe housing projects in St. Louis were emblematic of the arrogant, community-destroying, over-reaching that characterized all too much of the "bulldozer redevelopment" of the postwar decades. And many of the larger societal problems of cities that modern urban planning had hoped to alleviate – poverty, homelessness, racial discrimination, personal alienation – have proven impervious to the ministrations of the experts.

But while many of the failures of recent planning derive from the sometimes extravagant utopian goals of its founders, the great accomplishments of early city planning must not be overlooked or undervalued. The great urban parks still enhance the lives of millions and constitute an incalculable asset for the residents of great cities. The work of the early hygienists established public health programs that are simply essential to modern urban life. Both the elegant civic centers created by the city beautiful planners and the comfortable, sensitively designed garden suburbs built by the new town developers of the 1920s remain models for emulation today. And the many dedicated architects, landscape designers, legal experts, social reformers, environmental activists, and others who contributed to professionalization of modern urban planning deserve both the interest and respect of subsequent generations of urban specialists.

The future of urban planning

Thus, as the twentieth century comes to a close – and as the culture of modernism gives way to a widely felt and anxiously anticipated, if still poorly defined, postmodernism – a review of urban planning history discloses both continuities and discontinuities. The discontinuities are perhaps the most strongly felt, especially at a time when one great historical period is giving way to another. The collapse of the Soviet Union calls into question many of the postulates of command-economy socialist planning. The emergence of new economic and geo-political forces such as the Islamic world, a revivifying China, and a worldwide marketplace dominated by truly multinational corporations suggest the possible development of a new global system of cities. And new telecommunications technologies along with their associated information-based economies prepare the way for fundamentally new relationships between city and citizen, self and society, identity and community.

But even as it seems clear that the future will be defined by a whole new array of social, political-economic, cultural and technological norms, the continuities of human social existence persist. Men and women will need healthy housing and efficient transportation, meaningful jobs and life-enhancing amenities, access to community and access to nature, a sense of individual self-worth and a sense of collective participation. These have been the concerns of urban planning from the beginning of the modern period and will certainly continue to define both the practice and the purposes of urban planning in the future.

FREDERICK LAW OLMSTED

"Public Parks and the Enlargement of Towns"

American Social Science Association (1870)

Editors' introduction Frederick Law Olmsted (1822–1903) has been called "America's great pioneer landscape architect," and, during his lifetime, he was widely recognized as one of the most influential public figures in the nation. Along with his business partner, the English-born architect Calvert Vaux, he originated and dominated the urban parks movement, pioneered the development of planned suburbs, and laid out scores of public and private institutions. If Central Park in New York, illustrated as it looked in 1863 in Plate 23, remains his best known masterpiece, the designs for Riverside, Illinois (outside Chicago), the Boston park system, the Capitol grounds in Washington, DC, the 1893 World's Fair, and the campus of Stanford University in California are equally impressive contributions to the built environment.

Olmsted began his career practicing and writing about farming, then turned his talents to journalism and, in the 1850s, published a series of books describing the society and economy of the slave states of the American South (collected into one volume as *The Cotton Kingdom* in 1861). With this background, it is hardly surprising that Olmsted thoroughly imbued his art of landscape architecture with a wide variety of social and political, as well as cultural, concerns.

"Public Parks and the Enlargement of Towns" was originally presented as an address to the American Social Science Association, meeting at the Lowell Institute, Boston, in 1870. In it, Olmsted provides a number of specific guidelines for parks and parkways and suggests ways to overcome political resistance to public funding for parks and planned urban growth. Most importantly, however, he lays out the political and philosophical case for public parks in terms of three great moral imperatives: first, the need to improve public health by sanitation measures and the use of trees to combat air and water pollution; second, the need to combat urban vice and social degeneration, particularly among the children of the urban poor; and third, the need to advance the cause of civilization by the provision of urban amenities that would be democratically available to all.

Both as a practitioner and as a theorist, Olmsted anticipated many of the principal concerns of urban planning, both infrastructural and social, down to the present day. Indeed, behind the somewhat convoluted Victorianisms of his prose lies a strikingly modern mind. In the design of the Garden City, Ebenezer Howard (p. 321) borrowed directly from Olmsted, and even plans so fundamentally different as those of Frank Lloyd Wright (p. 344) and Le Corbusier (p. 336) owe a debt to Olmsted insofar as they recognize and address the central problem of the relationship between nature and the built urban environment. As the father of modern landscape architecture, Olmsted's work and thought invite comparison with all those, including J. B. Jackson (p. 162), who came after him in the profession, either as practitioners or critics.

A selection of Olmsted's most important writings may be found in S. B. Sutton (ed.), *Civilizing American Cities: Writings on City Landscapes by Frederick Law Olmsted*, (Cambridge, MA: MIT Press,

1971). Johns Hopkins University has published most of Olmsted's work in the multi-volume *Collected Papers of Frederick Law Olmsted* (Baltimore: Johns Hopkins University Press, 1977–1992).

Biographies of Olmsted and commentary on his work include Laura Wood Roper, *FLO: A Biography of Frederick Law Olmsted* (Baltimore: Johns Hopkins University Press, 1973), Elizabeth Stevenson, *Park Maker: A Life of Frederick Law Olmsted* (New York: Macmillan, 1977), and Charles E. Beveridge, Paul Rocheleau, and David Larkin, *Frederick Law Olmsted: Designing the American Landscape* (New York: Rizzoli, 1995).

Galen Cranz, *The Politics of Park Design: A History of Urban Parks in America* (Cambridge, MA: MIT Press, 1982) is a superb overview which places Olmsted's planning and landscape design achievements in the context of a larger movement for urban social reform.

We have reason to believe, then, that towns which of late have been increasing rapidly on account of their commercial advantages, are likely to be still more attractive to population in the future; that there will in consequence soon be larger towns than any the world has yet known, and that the further progress of civilization is to depend mainly upon the influences by which men's minds and characters will be affected while living in large towns.

Now, knowing that the average length of the life of mankind in towns has been much less than in the country, and that the average amount of disease and misery and of vice and crime has been much greater in towns, this would be a very dark prospect for civilization, if it were not that modern Science has beyond all question determined many of the causes of the special evils by which men are afflicted in towns, and placed means in our hands for guarding against them. It has shown, for example, that under ordinary circumstances, in the interior parts of large and closely built towns, a given quantity of air contains considerably less of the elements which we require to receive through the lungs than the air of the country or even of the outer and more open parts of a town, and that instead of them it carries into the lungs highly corrupt and irritating matters, the action of which tends strongly to vitiate all our sources of vigor – how strongly may perhaps be indicated in the shortest way by the statement that even metallic plates and statues corrode and wear away under the atmosphere influences which prevail in the midst of large towns, more rapidly than in the country.

The irritation and waste of the physical powers which result from the same cause, doubtless indirectly affect and very seriously affect the mind and the moral strength; but there is a general impression that a class of men are bred in towns whose peculiarities are not perhaps adequately accounted for in this way. We may understand these better if we consider that whenever we walk through the denser part of a town, to merely avoid collision with those we meet and pass upon the sidewalks, we have constantly to watch, to foresee, and to guard against their movements. This involves a consideration of their intentions, a calculation of their strength and weakness, which is not so much for their benefit as our own. Our minds are thus brought into close dealings with other minds without any friendly flowing toward them, but rather a drawing from them. Much of the intercourse between men when engaged in the pursuits of commerce has the same tendency – a tendency to regard others in a hard if not always hardening way. Each detail of observation and of the process of thought required in this kind of intercourse or contact of minds is so slight and so common in the experience of towns-people that they are seldom conscious of it. It certainly involves some expenditure nevertheless. People from the country are even conscious of the effect on their nerves and minds of the street contact – often complaining that they feel confused by it; and if we had no relief from it at all during our waking hours, we should all be conscious of suffering from it. It is upon our opportunities of relief from it, therefore, that not only our comfort in town life, but our ability to maintain a temperate, good-natured, and healthy state of mind, depends. This is one of many ways in which it happens that men who have been brought up, as the saying is, in the streets, who have been most directly and completely affected by town influences, so generally show, along with a remarkable quickness of apprehension, a peculiarly hard sort of selfishness. Every day of their lives they have seen thousands of their fellow-men, have met them face to face, have brushed against them, and yet have had no experience of anything in common with them.

[. . .]

It is practically certain that the Boston of to-day is the mere nucleus of the Boston that is to be. It is practically certain that it is to extend over many miles of country now thoroughly rural in character, in parts of which farmers are now laying out roads with a view to shortening the teaming distance between their wood-lots and a railway station, being governed in their courses by old property lines, which were first run simply with reference to the equitable division of heritages, and in other parts of which, perhaps, some wild speculators are having streets staked off from plans which they have formed with a rule and pencil in a broker's office, with a view, chiefly, to the impressions they would make when seen by other speculators on a lithographed map. And by this manner of planning, unless views of duty or of interest prevail that are not yet common, if Boston continues to grow at its present rate even for but a few generations longer, and then simply holds its own until it shall be as old as the Boston in Lincolnshire now is, more men, women, and children are to be seriously affected in health and morals than are now living on this Continent.

Is this a small matter – a mere matter of taste; a sentimental speculation?

It must be within the observation of most of us that where, in the city, wheel-ways originally twenty feet wide were with great difficulty and cost enlarged to thirty, the present width is already less nearly adequate to the present business than the former was to the former business; obstructions are more frequent, movements are slower and oftener arrested, and the liability to collision is greater. The same is true of sidewalks. Trees thus have been cut down, porches, bow-windows, and other encroachments removed, but every year the walk is less sufficient for the comfortable passing of those who wish to use it.

It is certain that as the distance from the interior to the circumference of towns shall increase with the enlargement of their population, the less sufficient relatively to the service to be performed will be any given space between buildings.

In like manner every evil to which men are specially liable when living in towns, is likely to be aggravated in the future, unless means are devised and adapted in advance to prevent it.

Let us proceed, then, to the question of means, and with a seriousness in some degree befitting a question, upon our dealing with which we know the misery or happiness of many millions of our fellow-beings will depend.

We will for the present set before our minds the two sources of wear and corruption which we have seen to be remediable and therefore preventible. We may admit that commerce requires that in some parts of a town there shall be an arrangement of buildings, and a character of streets and of traffic in them which will establish conditions of corruption and of irritation, physical and mental. But commerce does not require the same conditions to be maintained in all parts of a town.

Air is disinfected by sunlight and foliage. Foliage also acts mechanically to purify the air by screening it. Opportunity and inducement to escape at frequent intervals from the confined and vitiated air of the commercial quarter, and to supply the lungs with air screened and purified by trees, and recently acted upon by sunlight, together with opportunity and inducement to escape from conditions requiring vigilance, wariness, and activity toward other men, – if these could be supplied economically, our problem would be solved.

In the old days of walled towns all tradesmen lived under the roof of their shops, and their children and apprentices and servants sat together with them in the evening about the kitchen fire. But now that the dwelling is built by itself and there is greater room, the inmates have a parlor to spend their evening in; they spread carpets on the floor to gain in quiet, and hang drapery in their windows and papers on their walls to gain in seclusion and beauty. Now that our towns are built without walls, and we can have all the room that we like, is there any good reason why we should not make some similar difference between parts which are likely to be dwelt in, and those which will be required exclusively for commerce?

Would trees, for seclusion and shade and beauty, be out of place, for instance, by the side of certain of our streets? It will, perhaps, appear to you that it is hardly necessary to ask such a question, as throughout the United States trees are commonly planted at the sides of streets. Unfortunately they are seldom so planted as to have fairly settled the question of the desirableness of systematically maintaining trees under these circumstances. In the first place, the streets are planned, wherever they are, essentially alike. Trees are planted in the space assigned for sidewalks, where at first, while they are saplings and the vicinity is rural or suburban, they are not much in the way, but where, as they grow larger, and the vicinity becomes urban, they take up more and more space, while space is more and more required for passage. That is not all. Thousands and tens of thousands are planted every year in a manner and under conditions as nearly certain as possible either to kill them outright, or to so lessen their vitality as to prevent their natural and beautiful development, and to cause premature decrepitude. Often, too, as their lower limbs are found inconvenient, no space having been provided for trees in laying out the street, they are deformed by butcherly amputations. If by rare good fortune they are suffered to become beautiful, they still stand subject to be condemned to death at any time, as obstructions in the highway.

What I would ask is, whether we might not with economy make special provision in some of our streets – in a twentieth or a fiftieth part, if you please, of all – for trees to remain as a permanent furniture of the city? I mean, to make a place for them in which they would have room to grow naturally and gracefully. Even if the distance between the houses should have to be made half as much again as it is required to be in our commercial streets, could not the space be afforded? Out of town space is not costly when measures to secure it are taken early. The assessments for benefit where such streets were provided for, would, in nearly all cases, defray the cost of the land required. The strips of ground required for the trees, six, twelve, twenty feet wide, would cost nothing for paving or flagging.

The change both of scene and of air which would be obtained by people engaged for the most part in the necessarily confined interior commercial parts of the town, on passing into a street of this character after the trees have become stately and graceful, would be worth a good deal. If such streets were made still broader in some parts, with spacious malls, the advantage would be increased. If each of them were given the proper capacity, and laid out with laterals and connections in suitable directions to serve as a convenient trunk line of communication between two large districts of the town or the business centre and the suburbs, a very great number of people might thus be placed every day under influences counteracting those with which we desire to contend.

These, however, would be merely very simple improvements upon arrangements which are in common use in every considerable town. Their advantages would be incidental to the general uses of streets as they are. But people are willing very often to seek recreations as well as receive it by the way. Provisions may indeed be made expressly for public recreations, with certainty that if convenient they will be resorted to.

We come then to the question: what accommodations for recreation can we provide which shall be so agreeable and so accessible as to be efficiently attractive to the great body of citizens, and which, while giving decided gratification, shall also cause those who resort to them for pleasure to subject themselves, for the time being, to conditions strongly counteractive to the special, enervating conditions of the town?

In the study of this question all forms of recreation may, in the first place, be conveniently arranged under two general heads. One will include all of which the predominating influence is to stimulate exertion of any part or parts needing it; the other, all which cause us to receive pleasure without conscious exertion. Games chiefly of mental skill, as chess, or athletic sports, as baseball, are examples of means of recreation of the first class, which may be termed that of *exertive* recreation; music and the fine arts generally of the second or *receptive* division.

Considering the first by itself, much consideration will be needed in determining what classes of exercises may be advantageously provided for. In the Bois de Boulogne there is a race-course; in the Bois de Vincennes a ground for artillery target-practice. Military parades are held in Hyde Park. A few cricket clubs are accommodated in most of the London parks, and swimming is permitted in the lakes at certain hours. In the New York Park, on the other hand, none of these exercises are provided for or permitted, except that the boys of the public schools are given the use on holidays of certain large spaces for ball playing. It is considered that the advantage to individuals which would be gained in providing for them would not compensate for the general inconvenience and expense they would cause.

I do not propose to discuss this part of the subject at present, as it is only necessary to my immediate purpose to point out that if recreations requiring large spaces to be given up to the use of a comparatively small number, are not considered essential, numerous small grounds so distributed through a large town that some one of them could be easily reached by a short walk from every house, would be more desirable than a single area of great extent, however rich in landscape attractions it might be. Especially would this be the case if the numerous local grounds were connected and supplemented by a series of trunk-roads or boulevards such as has already been suggested.

Proceeding to the consideration of receptive recreations, it is necessary to ask you to adopt and bear in mind a further subdivision, under two heads, according to the degree in which the average enjoyment is greater when a large congregation assembles for a purpose of receptive recreation, or when the number coming together is small and the circumstances are favorable to the exercise of personal friendliness.

The first I shall term *gregarious*; the second, *neighborly*. Remembering that the immediate matter in hand is a study of fitting accommodations, you will, I trust, see the practical necessity of this classification.

Purely gregarious recreation seems to be generally looked upon in New England society as childish and savage, because, I suppose, there is so little of what we call intellectual gratification in it. We are inclined to engage in it indirectly, furtively, and with complication. Yet there are certain forms of recreation, a large share of the attraction of which must, I think, lie in the gratification of the gregarious inclination, and which, with those who can afford to indulge in them, are so popular as to establish the importance of the requirement.

If I ask myself where I have experienced the most complete gratification of this instinct in public and out of doors, among trees, I find that it has been in the promenade of the Champs-Élysées. As closely following it I should name other promenades of Europe, and our own upon the New York parks. I have studiously watched the latter for several years. I have several times seen fifty thousand people participating in them; and the more I have seen of them, the more highly have I been led to estimate their value as means of counteracting the evils of town life.

Consider that the New York Park and the Brooklyn Park are the only places in those associated cities where, in this eighteen hundred and seventieth year after Christ, you will find a body of Christians coming together, and with an evident glee in the prospect of coming together, all classes largely represented, with a common purpose, not at all intellectual, competitive with none, disposing to jealousy and spiritual or intellectual pride toward none, each individual adding by his mere presence to the pleasure of all others, all helping to the greater happiness of each. You may thus often see vast numbers of persons brought closely together, poor and rich, young and old, Jew and Gentile. I have seen a hundred thousand thus congregated, and I assure you that though there have been not a few that seemed a little dazed, as if they did not quite understand it, and were, perhaps, a little ashamed of it, I have looked studiously but vainly among them for a single face completely unsympathetic with the prevailing expression of good nature and light-heartedness.

Is it doubtful that it does men good to come together in this way in pure air and under the light of heaven, or that it must have an influence directly counteractive to that of the ordinary hard, hustling working hours of town life?

You will agree with me, I am sure, that it is not, and that opportunity, convenient, attractive opportunity, for such congregation, is a very good thing to provide for, in planning the extension of a town.

[. . .]

Think that the ordinary state of things to many is at this beginning of the town. The public is reading just now a little book in which some of your streets of which you are not proud are described. Go into one of those red cross streets any fine evening next summer, and ask how it is with their residents. Oftentimes you will see half a dozen sitting together on the door-steps or, all in a row, on the curb-stones, with their feet in the gutter; driven out of doors by the closeness within; mothers among them anxiously regarding their children who are dodging about at their play, among the noisy wheels on the pavement.

Again, consider how often you see young men in knots of perhaps half a dozen in lounging attitudes rudely obstructing the sidewalks, chiefly led in their little conversation by the suggestions given to their minds by what or whom they may see passing in the street, men, women, or children, whom they do not know and for whom they have no respect or sympathy. There is nothing among them or about them which is adapted to bring into play a spark of admiration, of delicacy, manliness, or tenderness. You see them presently descend in search of physical comfort to a brilliantly lighted basement, where they find others of their sort, see, hear, smell, drink, and eat all manner of vile things.

Whether on the curb-stones or in the dram-shops, these young men are all under the influence of the same impulse which some satisfy about the tea-table with neighbors and wives and mothers and children, and all things clean and wholesome, softening, and refining.

If the great city to arise here is to be laid out little by little, and chiefly to suit the views of land-owners, acting only individually, and thinking only of how what they do is to affect the value in the next week or the next year of the few lots that each may hold at the time, the opportunities of so obeying this inclination as at the same time to give the lungs a bath of pure sunny air, to give the mind a suggestion of rest from the devouring eagerness and intellectual strife of town life, will always be few to any, to many will amount to nothing.

But is it possible to make public provision for recreation of this class, essentially domestic and secluded as it is?

It is a question which can, of course, be conclusively answered only from experience. And from experience in some slight degree I shall answer it. There is one large American town, in which it may happen that a man of any class shall say to his wife, when he is going out in the morning: "My dear, when the children come home from school, put some bread and butter and salad in a basket, and go to the spring under the chestnut-tree where we found the Johnsons last week. I will join you there as soon as I can get away from the office. We will walk to the dairy-man's cottage and get some tea, and some fresh milk for the children, and take our supper by the brook-side"; and this shall be no joke, but the most refreshing earnest.

There will be room enough in the Brooklyn Park, when it is finished, for several thousand little family and neighborly parties to bivouac at frequent intervals through the summer, without discommoding one another, or interfering with any other purpose, to say nothing of those who can be drawn out to make a day of it, as many thousand were last year. And although the arrangements for the purpose were yet very incomplete, and but little ground was at all prepared for such use, besides these small parties, consisting of one or two families, there came also, in companies of from thirty to a hundred and fifty, somewhere near twenty thousand children with their parents, Sunday-school teachers, or other guides and friends, who spent the best part of a day under the trees and on the turf, in recreations of which the predominating element was of this neighborly receptive class. Often they would bring a fiddle, flute, and harp, or other music. Tables, seats, shade, turf, swings, cool spring-water, and a pleasing rural prospect, stretching off half a mile or more each way, unbroken by a carriage road or the slightest evidence of the vicinity of the town, were supplied them without charge and bread and milk and ice-cream at moderate fixed charges. In all my life I have never seen such joyous collections of people. I have, in fact, more than once observed tears of gratitude in the eyes of poor women, as they watched their children thus enjoying themselves.

The whole cost of such neighborly festivals, even when they include excursions by rail from the distant parts of the town, does not exceed for each person, on an average, a quarter of a dollar; and when the arrangements are complete, I see no reason why thousands should not come every day where hundreds come now to use them; and if so, who can measure the value, generation after generation, of such provisions for recreation to the over-wrought, much-confined people of the great town that is to be?

For this purpose neither of the forms of ground we have heretofore considered are at all suitable. We want a ground to which people may easily go after their day's work is done, and where they may stroll for an hour, seeing, hearing, and feeling nothing of the bustle and jar of the streets, where they shall, in effect, find the city put far away from them. We want the greatest possible contrast with the streets and the shops and the rooms of the town which will be consistent with convenience and the preservation of good order and neatness. We want, especially, the greatest possible contrast with the restraining and confining conditions of the town, those conditions which compel us to walk circumspectly, watchfully, jealously, which compel us to look closely upon others without sympathy. Practically, what we most want is a simple, broad, open space of clean greensward, with sufficient play of surface and a sufficient number of trees about it to supply a variety of light and shade. This we want as a central feature. We want depth of wood enough about it not only for comfort in hot weather, but to completely shut out the city from our landscapes.

The word *park*, in town nomenclature, should, I think, be reserved for grounds of the character and purpose thus described.

[. . .]

A park fairly well managed near a large town, will surely become a new center of that town. With the determination of location, size, and boundaries should therefore be associated the duty of arranging new trunk routes of communication between it and the distant parts of the town existing and forecasted.

These may be either narrow informal elongations of the park, varying say from two to five hundred feet in width, and radiating irregularly from it, or if, unfortunately, the town is already laid out in the unhappy way that New York and Brooklyn, San Francisco and Chicago, are, and, I am glad to say, Boston is not, on a plan made long years ago by a man who never saw a spring-carriage, and who had a conscientious dread of the Graces, then we must probably adopt formal Park-ways. They should be so planned and constructed as never to be noisy and seldom crowded, and so also that the straightforward movement of pleasure-car carriages need never be obstructed, unless at absolutely necessary crossings, by slow-going heavy vehicles used for commercial purposes. If possible, also, they should be branched or reticulated with other ways of a similar class, so that no part of the town should finally be many minutes' walk from some one of them; and they should be made interesting by a process of planting and decoration, so that in necessarily passing through them, whether in going to or from the park, or to and from business, some substantial recreative advantage may be incidentally gained. It is a common error to regard a park as something to be produced complete in itself, as a picture to be painted on canvas. It should rather be planned as one to be done in fresco, with constant consideration of exterior objects, some of them quite at a distance and even existing as yet only in the imagination of the painter.

I have thus barely indicated a few of the points from which we may perceive our duty to apply the means in our hands to ends far distant, with reference to this problem of public recreations. Large operations of construction may not soon be desirable, but I hope you will agree with me that there is little room for question, that reserves of ground for the purposes I have referred to should be fixed upon as soon as possible, before the difficulty of arranging them, which arises from private building, shall be greatly more formidable than now.

EBENEZER HOWARD

"Author's Introduction" and "The Town–Country Magnet"

from *Garden Cities of To-morrow* (1898)

Editors' introduction A stenographer by trade, Ebenezer Howard (1850–1928) was a quiet, modest, self-effacing man – "a man without credentials or connections," as one biographer put it – who nevertheless managed to change the world.

Born in London, Howard early experienced the pollution, congestion, and social dislocations of the modern industrial metropolis. After five years in America (one as a homesteader in Nebraska!), he returned to England in 1876 and became involved in political movements and discussion groups addressing what was then termed "the Social Question." Howard was influenced by a number of radical theorists and visionaries including the social reformer Robert Owen, the utopian novelist Edward Bellamy, and the "single tax" advocate Henry George. He published *To-morrow: A Peaceful Path to Real Reform* in 1898 (now better known under its 1902 title, *Garden Cities of To-morrow*) and methodically set about convincing people of the beauty and utility of "the Garden City idea."

Although Howard's plan may seem quaintly Victorian to the modern reader, the ideas he put forth were nothing short of revolutionary at the time. Indeed, Howard's ideas of urban decentralization, zoning for different uses, the integration of nature into cities, greenbelting, and the development of self-contained "new town" communities outside crowded central cities illustrated in Plate 25 laid the groundwork for the entire tradition of modern city planning. Unlike many other utopian dreamers, Howard lived to see his plans (if in a somewhat compromised form) actually put into action. In his own lifetime, the garden cities of Letchworth and Welwyn were built in England. Later, the Garden City idea spread to continental Europe, to America by way of the New Deal, and to much of the rest of the world.

Howard's argument begins with a protest against urban overcrowding; the one issue upon which, he writes, "men of all parties" are "well-nigh universally agreed." He then explains why "the people should continue to stream into the already over-crowded cities" by reference to "the town magnet," that combination of jobs and amenities that characterizes the modern metropolis. Arrayed against this urban magnetic force is "the country magnet," the appealing features of the more natural, but increasingly desolate, rural districts. Finally, Howard describes his own plan, a new kind of human community based on "the town–country magnet," which is the best of both worlds. As detailed in his famous concentric-ring diagram (which, he is careful to warn, is "a diagram only," not an actual site plan), the center of Garden City is to be a central park containing important public buildings and surrounded by a "Crystal Palace" ring of retail stores. The entire city of approximately 1,000 acres is to be encircled by a permanent agricultural greenbelt of some 5,000 acres, and the new cities are to be connected with central "Social Cities" (and each other) by a system of railroad lines.

Howard's ideas about the evils of overcrowding are similar to those of Friedrich Engels (p. 46), and his solution to the problem invites comparison with the very different solutions proposed by Le Corbusier

(p. 336) and Frank Lloyd Wright (p. 344). Direct followers of Howard include Patrick Geddes (p. 330), and Lewis Mumford (p. 92), who helped to spread the Garden City idea throughout Europe and America. More recently, Peter Calthorpe (p. 350) has effectively reinvented the Garden City idea in California in the form of greenbelted, suburban "pedestrian pockets" linked to central cities (and each other) by a network of light-rail transportation systems.

Garden Cities of To-morrow remains a readable and relevant book. It is available as the second volume of Richard T. LeGates and Frederic Stout (eds.), *Early Urban Planning, 1870–1940* (9 vols, London: Routledge/Thoemmes, 1998) and earlier editions by Attic Books (1985), Eastbourne (1985), MIT (1965), Faber & Faber (1960, 1951, and 1946), Sonnenschein (1902), and the original edition which appeared under the title *To-morrow; A Peaceful Path to Real Reform* (London: Swan Sonnenschein, 1898).

Biographies of Ebenezer Howard include Robert Beevers, *The Garden City Utopia: A Critical Biography of Ebenezer* (New York: St. Martin's, 1988), and Dugald Macfadyen, *Sir Ebenezer Howard and the Town Planning Movement* (Manchester: Manchester University Press, 1933; reprinted Cambridge, MA: MIT Press, 1970).

Excellent accounts of Howard and the Garden City movement may be found in Robert Fishman's *Urban Utopias in the Twentieth Century* (New York: Basic Books, 1977) and Peter Hall's *Cities of Tomorrow* (Oxford: Basil Blackwell, 1988).

Additional books about Ebenezer Howard and the Garden City movement include Standish Meacham, *Regaining Paradise: Englishness and the Early Garden City Movement* (New Haven, CT: Yale University Press, 1999), Peter Geoffrey Hall and Colin Ward, *Sociable Cities: The Legacy of Ebenezer Howard* (New York: John Wiley and Sons, 1998), Stephen V. Ward (ed.), *The Garden City: Past, Present and Future* (London and New York: E & FN Spon, 1992), and Stanley Buder, *Visionaries and Planners: The Garden City Movement and the Modern Community* (Oxford: Oxford University Press, 1990).

AUTHOR'S INTRODUCTION

In these days of strong party feeling and of keenly contested social and religious issues, it might perhaps be thought difficult to find a single question having a vital bearing upon national life and well-being on which all persons, no matter of what political party, or of what shade of sociological opinion, would be found to be fully and entirely agreed . . .

[. . .]

There is, however, a question in regard to which one can scarcely find any difference of opinion . . . It is wellnigh universally agreed by men of all parties, not only in England, but all over Europe and America and our colonies, that it is deeply to be deplored that the people should continue to stream into the already overcrowded cities, and should thus further deplete the country districts.

All . . . are agreed on the pressing nature of this problem, all are bent on its solution, and though it would doubtless be quite Utopian to expect a similar agreement as to the value of any remedy that may be proposed, it is at least of immense importance that, on a subject thus universally regarded as of supreme importance, we have such a consensus of opinion at the outset. This will be the more remarkable and the more hopeful sign when it is shown, as I believe will be conclusively shown in this work, that the answer to this, one of the most pressing questions of the day, makes of comparatively easy solution many other problems which have hitherto taxed the ingenuity of the greatest thinkers and reformers of our time. Yes, the key to the problem how to restore the people to the land – that beautiful land of ours, with its canopy of sky, the air that blows upon it, the sun that warms it, the rain and dew that moisten it – the very embodiment of Divine love for man – is indeed a *Master Key*, for it is the key to a portal through which, even when scarce ajar, will be seen to pour a flood of light on the problems of intemperance, of excessive toil, of restless anxiety, of grinding poverty – the true limits of Governmental interference, ay, and even the relations of man to the Supreme Power.

It may perhaps be thought that the first step to be taken towards the solution of this question – how to restore the people to the land – would involve a careful consideration of the very numerous causes which have hitherto led to their aggregation in large cities. Were this the case, a very prolonged enquiry would be necessary at the outset. Fortunately, alike for writer and for reader, such an analysis is not, however, here requisite, and for a very simple reason, which may be stated thus: Whatever may have been the causes which have operated in the past, and are operating now, to draw the people into the cities, those causes may all be summed up as "attractions"; and it is obvious, therefore, that no remedy can possibly be effective which will not present to the people, or at least to considerable portions of them, greater "attractions" than our cities now possess, so that the force of the old "attractions" shall be overcome by the force of new "attractions" which are to be created. Each city may be regarded as a magnet, each person as a needle; and, so viewed, it is at once seen that nothing short of the discovery of a method for constructing magnets of yet greater power than our cities possess can be effective for redistributing the population in a spontaneous and healthy manner.

So presented, the problem may appear at first sight to be difficult, if not impossible, of solution. "What", some may be disposed to ask, "can possibly be done to make the country more attractive to a workaday people than the town – to make wages, or at least the standard of physical comfort, higher in the country than in the town; to secure in the country equal possibilities of social intercourse, and to make the prospects of advancement for the average man or woman equal, not to say superior, to those enjoyed in our large cities?" The issue one constantly finds presented in a form very similar to that. The subject is treated continually in the public press, and in all forms of discussion, as though men, or at least working men, had not now, and never could have, any choice or alternative, but either, on the one hand, to stifle their love for human society – at least in wider relations than can be found in a straggling village – or, on the other hand, to forgo almost entirely all the keen and pure delights of the country. The question is universally considered as though it were now, and for ever must remain, quite impossible for working people to live in the country and yet be engaged in pursuits other than agricultural; as though crowded, unhealthy cities were the last word of economic science; and as if our present form of industry, in which sharp lines divide agricultural from industrial pursuits, were necessarily an enduring one. This fallacy is the very common one of ignoring altogether the possibility of alternatives other than those presented to the mind. There are in reality not only, as is so constantly assumed, two alternatives – town life and country life – but a third alternative, in which all the advantages of the most energetic and active town life, with all the beauty and delight of the country, may be secured in perfect combination; and the certainty of being able to live this life will be the magnet which will produce the effect for which we are all striving – the spontaneous movement of the people from our crowded cities to the bosom of our kindly mother earth, at once the source of life, of happiness, of wealth, and of power. The town and the country may, therefore, be regarded as two magnets, each striving to draw the people to itself – a rivalry which a new form of life, partaking of the nature of both, comes to take part in. This may be illustrated by a diagram (Figure 1) of "The Three Magnets", in which the chief advantages of the Town and of the Country are set forth with their corresponding drawbacks, while the advantages of the Town–Country are seen to be free from the disadvantages of either.

The Town magnet, it will be seen, offers, as compared with the Country magnet, the advantages of high wages, opportunities for employment, tempting prospects of advancement, but these are largely counterbalanced by high rents and prices. Its social opportunities and its places of amusement are very alluring, but excessive hours of toil, distance from work, and the "isolation of crowds" tend greatly to reduce the value of these good things. The well-lit streets are a great attraction, especially in winter, but the sunlight is being more and more shut out, while the air is so vitiated that the fine public buildings, like the sparrows, rapidly become covered with soot, and the very statues are in despair. Palatial edifices and fearful slums are the strange, complementary features of modern cities.

The Country magnet declares herself to be the source of all beauty and wealth; but the Town magnet mockingly reminds her that she is very dull for lack of society, and very sparing of her gifts for lack of capital. There are in the country beautiful vistas, lordly parks, violet-scented woods, fresh air, sounds of rippling water; but too often one sees those threatening words, "Trespassers will be prosecuted". Rents, if estimated by the acre, are certainly low, but such low rents are the natural fruit of low wages rather than a cause of substantial comfort; while long hours and lack of amusements forbid the bright sunshine and the pure air to gladden the hearts of the people. The one industry, agriculture, suffers frequently from excessive rainfalls; but this wondrous harvest of the clouds is seldom properly in-gathered, so that, in times of drought, there is frequently, even for drinking purposes, a most insufficient supply. Even the natural healthfulness of the country is largely lost for lack of proper drainage and other sanitary conditions, while, in parts almost deserted by the people, the few who remain are yet frequently huddled together as if in rivalry with the slums of our cities.

But neither the Town magnet nor the Country magnet represents the full plan and purpose of nature. Human society and the beauty of nature

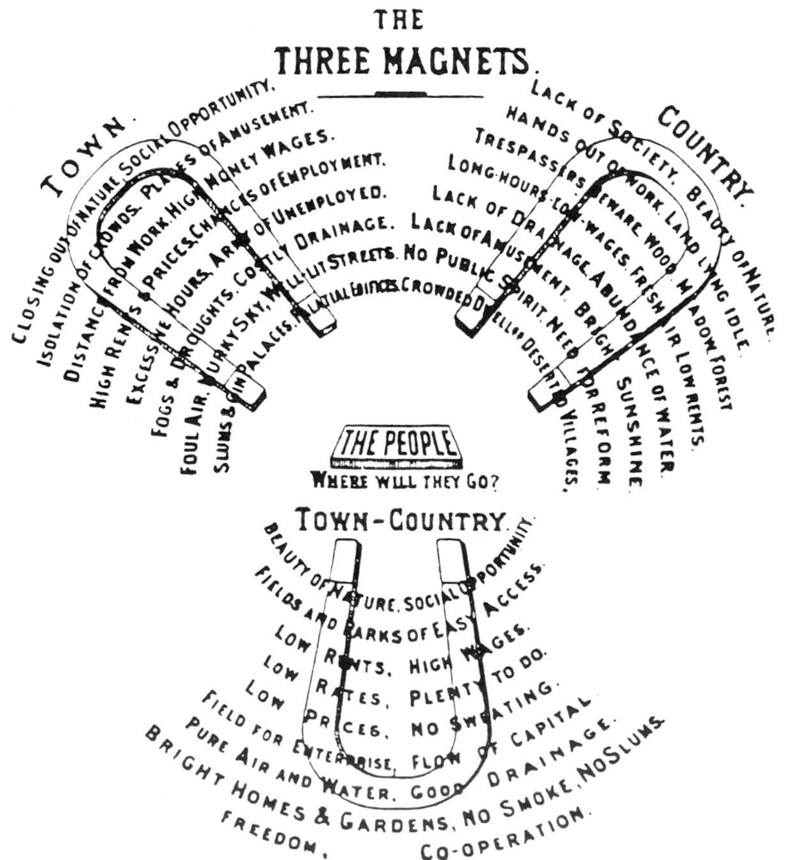

Figure 1

are meant to be enjoyed together. The two magnets must be made one. As man and woman by their varied gifts and faculties supplement each other, so should town and country. The town is the symbol of society – of mutual help and friendly co-operation, of fatherhood, motherhood, brotherhood, sisterhood, of wide relations between man and man – of broad, expanding sympathies – of science, art, culture, religion. And the country! The country is the symbol of God's love and care for man. All that we are and all that we have comes from it. Our bodies are formed of it; to it they return. We are fed by it, clothed by it, and by it are we warmed and sheltered. On its bosom we rest. Its beauty is the inspiration of art, of music, of poetry. Its forces propel all the wheels of industry. It is the source of all health, all wealth, all knowledge. But its fullness of joy and wisdom has not revealed itself to man. Nor can it ever, so long as this unholy, unnatural separation of society and nature endures. Town and country *must be married*, and out of this joyous union will spring a new hope, a new life, a new civilization. It is the purpose of this work to show how a first step can be taken in this direction by the construction of a Town–Country magnet; and I hope to convince the reader that this is practicable, here and now, and that on principles which are the very soundest, whether viewed from the ethical or the economic standpoint.

I will undertake, then, to show how in "Town–Country" equal, nay better, opportunities of social intercourse may be enjoyed than are enjoyed in any crowded city, while yet the beauties of nature may encompass and enfold each dweller therein; how higher wages are compatible with reduced rents and rates; how abundant opportunities for employment and bright prospects of advancement may be secured for all; how capital may be attracted and wealth created; how the most admirable sanitary conditions may be ensured; how beautiful homes and gardens may be seen on every hand; how the bounds of freedom may be widened, and yet all the best results of concert and co-operation gathered in by a happy people.

The construction of such a magnet, could it be effected, followed, as it would be, by the construction of many more, would certainly afford a solution of the burning question set before us by Sir John Gorst, "how to back the tide of migration of the people into the towns, and to get them back upon the land".

[. . .]

THE TOWN–COUNTRY MAGNET

The reader is asked to imagine an estate embracing an area of 6,000 acres, which is at present purely agricultural, and has been obtained by purchase in the open market at a cost of £40 an acre, or £240,000. The purchase money is supposed to have been raised on mortgage debentures, bearing interest at an average rate not exceeding 4 per cent. The estate is legally vested in the names of four gentlemen of responsible position and of undoubted probity and honour, who hold it in trust, first, as a security for the debenture-holders, and, secondly, in trust for the people of Garden City, the Town–Country magnet, which it is intended to build thereon. One essential feature of the plan is that all ground rents, which are to be based upon the annual value of the land, shall be paid to the trustees, who, after providing for interest and sinking fund, will hand the balance to the Central Council of the new municipality, to be employed by such Council in the creation and maintenance of all necessary public works – roads, schools, parks, etc. The objects of this land purchase may be stated in various ways, but it is sufficient here to say that some of the chief objects are these: To find for our industrial population work at wages of *higher purchasing power*, and to secure healthier surroundings and more regular employment. To enterprising manufacturers, co-operative societies, architects, engineers, builders, and mechanicians of all kinds, as well as to many engaged in various professions, it is intended to offer a means of securing new and better employment for their capital and talents, while to the agriculturists at present on the estate as well as to those who may migrate thither, it is designed to open a new market for their produce close to their doors. Its object is, in short, to raise the standard of health and comfort of all true workers of whatever grade – the means by which these objects are to be achieved being a healthy, natural, and economic combination of town and country life, and this on land owned by the municipality.

Garden City, which is to be built near the centre of the 6,000 acres, covers an area of 1,000 acres, or a sixth part of the 6,000 acres, and might be of circular form, 1,240 yards (or nearly three-quarters of a mile) from centre to circumference. (Figure 2 is a ground plan of the whole municipal area, showing the town in the centre; and Figure 3, which represents one section or ward of the town, will be useful in following the description of the town itself – *a description which is, however, merely suggestive, and will probably be much departed from . . .*)

Six magnificent boulevards – each 120 feet wide – traverse the city from centre to circumference, dividing it into six equal parts or wards. In the centre is a circular space containing about five and a half acres, laid out as a beautiful and well-watered garden; and, surrounding this garden, each standing in its own ample grounds, are the larger public buildings – town hall, principal concert and lecture hall, theatre, library, museum, picture-gallery, and hospital.

The rest of the large space encircled by the "Crystal Palace" is a public park, containing 145 acres, which includes ample recreation grounds within very easy access of all the people.

Running all round the Central Park (except where it is intersected by the boulevards) is a wide glass arcade called the "Crystal Palace", opening on to the park. This building is in wet weather one of the favourite resorts of the people, whilst the knowledge that its bright shelter is ever close at hand tempts people into Central Park, even in the most doubtful of weathers. Here manufactured goods are exposed for sale, and here most of that class of shopping which requires the joy of deliberation and selection is done. The space enclosed by the Crystal Palace is, however, a good deal larger than is required for these purposes, and a considerable part of it is used as a Winter Garden – the whole forming a permanent exhibition of a most attractive character, whilst its circular form brings it near to every dweller in the town – the furthest removed inhabitant being within 600 yards.

Passing out of the Crystal Palace on our way to the outer ring of the town, we cross Fifth Avenue – lined, as are all the roads of the town, with trees – fronting which, and looking on to the Crystal Palace, we find a ring of very excellently built houses, each standing in its own ample grounds; and, as we continue our walk, we observe that the houses are for the most part built either in concentric rings, facing the various avenues (as the circular roads are termed), or fronting the boulevards and roads which all converge to the centre of the town. Asking the friend who accompanies us on our journey what the population of this little city may be, we are told about 30,000 in the city itself, and about 2,000 in the agricultural estate, and that there are in the town 5,500 building lots of an *average* size of 20 feet × 130 feet – the minimum space allotted for the purpose being 20 × 100. Noticing the very varied architecture and design which the houses and groups of houses display – some having common gardens and co-operative kitchens – we learn that general observance of street line or harmonious departure from it are the chief points as to house building, over which the municipal authorities exercise control, for, though proper sanitary arrangements are strictly enforced, the fullest measure of individual taste and preference is encouraged.

Walking still toward the outskirts of the town, we come upon "Grand Avenue". This avenue is fully entitled to the name it bears, for it is 420 feet wide, and, forming a belt of green upwards of three miles long, divides that part of the town which lies outside Central Park into two belts. It really constitutes an additional park of 115 acres – a park which is within 240 yards of the furthest removed inhabitant. In this splendid avenue six sites, each of four acres, are occupied by public schools and their surrounding playgrounds and gardens, while other sites are reserved for churches, of such denominations as the religious beliefs of the people may determine, to be erected and maintained out of the funds of the worshippers and their friends. We observe that the houses fronting on Grand Avenue have departed (at least in one of the wards – that of which Figure 3 is a representation) – from the general plan of concentric rings, and, in order to ensure a longer line of frontage on Grand Avenue, are arranged in crescents – thus also to the eye yet further enlarging the already splendid width of Grand Avenue.

On the outer ring of the town are factories, warehouses, dairies, markets, coal yards, timber yards, etc., all fronting on the circle railway, which encompasses the whole town, and which

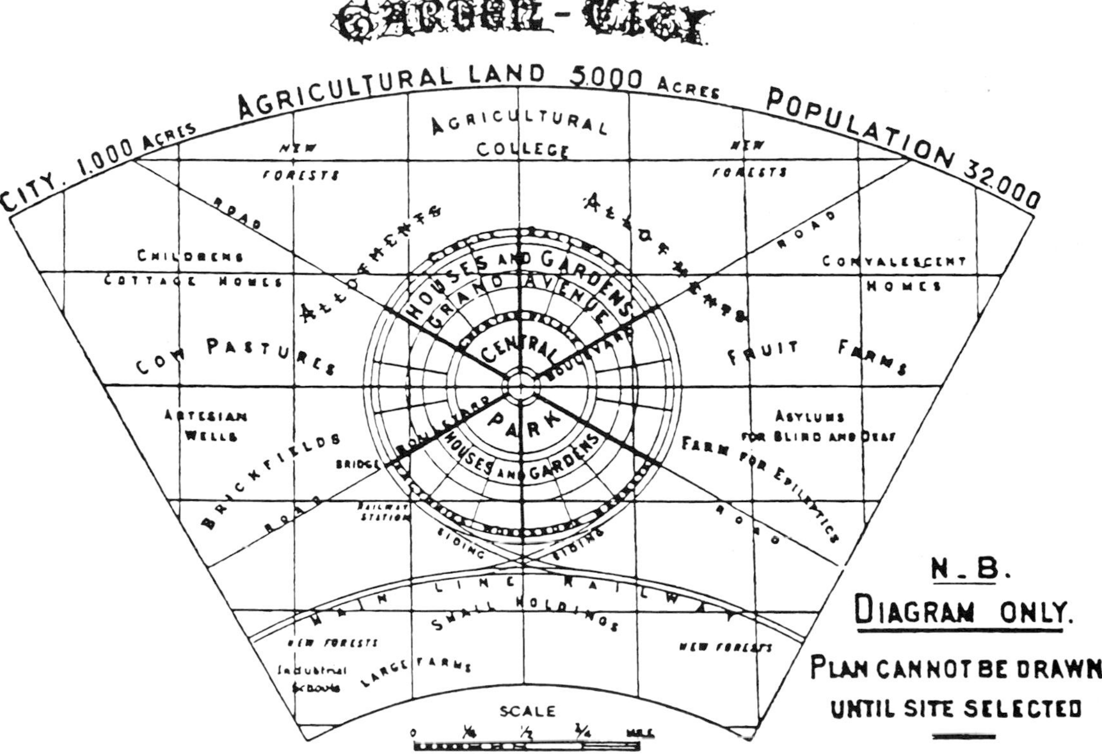

Figure 2

has sidings connecting it with a main line of railway which passes through the estate. This arrangement enables goods to be loaded direct into trucks from the warehouses and work shops, and so sent by railway to distant markets, or to be taken direct from the trucks into the warehouses or factories; thus not only effecting a very great saving in regard to packing and cartage, and reducing to a minimum loss from breakage, but also, by reducing the traffic on the roads of the town, lessening to a very marked extent the cost of their maintenance. The smoke fiend is kept well within bounds in Garden City; for all machinery is driven by electric energy, with the result that the cost of electricity for lighting and other purposes is greatly reduced.

The refuse of the town is utilized on the agricultural portions of the estate, which are held by various individuals in large farms, small holdings, allotments, cow pastures, etc.; the natural competition of these various methods of agriculture, tested by the willingness of occupiers to offer the highest rent to the municipality, tending to bring about the best system of husbandry, or, what is more probable, the best systems adapted for various purposes. Thus it is easily conceivable that it may prove advantageous to grow wheat in very large fields, involving united action under a capitalist farmer, or by a body of co-operators; while the cultivation of vegetables, fruits, and flowers, which requires closer and more personal care, and more of the artistic and inventive faculty, may possibly be best dealt with by individuals, or by small groups of individuals having a common belief in the efficacy and value of certain dressings, methods of culture, or artificial and natural surroundings.

This plan, or, if the reader be pleased to so term it, this absence of plan, avoids the dangers of stagnation or dead level, and, though encouraging individual initiative, permits of the fullest co-operation, while the increased rents which follow from this form of competition are common or municipal property, and by far the larger part of them are expended in permanent improvements.

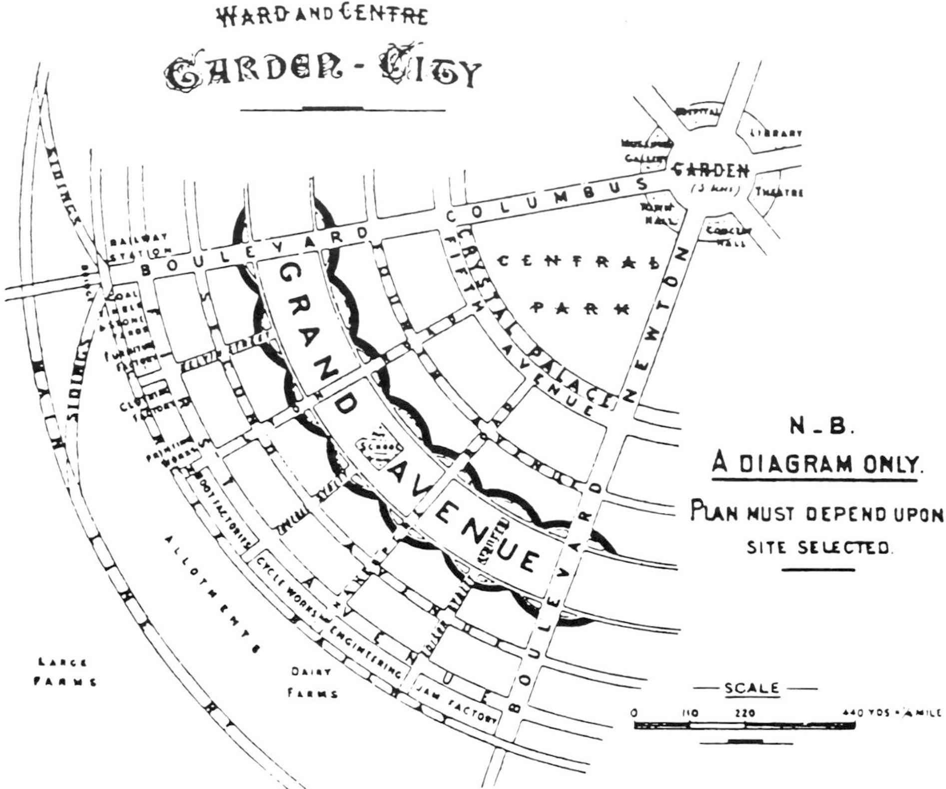

Figure 3

While the town proper, with its population engaged in various trades, callings, and professions, and with a store or depot in each ward, offers the most natural market to the people engaged on the agricultural estate, inasmuch as to the extent to which the towns-people demand their produce they escape altogether any railway rates and charges; yet the farmers and others are not by any means limited to the town as their only market, but have the fullest right to dispose of their produce to whomsoever they please. Here, as in every feature of the experiment, it will be seen that it is not the area of rights which is contracted, but the area of choice which is enlarged.

This principle of freedom holds good with regard to manufacturers and others who have established themselves in the town. These manage their affairs in their own way, subject, of course, to the general law of the land, and subject to the provision of sufficient space for workmen and reasonable sanitary conditions. Even in regard to such matters as water, lighting, and telephonic communication – which a municipality, if efficient and honest, is certainly the best and most natural body to supply – no rigid or absolute monopoly is sought; and if any private corporation or any body of individuals proved itself capable of supplying on more advantageous terms, either the whole town or a section of it, with these or any commodities the supply of which was taken up by the corporation, this would be allowed. No really sound system of *action* is in more need of artificial support than is any sound system of *thought*. The area of municipal and corporate action is probably destined to become greatly enlarged; but, if it is to be so, it will be because the people possess faith in such action, and that faith can be best shown by a wide extension of the area of freedom.

Dotted about the estate are seen various charitable and philanthropic institutions. These are not under the control of the municipality, but are supported and managed by various public-spirited people who have been invited by the municipality to establish these institutions in an open healthy district, and on land let to them at a pepper-corn rent, it occurring to the authorities that they can the better afford to be thus generous, as the spending power of these institutions greatly benefits the whole community. Besides, as those persons who migrate to the town are among its most energetic and resourceful members, it is but just and right that their more helpless brethren should be able to enjoy the benefits of an experiment which is designed for humanity at large.

PATRICK GEDDES

"City Survey for Town Planning Purposes, of Municipalities and Government"

from *Cities in Evolution* (1915)

Editors' introduction Patrick Geddes (1854–1932) was an eccentric, even enigmatic Scot who rarely published but who nonetheless became one of the most widely influential urban theorists and urban planning practitioners of the twentieth century. In Peter Hall's phrase, Geddes was "an unclassifiable polymath." Trained as a biological scientist, he was a professor of botany at University College, Dundee, in Scotland; self-trained as a sociologist and historian, he was later a professor of sociology and civics at the University of Bombay.

Through a network of close friendships with such notables as the anarchist philosopher Peter Kropotkin, Ebenezer Howard in England, and Lewis Mumford in America, Geddes was instrumental in the process of spreading Garden City ideas and town planning practices throughout the world. He coined the term "conurbation" to describe the emergent urban agglomerations that have become such a striking feature of the modern world. He pioneered modern conceptions of environmentalism by insisting that planning be regional in scope and firmly rooted in local geography (a radical idea at a time when urban planning was dominated by the primarily aesthetic notions of the City Beautiful movement). And he was an early advocate of community empowerment, having carried out an "unslumming" project in the Old Town district of Edinburgh that began with the distribution of whitewash and flower boxes to the slum residents themselves.

Geddes published *Cities in Evolution* in 1915. In it, he presented a macro-historical view of urban evolution from village, to civilization, and to eventual decadence. In reference to the then current state of urban affairs, he wrote: "slum, semi-slum, and super-slum, to this has come the evolution of cities." His visionary proposal was for the systematic planning of entire regions and for regarding each city's unique regional environment as the basis for a total reconstruction of social and political life.

In the selection here reprinted, Geddes calls for the completion of a complex city survey of local and regional conditions (including physical, social, cultural, and even historical investigations) that must precede any actual planning efforts by local government boards (LGBs). In addition, he argued strongly for a civic exhibition of the survey results so that the local citizenry could participate democratically in major planning decisions. One measure of the far-sighted brilliance of Patrick Geddes is that both ideas – "survey before plan" and citizen participation in the planning process – strike us today as simple common sense. But while inventories are routinely done before preparing city plans today, many fail to include the full range of concerns that Geddes wanted to see addressed. To him, a survey should examine not just the physical and infrastructural environment, but the historical and cultural background of a place as well.

As an early prophet of regionalism, Geddes had a direct influence on environmental planners like Ian McHarg and, through Mumford and others, on the course of much urban planning practice during the

New Deal period in America. For an insight into the origins of his thought, consult his early theoretical article – "The Twofold Aspect of the Industrial Age: Paleotechnic and Neotechnic," *Town Planning Review*, 31 (1912), and Carl Sussman (ed.), *Planning the Fourth Migration: The Neglected Vision of the Regional Planning Association of America* (Cambridge, MA: MIT Press, 1976), which reprints the 1 May, 1925 edition of *Survey* magazine devoted to popularizing Geddes's ideas in America and which includes articles by Mumford, Clarence Stein, Stuart Chase, and Benton MacKaye.

This selection is from Patrick Geddes, *Cities in Evolution* (London: Williams and Norgate, 1915). Geddes's other writings include *City Development: A Study of Parks, Gardens and Culture Institutes* (Edinburgh: Geddes and Co., 1904), essays in *Survey* magazine and a series of reports on planning cities in India.

Biographies of Geddes include Helen Meller, *Patrick Geddes: Social Evolutionist and City Planner* (London: Routledge, 1990) and Marshall Stalley, *Patrick Geddes: Spokesman for Man and the Environment* (New Brunswick, NJ: Rutgers University Press, 1972) which contains a reprint of most of *Cities in Evolution*.

[. . .]

We come now to the need of City Surveys and Local Exhibitions as preparatory to Town Planning Schemes. It may but bring our whole argument together, and in a way, we trust, practically convincing to municipal bodies, and appealing also to the Local Government Boards – which in each of the kingdoms have to supervise their schemes – if we here utilize with slight abbreviation, a memorandum prepared in the Sociological Society's Cities Committee, and addressed to the authorities concerned, local and central alike.

SUMMARY OF THE CITIES COMMITTEE'S WORK

We welcomed and highly appreciated the Town Planning Act, and we early decided that it was not necessary for this Committee to enter into its discussion in detail, or that of its proposed amendments. We have addressed ourselves essentially to the problem of Town Planning itself, as raised by the study of particular types of towns and districts involved; and to the nature and method of the City Survey which we are unanimously of opinion is necessary before the preparation of any Town Planning Scheme can be satisfactorily undertaken. Schemes, however, are in incubation, alike by municipal officials, by public utility associations, and by private individuals, expert or otherwise, which, whatever their particular merits, are not based upon any sufficient surveys of the past development and present conditions of their towns, nor upon adequate knowledge of good and bad town planning elsewhere. In such cases the natural order, that of town survey before town planning, is being reversed; and in this way individuals and public bodies are in danger of committing themselves to plans which would have been widely different with fuller knowledge; yet which, once produced, it will be too late to replace, and even difficult to modify.

We have, therefore, during the past few years addressed ourselves towards the initiation of a number of representative and typical City Surveys, leading towards Civic Exhibitions; and these we hope to see under municipal auspices, in conjunction with public museums and libraries, and with the co-operation of leading citizens representative of different interests and points of view. In Leicester and Saffron Walden, Lambeth, Woolwich, and Chelsea, Dundee, Edinburgh, Dublin, and other cities progress has already been made: and with the necessary skilled and clerical assistance, and moderate outlays, we should be able to assist such surveys in many other towns and cities. Our experience already shows that in this inspiring task, of surveying, usually for the first time, the whole situation and life of a community in past and present, and of thus preparing for the planning scheme which is to forecast, indeed largely decide, its material future, we have the beginnings of a new movement – one already characterized by an arousal of civic feeling, and the corresponding awakening of more enlightened and more generous citizenship.

RECOMMENDATION BY THE COMMITTEE

The preparation of a local and civic survey previous to the preparation of a Town Planning Scheme, though not actually specified in the Act, is fully within its spirit; and we are therefore most anxious that at least a strong recommendation to this effect should form part of the regulations for Town Planning Schemes provided for the guidance of local authorities by the Local Government Board. Without this, municipalities and others interested are in danger of taking the very opposite course, that of planning before survey. Our suggestion towards guarding against this is hence of the most definite kind, viz.:

Before proceeding to the preparation of a Town Planning Scheme, it is desirable to institute a Preliminary Local Survey – to include the collection and public exhibition of maps, plans, models, drawings, documents, statistics, etc., illustrative of Situation, Historic Development, Communications, Industry and Commerce, Population, Town Conditions and Requirements, etc.

We desire to bring this practical suggestion before local authorities, and also to ventilate it as far as may be in public opinion and through the press, and in communication to the many bodies whose interest in Town Planning Schemes from various points of view has been recognized in the Third Schedule of the Act, as lately amended by the Government in response to representations from our own and other societies.

DANGERS OF TOWN PLANNING BEFORE TOWN SURVEY

What will be the procedure of any community of which the local authorities have not as yet adequately recognized the need of the full previous consideration implied by our proposed inquiry, with its Survey and Exhibition? It is that the Town Council, or its Streets and Buildings Committee, may simply remit to its City Architect, if it has one, more usually to its Borough Surveyor or Engineer, to draw up the Town Planning Scheme.

This will be done after a fashion. But too few of these officials or of their committees have as yet had time or opportunity to follow the Town Planning movement even in its publications, much less to know it at first hand, from the successes and blunders of other cities. Nor do they always possess the many-sided preparation – geographic, economic, artistic, etc. – which is required for this most complex of architectural problems, one implying, moreover, innumerable social ones.

If the calling in of expert advice be moved for, the Finance Committee of the Town Council, the ratepayers also, will tend to discourage the employment of an external architect. Moreover, with exceptions, still comparatively rare, even the skilled architect, however distinguished as a designer of buildings, is usually as unfamiliar with town planning as can be the town officials; often, if possible, yet more so. For they have at least laid down the existing streets; he has merely had to accept them.

No doubt, if the plan thus individually prepared be so positively bad, in whole or in part, that its defects can be seen by those not specially acquainted with the particular town or with the quarter in question, the LGB [Local Government Board] can disapprove or modify. But even accepting what can be thus done at the distance of London, or even by the brief visit from an LGB advisory officer, the real danger remains. Not that of streets, etc., absurdly wrong perhaps; but that of the *low pass standard* – that of the mass of municipal art hitherto; despite exceptions, usually due to skilled individual initiative.

Town Planning Schemes produced under this too simple and too rapid procedure may thus escape rejection by the LGB rather than fulfil the spirit and aims of its Act; and they will thus commit their towns for a generation, or irreparably, to designs which the coming generation may deplore. Some individual designs will no doubt be excellent; but there are not as yet many skilled town planners among us. Even in

Germany, still more in America (despite all recent praise, much of which is justified), this new art is still in its infancy.

As a specific example of failures to recognize and utilize all but the most obvious features and opportunities of even the most commanding sites, the most favourable situations, Edinburgh may be chosen. For, despite its exceptional advantages, its admired examples of ancient and modern town planning, its relatively awakened architects, its comparatively high municipal and public interest in town amenity, Edinburgh notoriously presents many mistakes, disasters, and even vandalisms, of which some are recent ones. If such things happen in cities which largely depend upon their attractive aspect, and whose town council and inhabitants are relatively interested and appreciative, what of towns less favourably situated, less generally aroused to architectural interest, to local vigilance and civic pride? Even with real respect to the London County Council and the record of its individual members, past or present, it must be said that this is hardly a matter in which London can expect the provincial cities to look to her for much light and leading as a whole, while her few great and monumental improvements are naturally beyond their reach.

In short, *passable* Town Planning Schemes may be obtained without this preliminary Survey and Exhibition which we desire to see in each town and city; but the best *possible* cannot be expected. From the confused growth of the recent industrial past, we tend to be as yet easily contented with any improvement: this, however, will not long satisfy us, and still less our successors. This Act seeks to open a new and better era, and to render possible cities which may again be beautiful: it proceeds from Housing to Town (Extension) Planning, and it thus raises inevitably before each municipality the question of town planning at its best – in fact of city development and city design.

METHOD AND USES OF PRELIMINARY SURVEY

The needed preliminary inquiry is readily outlined. It is that of a City Survey. The whole topography of the town and its extensions must be taken into account, and this more fully than in the past, by the utilization not only of maps and plans of the usual kind, but of contour maps, and, if possible, even relief models. Of soil and geology, climate, rainfall, winds, etc., maps are also easily obtained, or compiled from existing sources.

For the development of the town in the past, historical material can usually be collected without undue difficulty. For the modern period, since the railway and industrial period have come in, it is easy to start with its map on the invaluable "Reform Bill Atlas of 1832," and compare with this its plans in successive periods up to the present.

By this study of the actual progress of town developments (which have often followed lines different from those laid down or anticipated at former periods) our present forecasts of future developments may usefully be aided and criticized.

Means of communication in past and present, and in possible future, of course need specially careful mapping.

In this way also appears the need of relating the given town not only to its immediate environs, but to the larger surrounding region. This idea, though as old as geographical science, and though expressed in such a term as "County Town," and implicit in "Port," "Cathedral City," etc., etc., is in our present time only too apt to be forgotten, for town and country interests are commonly treated separately with injury to both. The collaboration of rustic and urban points of view, of county and rural authorities, should thus as far as possible be secured, and will be found of the greatest value. The recent agricultural development in Ireland begins to bring forward the need of a more intelligent and practical co-operation of town and country than has yet been attempted; and towards this end surveys are beginning, and are being already found of value.

Social surveys of the fulness and detail of Mr Booth's well-known map of London may not be necessary; but such broader surveys as those of Councillor Marr in his *Survey of Manchester*, or of Miss Walker for Dundee, and the like, represent the very minimum wherever adequate civic betterment is not to be ignored.

The preparation of this survey of the town's Past and Present may usually be successfully undertaken in association with the town's library and museum, with such help as their curators can readily obtain from the town-house, from fellow-citizens acquainted with special departments, and, when desired, from the Sociological Society's Cities Committee.

Experience in various cities shows that such a Civic Exhibition can readily be put in preparation in this way, and without serious expense.

The urgent problem is, however, to secure a similar thoroughness of preparation of the Town Planning Scheme which is so largely to determine the future.

To the Exhibition of the City's Past and Present there therefore needs to be added a corresponding wall-space (a) to display good examples of town planning elsewhere; (b) to receive designs and suggestions towards the City's Future. These may be received from all quarters; some, it may be, invited by the municipality, but others independently offered, and from local or other sources, both professional and lay.

In this threefold Exhibition, then – of their Borough or City, Past, Present, and Possible – the municipality and the public would practically have the main outlines of the inquiry needful before the preparation of the Town Planning Scheme clearly before them; and the education of the public, and of their representatives and officials alike, may thus – and so far as yet suggested, thus only – be arranged for. Examples of town plans from other cities, especially those of kindred site or conditions, will here be of peculiarly great value, indeed are almost indispensable.

After this exhibition – with its individual contributions, its public and journalistic discussion, its general and expert criticism – the municipal authorities, their officials, and the public are naturally in a much more advanced position as regards knowledge and outlook from that which they occupy at present, or can occupy if the short and easy off-hand method above criticized be adopted, obeying only the minimum requirements of the Act. The preparation of a Town Planning Scheme as good as our present (still limited) lights allow, can then be proceeded with. This should utilize the best suggestions on every hand, selecting freely from designs submitted, and paying for so much as may be accepted on ordinary architectural rates.

As the scheme has to be approved by the LGB, their inspector will have the benefit of the mass of material collected in this exhibition, with corresponding economy of his time and gain to his efficiency. His inspection would essentially be on the spot; any critic who may be appointed would naturally require to do this. His suggestions and emendations could thus be more easily and fully made, and more cheerfully adopted. The selection of the best designs would be of immense stimulus to individual knowledge and invention in this field, and to a worthy civic rivalry also.

OUTLINE SCHEME FOR A CITY SURVEY AND EXHIBITION

The incipient surveys of towns and cities, above referred to, are already clearly bringing out their local individuality in many respects, in situation and history, in activities and in spirit. No single scheme of survey can therefore be drawn up so as to be equally applicable in detail to all towns alike. Yet unity of method is necessary for clearness, indispensable for comparison; and after the careful study of schemes prepared for particular towns and cities, a general outline has been drafted, applicable to all towns, and easily elaborated and adapted in detail to the individuality of each town or city. It is therefore appended, as suitable for general purposes, and primarily for that Preliminary Survey previous to the preparation of a Town Planning Scheme, which is the urgent recommendation of this Committee.

The survey necessary for the adequate preparation of a Town Planning Scheme involves the collection of detailed information upon the following heads. Such information should be as far as possible in graphic form, i.e. expressed in maps and plans illustrated by drawings, photographs, engravings, etc., with statistical summaries, and with the necessary descriptive text; and is thus suitable for exhibition in town-house, museum, or library; or, when possible, in the city's art galleries.

The following general outline of the main headings of such an inquiry admits of adaptation and extension to the individuality and special conditions of each town and city.

Situation, Topography, and Natural Advantages:

(a) Geology, Climate, Water Supply, etc.
(b) Soils, with Vegetation, Animal Life, etc.
(c) River or Sea Fisheries.
(d) Access to Nature (Sea Coast, etc.).

Means of Communication, Land and Water:

(a) Natural and Historic.
(b) Present State.
(c) Anticipated Developments.

Industries, Manufactures, and Commerce:

(a) Native Industries.
(b) Manufactures.
(c) Commerce, etc.
(d) Anticipated Developments.

Population:

(a) Movement.
(b) Occupations.
(c) Health.
(d) Density.
(e) Distribution of Well-being (Family Conditions, etc.).
(f) Education and Culture Agencies.
(g) Anticipated Requirements.

Town Conditions:

(a) Historical: Phase by Phase, from Origins onwards. Material Survivals and Associations, etc.
(b) Recent: Particularly since 1832 Survey, thus indicating Areas, Lines of Growth and Expansion, and Local Changes under Modern Conditions, e.g. of Streets, Open Spaces, Amenity, etc.
(c) Local Government Areas (Municipal, Parochial, etc.).
(d) Present: Existing Town Plans, in general and detail:
    Streets and Boulevards. Open Spaces, Parks, etc.
    Internal Communications, etc.
    Water, Drainage, Lighting, Electricity, etc.
    Housing and Sanitation (of localities in detail).
    Existing activities towards Civic Betterment, both Municipal and Private.

Town Planning: Suggestions and Designs:

(A) Example from other Towns and Cities, British and Foreign.
(B) Contributions and Suggestions towards Town Planning Scheme, as regards:
    (a) Areas.
    (b) Possibilities of Town Expansion (Suburbs, etc.).
    (c) Possibilities of City Improvement and Development.
    (d) Suggested Treatments of these in detail (alternatives when possible).

A fuller outline for city activities in detail would exceed our present limits; moreover, it will be found to arise more naturally in each city as its survey begins, and in course of the varied collaboration which this calls forth. The preparation of such more detailed surveys is in progress in some of the towns above mentioned, and is well advanced, for instance, in Edinburgh and Dublin: and though these surveys are as yet voluntary and unofficial, there are indications that they may before long be found worthy of municipal adoption. The recent example of the corporation of Newcastle upon Tyne, towards establishing a Civic Museum and Survey, may here again be cited as encouraging, and even predicted as likely before long to become typical.

The question is sometimes asked, How can we, in our town or city, more speedily set agoing this survey and exhibition without the delay of depending entirely on private and personal efforts? Here the services of the Cities and Town Planning Exhibition may be utilized, as notably in the case of Dublin. In this way the city's survey is initiated in consultation with the local experts of all kinds; and the broad outline thus prepared is capable of later local development in detail, with economy of time and convenience of comparison with other cities. The Exhibition, with its civic surveys from other places, is also suggestive and encouraging to local workers: while the variety of examples of town planning and design from all sources are of course helpful to all interested in the preparation of the best possible local schemes.

LE CORBUSIER (CHARLES-EDOUARD JEANNERET)

"A Contemporary City"

from *The City of Tomorrow and Its Planning* (1929)

Editors' introduction Le Corbusier (1887–1965) was one of the founding fathers of the Modernist movement and of what has come to be known as the International Style in architecture. Painter, architect, city planner, philosopher, author of revolutionary cultural manifestos – Le Corbusier exemplified the energy and efficiency of the Machine Age. His was the bold, nearly mystical rationality of a generation that was eager to accept the scientific spirit of the twentieth century on its own terms and to throw off all pre-existing ties – political, cultural, conceptual – with what it considered an exhausted, outmoded past.

Born Charles-Edouard Jeanneret, Le Corbusier grew up in the Swiss town of La Chaux-de-Fonds, noted for its watch-making industry. He took his famous pseudonym after he moved to Paris to pursue a career in art and architecture. From the first, his designs for modern houses – he called them 'machines for living' – were strikingly original, and many people were shocked by the spare cubist minimalism of his designs. The real shock, however, came in 1922 when Le Corbusier presented the public with his plan for "A Contemporary City of Three Million People." Laid out in a rigidly symmetrical grid pattern, the city consisted of neatly spaced rows of identical, strictly geometrical skyscrapers as illustrated in Plate 27. This was not the city of the future, Le Corbusier insisted, but the city of today. It was to be built on the Right Bank, after demolishing several hundred acres of the existing urban fabric of Paris!

The "Contemporary City" proposal certainly caught the attention of the public, but it did not win Le Corbusier many actual urban planning commissions. Throughout the 1920s, 1930s, and 1940s, he sought out potential patrons wherever he could find them – the industrial capitalists of the Voisin automobile company, the communist rulers of the Soviet Union, and the fascist Vichy government of occupied France – mostly without success. Le Corbusier's real impact came not from cities he designed and built himself but from cities that were built by others incorporating the planning principles that he pioneered. Most notable among these was the notion of "the skyscraper in the park," an idea that is today ubiquitous. Whether in relatively complete examples like Brasilia and Chandigarh, India (where new cities were built from scratch), or in partial examples such as the skyscraper parks and the high-rise housing blocks that have been built in cities worldwide, the Le Corbusier vision has truly transformed the global urban environment.

Le Corbusier's "Contemporary City" plan has often been contrasted to Frank Lloyd Wright's "Broadacre" (p. 344), and the comparison of a thoroughly centralized versus a thoroughly decentralized plan is indeed striking. Le Corbusier's boldness invites comparison with the original optimism of the post-World War II reconstruction and redevelopment efforts and even with the work of such visionary megastructuralists as Paolo Soleri (p. 540). Jane Jacobs (p. 106) may be counted as one of the severest critics and grassroots opponents of Corbusian city planning principles, and Allan Jacobs and Donald Appleyard's "Urban Design Manifesto" (p. 491) deliberately takes the form of a Le Corbusier pronouncement but rejects his program, opting instead for lively streets, participatory planning, and the integration of old buildings into the new urban fabric. Beneath all the sparkling clarity of Le Corbusier's urban designs are questions that must forever remain conjectural: How would democratic politics be practiced in a Corbusian city? What would social relationships be like amid the gleaming towers?

Le Corbusier's writings include *The City of Tomorrow and Its Planning* (New York: Dover, 1987, translated by Frederich Etchells from *Urbanisme*, 1929), *Concerning Town Planning* (New Haven, CT: Yale University Press, 1948, translated by Clive Entwistle from *Propos d'urbanisme*, 1946), and *L'Urbanisme des trois établissements humaines* (Paris: Editions de Minuit, 1959).

Excellent accounts of Le Corbusier's ideas may be found in Robert Fishman's *Urban Utopias in the Twentieth Century* (New York: Basic Books, 1977) and Peter Hall's *Cities of Tomorrow* (Oxford: Basil Blackwell, 1988).

The existing congestion in the centre must be eliminated.

The use of technical analysis and architectural synthesis enabled me to draw up my scheme for a contemporary city of three million inhabitants. The result of my work was shown in November 1922 at the Salon d'Automne in Paris. It was greeted with a sort of stupor; the shock of surprise caused rage in some quarters and enthusiasm in others. The solution I put forward was a rough one and completely uncompromising. There were no notes to accompany the plans, and, alas! not everybody can read a plan. I should have had to be constantly on the spot in order to reply to the fundamental questions which spring from the very depths of human feelings. Such questions are of profound interest and cannot remain unanswered. When at a later date it became necessary that this book should be written, a book in which I could formulate the new principles of Town Planning, I resolutely decided *first of all* to find answers to these fundamental questions. I have used two kinds of argument: first, those essentially human ones which start from the mind or the heart or the physiology of our sensations as a basis; secondly, historical and statistical arguments. Thus I could keep in touch with what is fundamental and at the same time be master of the environment in which all this takes place.

In this way I hope I shall have been able to help my reader to take a number of steps by means of which he can reach a sure and certain position. So that when I unroll my plans I can have the happy assurance that his astonishment will no longer be stupefaction nor his fears mere panic.

[. . .]

A CONTEMPORARY CITY OF THREE MILLION INHABITANTS

Proceeding in the manner of the investigator in his laboratory, I have avoided all special cases, and all that may be accidental, and I have assumed an ideal site to begin with. My object was not to overcome the existing state of things, but *by constructing a theoretically water-tight formula to arrive at the fundamental principles of modern town planning*. Such fundamental principles, if they are genuine, can serve as the skeleton of any system of modern town planning; being as it were the rules according to which development will take place. We shall then be in a position to take a special case, no matter what: whether it be Paris, London, Berlin, New York or some small town. Then, as a result of what we have learnt, we can take control and decide in what direction the forthcoming battle is to be waged. For the desire to rebuild any great city in a modern way is to engage in a formidable battle. Can you imagine people engaging in a battle without knowing their objectives? Yet that is exactly what is happening. The authorities are compelled to do something, so they give the police white sleeves or set them on horseback, they invent sound signals and light signals, they propose to put bridges over streets or moving pavements under the streets; more garden cities are suggested, or it is decided to suppress the tramways, and so on. And these decisions are reached in a sort of frantic haste in order, as it were, to hold a wild beast at bay. That beast is the great city. It is infinitely more powerful than all these devices. And it is just beginning to wake. What will to-morrow bring forth to cope with it?

We must have some rule of conduct.

We must have fundamental principles for modern town planning.

Site

A level site is the ideal site [for the contemporary city (Figure 1)]. In all those places where traffic becomes over-intensified the level site gives a chance of a normal solution to the problem. Where there is less traffic, differences in level matter less.

The river flows far away from the city. The river is a kind of liquid railway, a goods station and a sorting house. In a decent house the servants' stairs do not go through the drawing room – even if the maid is charming (or if the little boats delight the loiterer leaning on a bridge).

Population

This consists of the citizens proper; of suburban dwellers; and of those of a mixed kind.

(a) Citizens are of the city: those who work and live in it.

(b) Suburban dwellers are those who work in the outer industrial zone and who do not come into the city: they live in garden cities.

(c) The mixed sort are those who work in the business parts of the city but bring up their families in garden cities.

To classify these divisions (and so make possible the transmutation of these recognized types) is to attack the most important problem in town planning, for such a classification would define the areas to be allotted to these three sections and the delimitation of their boundaries. This would enable us to formulate and resolve the following problems:

1 The *City*, as a business and residential centre.
2 The *Industrial City* in relation to the *Garden Cities* (i.e. the question of transport).
3 The *Garden Cities* and the *daily transport* of the workers.

Our first requirement will be an organ that is compact, rapid, lively and concentrated: this is

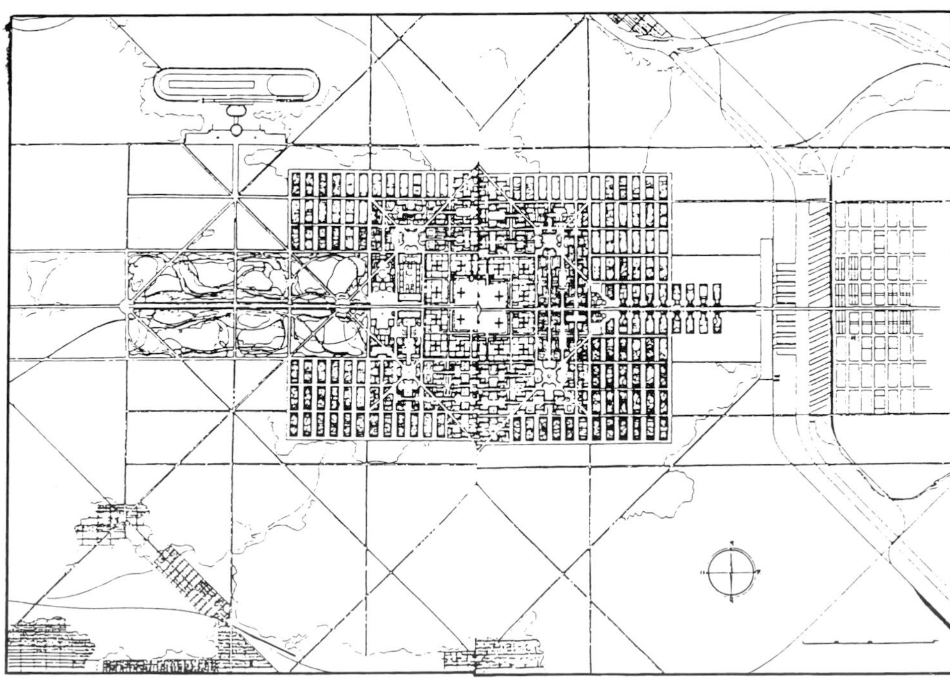

Figure 1

the City with its well organized centre. Our second requirement will be another organ, supple, extensive and elastic; this is the *Garden City* on the periphery. Lying between these two organs, we must *require the legal establishment* of that absolute necessity, a protective zone which allows of extension, *a reserved zone* of woods and fields, a fresh-air reserve.

Density of population

The more dense the population of a city is the less are the distances that have to be covered. The moral, therefore, is that we must *increase the density of the centres of our cities, where business affairs are carried on.*

Lungs

Work in our modern world becomes more intensified day by day, and its demands affect our nervous system in a way that grows more and more dangerous. Modern toil demands quiet and fresh air, not stale air.

The towns of to-day can only increase in density at the expense of the open spaces which are the lungs of a city.

We must *increase the open spaces and diminish the distances to be covered.* Therefore the centre of the city must be constructed *vertically.*

The city's residential quarters must no longer be built along "corridor-streets", full of noise and dust and deprived of light.

It is a simple matter to build urban dwellings away from the streets, without small internal courtyards and with the windows looking on to large parks; and this whether our housing schemes are of the type with "set-backs" or built on the "cellular" principle.

The street

The street of to-day is still the old bare ground which has been paved over, and under which a few tube railways have been run.

The modern street in the true sense of the word is a new type of organism, a sort of stretched-out workshop, a home for many complicated and delicate organs, such as gas, water and electric mains. It is contrary to all economy, to all security, and to all sense to bury these important service mains. They ought to be accessible throughout their length. The various storeys of this stretched-out workshop will each have their own particular functions. If this type of street, which I have called a "workshop", is to be realized, it becomes as much a matter of construction as are the houses with which it is customary to flank it, and the bridges which carry it over valleys and across rivers.

The modern street should be a masterpiece of civil engineering and no longer a job for navvies.

The "corridor-street" should be tolerated no longer, for it poisons the houses that border it and leads to the construction of small internal courts or "wells".

Traffic

Traffic can be classified more easily than other things.

To-day traffic is not classified – it is like dynamite flung at hazard into the street, killing pedestrians. Even so, *traffic does not fulfil its function.* This sacrifice of the pedestrian leads nowhere.

If we classify traffic we get:

(a) Heavy goods traffic.
(b) Lighter goods traffic, i.e vans, etc., which make short journeys in all directions.
(c) Fast traffic, which covers a large section of the town.

Three kinds of roads are needed, and in superimposed storeys:

(a) Below-ground there would be the street for heavy traffic. This storey of the houses would consist merely of concrete piles, and between them large open spaces which would form a sort of clearing-house where heavy goods traffic could load and unload.

(b) At the ground floor level of the buildings there would be the complicated and delicate network of the ordinary streets taking traffic in every desired direction.

(c) Running north and south, and east and west, and forming the two great axes of the city, there would be great *arterial roads for fast one-way traffic* built on immense reinforced concrete bridges 120 to 180 yards in width and approached every half-mile or so by subsidiary roads from ground level. These arterial roads could therefore be joined at any given point, so that even at the highest speeds the town can be traversed and the suburbs reached without having to negotiate any cross-roads.

The number of existing streets should be diminished by two-thirds. The number of crossings depends directly on the number of streets; and *cross-roads are an enemy to traffic*. The number of existing streets was fixed at a remote epoch in history. The perpetuation of the boundaries of properties has, almost without exception, preserved even the faintest tracks and footpaths of the old village and made streets of them, and sometimes even an avenue . . . The result is that we have cross-roads every fifty yards, even every twenty yards or ten yards. And this leads to the ridiculous traffic congestion we all know so well.

The distance between two bus stops or two tube stations gives us the necessary unit for the distance between streets, though this unit is conditional on the speed of vehicles and the walking capacity of pedestrians. So an average measure of about 400 yards would give the normal separation between streets, and make a standard for urban distances. My city is conceived on the gridiron system with streets every 400 yards, though occasionally these distances are subdivided to give streets every 200 yards.

This triple system of superimposed levels answers every need of motor traffic (lorries, private cars, taxis, buses) because it provides for rapid and *mobile* transit.

Traffic running on fixed rails is only justified if it is in the form of a convoy carrying an immense load; it then becomes a sort of extension of the underground system or of trains dealing with suburban traffic. *The tramway has no right to exist in the heart of the modern city.*

If the city thus consists of plots about 400 yards square, this will give us sections of about 40 acres in area, and the density of population will vary from 50,000 down to 6,000, according as the "lots" are developed for business or for residential purposes. The natural thing, therefore, would be to continue to apply our unit of distance as it exists in the Paris tubes to-day (namely, 400 yards) and to put a station in the middle of each plot.

Following the two great axes of the city, two "storeys" below the arterial roads for fast traffic, would run the tubes leading to the four furthest points of the garden city suburbs, and linking up with the metropolitan network . . . At a still lower level, and again following these two main axes, would run the one-way loop systems for suburban traffic, and below these again the four great main lines serving the provinces and running north, south, east and west. These main lines would end at the Central Station, or better still might be connected up by a loop system.

The station

There is only one station. The only place for the station is in the centre of the city. It is the natural place for it, and there is no reason for putting it anywhere else. The railway station is the hub of the wheel.

The station would be an essentially subterranean building. Its roof, which would be two storeys above the natural ground level of the city, would form the aerodrome for aero-taxis. This aerodrome (linked up with the main aerodrome in the protected zone) must be in close contact with the tubes, the suburban lines, the main lines, the main arteries and the administrative services connected with all these . . .

The plan of the city

The basic principles we must follow are these:

1 We must de-congest the centres of our cities.
2 We must augment their density.
3 We must increase the means for getting about.
4 We must increase parks and open spaces.

At the very centre we have the *station* with its landing stage for aero-taxis.

Running north and south, and east and west, we have the *main arteries* for fast traffic, forming elevated roadways 120 feet wide.

At the base of the sky-scrapers and all round them we have a great open space 2,400 yards by 1,500 yards, giving an area of 3,600,000 square yards, and occupied by gardens, parks and avenues. In these parks, at the foot of and round the sky-scrapers, would be the restaurants and cafes, the luxury shops, housed in buildings with receding terraces: here too would be the theatres, halls and so on; and here the parking places or garage shelters.

The sky-scrapers are designed purely for business purposes.

On the left we have the great public buildings, the museums, the municipal and administrative offices. Still further on the left we have the "Park" (which is available for further logical development of the heart of the city).

On the right, and traversed by one of the arms of the main arterial roads, we have the warehouses, and the industrial quarters with their goods stations.

All around the city is the *protected zone* of woods and green fields.

Further beyond are the *garden cities*, forming a wide encircling band.

Then, right in the midst of all these, we have the *Central Station*, made up of the following elements:

(a) The landing-platform; forming an aerodrome of 200,000 square yards in area.
(b) The entresol or mezzanine; at this level are the raised tracks for fast motor traffic: the only crossing being gyratory.
(c) The ground floor where are the entrance halls and booking offices for the tubes, suburban lines, main line and air traffic.
(d) The "basement": here are the tubes which serve the city and the main arteries.
(e) The "sub-basement": here are the suburban lines running on a one-way loop.
(f) The "sub-sub-basement": here are the main lines (going north, south, east and west).

The city

Here we have twenty-four sky-scrapers capable each of housing 10,000 to 50,000 employees; this is the business and hotel section, etc., and accounts for 400,000 to 600,000 inhabitants.

The residential blocks, of the two main types already mentioned, account for a further 600,000 inhabitants.

The garden cities give us a further 2,000,000 inhabitants, or more.

In the great central open space are the cafes, restaurants, luxury shops, halls of various kinds, a magnificent forum descending by stages down to the immense parks surrounding it, the whole arrangement providing a spectacle of order and vitality.

Density of population

(a) The sky-scraper: 1,200 inhabitants to the acre.
(b) The residential blocks with set-backs: 120 inhabitants to the acre. These are the luxury dwellings.
(c) The residential blocks on the "cellular" system, with a similar number of inhabitants.

This great density gives us our necessary shortening of distances and ensures rapid inter-communication.

Note. The average density to the acre of Paris in the heart of the town is 146, and of London 63; and of the over-crowded quarters of Paris 213, and of London 169.

Open spaces

Of the area (a), 95 per cent of the ground is open (squares, restaurants, theatres).

Of the area (b), 85 per cent of the ground is open (gardens, sports grounds).

Of the area (c), 48 per cent of the ground is open (gardens, sports grounds).

Educational and civic centres, universities, museums of art and industry, public services, county hall

The "Jardin anglais". (The city can extend here, if necessary.)

Sports grounds: Motor racing track, Racecourse, Stadium, Swimming baths, etc.

The protected zone (which will be the property of the city), with its aerodrome

A zone in which all building would be prohibited; reserved for the growth of the city as laid down by the municipality: it would consist of woods, fields, and sports grounds. The forming of a "protected zone" by continual purchase of small properties in the immediate vicinity of the city is one of the most essential and urgent tasks which a municipality can pursue. It would eventually represent a tenfold return on the capital invested.

Industrial quarters: types of buildings employed

For business: sky-scrapers sixty storeys high with no internal wells or courtyards . . .

Residential buildings with "set-backs", of six double storeys; again with no internal wells: the flats looking on either side on to immense parks.

Residential buildings on the "cellular" principle, with "hanging gardens", looking on to immense parks; again no internal wells. These are "service-flats" of the most modern kind.

Garden cities: their aesthetic, economy, perfection and modern outlook

A simple phrase suffices to express the necessities of tomorrow: WE MUST BUILD IN THE OPEN.

The lay-out must be of a purely geometrical kind, with all its many and delicate implications.

[. . .]

The city of to-day is a dying thing because it is not geometrical. To build in the open would be to replace our present haphazard arrangements, *which are all we have to-day*, by a *uniform* lay-out. Unless we do this *there is no salvation.*

The result of a true geometrical lay-out is *repetition*. The result of repetition is a *standard*, the perfect form (i.e. the creation of standard types). A geometrical lay-out means that mathematics play their part.

There is no first-rate human production but has geometry at its base. It is of the very essence of Architecture. To introduce uniformity into the building of the city we must *industrialize building*. Building is the one economic activity which has so far resisted industrialization. It has thus escaped the march of progress, with the result that the cost of building is still abnormally high.

The architect, from a professional point of view, has become a twisted sort of creature. He has grown to love irregular sites, claiming that they inspire him with original ideas for getting round them. Of course he is wrong. For nowadays the only building that can be undertaken must be either for the rich or built at a loss (as, for instance, in the case of municipal housing schemes), or else by jerry-building and so robbing the inhabitant of all amenities. A motor-car which is achieved by mass production is a masterpiece of comfort, precision, balance and good taste. A house built to order (on an "interesting" site) is a masterpiece of incongruity – a monstrous thing.

If the builder's yard were reorganized on the lines of standardization and mass production we might have gangs of workmen as keen and intelligent as mechanics.

The mechanic dates back only twenty years, yet already he forms the highest caste of the working world.

The mason dates . . . from time immemorial! He bangs away with feet and hammer. He smashes up everything round him, and the plant entrusted to him falls to pieces in a few months. The spirit of the mason must be disciplined by making him part of the severe and exact machinery of the industrialized builder's yard.

The cost of building would fall in the proportion of 10 to 2.

The wages of the labourers would fall into definite categories; to each according to his merits and service rendered.

The "interesting" or erratic site absorbs every creative faculty of the architect and wears him out. What results is equally erratic: lopsided abortions; a specialist's solution which can only please other specialists.

We must build *in the open*: both within the city and around it.

Then having worked through every necessary technical stage and using absolute ECONOMY, we shall be in a position to experience the intense joys of a creative art which is based on geometry.

THE CITY AND ITS AESTHETIC

(The plan of a city which is here presented is a direct consequence of purely geometric considerations.)

A new unit *on a large scale* (400 yards) inspires everything. Though the gridiron arrangement of the streets every 400 yards (sometimes only 200) is uniform (with a consequent ease in finding one's way about), no two streets are in any way alike. This is where, in a magnificent contrapuntal symphony, the forces of geometry come into play.

Suppose we are entering the city by way of the Great Park. Our fast car takes the special elevated motor track between the majestic sky-scrapers: as we approach nearer there is seen the repetition against the sky of the twenty-four sky-scrapers; to our left and right on the outskirts of each particular area are the municipal and administrative buildings; and enclosing the space are the museums and university buildings.

Then suddenly we find ourselves at the feet of the first sky-scrapers. But here we have, not the meager shaft of sunlight which so faintly illumines the dismal streets of New York, but an immensity of space. The whole city is a Park. The terraces stretch out over lawns and into groves. Low buildings of a horizontal kind lead the eye on to the foliage of the trees. Where are now the trivial *Procuracies?* Here is the *city* with its crowds living in peace and pure air, where noise is smothered under the foliage of green trees. The chaos of New York is overcome. Here, bathed in light, stands the modern city [Figure 2].

Our car has left the elevated track and has dropped its speed of sixty miles an hour to run gently through the residential quarters. The "set-backs" permit of vast architectural perspectives. There are gardens, games and sports grounds. And sky everywhere, as far as the eye can see. The square silhouettes of the terraced roofs stand clear against the sky, bordered with the verdure of the hanging gardens. The uniformity of the units that compose the picture throw into relief the firm lines on which the far-flung masses are constructed. Their outlines softened by distance, the sky-scrapers raise immense geometrical facades all of glass, and in them is reflected the blue glory of the sky. An overwhelming sensation. Immense but radiant prisms.

And in every direction we have a varying spectacle: our "gridiron" is based on a unit of 400 yards, but it is strangely modified by architectural devices! (The "set-backs" are in counterpoint, on a unit of 600 × 400.)

The traveller in his airplane, arriving from Constantinople or Pekin it may be, suddenly sees appearing through the wavering lines of rivers and patches of forests that clear imprint which marks a city which has grown in accordance with the spirit of man: the mark of the human brain at work.

As twilight falls the glass sky-scrapers seem to flame.

This is no dangerous futurism, a sort of literary dynamite flung violently at the spectator. It is a spectacle organized by an Architecture which uses plastic resources for the modulation of forms seen in light.

A city made for speed is made for success.

Figure 2 A contemporary city

FRANK LLOYD WRIGHT

"Broadacre City: A New Community Plan"

Architectural Record (1935)

Editors' introduction For more than half a century, the question "Who is the greatest American architect?" could have only one answer: Frank Lloyd Wright (1867–1959). First with his revolutionary "prairie houses" that seemed to grow directly out of the Midwest landscape with their long, low cantilevered rooflines, and later with such masterpieces as the Imperial Hotel in Tokyo, the Guggenheim Museum of Art in New York, and the breathtaking "Falling Water" in Western Pennsylvania, Wright became the spokesman for "organic architecture" and a style of building that expressed "the nature of the materials."

To many, Wright's architecture and "the architecture of American democracy" were synonymous. As an unabashed egotist and a pioneer in the field of media celebrity, Wright encouraged the popular identification of himself with the American spirit. He cultivated an imperious image of plain-speaking, anti-collectivist democracy and sought personally to embody the notion of radical individualism. As an artistic genius, Wright despised the popular philistinism of his day and attributed the observable decline of American popular culture to "the mobocracy" and to the unprincipled bankers and politicians who served its interests. By the 1920s and 1930s, Wright had become a social revolutionary but not, characteristically, of the socialist Left. Rather, Wright called for a radical transformation of American society to restore earlier Emersonian and Jeffersonian virtues. The physical embodiment of that utopian vision was Broadacre City.

Wright unveiled his model of Broadacre City, illustrated in Plate 29 at Rockefeller Center, New York, in 1935. The article reprinted here represents his first and clearest statement of the revolutionary proposal whereby every citizen of the United States would be given a minimum of one acre of land per person, with the family homestead being the basis of civilization, and with government reduced to nothing more than a county architect who would be in charge of directing land allotments and the construction of basic community facilities. Many at the time thought the idea was totally outlandish, but Broadacre (and the small, efficient "Usonian" house) proved to be prophetic as sprawling suburban regions transformed the American landscape during the second half of the twentieth century.

Wright believed that two inventions – the telephone and the automobile – made the old cities "no longer modern," and he fervently looked forward to the day when dense, crowded conglomerations like New York and Chicago would wither and decay. In their place, Americans would reinhabit the rural landscape (and re-acquire the rural virtues of individual freedom and self-reliance) with a "city" of independent homesteads in which people would be isolated enough from one another to insure family stability but connected enough, through modern telecommunications and transportation, to achieve a real sense of community. Borrowing an idea from the anarchist philosopher Kropotkin, Wright believed that the citizens of Broadacre would pursue a combination of manual and intellectual work every day, thus achieving a human wholeness that modern society and the modern city had destroyed. He also believed that a system of personal freedom and dignity through land ownership was the way to guarantee social harmony and avoid class struggle.

Broadacre City invites immediate comparison with the very different models of Ebenezer Howard's Garden City (p. 321) and Le Corbusier's cities based on towers in a park (p. 336). Intriguingly, the overall population density of Broadacre, on the one hand, and the Garden City and Corbusian visions, on the other, were not all that different, depending on the actual acreage of the surrounding parkland or greenbelt. And both Wright's and Le Corbusier's plans are wedded to the automobile, one vision seeing a centralizing, the other a decentralizing, effect. But the most revealing comparisons are with Robert Fishman's description of the now-emerging "technoburbs" (p. 77) and Melvin Webber's prediction of a "post-urban age" (p. 535). One cannot help but wonder whether what seemed impossible in 1935 may actually be realized, with the help of computer-based telecommunications and the possibility of "telecommuting" to work over the Internet, in the twenty-first century.

This selection is from *Architectural Record*, vol. 77 (April 1935). For more on Broadacre City see Robert Fishman's *Urban Utopias of the Twentieth Century* (New York: Basic Books, 1977).

For a biography of Wright see Meryle Secrest, *Frank Lloyd Wright: A Biography* (Chicago: University of Chicago, 1998). For good overviews of Wright's work see David Larkin and Bruce Brooks Pfeiffer (eds.), *Frank Lloyd Wright: The Masterworks* (New York: Rizzoli, 1993) and Neil Levine, *The Architecture of Frank Lloyd Wright* (Princeton, NJ: Princeton University Press, 1996). But the very best sources on Wright are Wright himself, although his writing style is often quirky and hyperbolic. Of particular interest are *When Democracy Builds* (Chicago: University of Chicago Press, 1945), *Genius and the Mobocracy* (New York: Duell, Sloan & Pearce, 1949), and *The Living City* (New York: Horizon, 1958).

Given the simple exercise of several inherently just rights of man, the freedom to decentralize, to redistribute and to correlate the properties of the life of man on earth to his birthright – the ground itself – and Broadacre City becomes reality.

As I see Architecture, the best architect is he who will devise forms nearest organic as features of human growth by way of changes natural to that growth. Civilization is itself inevitably a form but not, if democracy is sanity, is it necessarily the fixation called "academic." All regimentation is a form of death which may sometimes serve life but more often imposes upon it. In Broadacres all is symmetrical but it is seldom obviously and never academically so.

Whatever forms issue are capable of normal growth without destruction of such pattern as they may have. Nor is there much obvious repetition in the new city. Where regiment and row serve the general harmony of arrangement both are present, but generally, both are absent except where planting and cultivation are naturally a process or walls afford a desired seclusion. Rhythm is the substitute for such repetitions everywhere. Wherever repetition (standardization) enters, it has been modified by inner rhythms either by art or by nature as it must, to be of any lasting human value.

The three major inventions already at work building Broadacres, whether the powers that over-built the old cities otherwise like it or not, are:

1 The motor car: general mobilization of the human being.
2 Radio, telephone and telegraph: electrical intercommunication becoming complete.
3 Standardized machine-shop production: machine invention plus scientific discovery.

The price of the major three to America has been the exploitation we see everywhere around us in waste and in ugly scaffolding that may now be thrown away. The price has not been so great if by way of popular government we are able to exercise the use of three inherent rights of any man:

1 His social right to a direct medium of exchange in place of gold as a commodity: some form of social credit.
2 His social right to his place on the ground as he has had it in the sun and air: land to be held only by use and improvements.
3 His social right to the ideas by which and for which he lives: public ownership of invention and scientific discoveries that concern the life of the people.

The only assumption made by Broadacres as ideal is that these three rights will be the citizen's so soon as the folly of endeavoring to cheat him of their democratic values becomes apparent to those who hold (feudal survivors or survivals), as it is becoming apparent to the thinking people who are held blindly abject or subject against their will.

The landlord is no happier than the tenant. The speculator can no longer win much at a game about played out. The present success-ideal, placing, as it does, premiums upon the wolf, the fox and the rat in human affairs and above all, upon the parasite, is growing more evident every day as a falsity just as injurious to the "successful" as to the victims of such success. Well – sociologically, Broadacres is release from all that fatal "success" which is, after all, only excess. So I have called it a new freedom for living in America. It has thrown the scaffolding aside. It sets up a new ideal of success.

In Broadacres, by elimination of cities and towns the present curse of petty and minor officialdom, government, has been reduced to one minor government for each county. The waste motion, the back and forth haul, that today makes so much idle business is gone. Distribution becomes automatic and direct, taking place mostly in the region of origin. Methods of distribution of everything are simple and direct. From the maker to the consumer by the most direct route.

Coal (one-third the tonnage of the haul of our railways) is eliminated by burning it at the mines and transferring that power, making it easier to take over the great railroad rights of way; to take off the cumbersome rolling stock and put the right of way into general service as the great arterial on which truck traffic is concentrated on lower side lanes, many lanes of speed traffic above and monorail speed trains at the center, continuously running. Because traffic may take off or take on at any given point, these arterials are traffic not dated but fluescent. And the great arterial as well as all the highways become great architecture, automatically affording within their structure all necessary storage facilities of raw materials, the elimination of all unsightly piles of raw material.

In the hands of the state, but by way of the county, is all redistribution of land – a minimum of one acre going to the childless family and more to the larger family as effected by the state. The agent of the state in all matters of land allotment or improvement, or in matters affecting the harmony of the whole, is the architect. All building is subject to his sense of the whole as organic architecture. Here architecture is landscape and landscape takes on the character of architecture by way of the simple process of cultivation.

All public utilities are concentrated in the hands of the state and county government as are matters of administration, patrol, fire, post, banking, license and record, making politics a vital matter to everyone in the new city instead of the old case where hopeless indifference makes "politics" a grafter's profession.

In the buildings for Broadacres no distinction exists between much and little, more and less. Quality is in all, for all, alike. The thought entering into the first or last estate is of the best. What differs is only individuality and extent. There is nothing poor or mean in Broadacres.

Nor does Broadacres issue any dictum or see any finality in the matter either of pattern or style.

Organic character is style. Such style has myriad forms inherently good. Growth is possible to Broadacres as a fundamental form, not as mere accident of change but as integral pattern unfolding from within.

Here now may be seen the elemental units of our social structure [Figure 1]: the correlated farm, the factory – its smoke and gases eliminated by burning coal at places of origin, the decentralized school, the various conditions of residence, the home offices, safe traffic, simplified government. All common interests take place in a simple coordination wherein all are employed: *little* farms, *little* homes for industry, *little* factories, *little* schools, a *little* university going to the people mostly by way of their interest in the ground, *little* laboratories on their own ground for professional men. And the farm itself, notwithstanding its animals, becomes the most attractive unit of the city. The husbandry of animals at last is in decent association with them and with all else as well. True farm relief.

To build Broadacres as conceived would automatically end unemployment and all its evils forever. There would never be labor enough nor could under-consumption ever ensue. Whatever a man did would be done – obviously and directly – mostly by himself in his own interest under the most valuable inspiration and direction: under training, certainly, if necessary. Economic independence would be near, a subsistence certain; life varied and interesting.

Every kind of builder would be likely to have a jealous eye to the harmony of the whole within broad limits fixed by the county architect, an architect chosen by the county itself. Each county would thus naturally develop an individuality of its own. Architecture – in the broad sense – would thrive.

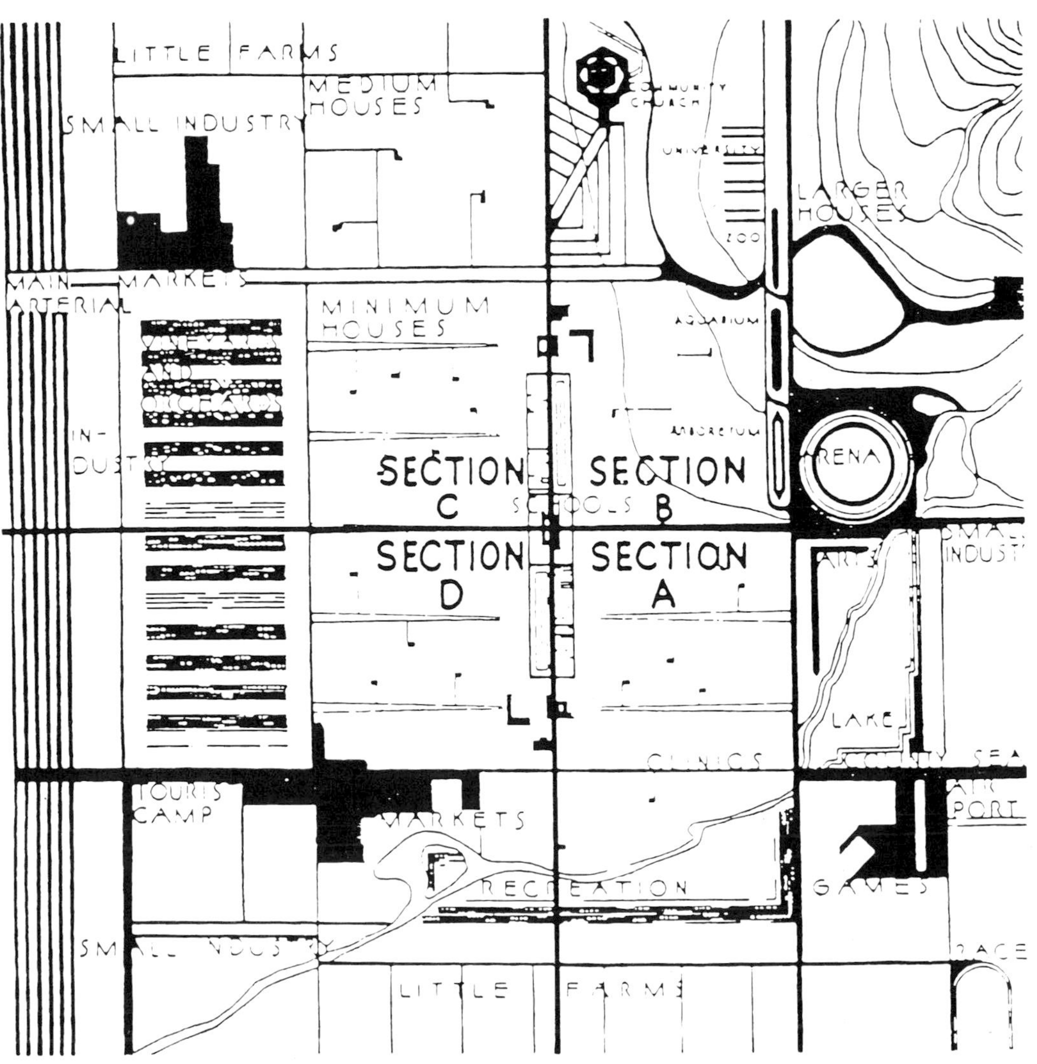

Figure 1

In an organic architecture the ground itself predetermines all features; the climate modifies them; available means limit them; function shapes them.

Form and function are one in Broadacres. But Broadacres is no finality! The model shows four square miles of a typical countryside developed on the acre as unit according to conditions in the temperate zone and accommodating some 1,400 families. It would swing north or swing south in type as conditions, climate and topography of the region changed.

In the model the emphasis has been placed upon diversity in unity, recognizing the necessity of cultivation as a need for formality in most of the planting. By a simple government subsidy certain specific acres or groups of acre units are, in every generation, planted to useful trees, meantime beautiful, giving privacy and various rural divisions. There are no rows of trees alongside the roads to shut out the view. Rows where they occur are perpendicular to the road or the trees are planted in groups. Useful trees like white pine, walnut, birch, beech, fir, would come to maturity as well as fruit and nut trees and they would come as a profitable crop meantime giving character, privacy and comfort to the whole city.The general park is a flowered meadow beside the stream and is bordered with ranks of trees, tiers gradually rising in height above the flowers at the ground level. A music-garden is sequestered from noise at one end. Much is made of general sports and festivals by way of the stadium, zoo, aquarium, arboretum and the arts.

The traffic problem has been given special attention, as the more mobilization is made a comfort and a facility the sooner will Broadacres arrive. Every Broadacre citizen has his own car. Multiple-lane highways make travel safe and enjoyable. There are no grade crossings nor left turns on grade. The road system and construction is such that no signals nor any lamp-posts need be seen. No ditches are alongside the roads. No curbs either. An inlaid purfling over which the car cannot come without damage to itself takes its place to protect the pedestrian.

In the affair of air transport Broadacres rejects the present airplane and substitutes the self-contained mechanical unit that is sure to come: an aerator capable of rising straight up and by reversible rotors able to travel in any given direction under radio control at a maximum speed of, say, 200 miles an hour, and able to descend safely into the hexacomb from which it arose or anywhere else. By a doorstep if desired.

The only fixed transport trains kept on the arterial are the long-distance monorail cars traveling at a speed (already established in Germany) of 220 miles per hour. All other traffic is by motor car on the twelve lane levels or the triple truck lanes on the lower levels which have on both sides the advantage of delivery direct to warehousing or from warehouses to consumer. Local trucks may get to warehouse-storage on lower levels under the main arterial itself. A local truck road parallels the swifter lanes.

Houses in the new city are varied: make much of fireproof synthetic materials, factory-fabricated units adapted to free assembly and varied arrangement, but do not neglect the older nature-materials wherever they are desired and available. House-holders' utilities are nearly all planned in prefabricated utility stacks or units, simplifying construction and reducing building costs to a certainty. There is the professional's house with its laboratory, the minimum house with its workshop, the medium house ditto, the larger house and the house of machine-age luxury. We might speak of them as a one-car house, a two-car house, a three-car house and a five-car house. Glass is extensively used as are roofless rooms. The roof is used often as a trellis or a garden. But where glass is extensively used it is usually for domestic purposes in the shadow of protecting overhangs.

Copper for roofs is indicated generally on the model as a permanent cover capable of being worked in many appropriate ways and giving a general harmonious color effect to the whole.

Electricity, oil and gas are the only popular fuels. Each land allotment has a pit near the public lighting fixture where access to the three and to water and sewer may be had without tearing up the pavements.

The school problem is solved by segregating a group of low buildings in the interior spaces of the city where the children can go without crossing traffic. The school building group includes galleries for loan collections from the museum, a concert and lecture hall, small gardens for the children in small groups and well-lighted cubicles for individual outdoor study: there is a small zoo, large pools and green playgrounds.

This group is at the very center of the model and contains at its center the higher school adapted to the segregation of the students into small groups.

This tract of four miles square, by way of such liberal general allotment determined by acreage and type of ground, including apartment buildings and hotel facilities, provides for about 1,400 families at, say, an average of five or more persons to the family.

To reiterate: the basis of the whole is general decentralization as an applied principle and architectural reintegration of all units into one fabric; free use of the ground held only by use and improvements; public utilities and government itself owned by the people of Broadacre City; privacy on one's own ground for all and a fair means of subsistence for all by way of their own work on their own ground or in their own laboratory or in common offices serving the life of the whole.

There are too many details involved in the model of Broadacres to permit complete explanation. Study of the model itself is necessary study. Most details are explained by way of collateral models of the various types of construction shown: highway construction, left turns, crossovers, underpasses and various houses and public buildings.

Anyone studying the model should bear in mind the thesis upon which the design has been built by the Taliesin Fellowship, built carefully not as a finality in any sense but as an interpretation of the changes inevitable to our growth as a people and a nation.

Individuality established on such terms must thrive. Unwholesome life would get no encouragement and the ghastly heritage left by over-crowding in overdone ultra-capitalistic centers would be likely to disappear in three or four generations. The old success ideals having no chance at all, new ones more natural to the best in man would be given a fresh opportunity to develop naturally.

PETER CALTHORPE

"The Pedestrian Pocket"

from Doug Kelbaugh (ed.), *The Pedestrian Pocket Book* (1989)

Editors' introduction Peter Calthorpe is an urban futurist very much rooted in both the realities of the present and the traditions of the past. As a practicing architect who also teaches at the University of California, Berkeley, Calthorpe is a successful designer/builder. As a leading proponent of ecology and environmentalism as applied to urban design, he is a prophet of a new kind of twenty-first century community, the "pedestrian pocket" that is descended directly from Ebenezer Howard's garden cities (p. 321), transit-oriented development (TOD), and the "new urbanism."

Working as a researcher for the California Energy Commission and the United States Department of Energy, Calthorpe wrote extensively and lectured around the world on the necessity of ecologically sensitive design and energy-efficient building based on the application of passive solar techniques. For a time, he was a partner of Sim Van der Ryn, the California state architect in the administration of progressive governor Jerry Brown and co-author of "An introduction to Ecological Design" (p. 519). Together, they published *Sustainable Communities* (1986), an influential volume that helped spread the ideas of solar energy, recycling, appropriate technology, and environmentalist approaches to urban planning and design.

In *The Pedestrian Pocket Book*, Calthorpe examines the "profound mismatch between the old suburban patterns . . . and the post-industrial culture in which we now find ourselves." The similarities between his solution and the one proposed by Howard in 1898 are striking. Both the garden city and the pedestrian pocket are surrounded by greenbelts of permanent agricultural land. Both are relatively dense developments, allowing residents to walk to the urban center in a short period of time. Both combine residential, commercial, and workplace elements; and where the garden city was served by a railroad connection, the pedestrian pocket avoids the typical suburban monoculture of the automobile by a system of light-rail transit connectors.

Although the source of the pedestrian pocket is the garden city, the real application of the Calthorpe plan will be – indeed, already is – in the blossoming new ring of suburban development currently springing up around the old metropolitan cores, the area that journalist Joel Garreau has dubbed "Edge City" and which Robert Fishman calls "technoburbs" (p. 77). In his most recent book, *The Next American Metropolis* (1993), Calthorpe has further refined and matured the pedestrian pocket idea to fit the emerging realities of Edge City technoburbia with what he calls TODs or "transit-oriented developments." Calthorpe is the most practical of urban visionaries because his visions represent "a response to a transformation that has already expressed itself: the transformation from the industrial forms of segregation and centralization to the decentralized and integrated forms of the post-industrial era." This, he writes, is the result of "a culture adjusting itself" to new realities.

Andres Duany and Elizabeth Plater-Zyberk, Jaime Correa, Steven Peterson and Barbara Littenberg, Mark Schimmenti, Daniel Solomon and a number of other architects and planners are designing human scale communities with design aspects similar to the ones Calthorpe espouses. Because their architecture draws on traditional small-town elements these architects are sometimes referred to as neotraditionalists, their work as "the New Urbanism."

This selection is from Doug Kelbaugh (ed.), *The Pedestrian Pocket Book* (New York: Princeton Architectural Press, 1989). Other books by Peter Calthorpe include, co-edited with Sym Van der Ryn, *Sustainable Communities* (San Francisco: Sierra Club Books, 1986), and *The Next American Metropolis* (New York: Princeton Architectural Press, 1993).

Peter Katz, *The New Urbanism: Towards an Architecture of Community* (New York: McGraw-Hill, 1994) provides an overview of designs by Peter Calthorpe and other of the "new urbanists." See also James Howard Kunstler's, *The Geography of Nowhere* (New York: Simon and Schuster, 1993) for a lively description of Calthorpe and other new urbanists' work.

There is a profound mismatch between the old suburban patterns of settlement we have evolved since World War II and the post-industrial culture in which we now find ourselves. This mismatch is generating traffic congestion, a dearth of affordable and appropriate housing, environmental stress, a loss of open space, and lifestyles that burden working families and isolate the elderly and singles living alone. This mismatch has two primary sources: a dramatic shift in the nature and location of our work place and a fundamental deviation in the character of our increasingly diverse households.

Traffic congestion in the suburbs signals a strong change in the structure of our culture. The computer and service industries have led to the decentralization of the work place, causing new traffic patterns and "suburban gridlock." Where downtown employment once dominated, suburb-to-suburb traffic now produces greater commuting distances and time. Throughout the country, over 40 percent of all commuting trips are now between suburbs. These new patterns have seriously eroded the quality of life in formerly quiet suburban towns. In the San Francisco Bay area, for example, 212 of the region's 812 miles of suburban freeway are regularly backed up during rush hours. That figure is projected to double within the next twelve years. As a result, recent polls have traffic continually heading the list as the primary regional problem, with the difficulty of finding good affordable housing running a close second.

Home ownership has become a troublesome – if not unattainable – goal, even with our double-income families. Affordable housing grows ever more elusive, and families have had to move to less expensive but more peripheral sites, consuming irreplaceable agricultural land and overloading roads. In 1970 about half of all families could meet the expense of a median-priced single-family home; today less than a quarter can.

Moreover, the basic criteria for housing have changed dramatically as single occupants, single parents, the elderly, and small double-income families redefine the traditional home. Our old suburbs were designed around a stereotypical household which is no longer prevalent. Over 73 percent of the new households in the 1980s lack at least one component of the traditional husband, wife and children model. Elderly people over 65 make up 23 percent of the total number of new homeowners, and single parents represent an astonishing 20 percent. Certainly the traditional three-bedroom, single-family residence is relevant to a decreasing segment of the population. The suburban dream becomes even more complicated when one considers the problem of affordability.

In addition to these dominant questions of traffic and housing, longer-range consequences of pollution, air quality, open-space preservation, the conversion of prime agricultural land, and growing infrastructure costs add to the crisis of post-industrial sprawl. These issues are manifested in a growing sense of frustration – placelessness – with the fractured quality of our suburban megacenters. The unique qualities of place are continually consumed by chain-store architecture, scaleless office parks and monotonous subdivisions.

THE SERVICE ECONOMY: DRIVING DECENTRALIZATION

As new jobs have shifted from blue collar to white and grey, the computer has allowed the decentralization of the new service industries into mammoth low-rise office parks on inexpensive and often remote sites. The shift is dramatic: from 1973 to 1985 five million blue-collar jobs were lost nationwide while the service and information fields gained from 82 to 110 million jobs. This translated directly into new office complexes, with 1.1 billion square feet of office space constructed. Nationwide, these complexes have moved outside the central cities, with the percentage of total office space in the suburbs shifting from 25 percent in 1970 to 57 percent in 1984.

Central to this shift is a phenomenon called the "back office," the new sweatshop of the post-industrial economy. The typical back office is large, often with a single floor area of one to two acres. About 80 percent of its employees are clerical, 12 percent supervisory and only 8 percent managerial. In a survey of criteria for back-office locations, forty-seven major Manhattan corporations ranked cost of space first, followed by the quality of the labor pool and site safety. These criteria lead directly to the suburbs where land is inexpensive, parking is easy, and (most importantly) the work force is supplemented by housewives – college-educated, poorly paid, nonunionized, and dependable.

This low-density office explosion has rejuvenated suburban growth just as urban "gentrification" has run its course. The young urban professional has recently made a family commitment and feels the draw of the suburbs. Most of the growth areas in the United States – office parks, shopping malls and single-family dwelling sub-divisions – have a suburban character. Although such growth continually seems to reach the limits of automobile congestion and building moratoriums, there are no readily available alternatives that will enrich the dialogue between growth and no-growth factions, between public benefit and private gain, between the environmentalist and the businessperson.

THE PEDESTRIAN POCKET: A POST-INDUSTRIAL SUBURB

Single-function land-use zoning at a scale and density that eliminates the pedestrian has been the norm for so long that Americans have forgotten that walking can be part of their daily lives. Certainly, the present suburban environment is not walkable, much to the detriment of children, their chauffeur parents, the elderly, and the general health of the population. Urban redevelopment is a strong and compelling alternative to the suburban world but does not seem to fit the character or aspirations of major parts of our population and of many businesses. Mixed-use New Towns are no alternative, as the political consensus needed to back the massive infrastructure investments is lacking. By default, growth is directed mainly by the location of new freeway systems, the economic strength of the region and standard single-use zoning practices. Environmental and local opposition to growth only seems to spread the problem, either transferring the congestion to the next county or creating lower and more auto-dependent densities.

Much smaller than a New Town, the Pedestrian Pocket is defined as a balanced, mixed-use area within a quarter-mile or a five-minute walking radius of a transit station. The functions within this 50- to 100-acre zone include housing, offices, retail, day care, recreation, and parks. Up to two thousand units of housing and one million square feet of office space can be located within three blocks of the transit station using typical residential densities and four-story office configurations.

The Pedestrian Pocket accommodates the car as well as transit and walking. Parking is provided for all housing and commercial space. The housing types are standard low-rise, high- density forms such as three-story walk-up apartments and two-story townhouses. Only the interrelationships and adjacent land use have changed. People have a choice: walk to work or to stores within the Pedestrian Pocket; take the light rail to work or to shop at another station; car pool on a dedicated right-of-way; drive on crowded freeways. In a small Pedestrian Pocket, homes are within walking distance of a neighborhood shopping center, several three-acre parks, day care, various services, and two thousand jobs. Within four stops of the light rail in either direction (ten minutes), employment is available for 16,000, or the amount of back-office growth equivalent to that of one of the nation's highest-growth suburbs over the last five years.

This mix of uses supports a variety of transportation means: walking, bus, light rail, car pool, and standard automobile. The goal is to create an environment that offers choices. Providing comfortable mid-day pedestrian access to retail, services, recreation, and civic functions is essential in order to encourage people to car pool. Similarly, the location of the station, whether bus or rail, near home or work and the realistic opportunity to handle errands without a car are tied to an individual's decision to use mass transit. A Pocket configuration that allows easy access by car to all commercial and residential development maintains the freedom of choice. The result is the best of both worlds.

The Pedestrian Pocket is located on a dedicated right-of-way which evolves with the development. Rather than bearing the large cost of a complete rail system as an initial expense, this right-of-way facilitates mass transit by providing exclusively for car pools, van pools, bikes, and buses. As the cluster matures, transit investments are made for light rail in the developed right-of-way. But the growth of this land-use pattern is not dependent on this investment; the system is designed to support many modes of traffic and to phase light rail into place when the population is great enough to support it.

The Pedestrian Pocket system would eventually act in concert with new light rail lines, reinforcing ridership and connecting existing employment centers, towns and neighborhoods with new development. Light rail lines are currently under construction in many suburban environments, such as, in California alone, Sacramento, San Jose, San Diego, Long Beach, and Orange County. They emphasize the economies of using existing right-of-ways and a simpler, more cost-effective technology than heavy rail. In creating a line of Pedestrian Pockets, the public sector's role is merely to organize the transit system and set new zoning guidelines, leaving development to the private sector. Much of the cost of the transit line can be covered by assessing the property owners benefiting from the increased densities.

The light rails in current use provide primarily a park-and-ride system to connect low-density sprawl with downtown commercial areas. In contrast, the Pedestrian Pocket system is decentralized, linking many nodes of high-density housing with many commercial destinations. Peak-hour traffic is multidirectional, reducing congestion and making the system more efficient. Bus systems, along with car-pool systems, can tie into the light rail. Several of the Pockets on a line have large parking facilities for park-and-ride access, allowing the existing suburban development to enjoy the services and opportunities of the Pockets. However, the location of the office, stores and services adjacent to the station and each other avoids the need for secondary mass transit or additional large parking areas.

The importance of the Pedestrian Pocket is that it provides balanced growth in jobs, housing, and services, while creating a healthy mass- transit alternative for the existing community. The key lies in the form and mix of the Pocket. The pedestrian path system must be carefully designed and form a primary order for the place. If this is configured to allow the pedestrian comfortable and safe access, up to 50 percent of a household's typical automobile trips can be replaced by walking, car pool and light rail journeys. Not only does this produce a better living environment within the Pocket, but the reduction of traffic in the region is significant and in many cases essential.

HOUSING: DIVERSITY IN NEEDS AND MEANS

Housing in the Pedestrian Pocket is planned to provide each of the primary household types with affordable homes that meet their needs. Families with children, single parents or couples need an environment in which kids can move safely, in which day care is integrated into the neighborhood, and in which commuting time is reduced. The townhouses and duplexes proposed for the Pedestrian Pocket allow these families to have all this as well as an attached garage, land ownership and a small private yard. These building types are more affordable to build and maintain than their detached counterparts yet still offer individual ownership and a private identity. The common open space, recreation, day care, and convenient shopping render these houses even more desirable. Group play areas are located off the townhouses' private yards and are connected to the central park and the commercial section by paths. One-third of the housing in a Pocket is of this type.

For singles and "empty-nesters," traditional two- and three-story apartment buildings or condominiums keep costs down while allowing access to the civic facilities, retail services and recreational amenities of the extended community. This segment of the population is traditionally more mobile and thus has an option of either rental or ownership housing. Elderly housing is located close to the parks, light rail and service retail; this eliminates some of the distance and alienation typical of housing facilities for the elderly. The housing is formed into courtyard clusters of two-story buildings which provide a private retreat area and the capacity for common facilities for dining and social activities. Residence in a pedestrian community allows the elderly to become a part of our everyday culture again and to enjoy the parks, stores, and restaurants close at hand.

Several parks double as paths to the station area, a route which is pleasant and free of automobile crossings. The housing overlooking the parks provides security surveillance and twenty-four-hour activity. Each Pocket offers a different arrangement of day-care buildings and general recreational facilities in its parks. Although the housing forms small clusters, the central park and facilities tend to unify the neighborhood, giving it an identity and commonality missing in most of our suburban tracts. An organization (much like a condominium homeowner's organization), which includes landlords, townhouse owners, tenants, office managers, and worker representatives, maintains the centers.

The goal of this tight mix of housing and open space is not just to provide more appropriate homes for the different users or to offer the convenience of walking but hopefully to reintegrate the currently separated age and social groups of our diverse culture. The shared common spaces and local stores may create a rebirth of our lost sense of community and place.

COMMERCE AND COMMUNITY

Jobs are the fuel of new growth, of which the service and high-technology fields are the spearhead. For example, the San Francisco Bay region has currently about 63 percent of all its jobs in these areas. That percentage is expected to increase in the next twenty years, adding about 200,000 new jobs in high technology and 370,000 new jobs in service. Retail activity and housing growth always follow in proportion to these primary income generators. The Pedestrian Pocket provides a framework that allows jobs and housing to grow in tandem.

The commercial buildings in the Pocket offer retail opportunities at their ground floor and offices above. The retail stores enjoy the local walk-in trade from offices and housing, as well as exposure to light rail and drive-in customers. All the stores face a "Main Street" on which the light rail line, the station and convenience parking for cars are mixed. This multiple exposure and access, along with the abundance of office workers, creates a strong market for the theaters, library, post office, food stores, and other convenience stores located in the one hundred thousand square feet of retail.

The offices above the retail stores provide space for small entrepreneurial businesses, start-up firms and local community services. Behind these offices, parking structures capable of accommodating one-half the workers in all the commercial space are located. Presumably, the other half of the employees walk, car pool, or arrive by light rail.

There is a 500,000 to 1,000,000 square foot potential in two to four office buildings per Pedestrian Pocket. These four-story buildings, with 60,000 square feet per floor, fit the size and cost criteria of most back-office employers. The buildings form a courtyard open to the station on one side and the park on the other. Office employees share day-care facilities and open space with the neighborhood.

The commercial mix attempts to balance housing with a desirable job market, stores, entertainment, and services. But the commercial facilities and the offices are not entirely financially dependent on the local housing; access by automobile from the existing neighborhoods and by light rail from other Pockets augments the market. Similarly the transportation system makes a pool of employees available from a twenty-mile range.

REGIONAL PLANNING AND THE PEDESTRIAN POCKET

Pedestrian Pockets are not meant to stand alone as developments; they are intended to form a network offering long-range growth within a region. They will vary considerably given the complexities of place and their internal makeup. Some may be larger than the sixty-acre model we've been using as an example: the quarter-mile walking radius actually encloses 120 acres. Pockets may offer different focuses, with one providing a regional shopping center, one, a cultural center, or a third, housing and recreation. Some may be used as redevelopment tools to provide economic incentives in a depressed area; others may rejuvenate an aging shopping area; the remaining Pockets may be located in new areas zoned for low-density sprawl and in this way save much of the land from more drastic development.

Pockets and their rail lines also connect to the existing assets of an area. The system links the major towns, office parks, shopping areas, and government facilities and allows those from earlier communities to gain access to the Pocket system. Many new light rail systems, built only to connect existing low-density development, are experiencing some resistance from people not wanting to leave their cars. The importance of rezoning for a comfortable walking distance from house to station is to ease people out of their cars, to give them an alternative which is convenient and pleasing. There is evidence that in time such planning will succeed: in a study of San Francisco's rail transit system, BART, it was discovered that fully 40 percent of those who lived and worked within a five-minute walk of the station used the train to get to work.

To test this regional planning concept I chose an area north of San Francisco, combining Marin and Sonoma counties. Many consider the area prime turf for new post-industrial sprawl. Sonoma is projected to have a 61 percent growth in employment in the next twenty years, the highest in the Bay region. Combined, these two counties are to grow by about 88,000 jobs and 63,000 households in the next fifteen years. Of the new jobs, around 60,000 will be in the service, high-technology and knowledge fields, the equivalent of twenty million square feet of office and light industrial space. With standard planning techniques, this growth will consume massive quantities of open space and necessitate a major expansion of the freeway system. The result will still involve frustrating traffic jams.

Instead, twenty Pedestrian Pockets along a new light rail line accommodate this office growth with matching retail facilities, businesses and approximately 30,000 new houses. Several additional pockets dedicated primarily to homes allow two-thirds of the area's housing demand to be met while linking the counties' main cities with a viable mass-transit system. A recently acquired, Northwestern Pacific railroad abandoned right-of-way, which would connect a San Francisco ferry terminal to the northernmost county seat, forms the spine for such a new pattern of growth.

SOCIAL AND ENVIRONMENTAL FORM

It is easy to talk quantitatively about the physical and environmental consequences of our new sprawl but very difficult to postulate their social implications. Many argue that there is no longer a causal relationship between the structure of our physical environment and that of our human well-being or social health. We are adaptable, they claim, and our communities form around interest groups and work rather than around any sense of place or group of individuals. Our center is abstract, not grounded in place, and our social forms are disconnected from home and neighborhood. Planners complicate the issue by polarizing urban and suburban forms. Some advocate a rigorous return to traditional city forms and almost preindustrial culture, while others praise the evolution of the suburban megalopolis as the inevitable and desirable expression of our new technologies and hyper-individualized culture. However rationalized, these new forms have a restless and hollow feel, reinforcing our mobile state and the instability of our families. Moving at a speed that allows only generic symbols to be recognized, we cannot wonder that the man-made environment seems trite and overstated.

In proposing the Pedestrian Pocket the practical comes first; the Pedestrian Pocket preserves land, energy and resources, reduces traffic, renders homes more affordable, allows children and the elderly more access to services, and decreases commuting time for working people. The social consequences are less quantitative but perhaps equally compelling. They have to do with the quality of our shared world.

Mobility and privacy have increasingly displaced the traditional commons, which once provided the connected quality of our towns and cities. Our shared public space has been given over to the car and its accommodation, while our private world has become bloated and isolated. As our private world grows in breadth, our public world becomes more remote and impersonal. As a result, our public space lacks identity and is largely anonymous, while our private space strains toward a narcissistic autonomy. Our communities are zoned black or white, private or public, my space or nobody's. The automobile destroys the urban street, the shopping center destroys the neighborhood store, and the depersonalization of public space grows with the scale of government. Inversely, private space is taxed by the necessity of providing for many activities that were once shared and is further burdened by the need to create identity in a sea of monotony. Although the connection between such social issues and development is elusive and complex, it must be addressed by any serious theory of growth.

In one way, Pedestrian Pockets are utopian – they involve the directed choice of an ideal rather than of laissez-faire planning, and they make certain assumptions about social well-being. But by not assuming a transformation of our society or its people, they avoid the full label, and its subsequent pitfalls, of most utopian schemes. They represent instead a response to a transformation that has already expressed itself: the transformation from the industrial forms of segregation and centralization to the decentralized and integrated forms of the post-industrial era. And perhaps, Pedestrian Pockets express the positive environmental and social results of a culture adjusting itself to this new reality.

PART 6

Urban Planning Theory and Practice

PART SIX

INTRODUCTION

Contemporary urban planning has come a long way from its nineteenth-century origins in the public health and Olmsted-inspired parks movement, the visionary plans of Howard and the garden cities that more or less followed his ideas, Burnham's monumental City Beautiful projects, and the plans of Geddes, Le Corbusier, Wright and a host of other imagined and implemented plans discussed in Part 5, Urban Planning History and Visions. Today city and regional planning (or town and country planning as it is called in the UK) has matured into an important profession with its own body of theory and set of professional practices.

Large local governments may employ dozens or even hundreds of professionally trained planners to carry out this planning work, and even most small and medium-sized cities and towns employ city planners and have explicit plans for their future development. In addition to professionals whose formal education is in city (or town) planning, planning staffs are likely to include architects and urban designers, engineers, geographers, economists, transportation experts, environmental professionals, quantitative analysts and others.

City plans are grounded in analysis of local conditions and contain visions of futures the citizens, local elected officials, planning staff and consultants consider desirable. They vary greatly in approach, content, sophistication, comprehensiveness, time frame, and format. Plans developed for the former London County Council or the Los Angeles City Planning Commission run to many volumes built on mountains of data and sophisticated analysis. The town plan for a provincial market town may contain a common sense description of the town's situation and some practical suggestions worked out by the residents under the direction of a part-time planning consultant.

The first two selections in this section pick up where Part 5 leaves off. Sir Peter Hall's "The City of Theory" (p. 362) discusses the evolution and current status of twentieth-century urban planning theory, and Edward J. Kaiser and David R. Godschalk's "Twentieth Century Land Use Planning" describes planning practice (p. 375). Both selections touch on planning in the first half of the twentieth century, but focus on how planning has evolved during the last half century and what it is like today. Read together with LeGates and Stout (p. 299), they give a good overview of twentieth-century urban planning theory and practice.

In the first half of the twentieth century, urban planning was elitist and paid little attention to implementation. According to Hall, planning theory in the first half of the twentieth century was preoccupied with how to create stable cities "geared to a static world." During this "golden age," the planner was "free from political interference" and "serenely sure of his technical capacities." He was also generally ineffective. Heavily influenced by the Garden City and City Beautiful movements, early planners worked out theoretically perfect physical plans in excruciating detail. This ivory tower approach was never very practical and was no longer defensible by the 1950s, as the pace of urban development and urban change began to accelerate to a superheated level. By the 1950s, according to Kaiser and Godschalk, there was a general consensus that urban general plans should be long-term, visionary documents, charting the desired physical form of the city.

In the last half century, urban planning theory has been buffeted by a whole series of conflicting approaches – Marxists and equity planners, systems analysts, pluralists, disjointed incrementalists, advocacy planners, probabalistic planners, and communicative action theorists. In practice the general plan "trunk" of the urban land use planning "tree" has branched into management and policy as well as design. Planning no longer takes place in ivory towers. As planning has become more relevant it has also become more significant – even conflictual – economically and politically.

Searching for "what works, what doesn't," Yale professor Alexander Garvin argues that to succeed, planning must consider private market forces. Markets, financing, and entrepreneurship are three key ingredients to success. Location, design, and time are the others. Garvin argues that planning should be designed to catalyze private development in urban areas. Strategic public investment in infrastructure and regulation can direct the forces which a good plan unleashes in positive directions. Planning that does not consider or misjudges market forces will not catalyze private investment. It will fail within the definition of successful planning that Garvin himself adopts. Other planners in this anthology would agree with many of the ingredients for success that Garvin identifies, but not all agree that market success is the best measure of a plan's success.

Cornell planning professor John Forester's description of "planning in the face of conflict" picks up many of Garvin's themes, but Forester does not use market success as the basis for judging a plan. He describes how planners actually work with private developers. Forester provides perhaps the best view we have of what current urban planning is really like as perceived by planners themselves. Based on interviews with dozens of practicing planners, Forester describes planners' day-to-day activities, the nature of and limitations on their power, and the strategies they employ to get things done. Planners negotiate, mediate, resolve conflict, and serve as diplomats. Forester is interested in equity planning and applauds the efforts of many planners to redirect market forces and to empower people and communities poorly served by the private market.

Paul Davidoff develops this theme still further. Davidoff envisioned a planning practice in which city planners would act as advocates for poor people and disenfranchised groups. Davidoff and many planners whom he inspired saw "advocacy planning" as one way to bring about non-violent social change. This concern with social justice is an enduring one as described in LeGates and Stout's history of early urban planning (p. 299). From the early efforts of nineteenth-century reformers advocating on behalf of slum residents in the new industrial slums, through Benjamin Marsh Clark and the New York congestion movement, to the New Deal regional planners, some progressives have always made a connection between urban planning and social justice. The most recent incarnation of this tradition is in equity planning, a subfield developed by Cleveland state professor and former Cleveland city planning director Norman Krumholz.

Another enduring value in urban planning has been a concern to harmonize the built environment with the natural environment. By the middle of the nineteenth century, park planners like Joseph Paxton in England and Frederick Law Olmsted in the United States were working hard to bring nature into crowded cities. In the early twentieth century, Patrick Geddes had worked out an elaborate scheme for regional planning that reflected different ecosystems. Similar thinking informed Ian McHarg's theories in his classic *Design with Nature* (1965). Today "sustainable" urban planning has moved to the fore. In the selection "Planning Sustainable and Livable Cities," Stephen Wheeler summarizes the current sustainable urban planning literature and provides his own definition of sustainable urban development, core aspects of sustainability, and principles to achieve it. He illustrates his ideas with actual examples of good sustainable planning practice.

The final selection in this section, by Leonie Sandercock and Ann Forsyth, introduces the important issue of gender to urban planning theory. Sandercock and Forsyth argue that as women are increasingly included in the planning profession and as planning practice becomes more sensitive to their needs, a body of planning theory is needed to guide the evolution of gender-sensitive planning practice. Drawing on a wide range of planning and nonplanning sources, the authors describe directions feminist urban planning might take.

One of the things that makes urban planning so fascinating is the tremendous variety of issues planning theory and practice must confront. Issues of aesthetics, design, economic feasibility, decision-making theory, equity, conflict resolution, advocacy, race and gender sensitivity, and sustainability are just some of the core concerns that planning must confront in the twenty-first century.

PETER HALL

"The City of Theory"

from *Cities of Tomorrow: An Intellectual History of Urban Planning and Design in the Twentieth Century* (1996)

Editors' introduction In this selection British geographer/planner Sir Peter Hall reviews the evolution of city planning theory in the United States and Europe during the last half century.

Hall points out that before World War II, city/town planning was defined as the craft of physical planning. Professors who had been educated as architects, landscape architects, and other design professionals dominated teaching and writing. They taught students to prepare self-contained, end-state physical plans like Raymond Unwin's plan for the garden city of Letchworth, England: architectural drawings extended to city scale. Once a well-crafted, aesthetically pleasing, functional plan was complete on paper the early planning theorists believed that a city could be built just as a house is built from architectural drawings.

But, as Hall notes, few planners today (or indeed in the past) actually get to plan whole new towns from scratch. Rather they have to decide how to tie new housing, streets, retail districts, industrial areas, parks, and infrastructure into existing cities. And city planning does not end the way house construction does with a finished product.

During the 1960s, systems analysts who had been educated as economists, computer scientists, mathematical geographers, and engineers developed a competing paradigm of what city planning should be. They argued that city planning should be a science, not a craft. They felt planners could and should use quantitative and exact methodologies to plan urban transportation, utility, or other "systems" on an ongoing basis. They were pioneers in using computers at a time when computers were new, costly, and extraordinarily hard to use. The systems theorists thought plans should be mathematical models rather than end-state architectural drawings. It is helpful to imagine a planner from the old school saying, "Here is my plan," and unrolling a series of large, carefully drafted blueprints and design sketches, confronting a systems planner saying, "Here's my plan," as she types an equation into a huge mainframe computer!

But planning for what kind of city and for whom? Normative theorists, shaken by 1960s racial and class conflict in cities, felt city planning was too important to leave to either elitist designers or technocratic systems analysts. They were more concerned with who the city was being built for and how planning might benefit poor and powerless people than with either design excellence or elegant mathematical modeling.

During the 1970s, Marxist theorists developed a particularly coherent body of urban theory in terms of class and race – but not in terms of gender as Leonie Sandercock and Ann Forsyth point out (p. 446). Marxist urban theory dominated much academic discourse through the 1970s.

Hall ends his tour of planning theory with some critical comments on the current divorce between planning theory and practice. He argues that as graduate programs in city and regional planning (often called town and country planning in Europe) have grown in number and size and as a formal body of planning theory has developed, too often academics today merely debate each other's academic theories with little attention to actual practice or the needs of planning practitioners on the front lines. Finding academic planning theory irrelevant, city planning practitioners concern themselves only with the nuts and bolts of planning practice without the deeper understanding relevant theory could offer. Hall calls for an improved, reciprocal relationship between the two: theory that is informed by and relevant to planning practice and planning practice informed and improved by theory.

Until assuming the Bartlett Professorship of City Planning at University College, London, in 1994, Peter Hall shuttled between the University of California, Berkeley's Department of City and Regional Planning and the University of Reading's Geography Department. He has traveled widely and written prolifically in urban geography and city planning. In addition to his academic accomplishments Peter Hall has worked with actual planning projects as varied as long-term plans for the London region and bringing high-speed bullet trains to California. His theory is informed by the experience of actual planning, and practicing planners as well as academics use his writings.

This selection is from *Cities of Tomorrow: An Intellectual History of Urban Planning and Design in the Twentieth Century*, 2nd edn. (Oxford: Blackwell, 1996). Other books by Peter Hall include *Cities in Civilization* (New York: Pantheon, 1998), with Colin Ward, *Sociable Cities: The Legacy of Ebenezer Howard* (London: John Wiley & Sons, 1998), *Urban and Regional Planning*, 3rd edn. (London: Routledge, 1992), *The World Cities*, 3rd edn. (London: Weidenfeld and Nicolson; New York: St. Martin's, 1984), and *Great Planning Disasters* (Berkeley: University of California Press, 1982).

Other overviews of planning theory include Seymour Mandelbaum (ed.), *Explorations in Planning Theory* (Rutgers, NJ: CUPR Press, 1996), John Friedmann, *Planning in the Public Domain: From Knowledge to Action* (Princeton, NJ: Princeton University Press, 1987), and Robert Burchell and George Sternlieb (eds.), *Planning Theory in the 1980s* (New Brunswick, NJ: Rutgers University Press, 1978).

PLANNING AND THE ACADEMY: PHILADELPHIA, MANCHESTER, CALIFORNIA, PARIS, 1955–1987

. . . about 1955 . . . city planning at last became legitimate; but in doing so, it began to sow the seeds of its own destruction. All too quickly, it split into two separate camps: the one, in the schools of planning, increasingly and exclusively obsessed with the theory of the subject; the other, in the offices of local authorities and consultants, concerned only with the everyday business of planning in the real world. That division was not at first evident; indeed, during the late 1950s and most of the 1960s, it seemed that at last a complete and satisfactory link had been forged between the world of theory and the world of practice. But all too soon, illusion was stripped aside: honeymoon was followed in quick succession during the 1970s by tiffs and temporary reconciliations, in the 1980s by divorce. And, in the process, planning lost much of its new-found legitimacy.

The prehistory of academic city planning 1930–1955

It was not that planning was innocent of academic influence before the 1950s. On the contrary: in virtually every urbanized nation, universities and polytechnics had created courses for the professional education of planners; professional bodies had come into existence to define and protect standards, and had forged links with the academic departments. Britain took an early lead when in 1909 . . . the soap magnate William Hesketh Lever, founder of Port Sunlight, won a libel action against a newspaper and used the proceeds to endow his local University of Liverpool with a Department of Civic Design. Stanley Adshead, the first professor, almost immediately created a new journal, the *Town Planning Review*, in which theory and good practice were to be firmly joined; its first editor was a young faculty recruit, Patrick Abercrombie, who was later to succeed Adshead in the chair first at Liverpool, then at Britain's second school of planning: University College London, founded in 1914. The Town Planning Institute – the Royal accolade was conferred only in 1959 – was founded in 1914 on the joint initiative of the Royal Institute of British Architects, the Institution of Civil Engineers and the Royal Institution of Chartered Surveyors; by the end of the 1930s, it had recognized seven schools whose examinations provided an entry to membership.

The United States was slower: though Harvard had established a planning course in 1909, neck and neck with Liverpool, it had no separate department until 1929. Nevertheless, by the 1930s America had schools also at MIT, Cornell, Columbia and Illinois, as well as courses taught in other departments at a great many universities across the country. And the American City Planning Institute, founded in 1917 as a breakaway from the National Conference on City Planning, ten years later became – mainly through the insistence of Thomas Adams – a full-fledged professional body on TPI lines, a status it retained when in 1938 it broadened to include regional planning and renamed itself the American Institute of Planners.

The important point about these, and other, initiatives was this: stemming as they did from professional needs, often through spin-offs from related professions like architecture and engineering, they were from the start heavily suffused with the professional styles of these design-based professions.

The job of the planners was to make plans, to develop codes to enforce these plans, and then to enforce those codes; relevant planning knowledge was what was needed for that job; planning education existed to convey that knowledge together with the necessary design skills. So, by 1950, the utopian age of planning . . . was over; planning was now institutionalized into comprehensive land-use planning. All this was strongly reflected in the curricula of the planning schools down to the mid-1950s, and often for years after that; and these in turn were reflected in the books and articles that academic planners wrote. Land-use planning, Keeble told his British audience in 1959 and Kent reminded the American counterpart in 1964, was a distinct and tightly bounded subject, quite different from social or economic planning. And these texts reflected the fact that "city planners early adopted the thoughtways and the analytical methods that engineers developed for the design of public works, and they then applied them to the design of cities."

The result, as Michael Batty has put it, was a subject that for the ordinary citizen was 'somewhat mystical' or arcane, as law or medicine were, but that was – in sharp contrast to education for these older professions – not based on any consistent body of theory; rather, in it, "scatterings of social science bolstered the traditional architectural determinism." Planners acquired a synthetic ability not through abstract thinking, but by doing real jobs; in them, they used first creative intuition, then reflection. Though they might draw on bits and pieces of theory about the city – the Chicago school's sociological differentiation of the city, the land economists' theory of urban land rent differentials, the geographers' concepts of the natural region – these were employed simply as snippets of useful knowledge. In the important distinction later made by a number of writers, there was some theory in planning but there was no theory of planning. The whole process was very direct, based on a single-shot approach: survey (the Geddesian approach) was followed by analysis (an implicit learning approach), followed immediately by design.

True, as Abercrombie's classic text of 1933 argued, the making of the plan was only half the planner's job; the other half consisted of planning, that is implementation, but it was nowhere assumed that some kind of continuous learning process was needed. True, too, the 1947 Act provided for plans – and the surveys on which they were based – to be quinquennially updated; the assumption was still that the result would be a fixed land-use plan. And, a decade after that, though Keeble's equally classic text referred to the planning process, by this he simply meant the need for a spatial hierarchy of related plans from the regional to the local, and the need at each scale for survey before plan. Nowhere is found a discussion of implementation or updating. Thus – apart from extremely generalized statements like Abercrombie's famous triad of 'beauty, health and convenience' – the goals were left implicit; the planner would develop them intuitively from his own values, which by definition were 'expert' and apolitical.

So, in the classic British land-use planning system created by the 1947 Town and Country Planning Act, no repeated learning process was involved, since the planner would get it right first time:

> The process was therefore not characterized by explicit feedback as the search "homed in" on the best plan, for the notion that the planner had to learn about the nature of the problem was in direct conflict with his assumed infallibility as an expert, a professional . . . The assumed certainty of the process was such that possible links back to the reality in the form of new surveys were rarely if ever considered . . . This certainty, based on the infallibility of the expert, reinforced the apolitical, technical nature of the process. The political environment was regarded as totally passive, indeed subservient to the 'advice' of the planners and in practice, this was largely the case.
>
> (Batty, 1979)

It was, as Batty calls it, the golden age of planning: the planner, free from political interference, serenely sure of his technical capacities, was left to get on with the job. And this was appropriate to the world outside, with which planning had to deal: a world of glacially slow change – stagnant population, depressed economy – in which major planning interventions would come only seldom and for a short time, as after a major war. Abercrombie, in the plan for the West Midlands he produced with Herbert Jackson in 1948, actually wrote that a major objective of the plan should be to slow down the rate of urban change, thus reducing the rate at which built structures became obsolescent: the ideal city would be a static, stable city:

> Let us assume . . . that a maximum population has been decided for a town, arrived at after consideration of all the factors appearing to be relevant . . . Allowance has been made for proper space for all conceivable purposes in the light of present facts and the town planner's experience and imagination. Accordingly, an envelope or green belt has been prescribed, outside which the land uses will be those involving little in the way of resident population. The town planner is now in the happy position for the first time of knowing the limits of his problem. He is able to address himself to the design of the whole and the parts in the light of a basic overall figure for population. The process will be difficult enough in itself, but at least he starts with one figure to reassure him.
>
> (Abercrombie and Jackson, 1948)

American planning was never quite like that. Kent's text of 1964 on the urban general plan, though it deals with the same kind of land-use planning, reminds its students of end-directions which are continually adjusted as time passes. And, because the planner's basic understanding of the interrelationship between socio-economic forces and the physical environment was largely intuitive and speculative, Kent warned his student readers,

> In most cases it is not possible to know with any certainty what physical design measures should be taken to bring about a given social or economic objective, or what social and economic consequences will result from a given physical-design proposal. Therefore, the city council and the city-planning commission, rather than professional city planners, should make the final value judgements upon which the plan is based.
>
> (Kent, 1964)

But even Kent was certain that, despite all this, it was still possible for the planner to produce some kind of optimal land use plan; the problem of objectives was just shunted off.

The systems revolution

It was a happy, almost dream-like, world. But increasingly, during the 1950s, it did not correspond to reality. Everything began to get out of hand. In every industrial country, there was an unexpected baby boom, to which the demographers reacted with surprise, the planners with alarm; only its timing varied from one country to another, and everywhere it created instant demands for maternity wards and child-care clinics, only slightly delayed needs for schools and playgrounds. In every one, almost simultaneously, the great postwar economic boom got under way, bringing pressures for new investment in factories and offices. And, as boom generated affluence, these countries soon passed into the realms of high mass-consumption societies, with unprecedented demands for durable consumer goods: most notable among these, land-hungry homes and cars. The result everywhere – in America, in Britain, in the whole of western Europe – was that the pace of urban development and urban change began to accelerate to an almost superheated level. The old planning system, geared to a static world, was overwhelmed.

These demands in themselves would force the system to change; but, almost coincidentally, there were changes on the supply side too. In the mid-1950s there occurred an intellectual revolution in the whole cluster of urban and regional social studies, which provided planners with much of their borrowed intellectual baggage. A few geographers and industrial economists discovered the works of German theorists of location, such as Johann Heinrich von Thünen (1826) on agriculture, Alfred Weber (1909) on industry, Walter Christaller (1933) on central places, and August Lösch (1940) on the general theory of location; they began to summarize and analyse these works, and where necessary to translate them. In the United States, academics coming from a variety of disciplines began to find regularities in many distributions, including spatial ones. Geographers, beginning to espouse the tenets of logical positivism, suggested that their subject should cease to be concerned with descriptions of the detailed differentiation of the earth's surface, and should instead begin to develop general hypotheses about spatial distributions, which could then be rigorously tested against reality: the very approach which these German pioneers of location theory had adopted. These ideas, together with the relevant literature, were brilliantly synthesized by an American economist, Walter Isard, in a text that became immediately influential. Between 1953 and 1957, there occurred an almost instant revolution in human geography and the creation, by Isard, of a new academic discipline uniting the new geography with the German tradition of locational economics. And, with official blessing – as in the important report of Britain's Schuster Committee of 1950, which recommended a greater social science content in planning education – the new locational analysis began to enter the curricula of the planning schools.

The consequences for planning were momentous: with only a short timelag, "the discipline of physical planning changed more in the 10 years from 1960 to 1970, than in the previous 100, possibly even 1000 years" (Batty, 1979).

The subject changed from a kind of craft, based on personal knowledge of a rudimentary collection of concepts about the city, into an apparently scientific activity in which vast amounts of precise information were garnered and processed in such a way that the planner could devise very sensitive systems of guidance and control, the effects of which could be monitored and if necessary modified. More precisely, cities and regions were viewed as complex systems – they were, indeed, only a particular spatially based subset of a whole general class of systems – while planning was seen as a continuous process of control and monitoring of these systems, derived from the then new science of cybernetics developed by Norbert Wiener.

There was thus, in the language later used in the celebrated work of Thomas Kuhn, a "paradigm shift." It affected city planning as it affected many other related areas of planning and design. Particularly, its main early applications – already in the mid 1950s – concerned defence and aerospace; for these were the Cold War years, when the United States was engaging in a crash programme to build new and complex electronically controlled missile systems. Soon, from that field, spun off another application. Already in 1954, Robert Mitchell and Chester Rapkin – colleagues of Isard at the University of Pennsylvania – had published a book suggesting that urban traffic patterns were a direct and measurable function of the pattern of activities – and thus land uses – that generated them. Coupled with earlier work on spatial interaction patterns, and using for the first time the data-processing powers of the computer, this produced a new science of urban transportation planning, which for the first time claimed to be able scientifically to predict future urban-traffic patterns. First applied in the historic Detroit Metropolitan Area transportation study of 1955, further developed in the Chicago study of 1956, it soon became a standardized methodology employed in literally hundreds of such studies, first across the United States, then across the world.

Heavily engineering-based in its approach, it adopted a fairly standardized sequence. First, explicit goals and objectives were set for the performance of the system. Then, inventories were taken of the existing state of the system: both the traffic flows, and the activities that gave rise to them. From this, models were derived which sought to establish these relationships in precise mathematical form. Then, forecasts were made of the future state of the system, based on the relationships obtained from the models. From this, alternative solutions could be designed and evaluated in order to choose a preferred option. Finally, once implemented the network would be continually monitored and the system modified as necessary.

At first, these relationships were seen as operating in one direction: activities and land uses were given; from these, the traffic patterns were derived. So the resulting methodology and techniques were part of a new field, transportation planning, which came to exist on one side of traditional city planning. Soon, however, American regional scientists suggested a crucial modification: the locational patterns of activities – commercial, industrial, residential – were in turn influenced by the available transportation opportunities; these relationships, too, could be precisely modelled and used for prediction; therefore the relationship was two-way, and there was a need to develop an interactive system of land-use–transportation planning for entire metropolitan or subregional areas. Now, for the first time, the engineering-based approach invaded the professional territory of the traditional land-use planner. Spatial interaction models, especially the Garin–Lowry model – which, given basic data about employment and transportation links, could generate a resulting pattern of activities and land uses – became part of the planner's stock in trade. As put in one of the classic systems texts:

> In this general process of planning we particularise in order to deal with more specific issues: that is, a specific real world system or subsystem must be represented by a specific conceptual system or subsystem within the general conceptual system. Such a particular representation of a system is called a *model* . . . the use of models is a means whereby the high variety of the real world is reduced to a level of variety appropriate to the channel capacities of the human being.
>
> (Chadwick, 1971)

This involved more than a knowledge of computer applications – novel as that seemed to the average planner of the 1960s. It meant also a fundamentally different concept of planning. Instead of the old master-plan or blueprint approach, which assumed that the objectives were fixed from the start, the new concept was of planning as a process, "whereby programmes are adapted during their implementation as and when incoming information requires such changes". And this planning process was independent of the thing that was planned; as Melvin Webber put it, it was "a special way of deciding and acting" which involved a constantly recycled series of logical steps: goal-setting, forecasting of change in the outside world, assessment of chains of consequences of alternative courses of action, appraisal of costs and benefits as a basis for action strategies, and continuous monitoring. This was the approach of the new British textbooks of systems planning, which started to emerge at the end of the 1960s, and which were particularly associated with a group of younger British graduates, many teaching or studying at the University of Manchester. It was also the approach of a whole generation of subregional studies, made for fast-growing metropolitan areas in Britain during that heroic period of growth and change, 1965–75: Leicester–Leicestershire, Nottinghamshire–Derbyshire, Coventry–Warwickshire–Solihull, South Hampshire. All were heavily suffused with the new approach and the new techniques; in several, the same key individuals – McLoughlin in Leicester, Batty in Notts–Derby – played a directing or a crucial consulting role.

But the revolution was less complete – at least, in its early stages – than its supporters liked to argue: many of these "systems" plans had a distinctly blueprint tint, in that they soon resulted in all-too-concrete proposals for fixed investments like freeway systems. Underlying this, furthermore, were some curious metaphysical assumptions, which the new systems planners shared with their blueprint elders: the planning system was seen as active, the city system as purely passive; the political system was regarded as benign and receptive to the planner's expert advice. In practice, the systems planner was involved in two very different kinds of activity: as a social scientist, he or she was passively observing and analysing reality; as a designer, the same planner was acting on reality to change it – an activity inherently less certain, and also inherently subject to objectives that could only be set through a complex, often messy, set of dealings between professionals, politicians and public.

The core of this problem was a logical paradox: despite the claims of the systems planners, the urban planning system was different from (say) a weapons system. In this latter kind of system, to which the 'systems approach' had originally and successfully been applied, the controls were inside the system; but here, the urban-regional system was inside its own system of control. Related to this were other crucial differences: in urban planning, there was not just one problem and one overriding objective, but many, perhaps contradictory; it was difficult to move from general goals to specific operational ones; not all were fully perceived; the systems to be analysed did not self-evidently exist, but had to be synthesized; most aspects were not deterministic, but probabilistic; costs and benefits were difficult to quantify. So the claims of the systems school to scientific objectivity could not readily be fulfilled. Increasingly, members of the school came to admit that in such 'open' systems, systematic analysis would need to play a subsidiary role to intuition and judgement; in other words, the traditional approach. By 1975 Britton Harris, perhaps the most celebrated of all the systems planners, could write that he no longer believed that the more difficult problems of planning could be solved by optimizing methods.

The search for a new paradigm

All this, in the late 1960s, came to focus in an attack from two very different directions, which together blew the ship of systems planning at least half out of the water. From the philosophical right came a series of theoretical and empirical studies from American political scientists, arguing that – at least in the United States – crucial urban decisions were made within a pluralist political structure in which no one individual or group had total knowledge or power, and in which, consequently, the decision-making process could best be described as "disjointed incrementalism" or "muddling through". Meyerson and Banfield's classic analysis of the Chicago Housing Authority concluded that it engaged in little real planning, and failed because it did not correctly identify the real power structure in the city; its elitist view of the public interest was totally opposed to the populist view of the ward politicians, which finally prevailed. Downs theorized about such a structure, suggesting that politicians buy votes by offering bundles of policies, rather as in a market. Lindblom contrasted the whole rational-comprehensive model of planning with what he found to be the actual process of policy development, which was characterized by a mixture of values and analysis, a confusion of ends and means, a failure to analyse alternatives, and an avoidance of theory. Altshuler's analysis of Minneapolis–St Paul suggested that the professional planner carried no clout against the political machine, which backed the highway-building engineers against him; they won by stressing expertise and concentrating on narrow goals, but theirs was a political game; the conclusion was that planners should recognize their own weakness, and devise strategies appropriate to that fact.

All these analyses arose from study of American urban politics, which is traditionally more populist, more pluralist, than most. Even there, Rabinowitz's study of New Jersey cities suggested that they varied greatly in style, from the highly fragmented to the very cohesive; while Etzioni, criticizing Lindblom, suggested that recent United States history showed several important examples of non-incremental decision-making, especially in defence. But, these reservations taken, the studies did at least suggest that planning in actuality was a very long way indeed from the cool, rational, Olympian style envisaged in the systems texts. Perhaps it might have been better if it had been closer; perhaps not. The worrisome point was that in practice, local democracy proved to be an infinitely messier business than the theory would have liked. Some theorists accordingly concluded that if this was the way planning was, this was the way it should be encouraged to be: partial, experimental, incremental, working on problems as they arose.

That emerged even more clearly, because – as so often seems to happen – in America the left-wing criticism was reaching closely similar conclusions. By the late 1960s, fuelled by the civil-rights movement and war on poverty, the protests against the Vietnam war and the campus free-speech movement, it was this wing that was making all the running. Underlying the general current of protest were three key themes, which proved fatal to the legitimacy of the systems planners. One was a widespread distrust of expert, top-down planning generally – whether for problems of peace and war, or for problems of the cities. Another, much more specific, was an increasing paranoia about the systems approach, which in its military applications was seen as employing pseudo-science and incomprehensible jargon to create a smokescreen, behind which ethically reprehensible policies could be pursued. And a third was triggered by the riots that tore through American cities starting with Birmingham, Alabama, in 1963 and ending with Detroit, in 1967. They seemed to prove the point: systems planning had done nothing to ameliorate the condition of the cities; rather, by assisting or at least conniving in the dismemberment of inner-city communities, it might actually have contributed to it. By 1967 one critic, Richard Bolan, could argue that systems planning was old-fashioned comprehensive planning, dressed up in fancier garb; both, alike, ignored political reality.

The immediate left-wing reaction was to call on the planners themselves to turn the tables, and to practise bottom-up planning by becoming advocate-planners. Particularly, in this way they would make explicit the debate about the setting of goals and objectives, which both the blueprint and systems approaches had bypassed by means of their comfortable shared assumption that this was the professional planner's job. Advocacy planners would intervene in a variety of ways, in a variety of groups; diversity should be their keynote. They would help to inform the public of alternatives; force public planning agencies to compete for support; help critics to generate plans that were superior to official ones; compel consideration of underlying values. The resulting structure was highly American: democratic, locally grounded, pluralistic, but also legalistic in being based on institutionalized conflict. But, interestingly, while demoting the planner in one respect, it enormously advanced his or her power in another: the planner was to take many of the functions that the locally elected official had previously exercised. And, in practice, it was not entirely clear how it would all work; particularly, how the process would resolve the very real conflicts of interest that could arise within communities, or how it could avoid the risk that the planners, once again, would become manipulators.

At any rate, there is more than a passing resemblance between the planner as a disjointed incrementalist, and the planner-advocate; and, indeed, between either of these and a third model set out in Bolan's paper of 1967, the planner as informal co-ordinator and catalyst, which in turn shades into a fourth: Melvin Webber's probabilistic planner, who uses new information systems to facilitate debate and improve decision-making. All are assumed to work within a pluralist world, with very many different competing groups and interests, where the planner has at most (and, further should have) only limited power or influence; all are based, at least implicitly, on continued acceptance of logical positivism. As Webber put it, at the conclusion of his long two-part paper of 1968–9:

> The burden of my argument is that city planning failed to adopt the planning method, choosing instead to impose input bundles, including regulatory constraints, on the basis of ideologically defined images of goodness. I am urging, as an alternative, that planning tries out the planning idea and the planning method.

In turn, Webber's view of planning – which flatly denied the possibility of a stable predictable future or agreed goals – provided some of the philosophical underpinnings of the Social Learning or New Humanist approach of the 1970s, which stressed the importance of learning systems in helping cope with a turbulent environment. But finally, this approach divorced itself from logical positivism, returning to a reliance on personal knowledge which was strangely akin to old-style blueprint planning; and, as developed by John Friedmann of the University of California at Los Angeles, it finally resulted in a demand for all political activity to be decomposed into decision by minute political groups: a return to the anarchist roots of planning, with a vengeance.

So these different approaches diverged, sometimes in detailed emphasis, sometimes more fundamentally. What they shared was the belief that – at any rate in the American political system – the planner did not have much power and did not deserve to have much either; within a decade, from 1965 to 1975, these approaches together neatly stripped the planner of whatever priestly clothing, and consequent mystique, s/he may have possessed. Needless to say, this view powerfully communicated itself to the professionals themselves. Even in countries with more centralized, top-down political systems, such as Great Britain, young graduating planners increasingly saw their roles as rather like barefoot doctors, helping the poor down on the streets of the inner city, working either for a politically acceptable local authority, or, failing that, for community organizations battling against a politically objectionable one.

Several historical factors, in addition to the demolition job on planning by the American theorists, contributed to this change: planners and politicians belatedly discovered the continued deprivation of the inner-city poor; then, it was seen that the areas where these people lived were suffering depopulation and deindustrialization; in consequence, planners progressively moved away from the merely physical, and into the social and the economic. The change can be caricatured thus: in 1955, the typical newly graduated planner was at the drawing board, producing a diagram of desired land uses; in 1965, s/he was analysing computer output of traffic patterns; in 1975, the same person was talking late into the night with community groups, in the attempt to organize against hostile forces in the world outside.

It was a remarkable inversion of roles. For what was wholly or partly lost, in that decade, was the claim to any unique and useful expertise, such as was possessed by the doctor or the lawyer. True, the planner could still offer specialized knowledge on planning laws and procedures, or on how to achieve a particular design solution; though often, given the nature of the context and the changed character of planning education, s/he might not have enough of either of these skills to be particularly useful. And, some critics were beginning to argue, this was because planning had extended so thinly over so wide an area that it became almost meaningless; in the title of Aaron Wildavsky's celebrated paper, "If Planning is Everything, Maybe it's Nothing".

The fact was that planning, as an academic discipline, had theorized about its own role to such an extent that it was denying its own claim to legitimacy. Planning, Faludi pointed out in his text of 1973, could be merely *functional,* in that the goals and objectives are taken as given; or *normative,* in that they are themselves the object of rational choice. The problem was whether planning was really capable of doing that latter job. As a result, by the mid-1970s planning had reached the stage of a "paradigm crisis"; it had been theoretically useful to distinguish the planning process as something separate from what is planned, yet this had meant a neglect of substantive theory, pushing it to the periphery of the whole subject. Consequently, new theory is needed which attempts to bridge current planning strategies and the urban physical and social systems to which strategies are applied.

The Marxist ascendancy

That became ever clearer in the following decade, when the logical positivists retreated from the intellectual field of battle and the Marxists took possession. As the whole world knows, the 1970s saw a remarkable resurgence – indeed a veritable explosion – of Marxist studies. This could not fail to affect the closely related worlds of urban geography, sociology, economics and planning. True, like the early neo-classical economists, Marx had been remarkably uninterested in questions of spatial location – even though Engels had made illuminating comments on the spatial distribution of classes in mid-Victorian Manchester. The disciples now reverently sought to extract from the holy texts, drop by drop, a distillation that could be used to brew the missing theoretical potion. At last, by the mid-1970s, it was ready; then came a flood of new work. It originated in various places and in various disciplines: in England and the United States the geographers David Harvey and Doreen Massey helped to explain urban growth and change in terms of the circulation of capital; in Paris, Manuel Castells and Henri Lefebvre developed sociologically based theories. In the endless debates that followed among the Marxists themselves, a critical question concerned the role of the state. In France, Lokjine and others argued that it was mainly concerned, through such devices as macroeconomic planning and related infrastructure investment, directly to underpin and aid the direct productive investments of private capital. Castells, in contrast, argued that its main function had been to provide collective consumption – as in public housing, or schools, or transportation – to help guarantee the reproduction of the labour force and to dampen class conflict, essential for the maintenance of the system. Clearly, planning might play a very large role in both these state functions; hence, by the mid-1970s French Marxist urbanists were engaging in major studies of this role in the industrialization of such major industrial areas as Dieppe.

At the same time, a specifically Marxian view of planning emerged in the English-speaking world. To describe it adequately would require a course in Marxist theory. But, in inadequate summary, it states that the structure of the capitalist city itself, including its land-use and activity patterns, is the result of capital in pursuit of profit. Because capitalism is doomed to recurrent crises, which deepen in the current stage of late capitalism, capital calls upon the state, as its agent, to assist it by remedying disorganization in commodity production, and by aiding the reproduction of the labour force. It thus tries to achieve certain necessary objectives: to facilitate continued capital accumulation, by ensuring rational allocation of resources; by assisting the reproduction of the labour force through provision of social services, thus maintaining a delicate balance between labour and capital and preventing social disintegration; and by guaranteeing and legitimating capitalist social and property relations. As Dear and Scott put it: "In summary, planning is an historically-specific and socially-necessary response to the self-disorganizing tendencies *of privatized capitalist* social and property relations as these appear in urban space." In particular, it seeks to guarantee collective provision of necessary infrastructure and certain basic urban services, and to reduce negative externalities whereby certain activities of capital cause losses to other parts of the system.

But, since capitalism also wishes to circumscribe state planning as far as possible, there is an inbuilt contradiction: planning, because of this inherent inadequacy, always solves one problem only by creating another. Thus, say the Marxists, nineteenth-century clearances in Paris created a working-class housing problem; American zoning limited the powers of industrialists to locate at the most profitable locations. And planning can never do more than modify some parameters of the land development process; it cannot change its intrinsic logic, and so cannot remove the contradiction between private accumulation and collective action. Further, the capitalist class is by no means homogenous; different fractions of capital may have divergent, even contradictory interests, and complex alliances may be formed in consequence; thus, latter-day Marxist explanations come close to being pluralist, albeit with a strong structural element. But in the process, "the more that the State intervenes in the urban system, the greater is the likelihood that different social groups and fractions will contest the legitimacy of its decisions. *Urban life as a whole becomes progressively invaded by political controversies and dilemmas.*"

Because traditional non-Marxian planning theory has ignored this essential basis of planning, so Marxian commentators argue, it is by definition vacuous: it seeks to define what planning ideally ought to be, devoid of all context; its function has been to depoliticize planning as an activity, and thus to legitimate it. It seeks to achieve this by representing itself as the force which produces the various facets of real-world planning. But in fact, its various claims – to develop abstract concepts that rationally represent real-world processes, to legitimate its own activity, to explain material processes as the outcome of ideas, to present planning goals as derived from generally shared values, and to abstract planning activity in terms of metaphors drawn from other fields like engineering – all these are both very large and quite unjustified. The reality, Marxists argue, is precisely the opposite: viewed objectively, planning theory is nothing other than a creation of the social forces that bring planning into existence.

It makes up a disturbing body of coherent criticism: yes, of course, planning cannot simply be an independent self-legitimating activity, as scientific inquiry may claim to be; yes, of course, it is a phenomenon that – like all phenomena –

represents the circumstances of its time. As Scott and Roweis put it:

> . . . there is a definite mismatch between the world of current planning theory, on the one hand, and the real world of practical planning intervention on the other hand. The one is the quintessence of order and reason in relation to the other which is full of disorder and unreason. Conventional theorists then set about resolving this mismatch between theory and reality by introducing the notion that planning theory is in any case not so much an attempt to explain the world as it is but as it ought to be. Planning theory then sets itself the task of rationalizing irrationalities, and seeks to materialize itself in social and historical reality (like Hegel's World Spirit) by bringing to bear upon the world a set of abstract, independent, and transcendent norms.
>
> (Scott and Roweis, 1977)

It was powerful criticism. But it left in turn a glaringly open question, both for the unfortunate planner – whose legitimacy is now totally torn from him, like the epaulette from the shoulder of a disgraced officer – and, equally, for the Marxist critic: what, then, is planning theory about? Has it any normative or prescriptive content whatsoever? The answer, logically, would appear to be no. One of the critics, Philip Cooke, is uncompromising:

> The main criticism that tends to have been made, justifiably, of planning is that it has remained stubbornly normative . . . in this book it will be argued that [planning theorists] should identify mechanisms which cause changes in the nature of planning to be brought about, rather than assuming such changes to be either the creative idealizations of individual minds, or mere regularities in observable events.
>
> (Cooke, 1983)

This is at least consistent: planning theory should avoid all prescription; it should stand right outside the planning process, and seek to analyse the subject – including traditional theory – for what it is, the reflection of historical forces. Scott and Roweis, a decade earlier, seem to be saying exactly the same thing: planning theory cannot be normative, it cannot assume "transcendent operational norms". But then, they stand their logic on its own head, saying that "a viable theory of urban planning should not only tell us what planning is, but also what we can, and must, do as progressive planners".

This, of course, is sheer rhetoric. But it nicely displays the agony of the dilemma. Either theory is about unravelling the historical logic of capitalism, or it is about prescription for action. Since the planner-theorist – however sophisticated – could never hope to divert the course of capitalist evolution by more than a millimetre or a millisecond, the logic would seem to demand that s/he sticks firmly to the first and abjures the second. In other words, the Marxian logic is strangely quietist; it suggests that the planner retreats from planning altogether into the academic ivory tower.

Some were acutely conscious of the dilemma. John Forester tried to resolve it by basing a whole theory of planning action on the work of Jürgen Habermas. Habermas, perhaps the leading German social theorist of the post-World War Two era, had argued that latter-day capitalism justified its own legitimacy by spinning around itself a complex set of distortions in communication, designed to obscure and prevent any rational understanding of its own workings. Thus, he argued, individuals became powerless to understand how and why they act, and so were excluded from all power to influence their own lives,

> as they are harangued, pacified, mislead [*sic*], and ultimately persuaded that inequality, poverty, and ill-health are either problems for which the victim is responsible or problems so "political" and "complex" that they can have nothing to say about them. Habermas argues that democratic politics or planning requires the consent that grows from processes of collective criticism, not from silence or a party line.
>
> (Forester, 1980)

But, Forester argues, Habermas's own proposals for communicative action provide a way for planners to improve their own practice:

> By recognizing planning practice as normatively role-structured communication action which distorts, covers up, or reveals to the public the prospects and possibilities they face, a critical theory of planning aids us practically and ethically as well. This is the contribution of critical theory to planning: pragmatics with vision – to reveal true alternatives, to correct false expectations, to counter cynicism, to foster

> inquiry, to spread political responsibility, engagement, and action. Critical planning practice, technically skilled and politically sensitive, is an organizing and democratizing practice.
>
> (Forester, 1980)

Fine. The problem is that – stripped of its Germanic philosophical basis, which is necessarily a huge oversimplification of a very dense analysis – the practical prescription all comes out as good old-fashioned democratic common sense, no more and no less than Davidoff's advocacy planning of fifteen years before: cultivate community networks, listen carefully to the people, involve the less-organized groups, educate the citizens in how to join in, supply information and make sure people know how to get it, develop skills in working with groups in conflict situations, emphasize the need to participate, compensate for external pressures. True, if in all this planners can sense that they have penetrated the mask of capitalism, that may help them to help others to act to change their environment and their lives; and, given the clear philosophical impasse of the late 1970s, such a massive metaphysical underpinning may be necessary.

The world outside the tower: practice retreats from theory

Meanwhile, if the theorists were retreating in one direction, the practitioners were certainly reciprocating. Whether baffled or bored by the increasingly scholastic character of the academic debate, they lapsed into an increasingly untheoretical, unreflective, pragmatic, even visceral style of planning. That was not entirely new: planning had come under a cloud before, as during the 1950s, and had soon reappeared in a clear blue sky. What was new, strange, and seemingly unique about the 1980s was the divorce between the Marxist theoreticians of academe – essentially academic spectators, taking grandstand seats at what they saw as one of capitalism's last games – and the anti-theoretical, anti-strategic, anti-intellectual style of the players on the field down below. The 1950s were never like that; then, the academics were the coaches, down there with the team.

The picture is of course exaggerated. Many academics did still try to teach real-life planning through simulation of real-world problems. The Royal Town Planning Institute enjoined them to become ever more practice-minded. The practitioners had not all shut their eyes and ears to what comes out of the academy; some even returned there for refresher courses. And if all this was true in Britain, it was even more so of America, where the divorce had never been so evident. Yet the picture does describe a clear and unmistakable trend; and it was likely to be more than a cyclical one.

The reason is simple: as professional education of any kind becomes more fully absorbed by the academy, as its teachers become more thoroughly socialized within it, as careers are seen to depend on academic peer judgements, then its norms and values – theoretical, intellectual, detached – will become ever more pervasive; and the gap between teaching and practice will progressively widen. One key illustration: of the huge output of books and papers from the planning schools in the 1980s, there were many – often, those most highly regarded within the academic community – that were simply irrelevant, even completely incomprehensible, to the average practitioner.

Perhaps, it might be argued, that was the practitioner's fault; perhaps too, we need fundamental science, with no apparent payoff, if we are later to enjoy its technological applications. The difficulty with that argument was to find convincing evidence that – not merely here, but in the social sciences generally – such payoff eventually comes. Hence the low esteem into which the social sciences had everywhere fallen, not least in Britain and the United States: hence too the diminished level of support for them, which – at any rate in Britain – had directly redounded on the planning schools. The relationship between planning and the academy had gone sour, and that is the major unresolved question that must now be addressed.

REFERENCES

Abercrombie, P. and Jackson H. (1948) *West Midlands Plan*. Interim Confidential Edition. 5 vols. London: Ministry of Town and Country Planning.

Batty, M. (1979) "On Planning Processes", in: B. Goodall and A. Kirby (eds) *Resources and Planning*. Oxford: Pergamon.

Chadwick, G. (1971) *A Systems View of Planning: Towards a Theory of the Urban and Regional Planning Process*. Oxford: Pergamon.

Cooke, P. N. (1983) *Theories of Planning and Spatial Development*. London: Hutchinson.

Forester, J. (1980) "Critical Theory and Planning Practice", *Journal of the American Planning Association*, 46, 275–86.

Kent, T. J. (1964) *The Urban General Plan*. San Francisco: Chandler.

Scott, A. J. and Roweis, S. T. (1977) "Urban Planning in Theory and Practice: An Appraisal", *Environment and Planning*, 9, 1097–19.

Webber, M. M. (1968–69) "Planning in an Environment of Change", *Town Planning Review*, 39, 179–95, 277–95.

EDWARD J. KAISER and DAVID R. GODSCHALK

"Twentieth Century Land Use Planning: A Stalwart Family Tree"

American Planning Association Journal (1995)

Editors' introduction Much of what city planning departments do is *physical planning* related to land use, transportation, capital improvements, and infrastructure. In the following selection, Edward J. Kaiser and David R. Godschalk describe how concepts of what land use plans should be like have evolved over time and what land use planning practice in the United States is like today.

Kaiser and Godschalk liken the development of land use planning to a tree. From disparate roots in planning theory and practice, a sturdy trunk based on the 1950s vision of a general plan for cities' long-term physical development has grown over time, with periodic branches, to a rich foliage of "hybrid" contemporary plans which typically blend aspects of design, policy, and management.

Kaiser and Godschalk begin their account of twentieth-century land use planning by describing the elitist, architecturally based, and often unrealistic plans that Peter Hall noted were common prior to World War II (p. 362). These have evolved into contemporary hybrid urban land use plans developed through participatory processes and which incorporate elements of policy and management that make them far more realistic than plans developed during planning's comfortable, but ineffective "golden age."

Kaiser and Godschalk trace the roots of their tree to mid-century. By that time, they say, there was a consensus that city plans should be focused on long-term physical development – what is called a "master plan" or "general plan."

At mid-century, city general plans tended to be elitist, inspirational, long-range visions developed with little attention to implementation. In contrast, Kaiser and Godschalk argue that today urban land use plans have become frameworks for community consensus on future growth supported by fiscally grounded actions to manage change. Plans are becoming more sensitive to the ecological design issues raised by Van der Ryn and Cowan (p. 519) and sustainable urban development as discussed by Wheeler (p. 434).

Most modern city plans still contain maps showing projected long range urban form for the city's land uses, transportation, community facilities, and other infrastructure. But today's plans do not consist of architecturally complete renderings of an entire static new town or section of a city like many pre-war plans. In addition to or in place of maps, the city plan's vision may be expressed in words and visual images.

At the turn of the millennium it appears that the "sturdy tree" of urban land use planning will continue to grow into the twenty-first century. Kaiser and Godschalk anticipate that the next generation of development plans will mature and adapt without abandoning the urban physical plan heritage.

Edward J. Kaiser and David R. Godschalk are professors of city planning at the University of North Carolina. They are co-authors (with Stuart Chapin) of *Urban Land Use Planning*, 4th edn. (Chicago: University of Illinois Press, 1995), a leading American text on land use planning.

Other important books on city planning in the United States include International City Management Association (ICMA), *The Practice of Local Government Planning*, 3rd edn. (Washington, DC: ICMA, 1999), J. Barry Cullingworth, *Planning in the USA: Policies, Issues, and Process* (London: Routledge, 1997), John M. Levy, *Contemporary Urban Planning* (New York: Prentice-Hall, 1996), Alexander Garvin, *The American City: What Works, What Doesn't* (New York: McGraw-Hill, 1995), Jonathan Bartlett, *the Fractured Metropolis* (New York: HarperCollins, 1996), Jay Stein (ed.), *Classic Readings in Urban Planning* (New York: McGraw-Hill, 1995), and Anthony J. Catanese and James C. Snyder, *Urban Planning*, 2nd edn. (New York: McGraw-Hill, 1988).

Leading books on European urban planning include Peter Hall, *Urban and Regional Planning*, 2nd edn. (London: Routledge, 1996), Patsy Healey and Stephen Graham (eds.), *Managing Cities* (London: Wiley, 1995) and Barry Cullingworth, *Town and Country Planning in England and Wales*, 11th edn. (London: Routledge, 1997). Cullingworth has written additional books which provide overviews of city planning in Canada, Britain, Scotland, and Wales.

> How a city's land is used defines its character, its potential for development, the role it can play within a regional economy and how it impacts the natural environment.
>
> Seattle Planning Commission, 1993

During the twentieth century, community physical development plans have evolved from elite, City Beautiful designs to participatory, broad-based strategies for managing urban change. A review of land use planning's intellectual and practice history shows the continuous incorporation of new ideas and techniques. The traditional mapped land use design has been enriched with innovations from policy plans, land classification plans, and development management plans. Thanks to this flexible adaptation, local governments can use contemporary land use planning to build consensus and support decisions on controversial issues about space, development, and infrastructure. If this evolution persists, local plans should continue to be mainstays of community development policy into the twenty-first century.

Unlike the more rigid, rule-oriented modern architecture, contemporary local planning does not appear destined for deconstruction by a postmodern revolution. Though critics of comprehensive physical planning have regularly predicted its demise, the evidence demonstrates that spatial planning is alive and well in hundreds of United States communities. A 1994 tabulation found 2,742 local comprehensive plans prepared under state growth management regulations in twelve states. (See table 1.) This figure of course significantly understates the overall nationwide total, which would include all those plans prepared in the other thirty-eight states and in the noncoastal areas of California and North Carolina. It is safe to assume that most, if not all, of these plans contain a mapped land use element. Not only do such plans help decision-makers to manage urban growth and change, they also provide a platform for the formation of community consensus about land use issues, now among the most controversial items on local government agendas.

This article looks back at the history of land use planning and forward to its future. It shows how planning ideas, growing from turn-of-the-century roots, culminated in a midcentury consensus on a general concept – the traditional land use design plan. That consensus was stretched as planning branched out to deal with public participation, environmental protection, growth management, fiscal responsibility, and effective implementation under turbulent conditions. To meet these new challenges, new types of plans arose: verbal policy plans, land classification plans, and growth management plans. These in turn became integrated into today's hybrid comprehensive plans, broadening and strengthening the traditional approach.

Future land use planning will continue to evolve in certain foreseeable directions, as well

Table 1 Local comprehensive plans in growth-managing states and coastal areas as of 1994

State	*Number of comprehensive plans* Cities/towns	Counties	Regions	Total	*Source*
California (coastal)	97	7	0	104	Coastal Commission
Florida	377	49	0	426	Department of Community Affairs
Georgia	298	94	0	392	Department of Community Affairs
Maine	270	0	0	270	Dept. of Economic and Community Development
Maryland	1	1	0	2	Planning Office
New Jersey	567	0	0	567	Community Affairs Department
North Carolina (coastal)	70	20	0	90	Division of Coastal Management
Oregon	241	36	1	278	Department of Local Community Development
Rhode Island	39	0	1	40	Department of Planning and Development
Vermont	235	0	10	245	Department of Housing and Community Affairs
Virginia	211	94	0	305	Department of Housing and Community Affairs
Washington	23	9	9	23	Office of Growth Management
Total	**2429**	**301**	**12**	**2742**	

Compiled from telephone survey of state sources.

as in ways unforeseen. Among the foreseeable developments are even more active participation by interest groups, calling for planners' skills at consensus building and managing conflict; increased use of computers and electronic media, calling for planners' skills in information management and communication; and continuing concerns over issues of diversity, sustainability, and quality of life, calling for planners' ability to analyze and seek creative solutions to complex and interdependent problems.

THE LAND USE PLANNING FAMILY TREE

We liken the evolution of the physical development plan to a family tree. The early genealogy is represented as the roots of the tree (figure 1). The general plan, constituting consensus practice at midcentury, is represented by the main trunk. Since the 1970s this traditional "land use design plan" has been joined by several branches – the verbal policy plan, the land classification plan, and the development management plan. These branches connect to the trunk although springing from different planning disciplines, in a way reminiscent of the complex structure of a Ficus tree. The branches combine into the contemporary, hybrid comprehensive plan integrating design, policy, classification, and management, represented by the foliage at the top of the tree.

As we discuss each of these parts of the family tree, we show how plans respond both to social climate changes and to "idea genes" from the literature. We also draw conclusions about the survival of the tree and the prospects for new branches in the future. The focus of the article is the plan prepared by a local government – a county, municipality, or urban region – for the long-term development and use of the land.

ROOTS OF THE FAMILY TREE: THE FIRST 50 YEARS

New World city plans certainly existed before this century. They included L'Enfant's plan for Washington, William Penn's plan for Philadelphia, and General Oglethorpe's plan for Savannah. These plans, however, were blueprints for undeveloped sites, commissioned by unitary authorities with power to implement them unilaterally.

In this century, perhaps the most influential

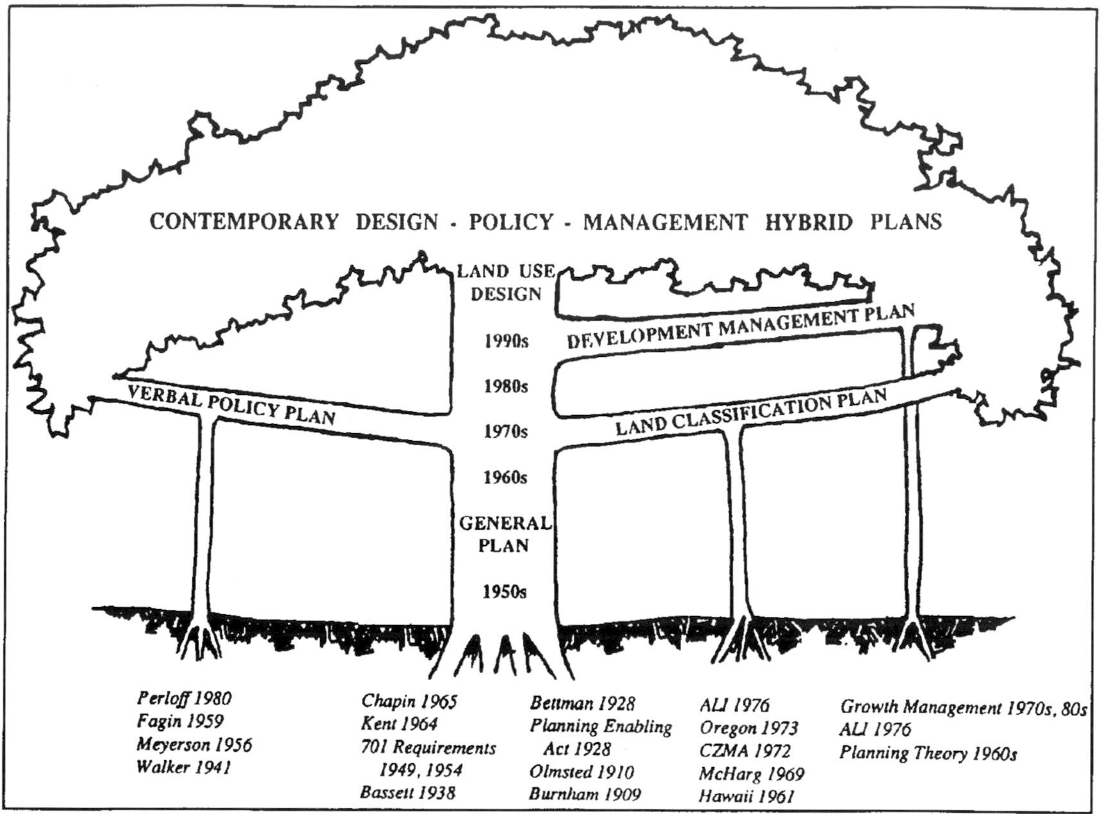

Figure 1 The family tree of the land use plan

early city plan was Daniel Burnham's plan for Chicago, published by the Commercial Club of Chicago (a civic, not a government entity) in 1909. The archetypical plan-as-inspirational-vision, it focuses only on design of public spaces as a City Beautiful effort.

The City Beautiful approach was soon broadened to a more comprehensive view. At the 1911 National Conference on City Planning, Frederick Law Olmsted, Jr., son of the famous landscape architect and in his own right one of the fathers of planning, defined a city plan as encompassing all uses of land, private property, public sites, and transportation. Alfred Bettman, speaking at the 1928 National Conference of City Planning, envisioned the plan as a master design for the physical development of the city's territory, including "the general location and extent of new public improvements . . . and in the case of private developments, the general distribution amongst various classes of land uses, such as residential, business, and industrial uses . . . designed for . . . the future, twenty-five to fifty years" (Black 1968, 352–3). Together, Olmsted and Bettman anticipated the development of the midcentury land use plan.

Another early influence, the federal Standard City Planning Enabling Act of 1928, shaped enabling acts passed by many states. However, the Act left many planners and public officials confused about the difference between a master plan and a zoning ordinance, so that hundreds of communities adopted "zoning plans" without having created comprehensive plans as the basis for zoning (Black 1968, 353). Because the Act also did not make clear the importance of comprehensiveness or define the essential elements of physical development, no consensus about the essential content of the plan existed.

Ten years later, Edward Bassett's book, *The Master Plan* (1938), spelled out the plan's subject matter and format – supplementing the 1928 Act, and consistent with it. He argued that the plan should have seven elements, all relating to land areas (not buildings) and capable of being shown on a map: streets, parks, sites for public buildings, public reservations, routes for public utilities, pier-head and bulkhead lines (all public facilities), and zoning districts for private lands. Bassett's views were incorporated in many state enabling laws.

The physical plans of the first half of the century were drawn by and for independent commissions, reflecting the profession's roots in the Progressive Reform movement, with its distrust of politics. The 1928 Act reinforced that perspective by making the planning commission, not the legislative body, the principal client of the plan, and purposely isolating the commission from politics. Bassett's book reinforced the reliance on an independent commission. He conceived of the plan as a "plastic" map, kept within the purview of the planning commission, capable of quick and easy change. The commission, not the plan, was intended to be the adviser to the local legislative body and to city departments.

By the 1940s, both the separation of the planning function from city government and the plan's focus on physical development were being challenged. Robert Walker, in *The Planning Function in Local Government,* argued that the "scope of city planning is properly as broad as the scope of city government". The central planning agency might not necessarily do all the planning, but it would coordinate departmental planning in the light of general policy considerations – creating a comprehensive plan but one without a physical focus. That idea was not widely accepted. Walker also argued that the independent planning commission should be replaced by a department or bureau attached to the office of mayor or city manager. That argument did take hold, and by the 1960s planning in most communities was the responsibility of an agency within local government, though planning boards still advised elected officials on planning matters.

This evolution of ideas over 50 years resulted at midcentury in a consensus concept of a plan as focused on long-term physical development; this focus was a legacy of the physical design professions. Planning staff worked both for the local government executive officer and with an appointed citizen planning board, an arrangement that was a legacy of the Progressive insistence on the public interest as an antidote to governmental corruption. The plan addressed both public and private uses of the land, but did not deal in detail with implementation.

THE PLAN AFTER MIDCENTURY: NEW GROWTH INFLUENCES

Local development planning grew rapidly in the 1950s, for several reasons. First, governments had to contend with the postwar surge of population and urban growth, as well as a need for the capital investment in infrastructure and community facilities that had been postponed during the depression and war years. Second, municipal legislators and managers became more interested in planning as it shifted from being the responsibility of an independent commission to being a function within local government. Third, and very important, Section 701 of the Housing Act of 1954 required local governments to adopt a long-range general plan in order to qualify for federal grants for urban renewal, housing, and other programs, and it also made money available for such comprehensive planning. The 701 program's double-barreled combination of requirements and financial support led to more urban planning in the United States in the latter half of the 1950s than at any previous time in history.

At the same time, the plan concept was pruned and shaped by two planning educators. T. J. Kent, Jr. was a professor at the University of California at Berkeley, a planning commissioner, and a city councilman in the 1950s. His book, *The Urban General Plan* (1964), clarified the policy role of the plan. F. Stuart Chapin, Jr. was a TVA planner and planning director in Greensboro, NC in the 1940s, before joining the planning faculty at the University of North Carolina at Chapel Hill in 1949. His contribution was to codify the methodology of land use planning in the various editions of his book, *Urban Land Use Planning* (1957, 1965).

What should the plan look like? What should it be about? What is its purpose (besides the cynical purpose of qualifying for federal grants)? The 701 program, Kent, and Chapin all offered answers.

The "701" program comprehensive plan guidelines

In order to qualify for federal urban renewal aid and, later, for other grants – a local government had to prepare a general plan that consisted of plans for physical development, programs for redevelopment, and administrative and regulatory measures for controlling and guiding development. The 701 program specified what the content of a comprehensive development plan should include:

- A land use plan, indicating the locations and amounts of land to be used for residential, commercial, industrial, transportation, and public purposes
- A plan for circulation facilities
- A plan for public utilities
- A plan for community facilities

T. J. Kent's urban general plan

Kent's view of the plan's focus was similar to that of the 701 guidelines: long-range physical development in terms of land use, circulation, and community facilities. In addition, the plan might include sections on civic design and utilities, and special areas, such as historic preservation or redevelopment areas. It covered the entire geographical jurisdiction of the community, and was in that sense comprehensive. The plan was a vision of the future, but not a blueprint; a policy statement, but not a program of action; a formulation of goals, but not schedules, priorities, or cost estimates. It was to be inspirational, uninhibited by short-term practical considerations.

Kent (1964, 65–89) believed the plan should emphasize policy, serving the following functions:

- Policy determination – to provide a process by which a community would debate and decide on its policy
- Policy communication – to inform those concerned with development (officials, developers, citizens, the courts, and others) and educate them about future possibilities
- Policy effectuation – to serve as a general reference for officials deciding on specific projects
- Conveyance of advice – to furnish legislators with the counsel of their advisors in a coherent, unified form

The format of Kent's proposed plan included a unified, comprehensive, but general physical design for the future, covering the whole community and represented by maps. (See figure 2.) It also contained goals and policies (generalized guides to conduct, and the most important ingredients of the plan), as well as summaries of background conditions, trends, issues, problems, and assumptions. (See figure 3.) So that the plan would be suitable for public debate, it was to be a complete, comprehensible document, containing factual data, assumptions, statements of issues, and goals, rather than merely conclusions and recommendations. The plan belonged to the legislative body and was intended to be consulted in decision-making during council meetings.

Kent recommended overall goals for the plan:

- Improve the physical environment of the community to make it more functional, beautiful, decent, healthful, interesting, and efficient
- Promote the overall public interest, rather than the interests of individuals or special groups within the community
- Effect political and technical coordination in community development
- Inject long-range considerations into the determination of short-range actions
- Bring professional and technical knowledge to bear on the making of political decisions about the physical development of the community

F. Stuart Chapin, Jr.'s urban land use plan

Chapin's ideas, though focusing more narrowly on the land use plan, were consistent with Kent's in both the 1957 and 1965 editions of *Urban Land Use Planning*, a widely used text and reference work for planners. Chapin's concept of the plan was of a generalized, but scaled, design for the future use of land, covering private land uses and public facilities, including the thoroughfare network.

Chapin conceived of the land use plan as the first step in preparing a general or comprehensive plan. Upon its completion, the land use plan served as a temporary general guide for decisions, until the comprehensive plan was developed. Later, the land use plan would become a cornerstone in the comprehensive plan, which also included plans for transportation, utilities, community facilities, and renewal, only the general rudiments of which are suggested in the land use plan. Purposes of the plan were to guide government decisions on public facilities, zoning, subdivision control, and urban renewal, and to inform private developers about the proposed future pattern of urban development.

The format of Chapin's land use plan included a statement of objectives, a description of existing conditions and future needs for space and services, and finally the mapped proposal for the future development of the community, together with a program for implementing the plan (customarily including zoning, subdivision control, a housing code, a public works expenditure program, an urban renewal program, and other regulations and development measures).

The typical general plan of the 1950s and 1960s

Influenced by the 701 program, Kent's policy vision, and Chapin's methods, the plans of the 1950s and 1960s were based on a clear and straightforward concept: The plans' purposes were to determine, communicate, and effectuate comprehensive policy for the private and public physical development and redevelopment of the city. The subject matter was long-range physical development, including private uses of the land, circulation, and community facilities. The standard format included a summary of existing and emerging conditions and needs; general goals; and a long-range urban form in map format, accompanied by consistent development policies. The coverage was comprehensive, in the sense of addressing both public and private development and covering the entire planning jurisdiction, but quite general. The tone was typically neither as "inspirational" as the Burnham plan for Chicago, nor as action-oriented as today's plans. Such was the well-defined trunk of the family tree in the 1950s and 1960s, in which today's contemporary plans have much of their origin.

CONTEMPORARY PLANS: INCORPORATING NEW BRANCHES

Planning concepts and practice have continued to evolve since midcentury, maturing in the process. By the 1970s, a number of new ideas had taken root. Referring back to the family tree in figure 1, we can see a trunk and several distinct branches:

- *The land use design*, a detailed mapping of future land use arrangements, is the most direct descendant of the 1950s plan. It still constitutes the trunk of the tree. However, today's version is more likely to be accompanied by action strategies, also mapped, and to include extensive policies.
- *The land classification plan*, a more general map of growth policy areas rather than a detailed land use pattern, is now also common, particularly for counties, metropolitan areas, and regions that want to encourage urban growth in designated development areas and to discourage it in conservation or rural areas. The roots of the land classification plan include McHarg's *Design With Nature* (1969), the 1976 American Law Institute (ALI) Model Land Development Code, the 1972 Coastal Zone Management Act, and the 1973 Oregon Land Use Law.
- *The verbal policy plan* de-emphasizes mapped policy or end-state visions and focuses on verbal action policy statements, usually quite detailed; sometimes called a strategic plan, it is rooted in Meyerson's middle-range bridge to comprehensive planning, Fagin's policies plan, and Perloff's strategies and policies general plan.
- *The development management plan* lays out a specific program of actions to guide development, such as a public investment program, a development code, and a program to extend infrastructure and services; and it assumes public sector initiative for influencing the location, type, and pace of growth. The roots of the development management plan are in the environmental movement, and the movements for state growth management and community growth control, as well as in ideas from Fagin (1959) and the ALI Code.

We looked for, but could not find, examples of land use plans that could be termed purely prototypical "strategic plans," in the sense of Bryson and Einsweller. Hence, rather than identifying strategic planning as a separate branch on the family tree of the land use plan, we see the influence of strategic planning showing up across a range of contemporary plans. We tend to agree with the planners surveyed by Kaufman and Jacobs that strategic planning differs from good comprehensive planning more in emphasis (shorter range, more realistically targeted, more market-oriented) than in kind.

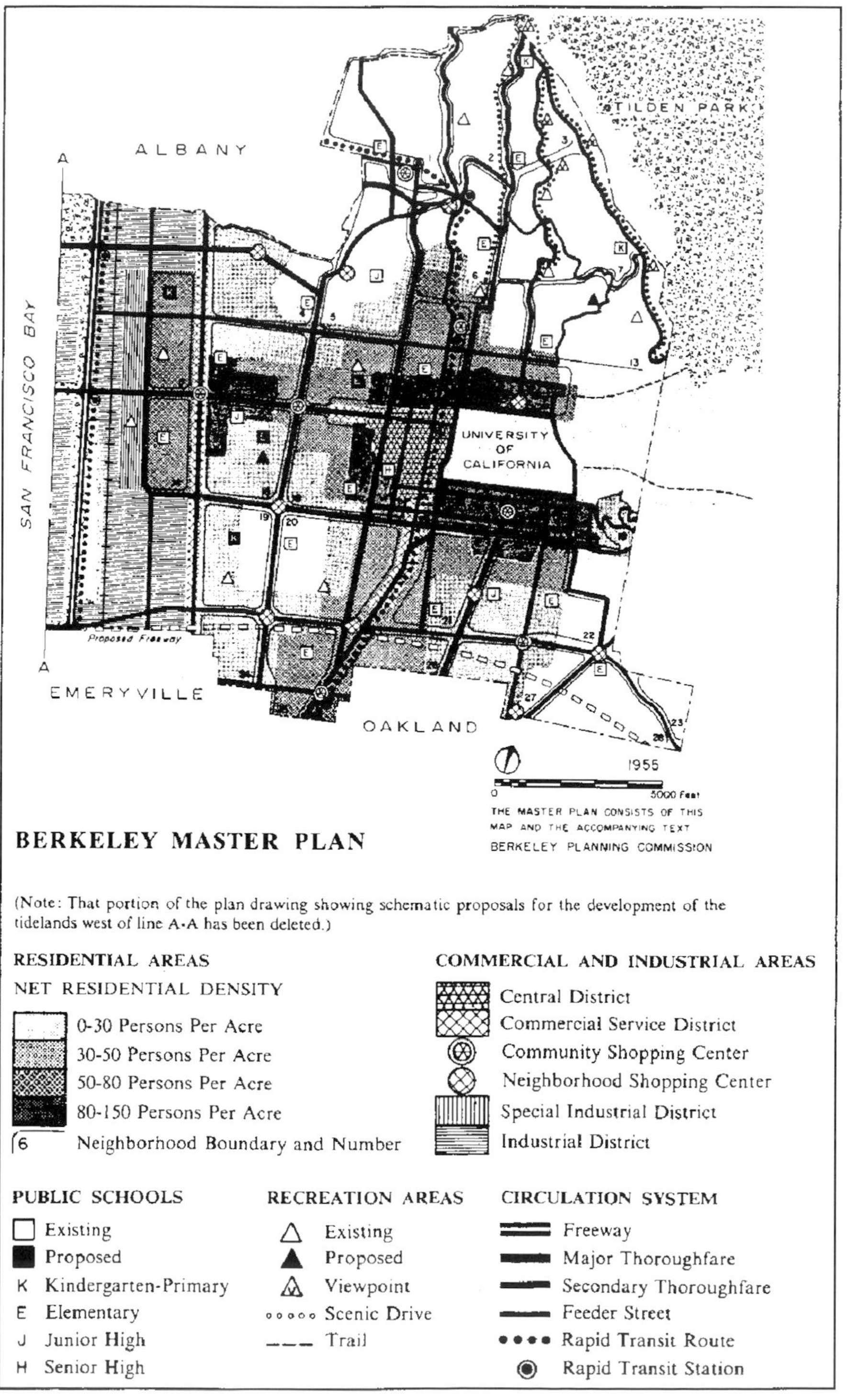

Figure 2 Example of the land use design map featured in the 1950s General Plan
Source: Kent 1991, 111

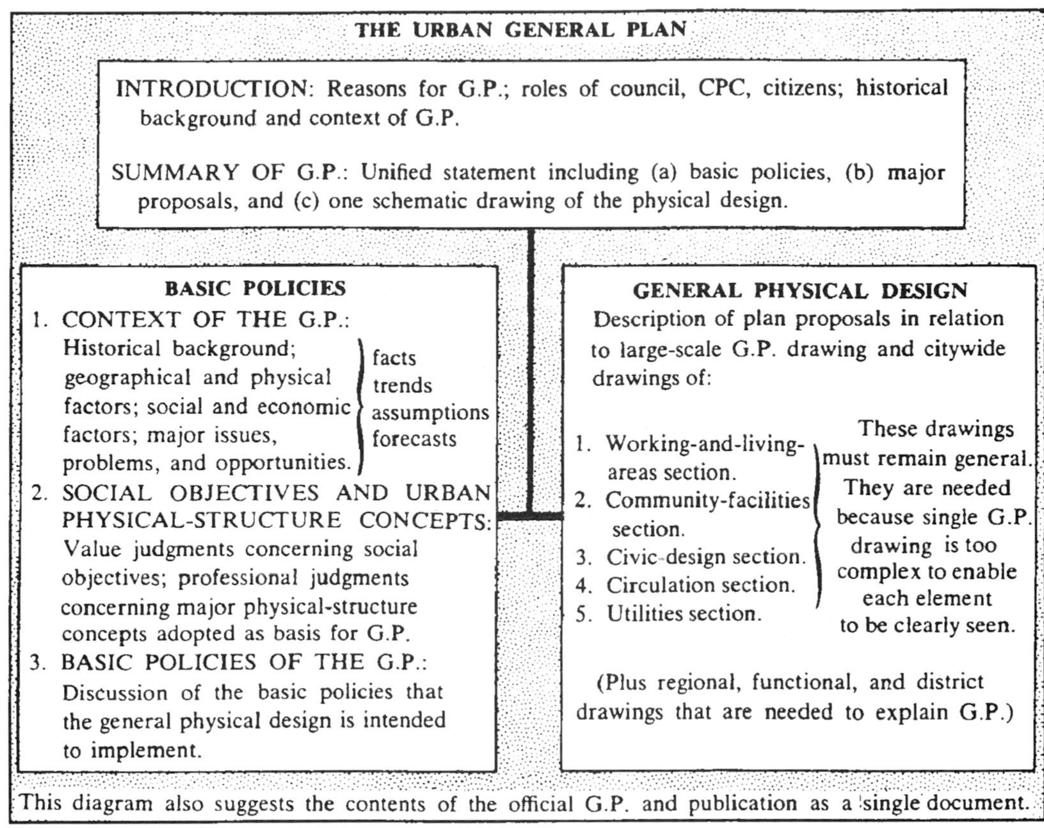

Figure 3 Components of the 1950s–1960s General Plan
Source: Kent 1964, 93

The land use design plan

The land use design plan is the most traditional of the four prototypes of contemporary plans and is the most direct descendent of the Kent–Chapin–701 plans of the 1950s and 1960s. It proposes a long-range future urban form as a pattern of retail, office, industrial, residential, and open spaces, and public land uses and a circulation system. Today's version, however, incorporates environmental processes, and sometimes agriculture and forestry, under the "open space" category of land use. Its land uses often include a "mixed use" category, honoring the neotraditional principle of closer mingling of residential, employment, and shopping areas. In addition, it may include a development strategy map, which is designed to bring about the future urban form and to link strategy to the community's financial capacity to provide infrastructure and services. The plans and strategies are often organized around strategic themes or around issues about growth, environment, economic development, transportation, or neighborhood/community scale change.

Like the other types of plans in vogue today, the land use design plan reflects recent societal issues, particularly the environmental crisis, the infrastructure crisis, and stresses on local government finance. Contemporary planners no longer view environmental factors as development constraints, but as valuable resources and processes to be conserved. They also may question assumptions about the desirability and inevitability of urban population and economic growth, particularly as such assumptions stimulate demand for expensive new roads, sewers, and schools. While at midcentury plans unquestioningly accommodated growth, today's plans cast the amount, pace, location, and costs of growth as policy choices to be determined in the planning process.

The 1990 Howard County (Maryland) General Plan, winner of an American Planning Association (APA) award in 1991 for outstanding comprehensive planning, exemplifies contemporary land use design. (See figure 4.) While clearly a direct descendent of the traditional general plan, the Howard County plan adds new types of goals, policies, and planning techniques. To enhance communication and public understanding, it is organized strategically around six themes/chapters (responsible regionalism, preservation of the rural area, balanced growth, working with nature, community enhancement, and phased growth), instead of the customary plan elements. Along with the traditional land use design, the plan includes a "policy map" (strategy map) for each theme and an overall policies map for the years 2000 and 2010. A planned service area boundary is used to contain urban growth within the eastern urbanized part of the county, home to the well-known Columbia New Town. The plan lays out specific next steps to be implemented over the next two years, and defines yardsticks for measuring success. An extensive public participation process for formulating the plan involved a 32-member General Plan Task Force, public opinion polling to discover citizen concerns, circulation of preplan issue papers on development impacts, and consideration of six alternative development scenarios.

The land classification plan

Land classification, or development priorities mapping, is a proactive effort by government to specify where and under what conditions growth will occur. Often, it also regulates the pace or timing of growth. Land classification addresses environmental protection by designating "non-development" areas in especially vulnerable locations. Like the land use design, the land classification plan is spatially specific and map-oriented. However, it is less specific about the pattern of land uses within areas specified for development, which results in a kind of silhouette of urban form. On the other hand, land classification is more specific about development strategy, including timing. Counties, metropolitan areas, and regional planning agencies are more likely than cities to use a land classification plan.

The land classification plan identifies areas where development will be encouraged (called urban, transition, or development areas) and areas where development will be discouraged (open space, rural, conservation, or critical environmental areas). For each designated area, policies about the type, timing, and density of allowable development, extension of infrastructure, and development incentives or constraints apply. The planning principle is to concentrate financial resources, utilities, and services within a limited, prespecified area suitable for development, and to relieve pressure on nondevelopment areas by withholding facilities that accommodate growth.

Ian McHarg's (1969) approach to land planning is an early example of the land classification concept. He divides planning regions into three categories: natural use, production, and urban. Natural use areas, those with valuable ecological functions, have the highest priority. Production areas, which include agriculture, forestry, and fishing uses, are next in priority. Urban areas have the lowest priority and are designated after allocating the land suitable to the two higher-priority uses. McHarg's approach in particular, and land classification generally, also reflect the emerging environmental consciousness of the 1960s and 1970s.

As early as 1961, Hawaii had incorporated the land classification approach into its state growth management system. The development framework plan of the Metropolitan Council of the Twin Cities Area defined "planning tiers," each intended for a different type and intensity of development. The concepts of the "urban service area," first used in 1958 in Lexington, Kentucky, and the "urban growth boundary," used throughout Oregon under its 1973 statewide planning act, classify land according to growth management policy. Typically, the size of an urban growth area is based on the amount of land necessary to accommodate development over a period of ten or twenty years.

Vision 2005: A Comprehensive Plan for Forsyth County, North Carolina exemplifies the contemporary approach to land classification plans. The plan, which won honorable mention from APA in 1989, employs a six category system of districts, plus a category for activity centers. It identifies both short- and long-range

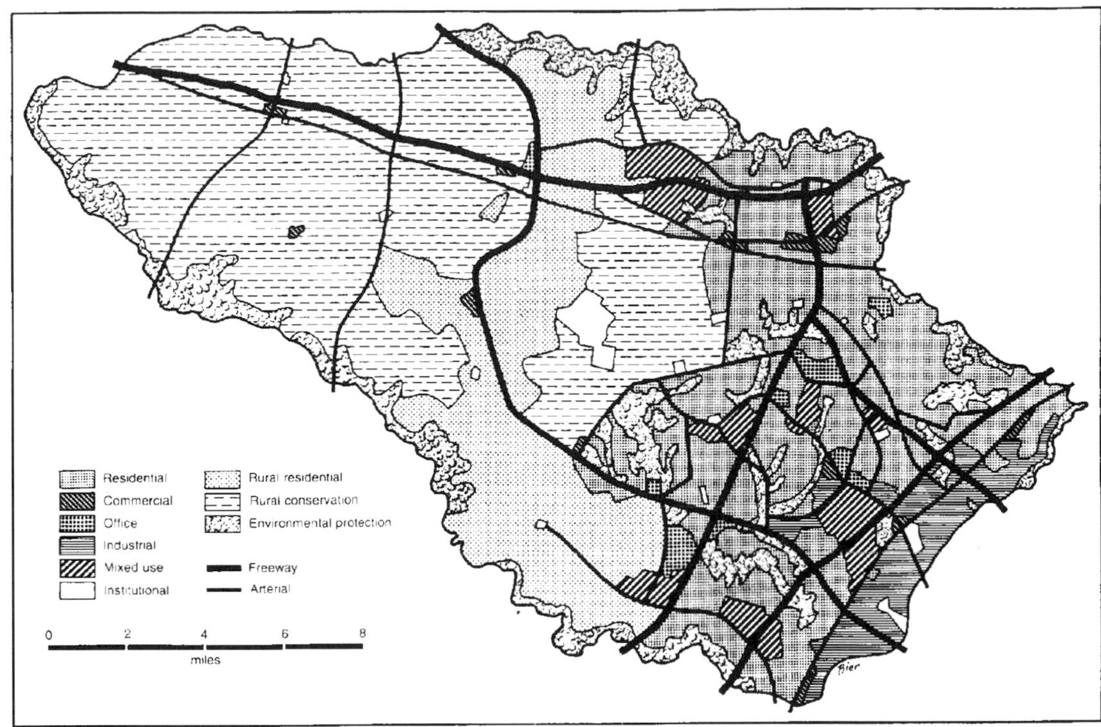

Figure 4 Howard County, Maryland, General Plan, Land Use 2010
Source: Adapted from Howard County 1990

growth areas (4A and 4B in figure 5). Policies applicable to each district are detailed in the plan.

The verbal policy plan: shedding the maps

The verbal policy plan focuses on written statements of goals and policy, without mapping specific land use patterns or implementation strategy. Sometimes called a policy framework plan, a verbal policy plan is more easily prepared and flexible than other types of plans, particularly for incorporating nonphysical development policy. Some claim that such a plan helps the planner to avoid relying too heavily on maps, which are difficult to keep up to date with the community's changes in policy. The verbal policy plan also avoids falsely representing general policy as applying to specific parcels of property. The skeptics, however, claim that verbal statements in the absence of maps provide too little spatial specificity to guide implementation decisions.

The verbal policy plan may be used at any level of government, but is especially common at the state level, whose scale is unsuited to land use maps. The plan usually contains goals, facts and projections, and general policies corresponding to its purposes – to understand current and emerging conditions and issues, to identify goals to be pursued and issues to be addressed, and to formulate general principles of action. Sometimes communities do a verbal policy plan as an interim plan or a first step in the planning process. Thus, verbal policies are included in most land use design plans, land classification plans, and development management plans.

The *Calvert County, MD, Comprehensive Plan* (Calvert County 1983), winner of a 1985 APA award, exemplifies the verbal policy plan. Its policies are concise, easy to grasp, and grouped in sections corresponding to the six divisions of county government responsible for implementation. It remains a policy plan, however, because it does not specify a program of specific actions for development management. Though the plan clearly addresses physical development and discusses specific spatial areas, it contains no land use map. (See figure 6 for an illustrative page from the Calvert County plan.)

The development management plan

The development management plan features a coordinated program of actions, supported by analyses and goals, for specific agencies of local government to undertake over a three-to-ten-year period. The program of actions usually specifies the content, geographic coverage, timing, assignment of responsibility, and coordination among the parts. Ideally, the plan includes most or all of the following components:

- Description of existing and emerging development conditions, with particular attention to development processes, the political-institutional context, and a critical review of the existing systems of development management
- Statement of goals and/or legislative intent, including management-oriented goals
- Program of actions – the heart of the plan – including:
  1 Outline of a proposed development code, with: (a) procedures for reviewing development permits; (b) standards for the type of development, density, allowable impacts and/or performance standards; (c) site plan, site engineering, and construction practice requirements; (d) exactions and impact fee provisions and other incentives/disincentives; and (e) delineation of districts where various development standards, procedures, exactions, fees, and incentives apply

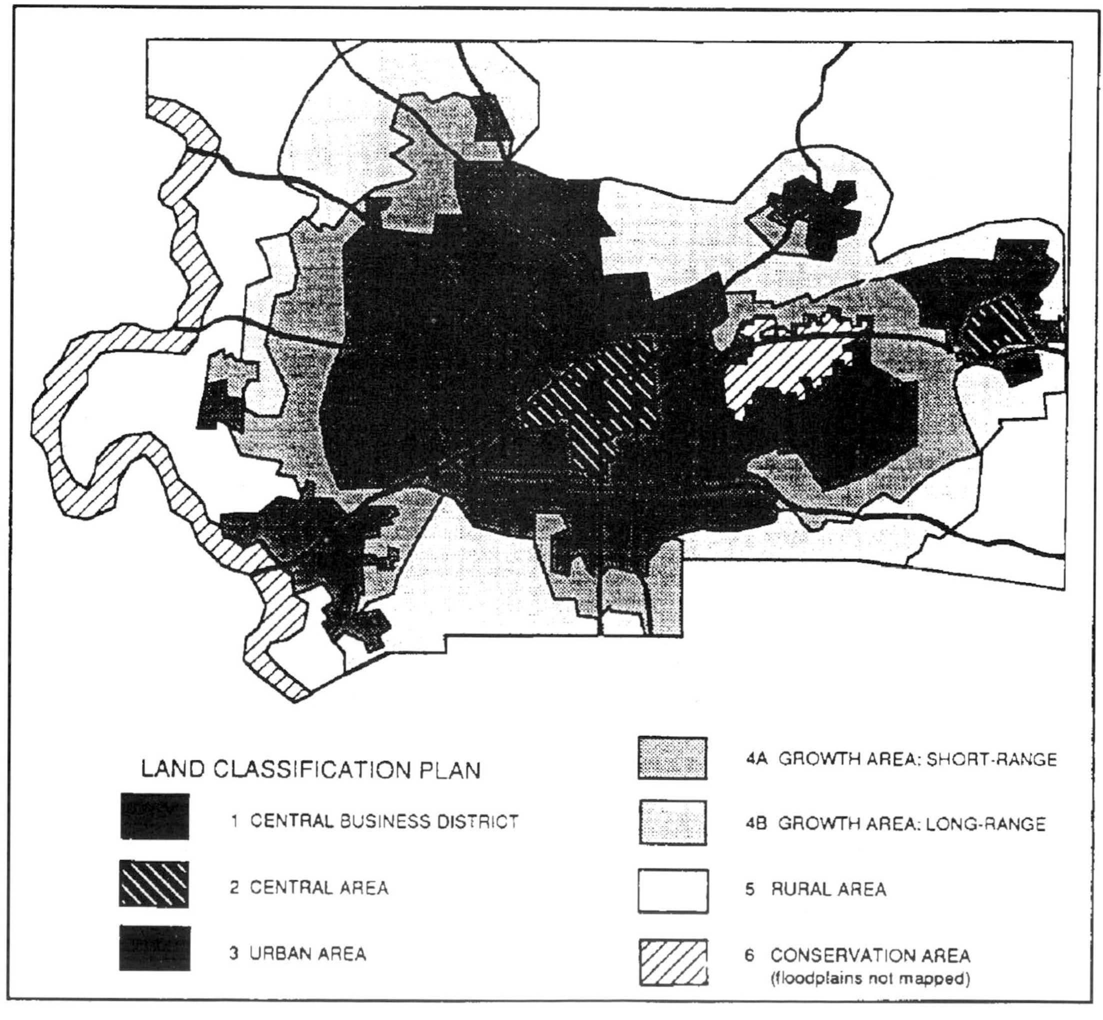

Figure 5 Example of a land classification plan
Source: Adapted from Forsyth County City–County Planning Board 1988

2. Program for the expansion of urban infrastructure and community facilities and their service areas
3. Capital improvement program
4. Property acquisition program
5. Other components, depending on the community situation, for example, a preferential taxation program, an urban revitalization program for specific built-up neighborhoods, or a historic preservation program

- Official maps, indicating legislative intent, which may be incorporated into ordinances, with force of law – among them, goal-form maps (e.g., land classification plan or land use design); maps of zoning districts, overlay districts, and other special areas for which development types, densities, and other requirements vary; maps of urban services areas; maps showing scheduled capital improvements; or other maps related to development management standards and procedures

The development management plan is a distinct type, emphasizing a specific course of action, nor general policy. At its extreme the management plan actually incorporates implementation measures, so that the plan becomes part of a regulative ordinance. Although the spatial specifications for regulations and other implementation measures are included, a land use map may not be.

One point of origin for development management plans is Henry Fagin's concept of the "policies plan," whose purpose was to coordinate the actions of line departments and provide a basis for evaluating their results, as well as to formulate, communicate, and implement policy (the traditional purpose). Such a plan's subject matter was as broad as the responsibilities of the local government, including but not limited to physical development. The format included a "state of the community" message, a physical plan, a financial plan, implementation measures, and detailed sections for each department of the government.

A more recent point of origin is *A Model Land Development Code* (American Law Institute 1976), intended to replace the 1928 Model Planning Enabling Act as a model for local planning and development management. The model plan consciously retains an emphasis on physical development (unlike Fagin's broader concept), but stresses a short-term program of action, rather than a long-term, mapped goal form. The ALI model plan contains a statement of conditions and problems; objectives, policies, and standards; and a short-term (from one to five years) program of specific public actions. It may also include land acquisition requirements, displacement impacts, development regulations, program costs and fund sources, and environmental, social, and economic consequences. More than other plan types, the development management plan is a "course of action" initiated by government to control the location and timing of development.

The *Sanibel, Florida, Comprehensive Land Use Plan* (1981) exemplifies the development management plan. The plan outlines the standards and procedures of regulations (i.e., the means of implementation), as well as the analyses, goals, and statements of intent normally presented in a plan. Thus, when the local legislature adopts the plan, it also adopts an ordinance for its implementation. Plan and implementation are merged into one instrument, as can be seen in the content of its articles:

Article 1: Preamble: including purposes and objectives, assumptions, coordination with surrounding areas, and implementation

Article 2: Elements of the Plan: Safety, Human Support Systems, Protection of Natural Environmental, Economic and Scenic Resources, Intergovernmental Coordination, and Land Use

Article 3: Development Regulations: Definitions, Maps, Requirements, Permitted Uses, Subdivisions, Mobile Home and Recreation Vehicles, Flood and Storm Proofing, Site Preparation, and Environmental Performance Standards

Article 4: Administrative Regulations (i.e., procedures): Standards, Short Form Permits, Development Permits, Completion Permits, Amendments to the Plan, and Notice, Hearing and Decision Procedures on Amendments

Figure 7 shows the Sanibel plan's map of permitted uses, which is more like a zoning plan than a land use design plan, because it shows where regulations apply, and boundaries are exact.

INDUSTRIAL DISTRICTS

Industrial Districts are intended to provide areas in the county which are suitable for the needs of industry. They should be located and designed to be compatible with the surrounding land uses, either due to existing natural features or through the application of standards.

RECOMMENDATIONS:

1. Identify general locations for potential industrial uses.
2. Permit retail sales as an accessory use in the Industrial District.

SINGLE-FAMILY RESIDENTIAL DISTRICTS

Single-Family Residential Districts are to be developed and promoted as neighborhoods free from any land usage which might adversely affect them.

RECOMMENDATIONS:

1. For new development, require buffering for controlling visual, noise, and activity impacts between residential and commercial uses.
2. Encourage single-family residential development to locate in the designated towns.
3. Allow duplexes, triplexes, and fourplexes as a conditional use in the "R-1" Residential Zone so long as the design is compatible with the single-family residential development.
4. Allow home occupations (professions and services, but not retail sales) by permitting the employment of one full-time equivalent individual not residing on the premises.

MULTIFAMILY RESIDENTIAL DISTRICTS

Multifamily Residential Districts provide for townhouses and multifamily apartment units. Areas designated in this category are those which are currently served or scheduled to be served by community or multi-use sewerage and water supply systems.

RECOMMENDATIONS:

1. Permit multifamily development in the Solomons, Prince Frederick, and Twin Beach Towns.
2. Require multifamily projects to provide adequate recreational facilities—equipment, structures, and play surfaces.
3. Evaluate the feasibility of increasing the dwelling unit density permitted in the multifamily Residential Zone (R-2).

Figure 6 An excerpt from a verbal policy plan
Source: Calvert County, Maryland 1983

THE CONTEMPORARY HYBRID PLAN: INTEGRATING DESIGN, POLICY, AND MANAGEMENT

The rationality of practice has integrated the useful parts of each of the separate prototypes reviewed here into contemporary hybrid plans that not only map and classify land use in both specific and general ways, but also propose policies and management measures. For example, Gresham, Oregon, combined land use design (specifying residential, commercial, and industrial areas, and community facilities and public lands) with an overlay of land classification districts (developed, developing, rural, and conservation), and also included standards and procedures for issuing development permits (i.e., a development code). Prepared with considerable participation by citizens and interest groups, such plans usually reflect animated political debates about the costs and benefits of land use alternatives.

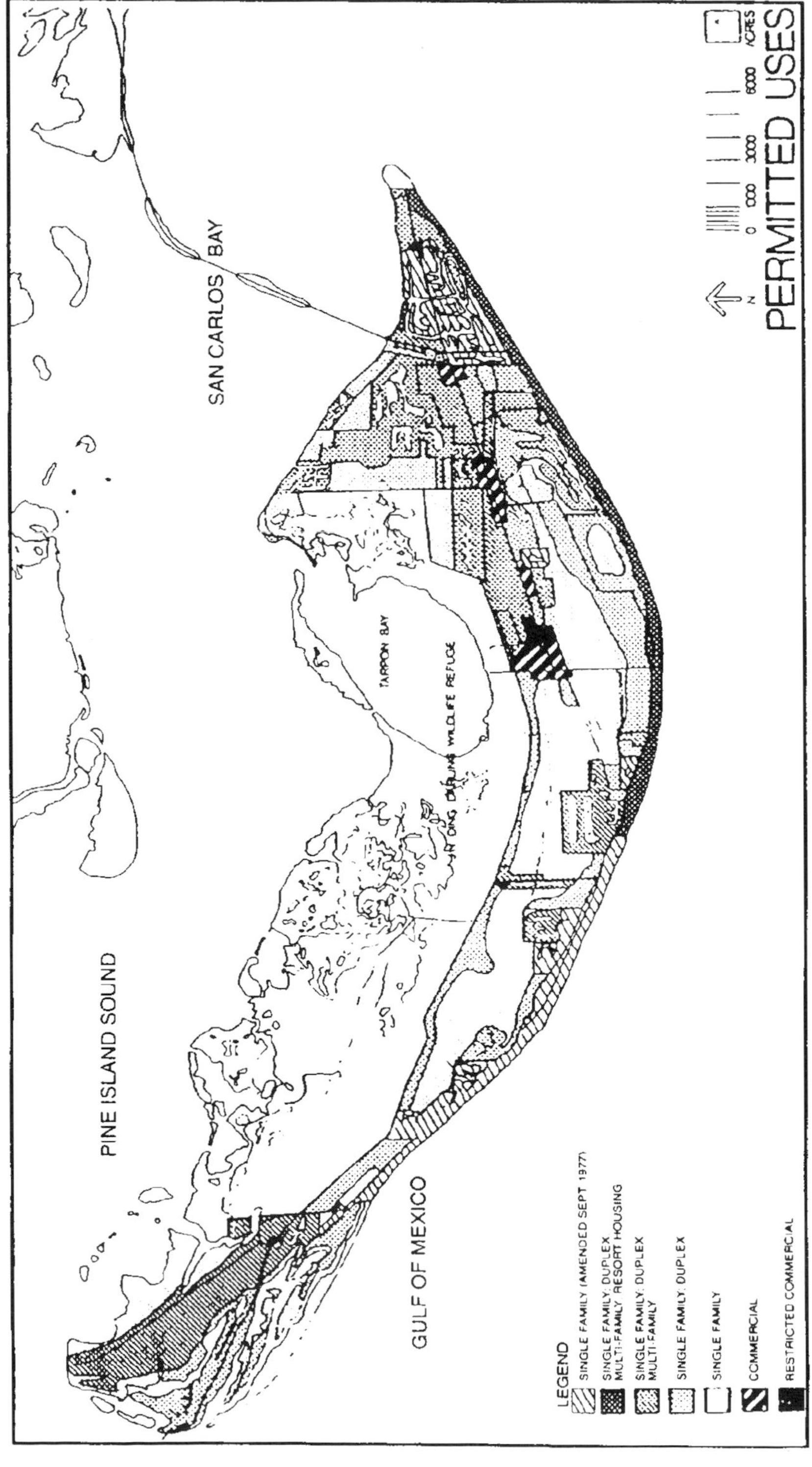

Figure 7 Map of permitted uses, Sanibel
Source: City of Sanibel 1981

The states that manage growth have created new land use governance systems whose influence has broadened the conceptual arsenals of local planners. DeGrove identifies the common elements of these systems:

- Consistency – intergovernmentally and internally (i.e., between plan and regulations)
- Concurrency – between infrastructure and new development
- Compactness – of new growth, to limit urban sprawl affordability – of new housing
- economic development, or "managing to grow"
- sustainability – of natural systems

DeGrove attributes the changes in planning under growth management systems to new hard-nosed concerns for measurable implementation and realistic funding mechanisms. For example, Florida local governments must adopt detailed capital improvement programs as part of their comprehensive plans, and substantial state grants may be withheld if their plans do not meet consistency and concurrency requirements.

Another important influence on contemporary plans is the renewed attention to community design. The neotraditional and transit-oriented design movements have inspired a number of proposals for mixed-use villages in land use plans. *Toward a Sustainable Seattle: A Plan for Managing Growth* (1994) exemplifies a city approach to the contemporary hybrid plan. Submitted as the Mayor's recommended comprehensive plan, it attempted to muster political support for its proposals. Three core values – social equity, environmental stewardship, and economic security and opportunity – underlie the plan's overall goal of sustainability. This goal is to be achieved by integrating plans for land use and transportation, healthy and affordable housing, and careful capital investment in a civic compact based on a shared vision. Citywide population and job growth targets, midway between growth completely by regional sprawl and growth completely by infill, are set forth within a 20-year time frame. The plan is designed to meet the requirements of the Washington State Growth Management Act.

The land use element designates urban center villages, hub urban villages, residential urban villages, neighborhood villages, and manufacturing/industrial centers, each with specific design guidelines (figure 8). The city's capacity for growth is identified, and then allocated according to the urban village strategy. Future development is directed to mixed-use neighborhoods, some of which are already established; existing single-family areas are protected. Growth is shaped to build community, promote pedestrian and transit use, protect natural amenities and existing residential and employment areas, and ensure diversity of people and activities. Detailed land use policies carry out the plan.

Loudoun County Choices and Changes: General Plan (1991), which won APA's 1994 award for comprehensive planning in small jurisdictions, exemplifies a county approach to the contemporary hybrid plan. Its goals are grouped into three categories:

1. Natural and cultural resources goals seek to protect fragile resources by limiting development or mitigating disturbances, while at the same time not unduly diminishing land values.
2. Growth management: goals seek to accommodate and manage the county's fair share of regional growth, guiding development into the urbanized eastern part of the county or existing western towns and their urban growth areas, and conserving agriculture and open space areas in the west. (See figure 9.)
3. Community design goals seek to concentrate growth in compact, urban nodes to create mixed-use communities with strong visual identities, human-scale street networks, and a range of housing and employment opportunities utilizing neotraditional design concepts (illustrated in figure 9).

Three time horizons are addressed: the "ultimate" vision through 2040, the 20-year, long-range development pattern, and the five-year, short-range development pattern. The plan uses the concept of community character areas as an organizing framework for land use management. Policies are proposed for the overall county, as well as for the eastern urban growth areas, town urban growth areas, rural areas, and existing rural village areas. Implementation tools include capital facility and transportation proffers by developers, density transfers, community design guidelines, annexation guidelines, and an action schedule of next steps.

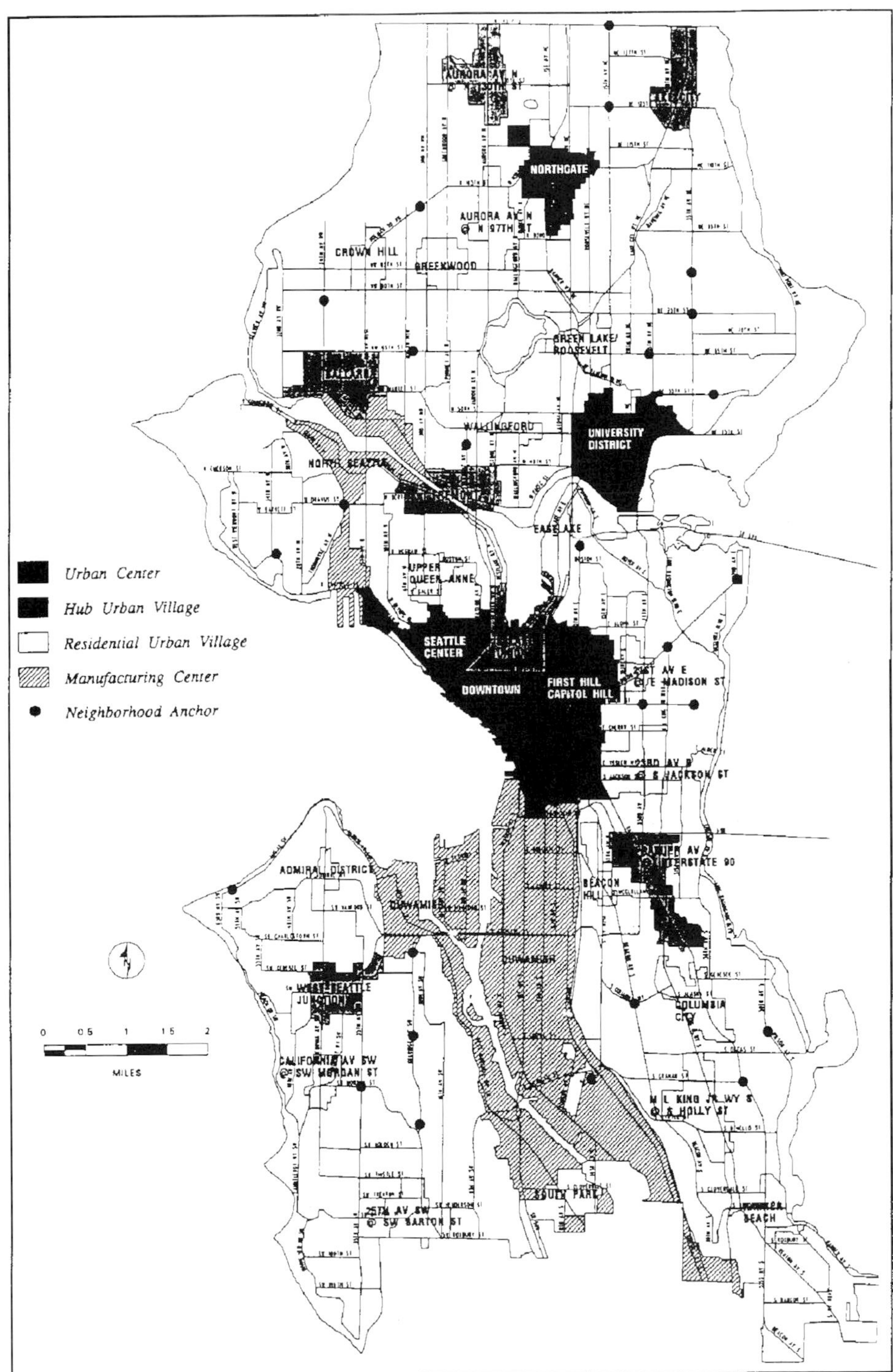

Figure 8 Seattle urban villages strategy
Source: Seattle Planning Department 1993

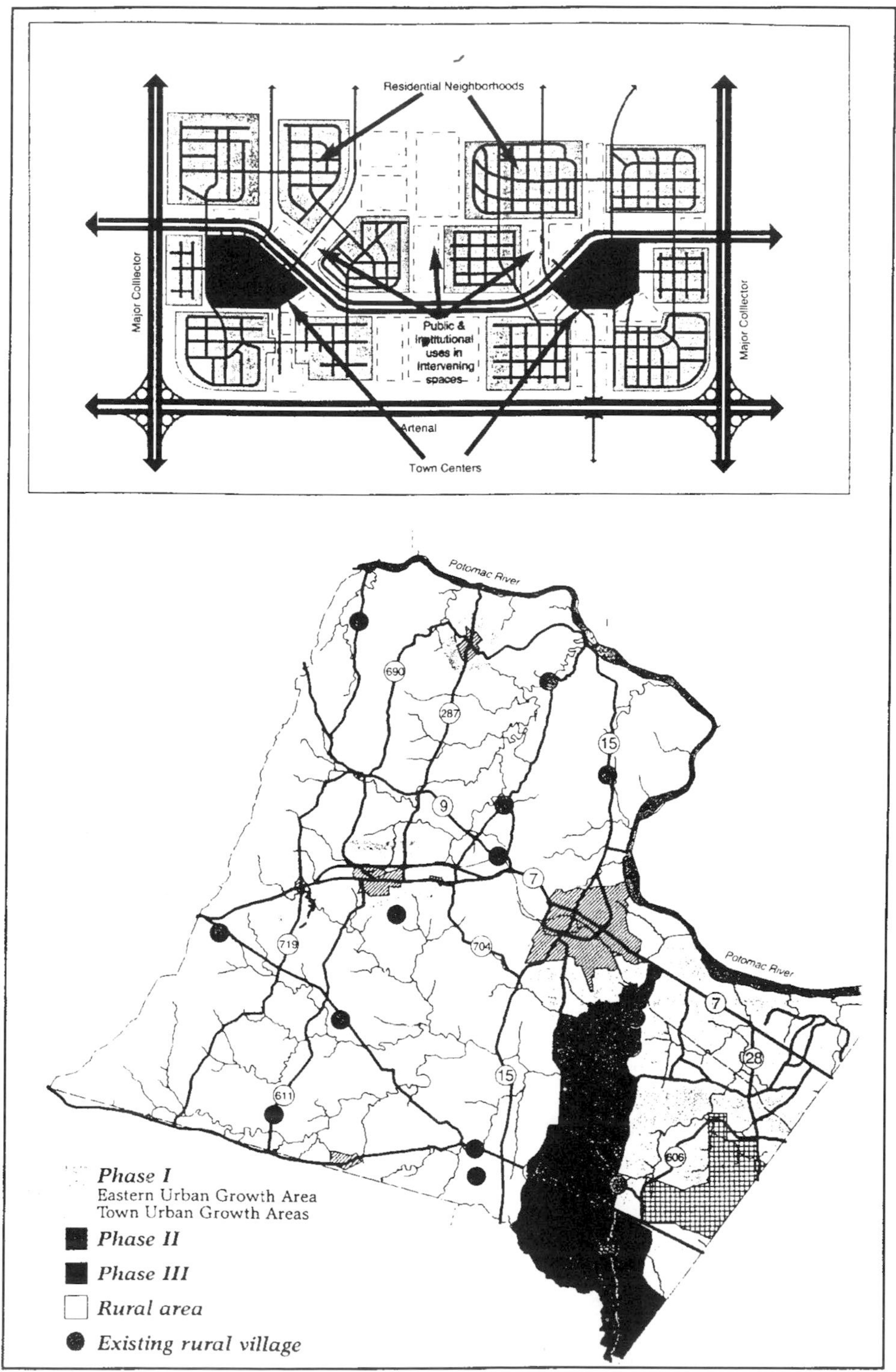

Figure 9 Neotraditional community schematic and generalized policy planning areas, Loudoun County, Virginia General Plan
Source: Planning 60, 3:10 (1994)

SUMMARY OF THE CONTEMPORARY SITUATION

Since midcentury, the nature of the plan has shifted from an elitist, inspirational, long-range vision that was based on fiscally innocent implementation advice, to a framework for community consensus on future growth that is supported by fiscally grounded actions to manage change. Subject matter has expanded to include the natural as well as the built environment. Format has shifted from simple policy statements and a single large-scale map of future land use, circulation, and community facilities, to a more complex combination of text, data, maps, and timetables. In a number of states' plans are required by state law, and their content is specified by state agencies. Table 2 compares the general plan of the 1950s–1960s with the four contemporary prototype plans and the new 1990s hybrid design-policy management plan, which combines aspects of the prototype plans.

Today's prototype land use design continues to emphasize long-range urban form for land uses, community facilities, and transportation systems as shown by a map; but the design is also expressed in general policies. Land use design is still a common form of development plan, especially in municipalities.

The land classification plan also still emphasizes mapping, but of development policy rather than policy about a pattern of urban land uses. Land classification is more specific about development management and environmental protection, but less specific about transportation, community facilities, and the internal arrangement of the future urban form. County and regional governments are more likely than are municipalities to use land classification plans.

The verbal policies plan eschews the spatial specificity of land use design and land classification plans and focuses less on physical development issues. It is more suited to regions and states, or may serve as an interim plan for a city or county while another type of plan is being prepared.

The development management plan represents the greatest shift from the traditional land use plan. It embodies a short-to-intermediate-range program of governmental actions for ongoing growth management rather than for long-range comprehensive planning.

In practice, these four types of plans are not mutually exclusive. Communities often combine aspects of them into a hybrid general plan that has policy sections covering environmental/social/economic/housing/infrastructure concerns, land classification maps defining spatial growth policy, land use design maps specifying locations of particular land uses, and development management programs laying out standards and procedures for guiding and paying for growth. Regardless of the type of plan used, the most progressive planning programs today regard the plan as but one part of a coordinated growth management program, rather than, as in the 1950s, the main planning product. Such a program incorporates a capital improvement program, land use controls, small area plans, functional plans, and other devices, as well as a general plan.

THE ENDURING LAND USE FAMILY TREE AND ITS FUTURE BRANCHES

For the first 50 years of this century, planning responded to concerns about progressive governmental reform, the City Beautiful, and the "City Efficient." Plans were advisory, specifying a future urban form, and were developed by and for an independent commission. By midcentury this type of plan, growing out of the design tradition, had become widespread in local practice. During the 1950s and 1960s the 701 program, T. J. Kent, and F. Stuart Chapin, Jr. further articulated the plan's content and methodology. Over the last 30 years, environmental and infrastructure issues have pushed planning toward growth management. As citizen activists and interest groups have taken more of a role, land use politics have become more heated. Planning theorists, too, have questioned the midcentury approach to planning, and have proposed changes in focus, process, subject matter, and format, sometimes challenging even the core idea of rational planning. As a result, practice has changed, though not to a monolithic extent and without entirely abandoning the traditional concept of a plan. Instead, at least four distinct types of plans have evolved, all descending more or less from the midcentury model, but advocating very different concepts of what a plan should be. With a kind of

Table 2 Comparison of plan types

	Contemporary prototype plans					
Features of plans	*1950s general plan*	*Land use design*	*Land classification plan*	*Verbal policy plan*	*Development management plan*	*1990s Hybrid design policy management plan*
Land use maps	Detailed	Detailed	General	No	By growth areas	General *and* area specific
Nature of recommendations	General community goals	Land use policies & objectives	Growth locations & incentives	Variety of community policies	Specific management actions	Policy *and* actions
Time horizon	Long range	Long range	Long range	Intermediate range	Short range	Short *and* long range
Link to implementation	Very weak	Weak to moderate	Moderate	Moderate	Strong	Moderate to strong
Public participation	Pro-forma	Active	Moderate	Moderate	Active	Active
Capital improvements	Advisory	Recommended	Recommended	Recommended	Required	Recommended to required
Land use/transportation linkage	Moderate	Strong	Weak	Varies	Strong	Strong
Environmental protection	Weak	Moderate	Strong	Varies	Varies	Strong
Social policy linkage	Weak	Weak	Weak	Moderate to strong	Weak	Moderate

self-correcting common sense, the plans of the 1990s have subsequently incorporated the useful parts of each of these prototypes to create today's hybrid design/policy/management plans.

To return to our analogy of the plan's family tree: Roots for the physical development plan became well established during the first half of this century. By 1950, a sturdy trunk concept had developed. Since then, new roots and branches have appeared – land classification plans, verbal policy plans, and development management plans. Meanwhile development of the main trunk of the tree – the land use design – has continued. Fortunately, the basic gene pool has been able to combine with new genes in order to survive as a more complex organism – the 1990s design-policy management hybrid plan. The present family tree of planning reflects both its heredity and its environment.

The next generation of physical development plans also should mature and adapt without abandoning their heritage. We expect that by the year 2000, plans will be more participatory, more electronically based, and concerned with increasingly complex issues. An increase in participation seems certain, bolstered by interest groups' as well as governments' use of expert systems and computer databases. A much broader consideration of alternative plans and scenarios, as well as a more flexible and responsive process of plan amendment will become possible. These changes will call upon planners to use new skills of consensus building and conflict management, as more groups articulate their positions on planning matters, and government plans and interest group plans compete, each backed by experts.

With the advent of the "information highway," plans are more likely to be drafted, communicated, and debated through electronic networks and virtual reality images. The appearance of plans on CD ROM and cable networks will allow more popular access and input, and better understanding of plans' three-dimensional consequences. It will be more important than ever for planners to compile information accurately and ensure it is fairly communicated. They will need to compile, analyze, and manage complex databases, as well as to translate abstract data into understandable impacts and images.

Plans will continue to be affected by dominant issues of the times: aging infrastructure and limited public capital, central city decline and suburban growth, ethnic and racial diversity, economic and environmental sustainability, global competition and interdependence, and land use/transportation/air quality spillovers. Many of these are unresolved issues from the last thirty years, now grown more complex and interrelated. Some are addressed by new programs like the Intermodal Surface Transportation Efficiency Act (ISTEA) and HUD's Empowerment Zones and Enterprise Communities. To cope with others, planners must develop new concepts and create new techniques.

One of the most troubling new issues is an attempt by conservative politicians (see the Private Property Protection Act of 1995 passed by the US House of Representatives) and "wise use" groups to reverse the precedence of the public interest over individual private property rights. These groups challenge the use of federal, state, and local regulations to implement land use plans and protect environmental resources when the result is any reduction in the economic value of affected private property. Should their challenge succeed and become widely adopted in federal and state law, growth management plans based on regulations could become toothless. Serious thinking by land use lawyers and planners would be urgently needed to create workable new implementation techniques, setting in motion yet another planning evolution.

We are optimistic, however, about the future of land use planning. Like democracy, it is not a perfect institution but works better than its alternatives. Because land use planning has adapted effectively to this century's turbulence and become stronger in the process, we believe that the twenty-first century will see it continuing as a mainstay of strategies to manage community change.

ALEXANDER GARVIN

"A Realistic Approach to City and Suburban Planning" and "Ingredients of Success"

from *The American City: What Works, What Doesn't* (1996)

Editors' introduction The following selection by Yale professor Alexander Garvin describes "what works and what doesn't" in American cities. Garvin summarizes the lessons he has drawn from a lifetime as an architect, urban planner, developer, public servant, and professor.

Garvin acknowledges widespread disillusionment about planning as a way of fixing cities. He agrees with Jane Jacobs that American city planning has been plagued with continuing mistakes. But among the dull housing projects, empty civic centers, tasteless commercial centers, and promenades to nowhere, Garvin has carefully sought out examples of what *does* work. And he concludes that much of what works best in American cities is the result of good city planning. He points to Chicago's magnificent parklands along Lake Michigan, which might still be dump and derelict land but for Daniel Burnham's 1909 Plan for Chicago, precious antebellum sections of Charleston, South Carolina, saved from destruction by the cities zoning laws and historic preservation efforts, Pittsburg, Pennsylvania, and Portland, Oregon's, thoughtfully planned and eminently successful downtowns, and many other examples.

Garvin emphasizes the primacy of private market forces and planning that does not consider them as doomed to failure. He defines urban planning as "*public action that will produce a sustained and widespread private market reaction.*" Cities can leverage billions of dollars of private investment, Garvin argues, if they have formulated plans that will work in the marketplace and use their own infrastructure investments strategically to leverage private capital. Municipal regulation can channel private market forces in positive directions. Alternatively, if they are poorly designed or ill thought out Garvin argues that regulations can actually make matters worse.

Garvin is an experienced real estate developer, and the first of six ingredients of success Garvin identifies is understanding and responding to the market. For example, a plan to redevelop a site in downtown New Haven for department stores, a mall, office building, and a garage in the 1950s could not succeed because suburban shopping malls could meet market demand cheaper and better.

Not all planners would agree with Garvin's assessment that without a market there is no reason to consider any planning prescription. To the contrary, liberal advocacy and equity planners feel that "market failure" is the main reason why planning is needed. Poor people may have *needs* for housing, transportation, open space, and other urban goods, but not the ability to pay for them.

Location is another of Garvin's ingredients of planning success. Garvin notes that the right planning solution in the wrong place simply won't work. Rehabilitation of a decaying, multi-storied brick chocolate factory in San Francisco's Ghirardelli Square was a planning and market success because the building was located on the San Francisco waterfront in a tourist area. Rehabilitation of similar structures in Kansas City's Riverside Quay flopped, according to Garvin, because the location was wrong.

Design – which Garvin is careful to distinguish from mere decoration – is the most misunderstood of his six elements of success. He considers it integral to the success or failure of any planning project. As will be further developed in Part 7, Perspectives on Urban Design, height, bulk, scale, orientation, access, views, materials, color, style, and other design features can make or break a city planning project.

Alexander Garvin is an architect, city planner, and real estate developer. He was formerly New York City's Deputy Commissioner of Housing and Director of Comprehensive Planning. The book from which this selection is taken grew out of his teaching at Yale University where he has taught the introductory course on "American Cities" for three decades.

This selection is from *The American City: What Works, What Doesn't* (New York: McGraw-Hill, 1996). Professor Garvin is also the author, with Gayle Berens, Christopher B. Leinberger, *et al.*, of *Urban Parks and Open Space* (Washington, DC: Urban Land Institute, 1997).

There are many books about what does *not* work in American urban planning. Among the best are Jane Jacobs's classic *The Death and Life of Great American Cities* (New York: Vintage Books, 1961), Peter Hall's *Great Planning Disasters* (Berkeley: University of California Press, 1982), Chester Hartman's *The Transformation of San Francisco* (Towota, NJ: Rowman and Allanheld, 1984), and James Howard Kunstler's *The Geography of Nowhere* (New York: Touchstone Books, 1994).

Other books discussing good city planning and what *does* work include Allan Jacobs, *Making City Planning Work* (Chicago: American Planning Association Press, 1978) and Kevin Lynch, *Good City Form* (Cambridge, MA: MIT Press, 1984).

A REALISTIC APPROACH TO CITY AND SUBURBAN PLANNING

There is agreement neither on what to do to improve our cities and suburbs nor on how to get the job done. Some believe the answers are a matter of money; others believe they involve politics, or racial and ethnic conflict, or some other factor. One thing most people share, though, is disillusionment with urban planning as a way of fixing the American city.

This disillusionment with urban planning is far from justified. There are dozens of projects that are triumphs of American city planning:

- Chicago would not have 23 miles of continuous park land along Lake Michigan if this land had not been included in the city's comprehensive plan of 1909.
- The glorious antebellum sections of Charleston, South Carolina, would not have survived if the city had not adopted zoning in 1931.
- Pittsburgh would not rank third in the nation as a major corporate headquarters center if it had not virtually rebuilt its downtown during the 1940s and 1950s.
- Portland, Oregon, would not be a lively retail and employment center if during the 1970s and 1980s it had not enriched its pedestrian environment, built a light-rail system, and reclaimed its riverfront.

Such triumphs are easy to overlook. Once a problem is solved it disappears and is forgotten. Even local excitement over a successful project rarely spills over into national publications other than those with a narrow group of readers (preservationists, environmentalists, realtors, lawyers, architects, bankers, or some other group that is intimately involved with specific sets of city problems).

Many people are disillusioned with urban

planning because so many of its promises are not kept. Usually these promises are made in good faith by city planners who believe that their job is to establish municipal goals and provide blueprints for a better city. Too often the efforts of these planners end without much consideration of how they will obtain political support for their proposals, who will execute them, or where the money to finance them will come from.

Disillusionment with urban planning also develops when physical improvements fail to solve deep-seated social problems. This is not the fault of urban planning. After all, fixing cities does not fix people. The disillusionment is the product of false expectations. Crime, delinquency, and poverty are afflictions of city residents, not of the cities themselves. Such problems can be found in suburban and rural areas as well.

We need more realistic expectations of what urban planning can accomplish. While it cannot change human nature and is therefore not a panacea for all urban ills, it surely can improve a city's physical plant and consequently affect the safety, utility, attractiveness, and character of city life. When Chicago began creating its waterfront parks, for example, large sections of the shoreline of Lake Michigan were being used as railyards and garbage dumps. Simply removing these uses reduced hazards and made neighboring property more attractive.

We also need a better understanding of how effective planning is translated into a better quality of life. It is not accomplished by planners operating in a vacuum. By themselves, urban planners cannot accomplish very much. Improving cities requires the active participation of property owners, bankers, developers, architects, lawyers, contractors, and all sorts of people involved with real estate. It also requires the sanction of community groups, civic organizations, elected and appointed public officials, and municipal employees. Together they provide the financial and political means of bringing plans to fruition. Without them even the best plans will remain irrelevant dreams.

Finally, the planning profession itself needs to improve its understanding of the way physical changes to a city can achieve a more smoothly functioning environment, a healthier economy, and a better quality of life. For example, the restoration of Charleston's historic district generated substantial tourist spending, just as the reconstruction of the bridges and highways leading into downtown Pittsburgh reduced the cost of doing business and initiated an era of major corporate investment. These and other successful planning strategies are too frequently ignored in the search for more innovative prescriptions.

[. . .]

Defining the planning process

Much of the nation's unsuccessful urban planning arises from the erroneous belief that project success equals urban planning success. Highways that are filled with automobiles, housing projects that are fully rented, and civic centers with plenty of busy bureaucrats may be successful on their own terms. The cities around them, however, may be completely unaffected. Worse, they may be in even greater trouble than they were prior to these projects.

Only when a project also has beneficial impact on the surrounding community can it be considered successful planning. Thus, urban planning should be defined as *public action that will produce a sustained and widespread private market reaction.* That is precisely what has occurred wherever urban planning has been successful.

- When Chicago transformed its lake shore into a continuous park and drive, the real estate industry responded by spending billions to make it a setting for tens of thousands of new apartments.
- When Charleston preserved its "old and historic" district, it retained an extraordinary physical asset that, decades later, would attract a growing population and provide the basis of a thriving economy.
- When Pittsburgh cleared its downtown of the clutter of railyards and warehouses; reduced air and water pollution; and built new highways, bridges, and downtown garages, businesses responded by rebuilding half the central business district.
- When Portland invested in a riverfront park, a light-rail system, and pedestrianized streets, the private sector responded by erecting office buildings, retail stores, hotels, and apartment houses.

The scope of urban planning must be

broadened. Over the past few decades, the areas of public concern and therefore of public action have expanded both substantively and geographically. Outraged citizens have demanded action to protect the natural environment, to preserve the national heritage, to provide a range of services that had never before been considered a public responsibility, and to deal with territory outside local political jurisdictions. The country should be deeply grateful to these activists for insisting that government fill important vacuums.

Too often, we have responded to their legitimate demands by creating a set of protected special interests that are excluded from competition with other equally legitimate public concerns. As a result, large geographic areas are removed from active use without consideration of the social consequences. Buildings are declared landmarks without reference to economic impact. Services are provided to socially impaired individuals without any thought of the effect on the surrounding community. The situation can be rectified simply by including these new areas of public concern within the scope of city planning and simultaneously including a far broader range of participants in the planning process.

The broad definition of urban planning suggested above highlights the fact that planning is about *change:* preventing undesirable change and encouraging desirable change. It may involve a tax incentive, a zoning regulation, or some other technical prescription, but only as a mechanism for instigating change. The important element is change itself. Planners obtain changes in safety, utility, and attractiveness of city life through strategic public investment, regulation, and incentives for private action.

Strategic public investment

Nineteenth-century planning was particularly enamored of strategic government investment. Just think of the many locations that were made more attractive for development by installing water mains, sewer pipes, or transit lines prior to development.

A more recent example is federal subsidization of the interstate highway system. It vastly increased the amount of land within commuting distance of cities and, in the process, increased the attractiveness of suburban locations. Developers eagerly purchased the newly accessible land and built houses, shopping malls, and office parks. In the process millions of consumers were given the opportunity of owning a house in the country, close to shopping facilities and sometimes also near their jobs.

There were adverse impacts as well. The interstate highway system, for example, attracted motorists away from traditional urban arterials, thereby reducing demand in the retail establishments that had previously catered to the large market of automobile-oriented consumers. For decades after the highways were built, cities were plagued with blighted retail streets, unable to replace the customers that previously had filled their no longer active stores.

The difference between routine capital spending and strategically planned investments lies in using these expenditures to spark further investment by private businesses, financial institutions, property owners, and developers. The revitalization of downtown Portland, Oregon, during the 1970s and 1980s provides a vivid demonstration of the effectiveness of such strategic capital investment. During this period the city rebuilt the streets and sidewalks of two parallel avenues, transforming them into handsome red-brick pedestrian transitways lined with bus shelters, artwork, fountains, and new street furniture. Portland also established a 27-stop light-rail system that starts downtown, moves along two parallel streets that cross the two transitways, and extends 15 miles into the suburbs. Eventually the city purchased the block where the pedestrian transitways and the light-rail system cross for a new public square.

All this public investment transformed the area into the most convenient spot in downtown Portland – perfect for retail shopping. The response from the private sector was both predictable and impressive. Nordstrom's built a new store facing the square. The Rouse Company acquired the nearby Olds & King department store and converted it into "The Galleria," a 75-foot-high atrium surrounded by a variety of restaurants, cafes, and retail stores. Saks Fifth Avenue, like Nordstrom's, opened a department store. The block next to Saks was rebuilt as Pioneer Place, a multistory, air-conditioned atrium with shops, restaurants, and tourist-oriented retail outlets. Few cities have been as effective in using capital expenditures to spur private investment or in obtaining the accompanying increase in retail sales, employment, and taxes.

Regulation

Regulation is most often used to alter the size and character of the market and the design of the physical environment. Perhaps the single most effective example occurred during the 1930s when the federal government restructured the banking system and in the process dramatically altered the housing market. Prior to that time few banks provided mortgage loans that covered more than half the cost of a house. These loans were extended for relatively short periods of time (two to five years) and involved little or no amortization.

The National Housing Act of 1934, which created the Federal Housing Administration (FHA), changed all that. It regulated the rate of interest and the terms of every mortgage that it insured. By 1938, a house could be bought for a cash down payment equal to 10 percent of the purchase price. The other 90 percent came in the form of a 25-year, self-amortizing, FHA-insured mortgage loan. These new mortgage lending practices greatly increased the number of people who could afford a down payment on a house as well as monthly debt service payments on a mortgage, and thereby also increased the size of the market for single-family houses. That is one of several reasons that the proportion of American households that owned their home increased from 44 percent in 1940 to 64 percent in 1990.

Not only did the FHA alter the size of the market, it also determined the design of the product. In order to be eligible for FHA mortgage insurance a house had to conform to published minimum property standards that included structure, materials, and room sizes. The effect of these regulations was to guarantee a minimum standard of quality on a national scale.

Regulation also can be used to alter the character of an entire area. This process usually begins with an attempt to prevent hazardous conditions. Local governments, for example, are usually interested in providing sufficient open land to permit natural drainage of rain and snow, to prevent waste from percolating through the ground to contaminate the water supply, and to ensure privacy. One way of achieving these objectives is to require a minimum lot size for any development (e.g., no more than one house per acre). The end result is the landscape of one-family houses on large lots that can be found throughout the nation.

As with strategic government investment, a regulation such as mandating minimum lot sizes also can produce an adverse impact. Since the amount of land in any community is finite, whenever a minimum lot size is adopted the future supply of house sites is reduced. This reciprocal relationship between the degree of regulation and the size of the market for the resulting product is inevitable. It was poignantly explained by Jacob Riis, who, in 1901, was already lamenting that the minimum construction requirements of "tenement house reform ... tended to make it impossible for anyone [not able] to pay $75 to live on Manhattan Island."

[. . .]

Incentives

Although the use of incentives is becoming more popular, the approach has been around a long time. One of the oldest examples is the incentive for people to own their homes. During the Civil War, Congress allowed taxpayers to deduct interest payments and local taxes from the income that formed the basis of federal tax payments. The same deduction of mortgage interest and taxes was reintroduced in 1913, when the federal income tax was adopted.

There can be no serious change either in cities or suburbs without a favorable investment climate. In many instances government need only guarantee two things: intelligent spending on capital improvements and regulatory policies that provide stability and encourage market demand. Only when investment and regulation are insufficient to do the job should incentives come into play.

New York City faced such a situation during the mid-1970s. The city's fiscal crisis precluded most capital spending. Political gridlock prevented serious regulatory reform. At the same time the rate of housing deterioration and abandonment had reached alarming proportions. The city administration had to develop a strategy that would prevent further deterioration. One technique seemed most likely to succeed: incentives that were sufficiently generous to induce private investment in the existing housing stock. Consequently, the Housing and Development Administration proposed to revise the city's little-known J-51 program. It provided a 12-year exemption from any increase in real estate tax assessment due to physical improvements and a deduction from annual real estate tax payments of a portion of the cost of those improvements.

The problem with the earlier J-51 Program was that it did not apply to three-quarters of the city's housing stock. Existing apartments that were not subject to rent control . . . could not obtain these benefits unless they became subject to rent control. Nonresidential structures that had been converted to residential use were completely ineligible. Without J-51 benefits, any major investment in improvements resulted in punishment – a major increase in the real estate tax assessment. This was especially burdensome to the 770,000 apartments then subject to rent stabilization, New York City's second rent regulatory system.

During 1976, the Beame Administration persuaded the state legislature and the New York City Council to smash the rent control barrier by extending eligibility to rent-stabilized apartments. J-51 benefits were also provided for cooperative and condominium apartments in newly rehabilitated residential structures and to rental apartments in buildings converted from nonresidential to residential use, provided that they would become subject to some form of rent regulation.

These tax incentives completely altered the climate for investment in existing buildings. Banks increased their lending for housing rehabilitation, building owners increased their investments in building improvements, and developers began purchasing vacant structures for conversion to residential use. In fiscal 1977–1978, the first year in which the full impact of these incentives could be measured, more than 48,000 apartments were granted J-51 benefits.

[. . .]

A new approach to urban planning

We need a new approach to urban planning that explicitly deals with both *public action* and the probable *private market reaction*. Such change-oriented planning requires general acceptance of the idea that while urban planners are in the change business, it is others who will make that change: civic leaders, interest groups, community organizations, property owners, developers, bankers, lawyers, architects, engineers, elected and appointed public officials – the list is endless.

Being entirely dependent on these other players, urban planners must concentrate on increasing the chances that everybody else's agenda will be successful. They may choose to do so by targeting public investment in infrastructure and community facilities, or by shaping the regulatory system, or by introducing incentives that will encourage market activity. But whatever they select, their role must be to initiate and shepherd often controversial expenditures and legislation. More important, the public will be able to hold them accountable by evaluating the cost effectiveness of the private market reaction to their programs.

Only when this approach to urban planning takes hold will we get beyond the technical studies, needs analyses, and visions of the good city that currently masquerade as urban planning and get on with the business of fixing the American city.

INGREDIENTS OF SUCCESS

There is no formula that guarantees a desirable private market reaction in response to public action. However, there are six ingredients that must be intelligently dealt with for any project to succeed. They are: market, location, design, financing, entrepreneurship, and time.

The need to consider these ingredients may seem obvious. Unfortunately, the proliferation of still-born projects reveals how little they are understood. Otherwise, why would there be housing for which there is no *market*, commercial centers that are in the wrong location, civic centers for which *financing* is not available, places whose *design* makes them unpleasant and unsafe areas in which to congregate, economic development projects whose completion is beyond the *entrepreneurship* of the responsible public agency, and public works whose *time* has passed but are still under way.

If any of these six ingredients is absent or if they are not combined in a mutually reinforcing fashion, the project will fail. Even when all the ingredients are properly combined, they may be insufficient to guarantee project success because city planners, unlike chefs, cannot keep unexpected ingredients from getting into the pot. Nevertheless, an intelligent mix of market, location, design, financing, entrepreneurship, and time is the key to success. Thus, an understanding of how these elements operate and interact will increase the likelihood of favorable results.

Market

The existence of a market for any urban planning prescription is primary, for without it there is no reason even to consider action. The word "market" is not synonymous with population. It means a specific population's desire for something and its ability and willingness to pay for it in the face of available alternatives. Nor is market synonymous with "need." Too often what one person calls a need is really a preference for what other people ought to have.

To be successful, an urban planning prescription must reflect both market demand and supply. The demand side requires a user population with enough money to purchase what it desires and the willingness to spend it. That means sufficient users to cover both capital cost and operating expenses. If it requires private action, there will have to be user charges; if it is a public project, the electorate will have to be willing to pay the necessary taxes.

The role that demand plays in determining the success of an urban planning prescription is illustrated by two neighborhood revitalization programs adopted for Savannah, Georgia. Both tried to preserve some of the nation's most attractive nineteenth-century buildings that, prior to these programs, had been vacant or dilapidated. The first neighborhood revitalization program began during the late 1960s and successfully revived the relatively small Pulaski Ward. It was followed by a second, similar effort that failed to restore the city's much larger Victorian District.

In both instances, concerned citizens established nonprofit institutions to salvage threatened historic structures. The mechanism they employed was a revolving fund that provided money to purchase vacant or deteriorating buildings. The fund was reimbursed from the proceeds of the resale of these buildings to responsible owners who agreed to restore and maintain them.

The 15-acre Pulaski Ward, initially settled in the 1840s, surrounds one of Savannah's charming original squares. In 1964, when the Historic Savannah Foundation chose it as a target area, Pulaski Ward had become a dilapidated neighborhood with many vacant (albeit historic) structures. Over the next 18 months, Historic Savannah acquired and resold 54 buildings, generating more than $1.5 million in privately financed renovation. Subsequently another dozen buildings were acquired and rehabilitated privately. Eventually owners renovated every building in the ward and even began filling in vacant lots with small-scale new construction.

The effort to revitalize Pulaski Ward was so successful that in the mid-1970s, preservationists decided to try the same strategy in the 150 blocks that make up Savannah's Victorian District. This time two separate revolving funds were set up. Federal subsidies were obtained to reduce the cost of rehabilitation to a level that was affordable for the area's low-income population. By 1990 more than 300 housing units had been rehabilitated and another 40 units built.

These efforts did not spark widespread investment in the area, which in 1992 remained riddled with vacant and deteriorating structures. Failure became inevitable when federal subsidy programs were curtailed during the early 1980s. Those who had conceived this preservation strategy based it on subsidies without which the area's low-income residents would be unable to afford debt service on a mortgage (covering the cost of acquisition and rehabilitation). Thus, when the federal government terminated its programs there was no way to pay for further renovation of the area's vacant but dilapidated buildings.

The revolving fund was successful in Pulaski Ward because there had been enough households who desired and could afford to live in charming, restored residences on the edge of downtown Savannah. All that had been necessary to tap that market was an initial investment in some of the area's vacant buildings. The same prescription failed in the Victorian District because without subsidies there was an insufficient market for the renovation of its no-less-charming historic structures.

[. . .]

Location

Location consists of two elements: a site's inherent characteristics and its proximity to other locations. Site characteristics alone may be sufficient to make it attractive. A spectacular view is an example. Another is an architecturally distinctive housing stock, such as the one that made renovation particularly inviting in the historic districts of Savannah.

Site conditions can also ruin an otherwise desirable location. During the first half of the twentieth century, air pollution in downtown Pittsburgh was so serious that street lights often remained on 24 hours a day. Raw sewage polluted both riverfronts. Daytime traffic congestion seriously restricted both circulation and business activity. In order to alter these inhibiting site conditions, the city obtained state legislation that allowed it to regulate air and water pollution, rebuild its highways and bridges, create more than 5000 parking spaces, and clear away the tangle of downtown railyards, dilapidated warehouses, and obsolete manufacturing lofts. Once these site conditions were eliminated, property owners invested hundreds of millions in redevelopment. Within a couple of decades, more than half of the business district had been rebuilt.

Proximity involves both time and space. The temporal dimension is shaped by technology and can be understood in terms of available means of conveyance. During the eighteenth century, when people were concerned with walking distances, cities had to be compact and densely built up. By the end of the twentieth century, when distance is measured in driving time, the resultant landscape is "spread city."

The spatial dimension of proximity involves interdependence with neighboring areas. An obvious example is the relationship between movie theaters, parking facilities, and eating places. On a larger scale, nineteenth-century warehouse and manufacturing districts often developed in close proximity to waterfront areas through which they received and shipped goods and materials.

Even before the end of World War II, most mercantile districts, especially in port cities and railroad towns, had begun a slow and steady decline. There was no longer the same need for large, multistory warehouses and manufacturing structures near the traffic-congested waterfront. Now merchandise could be stored in large prepackaged containers that were lifted by crane and shipped by truck along an increasingly convenient highway system. Containerports needed too much upland open space to be easily located along already built-up city waterfronts. Instead, they were being established along vacant shorefronts, nearer to major highways. Production was easier and cheaper in single-story, suburban factories that could provide extended horizontal production lines, easy parking for employees, and even easier highway access for trucks. Technological change had transformed proximity to the waterfront from an asset into a liability.

Recognition of changing demand for different locations is often quite slow. Most city officials only became aware of the decreasing importance of waterfront shipping from declining tax collections and increasing building vacancies. Recognition of the opportunities provided by declining but still attractive waterfront locations became apparent only after the success of Ghirardelli Square in San Francisco.

This project, conceived in 1962, converted into an urban marketplace 2.5 acres of factory and warehouse structures that had once housed a chocolate company. The design . . . established a charming combination of fashionable retail stores and restaurants in a physical setting redolent of old San Francisco. Ghirardelli Square became an instant tourist attraction. More important, it became an inspiration for similar projects in the surrounding Fisherman's Wharf section of San Francisco and throughout the country.

[. . .]

Design

The most misunderstood of the six ingredients of success is design. Too often, it is thought of as decoration that can be applied after the important decisions have been made. In fact, design is the physical manifestation of any prescription and, therefore, is integral to its success or failure from the time of inception.

Design is not just a matter of architectural style. Styles go in and out of fashion; successful planning has to survive for decades. Other more enduring aspects of design are more important. They include the arrangement of project components, the relative size of those components, and their character. Each element affects a project's utility, cost, and attractiveness. When they are organized in a mutually supportive manner, the result is an identifiable destination that provides an auspicious place for the activities occurring there. When arranged to fit the right combination of market, location, financing, entrepreneurship, and time, the result is a successful project.

The components of New Haven's Chapel Square, for example, are assembled in a manner that reduces utility to retail shoppers and, therefore, retail sales. Its two-story shopping mall, instead of being placed between the two department stores, is at one end of the scheme. The five-story parking garage is next to and provides direct access to the two department stores, but not to the shopping mall. As a result, none of the mall's retail facilities profits from purchases made by customers stopping in on their way to another intended destination.

If Chapel Square illustrates how the inept arrangement of the components of a design can exacerbate already poor market conditions, Ghirardelli Square illustrates how it can enhance a potentially wonderful location. At Ghirardelli Square the components are terraced in a manner that increases the utility of the site, reduces costs, and attracts customers. On this steeply sloping site, parking is fitted in under several levels of shopping without taking up otherwise rentable floor area. At the higher end of the site, the parking structure provides the foundation for retail stores. In the middle, its roof provides an outdoor pedestrian level in which retail shoppers can freely circulate among the stores. Only at the lowest end of the site is parking fully underground.

By including in the design formerly obsolete buildings (especially the factory building that now includes a display about the chocolate company that was its initial occupant) and by reserving for public use spots with panoramic views of the waterfront, the design attracts additional tourists. It is a profitable combination of utility, economy, and picturesque features. Today this arrangement seems obvious, but when Ghirardelli Square was conceived, nothing like it had ever been designed.

[. . .]

Financing

Every prescription for fixing cities requires financing. When this involves public action, as is the case with parks, the financing comes from taxes. Among the reasons that Minneapolis has the best designed and best maintained park system in America is that its elected Park and Recreation Board can levy taxes and issue bonds. As a result it has money to pay for acquisition, design, development, program delivery, and maintenance. Elsewhere, whenever cities face a period of budget stringency, they transfer money from the parks to other "more pressing priorities."

Financing is equally important when a planning proposal requires private sector activity. Privately financed projects need *capital* to cover start-up costs, a short-term *development loan* to pay expenses until it is operational, and a *permanent mortgage* to replace the other two when the project is complete and tenanted. The obvious place to obtain financing is a bank. Banks lend their depositors' money to developers whose projects pay a large enough return to keep depositors happy and contribute toward covering the costs of bank operations. In other words, developers pay banks for the use of their money. The price will depend on its assessment of the risk involved. If the deal looks too risky, the bank will not lend a penny.

Most banks will not lend enough to cover project cost. The rest of the money, the *equity* investment, usually comes from the developer and from investors who have confidence in the venture. Investors know that the bank has not lent enough money to complete the project. They know that if the venture fails, the bank may recoup its investment but they may not. They also know how much the bank is getting for its money. Consequently, developers have to pay investors a higher price for equity funds than they are paying for bank money. Developers will put up their own money if bank mortgages and investor equity do not cover all development costs. Typically, if something goes wrong and the investment has to be liquidated, the bank mortgage will be repaid first, then the equity investors, and finally the developer. Since the developer is taking the greatest risk, he or she will not go into the venture unless the return is better than that of the bank and the equity investors.

In other words, money is obtained at different prices, depending on risk and availability. Mortgage money is usually the least expensive. Equity money is more expensive. The developer's money is the most expensive. The greater the proportion of project costs that comes from other sources (the greater the *leverage* of the developer's cash investment), the more attractive the venture will appear to be to the developer. Government can increase the likelihood of project success by creating an investment climate in which bank financing is readily available and developers maximize leverage.

Congress has consistently tried to ensure adequate financing for housing construction by increasing the safety of residential mortgages. During the Great Depression, it restructured the banking industry and established mortgage insurance programs that eventually led to construction of millions of suburban houses. Title I of the Housing Act of 1949 (popularly known as the urban renewal program) provided two-thirds of the money needed to subsidize planning, start-up costs, property acquisition, demolition, and relocation for federally approved urban renewal projects. Local governments had to pay the remaining one-third. In 1954 Congress added federal insurance on mortgages for new or rehabilitated housing in urban renewal areas. This was followed by a series of mortgage-subsidy programs that reduced housing costs to a level that was affordable for low- and moderate-income families.

These programs reduced the cost of money and assured older cities of the financing needed to pay for major neighborhood reconstruction. Some cities (in particular, Philadelphia, New Haven, and New York) used the money for more than just wholesale clearance. They recognized that deteriorating neighborhoods also could be revived by cutting away scattered pockets of blight. Philadelphia's Washington Square East Urban Renewal Project (better known as Society Hill) is one of the earliest and most successful examples. The money to pay for planning, property acquisition, clearance, and site preparation came from the urban renewal program. Banks provided mortgage money for housing construction and rehabilitation because it was federally insured. Payments to cover ongoing operations came from the middle-income residents of the new and renovated housing. Where necessary financing was supplemented with further federal subsidies.

The revitalization of Society Hill was planned in the mid-1950s by Edmund Bacon, executive director of the Philadelphia City Planning Commission. He proposed selective clearance of only those structures that were beyond repair or that were incompatible with the rest of the neighborhood. These sites were to be filled in either with residential buildings sensitively fitted in between their older neighbors or with small "greenway" parks intended as both landscaped pedestrian paths and as small-scale recreation areas.

It was a perfect strategy for Society Hill. The bulk of the area's buildings were charming eighteenth- and nineteenth-century red-brick row houses. Once the blighting influence of neighboring properties had been eliminated, these row houses became extremely attractive to middle-class residents who wanted to live downtown. By 1970, owners had rehabilitated more than 600 of Society Hill's historic structures, property values had more than doubled, and the population had increased by a third.

[. . .]

Entrepreneurship

No prescription is self-implementing. Each requires talented public and private entrepreneurs. Without them a perfectly appropriate prescription will not get off the ground. Entrepreneurs conceive projects, often when others are unaware that there are any opportunities available. They assemble and coordinate the various players who will execute whatever needs to be done. Without the extra drive that entrepreneurs supply, these other players would be overwhelmed by the uncertainties of the marketplace.

Entrepreneurs do not appear automatically whenever there is unfulfilled demand for something. They have to believe that the risk of failure is minimal and the rewards that come with success are generous. Unless such favorable conditions are prevalent, entrepreneurs will exploit other, more attractive opportunities.

Public projects often fail because public officials ignore the role of entrepreneurship. They mistakenly believe that once a project has been assigned to a government agency, its role is purely administrative. In fact, public entrepreneurs are needed to assemble, coordinate, and inspire all the participants in the development process. Edmund Bacon performed that role in Society Hill. He successfully combined the activities of the bankers, bureaucrats, property owners, developers, architects, engineers, contractors, and countless other actors needed for the revitalization of the neighborhood. He also maintained public approval and bureaucratic momentum despite the uncertainty of acquiescence by property owners. He obtained timely approval by federal agencies and mortgage commitments from financial institutions. He stimulated developer interest in the project and political acceptance by Philadelphia's disparate civic and community groups. He sought and discovered opportunities for participation, funding, and implementation by previously uninvolved public agencies, nonprofit organizations, and private developers. Most important, Bacon implemented a strategy that had never been tried before: eliminating scattered pockets of blight, filling the resulting holes in the fabric of the neighborhood with new housing and parks.

While it is easier to understand the role of an entrepreneur in the private sector, it is essentially the same as that performed by public officials like Edmund Bacon. The role includes coordinating a plethora of participants, dealing with uncertainty, recognizing available opportunities that have not yet been exploited, and frequently accomplishing things in ways that have never been tried before. The difference between private and public entrepreneurial activity is only in the form of payment. The private entrepreneur is paid in hard currency; the public entrepreneur, in power. The sort of people capable of getting things done, however, will have to be extremely well paid in their respective coin.

In many cases private and public entrepreneurs work side by side. This is especially true in urban renewal projects like Society Hill where implementation is dependent on individual property owners and developers. When it enacted the Housing Act of 1949, Congress hoped to attract private developers into the business of redeveloping federally approved urban renewal areas by sharply reducing the risk of failure. This was accomplished by requiring the clearance of any blighted property that might affect the area. Local officials had to prepare a redevelopment plan that provided developers and financial institutions with certainty as to the future of every property within the area. Most important, Congress provided the subsidies needed to reduce land prices and site development costs to a marketable level.

Despite this reduced level of risk, few of the early renewal projects went into construction very quickly. Developers either were not willing to acquire approved urban renewal sites or, if they did acquire them, were unable to persuade financial institutions to provide the necessary financing. As a result, most cities initially generated government-subsidized clearance but were unable to find the proper combination of developer and financing to get very much built.

Not just developers, but banks and insurance companies were afraid of investing in officially designated "blighted areas." Without institutional financing, developers would have had to invest substantial amounts of equity capital. Initially, neither lenders nor equity investors perceived a return commensurate with their risk.

In 1954 Congress made the changes that were needed to interest private entrepreneurs in carrying out approved redevelopment projects. The vehicle it chose was federal mortgage insurance that covered up to 95 percent of the cost of new and rehabilitated housing in urban renewal areas. Since financing now could be insured, banks were ready to issue mortgages on most approved urban renewal projects. For the first time, risk was minimal, equity capital requirements extremely low, and profits entirely a matter of entrepreneurial skill. Naturally, all sorts of business people were eager to get involved.

[. . .]

Time

There are three time sequences that affect success. The first is relatively brief: the period during which a person passes through an area. The second takes into account what will occur 24 hours a day, 7 days a week. The third may take decades, during which political and financial climates will certainly change many times.

Developers of retail shopping facilities are perhaps the most skilled in predicting a person's activity pattern within an area. They have to be skilled in dealing with this brief time period because their tenants' profits are dependent on transient customer activity and their own profits are dependent on tenant success.

At Ghirardelli Square, for example, visitors come by foot or motor vehicle. In either case, when they arrive they are quickly faced with a wide variety of attractions. Passing from one to the next, these visitors invariably stop to look at or purchase something. The result is plenty of activity, a high volume of sales, and therefore high rents per square foot.

The movements of a single individual, on the other hand, are irrelevant in planning for a 24-hour day and a 7-day week. Such planning requires providing a suitable environment for a wide variety of users on a continuing basis. Thus, the crucial questions are who is likely to be in an area over a 7-day period, what will they want to do, how many people are needed to support those activities, and in what ways should the environment be organized to accommodate satisfactorily those people and activities.

[. . .]

Jane Jacobs calls for districts with a "diversity of uses that give each other constant mutual support both economically and socially." Rather than complete neighborhood units, she recommends districts that contain apartment houses with residents who leave for work every day, office buildings with daytime workers, performance halls that accommodate primarily nighttime customers, as well as bars, restaurants, retail stores, and all manner of service establishments. Together they constitute a district that is alive with people 24 hours a day, 7 days a week. Such a district surrounds and includes New York's Lincoln Center. It attracts people for different purposes at different times of the day, 52 weeks a year.

Successful planning also requires a strategy that will remain appropriate over long periods of time. Of all the strategies for fixing urban/suburban America, the planned "new community" is among the most sensitive to long-term cycles. During the decades required to plan, build, and market a new community, it will experience continually changing economic conditions, political trends, migration patterns, and consumer demand. Because of these inevitably changing market pressures, cash flow can vary substantially from year to year. However, to survive to completion, every planned new community must continue making debt-service payments on a massive, frontloaded investment in land, streets, sidewalks, sewers, water mains, and all the required infrastructure and community facilities. This requires access to plenty of capital and investors who are willing to wait for years before seeing profits.

Radburn, New Jersey, perhaps the best designed and most influential planned community in America, was never completed because it could not ride out these pressures. Radburn was developed by the City Housing Corporation, a limited-dividend company expressly created to demonstrate the efficacy of developing carefully planned new communities. In 1927, it purchased 1350 acres in Fair Lawn, New Jersey, 10 miles from the George Washington Bridge, where it intended to create "a new town for the motor age" with a projected population of 25,000.

Clarence Stein and Henry Wright, Radburn's architects, devised a unique plan in which you drove to your home, parked, and entered the rear of the house. The house itself was turned around so that it faced a private yard that fronted on a landscaped pedestrian walk. These pedestrian walks opened onto beautifully landscaped common open spaces, large enough for children to play ball. They were, in turn, connected by an underpass to Radburn's school, swimming pool, and community facilities.

Radburn quickly became famous among city planners. Photographs of its underpass were printed in books and articles all over the world. Architects and planners, particularly in Europe, began copying what they called "the Radburn idea." Ironically, while giving new life to the idea of building planned new communities, Radburn itself failed. During the Depression, few families could afford to purchase a new house. Sales were insufficient for the City Housing Corporation to service the debt it had incurred to pay for land, infrastructure, and community facilities. Its financial backers were not willing to continue the venture without receiving a return on their investment. So, in 1935, after completing about 300 houses, the City Housing Corporation declared bankruptcy. Nevertheless, more than half a century since its financial collapse, Radburn remains one of the world's most beautiful and important planned new communities.

[. . .]

Manipulating the ingredients of success to obtain desirable private-market reaction

City and suburban planning . . . must be evaluated in terms of the cost-effectiveness of the induced private-market reaction. That reaction is determined by the same ingredients that determine the community impact of profit-motivated projects.

While private developers rarely seek to generate and sustain a widespread private-market reaction, some of their projects make profound changes to surrounding communities. Ghirardelli Square, for example, altered the character of Fisherman's Wharf and shifted a substantial amount of San Francisco's tourism to the waterfront . . .

The only way to ensure that market demand will spill over into the surrounding area is to plan *not* to satisfy that market within the project. Then there will be a reason for people to go elsewhere. The 71 stores and restaurants that first opened at Ghirardelli Square could never satisfy all the demands of the customers who were attracted to the San Francisco waterfront. Nor was Ghirardelli Square conceived as a retail facility that would supply everything of interest to its visitors. In fact, its developers hoped to attract customers headed to other Fisherman's Wharf destinations.

[. . .]

The role of government

Government can play a major role in fostering desirable interaction between proposed real estate developments and their neighbors. By subsidizing housing construction in Society Hill, the city government increased the number of customers in walking distance of the downtown stores and restaurants. The additional consumer traffic allowed shops and restaurants to remain in operation for longer hours, and in the process increased the safety and attractiveness of downtown streets during the early evening. Some cities have enacted zoning ordinances that allow parking requirements to be satisfied at off-site locations. This increases pedestrian traffic between those parking facilities and the consumer's ultimate destination. Other cities offer a bonus of additional rentable space to developers who provide suitably designed open space, thereby increasing pedestrian traffic to and from more congested nearby locations.

These examples involve the use of *investment* (housing subsidies), *regulation* (parking requirements), or *incentives* (a zoning bonus) to alter four of the ingredients of success (market, location, design, time of operation). Success in generating further market activity may also require the other two ingredients: financing and entrepreneurship . . .

While private real estate ventures may be more likely to succeed when they interact with surrounding areas, publicly assisted projects are increasingly dependent on that interaction for their very existence. Think of the cities that want a plentiful supply of electricity but are unwilling to permit a power plant within their boundaries, or the neighborhoods that want clean streets but bitterly resist a sanitation garage in their neighborhood, or the homeowners who want a convenient school for their children but oppose locating it across the street. Public projects of this sort are regularly defeated by citizens who do not want them in their backyard and have little confidence that anything will mitigate their negative impact.

We can overcome citizen opposition and ensure project feasibility if we stop thinking solely in terms of individual projects. Instead, we must make decisions that are also based on the probability of a desired private-market reaction. Then the public dialogue will shift from consideration of the project itself to the ways in which its market, location, design, financing, entrepreneurs, and times of operation will benefit the surrounding community. More important, we will increase financial and political feasibility while simultaneously increasing the likelihood of the desirable, sustained, and widespread market reaction that is characteristic of good city and suburban planning.

JOHN FORESTER

"Planning in the Face of Conflict"

American Planning Association Journal (1987)

Editors' introduction While good city planning needs to be inspired by a vision of the end results, and should be informed by theory, in democracies planning is never achieved without conflict. Planners, citizens, local elected officials, developers, and others invariably have different views on what a city *should* be like and *how* to build. Passions run high at important city planning commission meetings.

John Forester, the chair of the Department of City and Regional Planning at Cornell University, got down in the trenches with practicing city planners and others involved in city development to study what the practice of city planning is really like in the face of conflict.

The following selection summarizes what he learned about the process and his ideas on how planners can be effective in the face of conflict. It is valuable for the practical lessons Forester learned. But equally important Forester helps point the way out of the impasse, as described by Peter Hall, that exists in much academic planning theory today. Unlike some ivory tower theorists, Forester has listened carefully to practicing city planners and learned from them. His work then synthesizes what he has learned and develops theoretical concepts which are highly relevant to planning practice.

Forester found that city planners need to guide both developers and neighborhood residents through the complexities of the planning process. They have to be attentive to timing. Successful planners handle conflicts both through formal channels and informally. They must respond to complex and contradictory duties – tugged this way by local politicians, that way by legal mandates, and yet another way by citizen demands. Through all of this, successful city planners must be true to professional norms and hold fast to their own visions of high-quality city development. City planners who retreat to their planning offices to draw beautiful plans or create elegant computerized models of how cities will develop are doomed to frustration. They need to wed these professional skills to the rough and tumble of conflicting local values. Ultimately that is what makes city planning such a fascinating profession.

There are many lessons in Forester's work. People who want to be effective in translating city plans into action need to expect opposition and should not be surprised or worn down by what often seems an endless and frustrating process. They need to be aware of their own power and also its limitations. They have to be sensitive to and understand the interests of other actors in the city development process, and form alliances. Finally, Forester argues, city planners can self-consciously follow any of a number of strategies to keep projects on track and achieve success – as rule-enforcers, negotiators and mediators, resource people, or shuttle diplomats.

Consider the kind of conflicts you would expect if you were trying to make city planning work according to Alexander Garvin's principles (p. 396) or the types of plans that Edward Kaiser and David Godschalk describe (p. 375). Compare Forester's insights with John Mollenkopf's observations on how to study urban power (p. 215).

A related article by John Forester is "Planning in the Face of Power," *American Planning Association Journal* 48, 1 (1982). Forester teamed up with former Cleveland city planning director Norman Krumholz to write an account of Krumholz's experience implementing socially responsible planning in *Making Equity Planning Work* (Philadelphia: Temple University Press, 1990). He is the author of *Critical Theory, Public Choice, and Planning Practice: Towards a Critical Pragmatism* (Buffalo: State University of New York Press, 1993).

Other books on what city planning is actually like include Allan Jacobs, *Making City Planning Work* (Chicago: American Society of Planning Officials, 1976), and Bruce W. McClendon (ed.), *Planners on Planning* (San Francisco: Jossey-Bass, 1996).

In the face of local land-use conflicts, how can planners mediate between conflicting parties and at the same time negotiate as interested parties themselves? To address that question, this article explores planners' strategies to deal with conflicts that arise in local processes of zoning appeals, subdivision approvals, special permit applications, and design reviews.

Local planners often have complex and contradictory duties. They may seek to serve political officials, legal mandates, professional visions, and the specific requests of citizens' groups, all at the same time. They typically work in situations of uncertainty, of great imbalances of power, and of multiple, ambiguous, and conflicting political goals. Many local planners, therefore, may seek ways both to negotiate effectively, as they try to satisfy particular interests, and to mediate practically, as they try to resolve conflicts through a semblance of a participatory planning process.

But these tasks – negotiating and mediating – appear to conflict in two fundamental ways. First, the negotiator's interest in the subject threatens the independence and the presumed neutrality of a mediating role. Second, although a negotiating role may allow planners to protect less powerful interests, a mediating role threatens to undercut this possibility and thus to leave existing inequalities of power intact. How can local planners deal with these problems? I discuss their strategies in detail below.

This article first presents local planners' own accounts of the challenges they face as simultaneous negotiators and mediators in local land-use permitting processes. Planning directors and staff in New England cities and towns, urban and suburban, shared their viewpoints with me during extensive open-ended interviews. The evidence reported here, therefore, is qualitative, and the argument that follows seeks not generalizability but strong plausibility across a range of planning settings.

The article next explores a repertoire of mediated negotiation strategies that planners use as they deal with local land-use permitting conflicts. It assesses the emotional complexity of mediating roles and asks: What skills are called for? Why do planners often seem reticent to adopt face-to-face mediating roles?

Finally, the article turns to the implications of these discussions. How might local planning organizations encourage both effective negotiation and equitable, efficient mediation? How might mediated-negotiation strategies empower the relatively powerless instead of simply perpetuating existing inequalities of power?

ELEMENTS OF LOCAL LAND-USE CONFLICTS

Consider first the settings in which planners face local permitting conflicts. Private developers typically propose projects. Formal municipal boards – typically planning boards and boards of zoning appeals – have decision-making authority to grant variances, special permits, or design approvals. Affected residents often have a say – but sometimes little influence – in formal public hearings before these boards. Planning staff report to these boards with analyses of specific proposals. When the reports are positive, they often recommend conditions to attach to a permit or suggest design changes to improve the final project. When the reports are negative, there are arguments to be made, reasons to be given.

Some municipalities have elected permit-granting boards; some have appointed boards. Some municipal ordinances mandate design review; others do not. Some local by-laws call for more than one planning board hearing on "substantial" projects, but others do not. Nevertheless, for several reasons, planners' roles in these different settings may be more similar than dissimilar.

Common planning responsibilities

First, planners must help both developers and neighborhood residents to navigate a potentially complex review process; clarity and predictability are valued goods. Second, the planners need to be concerned with timing. When a developer or neighborhood resident is told about an issue may be even more important than the issue itself. Third, planners typically need to deal with conflicts between project developers and affected neighborhood residents that usually concern several issues at once: scale, the income of tenants, new traffic, existing congestion, the character of a street, and so on. Such conflicts simultaneously involve questions of design, social policy, safety, transportation, and neighborhood character as well. Fourth, how much planners can do in the face of such conflicts depends not only upon their formal responsibilities, but also upon their informal initiatives. A zoning by-law, for example, can specify a time by which a planning board is to hold a public hearing, but it usually will not tell a planner how much information to give a developer or a neighbor, when to hold informal meetings with either or both, how to do it, just whom to invite, or how to negotiate with either party. So within the formal guidelines of zoning appeals, special permit applications, site plan and design reviews, planning staff can exercise substantial discretion and exert important influence as a result.

Planners' influence

The complexity of permitting processes is a source of influence for planning staff. Complexity creates uncertainties for everyone involved. Some planners eagerly use the resulting leverage, as an associate planning director explains, beginning with a truism but then elaborating:

> Time is money for developers. Once the money is in, the clock is ticking. Here we have some influence. We may not be able to stop a project that we have problems with, but we can look at things in more or less detail, and slow them down. Getting back to [the developers] can take two days or two months, but we try to make it clear, "We're people you can get along with." So many developers will say, "Let's get along with these people and listen to their concerns . . ."

He continues,

> But we have influence in other ways too. There are various ways to interpret the ordinance, for example. Or I can influence the building commissioner. He used to work in this office and we have a good relationship . . . his staff may call us about a project they're looking over and ask, "Hey, do you want this project or not?"

Planners think strategically about timing not only to discourage certain projects but to encourage or capture others. The associate director explains,

> On another project, we waited before pushing for changes. We wanted to let the developer get fully committed to it; then we'd push. If we'd pushed earlier, he might have walked away . . .

A director in another municipality echoes the point:

> Take an initial meeting with the developer, the mayor, and me. Depending on the benefits involved – fiscal or physical – the mayor might kick me under the table; "Not now," he's telling me. He doesn't want to discourage the project . . . and so I'll be able to work on the problems later . . .

For the astute, it seems, the complexity of the planning process creates more opportunities than headaches. For the novice, no doubt, the balance shifts the other way.

But isn't everything, in the last analysis, all written down in publicly available documents for everyone to see? Hardly. Could all the procedures ever be made entirely clear? Consider the experience of an architect planner who grappled with these problems in several planning positions. The following conversation took place toward the end of my interview with this planner. The planner pulled a diagram from a folder and said, "Here's the new flow-chart I just drew up that shows how our design review process works. If you have any questions, let's talk. I think it's still pretty cryptic."

"If you think it's cryptic," interjected the zoning appeals planner, who was standing nearby and had overheard this, "just think what developers and neighborhood people will think!"

Both planners shook their heads and laughed, since the problem was all too plain: the arrows on the design review flow chart seemed to run everywhere. The chart was no doubt correct, but it did look complicated.

I recalled my first interview with the zoning appeals planner. Probing with a deliberately leading question, I had asked, "But what influence can you have in the process if everything's written down as public information, if it's all clear there on the page?"

The zoning appeals planner had grinned: "But that's just it! The process is not clear! And that's where I come in . . ." The architect-planner developed the point further:

> Where I worked before, the planning director wanted to adopt a new "policy and procedures" document that would have every last item defined. We were going to get it all clear. The whole staff spent a lot of time writing that, trying to get all the elements and subsections and so on clearly defined . . . But it was chaos. Once we had the document, everyone fought about what each item meant . . .

So clarity, apparently, has its limits!

Different actors, different strategies

Planning staff point almost poignantly to the different issues that arise as they work with developers and neighborhood residents. The candor of one planning director is worth quoting at length:

> It's easy to sit down with developers or their lawyers. They're a known quantity. They want to meet. There's a common language – say, of zoning – and they know it, along with the technical issues. And they speak with one voice (although that's not to say that we don't play off the architect and the developer at times – we'll push the developer, for example, and the architect is happy because he agrees with us) . . .
>
> But then there's the community. With the neighbors, there's no consistency. One week one group comes in, and the next week it's another. It's hard if there's no consistent view. One group's worried about traffic; the other group's not worried about traffic but about shadows. There isn't one point of view there. They also don't know the process (though there are cases where there are too many experts).
>
> So at the staff level (as opposed to planning board meetings) we usually don't deal with both developers and neighbors simultaneously.

Although these comments may distress advocates of neighborhood power, they say much about the practical situation in which the director finds himself.

All people may be created equal, but when they walk into the planning department, they are simply not all the same. This director suggests that getting all the involved parties together around the table in the planners' conference room is not an obviously good idea, for several reasons. (It is, however, an idea we shall consider more closely below.)

First, the director suggests, planners generally know what to expect from developers; the developers' interests are often clearer than the neighbors', and project proponents may actually want to meet with the staff. Neighborhood residents may be less likely to treat planners as potential allies; after all, the planners are not the decision makers, and the decision makers can often easily ignore the planners' recommendations. Because developers may cultivate good relations with planning staff (this is in part their business, after all), while neighborhood groups do not, local planning staff may find meetings with developers relatively cordial and familiar, but meetings with neighborhood activists more guarded and uncertain.

Second, the planning director suggests that planners and developers often share a common professional language. They can pinpoint technical and regulatory issues and know that both sides understand what is being said. But on any given project, he implies, he may need to teach the special terms of the local zoning code to affected neighbors before they can really get to the issues at hand.

The planning director makes a third point. Developers speak with one voice; neighbors do not. When planners listen to developers talk, they know whom they're listening to, and they know what they're likely to hear repeated, elaborated, defended, or qualified next week. When planners listen to neighborhood residents, though, this director suggests, they can't be so sure how strongly to trust what they hear. "Who really speaks for the neighborhood?" the director wonders.

Planners must make practical judgments about who represents affected residents and about how to interpret their concerns. This director implies, therefore, that until planners find a way to identify "the neighborhood's voice," the problems of conducting joint mediated negotiations between developers and neighbors are likely to seem insurmountable. We return to this issue of representation below.

Inequalities of information, expertise, and financing

What about imbalances of power? Developers, typically, initiate site developments. Planners respond. Neighbors, if they are involved at all, then try to respond to both. Developers have financing and capital to invest; neighbors have voluntary associations and not capital, but lungs. Developers hire expertise; neighborhood groups borrow it. Developers typically have economic resources; neighbors often have time, but not always the staying power to turn that time into real negotiating power.

Where power relations are unbalanced, must mediated negotiation simply lead to coopting the weaker party? No, because, as we shall see below, mediated negotiation is not a gimmick or a recipe; it is a practical and political strategy to be applied in ways that address the specific relations of power at hand.

When either developers or neighborhood groups are so strong that they need not negotiate, mediated negotiation is irrelevant, and other political strategies are more appropriate. But when both developers and neighbors want to negotiate, planners can act both as mediators, assisting the negotiations, and as interested negotiators themselves. But how is this possible? What strategies can planners use?

PLANNERS' STRATEGIES: SIX WAYS TO MEDIATE LOCAL LAND-USE CONFLICTS

Consider the following six mediated-negotiation strategies that planning staff can utilize in the face of local land-use conflicts. They are *mediated* strategies because planners employ them to assure that the interests of the major parties legitimately come into play. They are *negotiation* strategies because (except for the first) they focus attention on the informal negotiations that may produce viable agreements even before formal decision-making boards meet.

Strategy one: The facts! The rules! (The planner as regulator)

The first strategy is a traditional response, pristine in its simplicity, but obviously more complex in practice. A young planner who handles zoning appeals and design review says:

> I see my role often as a fact finder so that the planning board can evaluate this project and form a recommendation; whether it's design review, special permits, or variances, you still need lots of facts . . .

Here of course is the clearest echo of the planner as technician and bureaucrat; the planner processes information and someone else takes responsibility for making decisions. But the echo quickly fades. A moment later, this planner continues,

> Our role is to listen to the neighbors, to be able to say to the board, "Okay, this project meets the technical requirements but there will be impacts . . ." The relief will usually then be granted, but with conditions . . . We'll ask for as much in the way of conditions as we think necessary for the legitimate protection of the neighborhood. The question is, is there a legitimate basis for complaint? And it's not just a matter of complaint, but of the merits.

This planner's role is much more complex than that of fact finder; it is virtually judicial in character. He implies, essentially, "I'm not just a bureaucrat, I'm a professional. I need to think not only about the technical requirements, but about what's legitimate protection for the neighbors. Now I have to think about the merits!" Thinking about the merits, though, does not yet mean thinking about politics, the feelings of other agencies, the chaos at community meetings – it means making professional judgments and then recommending to the planning board the conditions that should be attached to the permits.

Consider now a slightly more complex strategy.

Strategy two: Premediate and negotiate – representing concerns

When developers meet with planners to discuss project proposals, neighborhood representatives rarely join them. Yet planners might nevertheless speak *for* neighborhood concerns as well as *about* them. A planning director in a municipality where neighborhood groups are well-organized, vocal, and influential notes,

> We temper our recommendations to developers. While we might accept A, the neighbors want D, and so we'll tell the developers to think about something in the middle – if they can make it work.

Here, the planner anticipates the concerns of affected residents and changes the informal staff recommendation accordingly to search for an acceptable compromise with the developers. He explains,

> What we do is premediate rather than mediate after the fact. We project people's concerns and then raise them; so we do more before [explicit conflict arises] . . . The only other way we step in and mediate, later, is when we support changes to be made in a project, changes that consider the neighbors' views; but that's later, after the public hearing . . .

Unlike the planner-regulator quoted above, this planning director relies on far more than his professional judgment when he meets with a developer. He will negotiate to reach project outcomes that satisfy local statutes, professional standards, and the interests of affected residents as well. His calculation is not only judicial, but explicitly political. He anticipates the concerns of interested community members. So he seeks to represent neighborhood interests – without neighborhood representatives.

Such premediation – articulating others' concerns well before they can erupt into overt conflict – involves a host of political, strategic, and ethical issues. What relationships does the planner have with neighborhood groups? In what senses can the planner "know what the community wants"? To which "key actors" might the planner "steer" the developer? How much information and how much advice should the planner give, or withhold?

Such questions arise whether or not project developers ever meet with neighbors. In many cases, where "neighborhoods" are sprawling residential areas, and where "the interests of the neighbors" seem most difficult to represent through actual neighborhood representatives, the planners' premediation may be the only mediation that takes place.

Strategy three: Let them meet – the planner as a resource

The planner's influence might be used in still other ways. The director continues:

> Regardless of how our first meeting with a developer goes, we recommend to them that they meet with neighbors and the neighbors' representatives [on the permit-granting board]. We usually can give the developer a good inkling about what to expect both professionally and politically. The same elected representative might say that a project is "okay" professionally, but not "okay" for them in their elected capacity. We try to encourage back and forth meetings . . .

The director, then, regularly takes the pulse of neighborhood groups and elected representatives. Working in city hall has its advantages: "We'll discuss a project with the representatives; we see them so much here, just in the halls, and they ask us to let them know what's happening in their parts of the city." So the director listens to the developers, listens to the neighbors, and "encourages back and forth meetings."

A planning director who seldom met jointly with neighbors and developers had an acute sense of other strategies he used:

> We . . . urge the applicants, the developers, to deal directly with the neighborhood for several reasons: First, if the neighbors are confronted at a hearing with glossy plans, they'll think it's all a *fait accompli*; so they'll just adopt the "guns blazing, full charge ahead" strategy, since they think it'll just be a "yea" or "nay" decision. Second, we tell

them to talk to the neighbors since if they can come up with something that the neighbors will "okay," it'll be easier at the board of appeals. Third, we try to get them to meet one on one, or maybe as a group, but in as deinstitutionalized a way as possible, informally. We try to get the developers to sell their case that way; it'll get a much better hearing than at the big formal public hearings.

But why should planners be reluctant to convene joint negotiating sessions between developers and neighbors, yet still be willing to encourage both parties to meet on their own? Why don't these planners embrace opportunities to mediate local land-use conflicts face to face? One planner could hardly imagine such a mediating role:

> Work as a neutral between developers and neighbors? I don't know how I'd approach it. I'd just answer questions, suggest what could be done, and so on. That's what our role should be – although we should reach compromises between developers and neighbors. But we have to work within the rules – that's my reference point – to say what the rules of the game are; that's the job.

This planner's image of a "neutral" between disputing parties is less that of a mediator facilitating agreement than it is of a referee in a boxing match. The referee assures that the rules are followed, but the antagonists might still kill each other. No wonder planners might find this image of mediation unattractive!

A senior planner envisions further complications:

> If I could be assured I could be wholly independent, then I could mediate – but I still have to pay my bills ... The planning department always has some vested interests, as much as we try to stay objective, independent ... I work for a mayor, for the elected representatives, for 14 committees ... So there's always the question of compromise on my part: if the mayor says, "Tell me how to make this project work," for example. It took me a long time before I was able to say, "I'm going to have to say no." We have a very strong mayor ...

Strategy four: Perform shuttle diplomacy – probe and advise both sides

A planning director proposes another way to facilitate developer–neighbor negotiations:

> I feel more comfortable in shuttle diplomacy, if you will; trying to get the neighbors' concerns on the table, to get the developers to deal with them ... I'd rather bounce ideas off each side individually than be caught in the middle if they're both there. If both sides are there, I'm less likely to give my own ideas than if I'm alone with each of them.

Shuttle diplomacy, this director suggests, allows planners to address the concerns of each party in a professionally effective way. He explains:

> If I'm with the developer, I feel I can make a much more extreme proposal – "knock off three stories" – but I wouldn't dare say that if neighbors were there. The neighbors would be likely to pick up and run with it, and it could damage the negotiations rather than help them ... I'm willing to back off on an issue if the developer has a good argument, but the neighborhood might not, and then they might use my point as a club to hit the developer with: "Well, the planning director suggested that; it must be a good idea" ... and then I can't unsay it ...

This planning director is as concerned about how his suggestions, proposals, queries, and arguments will be understood and used as he is about what ought to be altered in the project at hand. He recognizes clearly that when he talks he acts politically and inevitably fuels one argument or another. He not only conveys information in talking, but he acts practically, influentially. He focuses attention on specific problems, shapes future agendas, legitimates a point of view, and suggests lines of further argument.

The director continued,

> I might not want to concede to a developer that there won't be a traffic problem, because I want to push him to relieve a problem or a perceived problem ... but I could say to the neighbors aside, "Look, this will be no big deal; it'll be five trips, not fifty." I can say that in a private meeting, but in a public meeting if I say it to a neighborhood representative I'm insulting him, even if the developer snickers silently ... So I lose my ability to be frank with both sides if we're all together. Not that this should be completely shuttle diplomacy, but it has its place.

These comments suggest that planning staff can certainly mediate conflicts in local permitting processes, if not in ways that mediators are thought typically to act. The planners may not be independent third parties who assist developers and neighbors in face-to-face meetings to reach development agreements – but they might still mediate such conflicts as "shuttle diplomats."

Strategy five: Active and interested mediation – thriving as a non-neutral

We can consider a case that involves not a zoning appeal but a rezoning proposal. One planner, who had earlier worked as a community organizer, had convened a working group of five community representatives and five local business representatives to draft a rezoning proposal for a large stretch of the major arterial street in their municipality. She considers her work on that project a kind of mediation and reflects about how she as a planner acts as a mediator, dealing with substantive and affective issues alike:

> Am I in a position of having to think about everyone's interest and yet being trusted by no one? Sure, all the time. But I've been in this job for seven years, and I have a reputation that's good, fortunately . . . Trust is an issue of your integrity and planning process. I talk to people a lot; communication is a big part of it . . . My approach is to let people let off steam – let them say negative things about other people to me, and then in a different conversation at another time, I'll be sure to say something positive about that person – to try to let them feel that they can say whatever they want to me, and to try to confront them with the fact that the other person isn't just out to ruin the process. But I'd do that in another conversation; I let them let off steam if they're angry.

This planner is well aware that distrust on all sides is an abiding issue, so she tries to build trust as she works. She works to assure others that she will listen to them and more; that she will acknowledge and respect their thoughts and feelings, whatever they have to say. She pays attention first to the person, then to the words. Then, as she establishes trust with her committee members and with others, she can also make sure, carefully, that real evidence is not ignored.

She realizes that anger makes its own demands, so she responds with an interested patience. She seeks throughout to mediate the conflicting interests of the groups with whom she's working:

> I also make a point to tell each side the other's concerns – categorically, not with names, but all the other sides' concerns . . . Why's that important? I like to let people anticipate the arguments and prepare a defense, either to stand or fall on its own merits. For people to be surprised is unfortunate. It's better to let people know what's coming so they can build a case. They can hear an objection, if you can retain credibility, and absorb it; but in another setting they might not be able to hear it . . . If they hear an objection first as a surprise, you're likely to get blamed for it. If concerns are raised in an emotional setting, people concentrate more on the emotion than on the substance. This is a concern of mine. In emotional settings, lots gets thrown out, and lots is peripheral, but possibly also central later . . .

This planner is keenly aware that emotion and substance are interwoven, and that planners who focus only upon substance and try to ignore or wish away emotion do so at their own practical peril. Yet she is saying even more.

She knows that in some settings disputing groups can hear objections, understand the points at stake, and address them, while in other settings those points may be lost. She tries to present each side's concerns to the other so that they can be understood and addressed. Anticipating issues is central; learning of important objections late in the process will be mostly emotionally and financially, and planning staff are likely to share the blame. "Why didn't you tell us sooner . . . ?" the refrain is likely to sound.

Consider next, then, this planner-mediator's thoughts about the sort of mediation role she is performing. She continues,

> But what I do is different from the independent mediator model. In a job like mine, you have an on-going relationship with parties in the city. You have more information than a mediator does about the history of various individuals, about participating organizations, about the political history of city agencies, and so on. You also have a vested interest in what happens. You want the process to be credible. You want the product to be successful; in my case I want the city council to adopt the committee's proposal. And you're invested . . . both professionally and emotionally. And then you have an opinion about particular proposals; you're a professional, you should have one, you should

be able to look at a proposal and have an opinion.

Thus, she suggests, mediation has its place in local land-use conflicts, but the "rules of the game" will not be those that labor mediators follow. Indeed, planners who now mediate local land-use conflicts are not waiting for someone else to write the rules of the game, they are writing them themselves.

Strategy six: Split the job – you mediate, I'll negotiate

Consider finally a planning strategy that promotes face-to-face mediation with planning staff at the table – but as negotiators or advisors, not as mediators. A planning director explains:

> There's another way we deal with these conflicts; we might involve a local planning board member. For example, if there's a sophisticated neighborhood group that's well organized, we've brought in an architect from the board who's as good with words as he is with his pencil. . . . The chair of the board might ask the board member to be a liaison to the neighborhood, say, and sometimes he'll talk just to the neighbors, sometimes with both . . .

Here the "process manager" comes from the planning board with highly developed "communications skills." How does the planner feel in these situations?

> It's more comfortable from my point of view, and the citizens', to have a board member in the convening role. I'm still a hired hand. It seems more appropriate in a negotiating situation to have a citizen in that role and not an employee . . . Since they've come from the neighborhoods, a board member is in a better position to bring neighbors and developers together – if they behave properly. Some board members are good communicators; some are more dynamic than others in pressing for specific solutions.

This planner identifies so strongly with the professional and political mandate of his position that he cannot imagine a role as neutral convener or mediator of neighborhood–developer negotiations. But that does not prevent mediation; it means rather that the planner retains a substantively interested posture while another party, here a planning board member, convenes informal, but organized, project negotiations between developers and neighbors. This planner's example makes the point:

> Take the example of the Mayfair Hospital site. The hospital was going to close, and the neighbors and the planning board were concerned about what might happen with the site. So Jan from the planning board got involved with the hospital and the neighborhood to look at the possibilities. Both the neighbors and the hospital set up re-use committees, and Jan and I went to the meetings. There was widespread agreement that the best use of the site would be residential – the neighbors definitely preferred that to an institutional use – but then there was a lot of haggling over scale, density, and so on. Ultimately, a special zoning district was proposed that included the site; the neighbors supported it, and it went to [the elected representatives] where they voted to rezone the several acres involved . . .

When local planners feel they cannot mediate disputes themselves, then one strategy may be to search for informal, most likely volunteer, mediators. These ad hoc mediators might be "borrowed" from respected local institutions, and their facilitation of meetings between disputing parties might allow planning staff to participate as professionally interested parties concerned with the site in question.

Table 1 summarizes the six approaches presented. Together, these approaches form a repertoire of strategies that land-use planners can use to encourage mediated negotiations in the face of conflicts in local zoning, special permit, and design review processes. To refine these strategies, local planning staff can build upon several basic theories and techniques of conflict resolution. Consider now the distinctive competences and sensitivities required by these strategies.

THE EMOTIONAL COMPLEXITY OF MEDIATED-NEGOTIATION STRATEGIES

More than a lack of independence keeps planners from easily adopting roles as mediators. The emotional complexity of the mediating role makes quite different demands upon planners than those that they have traditionally been prepared to meet. The community-organizer-turned-planner makes the point brilliantly:

> In the middle, you get all the flak. You're the release valve. You're seen as having some power, and you do have some . . . Look, if you have a financial interest in a project, or an emotional one, you want the person in the middle to care about your point of view, and if you don't think they do, you'll be angry!
>
> ["So when planners try to be professional by appearing detached, objective, does it get people angry at them?" I asked.]
>
> Sure!

This comment cuts to the heart of planners' professional identities. Must "professional," "objective," and "detached" be synonymous? If so, this planner suggests, then planners' own striving for an independent professionalism will fuel the anger, resentment, and suspicion of the same people those planners presume to serve!

Thus we can understand the caution with which a planner speaks of his way of handling emotional participants in public hearings:

> How do I deal with people's anger? I try to keep cool, but occasionally I get irritated. But that's how we're expected to behave, to be rational. It's all right for citizens to be irrational, but not the staff!

How does one keep cool, be rational, and still respond to the claims of an emotional public at formal hearings? This planner elaborates:

> It's one thing to begin the discussion of a project [to present our analysis] and anticipate problems. But it's another thing to *rebut* a neighborhood resident in public in a gentle way . . . Part of the problem is that if you antagonize people it'll haunt you in the future . . . We're here for the long haul, and we have to try to maintain our credibility . . .

The planner's problem here is precisely *not* the facts of the case: the facts themselves may be clear enough. But how should the planner present the analysis that he feels must be made and how should he decide which arguments to make and which to hold back at a given time?

> The biggest problem I have in the board meetings is when to respond and when to keep quiet. In a hearing, for example, I can't possibly respond to all the accusations and issues that come up. So I have to pick a direction, to deal with a generally felt concern. It's just not effective to enter into a debate on each point in turn; it's better to clarify things, to explain what's misunderstood . . .

This planner does much more than simply recapitulate facts. He tries to avoid an adversarial posture, even when he feels the situation is quite conflictual. He listens as much to the individuals and their concern as he does to each point. He knows that points and demands and positions may change as issues are clarified, but that if he cannot respond to people's concerns, he's in some trouble. Because he and his staff are there "for the long haul," he wants to be able to work with neighbors, community leaders, and elected representatives alike not just now but in the future as well. How he relates to the parties involved in local disputes, he suggests, is as important as what he has to say.

Another planner points to the skills involved:

> Whom would I try to hire to deal with such conflicts? I'd look for someone who's a careful listener, someone who's good at explaining a position coherently, succinctly, quietly, in a calm tone . . . someone who could hear a point, understand it whether he or she agreed with it or not, and then verbalize a clear, concise response. Most people though – myself included – try to jump the gun and answer before it's appropriate. So I want someone who's able to stay cool and stay on the issues . . .

A community development director first

Table 1 A repertoire of mediated-negotiation strategies used by local land-use planners in permitting processes

1	The Facts! The Rules! (The Planner as Regulator)
2	Pre-Mediate and Negotiate: Representing Concerns
3	Let Them Meet: The Planner as a Resource . . .
4	Perform Shuttle Diplomacy: Probe and Advise Both Sides
5	Active and Interested Mediation – Thriving as a None-neutral
6	Split the Job: You Mediate, I'll Negotiate

mentions "a good listener" and then elaborates:

> [To deal with these conflicting situations I'd want to hire staff] who won't say, "I know best," who won't get people's backs up just by their style. I'd want someone with some openness, with a sense of how things work who won't accept everything, but who won't offend people. They have to have critical judgment – to leave doors open, to give people a sense of involvement and a sense of the feasible – [someone who] can't be convinced of something that's not likely to work, just for the sake of getting agreement . . .

This planning director also points to the balance necessary between what planners say and how they say it. The "how" counts; he doesn't want staff who will "get people's backs up," "offend people," and not communicate an openness to others' concerns. Nor does he want someone who will sacrifice project viability for the temporary comfort of agreement. He asks for substantive judgment and the skills to manage a process.

Referring to the demands of working and negotiating with developers as they navigate the approval process, the director stresses the role of diplomacy:

> We [planners] have access to information, to resources, to skills . . . so developers usually want to work with us. They have certain problems getting through the process . . . so we'll go to them and ask, "What do you want?" and we'll start a process of meetings . . . It's diplomacy; that's the real work. You have to have the technical skill . . . but that's the first 25 percent. The next 75 percent is diplomacy, working through the process.

Percentages aside, the point remains. To the extent that planning practitioners and educators focus predominantly upon facts, rules, likely consequences, and mitigation measures, they may fail to attend to the pressing emotional and communicative dimensions of local land-use conflicts. Because the planning profession has not traditionally embraced the diplomat's skills, it should surprise no one that practicing planners envision mediating roles with more reticence than relish.

In the next section, we turn to administrative and political questions. What, initially, can be done in planning organizations to improve planners' abilities to mediate local land-use negotiations successfully? What about imbalances of power?

ADMINISTRATIVE IMPLICATIONS FOR PLANNING ORGANIZATIONS

What does this analysis imply for policymakers and planners who wish to build options for mediation into local review processes? Mediation may offer several opportunities, under conditions of interdependent power: a shift from adversarial to collaborative problem-solving; voluntary development controls and agreements; improved city–developer–neighborhood relationships enabling early and effective reviews of future projects; more effective neighborhood voice; and joint gains ("both gain" outcomes) for the municipality, neighbors, and developers alike. Such opportunities present themselves *only* when no single party is so dominant that it need not negotiate at all, that it is likely simply to get what it wants in any case.

Planners already use the strategies reviewed in diverse settings. Which strategy a planner uses, and at which times, depends largely on practical judgment: What skills does the planner have? How willing are developers or neighbors, or other agency staff, to meet jointly? Does enough time exist to allow early, joint meetings? Are the practical and political alternatives of any one party so attractive that they see no point in mediated negotiations?

No strategy is likely to be desirable in all circumstances, so no one approach will provide the model to formalize into new zoning or permitting procedures. But to say that we should not formalize these strategies does not mean that we cannot regularly use them. How, then, can planners apply the mediated-negotiation strategies in local zoning, permitting, and design review processes?

First, planning staff must distinguish clearly the two complementary but distinct mandates they typically must serve: to press professionally, and thus to negotiate, for particular substantive goals (design quality or affordable housing, for example), and to enable a participatory process that gives voice to affected parties; thus, like mediators, to facilitate negotiations between disputants.

Second, planning staff need to adopt, administratively if not formally, a goal of supplementing (not substituting for) formal permitting processes with mediated negotiations: attempting to craft workable and voluntary tentative agreements before formal hearing dates.

Third, planning staff should examine each of the strategies reviewed here. They need to determine how each could work, given the size of their agency, their zoning and related by-laws, the political and institutional history of elected officials, neighborhood groups, and other agencies. Planning staff must ask which skills and competencies they need to develop to employ each of these strategies appropriately.

Fourth, planning staff must be able to show others – developers, neighborhood groups, public works department staff, elected and appointed officials – how and when mediated negotiations can lead to "both gain" outcomes and so improve the local land-use planning and development process. Planners also have to be clear about what mediated negotiation will not do: it will not solve problems of radically unbalanced power, for example. It can, however, refine an adversarial process into a partially collaborative one. It will not solve problems of basic rights, but it can often expand the range of affected parties' interests that developers will take into account. Mediated negotiations will neither necessarily co-opt project opponents (as skeptical neighborhood residents might suspect) nor stall proposals and projects (as skeptical developers and builders might suspect). Yet when each side can effectively threaten the other, when each side's interests depend upon the other's actions, then mediated negotiations may enable voluntary agreements, incorporate measures of control on both sides, allow "both gain" trades to be achieved, and do so more efficiently for all sides than pursuing alternative strategies (e.g. going to court or, sometimes, community organizing).

Fifth, planners need administratively to create an organized process to match incoming projects with one or more of the mediated-negotiation strategies and to review their progress as they go along. With staff training in negotiation and mediation principles and techniques, planning departments would be better able to carry out these strategies effectively once they have organized administratively to promote them.

DEALING WITH POWER IMBALANCES: CAN THE SIX STRATEGIES MAKE A DIFFERENCE?

The six strategies we have considered are hardly "neutral." Planners who adopt them inevitably either perpetuate or challenge existing inequalities of information, expertise, political access, and opportunity. Consider each approach, briefly, in turn.

To provide only the facts, or information about procedures, to whomever asks for them seems to treat everyone equally. Yet where severe inequalities exist, to treat the strong and the weak alike only ensures that the strong remain strong, the weak remain weak. The planner who pretends to act as a neutral regulator may sound egalitarian but nevertheless act, ironically, to perpetuate and ignore existing inequalities.

The premediation strategy can involve substantial discretion on the part of the planning staff. If the staff fail to put the interests of weaker parties "on the negotiating table," then here, too, inequalities will be perpetuated, not mitigated. If the staff do defend neighborhood interests in the development negotiations, they may challenge existing inequalities. But which "neighborhood interests" should the planning staff identify? How should neighborhoods – especially weakly organized ones – be represented? These questions are both practical and theoretical and they have no purely technical, "recipe"-like answers.

At first glance, the strategy of letting developers and neighbors meet without an active staff presence seems only to reproduce the initial strengths of the parties. Yet depending on how the planning staff intervene, one party or another may be strengthened or weakened. At times planners have helped developers anticipate and ultimately evade the concerns of citizens who opposed projects. Yet planners may also provide expertise, access, information, and so on to strengthen weaker citizens' positions.

The same discretion exists for planning staff who act as shuttle diplomats. Here a planner may counsel weaker parties to help them both before and during actual negotiations by identifying concerns that might effectively be raised, experts or other influentials who might be called upon, prenegotiation strategies and tactics to be employed, and so on. The shuttle diplomat need not appear neutral to all parties but he or she does need to appear useful to, or needed by, those parties. Planners who act as "interested mediators" face many of the same problems and opportunities that shuttle diplomats confront. In addition, though, the activist mediator may risk being perceived by planning board members, officials, or elected representatives as making deals that preempt their own formal authority. Thus the invisibility of the shuttle diplomat has its advantages; the planners can give counsel discreetly, suggesting packages and "deals" but avoiding the glare – and the heat – of the limelight.

Finally, the strategy of separating mediation and negotiation functions also involves substantial staff discretion. Here, too, the ways that mediators and negotiators consider the interests and enable the voice of weaker parties will affect existing power imbalances.

Because negotiations always involve questions of relative power, they depend heavily upon the parties' *prenegotiation* work of marshalling resources, developing options, and organizing support. Thus politically astute planners need both organizing and mediated-negotiation skills if conflicts are to be addressed without pretending that structural power imbalances just do not exist. Finally, note that a planner who explicitly calls everyone's attention to class-based power imbalances, for example, may not obviously do better in any practical sense of the word than an activist mediator who knows the same thing and acts on it in just the same ways without explicitly framing the planning negotiations in those terms.

CONCLUSION

The repertoire of mediated-negotiation strategies inevitably requires that planners exercise practical judgment, both politically and ethically. These judgments involve who is and who is not invited to meetings; where, when, and which meetings are held; what issues should and should not appear on agendas; whose concerns are and are not acknowledged; how interventionist the planner's role is; and so on.

In local planning processes, then, planners often have the administrative discretion not only to mediate among conflicting parties, but to negotiate as interested parties themselves. Planning staff can routinely engage in the complementary tasks of supporting organizing efforts, negotiating, and mediating. In these ways, local planners can use a range of mediated-negotiation strategies to address practically existing power imbalances of access, information, class, and expertise that perpetually threaten the quality of local planning outcomes.

Mediated negotiations in local permitting processes will, of course, not resolve the structural problems of our society. Yet when local conflicts involve multiple issues, when differences in interests can be exploited by trading to achieve joint gains, and when diverse interests rather than fundamental rights are at stake, mediated-negotiation strategies for planners make good sense, politically, ethically, and practically.

PAUL DAVIDOFF

"Advocacy and Pluralism in Planning"

***Journal of the American Institute of Planners* (1965)**

Editors' introduction Each year at the annual meeting of the Association of Collegiate Schools of Planning (ACSP), professors of urban planning present the Paul Davidoff Award to a city planning scholar whose work exemplifies the practice and ideals of the professor/lawyer/activist who is the author of this selection. It is an honor to receive the Davidoff award, because Davidoff exemplified professional commitment to vigorous advocacy on behalf of the less fortunate members of society.

During the 1960s, Davidoff, a lawyer and city planner, taught city planning students at Hunter College and simultaneously fought successfully to get racially integrated low-income housing built in exclusive white suburbs. This experience as an advocate for low-income minority residents shaped his view of what city planning should be like.

Unlike the "advocacy planning" Davidoff proposes, most city or town planning is performed by a single local government agency which develops plans which, it feels, will best serve the welfare of the whole community, not of individual interest groups such as organizations of homeless people, merchants, environmentalists or bicycle enthusiasts. While city planning commissions may explore many alternatives and consider conflicting interest group demands before finalizing plans, generally they end up with a single unitary plan.

Davidoff's vision for how planning might be structured is quite different. He argues that different groups in society have different interests, which would result in fundamentally different plans if they were recognized. Business elites and other articulate, wealthy, and powerful groups have the skill and resources to shape city plans to serve their interests. But what about the poor and powerless? Davidoff argued that there should be planners acting as *advocates* articulating the interests of these and other groups much as a lawyer represents a client. For example, a planner might develop and advocate for a plan that would meet the needs of poor West Indian residents of London's Brixton neighborhood. Another planner might develop a different plan representing the point of view of shopkeepers in the same area. And yet another might work with Brixton environmentalists to develop and advocate for a plan for the Brixton area incorporating Sim Van der Ryn and Stuart Cowan's ecological design principals (p. 519) or the kind of sustainable urban development urged by Stephen Wheeler (p. 434). Confronted with these different "visions" (and the empirical data which makes the most compelling case for them), the local planning commission could weigh the merits of the competing plans much as a court weighs evidence and conflicting characterizations of a legal case by competing lawyers. Davidoff believed that the plan that would emerge from such a process would be better than a plan prepared by planning department staff without the interplay of competing advocate planners. And, Davidoff reasoned, the needs of the poor and powerless would be better met in city plans if – a big if – they were adequately represented by advocacy planners speaking on their behalf.

Davidoff's view of planning profoundly influenced activist planners of the 1960s and 1970s, many of whom defined themselves as advocacy planners and developed plans to meet the needs and interests of underrepresented groups, with some notable successes. "Equity planners" today continue this tradition.

Compare Davidoff's humanistic, grassroots, pluralistic approach to city planning with Le Corbusier's brilliant, but elitist vision of a cadre of CIAM architects to impose on the fabric of cities the forms they felt modern machine culture demanded (p. 336). Compare Davidoff's views with Forester's comments on how planners working within the system can use their influence to empower stakeholders in the planning process (p. 410). Compare the advocacy planning approach to strategies to empower communities to reach the highest possible level on the "ladder" of citizen participation that Sherry Arnstein has developed (p. 240). Review John Mollenkopf's description of how urban political power is actually exercised to understand how stakeholders represented by advocacy planners might better fit into pluralist local decision-making structures and urban regimes.

Norman Krumholz and John Forester describe Krumholz's experience as the planning director of Cleveland, Ohio, in *Making Equity Planning Work* (Philadelphia: Temple University Press, 1990). Krumholz worked hard to make city planning responsive to the needs of the poor and powerless. As a planning professor at Cleveland State University he continues to develop the theory and practice of equity planning. Chester Hartman's *The Transformation of San Francisco* (Totowa, NJ: Rowman and Allanheld, 1984) describes how advocacy planners and lawyers fought to make urban renewal more responsive to very low income residents. For an application of advocacy planning to women, see Jacqueline Levitt, "Feminist Advocacy Planning in the 1980s" in Barry Checkoway (ed.), *Strategic Perspectives in Planning Practice* (Lexington, MA: Lexington Books, 1986). For a radical critique of advocacy planning, see Frances Fox Piven, "Whom does the Advocate Planner Serve?" in Richard A. Cloward and Frances Fox Piven, *The Politics of Turmoil* (New York: Vintage, 1965). Piven sees advocacy planners as unwitting dupes of the system. She argues that angry and potentially violent groups will obtain more political leverage bargaining directly for themselves without professional intermediaries. She feels they need power, not plans.

The present can become an epoch in which the dreams of the past for an enlightened and just democracy are turned into a reality. The massing of voices protesting racial discrimination have roused this nation to the need to rectify racial and other social injustices. The adoption by Congress of a host of welfare measures and the Supreme Court's specification of the meaning of equal protection by law both reveal the response to protest and open the way for the vast changes still required.

The just demand for political and social equality on the part of the Negro and the impoverished requires the public to establish the bases for a society affording equal opportunity to all citizens. The compelling need for intelligent planning, for specification of new social goals and the means for achieving them, is manifest. The society of the future will be an urban one, and city planners will help to give it shape and content.

The prospect for future planning is that of a practice which openly invites political and social values to be examined and debated. Acceptance of this position means rejection of prescriptions for planning which would have the planner act solely as a technician. It has been argued that technical studies to enlarge the information available to decision makers must take precedence over statements of goals and ideals:

> We have suggested that, at least in part, the city planner is better advised to start from research into the functional aspects of cities than from his own estimation of the values which he is

attempting to maximize. This suggestion springs from a conviction that at this juncture the implications of many planning decisions are poorly understood, and that no certain means are at hand by which values can be measured, ranked, and translated into the design of a metropolitan system.

While acknowledging the need for humility and openness in the adoption of social goals, this statement amounts to an attempt to eliminate, or sharply reduce, the unique contribution planning can make: understanding the functional aspects of the city and recommending appropriate future action to improve the urban condition.

Another argument that attempts to reduce the importance of attitudes and values in planning and other policy sciences is that the major public questions are themselves matters of choice between technical methods of solution. Dahl and Lindblom put forth this position at the beginning of their important textbook *Politics, Economics, and Welfare*:

> In economic organization and reform, the "great issues" are no longer the great issues, if they ever were. It has become increasingly difficult for thoughtful men to find meaningful alternatives posed in the traditional choices between socialism and capitalism, planning and the free market, regulation and laissez faire, for they find their actual choices neither so simple nor so grand. Not so simple, because economic organization poses knotty problems that can only be solved by painstaking attention to technical details – how else, for example, can inflation be controlled? Nor so grand, because, at least in the Western world, most people neither can nor wish to experiment with the whole pattern of socio-economic organization to attain goals more easily won. If, for example, taxation will serve the purpose, why "abolish the wages system" to ameliorate income inequality?

These words were written in the early 1950s and express the spirit of that decade more than that of the 1960s. They suggest that the major battles have been fought. But the "great issues" in economic organization, those revolving around the central issue of the nature of distributive justice, have yet to be settled. The world is still in turmoil over the way in which the resources of nations are to be distributed. The justice of the present social allocation of wealth, knowledge, skill, and other social goods is clearly in debate. Solutions to questions about the share of wealth and other social commodities that should go to different classes cannot be technically derived; they must arise from social attitudes.

Appropriate planning action cannot be prescribed from a position of value neutrality, for prescriptions are based on desired objectives. One conclusion drawn from this assertion is that "values are inescapable elements of any rational decision-making process" and that values held by the planner should be made clear. The implications of that conclusion for planning have been described elsewhere and will not be considered in this article. Here I will say that the planner should do more than explicate the values underlying his prescriptions for courses of action; he should affirm them; he should be an advocate for what he deems proper.

Determinations of what serves the public interest, in a society containing many diverse interest groups, are almost always of a highly contentious nature. In performing its role of prescribing courses of action leading to future desired states, the planning profession must engage itself thoroughly and openly in the contention surrounding political determination. Moreover, planners should be able to engage in the political process as advocates of the interests both of government and of such other groups, organizations, or individuals who are concerned with proposing policies for the future development of the community.

The recommendation that city planners represent and plead the plans of many interest groups is founded upon the need to establish an effective urban democracy, one in which citizens may be able to play an active role in the process of deciding public policy. Appropriate policy in democracy is determined through a process of political debate. The right course of action is always a matter of choice, never of fact. In a bureaucratic age great care must be taken that choices remain in the area of public view and participation.

Urban politics, in an era of increasing government activity in planning and welfare, must balance the demands for ever-increasing central bureaucratic control against the demands for increased concern for the unique requirements of local, specialized interests. The welfare of all and the welfare of minorities are both deserving of support; planning must be so structured and so practiced as to account for this unavoidable bifurcation of the public interest.

The idealized political process in a democracy serves the search for truth in much the same manner as due process in law. Fair notice and hearings, production of supporting evidence, cross-examination, reasoned decision are all means employed to arrive at relative truth: a just decision. Due process and two- (or more) party political contention both rely heavily upon strong advocacy by a professional. The advocate represents an individual, group, or organization. He affirms their position in language understandable to his client and to the decision makers he seeks to convince.

If the planning process is to encourage democratic urban government then it must operate so as to include rather than exclude citizens from participating in the process. "Inclusion" means not only permitting the citizen to be heard. It also means that he be able to become well informed about the underlying reasons for planning proposals, and be able to respond to them in the technical language of professional planners.

A practice that has discouraged full participation by citizens in plan making in the past has been based on what might be called the "*unitary plan*." This is the idea that only one agency in a community should prepare a comprehensive plan; that agency is the city planning commission or department. Why is it that no other organization within a community prepares a plan? Why is only one agency concerned with establishing both general and specific goals for community development, and with proposing the strategies and costs required to effect the goals? Why are there not plural plans?

If the social, economic, and political ramifications of a plan are politically contentious, then why is it that those in opposition to the agency plan do not prepare one of their own? It is interesting to observe that "rational" theories of planning have called for consideration of alternative courses of action by planning agencies. As a matter of rationality it has been argued that all of the alternative choices open as means to the ends ought be examined. But those, including myself, who have recommended agency consideration of alternatives have placed upon the agency planner the burden of inventing "a few representative alternatives." The agency planner has been given the duty of constructing a model of the political spectrum, and charged with sorting out what he conceives to be worthy alternatives. This duty has placed too great a burden on the agency planner, and has failed to provide for the formulation of alternatives by the interest groups who will eventually be affected by the completed plans.

Whereas in a large part of our national and local political practice contention is viewed as healthy, in city planning where a large proportion of the professionals are public employees, contentious criticism has not always been viewed as legitimate. Further, where only government prepares plans, and no minority plans are developed, pressure is often applied to bring all professionals to work for the ends espoused by a public agency. For example, last year a Federal official complained to a meeting of planning professors that the academic planners were not giving enough support to Federal programs. He assumed that every planner should be on the side of the Federal renewal program. Of course government administrators will seek to gain the support of professionals outside of government, but such support should not be expected as a matter of loyalty. In a democratic system opposition to a public agency should be just as normal and appropriate as support. The agency, despite the fact that it is concerned with planning, may be serving undesired ends.

In presenting a plea for plural planning I do not mean to minimize the importance of the obligation of the public planning agency. It must decide upon appropriate future courses of action for the community. But being isolated as the only plan maker in the community, public agencies as well as the public itself may have suffered from incomplete and shallow analysis of potential directions. Lively political dispute aided by plural plans could do much to improve the level of rationality in the process of preparing the public plan.

The advocacy of alternative plans by interest groups outside of government would stimulate city planning in a number of ways. First, it would serve as a means of better informing the public of the alternative choices open, *alternatives strongly supported by their proponents*. In current practice those few agencies which have portrayed alternatives have not been equally enthusiastic about each. A standard reaction to rationalists' prescription for consideration of alternative courses of action has been "it can't be done; how can you expect planners to present alternatives which they don't approve?" The appropriate answer to that question has been that planners, like lawyers, may have a professional obligation to defend positions they oppose. However, in a system of plural planning, the public agency would be relieved of at least some of the burden of presenting alternatives. In plural planning the alternatives would be presented by interest groups differing with the public agency's plan. Such alternatives would represent the deep-seated convictions of their proponents and not just the mental exercises of rational planners seeking to portray the range of choice.

A second way in which advocacy and plural planning would improve planning practice would be in forcing the public agency to compete with other planning groups to win political support. In the absence of opposition or alternative plans presented by interest groups the public agencies have had little incentive to improve the quality of their work or the rate of production of plans. The political consumer has been offered a yes–no ballot in regard to the comprehensive plan; either the public agency's plan was to be adopted or no plan would be adopted.

A third improvement in planning practice which might follow from plural planning would be to force those who have been critical of "establishment" plans to produce superior plans, rather than only to carry out the very essential obligation of criticizing plans deemed improper.

THE PLANNER AS ADVOCATE

Where plural planning is practiced, advocacy becomes the means of professional support for competing claims about how the community should develop. Pluralism in support of political contention describes the process; advocacy describes the role performed by the professional in the process. Where unitary planning prevails, advocacy is not of paramount importance, for there is little or no competition for the plan prepared by the public agency. The concept of advocacy as taken from legal practice implies the opposition of at least two contending viewpoints in an adversary proceeding.

The legal advocate must plead for his own and his client's sense of legal propriety or justice. The planner as advocate would plead for his own and his client's view of the good society. The advocate planner would be more than a provider of information, an analyst of current trends, a simulator of future conditions, and a detailer of means. In addition to carrying out these necessary parts of planning, he would be a *proponent* of specific substantive solutions.

The advocate planner would be responsible to his client and would seek to express his client's views. This does not mean that the planner could not seek to persuade his client. In some situations persuasion might not be necessary, for the planner would have sought out an employer with whom he shared common views about desired social conditions and the means toward them. In fact one of the benefits of advocate planning is the possibility it creates for a planner to find employment with agencies holding values close to his own. Today the agency planner may be dismayed by the positions affirmed by his agency, but there may be no alternative employer.

The advocate planner would be above all a planner. He would be responsible to his client for preparing plans and for all of the other elements comprising the planning process. Whether working for the public agency or for some private organization, the planner would have to prepare plans that take account of the arguments made in other plans. Thus the advocate's plan might have some of the characteristics of a legal brief. It would be a document presenting the facts and reasons for supporting one set of proposals, and facts and reasons indicating the inferiority of counter-proposals. The adversary nature of plural planning might, then, have the beneficial effect of upsetting the tradition of writing plan proposals in terminology which makes them appear self-evident.

A troublesome issue in contemporary planning is that of finding techniques for evaluating alternative plans. Technical devices such as cost–benefit analysis by themselves are of little assistance without the use of means for appraising the values underlying plans. Advocate planning, by making more apparent the values underlying plans, and by making definitions of social costs and benefits more explicit, should greatly assist the process of plan evaluation. Further, it would become clear (as it is not at present) that there are no neutral grounds for evaluating a plan; there are as many evaluative systems as there are value systems.

The adversary nature of plural planning might also have a good effect on the uses of information and research in planning. One of the tasks of the advocate planner in discussing the plans prepared in opposition to his would be to point out the nature of the bias underlying information presented in other plans. In this way, as critic of opposition plans, he would be performing a task similar to the legal technique of cross-examination. While painful to the planner whose bias is exposed (and no planner can be entirely free of bias) the net effect of confrontation between advocates of alternative plans would be more careful and precise research.

Not all the work of an advocate planner would be of an adversary nature. Much of it would be educational. The advocate would have the job of informing other groups, including public agencies, of the conditions, problems, and outlook of the group he represented. Another major educational job would be that of informing his clients of their rights under planning and renewal laws, about the general operations of city government, and of particular programs likely to affect them.

The advocate planner would devote much attention to assisting the client organization to clarify its ideas and to give expression to them. In order to make his client more powerful politically the advocate might also become engaged in expanding the size and scope of his client organization. But the advocate's most important function would be to carry out the planning process for the organization and to argue persuasively in favor of its planning proposals.

Advocacy in planning has already begun to emerge as planning and renewal affect the lives of more and more people. The critics of urban renewal have forced response from the renewal agencies, and the ongoing debate has stimulated needed self-evaluation by public agencies. Much work along the lines of advocate planning has already taken place, but little of it by professional planners. More often the work has been conducted by trained community organizers or by student groups. In at least one instance, however, a planner's professional aid led to the development of an alternative renewal approach, one which will result in the dislocation of far fewer families than originally contemplated.

Pluralism and advocacy are means for stimulating consideration of future conditions by all groups in society. But there is one social group which at present is particularly in need of the assistance of planners. This group includes organizations representing low-income families. At a time when concern for the condition of the poor finds institutionalization in community action programs, it would be appropriate for planners concerned with such groups to find means to plan with them. The plans prepared for these groups would seek to combat poverty and would propose programs affording new and better opportunities to the members of the organization and to families similarly situated. The difficulty in providing adequate planning assistance to organizations representing low-income families may in part be overcome by funds allocated to local antipoverty councils. But these councils are not the only representatives of the poor; other organizations exist and seek help. How can this type of assistance be financed? This question will be examined below, when attention is turned to the means for institutionalizing plural planning.

THE STRUCTURE OF PLANNING

Planning by special interest groups

The local planning process typically includes one or more "citizens'" organizations concerned with the nature of planning in the community. The Workable Program requirement for "citizen participation" has enforced this tradition and brought it to most large communities. The difficulty with current citizen participation programs is that citizens are more often *reacting* to agency programs than proposing their concepts of appropriate goals and future action.

The fact that citizens' organizations have not played a positive role in formulating plans is to some extent a result of both the enlarged role in society played by government bureaucracies and the historic weakness of municipal party politics.

There is something very shameful to our society in the necessity to have organized "citizen participation." Such participation should be the norm in an enlightened democracy. The formalization of citizen participation as a required practice in localities is similar in many respects to totalitarian shows of loyalty to the state by citizen parades.

Will a private group interested in preparing a recommendation for community development be required to carry out its own survey and analysis of the community? The answer would depend upon the quality of the work prepared by the public agency, work which should be public information. In some instances the public agency may not have surveyed or analyzed aspects the private group thinks important; or the public agency's work may reveal strong biases unacceptable to the private group. In any event, the production of a useful plan proposal will require much information concerning the present and predicted conditions in the community. There will be some costs associated with gathering that information, even if it is taken from the public agency. The major cost involved in the preparation of a plan by a private agency would probably be the employment of one or more professional planners.

What organizations might be expected to engage in the plural planning process? The first type that comes to mind are the political parties; but this is clearly an aspirational thought. There is very little evidence that local political organizations have the interest, ability, or concern to establish well-developed programs for their communities. Not all the fault, though, should be placed upon the professional politicians, for the registered members of political parties have not demanded very much, if anything, from them as agents.

Despite the unreality of the wish, the desirability for active participation in the process of planning by the political parties is strong. In an ideal situation local parties would establish political platforms which would contain master plans for community growth and both the majority and minority parties in the legislative branch of government would use such plans as one basis for appraising individual legislative proposals. Further, the local administration would use its planning agency to carry out the plans it proposed to the electorate. This dream will not turn to reality for a long time. In the interim other interest groups must be sought to fill the gap caused by the present inability of political organizations.

The second set of organizations which might be interested in preparing plans for community development are those that represent special interest groups having established views in regard to proper public policy. Such organizations as chambers of commerce, real estate boards, labor organizations, pro- and anti-civil rights groups, and anti-poverty councils come to mind. Groups of this nature have often played parts in the development of community plans, but only in a very few instances have they proposed their own plans.

It must be recognized that there is strong reason operating against commitment to a plan by these organizations. In fact it is the same reason that in part limits the interests of politicians and which limits the potential for planning in our society. The expressed commitment to a particular plan may make it difficult for groups to find means for accommodating their various interests. In other terms, it may be simpler for professionals, politicians, or lobbyists to make deals if they have not laid their cards on the table.

There is a third set of organizations that might be looked to as proponents of plans and to whom the foregoing comments might not apply. These are the ad hoc protest associations which may form in opposition to some proposed policy. An example of such a group is a neighborhood association formed to combat a renewal plan, a zoning change, or the proposed location of a public facility. Such organizations may seek to develop alternative plans, plans which would, if effected, better serve their interests.

From the point of view of effective and rational planning it might be desirable to commence plural planning at the level of city-wide organizations, but a more realistic view is that it will start at the neighborhood level. Certain advantages of this outcome should be noted. Mention was made earlier of tension in government between centralizing and decentralizing forces. The contention aroused by conflict between the central planning agency and the neighborhood organization may indeed be healthy, leading to clearer definition of welfare policies and their relation to the rights of individuals or minority groups.

Who will pay for plural planning? Some organizations have the resources to sponsor the development of a plan. Many groups lack the means. The plight of the relatively indigent association seeking to propose a plan might be analogous to that of the indigent client in search of legal aid. If the idea of plural planning makes sense, then support may be found from foundations or from government. In the beginning it is more likely that some foundation might be willing to experiment with plural planning as a means of making city planning more effective and more democratic. Or the Federal Government might see plural planning, if carried out by local anti-poverty councils, as a strong means of generating local interest in community affairs.

Federal sponsorship of plural planning might be seen as a more effective tool for stimulating involvement of the citizen in the future of his community than are the present types of citizen participation programs. Federal support could only be expected if plural planning were seen, not as a means of combating renewal plans, but as an incentive to local renewal agencies to prepare better plans.

The public planning agency

A major drawback to effective democratic planning practice is the continuation of that non-responsible vestigial institution, the planning commission. If it is agreed that the establishment of both general policies and implementation policies are questions affecting the public interest and that public interest questions should be decided in accord with established democratic practices for decision making, then it is indeed difficult to find convincing reasons for continuing to permit independent commissions to make planning decisions. At an earlier stage in planning the strong arguments of John T. Howard and others in support of commissions may have been persuasive. But it is now more than a decade since Howard made his defense against Robert Walker's position favoring planning as a staff function under the mayor. With the increasing effect planning decisions have upon the lives of citizens the Walker proposal assumes great urgency.

Aside from important questions regarding the propriety of independent agencies which are far removed from public control determining public policy, the failure to place planning decision choices in the hands of elected officials has weakened the ability of professional planners to have their proposals effected. Separating planning from local politics has made it difficult for independent commissions to garner influential political support. The commissions are not responsible directly to the electorate and in turn the electorate is, at best, often indifferent to the planning commission.

During the last decade, in many cities power to alter community development has slipped out of the hands of city planning commissions, assuming they ever held it, and has been transferred to development coordinators. This has weakened the professional planner. Perhaps planners unknowingly contributed to this by their refusal to take concerted action in opposition to the perpetuation of commissions.

Planning commissions are products of the conservative reform movement of the early part of this century. The movement was essentially anti-populist and pro-aristocracy. Politics was viewed as dirty business. The commissions are relics of a not-too-distant past when it was believed that if men of good will discussed a problem thoroughly, certainly the right solution would be forthcoming. We know today, and perhaps it was always known, that there are no right solutions. Proper policy is that which the decision-making unit declares to be proper.

Planning commissions are responsible to no constituency. The members of the commissions, except for their chairman, are seldom known to the public. In general the individual members fail to expose their personal views about policy and prefer to immerse them in group decision. If the members wrote concurring and dissenting opinions, then at least the commissions might stimulate thought about planning issues. It is difficult to comprehend why this aristocratic and undemocratic form of decision making should be continued. The public planning function should be carried out in the executive or legislative office and perhaps in both. There has been some question about which of these branches of government would provide the best home, but there is much reason to believe that both branches would be made more cognizant of planning issues if they were each informed by their own planning staffs. To carry this division further, it would probably be advisable to establish minority and majority planning staffs in the legislative branch.

At the root of my last suggestion is the belief that there is or should be a Republican and Democratic way of viewing city development; that there should be conservative and liberal plans, plans to support the private market, and plans to support greater government control. There are many possible roads for a community to travel and many plans should show them. Explication is required of many alternative futures presented by those sympathetic to the construction of each such future. As indicated earlier, such alternatives are not presented to the public now. Those few reports which do include alternative futures do not speak in terms of interest to the average citizen. They are filled with professional jargon and present sham alternatives. These plans have expressed technical land use alternatives rather than social, economic, or political value alternatives. Both the traditional unitary plans and the new ones that present technical alternatives have limited the public's exposure to the future states that might be achieved. Instead of arousing healthy political contention as diverse comprehensive plans might, these plans have deflated interest.

The independent planning commission and unitary plan practice certainly should not co-exist. Separately they dull the possibility for enlightened political debate; in combination they have made it yet more difficult. But when still another hoary concept of city planning is added to them, such debate becomes practically impossible. This third of a trinity of worn-out notions is that city planning should focus only upon the physical aspects of city development.

AN INCLUSIVE DEFINITION OF THE SCOPE OF PLANNING

The view that equates physical planning with city planning is myopic. It may have had some historic justification, but it is clearly out of place at a time when it is necessary to integrate knowledge and techniques in order to wrestle effectively with the myriad of problems afflicting urban populations.

The city planning profession's historic concern with the physical environment has warped its ability to see physical structures and land as servants to those who use them. Physical relations and conditions have no meaning or quality apart from the way they serve their users. But this is forgotten every time a physical condition is described as good or bad without relation to a specified group of users. High density, low density, green belts, mixed uses, cluster developments, centralized or decentralized business centers are per se neither good nor bad. They describe physical relations or conditions, but take on value only when seen in terms of their social, economic, psychological, physiological, or aesthetic effects upon different users.

The profession's experience with renewal over the past decade has shown the high costs of exclusive concern with physical conditions. It has been found that the allocation of funds for removal of physical blight may not necessarily improve the overall physical condition of a community and may engender such harsh social repercussions as to severely damage both social and economic institutions. Another example of the deficiencies of the physical bias is the assumption of city planners that they could deal with the capital budget as if the physical attributes of a facility could be understood apart from the philosophy and practice of the service conducted within the physical structure. This assumption is open to question. The size, shape, and location of a facility greatly interact with the purpose of the activity the facility houses. Clear examples of this can be seen in public education and in the provision of low cost housing. The racial and other socioeconomic consequences of "physical decisions" such as location of schools and housing projects have been immense, but city planners, while acknowledging the existence of such consequences, have not sought or trained themselves to understand socioeconomic problems, their causes or solutions.

The city planning profession's limited scope has tended to bias strongly many of its recommendations toward perpetuation of existing social and economic practices. Here I am not opposing the outcomes, but the way in which they are developed. Relative ignorance of social and economic methods of analysis has caused planners to propose solutions in the absence of sufficient knowledge of the costs and benefits of proposals upon different sections of the population.

Large expenditures have been made on planning studies of regional transportation needs, for example, but these studies have been conducted in a manner suggesting that different social and economic classes of the population did not have different needs and different abilities to meet them. In the field of housing, to take another example, planners have been hesitant to question the consequences of locating public housing in slum areas. In the field of industrial development, planners have seldom examined the types of jobs the community needs; it has been assumed that one job was about as useful as another. But this may not be the case where a significant sector of the population finds it difficult to get employment.

"Who gets what, when, where, why, and how" are the basic political questions which need to be raised about every allocation of public resources. The questions cannot be answered adequately if land use criteria are the sole or major standards for judgment.

The need to see an element of city development, land use, in broad perspective applies equally well to every other element, such as health, welfare, and recreation. The governing of a city requires an adequate plan for its future. Such a plan loses guiding force and rational basis to the degree that it deals with less than the whole that is of concern to the public.

The implications of the foregoing comments for the practice of city planning are these. First, state planning enabling legislation should be amended to permit planning departments to study and to prepare plans related to any area of public concern. Second, planning education must be redirected so as to provide channels of specialization in different parts of public planning and a core focused upon the planning process. Third, the professional planning association should enlarge its scope so as to not exclude city planners not specializing in physical planning.

A year ago at the AIP convention it was suggested that the AIP Constitution be amended to permit city planning to enlarge its scope to all matters of public concern. Members of the Institute in agreement with this proposal should seek to develop support for it at both the chapter and national level. The Constitution at present states that the Institute's "particular sphere of activity shall be the planning of the unified development of urban communities and their environs and of states, regions and the nation *as expressed through determination of the comprehensive arrangement of land and land occupancy and regulation thereof*." It is time that the AIP delete the words in my italics from its Constitution. The planner limited to such concerns is not a city planner, he is a land planner or a physical planner. A city is its people, their practices, and their political, social, cultural and economic institutions as well as other things. The city planner must comprehend and deal with all these factors.

The new city planner will be concerned with physical planning, economic planning, and social planning. The scope of his work will be no wider than that presently demanded of a mayor or a city councilman. Thus, we cannot argue against an enlarged planning function on grounds that it is too large to handle. The mayor needs assistance; in particular he needs the assistance of a planner, one trained to examine needs and aspirations in terms of both short- and long-term perspectives. In observing the early stages of development of Community Action Programs, it is apparent that our cities are in desperate need of the type of assistance trained planners could offer. Our cities require for their social and economic programs the type of long-range thought and information that have been brought forward in the realm of physical planning. Potential resources must be examined and priorities set.

What I have just proposed does not imply the termination of physical planning, but it does mean that physical planning be seen as part of city planning. Uninhibited by limitations on his work, the city planner will be able to add his expertise to the task of coordinating the operating and capital budgets and to the job of relating effects of each city program upon the others and upon the social, political, and economic resources of the community.

An expanded scope reaching all matters of public concern will make planning not only a more effective administrative tool of local government but it will also bring planning practice closer to the issues of real concern to the citizens. A system of plural city planning probably has a much greater chance for operational success where the focus is on live social and economic questions instead of rather esoteric issues relating to physical norms.

THE EDUCATION OF PLANNERS

Widening the scope of planning to include all areas of concern to government would suggest that city planners must possess a broader knowledge of the structure and forces affecting urban development. In general this would be true. But at present many city planners are specialists in only one or more of the functions of city government. Broadening the scope of planning would require some additional planners who specialize in one or more of the services entailed by the new focus.

A prime purpose of city planning is the coordination of many separate functions. This coordination calls for men holding general knowledge of the many elements comprising the urban community. Educating a man for performing the coordinative role is a difficult job, one not well satisfied by the present tradition of two years of graduate study. Training of urban planners with the skills called for in this article may require both longer graduate study and development of a liberal arts undergraduate program affording an opportunity for holistic understanding of both urban conditions and techniques for analyzing and solving urban problems.

The practice of plural planning requires educating planners who would be able to engage as professional advocates in the contentious work of forming social policy. The person able to do this would be one deeply committed to both the process of planning and to particular substantive ideas. Recognizing that ideological commitments will separate planners, there is tremendous need to train professionals who are competent to express their social objectives.

The great advances in analytic skills, demonstrated in the recent May issue of this journal [*Journal of the American Institute of Planners*] dedicated to techniques of simulating urban growth processes, portend a time when planners and the public will be better able to predict the consequences of proposed courses of action. But these advances will be of little social advantage if the proposals themselves do not have substance. The contemporary thoughts of planners about the nature of man in society are often mundane, unexciting or gimmicky. When asked to point out to students the planners who have a developed sense of history and philosophy concerning man's situation in the urban world one is hard put to come up with a name. Sometimes Goodman or Mumford might be mentioned. But planners seldom go deeper than acknowledging the goodness of green space and the soundness of proximity of linked activities. We cope with the problems of the alienated man with a recommendation for reducing the time of the journey to work.

CONCLUSION

The urban community is a system comprised of interrelated elements, but little is known about how the elements do, will, or should interrelate. The type of knowledge required by the new comprehensive city planner demands that the planning profession be comprised of groups of men well versed in contemporary philosophy, social work, law, the social sciences, and civic design. Not every planner must be knowledgeable in all these areas, but each planner must have a deep understanding of one or more of these areas and he must be able to give persuasive expression to his understanding. As a profession charged with making urban life more beautiful, exciting, and creative, and more just, we have had little to say. Our task is to train a future generation of planners to go well beyond us in its ability to prescribe the future urban life.

STEPHEN WHEELER

"Planning Sustainable and Livable Cities"

(1998)

Editors' introduction The enormous and exponential growth of the world's population described by Kingsley Davis in the prologue to this anthology (p. 3) has had profound and often catastrophic effects on the natural environment of planet Earth. Non-renewable energy sources have been consumed, forests cleared, buffalo slaughtered, species extinguished. Rainforests are disappearing. Today, many argue that global warming caused by population growth, urbanization, and irresponsible consumption threatens to inundate low-lying cities and irreparably damage the climate of the entire world.

In the following selection Stephen Wheeler reviews the evolution of sustainable urban development thinking and ties it to the new concern with urban livability. He defines sustainable development as "development that improves long-term health of human and ecological systems." He dismisses as inadequate recent debates about "needs" (which are hard to distinguish from wants), the "carrying capacity" of areas (which is tough to pin down, particularly for people), and "sustainable end states" (since it is virtually impossible to decide on end states).

Wheeler lays out a helpful compendium of core themes in the sustainable urban development literature, a list of specific approaches that can guide planning practice, and numerous examples to learn from. Like Forester's description of how city planners actually make decisions (p. 410), Wheeler's theoretical framework provides real guidance to planners about how to carry out plans to foster sustainable development.

Whatever their specifics, Wheeler argues, sustainable urban development strategies need to be long-term – plans for 20, 50, 100 years or longer rather than year-by-year plans that optimize short term present enjoyment at the expense of future welfare. Wheeler emphasizes that a core theme in urban sustainability planning must be attention to the natural environment. And finally urban sustainability planning requires holistic and interdisciplinary approaches connecting the insights of biologists and transit planners, agronomists, economists, and many other disciplines. Sustainable urban development planning requires that land use, transportation, housing, community development, economic development, and environmental planning all be woven together.

Wheeler outlines nine directions sustainable urban development should take to move towards his definition of sustainable urban development. He illustrates each suggested direction with specific, concrete examples of actual good work some cities have done. In contrast to unplanned sprawl that consumes land and leads to costly and inefficient infrastructure, for example, Wheeler urges planners to move in the direction of compact, efficient land use. Wheeler lists each of his other eight proposed new directions – less automobile use, more efficient resource use, restoration of natural systems, good housing and living environments, a healthy social ecology, a sustainable economics, community participation and involvement, and preservation of local culture and wisdom.

Compare Wheeler's ideas on planning sustainable and livable cities to Sim Van der Ryn and Stuart Cowan's ideas about ecological design (p. 519) and David Clark's vision for the future urban world (p. 579). For ingenious and influential ideas on how to combine land use, housing, and transit design in environmentally sensitive ways, see Peter Calthorpe's suggestions for pedestrian pockets and transit oriented design (p. 350).

Stephen Wheeler is an environmental activist and lobbyist completing a Ph.D. related to sustainable urban development at the University of California, Berkeley, Department of City and Regional Planning. He is a board member of Urban Ecology and for eight years edited their publication *The Urban Ecologist*. Mr. Wheeler is active in Bay Area transportation issues and is the co-founder of the Bay Area Transportation Choices Forum.

Among the literature on the destructive effects of urbanization on the natural environment see particularly William Cronon, *Nature's Metropolis* (New York: W. W. Norton, 1992) and Mark Reisner, *Cadillac Desert* (New York: Penguin, 1993). An early and influential report on sustainability is the World Commission on Environment and Development, *Our Common Future* (New York: Oxford University Press, 1987), commonly referred to as the Brundtland Report. Other writings on sustainable urban development include Graham Haughton and Colin Hunter, *Sustainable Cities* (London: Regional Studies Association, 1994), the President's Council on Sustainable Development, *Sustainable America* (Washington, DC: Government Printing Office, 1996), and William Rees, *Our Ecological Footprint: Reducing Human Impact on Earth* (Philadelphia: New Society Publishers, 1996). Ian McHarg's classic book *Design With Nature* (Garden City, NY: Doubleday & Company, 1969) and Patrick Geddes, *Cities in Evolution* (London: Williams & Norgate, 1915), reprinted in Richard LeGates and Frederic Stout, *Early Urban Planning 1870–1940* (London: Routledge/Thoemmes, 1998) are classic works that anticipate today's sustainable urban development, regional planning, and ecological design debates.

INTRODUCTION

The term "sustainable" is now widely used to describe a world in which both human and natural systems can continue to exist long into the future. The concept of "sustainable development" is used to refer to alternatives to traditional patterns of physical, social and economic development that can avoid problems such as exhaustion of natural resources, ecosystem destruction, pollution, overpopulation, growing inequality, and the degradation of human living conditions.

However, the notion of sustainability is still very recent, and understandings of what it means and how to apply it are still evolving. Many questions surround this concept. Does it in fact express a coherent and meaningful philosophy? Can it be defined in ways that the general public can relate to? Can it be used to generate consensus around specific urban planning directions? Can it avoid cooptation by existing political forces?

In the following pages I will propose a framework for thinking about sustainable development in the metropolitan context. After examining the origins of the sustainability concept, different definitions, and key themes, I will outline implications for urban and regional planning, and suggest processes through which sustainable development planning might be achieved. I will also consider the related concept of "livability," another broad notion that is often used in the same breath as "sustainability." These terms are closely related, in that they both promote urban planning that enhances long-term community well-being. It is no accident that these concepts have come to the fore in urban planning discussions recently, because they address important unmet needs arising from the nature of twentieth-century urban development.

ORIGINS OF THE "SUSTAINABILITY" CONCEPT

The verb "sustain" has been used in English since the year 1290 or before, and comes from the Latin roots "sub" + "tenere," meaning "to uphold" or "to keep". The *Oxford English Dictionary* traces the adjective "sustenable" to around 1400 and the modern form "sustainable" to 1611. However, this term appears to have been used mainly in legal contexts until recently, as in "The Defendant has taken several technical objections to the order, none of which . . . are sustainable" (1884). Many other variants of "sustain" have existed for centuries, but only in the past several decades has the word "sustainable" emerged with its current meaning, perhaps most simply defined as "that which can be maintained into the future."

It is far from clear who was the first to use the term "sustainable development" in its current sense. Rather, it seems one of those inevitable expressions that so neatly encapsulates what many people are thinking that once the words are mentioned they quickly become ubiquitous. The birth of the sustainability concept in the 1970s can be seen as the logical outgrowth of a new consciousness about global problems related to environment and development, fueled in part by 1960s environmentalism, publications such as *The Limits to Growth*, and the first United Nations Conference on Environment and Development held in Stockholm in 1972. However, the notion of sustainability is also rooted in older environmental traditions, particularly "sustained yield" techniques of forest management developed by nineteenth-century German foresters. These concepts influenced American policy makers such as Gifford Pinchot, Theodore Roosevelt's chief forester, and natural resource scientists such as Aldo Leopold. Leopold's notion of a "land ethic" – a human responsibility to care for particular lands and ecosystems, discussed most fully in *A Sand County Almanac* (1948) – represents a fundamental shift from the view that natural resources should be seen in terms of their utility for human beings, toward the perspective that species and ecosystems have intrinsic value in their own right and should be stewarded and sustained indefinitely into the future.

In the post-World War II period a long line of environmentalist works such as William Vogt's best-selling *Road to Survival* (1948), Fairfield Osborn's *Our Plundered Planet* (1948), Rachel Carson's *Silent Spring* (1962), and Barry Commoner's *The Closing Circle* (1971) sounded a note of alarm about the global ecological situation, and helped tie the rise of ecological problems to ongoing patterns of industrial development. Particular events also helped change consciousness around development issues, such as the 1973 oil embargo during which millions of people suddenly realized that their fossil fuel use could not continue to expand forever. Social critics, futurists, feminists, and new age writers further prepared the way for discussions of sustainability by critiquing existing notions of development and proposing alternative paradigms that would emphasize the spiritual, the natural, and the human over values of profit and economic progress as traditionally conceived. At the same time humanistic and transpersonal psychologists pointed out ways in which human potential is shaped by the surrounding environment, and ways in which it can perhaps be shaped in healthier directions in the future. The implication of such work is that people and perhaps entire societies can evolve towards more conscious, compassionate, and sustainable modes of existence, given the right conditions.

The earliest specific reference to sustainability that I have been able to document occurs in the 1972 book *Limits to Growth*, in which Donella Meadows and other MIT researchers describe computer models showing a collapse of global systems in the mid-twenty-first century, but state optimistically that "It is possible to alter these growth trends and to establish a condition of ecological and economic stability that is sustainable far into the future." A 1974 conference of the World Council of Churches then issued a call for a "sustainable society," and the earliest book specifically discussing sustainability appeared in 1976, a volume entitled *The Sustainable Society: Ethics and Economic Growth* by Lutheran theologian Robert L. Stivers.

In the late 1970s the number of writings on sustainability grew rapidly. The sustainability literature got one of its strongest pushes from Lester Brown and others at the Worldwatch Institute, a group which began publishing an extensive series of papers and books related to global sustainability, including annual State of the World reports. The tide of literature expanded in the 1980s with the International Union for the Conservation of Nature's influential World Conservation Strategy (1980), the President's Council on Environmental Quality's Global 2000 Report (1981), and above all the 1987 report of the World Commission on Environment and Development, chaired by Norwegian Prime Minister Gro Harlem Brundtland. These documents warned about global environmental problems and critiqued notions of "development," although generally accepting the desirability of continued economic growth. The influence of the IUCN and Brundtland reports in particular stemmed from the broad participation of mainstream governmental officials and academics within these bodies, which gave their findings an air of authority going beyond the "alarmist" reports of the *Limits to Growth* researchers, Global 2000, or the Worldwatch Institute. The Brundtland Commission in particular received input from literally thousands of individuals and organizations from around the world. Initiated at the request of the United Nations Secretary-General, it followed in the footsteps of two other highly respected U.N.-sponsored commissions, the Brandt Commission on North–South Issues and the Palme Commission on Security and Disarmament Issues. A more authoritative body to explore the topic would have been hard to find.

With the release of the Brundtland Commission report *Our Common Future* in 1987 and the United Nations "Earth Summit" conference in 1991, calls for sustainable development entered the mainstream internationally. "Sustainable city" programs emerged in many parts of the world, some resulting from grassroots activism, some based on municipal initiative, some benefiting from the support of national governments, and some facilitated by multilateral entities such as the European Community, the World Bank, and the U.N. The 1996 U.N. Habitat II "City Summit" in Istanbul took slow but significant steps toward establishing global consensus on how the sustainability agenda can be applied to urban planning. National reports such as that of the President's Commission on Sustainable Development (PCSD) in 1996 attempted to establish sustainable development directions for particular countries. As the 1990s went on academics in fields such as urban planning began to delve into the subject. Although actual implementation of sustainability programs remains difficult, the persistence and spread of the concept over three decades indicates that sustainability is a notion of lasting importance.

DEFINITIONS

Unfortunately, no perfect definition of sustainable development has emerged. The most widely used is that of the Brundtland Commission: "development that meets the needs of the present without compromising the ability of future generations to meet their own needs." However, this formulation is open to criticism for being anthropocentric and for raising the difficult-to-define concept of needs. (Does every household really need two cars? A VCR? A 2,000-square-foot house on a 5,000-square-foot lot? What happens if every household worldwide has these things?)

Other definitions include that given by the World Conservation Union in 1991: "improving the quality of human life while living within the carrying capacity of supporting ecosystems." This version raises the problematic notion of "carrying capacity," which is useful to think about for educational purposes but extremely hard to pin down in practice. It is one thing to say that the carrying capacity of a given watershed is a certain number of white-tailed deer; it is far more difficult to say that it is a certain number of human beings, when humans readily transport themselves and the resources they use over vast distances, and can substitute some resources for others if necessary. Although William Rees at the University of British Columbia argues that it is useful to try to calculate the "ecological footprint" of cities in terms of their resource use, my own belief is that attempts to actually define "carrying capacity" are best avoided.

Still other writers prefer to define sustainability in terms of preserving existing stocks of "ecological capital" and "social capital." This approach builds on the economic wisdom of living on the interest of an investment

– in this case the earth's stock of natural resources – rather than the principal. For example, British economist David Pearce argues that "Sustainability requires at least a constant stock of natural capital, construed as the set of all environmental assets."

Most sustainability advocates throw up their hands when faced with the definition question, and fall back on Brundtland. However, my own preference is to move instead towards a relatively simple, process-oriented definition emphasizing long-term systemic welfare: "Sustainable development is development that improves the long-term health of human and ecological systems." This strategy avoids fruitless debates over "carrying capacity," "needs," or sustainable end states, while emphasizing the process of continually moving towards healthier human and natural communities. In theory the directions of this process can be agreed upon through participatory processes in which all relevant stakeholders are represented, and progress can be measured by means of various performance indicators.

CORE THEMES

The widespread use of the sustainability concept testifies to the strength and relevance of its underlying themes, both for urban planning and other fields. Foremost among these is a concern for the long-term perspective. Though it seems only common sense that planning and building should be for the long-term, this is manifestly often not the case in practice. In particular, long-term patterns of metropolitan growth, land use, resource use, and infrastructure development demand attention, giving new impetus to old quests such as halting suburban sprawl. From this perspective it is very important to think about expanding planning horizons from a year-by-year approach or even a 20-year horizon, to think instead about the effects of urban development over 50 years, 100 years, or longer.

A second main theme is concern about the earth's natural environment. As widespread as this sentiment is these days, it is remarkably recent within industrial society and should not be taken for granted. Also new is a recognition that current development patterns are leading to ecological and social problems on a global scale. Problems such as the greenhouse effect and damage to the earth's ozone layer were only taken seriously starting in the late 1980s. In its simplest formulation, concern about environmental problems often focuses on worries about global collapse or large-scale disaster. In more complex, systems-oriented formulations, environmental and social problems can be seen not so much as leading to specific disasters in specific time frames, but as contributing to an increasingly unstable and unhealthy global system, which could be plagued by any number of unpredictable catastrophes as well as by a generally increased level of suffering on the part of human beings and natural ecosystems. Either way, the environmental costs and risks of current development patterns are viewed as unacceptable by increasing numbers of observers. Hence the search for "sustainable" alternatives.

Lastly, the sustainable development discourse can be seen as based on a new recognition of the complex web of interconnections between different issues, fields, disciplines, and actors. This holistic and interdisciplinary perspective, based on the ecological metaphor of the world as an organic system, has huge implications for planning. Among other things, it means that different specialties having to do with transportation, land use, housing, community development, economic development, and environmental protection should not be handled in isolation from one another, but should be integrated to the extent possible even while specific tasks are carried out. Coordination of economic, environmental, and social goals within planning is also necessary. Indeed, it is widely believed that social dimensions of sustainable development should be given equal weight to environmental goals.

In some ways the related concept of "livability" is a bit simpler, since it focuses less on abstract themes and more on specific human needs and people's subjective reactions to places. Dictionary definitions such as "fit to live in" or "conducive to comfortable living" actually work fairly well. Of course human needs are to a large extent culturally determined and are open to extensive debate. However, there is widespread agreement on basic elements that make cities and towns livable – a healthy environment, decent housing, safe public places, uncongested roads, parks and recreational opportunities, vibrant social interaction, and so on. Such elements obviously contribute to sustainability as well.

Livability themes are becoming more and more important to modern societies in which the basic problems of food, shelter, public health and sanitation that plagued nineteenth century cities have long since been solved, at least for most people. Instead, in the postindustrial world the emphasis is increasingly on "quality of life." It is no longer enough just to throw up cities and suburbs that are ugly, uncoordinated, automobile-dominated, and lacking in parks, sidewalks, local shops, community vitality, and sense of place. The question becomes How do we make these places green, safe, convenient, and human-oriented? How do we turn mass-produced urban landscapes into places that have character and nurture community? How do we make cities attractive and comfortable to all groups within society, including women, children, the elderly, and minorities? All of these concerns touch upon the task of making urban places "livable" in the long run.

IMPLICATIONS FOR URBAN DEVELOPMENT

Exactly what constitutes a "sustainable city" is impossible to determine, given the extent to which cities are embedded in the global context. To be absolutely self-sustaining, an urban region would need to wall itself off from the rest of the world and produce all food, energy, and materials locally. Such an autarkic model is generally infeasible and would be seen as undesirable by most residents.

Rather, it is more useful to speak of cities as moving *toward* sustainability. A metropolitan area might seek to move toward greater resource efficiency, environmental quality, social equity, and community vitality, while moving away from automobile dependency, non-renewable resource consumption, hazardous waste generation, and inequity. While local self-sufficiency may indeed offer many benefits for sustainability, this can be set forth as a value to be enhanced rather than as an absolute goal.

Until the early 1990s very little of the sustainable development literature focused on cities or patterns of urban development. Instead, writers addressed topics such as the global environmental crisis, ecological economics, critiques of prevailing models of international development, and the need for a transformation of values and mindsets. However, in recent years architects and planners have begun looking more specifically at what sustainability means for patterns of metropolitan development. Some authors have emphasized urban design and physical planning. Others have focused on environmental planning concerns having to do with the quality of air, water, and natural ecosystems. A number have stressed the need to address social problems and inequities within the urban community, and emphasize that environmental and social issues are inextricably linked. In all of these categories, urban sustainability advocates can be seen as building on the work of past planning visionaries such as Patrick Geddes, Ebenezer Howard, Lewis Mumford, Jane Jacobs, and Ian McHarg.

Although authors may have different emphases, there is substantial agreement on many dimensions of sustainable urban development. Such directions have been endorsed by documents such as U.N.'s Agenda 21 and Habitat Agenda, professional manifestos such as the Charter of the Congress for the New Urbanism and the Local Government Commission's Ahwahnee Principles, and publications of the European Community and the President's Council on Sustainable Development.

"Sustainable urban development" might be defined as "development that improves the long-term social and ecological health of cities and towns". Based on this definition and sources such as those mentioned above, main directions for urban sustainability can be seen to include the following:

1. Compact, efficient land use. Land is perhaps our most important limited resource, and current urban development patterns are clearly consuming the landscape in unsustainable ways. Land is also often divided very inequitably, and in many parts of the world those inequities are increasing. A wide range of devices can help lead to more sustainable land use. For example, urban growth boundaries (UGBs) have been adopted by Portland, Oregon, and eleven San Francisco Bay Area cities to restrain sprawl. To be effective in the long run, UGBs need to be coupled with policies to increase the efficiency of land use within already built-up areas, and to make these places more green, safe, attractive, and livable. Other types of land use controls can help preserve farm land, ecological habitat, and open space near cities. Meanwhile, urban park systems can be expanded and property tax systems changed to promote equity. Beyond specific land use changes, sustainable patterns of development are likely to involve alterations to the relationship between people and the land. In particular, a new balance between private property rights and human responsibilities toward the land is needed, as Leopold urged. The view of land as a commodity for human use and profit needs to shift towards a respect for the landscape as a thing of value in its own right, and a renewed sense of connection between human beings and the land they live on.

2. Less automobile use, better access. Current transportation systems contribute to a complex web of urban problems such as air pollution, congestion, blight, suburban sprawl, ecosystem destruction, and social fragmentation. Transport in more sustainable cities will most likely be based on several key principles: access by proximity, an inversion of the current transportation hierarchy, and demand reduction. Together these are likely to reduce the total amount that people need to travel, while allowing them to travel by far cleaner, more resource efficient, and more community-enhancing modes.

"Access by proximity" means solving transportation problems by bringing people closer to the places they need to go everyday. This is done primarily through land use changes, for example by promoting mixed-use development and the creation of neighborhood centers and "urban villages" which contain homes, workplaces, shops, and recreational facilities in close proximity to one another. Not coincidentally, fine-grained, mixed-use development also tends to create more interesting and livable places. "Inverting the transportation hierarchy" means placing the heaviest emphasis on the pedestrian, who represents the most energy efficient form of transportation and adds a much-needed human presence to the city. Bicycle planning should also be near the top of the priority list, followed by public transit. The automobile should be given lowest priority in the new hierarchy, and existing automobile subsidies reduced, although it is enormously difficult to do this politically. Finally, efforts to develop more sustainable transportation systems should also include looking at the "demand side" of the equation. By providing incentives to reduce the amount that people travel (the "demand side"), congestion problems can be solved and quality of life improved without building new roads or other infrastructure (the "supply side"). Pricing mechanisms are particularly useful to do this, such as higher parking charges, gas taxes, tolls, and vehicle registration fees.

3. Efficient resource use, less pollution and waste. Moving toward sustainability means paying greater attention to flows of energy and materials through human society, and planning for wiser use of resources. The overall challenge can be seen as one of moving from open-ended resource flows, in which nonrenewable resources are harvested, used once by human systems, and then discarded (often creating pollution and toxic waste problems in the process), toward closed-loop flows in which resources are reused and recycled.

A great number of mechanisms are available to urban and regional authorities to encourage this transition toward more sustainable resource flows. Energy conservation and materials recycling are two areas in which ordinary citizens can most directly take action through small daily initiatives, and so are good subjects for public involvement efforts. Municipal recycling programs are one of the most obvious areas in which cities can demonstrate their commitment to sustainable resource use. As in the transportation field, demand-side management programs offer great potential to reduce resource consumption. Since the late 1970s, for example, many utility companies around the U.S. have offered consumers free or reduced-price compact fluorescent light bulbs, which typically use about one-fifth the electricity of incandescent models of similar wattage, as well as rebates on energy-efficient refrigerators, air conditioners, and water heaters. This saves both the consumer and the utility money, while reducing energy consumption. Unfortunately such programs were de-emphasized in the 1990s as many states shifted their attention to deregulating the utility industry.

Tougher energy conservation codes in building construction have also produced large energy savings in many cities and states. Pollution prevention programs are being developed to try to eliminate pollution before it is created rather than cleaning it up afterwards. "Industrial ecosystem" projects try to take a systematic look at manufacturing processes to see if the wastes from one industry can be used as inputs to another. And in the economic realm, attempts are being made to shift the costs of pollution from society as a whole onto the individual or group who produces it – what has become known as the "polluter pays principle."

4. Restoration of natural systems. Even though many urban areas may seem entirely artificial – full of pavement and buildings, and often landscaped with non-native plants – still in almost every location there are many elements of the original ecosystem that can be reclaimed. Such restoration efforts add to the livability, ecological health, and overall sustainability of the urban region. Creek restoration, for example, is an idea that is catching on rapidly in many parts of the U.S. as well as overseas. Restoring a natural watercourse provides corridors and habitat for wildlife as well as walkways and open space for people. It also helps reconnect urban dwellers to the bioregion, reminding them that they live in a natural world with cycles of rainfall and waterflow. Existing urban parks and areas of open space can benefit from restoration activities as well. Volunteer site restoration programs and stewardship approaches to watershed management can help this happen. Urban agriculture is another area in which nature is being brought back into the city and urban sustainability enhanced. Biointensive methods make it possible for urbanites to grow substantial amounts of food on very small areas of land. In both real and symbolic ways, urban gardening helps reconnect city dwellers with the earth. Finally, ecological restoration is urgently needed in many inner city areas which are often home to lower-income neighborhoods and communities of color. Abandoned or contaminated industrial land can be reclaimed and restored, while vacant lots can be recycled into parks, housing, and community gardens.

Cities are often seen as unlivable because they have lost any sense of connection between people and the natural world. People move to suburbs in search of trees and nature, and there is a widespread belief that densely settled areas cannot also be green. Restoring urban ecosystems can lead to healthier and more livable cities, while providing important amenities that can help entice residents back from suburbia.

5. Good housing and living environments. One of the main purposes of cities and towns is to create decent places for people to live, and if these do not exist or are not affordable, the urban system is bound to suffer. Housing affordability is a recurrent crisis in many North American cities and suburbs. Steps to address the affordability problem include active government construction of housing – which has a checkered history in the U.S., although a somewhat better record in European countries – support of non-profit housing developers, tenant subsidies, and requirements that developers include a certain number of affordable units in any market-rate project. Also, many urban areas are characterized by ugly, homogenous, pedestrian-unfriendly development which writer James Howard Kunstler has called a "geography of nowhere." The design of housing and neighborhoods needs to be rethought in many cases to ensure that people have access to open space, meeting areas, shared facilities, shops, offices, public transportation, child care facilities, and other essentials which can make urban communities more livable.

6. A healthy social ecology. The health of human communities within an urban region is more difficult to grasp than the natural ecology. Certain social problems such as homelessness are quite obvious to anyone who walks down an urban street. Other deeply entrenched social problems, which help to decrease overall sustainability and livability in the long run, are often more hidden. Racism, for example, has been an enormous factor shaping American cities for many decades. Expressed in particular through the denial of housing, financing, insurance, or other necessities to persons of color, this factor has done as much as any other to contribute to the decline of many central city areas.

Promoting a healthy and sustainable social ecology means looking for every opportunity to enhance human community, opportunity and empowerment. It requires planners in particular to advocate on behalf of those groups who do not have access to power or expertise, and to fight for equity and justice. On a personal level, it requires an ability to put oneself in the shoes of any resident of an urban region and ask, What are the opportunities available to him or her? What is the environment like in which he or she lives? What public policies, design improvements, and social programs could help improve this environment?

7. A sustainable economics. Developing an economy that values the long-term health of human and natural systems is one of the biggest challenges related to sustainability. It would be a mistake to say that any one economic model holds all the answers, but in general a sustainable regional economy is likely be oriented around three principles. First, it is likely to be what Paul Hawken terms a "restoration economy" – one which helps restore the environmental and social damage done in the past, and that prevents new problems from occurring. Second, it is likely to be a "human-centered economy," one which meets real human needs and provides meaningful work to people at decent pay. Third, it is likely to be a locally-oriented economy, one which emphasizes local ownership, local control, local investment, use of local resources, production for local markets. This does not mean that economic development policies should totally downplay export-oriented industries, but that they should encourage as much economic activity as possible to be rooted in particular communities and regions.

The sustainable economy is likely to meet these goals through a mixture of market mechanisms, government action, and incentives for social and environmental responsibility in economic decision-making. One important step toward a more sustainable economy will be to phase out industries or processes that consume large amounts of nonrenewable resources and produce large quantities of pollutants and toxics. In addition, economic sectors based on large-scale extraction of natural resources such as minerals and oil are unlikely to be sustainable in the long run, since these resources will eventually run out after enduring various price and supply shocks along the way. Likewise, sections of the regional economy based on government subsidy – such as military industries – are not particularly sustainable either, since these hand-outs may cease. Somewhat more arguably, sectors of the economy which support the automobile should not be considered sustainable, since cars and car-dependent patterns of suburban sprawl cause a great deal of ecological and social damage. In contrast businesses engaged in environmental cleanup, recycling, public transportation, affordable housing, organic food production and the like contribute to sustainability, in that these improve the social and environmental health of the region. But while some industries may dwindle in a sustainable economy and others grow, many will simply find ways of doing the same things better. This can have advantages for businesses as well as the environment.

Many writers have argued that a cooperative, locally-oriented economics, emphasizing worker, producer and consumer co-ops and small, locally-owned businesses, is healthiest for local communities. Such a system promotes economic democracy, local control, diversity of ownership, and social responsibility, and offers an alternative model to the global market economy envisioned under free trade agreements such as GATT. Such a free-market global economy led by large corporations has many negative effects on cities and towns, in that it tends to undermine local ownership and control, replace a diversity of small retailers with a few standardized chains, and export capital from local communities to financial centers in other parts of the world.

8. Community participation and involvement. One of the most important components of urban sustainability will be creation of more functional local and regional democracy, which in turn can bring about other positive changes. There is no single best way to promote this. But a package of policies aimed at opening up local political processes, insulating them from money and special interests, producing an educated and informed electorate, and promoting responsible local decision-making can help. Community participation in local planning and design is important, as is the broad-minded leadership of officials at local, state and federal levels of government, who must demonstrate that it is possible at every level to make decisions with global, regional, and local sustainability in mind.

9. Preservation of local culture and wisdom. Much of the strength of any particular urban region lies in its cultural traditions and the unique relationships that its residents develop with each other and with the land. This uniqueness gives a region vitality, helps it take advantage of particular local contexts, and makes it an interesting place to live. Local culture, history and wisdom can add to sustainability, and their best aspects should be preserved. Such preservation will often take conscious action by governments to encourage traditional crafts, languages, rituals, cultural practices, and building techniques; to protect important local products from mass-produced imports; to protect local farmland and resource stocks; and to integrate vernacular architecture and materials into local development.

HOW CAN SUSTAINABLE AND LIVABLE CITIES BE CREATED?

How can sustainability goals be put into practice for cities and towns? On the one hand, progress will depend on sustainability themes diffusing into all existing planning and development processes. On the other hand, specific planning efforts and changes to planning procedures may be necessary as well. Sustainability-oriented planning processes could focus on particular urban issues or problems, such as air quality or watershed management (as long as these are undertaken with an awareness of other related issues). Or city leaders and public participants could take a more comprehensive look at the sustainability of a city or region. Cities such as San Francisco, Seattle, Santa Monica, and Leicester, England, have taken this latter approach.

In this age of entrenched economic and political forces opposing sustainability, no single planning effort is going to set cities on a path towards a healthy long-term future. Rather, the need is for a long-term strategy emphasizing consensus processes, public education, political organizing, policy tools such as indicators and performance standards, development of vision documents and "best practices" examples, and the creation of institutions that can more effectively address physical planning and equity issues. Together, such efforts can develop the knowledge, political will, and institutional capacity to bring about change.

A systematic planning approach to promote urban sustainability might first seek to get a wide range of parties involved in efforts to improve the long-term health of a particular city or region. Participants would then attempt to reach agreement on particular values and goals that might move the community in more sustainable directions. Vision statements and review of "best practices" examples worldwide are often helpful in giving stakeholders ideas about the range of possible approaches. Performance standards and sustainability indicators could be developed to help measure whether or not the community is making progress toward long-term goals, and to allow policies and programs to be revised to better achieve their objectives.

The most difficult challenge comes in implementing sustainability visions, policies, and programs, and in modifying institutions so as to be able to do this. This process depends on effective political organizing, and on the development of a coalition of interests supporting common objectives around sustainability and quality of life. Participants should expect the process to be a long one. Most cities and towns contain entrenched political and economic forces with an interest in continuing unsustainable patterns of development, and so progress will be slow.

In the long run, sustainable development will require systemic cultural change that builds democracy and social capital (accumulations of trust and cooperation between people). The problems created by concentrations of economic power must be addressed, as well as the tendency of capitalist systems to reinforce values oriented toward short-term private profit rather than long-term social or ecological well-being. Ultimately, moving towards sustainable cities will require a different mix of values than dominates urban development at present. There will still be opportunities for individuals to make a profit. However, the emphasis must be instead on caring deeply about the community, the region, and each other. Each development or planning decision must be evaluated in terms of its effects on the health of human and ecological communities. In *A New Theory of Urban Design* (1987), architect Christopher Alexander sets forth a single overriding rule of city development: "Every increment of construction must be made in such a way as to heal the city." This is not a bad guideline for sustainable development planning.

CONCLUSION

Planning for urban sustainability is still in the very early stages. As of now little progress has been made on turning today's huge, resource-consumptive metropolises and sprawling, even more resource-intensive suburbs into sustainable communities. But the seeds are being laid for future change, in terms of emerging consensus on what more sustainable and livable cities would be like, and on some processes and institutions that can help implement these planning directions.

Different approaches will have to be found for different cities. The Third World's rapidly growing megacities face the challenge of providing basic water, sewer, utility, and transportation services for their residents in ways that will be sustainable in the long run and that will avoid some of the problems that industrialized cities in the "developed" world have gotten into. First World cities face a different task: redeveloping urban areas that have plenty of infrastructure but fail to provide an ecologically or socially healthy urban environment. Sometimes the needs of cities in different parts of the world will be exactly opposite. For example, planners in the U.S. often seek higher residential densities to support transit, conserve open land, and promote community interaction. Yet in high-rise Hong Kong – the world's densest city – efficient land use is not an issue, and densities may fall without affecting sustainability.

It is important to remember that the current city is very recent. Its form and environment are heavily determined by technological innovations such as the automobile and the elevator which have only existed since the late nineteenth century. The creation of megacities of more than ten million people is an even more recent phenomenon, occurring only in the latter part of the twentieth century. Just as recent patterns of suburban development are now layered on top of nineteenth-century streetcar grids and eighteenth-century walking cities, so new and more ecological patterns of development may someday be layered on top of these, gradually bringing cities back into a better balance with the ecological limits of regions and the planet as a whole.

Twentieth-century suburbanization was in large part a reaction against the dirty, crowded, unhealthy cities of the industrial revolution, in which people were crammed into terrible housing virtually without amenities, to serve the needs of ruthless early forms of industrial capitalism. In a similar manner sustainable city initiatives of the next century may form a reaction against the excesses of twentieth-century culture, which is dominated by economic rather than environmental or social values. The transition toward more sustainable cities will not happen overnight. But through a growing ecological and social consciousness, the development of innovative models and examples, and better understandings of the policies, programs and designs appropriate to urban sustainability, new, more sustainable forms of urban development can come about.

RECOMMENDED READINGS

Beatley, Timothy and Kristy Manning, *The Ecology of Place: Planning for Environment, Economy, and Community*, Island Press, Washington, D.C., 1997.

Braidotti, Rosi *et al.*, *Women, the Environment and Sustainable Development: Towards a Theoretical Synthesis*, Zed Books, London, 1994.

Brown, Lester R., *Building a Sustainable Society*, Norton, New York, 1981.

Calthorpe, Peter, *The Next American Metropolis*, Princeton Architectural Press, Princeton, 1993.

Elkin, Tim and Duncan McLaren, with Mayer Hillman, *Reviving the City: Toward Sustainable Urban Development*, Friends of the Earth, London, 1990.

Girardet, Herbert, *The GAIA Atlas of Cities: New Directions in Sustainable Urban Living*, Anchor Books/Doubleday, New York, 1992.

Haughton, Graham and Colin Hunter, *Sustainable Cities*, Regional Studies Association, London, 1994.

Hawken, Paul, *The Ecology of Commerce*, HarperCollins, New York, 1993.

Holmberg, Johan, ed., *Making Development Sustainable: Redefining Institutions, Policy, and Economics*, Island Press, Washington, D.C., 1992.

Kelbaugh, Douglas, *Common Place: Toward Neighborhood and Regional Design,* University of Washington Press, Seattle, 1997.

Local Government Commission, *Land Use Strategies for More Livable Places*, Sacramento, 1992.

Lowe, Marcia, *Shaping Cities: The Environmental and Human Dimensions*, Worldwatch Paper

105, The Worldwatch Institute, Washington, D.C., 1991.

Lyle, John Tillman, *Regenerative Design for Sustainable Development*, John Wiley & Sons, New York, 1994.

Maclaren, Virginia W., "Urban Sustainability Reporting," *Journal of the American Planning Association*, Vol. 62, No. 2, Spring 1996, 184–202.

Mitlin, Diana, "Sustainable Development: A Guide to the Literature," *Environment and Urbanization*, Vol. 4, No. 1, April 1992.

Norgaard, Richard B., *Development Betrayed: The End of Progress and a Coevolutionary Revisioning of the Future*, Routledge, New York, 1994.

Pearce, David, Edward Barbier, and Anil Markandya, *Blueprint for a Green Economy*, Earthscan Publications, London, 1989.

President's Council on Sustainable Development, *Sustainable America*, Government Printing Office, Washington, D.C. 1996.

Rees, William, *Our Ecological Footprint: Reducing Human Impact on Earth*, New Society Publishers, Philadelphia, 1996

Stren, Richard, Rodney White, and Joseph Whitney, eds., *Sustainable Cities: Urbanization and the Environment in International Perspective*, Westview Press, Boulder CO, 1992.

United Nations, *Agenda 21: Program of Action for Sustainable Development*, New York, 1992.

Van der Ryn, Sim and Stuart Cowan, *Ecological Design*, Island Press, Washington, D.C., 1995.

World Commission on Environment and Development, *Our Common Future*, Oxford University Press, New York, 1987.

Zuckerman, Wolfgang, *End of the Road: The World Car Crisis and How We Can Solve It*, Chelsea Green Publishing, Post Mills, VT, 1991.

LEONIE SANDERCOCK AND ANN FORSYTH

"A Gender Agenda: New Directions for Planning Theory"

American Planning Association Journal (1992)

Editors' introduction In the twentieth century there have been a succession of "waves" of planning theory and many cross-currents, as described by Peter Hall (p. 362) . One of the newest waves is theory about gender and planning. In the following selection Australian planners Leonie Sandercock and Ann Forsyth synthesize recent work on gender and city planning and sketch out "a gender agenda" for the future.

Leonie Sandercock is the director of the Urban Planning Program at the Royal Melbourne Institute of Technology in Melbourne. Her writings on city planning and the operation of the private real estate market in Australia have been influential both in local policy-making and in the understanding of cities worldwide. Ann Forsyth, also originally from Australia, is an associate professor of Landscape Architecture at the University of Massachusetts, Amherst.

Until recently, urbanists like Jane Jacobs (p. 106), Dolores Hayden (p. 503), and William Whyte (p. 483) who were concerned with gender issues in urban planning were extremely rare. The global rise of feminism has changed that. A recent explosion of writing about gender and cities by women focuses on the uses of cities and the built environment by women, as distinct from men. This literature argues that male-planned cities reflect current (unequal) gender relations and often serve women poorly. Female homemakers may be isolated in single-family suburban homes, in an economically inferior status, unserved by public transit related to work sites or daycare (if it exists at all), in buildings designed by male architects that reinforce patriarchal gender roles. Public spaces, the feminist planning literature argues, are generally designed by males and reflect predominantly male values and concerns.

More and more women are entering the planning profession, and women are increasingly joining the ranks of city planning school faculty. There is an increasing body of writings about how city planners can build new environments or retrofit old ones to better respond to working women, female headed households, two wage-earner nuclear families who share child rearing, lesbian lifestyles and other gender issues. If gender can no longer be ignored in planning practice, Sandercock and Forsyth argue that it should not be ignored in planning theory.

The first part of this selection describes currents in feminist thought which the authors argue should enrich planning theory. These include feminist theories of language and communication, critiques of epistemology and methodology which call for "connected knowing" (emphasizing both scientific reason and passion), feminist ethics, and gender-sensitive definitions of the public and private domain, and other themes.

The second part of this selection lays out a "gender agenda" for planning education, research, and theory building. Sandercock and Forsyth would like to see initial efforts at rewriting the history of city planning broadened to provide a full account of women, not just as victims of male-dominated city planning, but as actors who have made a contribution. They argue that feminist theory can enrich planning theory. They feel that research and theory building can help change the culture of planning so that gender considerations become an accepted part of practice by all planners, male and female.

Note the connections between themes Peter Hall (p. 362) develops and the authors' account of emerging feminist planning theory. Women designers and architects in the planning-as-craft tradition are generating designs to meet women's needs; socialist feminists like Dolores Hayden are exploring the relationships between gender, class, and power; and systems-oriented women theorists are writing about how transit, daycare, and other urban systems can better serve women's needs.

Leonie Sandercock's other books include *Towards Cosmopolis* (New York: John Wiley and Sons, 1998), *Making the Invisible Visible: A Multicultural History of Urban Planning* (Berkeley: University of California Press, 1998), *Property, Politics, and Urban Planning: A History of Australian City Planning, 1890–1990* (New Brunswick, NJ: Transaction Press, 1990), with Michael Berry, *Urban Political Economy: The Australian Case* (Sydney and Boston: G. Allen & Unwin, 1983) and *Cities For Sale* (London: Heinemann, 1976).

Other books on women in cities and city planning include Catherine Stimpson, Elsa Dixler, Martha Nelson, and Kathryn Yatrakis (eds.), *Women and the American City* (Chicago: University of Chicago Press, 1981), Caroline Andrew and Beth Moore Milroy (eds.), *Life Spaces: Gender, Household, Employment* (Vancouver: University of British Columbia Press, 1988), Jo Little, Linda Peake, and Pat Richardson (eds.), *Women in Cities: Gender and the Urban Environment* (Basingstoke, UK: Macmillan, 1988), Dolores Hayden, *Redesigning the American Dream* (New York: W. W. Norton, 1984), and Daphne Spain, *Gendered Space* (Chapel Hill: University of North Carolina Press, 1992).

"Women face problems of such significance in cities and society that gender can no longer be ignored in planning practice," says Leavitt (1986: 181). In "Toward a Woman-Centered University," Adrienne Rich speaks of the need to change the center of gravity within academia to encompass women's knowledge and experience (1979). Planners also must work to change the center of gravity within their field. Leavitt (1986), Wekerle (1980), Hayden (1981, 1984), Cooper Marcus and Sarkissian (1986), and Stimpson *et al.* (1981) write of the importance of gender as a focus in planning practice. The crucial connections between theory and practice are, however, still rare and tentative.

With the new wave of feminist thinking in the 1970s came a spate of research on women and the urban environment, but the integration of that rapidly growing body of work with theory and paradigms to explain women's urban experience was "still far in the future" (Wekerle, 1980). The 1980s witnessed some nourishing of attention to gender in policy questions in the "women and ..." literature (women and housing, women and transportation, women and economic development). But in the developed countries, of all of the subfields within planning, theory remains the most male dominated and the least influenced by any awareness of the importance of gender. (By contrast, for developing countries see Moser and Levi (1986) and Moser (1989).) The works of Hayden (1981, 1984) and Leavitt and Saegert (1989), as well as the literature on gender issues in international development, are path breaking and inspirational, but they are marginalized or ignored by most of the rest of planning theory. If gender can no longer be ignored in planning practice, how can the theoretical debates continue to be silent on the subject?

Of course, much depends on how we define planning theory. There is as little agreement within planning as to what constitutes planning theory, as there is within feminism as to what constitutes feminist theory. Not simply a semantic difficulty, it is a question of contested terrain. It is a political question. Just as feminists use competing theories to understand or explain the oppression and subordination of women, planners use competing theories to explain the role, practice, and effects of planning. Even more fundamentally, disagreement abounds as to the proper theoretical object of planning theory.

Planning theory can be delineated into three different emphases: planning practice, political economy, and metatheory (Sandercock and Forsyth, 1990). At one level are those authors who theorize about planning practice, both its processes and outcomes. In general, theories of planning practice involve analysis of the procedures, actions, and behavior of planners. They may also include an analysis of the context or concrete situation in which planners are working.

The political economy approach examines the nature and meaning of urban planning in capitalist society. This approach might encompass speculations about the relationships among capitalism, democracy, and reform. Generally this approach is disinterested in planning practice. Rather, this work begins with a general theory – most commonly some version of Marxism – and uses case studies from the planning arena to illustrate the prechosen theory.

The metatheory approach involves work that asks fundamental epistemological and methodological questions about planning. Its theoretical object is an abstract, general notion of planning as a rational human activity that involves the translation of knowledge into action. At this level, theorists are no longer necessarily talking specifically about urban or regional planning, but about planning as a generic activity and as a historical legacy of the Enlightenment.

Gender issues emerge in each of the three approaches and take the form of such themes as the economic status of women, the location and movement of women through the built environment, the connections between capitalist production and patriarchal relationships and between public and domestic life, how women know about the world and about what is good, and the forms of communication with which women are most comfortable or by which they are most threatened. An awareness of these issues is lacking in planning theory. The objective of this paper is not to present a singular feminist theory of planning practice. Rather it examines those aspects of feminist theory that seem to have the most to offer planning theory.

SPATIAL, ECONOMIC, AND SOCIAL RELATIONSHIPS

Contemporary Western feminism emerged from a particular urban form – the mid-twentieth-century capitalist city, "which expressed and reinforced differentiated gender roles" (Mackenzie, 1989: 110). As more women have become wage earners the physical constraints of this type of city have become apparent. Child care is rarely close to employment centers. When [it is] unavailable, women are severely constrained by the difficult decision between not having children and paying for child care in lost wages or lost time. Similarly, mass transit is scheduled for rational commutes to work rather than the erratic movements of women responsible for both domestic duties and paid work (Palm and Pred, 1976; Pickup, 1984). Theoretical accounts of these issues and the links among them emerge infrequently and only recently in the field of urban planning.

Feminist theory, however, has examined these issues. In a pioneering article, Ann Markusen (1980) argued that women's household work had been ignored by both Marxist and neoclassical economists, even though this work has a large impact on the use of cities. She examines these issues in relation to capitalism and patriarchy. Other feminist scholars are working on the relationship among capitalist urbanization, the built environment, and gender (Huxley, 1988; Mackenzie, 1989), or among household, community, and city (Leavitt and Saegert, 1989; Mackenzie, 1988). Some of this work has grown from attempts to develop a feminist Marxism.

Other feminists, dissatisfied with mainstream theories that define human relations primarily in terms of capitalist production in the official economy, have responded with subjectivist, communitarian, or hermeneutical approaches, and have emphasized the traditional, life-sustaining work of women. When the object of this work is to create new theories to better understand the context of planning, then it fits into the political economy approach to planning theory. When the primary object is to generate strategies and programs for change, then the work belongs with theories of feminist planning practice.

Dolores Hayden's article "What Would a Non-sexist City Be Like?" (1980) and her book *Redesigning the American Dream* (1984) provide the best-known and broadest theories and visions of feminist planning practice in developed nations. Theories that are broad in scope link different activities and scales of planning – home and transport, household sexual politics, work places, and the environment, for example – rather than just concentrating on one activity.

Hayden describes a diversity of women – single parents, poor women, battered wives, and so forth – and their different needs. This sense of women being at the same time a whole and also a collection of smaller populations grew during the 1980s, particularly as minority women began to speak out on women's issues (King, 1988). Women are divided by geographical, political, religious, class, and cultural boundaries. Yet the internationalized economy exacerbates the vulnerability of women, who continue to undertake the bulk of unpaid domestic work and are engaged in low-wage work and unorganized informal markets. Women are linked to each other more than ever by an international network of decisions. Immigrant workers, or nonmigrants working for mobile firms, exemplify this connection (Sassen-Koob, 1984).

Feminist theory is currently grappling with differences among women. The book title *All the Women Are White, All the Blacks Are Men, But Some of Us Are Brave* (Hull *et al.*, 1982) reflects the vigorous exchange about whether feminist theory is ethnocentric, grounded only in the experience of white women, and whether adding in minority women is enough. Some feminists hold that theoretical categories need to be reformed in light of the distinct experiences of minority women (Barrett and McIntosh, 1985; Bhavnani and Coulson, 1986; Lugones and Spelman, 1983; hooks, 1984). Planning theory must treat this diversity seriously. Theorists must also be able to determine when it is appropriate to distinguish between specific categories and when the experiences among women of different classes, races, and other backgrounds are actually congruent (Collins, 1990: 217–19).

Theorizing within this multiplicity of voices is a complex task, but not doing so can make "woman" as oppressive a category as "man" (Harding, 1986b). As yet, planning theory literature deals hardly at all with multiple oppressions by race, sexual preference, culture, and gender. Leavitt and Saegert's (1989) work on gender, race, and age among poor people is a notable exception. Outside planning there are more attempts.

LANGUAGE AND COMMUNICATION

Planning theorists are currently involved in debates over the types and uses of rational communication, the use of language as a means of empowerment, the construction of meaning (Marris, 1987), and microanalyses of communication as action and of listening as a crucial tool of social policy (Forester, 1989). Recent feminist scholarship has extended the scope of this important work.

Feminist theories of language often start by showing how language forms one's sense of reality, order, and place in the community. As such, language can be limiting as well as empowering (Spender, 1985; Collins, 1987). Feminists have pointed to inequalities in the use of language; for example, to how men interrupt women more often than women interrupt men and to how men listen less intently to women than women listen to men (Spender, 1985: 41–50, 121–9). Important empirical studies are being conducted on how through language women come to know their world differently from each other and from men (Belenky *et al.*, 1986). Minority women have pointed to their distinct use and experience of language (hooks, 1984; Williams, 1988; Collins, 1990). Empowering language and dominant forms of communications are frequently acquired through formal education. Where education is unequally distributed, inequalities in communication will be accentuated. The upbringing and life experience of many women have actively discouraged them from speaking out or speaking up for their own needs. And when women do speak, they are more ambivalent than men about speaking assertively and with authority and are less comfortable than men with the dominant rational, scientific modes of thought (Okin, 1989: 72).

Evidence of communication inequalities emerges in such areas as citizen participation. Professional jargon and argumentative speaking styles can alienate, confuse, or render women speechless. Although in practice residents and planners are likely to be somewhat "multilingual" (many planners are women, after all, and many men are sensitive to these issues), theory should address this need for appropriate styles of communication.

Theory needs to consider the assumption, implicit in pluralist political theory, that, if given the chance, all interest groups will articulate their demands in a roughly equivalent manner. Given the current socialization of women, particularly women who suffer multiple disadvantages because of class, race, education, health, and self-esteem, this simply may not be the case.

A feminist planner, experienced in neighborhood consultation and participatory planning, described her difficulties in encouraging people at public meetings to contribute equally, particularly when many women are socialized to believe they have nothing valuable to say. She responded to this problem in one large community meeting by asking people to sit in small groups and tell a story or anecdote about their neighborhood. People then had no trouble speaking out about their lives and their community. Previously silent or hesitant participants found that they too possessed knowledge. For example, women who were stuck in the suburbs all day talked about the problems of public transport for themselves and their family. The storytelling format gave a variety of people the courage to be more involved (Sarkissian, 1990).

Theories of professional communication and citizen representation and participation need to be developed to understand these complex inequalities in planning and to develop strategies to bring women out of silence. Balancing equality and special treatment is always a complicated task, but ignoring gender is a false equality.

METHODOLOGY AND EPISTEMOLOGY

The case for a feminist perspective on epistemology and methodology in planning is grounded in feminist critiques of content, theory, and method in the social sciences. The tendency in the social sciences has been to validate only scientific and technical knowledge and dismiss all other kinds of knowledge. Feminists are increasingly critical of the traditional dualism that pits reason against passion and rationality against politics, as if reason excludes passion, as if politics, by definition, were irrational. Instead, feminists argue for what Belenky *et al.* (1986) call "connected knowing," which emphasizes relationship, rather than separation between the self and the object of research, and for discussion of the politics of theory and method and of the origins and implications of theoretical hierarchies.

In her paper criticizing theory and method in geography, "On Being outside 'The Project,'" Christopherson (1989) notes recent work by feminists and other critical theorists in jurisprudence, history, philosophy, and aesthetics which discusses the relationship between theory construction and power. These feminists insist that theorists must identify their personal position relative to the theoretical object. By way of example, Collins (1990) elaborates four elements that shape her articulation of an Afrocentric feminist epistemology: concrete experience as a criterion of meaning; the use of dialogue in assessing knowledge claims; an ethic of caring that stresses a capacity for empathy and the appropriateness of emotions in dialogue; and an ethic of personal accountability. Feminists are certainly not alone in their critiques of positivist epistemology (Kuhn, 1962; Polanyi, 1958; Feyerabend, 1975), but their work originates in response to an alienation from the methods of research and definitions of knowledge that denigrate or ignore women's experiences and that refuse to consider the political content of knowledge creation.

John Friedmann's recent synthesis of planning theory, *Planning in the Public Domain: From Knowledge to Action*, is an example of planning theory's uncertainty about its knowledge base. Initially he defines his theoretical object – planning – as the linking of *scientific and technical knowledge* to action in the public domain. But in conclusion he turns away from purely technocratic planning and embraces subjective knowledge as the foundation of a radical planning approach – a stance more sympathetic to feminist critiques (Friedmann, 1987: 413–15).

A distinctively feminist epistemology would be controversial (Sandercock and Forsyth, 1992). Feminist insights, however, would expand the planner's perspective beyond scientific and technical knowledge to other ways of knowing. First, planners would accept that knowledge is gained through talking, especially through oral traditions and gossip, which Belenky *et al.* define as conversations among intimates, talk about feelings, about the personal, the particular, the petty, but not necessarily the trivial. "Gossip, like poetry and fiction, penetrates to the truth of things," say Belenky *et al.* It is a "special mode of knowing," which moves back and forth between large and small, between particular and general (Belenky *et al.* 1986: 116). Second, knowledge is gained through listening, which Forester (1989) insightfully describes as "the social policy of everyday life," and indispensable to those working in planning. Third, knowing is also tacit or intuitive (Polanyi, 1958). As microbiologist Barbara McLintock has argued, "Reason is not by itself adequate to describe and understand the vast complexity, indeed mystery, of living forms" (Keller, 1983: 199). Fourth, creating symbolic forms through painting, music, or poetry is a more important way of knowing and communicating than planners have yet been prepared to contemplate. (For example, graffiti, murals, and folk and rap songs are ways in which minorities express themselves.) And acting and reflecting on the meaning of action yields information about the world in a way that is unavailable through technical books and reports. This is the heart of the philosophy of learning by doing, practiced by Jane Addams in her community work in turn-of-the-century Chicago at Hull House (Addams, 1910), taken up by philosopher John Dewey who was a frequent visitor to Hull House (Dewey, 1980), and developed later in planning in the work of Donald Schon in his discussion of reflective practice (Schon, 1983).

All of these ways of knowing are inseparable from the subject who is doing the talking, listening, or acting. Knowledge thus is partially autobiographical, and, therefore, is gender based. Moreover, knowledge is a social construction. Different kinds of knowledge, including scientific and technical forms, must be shared through communication to construct meaning. The construction of meaning involves communication, politics, and passion. Knowledge is, therefore, an ongoing and unfinished business.

Expanding the ways of knowing leads to a rethinking of other methodological issues, such as how to go about research in planning. Again, the feminist social scientists can assist planners with these issues. Sociologists Judith Cook and Mary Fonow (1986) have outlined five basic principles of a feminist methodology:

1 to continuously and reflexively attend to the significance of gender and gender asymmetry as a basic feature of all social life, including the conduct of research;
2 to accept the centrality of consciousness raising as a specific methodological tool and as a general orientation, or way of seeing;
3 to challenge the norm of objectivity that assumes that the subject and object of research can be separated and that personal experiences are unscientific;
4 to be concerned with the ethical implications of feminist research, and recognition of the exploitation of women as objects of knowledge; and
5 to focus on the empowerment of women and transformation of patriarchal social institutions through research.

While a distinctive feminist method of research or a distinctive feminist epistemology would be unbalanced, it must be recognized with Westkott (1979) that knowledge is inherently dialectical and that feminist inquiry has emancipatory as well as critical power.

ETHICS IN PLANNING

Recently feminist attention has focused on ethics in response to the influential and controversial work of Carol Gilligan (1982), who with a group of colleagues published a series of studies critical of the work of Lawrence Kohlberg, a psychologist who advanced the theory that humans develop in a morally autonomous fashion, skilled in reasoning about rights and justice. Gilligan noticed that women rarely did as well as men in Kohlberg's studies, and proposed that this was not because of inferior moral development but rather because of their different development. She suggested that women tend to develop a morality of responsibility and care, based on relationships with loved ones in stark contrast to Kohlberg's prototypical liberal individuals with their focus on abstract reasoning about rights and justice (Gilligan, 1982; Gilligan *et al.*, 1988; also Chodorow, 1978).

Although Gilligan originally suggested that these two moral orientations were mutually exclusive, her later work indicates that many people use both when finding the best solution to a problem (Johnston, 1988). Other studies have revealed that moralities of responsibility and care can be ascribed not only to women but to other disadvantaged and oppressed groups (Tronto, 1987; Collins, 1990). Other feminist philosophers have posed the possibility of an ethic based on "maternal thinking" (Ruddick, 1983) and have supported findings of altruistic tendencies in the general population (Mansbridge, 1988). Gilligan's work has come under attack for valorizing the consequences of women's oppression, pointing out that caring can lead to prejudice as well as altruism. Her insights remain an important empirical finding, while leaving unanswered the question of the origin of gender differences in the approach to ethical debates.

For planners, this expanded feminist ethic is a companion, although sometimes an uneasy one, to ideas of community. The new communitarian theorists offer sophisticated critiques of liberalism and offer alternatives with obvious links to feminist theories of care. They tend, however, to be complacent about traditional structures, such as the family or nation, which are hierarchically organized and oppressive to women (Sandel, 1982; MacIntyre, 1981). Feminists have proposed instead alternative models based on such communities as trade unions and political and self-help groups (Friedman, 1989).

Planning theory's sensitive analyses of community have often included critiques of romanticism (Jacobs, 1961; Bell and Newby, 1976; Heskin, 1991; Marris, 1987). Leavitt and Saegert's (1989) study of the residents of landlord-abandoned buildings in Harlem, a population predominantly black and female, is a pioneering empirical feminist work. They discovered that many of the residents had formed "community households," which shared economic and administrative burdens and drew on reciprocal social relations and attachment to place and the historical community of Harlem. Saegert and Leavitt describe the sensitivity required of planners to respect this sense of connection and care, rather than rely solely on economic criteria and formal democratic processes. Black feminist theorist Collins also discusses the centrality of an ethic of caring in African-American women's culture, but notes that institutional supports validating this ethic are virtually nonexistent (Collins, 1990: 215–17).

THE PUBLIC DOMAIN

The place of women in the public domain is a complex issue in planning. Beginning in the Victorian era and culminating in the progressive era, "the city of separate spheres" emerged in which a woman's proper place was perceived to be in the home (Wright, 1980; Brown, 1990). For domestics, however, who were usually immigrants and black women, that home was someone else's (Collins, 1990: 55). Twentieth-century metropolitan spatial form, with its "masculine cities and feminine suburbs" (Saegert, 1980), has reinforced the notion of separate spheres.

While sociologist Richard Sennett has discussed "the fall of public man" (1977), feminists in the past few decades have campaigned for the rise of public woman. The feminist political struggle in recent decades has had three components: (1) claiming women's right to be actors in the public domain and to work and participate fully in the life of the city; (2) carving out and protecting public space for women; and (3) redefining the nature and extent of the public domain. Some feminists argue that dramatic changes in metropolitan spatial structures and improvements in social and transportation policy are required to improve the opportunities for women who are also primary care-givers to participate in the political and economic life of the city. Second, feminist planners are still struggling to incorporate the issue of women's safety into land-use planning. Third, in challenging the definition of the public domain in liberal theory, feminists have shown that liberal theory has ignored the political nature of personal life, the interconnections between gender relations in the family and the paid workplace, and the fact that socialization for citizenship occurs in the domestic realm

(Pateman, 1983; Okin, 1989; Hirschmann, 1989).

The feminist struggle has led to a variety of activist responses. One feminist stance is to say that the personal realm is political and that issues like domestic violence, which are traditionally seen as private issues, are actually public. A second strategy is to make a private issue, like sexuality or abortion, public until oppressive policies, programs, and plans are eliminated. A third strategy, often conducted by these same feminists, is to make private some actions and behaviors that have traditionally been seen as part of the public domain of planning. These include lobbying for the removal from public scrutiny of the family structure of households in residential areas or the sexual relationships of public housing tenants (information required for housing allocation and rental payments). The thrust of all these strategies is to redefine the meaning of public and private. While abolishing all divisions between the private realm and the larger world would be undesirable, feminists indicate that in the arena of urban planning the line between public and private or domestic life has been drawn to men's advantage. Thus the public domain is a physical construct that by definition represents a whole set of contested political and economic issues within planning.

Feminist analysis of the state can productively interact with planning theory. Feminist theory often characterizes the state as a kind of public patriarchy. Unquestionably its employees are divided by gender, with women concentrated in secretarial and clerical work and, at the senior level, primarily in human services. State policies about marriage, the family, legitimate violence, industrial subsidies, and schools tend to reproduce and form gender roles and relationships (Connell, 1990). Planners assist in this process when they create zoning policies that restrict cohabitation to only related individuals, forcing out or apart gay couples and communal households (Ritzdorf, 1989), or policies that attract industries with gender-segmented workforces to enterprise zones, thereby reinforcing different job options for men and women.

The involvement of planners in current moves to privatize public services also has a direct though complex effect on women. Women are more likely than men to receive public assistance, as single parents, as the dominant elderly population group, as residents in public housing, or as the majority users of public transport. This assistance has given women more choices, relieving them from some of the responsibilities formerly considered private or domestic, such as caring for children or older relatives, and giving them enough material resources to achieve some measure of independence.

The form privatization takes – corporate or community-based nonprofit group ownership, continued or cut resources – affects this trend toward greater independence. A limited-equity housing cooperative (like the ones studied by Leavitt and Saegert) is a far different form of privatization than ownership by an absentee corporate or unrestricted individual landlord. Feminists and planners need to consider this issue in all its complexity, which will mean dealing realistically and at many levels with issues of power and control in women's lives.

A GENDER RESEARCH AGENDA

A gender research agenda for planning theory concentrates on areas where feminist theory has had little to say: case studies of planning practice, practical and strategic gender interests, gender in the internal culture of planners, a gender-conscious reform of planning education, and the balance between multiple differences and equality.

STUDIES OF PLANNING PRACTICE

Feminist theory, unlike more academic theories, is related to and grows out of feminist practice. Studies of both feminist planning practice and the relationship of feminist activism to planning are needed. The case studies of traditional planning, however, seldom consider gender issues. Feminist planning needs what Krumholz and Forester (1990) have done for equity-based planning: an account of attempts to politicize gender issues in planning, followed by a theorizing of the successes and failures. Some international examples serve to inspire us: the Women's Committee of the Greater London Council (Brown, 1990); women's planning initiatives in Canada (Modlich, 1986); and the Dutch women movement's campaign against clustered deconcentration and their incorporation of social safety into city planning in Amsterdam and Eindhoven (Brown, 1990: 206–60). Surely there must be some homegrown equivalents.

Arguably the history of city planning should be rewritten, incorporating gender as a category of analysis. Feminist historiography has challenged the notion that the history of women is always the same as the history of men or that significant turning points in history have the same impact on both sexes (Lerner, 1979; Kelly, 1984; Scott, 1986). A gender-conscious approach to the writing of history produces a new set of questions about the history of city planning ideas and practice (Sandercock, 1990: 21–33) and develops a different sense of historical change. In the history of planning women have often suffered and been discriminated against because men or patriarchal capitalism have controlled their lives, but this is by no means the full story. Women have not simply been victims, they too have been actors, and recent work has begun to uncover their contributions (Hayden, 1981; Birch, 1983; Davis, 1983; Wirka, 1989).

PRACTICAL AND STRATEGIC GENDER INTERESTS

Feminist planners in developing countries have drawn a distinction between practical and strategic gender interests (Molyneux, 1985; Moser, 1989). Practical interests are derived directly from women's experiences in their gender relations and their interest in survival, given that context. The practical approach does not challenge current gender relations. Strategic gender interests are derived from a more theoretical or feminist analysis of women's subordination to men, and aim to alter those relationships. This theoretical construction seems worth exploring for its usefulness in planning in developed countries. It promises to provide a framework for linking the descriptive "women and . . ." literature with explanations of why gender oppression occurs and with programs for fundamental change.

GENDER AND THE CULTURE OF PLANNERS

With more women entering the planning profession gender inequality is not merely an issue of the numerical dominance of men. Rather it is male dominance in the theories, standards, and ideologies used to guide planners' work – that is, in the internal culture of planners. By the late 1980s most planning schools were admitting roughly equal proportions of male and female students, but there nevertheless remain considerable structural inequalities between men and women in the planning profession. There are very few women running or even in the senior ranks of planning agencies. Women are concentrated in human services and social planning, professional areas with small and vulnerable budgets and relatively little prestige and power compared with development control, metropolitan strategy, or transportation planning. In essence, despite their growing numbers, women are still on the periphery rather than at the center of planning practice. Perhaps this will change over time, as women move up through the ranks in the next decade. Or are there structural impediments embedded in the culture of planners that need to be addressed (as there are for women in other professions)? Are women treated differently (from and by men) in the planning workplace? Do they experience difficulties in being heard, in being taken seriously, in being drawn into the confidences or information sharing that constitute the informal web of daily life in a planning office? Are women planners punished by their male peers if they speak out on women's issues? Do women planners simply not speak out on such issues from fear or from a perception that they would be marginalized in some way for doing so? In other words, is there a dominant male definition of the key issues and roles in the planning workplace that could be at once progressive in class terms and yet gender blind?

Two anecdotes suffice. A group of women planners in an Australian capital city, when asked whether they thought that the notion that planning policies are *not* gender neutral had percolated through the male ranks of the profession and become built-in to their daily practice and discussion, simply laughed at the apparent naivete of the question, at the hopelessness of the situation, and perhaps, too, at their own tendency to avoid the issue because of the discomfort it inevitably causes.

A feminist planner became the manager of community services for a large suburban municipality in Australia. The managers of all the other planning departments within that council were male. She knew that they all met together at the local pub at the end of the week. She suspected that important informal information exchange and power plays took place at these gatherings. She was not invited. Women traditionally have not been part of pub culture in Australian life. This planner was not a drinker and did not like the pub atmosphere. Yet she felt excluded and debated raising the matter with the boys. The issue likely has no solution; if she raised it and was invited to join the men, their conversation would no doubt be constrained by her presence. This problem is an example of how the internal culture of planners reflects the biases of the wider masculine culture, and poses dilemmas for professional women about whether to adjust their behavior accordingly, or whether to try to introduce more female ways of socializing into the workplace.

Research based on in-depth interviews could be done about the experience of women in the planning workplace to assess whether and to what extent the gender inequalities and biases of the wider society are being reinforced or challenged. (This is an omission in Forester's otherwise very perceptive 1989 work on the internal culture of planners.)

REFORM OF PLANNING EDUCATION

During a 1990 Australian government review of metropolitan strategy, a group of women planners were asked about gender issues in the local planning scene. The response was that consciousness of these issues was very low. The respondents added that recent women graduates of the local planning program were actively antifeminist. These new graduates, it seems, are afraid of being stereotyped and dismissed by male colleagues as "noisy feminists." Further questioning revealed that the planning school has no women on its faculty and that gender issues are not in the syllabus.

A recent introductory undergraduate planning course at the University of California at Berkeley had eighty-five students, male and female almost equally represented. Three of thirty lectures dealt specifically with the question of gender in planning history, theory, and practice, while the importance of gender was integrated into the rest of the subject matter. In student evaluations at the end of the course, 10 percent of the students, when asked "What was the worst thing about this class?", replied, "The emphasis on gender." One student complained that the course should have been titled "Feminist City Planning." This group ranked the instructor and the course at the lowest possible grade. On the other hand, some 30 percent noted the exploration of gender issues as one of the best things about the class, and ranked both course and instructor at the highest possible grade. This dramatic polarization reveals both the need for and the resistance to a gender-conscious approach to the teaching of planning. Some feminist planners indicate that attempts to introduce a more gender-sensitive curriculum in planning programs continue to be met with resistance and incomprehension by male-dominated faculties.

BALANCING DIFFERENCES AND EQUALITY

Feminism starts from an experience of difference – the differences between men and women, differences that in some way cause women to be disadvantaged. Recently, however, the focus in some feminist work has shifted to considerations of differences among women, which raises new and important questions. How do different groups of women use and experience cities? Do public spaces hold the same intimidation for middle-class and poor women, for African-Americans, mothers, Chicanas, Jewish women, or lesbians? How are experiences of lone parenting different for women from different communities? Taking account of the systematic differences among women, as well as the systematic differences among women and the men in their various communities, is an important task for gender-conscious planning.

THE ONGOING DEBATE

In "The 'Thereness' of Women: A Selective Review of Urban Sociology," Lyn Lofland asserts that in empirical and theoretical urban sociology, women are perceived as being part of the scene but not part of the action.

> [Women] are part of the locale or neighborhood or area described like other important aspects of the setting such as income, ecology or demography – but largely irrelevant to the analytic *action*. They reflect a group's social organization and culture, but they never seem to be in the process of creating it.
>
> (Lofland, 1975: 145)

In mainstream planning theory women have scarcely even been seen as subjects of theory. The problem, however, is far more subtle and complex than a simple tradition of exclusion. The paradigms on which planning and theorizing about it have been based are informed by characteristics traditionally associated with the masculine in our society. There is a need to rethink the foundations of the discipline, its epistemology, and its various methodologies. Feminist critiques and feminist literature need to be incorporated into the debates within planning theory.

REFERENCES AND FURTHER READING

Addams, Jane (1910) *Twenty Years at Hull House: With Autobiographical Notes*. New York: Macmillan.

Andrew, Caroline and Milroy, Beth Moore (eds.) (1988) *Life Spaces: Gender, Household, Employment*. Vancouver: University of British Columbia Press.

Barrett, Michele and McIntosh, Mary (1985) "Ethnocentrism and Socialist Feminist Theory," *Feminist Review* 20: 23–48.

Belenky, Mary, Clinchy, Blythe, Goldberger, Nancy and Tarule, Jill (1986) *Women's Ways of Knowing: The Development of Self, Voice, and Mind*. New York: Basic Books.

Bell, Colin, and Newby, Howard (1976) "Community, Communion, Class and Community Action: The Social Sources of the New Urban Politics" in D. T. Herbert and R. J. Johnson (eds.) *Spatial Perspectives on Problems and Policies*, vol. 2 of *Social Areas in Cities*. London: John Wiley.

Bell, Derrick (1985) *And We Are Not Saved: The Elusive Request for Racial Justice*. New York: Basic Books.

Bhavnani, Kum-Kum and Coulson, Margaret (1986) "Transforming Socialist Feminism: The Challenge of Racism," *Feminist Review* 23: 81–92.

Birch, Eugenie (1983) "From Civic Worker to City Planner: Women and Planning, 1890–1980" in Donald Krueckeberg (ed.) *The American Planner: Biographies and Reflections*. New York: Methuen.

Brown, Karen Rebecca (1990) "Rethinking the Man-Made Metropolis: The Relationship between Constructs of Gender and Planned Urban Form," B.A. honors thesis, Women's Studies, Brown University.

Caine, Barbara, Grosz, E. A. and de Lepervanche, Marie (1988) *Crossing Boundaries: Feminisms and the Critique of Knowledges*. Sydney: Allen & Unwin.

Castells, Manuel (1977) *The Urban Question*. London: Edward Arnold.

Chodorow, Nancy (1978) *The Reproduction of Mothering*. Berkeley: University of California Press.

Christopherson, Susan (1989) "On Being outside 'The Project'," *Antipode* 21(2): 83–9.

Clavel, Pierre (1986) *The Progressive City: Planning and Participation, 1969–1984*. New Brunswick, N.J.: Rutgers University Press.

Collins, Patricia (1990) *Black Feminist Thought: Knowledge, Consciousness and the Politics of Empowerment*. Boston: Unwin Hyman.

Connell, R. W. (1987) *Gender and Power: Society, the Person and Sexual Politics*. Stanford: Stanford University Press.

——— (1990) "The State, Gender and Sexual Politics: Theory and Appraisal," *Theory and Society* 19: 507–44.

Cook, Judith and Fonow, Mary (1986) "Knowledge and Women's Interests: Issues of Epistemology and Methodology in Feminist Sociological Research," *Sociological Inquiry* 56(1): 2–29.

Cooper Marcus, Clare and Sarkissian, Wendy (1986) *Housing as if People Mattered: Site Design Guidelines for Medium-Density Family Housing*. Berkeley: University of California Press.

Davis, Allan (1983) "Playgrounds, Housing and City Planning" in Donald Krueckeberg (ed.) *Introduction to Planning History in the United States*. New Brunswick, N.J.: Center for Urban Policy Research, Rutgers University.

Delgado, Richard (1989) "Story Telling for Oppositionists and Others: A Plea for Narrative," *Michigan Law Review* 87: 2410–41.

Dewey, John [1929] (1980) *The Quest for Certainty: A Study of the Relation of Knowledge to Action*. New York: Perigree Books.

Fainstein, S. and Fainstein, N. (1982) *Urban Policy under Capitalism*. Beverly Hills, Calif.: Sage.

Faludi, Andreas (1986) *Critical Rationalism and Planning Methodology*. London: Pion.

Feyerabend, Paul (1975) *Against Method: Outline of an Anarchistic Theory of Knowledge*. London: New Left Books.

Fogelsong, Richard (1986) *Planning the Capitalist City: The Colonial Era to the 1920s*. Princeton, N.J.: Princeton University Press.

Forester, John (1989) *Planning in the Face of Power*. Berkeley: University of California Press.

Friedman, Marilyn (1989) "Feminism and Modern Friendship: Dislocating the Community," *Ethics* 99(1): 275–90.

Friedmann, John (1987) *Planning in the Public Domain: From Knowledge to Action*. Princeton: Princeton University Press.

Gilligan, Carol (1982) *In a Different Voice: Psychological Theory and Women's Development*. Cambridge, Mass.: Harvard University Press.

Gilligan, Carol, Ward, Janie and Taylor, Jill (eds.) (1988) *Mapping the Moral Domain: A Contribution of Women's Thinking to Psychological Theory and Education*. Cambridge, Mass.: Center for the Study of Gender, Education and Human Development, Harvard University.

Harding, Sandra (1986a) *The Science Question in Feminism*. Ithaca, N.Y.: Cornell University Press.

——— (1986b) "The Instability of the Analytical Categories of Feminist Theory," *Signs* 11(4): 645–64.

Harding, Sandra and Hintikka, Merrill (eds.) (1983) *Discovering Reality: Feminist Perspectives on Epistemology, Metaphysics, Methodology, and Philosophy of Science*. Dortrecht: D. Reidel.

Harvey, David (1978a) "The Urban Process under Capitalism: A Framework for Analysis," *International Journal for Urban and Regional Research* 2(1): 101–31.

——— (1978b) "Labour, Capital and Class Struggle around the Built Environment in Advanced Capitalist Societies" in Kevin Cox (ed.) *Urbanization and Conflict in Market Societies*. London: Methuen.

Hayden, Dolores (1980) "What Would a Non-sexist City Be Like? Speculations on Housing, Urban Design, and Human Work," *Signs* 5(3) (Supplement): S170–87.

——— (1981) *The Grand Domestic Revolution: A History of Feminist Designs for American Homes, Neighborhoods and Cities*. Cambridge, Mass.: MIT Press.

——— (1984) *Redesigning the American Dream: The Future of Housing, Work, and Family Life*. New York: W. W. Norton.

Heskin, Allan (1991) *The Struggle for Community*. Boulder, Colo.: Westview Press.

Hirschmann, Nancy (1989) "Freedom, Recognition, and Obligation: A Feminist Approach to Political Theory," *American Political Science Review* 83(4): 1227–44.

hooks, bell (1984) *Feminist Theory: From Margin to Center*. Boston: South End Press.

Hull, Gloria, Scott, Patricia Bell and Smith, Barbara (eds.) (1982) *All the Women Are White, All the Blacks Are Men, But Some of Us Are Brave: Black Women's Studies*. New York: Feminist Press.

Huxley, Margo (1988) "Feminist Urban Theory: Gender, Class and the Built Environment," *Transition* (Winter): 39–43.

Jacobs, Jane (1961) *The Death and Life of Great American Cities*. New York: Random House.

Jaggar, Alison and Bardo, Susan (eds.) (1989) *Gender/Body/Knowledge: Feminist Reconstructions of Being and Knowing*. New Brunswick, N.J.: Rutgers University Press.

Jaggar, Alison and Rothenberg, Paula (eds.) (1984) *Feminist Frameworks: Alternative Theoretical Accounts of the Relations Between Women and Men*, 2nd edn. New York: McGraw-Hill.

Keller, Evelyn Fox (1983) *A Feeling for the Organism: The Life and Work of Barbara McClintock*. San Francisco: W. H. Freeman.

Keller, Suzanne (ed.) (1981) *Building for Women*. Lexington, M.A.: Lexington Books.

Kelly, Joan Gadol (1984) *Woman, History and Theory: The Essays of Joan Kelly*. Chicago: University of Chicago Press.

King, Deborah (1988) "Multiple Jeopardy, Multiple Consciousness: The Context of a Black Feminist Ideology," *Signs* 14(1): 42–72.

Kohlberg, Lawrence (1981) *The Philosophy of Moral Development*. San Francisco: Harper & Row.

Krieger, Martin (1989) *Marginalization and Discontinuity: Tools for the Crafts of Knowledge and Decision*. New York: Russell Sage Foundation.

Krumholz, Norman and Forester, John (1990) *Making Equity Planning Work: Leadership in the Public Sector*. Philadelphia: Temple University Press.

Kuhn, Thomas (1962) *The Structure of Scientific Revolutions*. Chicago: University of Chicago Press.

Leavitt, Jacqueline (1986) "Feminist Advocacy Planning in the 1980s" in Barry Checkoway (ed.) *Strategic Perspectives in Planning Practice*. Lexington, Mass.: Lexington Books.

Leavitt, Jacqueline and Saegert, Susan (1989) *From Abandonment to Hope: Community-Households in Harlem*. New York: Columbia University Press.

Lerner, Gerda (1979) *The Majority Finds Its Past: Placing Women in History*. New York: Oxford University Press.

Lindblom, Charles (1990) *Inquiry and Change: The Troubled Attempt to Understand and Shape Society*. New Haven, Conn.: Yale University Press.

Lofland, Lyn (1975) "The 'Thereness' of Women: A Selective Review of Urban Sociology" in Marcia Millman and Rosabeth Moss Kanter (eds.) *Another Voice: Feminist Perspectives on Social Life and Social Science*. New York: Anchor Books.

Lugones, Maria and Spelman, Elizabeth (1983) "Have We Got Theory For You: Feminist Theory, Cultural

Imperialism and the Demand for 'The Woman's Voice'," *Women's Studies International Forum* 6(6): 573–81.

MacIntyre, Alasdair (1981) *After Virtue: A Study in Moral Theory*. London: Duckworth.

Mackenzie, Suzanne (1988) "Building Women, Building Cities: Toward Gender Sensitive Theory in Environmental Disciplines" in Carolyn Andrew and Beth Moore Milroy (eds.) *Life Spaces: Gender, Household, Employment*. Vancouver: University of British Columbia Press.

——— (1989) "Women in the City" in Richard Peet and Nigel Thrift (eds.) *New Models in Geography*. London: Routledge.

Majone, Giandomenico and Quade, Edward (1980) *Pitfalls of Analysis*. New York: John Wiley.

Mansbridge, Jane (ed.) (1988) *Beyond Self-Interest*. Chicago: University of Chicago Press.

Markusen, Ann (1980) "City Spatial Structure, Women's Household Work, and National Urban Policy," *Signs* 5(3) (Supplement): S23–44.

Marris, Peter [1982] (1987) *Meaning and Action: Community Planning and Conceptions of Change*, 2nd edn. London: Routledge & Kegan Paul.

Matrix (1984) *Making Space: Women in the Man-Made Environment*. London: Pluto Press.

Matsuda, Mari (1989) "Public Response to Racist Speech: Considering the Victim's Story," *Michigan Law Review* 87: 2320–81.

Millman, Marcia and Moss Kanter, Rosabeth (eds.) (1975) *Another Voice: Feminist Perspectives on Social Life and Social Science*. New York: Anchor Books.

Modlich, Regula (1986) "Women Plan Toronto," *Women and Environments* 8(1).

Molyneux, Maxine (1985) "Mobilization without Emancipation? Women's Interests, State, and Revolution in Nicaragua," *Feminist Studies* 11(2): 227–54.

Moser, Carolyn (1989) "Gender Planning in the Third World: Meeting Practical and Strategic Gender Needs," *World Development* 17(1): 1799–825.

Moser, Carolyn, and Levi, C. (1986) *A Theory and Methodology of Gender Planning: Meeting Practical and Strategic Gender Needs*. Gender and Planning Working Paper 11. London: Development Planning Unit, University College London.

Nicholson, Linda (ed.) (1990) *Feminism/Postmodernism*. New York: Routledge.

Okin, Susan Moller (1989) *Justice, Gender and the Family*. New York: Basic Books.

Palm, Risa and Pred, Alan (1976) *A Time-Geographic Perspective on the Problems of Inequality for Women*. Working Paper 236. Berkeley: Institute of Urban and Regional Development, University of California.

Paris, Chris (ed.) (1983) *Critical Readings in Planning Theory*. Oxford: Pergamon.

Pateman, Carole (1983) "Feminist Critiques of the Public/Private Dichotomy" in S. I. Benn and G. F. Gauss (eds.) *Public and Private in Social Life*. London: Croom Helm.

Pateman, Carole and Gross, Elizabeth (1986) *Feminist Challenges: Social and Political Theory*. Boston: Northeastern University Press.

Phillips, Anne (1987) *Divided Loyalties: Dilemmas of Sex and Class*. London: Virago.

Pickup, Laurie (1984) "Women's Gender-Role and Its Influence on Travel Behavior," *Built Environment* 10(1): 61–8.

Polanyi, Michael (1958) *Personal Knowledge: Towards a Post-critical Philosophy*. New York: Harper & Row.

Rich, Adrienne (1979) "Toward a Woman-Centered University" in *On Lies, Secrets and Silence*. New York: W. W. Norton.

Ritzdorf, Marsha (1989) "Regulating Separate Spheres: Municipal Land-Use Planning and the Changing Lives of Women," paper presented at the Conference of the Society for American City and Regional Planning History, 29 November.

Ruddick, Sara (1983) "Maternal Thinking" in Joyce Treblicot (ed.) *Mothering: Essays in Feminist Theory*. Savage, Md.: Rowman & Littlefield.

Saegert, Susan (1980) "Masculine Cities and Feminine Suburbs: Polarized Ideas, Contradictory Realities," *Signs* 5(3) (Supplement): S93–108.

Sandel, Michael (1982) *Liberalism and the Limits of Justice*. Cambridge: Cambridge University Press.

Sandercock, Leonie (1990) *Property, Politics and Urban Planning: A Political History of Australian City Planning 1890–1990*. New Brunswick, N.J.: Transaction Press.

Sandercock, Leonie and Forsyth, Ann (1990) *Gender: A New Agenda for Planning Theory*. Working Paper 521. Berkeley: Institute of Urban and Regional Development, University of California.

——— (1992) "Feminist Theory and Planning Theory: The Epistemological Links," *Planning Theory Newsletter* (Winter). Turin: Dipartimento Interateneo Territorio.

Sarkissian, Wendy (1990) Personal communication.

Sassen-Koob, Saskia (1984) "From Household to Workplace: Theories and Survey Research on Migrant Women in the Labor Market," *International Migration Review* 18(4): 1114–67.

Schon, Donald (1983) *The Reflective Practitioner: How Professionals Think in Action*. New York: Basic Books.

Scott, Joan (1986) "Gender: A Useful Category in Historical Analysis," *American Historical Review* (December): 1053–75.

Sennett, Richard (1977) *The Fall of Public Man*. New York: Knopf.

Soja, Edward (1989) *Postmodern Geographies: The Reassertion of Space in Critical Social Theory*. New York: Verso.

Spender, Dale (1985) *Man Made Language*, 2nd edn. London: Routledge & Kegan Paul.

Stimpson, Catharine, Dixler, Elsa, Nelson, Martha and Yatrakis, Kathryn (eds.) (1981) *Women and the*

American City. Chicago: University of Chicago Press. (Originally published in 1980 as a supplement to Signs 5(3).)

Tabb, W. K. and Sawers, L. (1978) *Marxism and the Metropolis*. Oxford: Oxford University Press.

Tronto, Joan (1987) "Beyond Gender Difference to a Theory of Care," *Signs* 12(4): 644–63.

Watson, Sophie (1986) "Women and Housing or Feminist Housing Analysis?" *Housing Studies* 1(1): 1–10.

——— (1988) *Accommodating Inequality: Gender and Housing*. Sydney: Allen & Unwin.

Wekerle, Gerda (1980) "Women in the Urban Environment," *Signs* 5(3) (Supplement): S188–214.

Wekerle, Gerda, Peterson, Rebecca and Morley, David (eds.) (1980) *New Space for Women*. Boulder, Colo.: Westview Press.

Westkott, Nancy (1979) "Feminist Criticism of the Social Sciences," *Harvard Educational Review* 49(4): 422–30.

Williams, Patricia (1988) "On Being the Object of Property," *Signs* 14(1): 5–24.

——— (1989) "The Obliging Shell: An Informal Essay on Formal Equal Opportunity," *Michigan Law Review* 87: 2128–59.

Wirka, Susan (1989) "Mary Kingsbury Simkhovitch and Neighborhood Planning in New York City, 1897–1909," Master's thesis, University of California, Los Angeles.

Wright, Gwendolyn (1980) *Moralism and the Model Home: Domestic Architecture and Cultural Conflict in Chicago, 1873–1913*. Chicago: University of Chicago Press.

Plate 23 Central Park, New York, 1863. Frederick Law Olmsted and his partner Calvert Vaux conceived and executed a park, which was not only a masterpiece of design excellence, but also articulated a philosophy about what urban parks were for. Central Park provided the illusion of nature in the city. It was an oasis of calm and an intended meeting place for different classes. Central Park provided areas for quiet contemplation, boating, strolling, riding, baseball, Sunday school picnics, and countless other activities. (© Museum of the City of New York.)

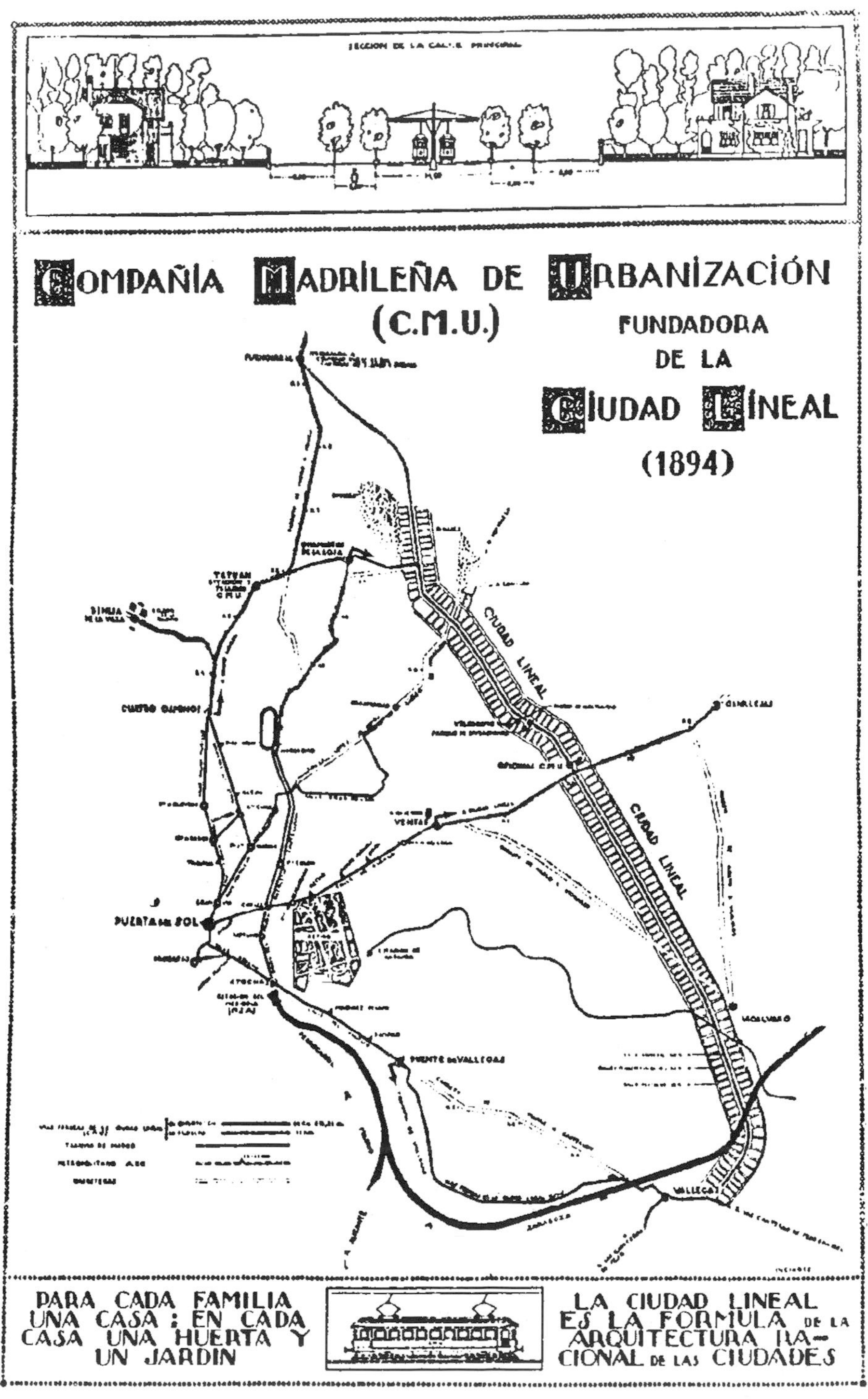

Plate 24 Arturo Soria y Mata's plan for a linear city around Madrid, 1894. Spanish engineer/planner Arturo Soria y Mata envisaged the liberating force of new transportation technology. He conceived of "linear cities" built along electric streetcar lines which would provide for access to nature, large and relatively inexpensive lots, quick transportation, and efficient provision of infrastructure. Portions of the linear city around Madrid pictured here were built. From Arturo Soria y Mata, *The Linear City*.

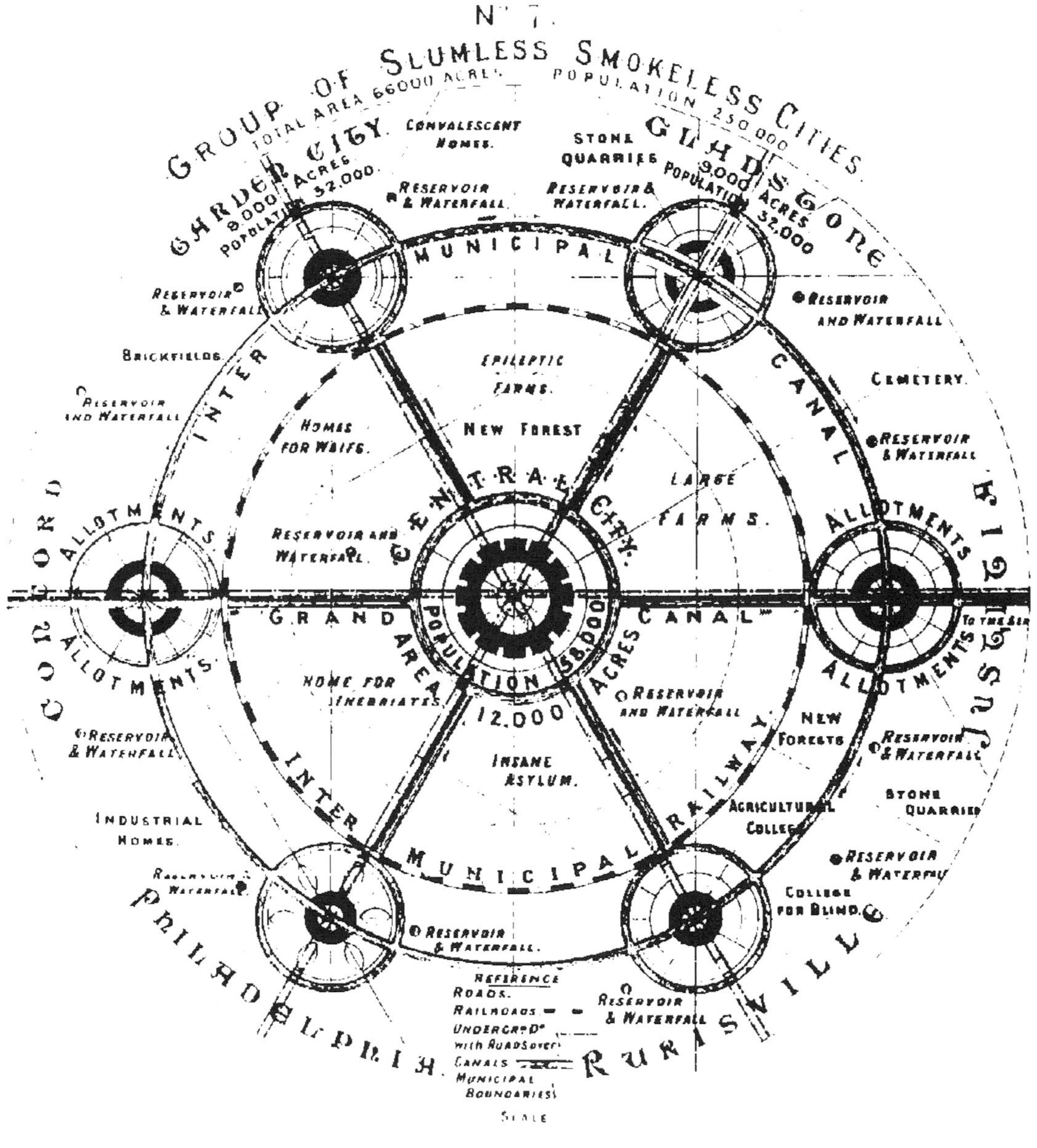

Plate 25 Ebenezer Howard's plan for a Garden City, 1898. From Ebenezer Howard, *To-morrow: A Peaceful Path to Real Reform* (London: Swan Sonnenschein, 1898). Reacting to the squalor of the nineteenth-century city, Ebenezer Howard proposed self-sufficient "garden cities" of about 32,000 people, carefully planned and surrounded by a permanent green belt. The Garden City movement spread worldwide and continues to inspire city planners.

Plate 26 Plan for Welwyn Garden City, 1909. Welwyn, the second of Britain's garden cities, closely followed Howard's vision. Howard lived to see Welwyn built and spent the last years of his life there.

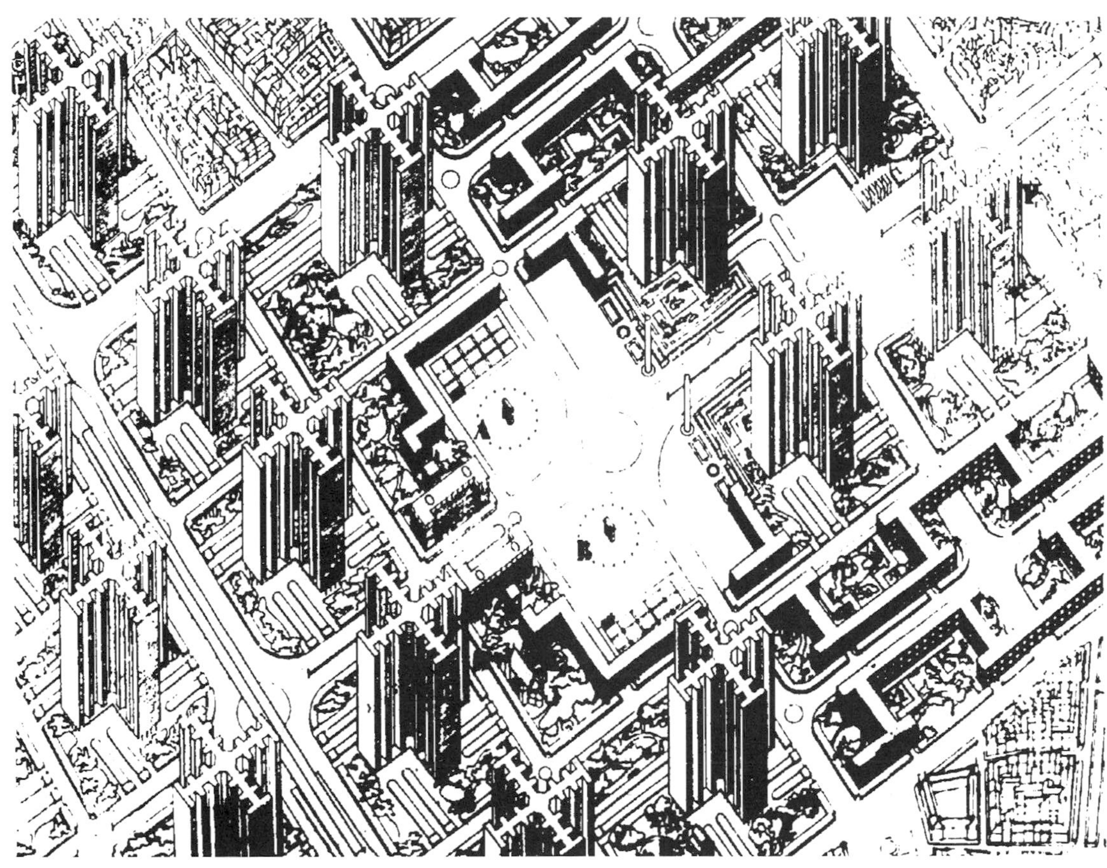

Plate 27 Le Corbusier's "Plan Voisin" for a hypothetical city of three million people, 1925. Visionary modernist Le Corbusier planned huge new cities of steel and concrete dominated by highways and large modern buildings in park-like settings. This 1925 plan contemplates an entire new city of three million people built on these principles.

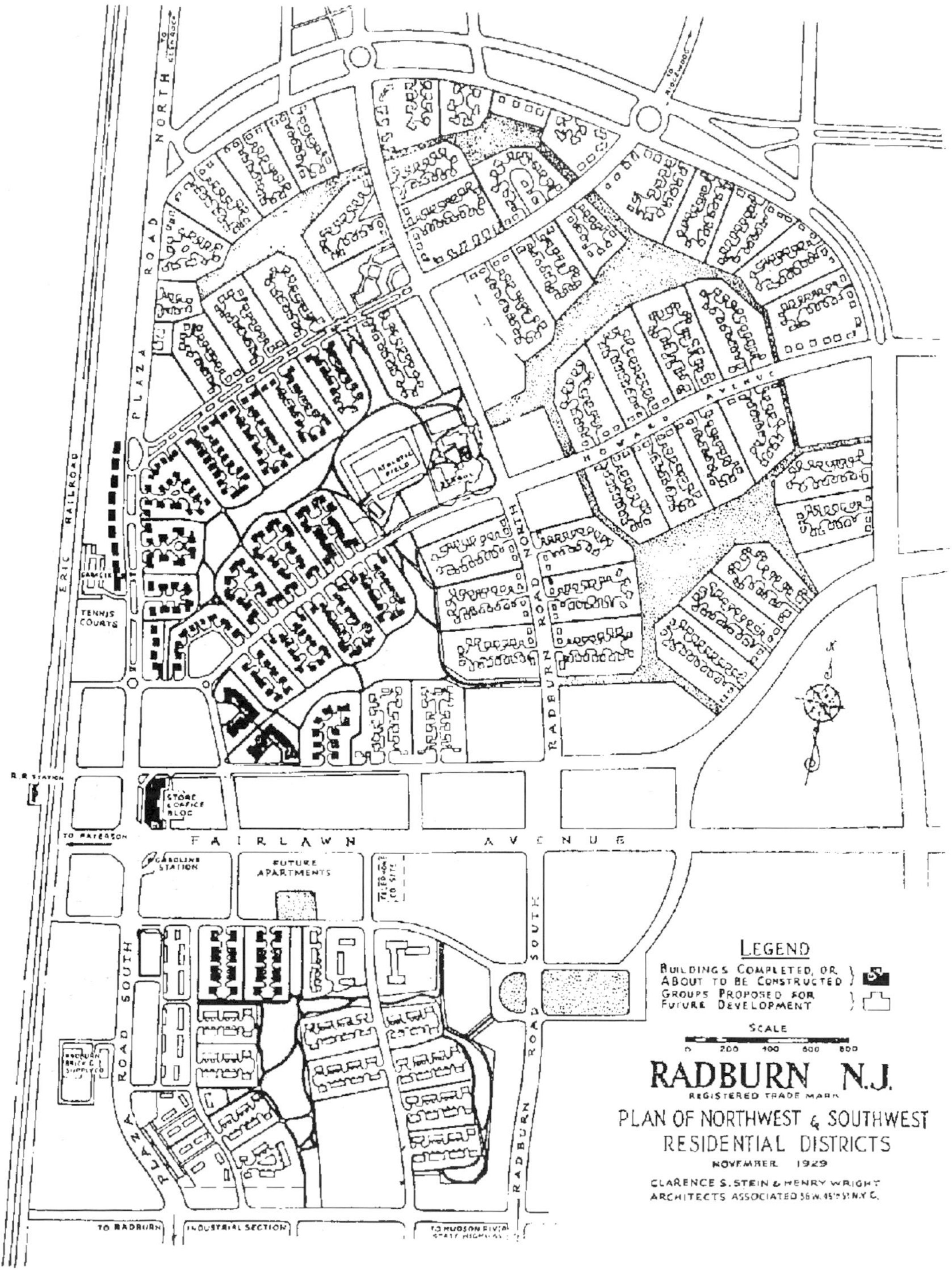

Plate 28 Plan for Radburn, New Jersey, 1929. Architects Clarence Stein and Henry Wright responded to the automobile by designing cities with a separation between pedestrian and traffic arteries, superblocks, and separate city areas with strong neighborhood identity. Their plan for Radburn, New Jersey, has inspired generations of planners.

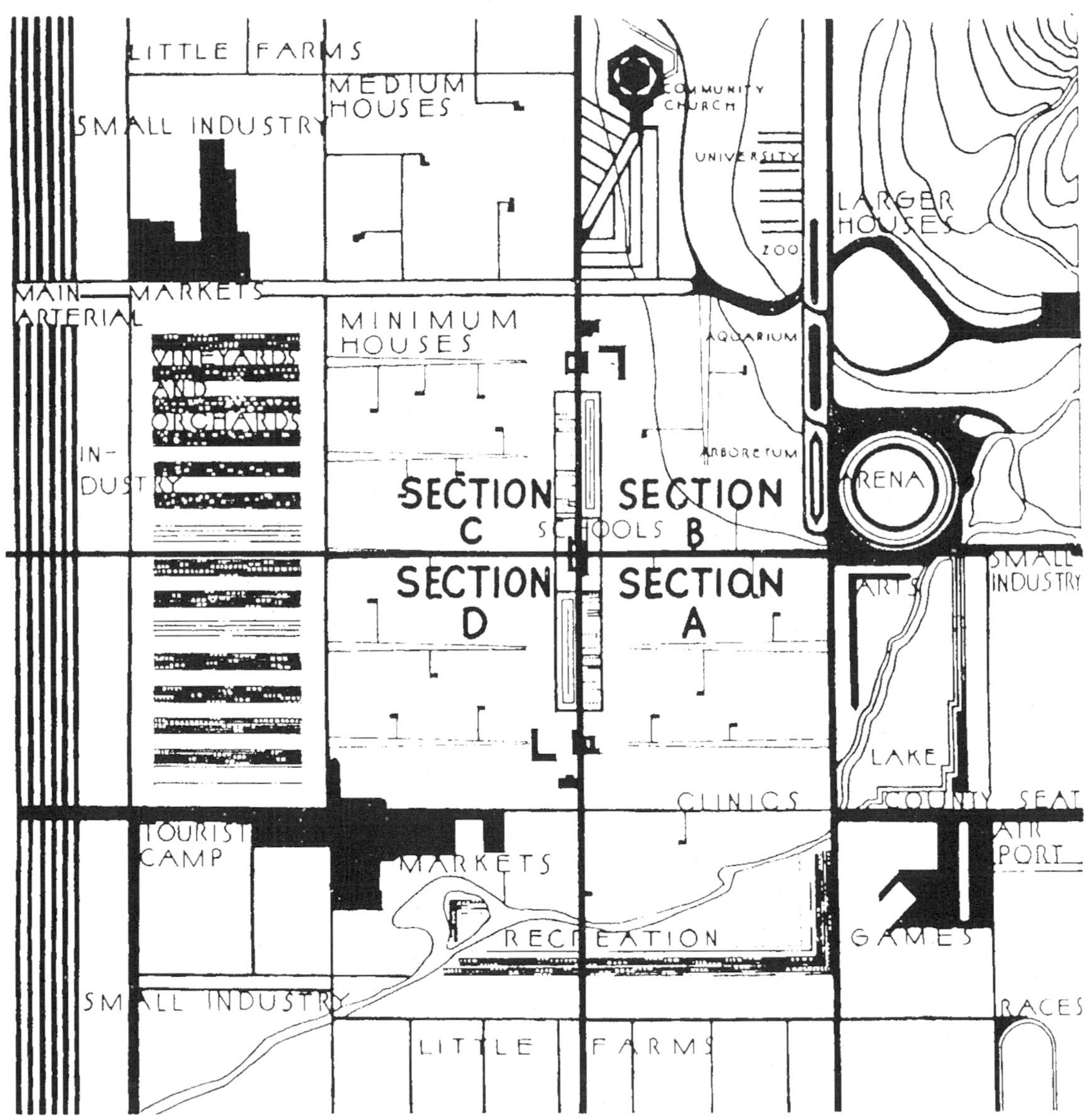

Plate 29 Frank Lloyd Wright's plan for Broadacre City, 1935. Frank Lloyd Wright's extreme decentralist vision of the modern city contemplated a new form of Jeffersonian democracy with decentralized households living with at least one acre of land for each person.

Plate 30 Paseo del Rio, San Antonio, Texas. Good city planning can turn the most mundane landscape into a good urban space. Through adopting a bold vision and seeing it through to completion, San Antonio, Texas, turned a blighted riverfront into a magnificent area of parkland, water-oriented activities, and retail shopping. (© Alexander Garvin.)

Plate 31 Quincy Market, Boston, Massachusetts. Suburban malls and shopping centers may be "the new downtown" in many metropolitan areas, but some cities like Boston, Massachusetts are rebuilding their urban core into attractive shopping and entertainment areas. Boston's Quincy Market has been transformed from a seedy and economically marginal market area into a bustling and successful commercial area. (© Alexander Garvin.)

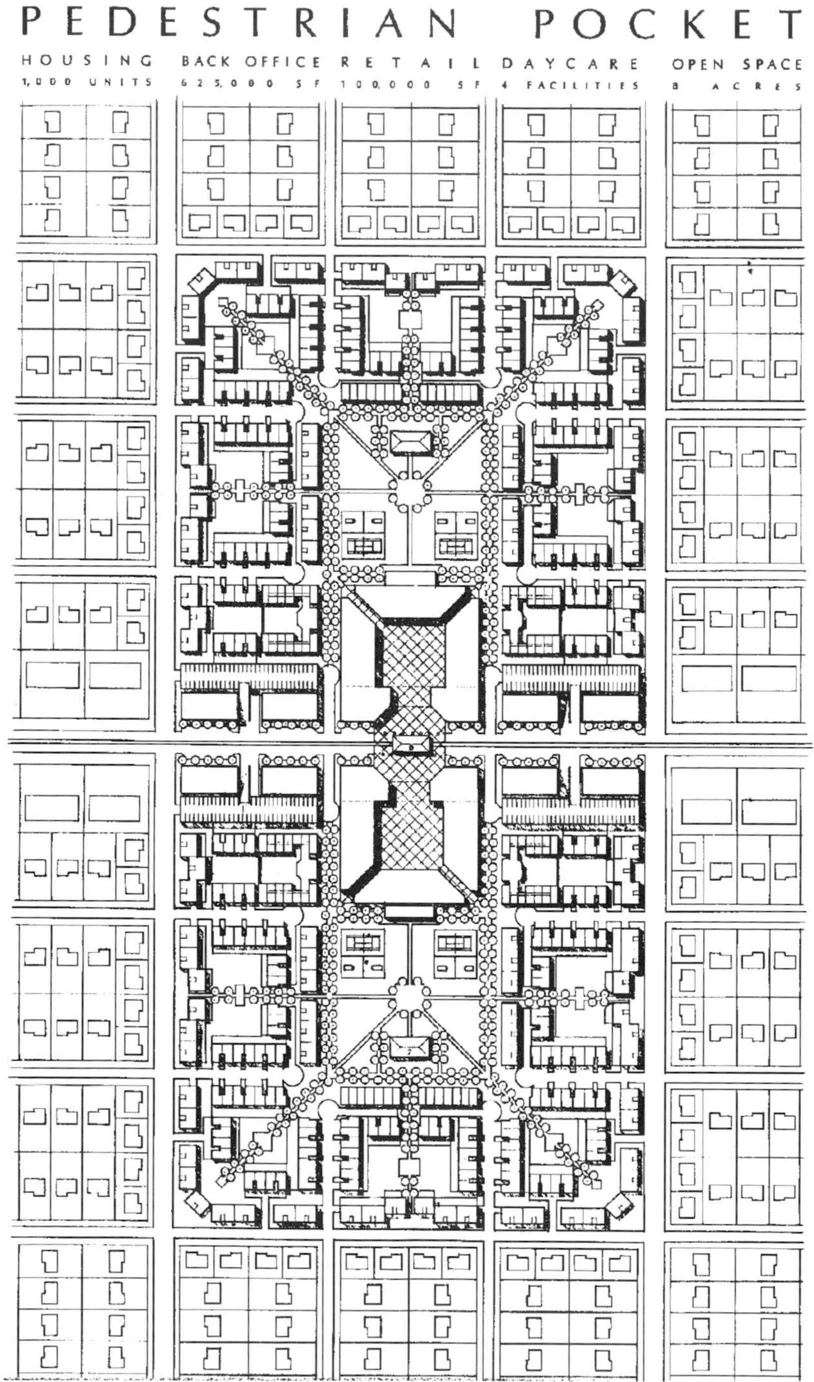

Plate 32 Peter Calthorpe's plan for a pedestrian pocket. Some modern architects and planners are incorporating elements from earlier small towns into their plans. California architect Peter Calthorpe advocates and has designed a series of "pedestrian pockets" which link land use and transportation. Calthorpe's pedestrian pockets are dense enough to support light rail systems, environmentally sensitive, and pedestrian friendly. Calthorpe's designs emphasize front porches and de-emphasize garages. (© Peter Calthorpe Associates.)

PART 7

Perspectives on Urban Design

PART SEVEN

INTRODUCTION

This section focuses on urban *design* – the way in which humans actually shape and structure the built environment. Urban designers are usually trained as architects with further training in urban design or city and regional planning. They focus on the design of sites larger than individual buildings – neighborhoods, park systems, highway corridors or even entire new towns. Professionals from the related field of landscape architecture are trained to provide advice on the relationship of the natural environment to the built environment. As in other areas where academics study cities or professionals work to build cities, material from many disciplines and professional fields is relevant.

The selections in this section by architects Camillo Sitte, Sim Van der Ryn and Stuart Cowan, urban designer Kevin Lynch, sociologist William Whyte, city planners Allan Jacobs and Donald Appleyard, and historian/architect/planner Dolores Hayden are examples of the way different approaches and perspectives enrich the field of urban design.

Urban designers may disagree on what makes for a good design, but they share a belief in the value of design itself, consciously thinking about physical relationships in the creation of urban space. Design is often expressed through drawings – generated by hand or computer – but design ideas may be expressed in many other ways.

Good urban design usually begins with intensive observation. Camillo Sitte (p. 466) observed squares and plazas in cities all over Europe in order to develop his principles for designing cities. William Whyte (p. 483) and his students spent hundreds of hours watching people use New York City parks and plazas – filming them, analyzing the films, and quantifying their behavior in order to reach principles for good park design. Kevin Lynch and his students (p. 478) surveyed people to see how they perceived the cities where they lived and asked them to draw maps to help understand what parts were and were not clear to them. Allan Jacobs, and Donald Appleyard's "urban design manifesto" grew out of many hours surveying, sketching, counting, mapping, photographing, measuring, and simply walking and looking at San Francisco.

Urban designers often criticize how badly the built environment fits human needs. Ugly, impersonal, dirty, dangerous, dysfunctional, race- and gender-segregated areas dominate many large cities today. Designers may have different priorities with respect to the value of improving traffic flow versus making pedestrian-friendly streets, economically revitalizing an area versus retaining historic buildings, or protecting the natural environment versus keeping the city competitive in the global economy, but urban design is never value free. The design of the built environment may not *determine* human behavior, but implicit in the following selections is the notion that bad design can numb the human spirit and good design can have powerful, positive influences on human beings.

The section begins with Austrian architect Camillo Sitte's theories on "the art of building cities," written at the end of the nineteenth century. Sitte is often viewed as the father of modern urban design. Dissatisfied with the impersonality of building projects in his own city of Vienna, Sitte set out to rediscover the artistic principles that guided classical Greek and Roman city builders and their mediaeval and renaissance successors. Sitte emphasizes the aesthetic, artistic character of city design. Underlying Sitte's theories is a profound belief in the importance of public places as venues to celebrate civic life. He values human scale development, the retention of historical elements, and the joys of irregularity. Sitte's writings ran counter to the grand building schemes the new industrial wealth in Europe was creating during his own day. But he touched a responsive chord, and he quickly developed followers throughout the world. Modernists enamoured of technology and machines dismissed him. But in today's postmodern era his ideas are enjoying a resurgence of interest. Sitte's perspective – the aesthetic character of cities – did not deal with many of the difficult and important issues raised in prior sections of this reader. Issues of class and race, economics and governance and how to implement city plans are nowhere discussed in his work. But within the sphere of aesthetics and design he made an important contribution.

Seventy years after *The Art of Building Cities* was published another slim volume on urban design appeared. MIT planning professor Kevin Lynch's *The Image of the City* quickly established itself as the foundation of much contemporary urban design. Lynch asked basic questions: How do people perceive the built environment? What are underlying elements common to human perception of the city? Armed with a better understanding of how people perceive the city image, what can urban designers actually do to design cities better? The selection by Lynch reprinted here describes the city image and the characteristics of five elements of urban form that Lynch considered to be fundamental. It is rich in suggestions about how these findings can shape better urban design.

Sociologist William Whyte's writing on the design of spaces summarizes ideas he developed studying the way in which New Yorkers use urban parks and plazas. Whyte was a sharp observer, fine writer, and exemplary of the way in which social scientists can link understanding of human behavior to improving urban design. His work also illustrates how urban research can lead directly to changes in city policy. The New York City Planning department and organizations in New York involved in planning and managing parks have incorporated Whyte's ideas directly into policy.

Allan Jacobs and Donald Appleyard also illustrate the connection between theory and practice. Jacobs is a professor of city planning at the University of California, Berkeley, where Appleyard also taught until his death. Both were students of Kevin Lynch. Jacobs served as the director of the San Francisco City Planning Department, and Appleyard worked with him on notable studies of street livability and urban design. Under Jacobs's leadership, the San Francisco City Planning Department produced an award-winning urban design plan, which draws heavily on Lynch's ideas and the insights of Appleyard's studies. That work has profoundly shaped the development of San Francisco over the last quarter century.

The final two selections in this section raise important questions about the relationship between design and gender and work roles and the relationship between design of the natural and built environments. Dolores Hayden condemns the isolated, wasteful, sexist world that male architects and planners have built reflecting outdated gender and work roles. She asks what a non-sexist city would be like and imagines groups reflecting the gender, age, and occupational structure of America sharing physical space, appliances, child rearing, and wage earning.

Architects Sim Van der Ryn and Stuart Cowan's concerns are environmental. They advocate a new process of "ecological design" and believe that architects, planners, and others should study natural ecological processes and learn from them. By using environmentally sensitive materials and building processes that respect natural systems and by pursuing strategies of conservation, restoration, and stewardship, Van der Ryn and Cowan believe it will be possible to heal much environmental damage and move towards sustainable urban development.

In sum, designing the urban environment requires sensitivity to both the natural environment and human needs. Urban planners, architects, landscape architects and other design professionals must weave together the natural environment and the urban fabric. To do that they must be sensitive to biological and other natural systems, physical form and function, aesthetics, and the way in which people use urban space. Each of the major subdivisions of this anthology can contribute to sensitive urban design.

CAMILLO SITTE

"Author's Introduction," "The Relationship Between Buildings, Monuments, and Public Squares," and "The Enclosed Character of the Public Square"

from *The Art of Building Cities* (1889)

Editors' introduction Camillo Sitte (1843–1903), an Austrian architect, felt that much was lost in the transformation of his native city of Vienna in the latter part of the nineteenth century. Sitte witnessed the old city walls – no longer useful against modern artillery – torn down, a "Ringstrasse" with new electric streetcars built to encircle the city, and old areas in the city leveled for monumental boulevards and impressive new buildings. Sitte felt nostalgia for the oddly shaped cathedral squares and narrow streets of Vienna, Salzburg, and other European cities that had evolved over time. He mourned the loss of structures built to human scale and public spaces embellished with statues, fountains, and other "municipal art" that adorned cites in classical Greece and Rome, the Middle Ages and the Renaissance.

Sitte knew instinctively that he enjoyed qualities in cities he had visited that had been built before the modern era. He also admired the urban form of Greek and Roman cities of classical antiquity discernable in ruins and fragments of existing cities. But what exactly were these qualities? And how might modern day designers, architects, and city planners incorporate the principles that Sitte and so many others enjoyed in new building projects? To answer these questions, Sitte embarked on a careful study of the built environment of notable European cities. Armed with a sketchbook he visited Athens, Rome, Florence, Venice, Paris, Pisa, Salzburg, Rothenburg on the Tauber, Dresden and dozens of other European cities. Everywhere he went, Sitte carefully sketched the physical form of squares and plazas, the outline of cathedrals and public buildings, the location of statues and fountains. He thought about scale and building materials, views and elevations, the integration of ornamental features with functional buildings. He imagined what civic life in these urban spaces must have been like at the time of Pericles and Julius Caesar, of Lorenzo the Magnficent, and Louis XI of France. Sitte reflected on how the architects and city planners designed aesthetically pleasing spaces that reinforced civic culture. The result was a masterful little book that set in motion the modern study of urban design. Sitte produced a volume that was passionate in its advocacy of human scale building and consideration of "artistic principles" in city building.

Sitte celebrated public space and was particularly enamoured of public squares and plazas. He applauded the practice in ancient Greece and Rome and during the Italian Renaissance of concentrating outstanding buildings around a single public square or plaza and ornamenting this center of community life with fountains, monuments, and statues.

Sitte had limited influence on the rebuilding of his native Vienna, but enormous and continuing impact elsewhere. "Sitte Schülen" (Sitte schools) sprang up all over Europe as young architects and planners read his book and discussed how to implement his ideas. *The Art of Building Cities* was translated into other languages. In the United States it was the bible of the turn of the century municipal arts movement described in LeGates and Stout above (p. 299). Dozens of local committees inspired by Sitte and the American writer Charles Mulford Robinson formed to "embellish and adorn" American cities.

Sitte fell out of favor in the interwar period when Le Corbusier and the insurgent young architects of the Congres International de Architecture Moderne (CIAM) developed plans to raze and rebuild what they saw as obsolete cities using modern materials, monumental scale, and designs inspired by industrial society (p. 336). Today there is a renewed interest in human scale postmodernist designs, and "new urbanist" architects and planners are rediscovering Sitte's writings and find much of value in the principles he developed more than a century ago.

Note the connections between Sitte's celebration of plazas and public squares in classical Greece and Rome and in medieval and renaissance European cities with Lewis Mumford's notion that cities above all should be theatres in which humans could display their culture (p. 92) and William Whyte's views on how urban parks and plazas contribute to urban life (p. 483).

Camillo Sitte's treatise is available in many languages and editions. The most recent English language edition is Camillo Sitte, *The Art of Building Cities: City Building According to its Artistic Fundamentals* (Westport, CT: Hyperion Press, 1979).

An exhaustive study of the form of European (and non-European) cities before the Industrial Revolution is A. E. J. Morris, *The History of Urban Form before the Industrial Revolutions* (New York: John Wiley, 1994). Another study of public spaces in European cities is Paul Zucker's *Town and Square from the Agora to the Village Green*, 2nd edn. (Cambridge, MA: MIT Press, 1959). Erwin Anton Gutkind's eight volume *International History of City Development* (New York: Free Press, 1964) has hundreds of illustrations of public spaces in European cities.

Spiro Kostoff's monumental studies of urban form, *The City Shaped* (New York: Little, Brown, 1991), and *The City Assembled* (New York: Little, Brown, 1992), contain illustrations of many of the features Sitte discusses. An account of the transformation of Vienna at the time that Sitte lived and wrote is contained in Carl Shorske, *Fin-De-Siècle Vienna: Politics and Culture* (New York: Vintage, 1981).

A biography and study of Sitte's work is George and Christiane Collins, *Camillo Sitte: The Birth of Modern City Planning* (New York: Rizzoli, 1981).

AUTHOR'S INTRODUCTION

Memory of travel is the stuff of our fairest dreams. Splendid cities, plazas, monuments, and landscapes thus pass before our eyes, and we enjoy again the charming and impressive spectacles that we have formerly experienced. If we could but stop again at those places where beauty never satiates, we could bear many dreary hours with a light heart and pursue life's long struggle with new energies. Assuredly the imperturbable lightheartedness of the South, on the Hellenic coast, in lower Italy and other favored climes, is above all a gift of nature. And the old cities of these countries, built after the beauty of nature itself, continue to augment nature's gentle and irresistible influence upon the soul of man. Only the person who has never understood the beauty of an ancient city could contradict this assertion.

Let him go ramble on the ruins of Pompeii to convince himself of it. If, after a day of patient investigation there, he walks across the bare Forum, he will be drawn, in spite of himself, to the summit of the monumental staircase toward the terrace of Jupiter's temple. On this platform, which dominates the entire place, he will sense, rising within him, waves of harmony like the pure, full tones of sublime music. Under this influence he will truly understand the words of Aristotle, who thus summarized all principles of city building: "A city should be built to give its inhabitants security and happiness."

The science of the technician will not suffice to accomplish this. We need, in addition, the talent of the artist. Thus it was in ancient times, in the Middle Ages, and in the Renaissance, wherever fine arts were held in esteem. It is only in our mathematical century that the construction and extension of cities has become a purely technical matter. Perhaps, then, it is not beside the point to recall that these problems have diverse aspects, and that he who has been given the least attention in our time is perhaps not the least important.

The object of this study, then, is clear. It is not our purpose to republish ancient and trite ideas, nor to reopen sterile complaints against the already proverbial banality of modern streets. It is useless to hurl general condemnations and to put everything that has been done in our time and place once more to the pillory. That kind of purely negative effort should be left to the critic who is never satisfied and who can only contradict. Those who have enough enthusiasm and faith in good causes should be convinced that our own era can create works of beauty and worth. We shall examine the plan of a number of cities, but neither as historian nor as critic. We wish to seek out, as technician and artist, the elements of composition which formerly produced such harmonious effects, and those which today produce only loose and dull results. Perhaps this study will permit us to find the means of satisfying the three principal requirements of practical city building: to rid the modern system of blocks and regularly aligned houses; to save as much as possible of that which remains from ancient cities; and in our creation to approach more closely the ideal of the ancient models.

This standpoint of practical art will lead us to consider especially the cities of the Middle Ages and the Renaissance. We shall be content, on recalling examples from Greek and Roman conceptions, either to explain the creations of following epochs, or to support the ideas that we propose to develop. For the principal architectural elements of cities have greatly changed since antiquity. Public squares (Forum, market, etc.) are used in our times not so much for great popular festivals or for the daily needs of our life. The sole reason for their existence is to provide more air and light, and to break the monotony of oceans of houses. At times they also enhance a monumental edifice by freeing its walls. It was quite different in ancient times. Public squares, or plazas, were then of prime necessity, for they were theaters for the principal scenes of public life, which today take place in enclosed halls. Under the open sky, on the agora, the council of the ancient Greeks gathered.

The market place, a further center of activity for our ancestors, has persisted, it is true, to the present time, but more and more it is being replaced by vast enclosed halls. And how many other scenes of public life have totally disappeared? Sacrifices before the temples, games, and theatrical presentations of all kinds. The temples themselves were scarcely covered, and the principal part of dwellings, around which were grouped large and small rooms, consisted of an open court. In a word, the distinction between the public square and other structures was so slight that it is amazing to our modern minds, accustomed to a very different state of things.

A review of the writings of the period proves to us that the ancients themselves sensed this similarity. Thus Vitruvius does not discuss the Forum in connection with the placement of public buildings or the arrangement of streets in his account of Dinocrates and his plan of Alexandria. But he does mention it in the same chapter which discusses the Basilica, and in the same book (1, 5) he deals with the theaters, palaces, the circus, and the baths. That is to say, all gathering places under the open sky constituted architectural works. The ancient Forum corresponds exactly to this definition, and Vitruvius logically places it in this group. This close relationship between the Forum and a public hall enhanced architecturally by statues and paintings is brought out clearly by the Latin writer's description, and more clearly still by an examination of the Forum of Pompeii. Vitruvius writes again on this subject:

> The Greeks arrange their market places in the form of a square and surround them by vast

double column supporting stone or marble architraves above which run the promenades. In Italian cities the Forum takes another aspect, for from time immemorial it has been the theater of gladiatorial combats. The columns, therefore, must be less densely grouped. They shelter the stalls of the silversmiths, and their upper floors have projections in the form of balconies which are advantageously placed for frequent use and for public revenue.

This description illustrates well the correspondence between theater and Forum. This relationship appears still more striking when we examine the plan of the Forum of Pompeii (Figure 1). The square is surrounded on all sides by public buildings. The temple of Jupiter alone rises in isolation. And the two-story colonnade which surrounds the entire space is interrupted only by the peristyle of the temple of the household gods, which makes a greater projection than the other buildings. The center of the Forum remains free, but its periphery is occupied by numerous monuments, the pedestals of which, covered with inscriptions, are still visible.

What a grandiose impression this place must have made! To our modern point of view its effect is like that of a great concert hall without ceiling. In every direction the eye fell upon edifices which in no respect resembled our files of modern houses, and there were far fewer streets opening directly on the plaza. Streets ran behind buildings III, IV, and V, but they did not extend as far as the Forum. Streets C, D, E, and F were closed by grilles, and even those on the north side passed under the monumental portals, A and B.

Forum Romanum (Figure 2) was conceived according to the same principles. It is surrounded, of course, by buildings more varied in type but all monumental. The streets which open onto it were arranged to avoid too frequent openings in the frame of the plaza. Monuments are located around its sides rather than in its center. In brief, the place of the forum in cities corresponds to that of the principal room of a house. It is to the city, so to speak, the principal hall, as well arranged as it is richly furnished. There stand assembled in immense bulk the columns, the statues, the monuments, and everything that can contribute to the splendor of the place. The art treasures of some of them were said to be numbered in hundreds and thousands. As they did not encumber the midst of the plaza, but were always located at the periphery, it was possible to encompass them all with a single glance; and the spectacle must have been imposing. This concentration of plastic and architectural masterpieces at a single point was a stroke of genius. Aristotle had taught it. He advocated grouping the temples of the gods with public buildings. Pausanias wrote similarly, "A city without public edifices and squares is not worthy of its name."

The market place of Athens is arranged in its principal features according to the same rules, as well as may be judged from the restoration projects. They are applied on a still grander scale in the consecrated cities of Hellenic antiquity (Olympia, Delphi, Eleusis) (Figure 3). Masterpieces of architecture, painting, and sculpture are found there in a superb and imposing union capable of rivaling the most powerful tragedies and the most majestic symphonies. The Acropolis of Athens (Figure 4) is the most finished creation of this character. A high plateau surrounded by high walls is the base of it. The lower entrance portal, the enormous flight of steps, and monumental vestibules constitute the first phrase of this symphony in marble, gold, ivory, bronze, and color. The interior temples and monuments are the stone myths of the Greek people. The highest poetry and thought are embodied in them. It is truly the center of a considerable city, an expression of the feelings of a great people. It is no longer a simple square in the ordinary sense of the term, but the work of several centuries grown to the maturity of pure art.

It is impossible to establish a higher aim in this style, and it is difficult to imitate successfully this splendid model, but it should always remain before our eyes in all our works as the most sublime ideal to attain. In the progress of our study we shall see that the principles which have inspired such building are not entirely lost, but that they remain to us.

THE RELATIONSHIP BETWEEN BUILDINGS, MONUMENTS, AND PUBLIC SQUARES

In the South of Europe, and especially in Italy, where ancient cities and ancient public customs have remained alive for ages, even to the present

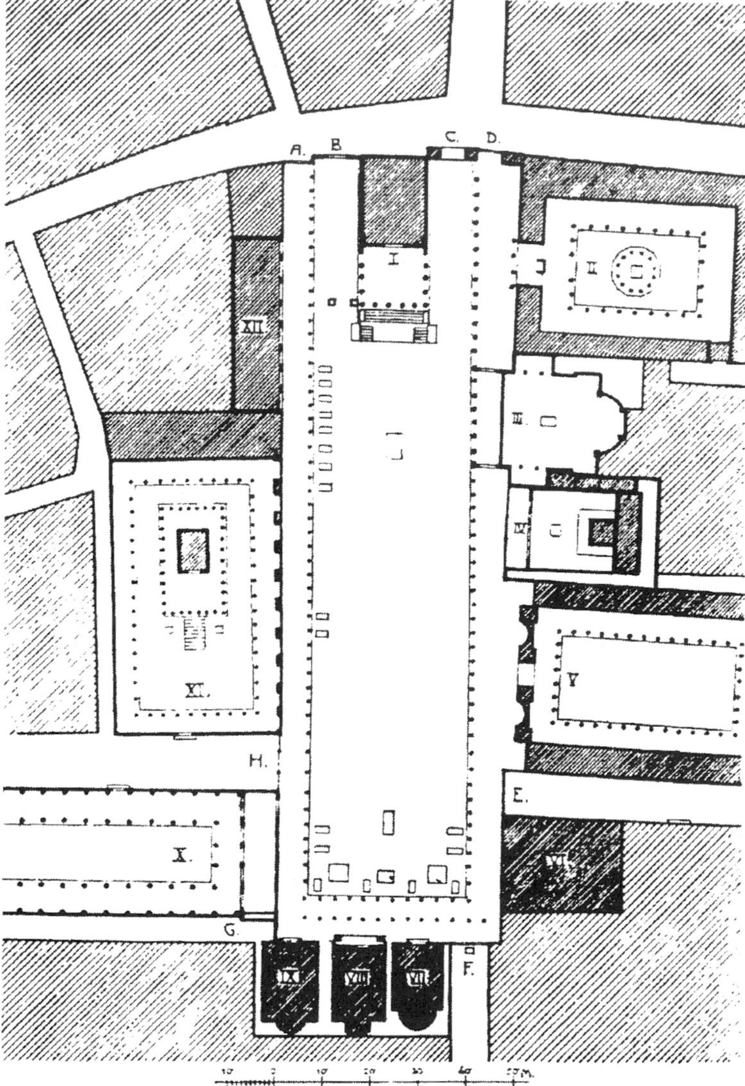

Figure 1 Forum of Pompeii
I Temple of Jupiter, II Enclosed Market, III Temple of Household Gods, IV Temple of Vespasianus, V Eumachia, VI Comitium, VII–IX Public Buildings, X Basilica, XI Temple of Apollo, XII Market Hall.

in some places, public squares still follow the type of the ancient forum. They have preserved their role in public life. Their natural relationships with the buildings which enclose them may still be readily discerned. The distinction between the forum, or agora, and the market place also remains. As before, we find the tendency to concentrate outstanding buildings at a single place, and to ornament this center of community life with fountains, monuments, and statues which can bring back historical memories and which, during the Middle Ages and the Renaissance, constituted the glory and pride of each city.

It was there that traffic was most intense. That is where public festivals and theatrical presentations were held. There it was that official ceremonies were conducted and laws promulgated. In Italy, according to varying circumstances, two or three public places, rarely

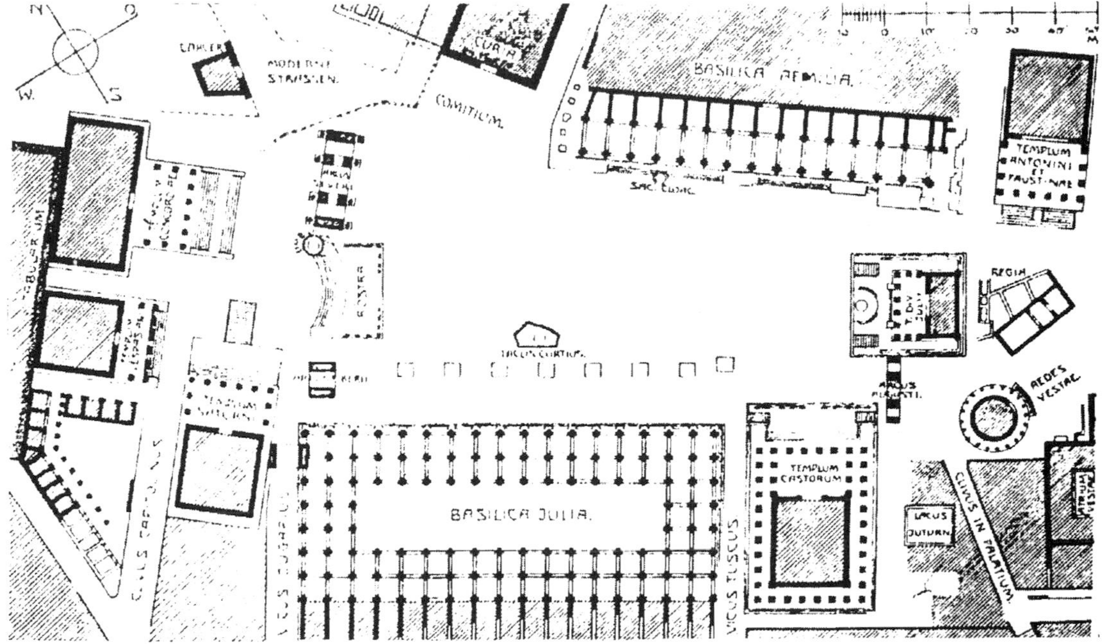

Figure 2 Forum Romanum

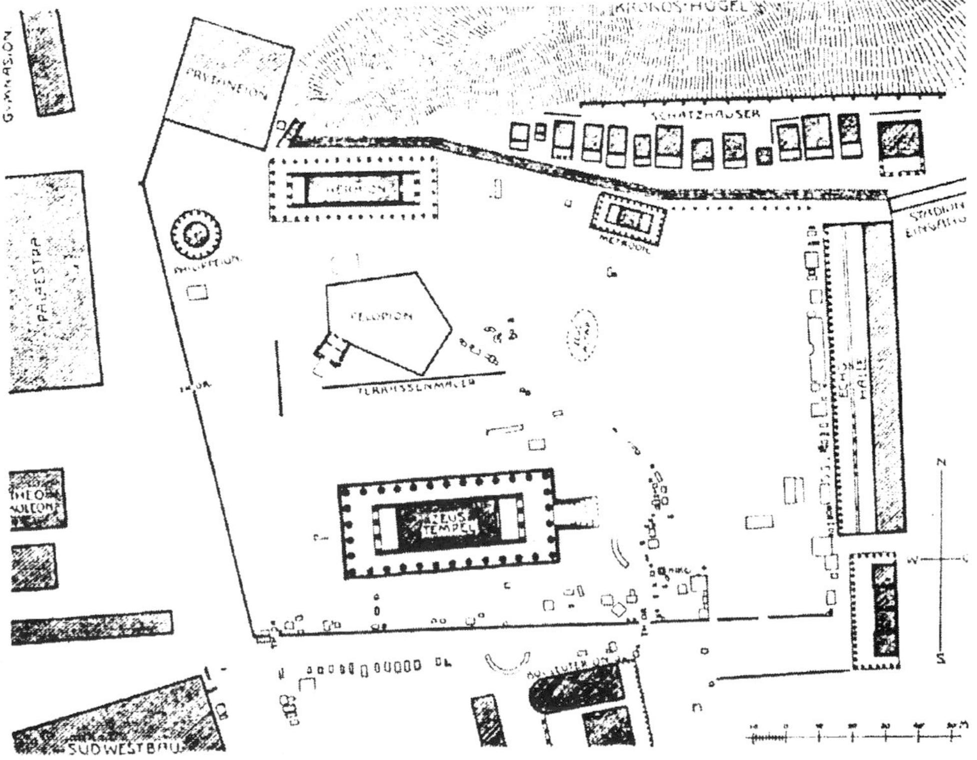

Figure 3 The Temple of Zeus and Plaza of Olympia

a single one, served these practical purposes.

The existence of two powers, temporal and spiritual, required two distinct centers: one, the cathedral square (Figure 5) dominated by the campanile, the baptistry, and the palace of the bishop; the other, the Signoria, or manor place, which is a kind of vestibule to a royal residence. It is enclosed by houses of the country's great and adorned with monuments. Sometimes we see there a loggia, or open gallery, used by a military guard, or a high terrace from which laws and public statements were promulgated. The Signoria of Florence (Figure 6) is the finest example of this. The market square, rarely lacking even in cities of northern Europe, is the meeting place of the citizens. There stand the City Hall and the more or less richly decorated traditional fountain, the sole vestige of the past that has been conserved since the lively activity of merchants and traders has been moved within to iron cages and glass market places.

The important function of the public square in the community life of past ages is evident. The period of the Renaissance saw the birth of masterpieces in the manner of the Acropolis of Athens, where everything concurred to produce a finished artistic effect. The cathedral place at Pisa, an Acropolis of Pisa (Figure 5), is the proof of this. It includes everything that the people of the City have been able to create in building religious edifices of unparalleled richness and grandeur. The splendid cathedral, the campanile, the baptistry, the incomparable Campo Santo are not depreciated by profane or banal surroundings of any kind. The effect produced by such a place, removed from the world of baseness while rich in the noblest works of the human spirit, is overpowering. Even those with a poorly developed sensitiveness to art are unable to escape the power of this impression. There is nothing there to distract our thoughts or to intrude our daily affairs. The esthetic enjoyment of those who look upon the noble facade of the Cathedral is not spoiled by the sight of a modern haberdashery, by the cries of drivers and porters, or by the tumult of a cafe. Peace reigns over the place. It is thus possible to give full attention to the artwork assembled there.

This situation is almost unique, although that of Saint Francis of Assisi and the arrangement of the Certosa de Pavia closely approach it. In general, the modern period does not encourage the formation of such perfect groupings. Cities, even in the fatherland of art, undergo the fate of palaces and dwellings. They no longer have distinct character. They present a mixture of motifs borrowed as much from the architecture of the north as from that of the southern countries. Ideas and tastes have been mingled as the people themselves have been interchanged. Local

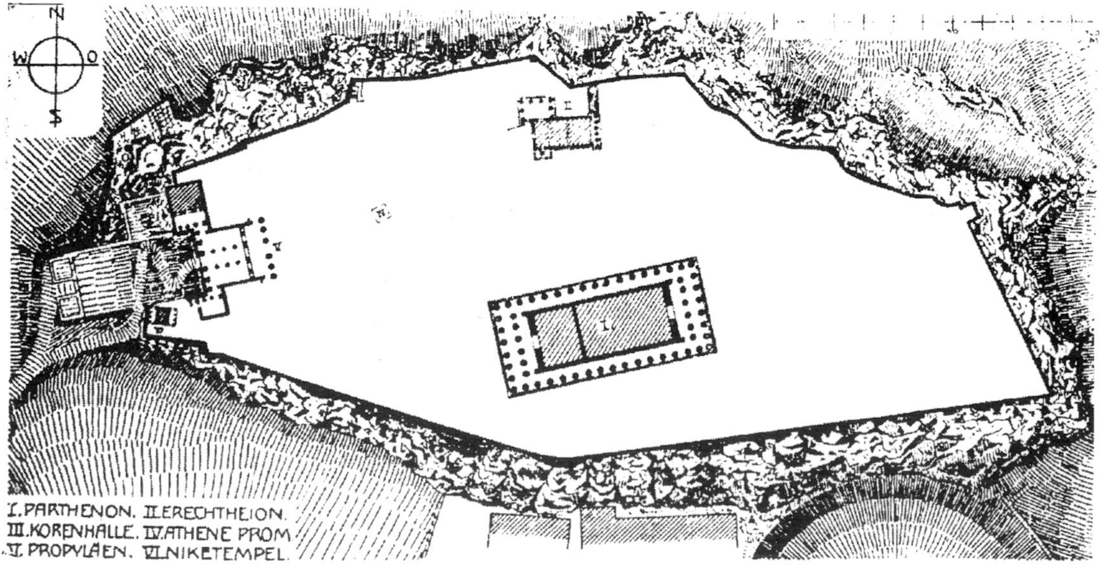

Figure 4 The Acropolis of Athens in the age of Pericles

characteristics are gradually disappearing. The market place alone, with its City Hall and fountain, has here and there remained intact.

In passing we should like to remark that our intention is not to suggest a sterile imitation of the beauties spoken of as "picturesque" in the ancient cities for our present needs. The proverb, "Necessity breaks even iron," is fully applicable here. Changes made necessary by hygiene or other requirements must be carried out, even if the picturesque suffers from it. But that does not prevent us from examining the work of our forebears at close range to determine how much of it may be adapted to modern conditions. In this way alone can we resolve the esthetic part of the practical problem of city building, and determine what can be saved from the heritage of our ancestors.

Before determining the question in a positive manner, we state the principle that during the Middle Ages and Renaissance public squares were often used for practical purposes, and that they formed an entirety with the buildings which enclosed them. Today they serve at best as places for stationing vehicles, and they have no relation to the buildings which dominate them. Our parliament buildings have no agora enclosed by columns. Our universities and cathedrals have lost their atmosphere of peace. Surging throngs no longer circulate on market days before our City Halls. In brief, activity is lacking precisely in those places where, in ancient times, it was most intense – near public structures. Thus, to a great extent, we have lost that which contributed to the splendor of public squares.

And the fabric of their very splendor, the numerous statues, is almost entirely lacking today. What have we to compare to the richness of ancient forums and to works of majestic style like the Signoria of Florence and its Loggia dei Lanzi?

A few years ago there flourished at Vienna a remarkable school of sculpture whose works of merit cannot be scorned. They were generally used to adorn buildings. In only a few exceptional cases were their works used in public squares. Statues adorn the two museums, the palace of Parliament, the two Court theaters, the City Hall, the new university, the Votive Church. But there is no interest in adorning public open spaces. And that is true not only in Vienna, but nearly everywhere.

Buildings lay claim to so many statues that commissions are needed to find new subjects to be represented. It is often necessary to wait for years to find a suitable place for a statue although many appropriate places remain empty in the meantime. After long efforts we have reconciled ourselves to modern public squares as vast as they are deserted, and the monument, without a place of refuge, becomes stranded on some small and ancient space. That is even more strange, yet true. After much groping about, this fortunate result occurs, for it is thus that a work of art derives its value and produces a more powerful impression. Indifferent artists who

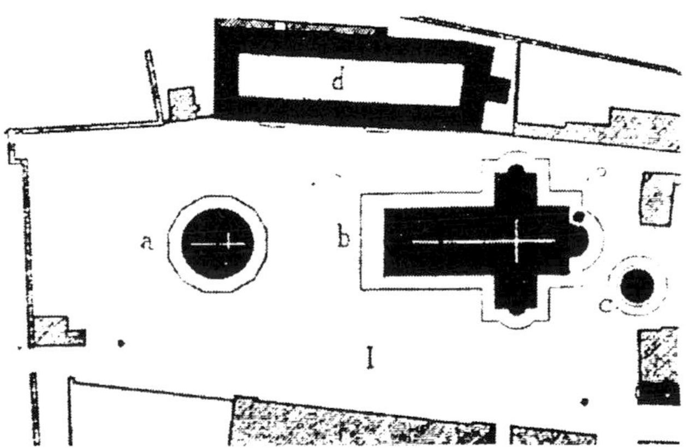

Figure 5 Pisa: Cathedral Square
a. Saint-Jean b. Cathedral
c. Campanile d. Campo Santo (Cemetery)

Figure 6 Florence, Piazza of the Signoria

neglect to provide for such effects must bear the entire responsibility of it.

The story of Michelangelo's *David* at Florence shows how mistakes of this kind are perpetrated in modern times. This gigantic marble statue stands close to the walls of the Palazzo Vecchio, to the left of its principal entrance, in the exact place chosen by Michelangelo. The idea of erecting a statue on this place of ordinary appearance would have appeared to moderns as absurd if not insane. Michelangelo chose it, however, and without doubt deliberately; for all those who have seen the masterpiece in this place testify to the extraordinary impression that it makes. In contrast to the relative scantiness of the place, affording an easy comparison with human stature, the enormous statue seems to swell even beyond its actual dimensions. The sombre and uniform, but powerful, walls of the palace provide a background on which we could not wish to improve to make all the lines of the figure stand out.

Today the David is moved into one of the academy's halls under a glass cupola in the midst of plaster reproductions, photographs, and engravings. It serves as a model for study and an object of research for historians and critics. A special mental preparation is needed now to resist the morbid influences of an art prison that we call a museum, and to have the ability to enjoy the imposing work. Moreover, the spirit of the times, which believed that it was perfecting art, and which was still not satisfied with this innovation, had a bronze cast made of the David in its original grandeur and put it up on a vast plaza (naturally in its mathematical center) far from Florence at the Via dei Colli. It has a superb horizon before it; behind it, cafes; on one side, a carriage station, a corso; and from all sides the murmurs of Baedeker readers ascend to it. In this setting the statue produces no effect at all. The opinion that its dimensions do not exceed human stature is often heard. Michelangelo thus understood best the kind of placement that would be suitable for his work, and, in general, the ancients were abler than we are in these matters.

The fundamental difference between the procedures of former times and those of today rests in the fact that we constantly seek the largest possible space for each little statue. Thus we diminish the effect that it could produce, instead of augmenting it with the assistance of a neutral background such as painters have used in their portraits.

This explains why the ancients erected their monuments by the sides of public places, as is shown in the view of the Signoria of Florence. In this way, the number of statues could increase indefinitely without obstructing the circulation of traffic, and each of them had a fortunate background. Contrary to this, we hold the middle of a public place as the sole spot worthy to receive a monument. Thus no esplanade, however magnificent, can have more than one. If by misfortune it is irregular and if its center cannot be located geometrically we become confused and allow the space to remain empty for eternity.

THE ENCLOSED CHARACTER OF THE PUBLIC SQUARE

The old practice of setting churches and palaces back against other buildings brings to mind the ancient forum and its unbroken frame of public buildings. In examining the public squares that came into being during the Middle Ages and the Renaissance, especially in Italy, it is seen that this pattern has been retained for ages by tradition. The old plazas produce a collective harmonious effect because they are uniformly enclosed. In fact, the public square owes its name to this characteristic in an expanse at the center of a city. It is true that we now use the term to indicate any parcel of land bounded by four streets on which all construction has been renounced.

That can satisfy the public health officer and the technician, but for the artist these few acres of ground are not yet a public square. Many things must be done to embellish the area to give it character and importance. For just as there are furnished and unfurnished rooms, we could speak of complete and incomplete squares. The essential thing of both room and square is the quality of enclosed space. It is the most essential condition of any artistic effect, although it is ignored by those who are now elaborating on city plans.

The ancients, on the contrary, employed the most diverse methods of fulfilling this condition under the most diverse circumstances. They were, it is true, supported by tradition and favored by the usual narrowness of streets and less active traffic movement. But it is precisely in cases where these aids were lacking that their talent and artistic feeling is displayed most conspicuously.

A few examples will assist in accounting for this. The following is the simplest. Directly facing a monumental building a large gap was made in a mass of masonry, and the square thus created, completely surrounded by buildings, produced a happy effect. Such is the Piazza S. Giovanni at Brescia. Often a second street opens on to a small square, in which case care is taken to avoid an excessive breach in the border, so that the principal building will remain well enclosed. The methods used by the ancients to accomplish this were so greatly varied that chance alone could not have guided them. Undoubtedly they were often assisted by circumstance, but they also knew how to use circumstances admirably.

Today in such cases all obstructions would be taken down and large breaches in the border of the public place would be opened, as is done when we decide to "modernize" a city. Ancient streets would be found to open on the square in a manner precisely contrary to the methods of modern city builders, and mere chance would not account for this. Today the practice is to join two streets that intersect at right angles at each corner of the square, probably to enlarge as much as possible the opening made in the enclosure and to destroy every impression of cohesion. Formerly the procedure was entirely different. There was an effort to have only one street at each angle of the square. If a second artery was needed in a direction at right angles to the first it was designed to terminate at a sufficient distance from the square to remain out of view from the square. And better still, the three or four streets which came in at the corners each ran in a different direction. This interesting arrangement was reproduced so frequently, and more or less completely, that it can be considered as one of the conscious or subconscious principles of ancient city building.

Careful study shows that there are many advantages to an arrangement of street openings in the form of turbine arms. From any part of the square there is but one exit on the streets opening into it, and the enclosure of buildings is not broken. It even seems to enclose the square completely, for the buildings set at an angle conceal each other, thanks to perspective, and unsightly impressions which might be made by openings are avoided. The secret of this is in having streets enter the square at right angles to the visual lines instead of parallel to them. Joiners and carpenters have followed this principle since the Middle Ages when, with subtle art, they sought to make joints of wood and stone inconspicuous if not invisible.

The Cathedral Square at Ravenna shows the purest type of the arrangement just described. The square of Pistoia (Figure 7) is in the same

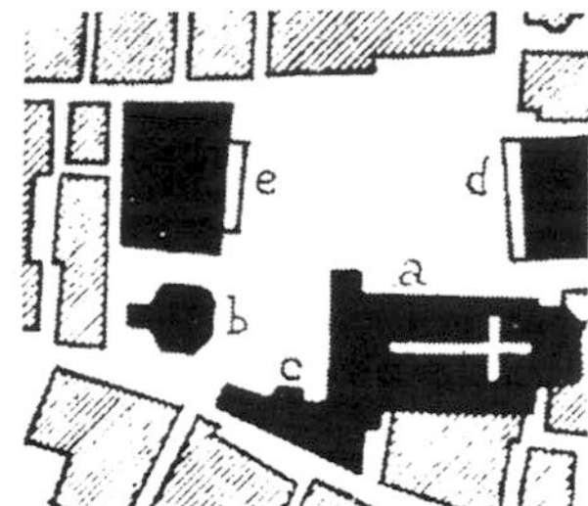

Figure 7 Pistoia: Cathedral Square
a. Cathedral b. Baptistry
c. Residence d. Palais de la Commune
e. Palais du Podestat

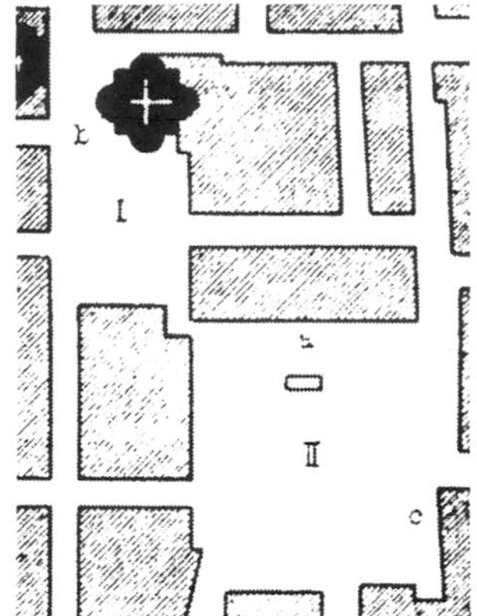

Figure 8 Parma
a. Pal. del Commune b. Madonna della Steccata
c. Pal. della Podesteria
I. Piazza d. Steccata II. Piazza Grande

manner; as is . . . the Piazza Grande at Parma (Figure 8).

The ancients had recourse to still other means of closing in their squares. Often they broke the infinite perspective of a street by a monumental portal or by several arcades of which the size and number were determined by the intensity of traffic circulation. This splendid architectural pattern has almost entirely disappeared, or, more accurately, it has been suppressed. Again Florence gives us one of the best examples in the portico of the Uffizi with its view of the Arno in the distance. Every Italian city of average importance has its portico, and this is also true north of the Alps. We mention only the Langasser Thor at Danzig, the entrance portal of the City Hall and Chancellery at Bruges, the Kerkboog at Nimeguen, the great Bell Tower at Rouen, the monumental Portals of Nancy, and the windows of the Louvre.

More or less ornate portals like those that simply but effectively frame the Piazza dei Signori at Verona (Figure 9) are to be found in all the royal residences, in the chateaux and city halls, and they are used as much for vehicular traffic as by pedestrians. While ancient architects used this pattern wherever possible with infinite

Figure 9 Verona: Piazza dei Signori

variations, our modern builders seem to ignore its existence. Let us recall, to demonstrate again the persistence of ancient traditions, that at Pompeii, too, there is an Arc de Triomphe at the entrance to the Forum.

Columns were used with porticos to form enclosures for public squares. Saint Peter's in Rome is the best example of this . . .

Arcades were used to embellish monumental buildings more frequently in former times than at present, either on the higher stories, as in the City Halls of Halle (1548) and Cologne (1568), or on the ground level . . .

All of these above-mentioned architectural forms in former times made up a complete system of enclosing public squares. Today there is a contrary tendency to open them on all sides. It is easy to describe the results that have come about. It has tended to destroy completely the old public squares. Wherever these openings have been made the cohesive effect of the square has been completely nullified.

KEVIN LYNCH

"The City Image and its Elements"

from *The Image of the City* (1960)

Editors' introduction If Camillo Sitte originated modern urban design at the end of the nineteenth century, Kevin Lynch (1918–1989) is the towering figure of twentieth century urban design. Lynch was a professor of urban studies and planning at the Massachusetts Institute of Technology where he taught courses in urban design. As a young student Kevin Lynch apprenticed himself to Frank Lloyd Wright. Drawing widely on material from psychology and the humanities, Lynch sought to understand how people perceive their environments and how design professionals can respond to the deepest human needs. Lynch's rambling, profoundly humane writings weave together a unique blend of theory and practical design suggestions drawn from his voluminous reading in history, anthropology, architecture, art, literature, and a host of other areas.

This influential chapter from *The Image of the City* presents Lynch's best known concepts on how people perceive cities. Lynch argues that people perceive cities as consisting of underlying city form "elements" such as *paths* (along which movement flows) and *edges* (which differentiate one part of the urban fabric from another). If they understand how people perceive these elements and design to make cities more imageable, Lynch argues, designers can create more psychologically satisfying urban environments.

Urban designers throughout the world today sketch out the elements of cities or parts of cities they are designing as paths, edges, nodes, landmarks, and districts – the underlying elements of city form that Lynch identified – and draw on his theories and practical suggestions to strengthen the city image. Planners in cities as diverse as San Francisco, Cairo, Ciudad Guyana in Venezuela, and Havana have used Lynch's concepts to inform their urban planning and design strategies.

Compare Lynch's ideas about what people find psychologically satisfying and aesthetically appealing about cities with Camillo Sitte's ideas (p. 478). Contrast his practical suggestions with William Whyte's applied principles and standards for park and plaza design (p. 483), and Frederick Law Olmsted's vision of urban parks (p. 314).

In addition to *The Image of the City* (Cambridge, MA: MIT Press, 1960), Lynch's many books include a textbook on site design co-authored with Gary Hack, *Site Planning*, 2nd edn. (Cambridge, MA: MIT Press, 1971), *What Time Is This Place* (Cambridge, MA: MIT Press, 1979), a book on historic preservation, *Managing the Sense of a Region* (Cambridge, MA: MIT Press, 1976), and his magnum opus, *Good City Form* (Cambridge, MA: MIT Press, 1991). Other of Lynch's writings are contained in Kevin Lynch, Tridib Banerjee, and Michael Southworth (eds.), *City Sense and City Design: Writings and Projects of Kevin Lynch* (Cambridge, MA: MIT Press, 1995).

Other books on the way in which people perceive the built environment and on urban design include Mike Greenberg, *The Poetics of Cities: Designing Neighborhoods that Work* (Columbus: Ohio State University Press, 1995) and Doug Kelbaugh, *Common Place: Toward Neighborhood and Regional*

Design (Seattle: University of Washington Press, 1997), Anthony Hiss, *The Experience of Place* (New York: Knopf, 1990), and Robert Sommer, *Personal Space* (Englewood Cliffs, NJ: Prentice-Hall, 1969).

There seems to be a public image of any given city which is the overlap of many individual images. Or perhaps there is a series of public images each held by some significant number of citizens. Such group images are necessary if an individual is to operate successfully within his environment and to cooperate with his fellows. Each individual picture is unique, with some content that is rarely or never communicated, yet it approximates the public image, which in different environments is more or less compelling, more or less embracing.

[. . .]

The contents of the city images so far studied, which are referable to physical forms, can conveniently be classified into five types of elements: paths, edges, districts, nodes, and landmarks . . . These elements may be defined as follows:

1 *Paths*. Paths are the channels along which the observer customarily, occasionally, or potentially moves. They may be streets, walkways, transit lines, canals, railroads. For many people, these are the predominant elements in their image. People observe the city while moving through it, and along these paths the other environmental elements are arranged and related.

2 *Edges*. Edges are the linear elements not used or considered as paths by the observer. They are the boundaries between two phases, linear breaks in continuity: shores, railroad cuts, edges of development, walls. They are lateral references rather than coordinate axes. Such edges may be barriers, more or less penetrable, which close one region off from another; or they may be seams, lines along which two regions are related and joined together. These edge elements, although probably not as dominant as paths, are for many people important organizing features, particularly in the role of holding together generalized areas, as in the outline of a city by water or wall.

3 *Districts*. Districts are the medium-to-large sections of the city, conceived of as having two-dimensional extent, which the observer mentally enters "inside of," and which are recognizable as having some common, identifying character. Always identifiable from the inside, they are also used for exterior reference if visible from the outside. Most people structure their city to some extent in this way, with individual differences as to whether paths or districts are the dominant elements. It seems to depend not only upon the individual but also upon the given city.

4 *Nodes*. Nodes are points, the strategic spots in a city into which an observer can enter, and which are the intensive foci to and from which he is traveling. They may be primarily junctions, places of a break in transportation, a crossing or convergence of paths, moments of shift from one structure to another. Or the nodes may be simply concentrations, which gain their importance from being the condensation of some use or physical character, as a street-corner hangout or an enclosed square. Some of these concentration nodes are the focus and epitome of a district, over which their influence radiates and of which they stand as a symbol. They may be called cores. Many nodes, of course, partake of the nature of both junctions and concentrations. The concept of node is related to the concept of path, since junctions are typically the convergence of paths, events on the journey. It is similarly related to the concept of district, since cores are typically the intensive foci of districts, their polarizing center. In any event, some nodal points are to be found in almost every image, and in certain cases they may be the dominant feature.

5 *Landmarks*. Landmarks are another type of point-reference, but in this case the observer does not enter within them, they are external. They are usually a rather simply defined physical object: building, sign, store, or mountain. Their use involves the singling out of one element from a host of possibilities. Some landmarks are

distant ones, typically seen from many angles and distances, over the tops of smaller elements, and used as radial references. They may be within the city or at such a distance that for all practical purposes they symbolize a constant direction. Such are isolated towers, golden domes, great hills. Even a mobile point, like the sun, whose motion is sufficiently slow and regular, may be employed. Other landmarks are primarily local, being visible only in restricted localities and from certain approaches. These are the innumerable signs, store fronts, trees, doorknobs, and other urban detail, which fill in the image of most observers. They are frequently used clues of identity and even of structure, and seem to be increasingly relied upon as a journey becomes more and more familiar.

[. . .]

PATHS

For most people interviewed, paths were the predominant city elements, although their importance varied according to the degree of familiarity with the city. People with the least knowledge of Boston tended to think of the city in terms of topography, large regions, generalized characteristics, and broad directional relationships. Subjects who knew the city better had usually mastered parts of the path structure; these people thought more in terms of specific paths and their interrelationships. A tendency also appeared for the people who knew the city best of all to rely more upon small landmarks and less upon either regions or paths.

The potential drama and identification in the highway system should not be underestimated. One Jersey City subject, who can find little worth describing in her surroundings, suddenly lit up when she described the Holland Tunnel. Another recounted her pleasure:

> You cross Baldwin Avenue, you see all of New York in front of you, you see the terrific drop of land [the Palisades] . . . and here's this open panorama of Lower Jersey City in front of you and you're going downhill, and there you know: there's the tunnel, there's the Hudson River and everything . . . I always look to the right to see if I can see the . . . Statue of Liberty . . . Then I always look up to see the Empire State Building, see how the weather is . . . I have a real feeling of happiness because I'm going someplace, and I love to go places.

[. . .]

Concentration of special use or activity along a street may give it prominence in the minds of observers. Washington Street in Boston is the outstanding Boston example: subjects consistently associated it with shopping and theaters . . . People seemed to be sensitive to variations in the amount of activity they encountered and sometimes guided themselves largely by following the main stream of traffic. Los Angeles' Broadway was recognized by its crowds and its street cars; Washington Street in Boston was marked by a torrent of pedestrians. Other kinds of activity at ground level also seemed to make places memorable, such as construction work near South Station, or the bustle of the food markets.

Characteristic spatial qualities were able to strengthen the image of particular paths. In the simplest sense, streets that suggest extremes of either width or narrowness attracted attention . . .

[. . .]

Where major paths lacked identity, or were easily confused one for the other, the entire city image was in difficulty . . . Boston's Longfellow Bridge was not infrequently confused with the Charles River Dam, probably since both carry transit lines and terminate in traffic nodes . . .

[. . .]

People tended to think of path destinations and origin points: they liked to know where paths came from and where they led. Paths with clear and well-known origins and destinations had stronger identities, helped tie the city together, and gave the observer a sense of his bearings whenever he crossed them. Some subjects thought of general destinations for paths, to a section of the city, for example, while others thought of specific places. One person, who made rather high demands for intelligibility upon the city environment, was troubled because he saw a set of railroad tracks, and did not know the destination of trains using them.

[. . .]

EDGES

Edges are the linear elements not considered as paths: they are usually, but not quite always, the boundaries between two kinds of areas. They act as lateral references. They are strong in Boston and Jersey City but weaker in Los Angeles. Those edges seem strongest which are not only visually prominent, but also continuous in form and impenetrable to cross movement. The Charles River in Boston is the best example and has all of these qualities . . .

[. . .]

It is difficult to think of Chicago without picturing Lake Michigan. It would be interesting to see how many Chicagoans would begin to draw a map of their city by putting down something other than the line of the lake shore. Here is a magnificent example of a visible edge, gigantic in scale, that exposes an entire metropolis to view. Great buildings, parks, and tiny private beaches all come down to the water's edge, which throughout most of its length is accessible and visible to all. The contrast, the differentiation of events along the line, and the lateral breadth are all very strong. The effect is reinforced by the concentration of paths and activities along its extent. The scale is perhaps unrelievedly large and coarse, and too much open space is at times interposed between city and water, as at the Loop. Yet the facade of Chicago on the Lake is an unforgettable sight.

DISTRICTS

Districts are the relatively large city areas which the observer can mentally go inside of, and which have some common character. They can be recognized internally, and occasionally can be used as external reference as a person goes by or toward them. Many persons interviewed took care to point out that Boston, while confusing in its path pattern even to the experienced inhabitant, has, in the number and vividness of its differentiated districts, a quality that quite makes up for it. As one person put it: "Each part of Boston is different from the other. You can tell pretty much what area you're in."

[. . .]

Subjects, when asked which city they felt to be a well-oriented one, mentioned several, but New York (meaning Manhattan) was unanimously cited. And this city was cited not so much for its grid, which Los Angeles has as well, but because it has a number of well-defined characteristic districts, set in an ordered frame of rivers and streets. Two Los Angeles subjects even referred to Manhattan as being "small" in comparison to their central area! Concepts of size may depend in part on how well a structure can be grasped.

[. . .]

Usually the typical features were imaged and recognized in a characteristic cluster, the thematic unit. The Beacon Hill image, for example, included steep narrow streets; old brick row houses of intimate scale; inset, highly maintained, white doorways; black trim; cobblestones and brick walks; quiet; and upper-class pedestrians. The resulting thematic unit was distinctive by contrast to the rest of the city and could be recognized immediately . . .

NODES

Nodes are the strategic foci into which the observer can enter, typically either junctions of paths, or concentrations of some characteristic. But although conceptually they are small points in the city image, they may in reality be large squares, or somewhat extended linear shapes, or even entire central districts when the city is being considered at a large enough level. Indeed, when conceiving the environment at a national or international level, then the whole city itself may become a node.

The junction, or place of a break in transportation, has compelling importance for the city observer. Because decisions must be made at junctions, people heighten their attention at such places and perceive nearby elements with more than normal clarity. This tendency was confirmed so repeatedly that elements located at junctions may automatically be assumed to derive special prominence from their location. The perceptual importance of such locations shows in another way as well. When subjects were asked where on a habitual trip they first felt a sense of arrival in downtown Boston, a large number of people singled out break-points of transportation as the key places . . .

LANDMARKS

Landmarks, the point reference considered to be external to the observer, are simple physical elements which may vary widely in scale. There seemed to be a tendency for those more familiar with a city to rely increasingly on systems of landmarks for their guides – to enjoy uniqueness and specialization, in place of the continuities used earlier.

Since the use of landmarks involves the singling out of one element from a host of possibilities, the key physical characteristic of this class is singularity, some aspect that is unique or memorable in the context. Landmarks become more easily identifiable, more likely to be chosen as significant, if they have a clear form; if they contrast with the background; and if there is some prominence of spatial location. Figure–background contrast seems to be the principal factor. The background against which an element stands out need not be limited to immediate surroundings: the grasshopper weathervane of Faneuil Hall, the gold dome of the State House, or the peak of the Los Angeles City Hall are landmarks that are unique against the background of the entire city.

[. . .]

WILLIAM H. WHYTE

"The Design of Spaces"

from *City: Rediscovering the Center* (1988)

Editors' introduction Puzzled by why some of New York's parks and plazas were well used while others were almost empty, the New York City Planning Commission asked sociologist William Whyte (1918–1999) to study park and plaza use and help draft a comprehensive design plan for the city.

Whyte's lucid writing on planning and design gave him great credibility. Hunter College appointed him a Distinguished Professor, and the National Geographic Society awarded him the first domestic "expedition grant" they had ever made.

Whyte worked with bright young designers and planners at the New York City Planning Department, Hunter College sociology students, and other talented people he drew to "The Street Life Project." This team produced an exceptional study of how people use urban space and a set of urban design guidelines for New York which have been widely praised and used in New York and many other cities.

The Street Life Project is an excellent example of how to do urban research. Whyte formed hypotheses about how people use urban space. Then he tested each hypothesis by filming people using many different plazas and parks in New York City and carefully analyzing the films. His results were often startling. Some initial hypotheses were validated, but Whyte found that he had to reject or modify many others that had seemed intuitively obvious. For example, Whyte hypothesized that the number of people using a plaza would be related to the *amount* of space or its *shape*. Big parks should have more people than little ones. A long skinny strip park should have fewer people than a rectangular one. But Chart 1 shows that is not the case. New Yorkers use tiny Greenacre Park much more than the much larger J. C. Penney Park. One of New York's most popular parks is just a long, narrow indentation in a building. Whyte eventually concluded that the amount of *sittable* space in a park or plaza was much more important than either the total space or its shape.

Most writing on urban design ignores gender differences or is written from a male perspective with a separate section on design implications for women. Whyte is one of the few authors to notice gender and to weave its significance into the fabric of his study. He noticed that women are more discriminating than men as to where they will sit and are more sensitive to annoyances. He concluded that if a plaza has a high proportion of women, it is probably a good and well-managed one.

William Whyte's ideas have had a wide impact. The New York City Planning Commission held hearings on his recommendations and, after much debate, adopted many of his suggestions as requirements or guidelines for new development. Sharon Zukin (p. 131) describes how a public/private partnership inspired by Whyte's ideas tore down walls isolating Bryant Park from the street, put in sittable space and food and transformed a dangerous, little-used park into one bustling with life, while at the same time evicting many homeless park "residents" and raising troubling questions about who should control public space and prescribe how it is to be used.

Note the similarity between Whyte's description of Seagram's Plaza as "the best of stages" and Lewis

Mumford's emphasis on the city as theater (p. 92). Mike Davis (p. 193) found designers in Los Angeles consciously design public spaces to keep people *away*. As Whyte comments, "it takes real work to create a lousy place." Note the importance of good public spaces in Camillo Sitte's work on *The Art of Building Cities* (p. 466) and in Kitto's description of the Greek polis (p. 31).

This selection is from *City: Rediscovering The Center* (New York: Anchor Books, 1988). In other chapters, Whyte explores water, wind, trees, light, steps and entrances, undesirables, walls, sun and shadows, and many other factors. Whyte produced a delightful film titled *Public Spaces/Human Places* based on his research (available from Direct Cinema Limited in Los Angeles).

Whyte's other writings on urban society and planning include *The Organization Man* (New York: Simon and Schuster, 1956) and *The Last Landscape* (Garden City, NY: Doubleday, 1968).

There are many books concerning human aspects of design. Spiro Kostoff's monumental *The City Shaped: Urban Patterns and Meanings through History* (London: Thames and Hudson, 1991) and *The City Assembled* (London: Thames and Hudson, 1992) synthesize a vast amount of material from around the world and contain excellent illustrations. Oscar Newman's *Defensible Space* (New York: Macmillan, 1972) is a study of the way in which low-rent public housing project residents use space with suggestions to architects and planners on how to meet their security concerns. Clare Cooper and Wendy Sarkissien's *Housing as if People Mattered* (Berkeley: University of California Press, 1986) provides practical suggestions for designing housing responsive to the needs of all its residents, particularly working women and children. Allan Jacobs's *Looking at Cities* (Cambridge, MA: MIT Press, 1985) provides a stimulating discussion of how close observation like that which Whyte undertook can inform city planning and how to do it.

. . . Since 1961 New York City had been giving incentive bonuses to developers who would provide plazas . . . Every new office building qualified for the bonus by providing a plaza or comparable space; in total, by 1972 some twenty acres of the world's most expensive open space.

Some plazas attracted lots of people . . .

But on most plazas there were few people. In the middle of the lunch hour on a beautiful day the number of people sitting on plazas averaged four per thousand square feet of space – an extraordinarily low figure for so dense a center . . .

. . . The city was being had. For the millions of dollars of extra floor space it was handing out to developers, it had every right to demand much better spaces in return.

I put the question to the chairman of the city planning commission, Donald Elliott . . . He felt tougher zoning was in order. If we could find out why the good places worked and the bad ones didn't and come up with tight guidelines, there could be a new code . . .

We set to work. We began studying a cross section of spaces – in all, sixteen plazas, three small parks, and a number of odds and ends of space . . .

[. . .]

We started by charting how people used plazas. We mounted time-lapse cameras at spots overlooking the plaza . . . and recorded the dawn-to-dusk patterns. We made periodic circuits of the plazas and noted on sighting maps where people were sitting, their gender, and whether they were alone or with others . . . We also interviewed people and found where they worked, how frequently they used the plaza, and what they thought of it. But mostly we watched what they did.

Most of them were young office workers from nearby buildings. Often there would be relatively few from the plaza's own building. As some secretaries confided, they would just as soon put a little distance between themselves and the boss come lunchtime. In most cases the plaza users came from a building within a three-block radius. Small parks, such as Paley and Greenacre, had a somewhat more varied mix of people – with more upper-income older people – but even here office workers predominated.

This uncomplicated demography underscores an elemental point about good spaces: supply creates demand. A good new space builds a new constituency. It gets people into new habits – such as alfresco lunches – and induces them to use new paths . . .

The best-used plazas are sociable places, with a higher proportion of couples and groups than you will find in less-used places. At the plazas in New York, the proportion of people in twos or more runs about 50–62 percent; in the least-used, 25–30 percent. A high proportion is an index of selectivity. If people go to a place in a group or rendezvous there, it is most often because they decided to beforehand. Nor are these places less congenial to the individual. In absolute numbers, they attract more individuals than do the less-used spaces. If you are alone, a lively place can be the best place to be.

The best-used places also tend to have a higher than average proportion of women. The male–female ratio of a plaza reflects the composition of the work force and this varies from area to area. In midtown New York it runs about 60 percent male, 40 percent female. Women are more discriminating than men as to where they will sit, they are more sensitive to annoyances, and they spend more time casing a place. They are also more likely to dust off a ledge with their handkerchief.

The male–female ratio is one to watch. If a plaza has a markedly low proportion of women, something is wrong. Conversely, if it has a high proportion, the plaza is probably a good and well-managed one and has been chosen as such.

The rhythms of plaza life are much alike from place to place. In the morning hours, patronage will be sporadic . . .

Around noon the main clientele begins to arrive. Soon activity will be near peak and will stay there until a little before two . . .

Some 80 percent of the people activity on plazas comes during the lunchtime, and there is very little of any kind after five-thirty . . .

During the lunch period, people will distribute themselves over space with considerable consistency, with some sectors getting heavy use day in and day out, others much less so. We also found that off-peak use often gives the best clues to people's preferences. When a place is jammed, people sit where they can; this may or may not be where they most want to. After the main crowd has left, however, the choices can be significant. Some parts of the plaza become empty; others continue to be used . . .

Men show a tendency to take the front row seats and if there is a kind of gate they will be the guardians of it. Women tend to favor places slightly secluded. If there are double-sided ledges parallel to the street, the inner side will usually have a higher proportion of women; the outer, of men.

Of the men up front the most conspicuous are the girl watchers. As I have noted, they put on such a show of girl watching as to indicate that their real interest is not so much the girls as the show. It is all machismo. Even in the Wall Street area, where girl watchers are especially demonstrative you will hardly ever see one attempt to pick up a girl.

Plazas are not ideal places for striking up acquaintances. Much better is a very crowded street with lots of eating and quaffing going on. An outstanding example is the central runway of the South Street Seaport. At lunch sometimes, one can hardly move for the crush. As in musical chairs, this can lead to interesting combinations. On most plazas, however, there isn't much mixing. If there are, say, two smashing blondes on a ledge, the men nearby will usually put on an elaborate show of disregard. Look closely, however, and you will see them giving away the pose with covert glances.

Lovers are to be found on plazas, but not where you would expect them. When we first started interviewing, people would tell us to be sure to see the lovers in the rear places. But they weren't usually there. They would be out front. The most fervent embracing we've recorded on film has taken place in the most visible of locations, with the couple oblivious of the crowd. (In a long clutch, however, I have noted that one of the lovers may sneak a glance at a wristwatch.)

Certain locations become rendezvous points for groups of various kinds. The south wall of the Chase Manhattan Plaza was, for a while, a gathering point for camera bugs, the kind who are always buying new lenses and talking about them. Patterns of this sort may last no more than a season – or persist for years. A black civic leader in Cincinnati told me that when he wants to make contact, casually, with someone, he usually knows just where to look at Fountain Square . . .

Standing patterns on the plazas are fairly regular. When people stop to talk they will generally do so athwart one of the main traffic flows, as they do on streets. They also show an inclination to station themselves near objects, such as a flagpole or a piece of sculpture. They like well-defined places, such as steps or the border of a pool. What they rarely choose is the middle of a large space.

There are a number of explanations. The preference might be ascribed to some primeval instinct: you have a full view of all comers but your rear is covered. But this doesn't explain the inclination men have for lining up at the curb. Typically, they face inward, with their backs exposed to the vehicle traffic of the street.

Whatever their cause, people's movements are one of the great spectacles of a plaza. You do not see this in architectural photographs, which are usually devoid of human beings and are taken from a perspective that few people share. It is a misleading one. Looking down on a bare plaza, one sees a display of geometry, done almost in monochrome. Down at eye level the scene comes alive with movement and color – people walking quickly, walking slowly, skipping up steps, weaving in and out on crossing patterns, accelerating and retarding to match the moves of others. Even if the paving and the walls are gray, there will be vivid splashes of color – in winter especially, thanks to women's fondness for red coats and colored umbrellas.

There is a beauty that is beguiling to watch, and one senses that the players are quite aware of this themselves. You can see this in the way they arrange themselves on ledges and steps. They often do so with a grace that they must appreciate themselves. With its brown-gray setting, Seagram is the best of stages – in the rain, too, when an umbrella or two puts spots of color in the right places, like Corot's red dots.

Let us turn to the factors that make for such places. The most basic one is so obvious it is often overlooked: people. To draw them, a space should tap a strong flow of them. This means location, and, as the old adage has it, location and location. The space should be in the heart of downtown, close to the 100 percent corner – preferably right on top of it.

Because land is cheaper further out, there is a temptation to pick sites away from the center. There may also be some land for the asking – some underused spaces, for example, left over from an ill-advised civic center campus of urban renewal days. They will be poor bargains. A space that is only a few blocks too far might as well be ten blocks for all the people who will venture to walk to it.

People *ought* to walk to it, perhaps; the exercise would do them good. But they don't. Even within the core of downtown the effective radius of a good place is about three blocks. About 80 percent of the users will have come from a place within that area. This does indicate a laziness on the part of pedestrians and this may change a bit, just as the insistence on close-in parking may. But there is a good side to the constrained radius. Since usage is so highly localized, the addition of other good open spaces will not saturate demand. They will increase it.

Given a fine location, it is difficult to design a space that will not attract people. What is remarkable is how often this has been accomplished. Our initial study made it clear that while location is a prerequisite for success, it in no way assures it. Some of the worst plazas are in the best spots . . .

All of the plazas and small parks that we studied had good locations; most were on the major avenues, some on attractive side-streets. All were close to bus-stops or subway stations and had strong pedestrian flows on the sidewalks beside them. Yet when we rated them according to the number of people sitting at peak time, there was a wide range: from 160 people at 77 Water Street to 17 at 280 Park Avenue.

How come? The first factor we studied was the sun. We thought it might well be the critical one, and our first time-lapse studies seemed to bear this out. Subsequent studies did not. As I will note later they show that sun was important but did not explain the differences in popularity of plazas.

Nor did aesthetics . . . The elegance and purity of a complex's design, we had to conclude, had little relationship to the usage of the spaces around it.

[. . .]

Another factor we considered was the shape of spaces. Members of the commission's urban design group believed this was very important and hoped our findings would support tight criteria for proportions and placement. They were particularly anxious to rule out strip plazas: long, narrow spaces that were little more than enlarged sidewalks, and empty of people more times than not . . .

Our data did not support such criteria. While it was true that most strip plazas were little used, it did not follow that their shape was the reason. Some squarish plazas were little used too, and, conversely, several of the most heavily used spaces were in fact long, narrow strips. One of the five most popular sitting places in New York is essentially an indentation in a building, long and narrow. Our research did not prove shape unimportant or designers' instincts misguided. As with the sun, however, it proved that other factors were more critical.

If not the shape of the space, what about the *amount* of it? Some conservationists believed this would be the key factor. In their view, people seek open space as a relief from overcrowding and it would follow that places with the greatest sense of space and light and air would draw the best. If we ranked plazas by the amount of space they provided, there surely would be a positive correlation between space and people.

Once again we found no clear relationship. Several of the smallest spaces had the largest number of people, and several of the largest spaces had the least number of people . . .

What about the amount of *sittable* space? Here we began to get close. As we tallied the number of linear feet of sitting space, we could see that the plazas with the most tended to be among the most popular . . .

. . . No matter how many other variables we checked, one basic point kept coming through. We at last recognized that it was the major one.

People tend to sit most where there are places to sit.

This may not strike the reader as an intellectual bombshell, and now that I look back on our study I wonder why it was not more apparent to us from the beginning . . . Whatever the attractions of a space, it cannot induce people to come and sit if there is no place to sit.

INTEGRAL SEATING

The most basic kind of seating is the kind that is built into a place, such as steps and ledges. Half the battle is seeing to it that these features are usable by people. And there is a battle. Another force has been diligently at work finding ways to deny these spaces. Here are some of the ways:

Horizontal metal strip with sawtooth points.
Jagged rocks set in concrete (Southbridge House, New York City).
Spikes imbedded in ledges (Peachtree Plaza Hotel).
Railing placed to hit you in small of back (GM Plaza, New York City).
Canted ledges of slippery marble (Celanese Building, New York City).

It takes real work to create a lousy place. In addition to spikes and metal objects, there are steps to be made steep, additional surveillance cameras to be mounted, walls to be raised high. Just not doing such things can produce a lot of sitting space.

It won't be the most comfortable kind but it will have the great advantage of enlarging choice. The more sittable the inherent features are made, the more freedom people have to sit up front, in the back, to the side, in the sun, or out of it. This means designing ledges and parapets and other flat surfaces so they can do double duty as seating, tables, and shelves. Since most building sites have some slope in them, there are bound to be opportunities for such features, and it is no more trouble to leave them sittable than not.

[. . .]

SITTING HEIGHT

One guideline we thought would be easy to establish was for sitting heights. It seemed obvious enough that somewhere around sixteen to seventeen inches would probably be the optimum. But how much higher or lower could a surface be and still be sittable? Thanks to slopes, several of the most popular ledges provided a

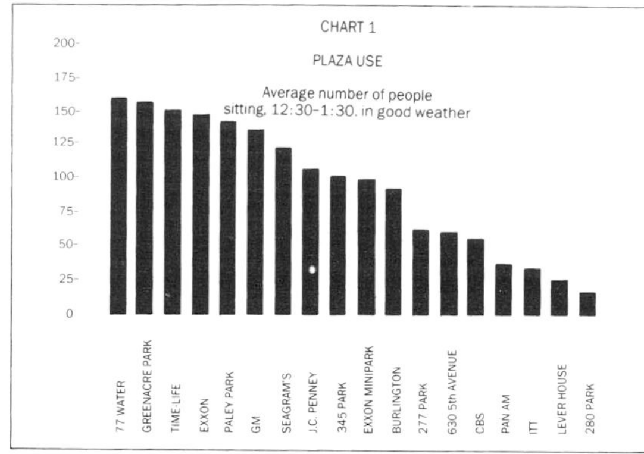

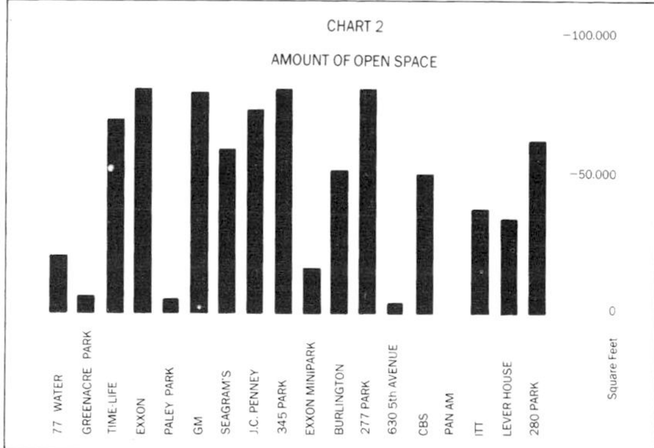

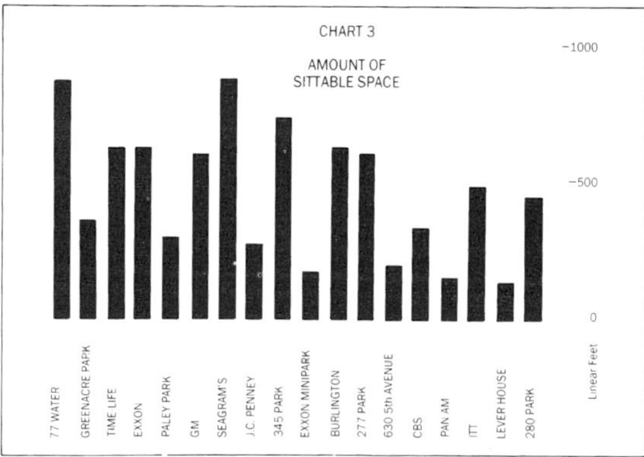

Figure 1

range of continuously variable heights. The front ledge at Seagram, for example started at seven inches at one corner and rose to forty-four inches at the other. Here was an opportunity for a definitive study, we thought; by recording over time how many people sat at what heights, we would get a statistical measure of preferences.

We didn't . . . We had to conclude that people will sit almost anywhere between a height of one foot and three, and this was the range that was to be specified in the zoning. People will sit lower or higher, of course, but there are apt to be special conditions – a wall too high for most adults to mount but just right for teenagers.

A dimension that is truly important is the human backside. It is a dimension many architects ignore. Rarely will you find a ledge or bench that is deep enough to be sittable on both sides . . . Most frustrating are the ledges just deep enough to tempt people to sit on both sides, but too shallow to let them do so comfortably. At peak times people may sit on both sides but they won't be comfortable doing it. They will be sitting on the forward edge, awkwardly.

Thus to another of our startling findings: ledges and spaces two backsides deep seat more people than those that are not as deep . . .

[. . .]

Steps work for the same reason. They afford an infinity of possible groupings, and the excellent sight lines make all the seating great for watching the theatre of the street . . .

[. . .]

Circulation and sitting, in sum, are not antithetical but complementary. I stress this because a good many planners think that the two should be kept separate. More to the point, so do some zoning codes. New York's called for "pedestrian circulation areas" separate from "activity areas" for sitting. People ignore such boundaries.

We felt that pedestrian circulation through and within plazas should be encouraged. Plazas that are sunken or elevated tend to attract low flows, and for that reason the zoning specifies that plazas be not more than three feet below street level or above it. The easier the flow between street and plaza the more likely they are to come in and tarry and sit.

This is true of the handicapped also. If a place is planned with their needs in mind, the place is apt to function more easily for everyone. Drinking fountains that are low enough for wheelchair users are low enough for children. Walkways that are made easier for the handicapped by ramps, handrails, and steps of gentle pitch are easier for all. The guidelines make such amenities mandatory . . . For the benefit of the handicapped, it is required that at least 5 percent of the seating spaces have backrests. These are not segregated for the handicapped. No facilities are segregated. The idea is to make all of the place useful for everyone.

BENCHES

Benches are design artifacts the purpose of which is to punctuate architectural photographs. They are most often sited in modular form, spaced equidistant from one another in a symmetry that is pleasing in plan view. They are not very good, however, for sitting. There are usually too few of them; they are too short and too narrow; they are isolated from other benches and from what action there is to look at.

[. . .]

Watch how benches fill up. The first arrival will usually take the end of a bench, not the middle. The next arrival will take the end of another bench. Subsequent arrivals head for whatever end spots are not taken. Only when there are few other places left will people sit in the middle of the bench, and some will elect to stand.

Since it's the ends of the benches that do most of the work, it could be argued that benches ought to be shortened so they're all end and no middle. But the unused middles are functional for not being used. They provide buffer space. They also provide choice, and if it is the least popular choice, that does not negate its utility.

[. . .]

CHAIRS

We come now to a wonderful invention: the movable chair. Having a back, it is comfortable, and even more so if it has armrests as well. But the big asset is movability. Chairs enlarge choice: to move into the sun, out of it; to move closer to someone, further away from another.

The possibility of choice is as important as the exercise of it. If you know you can move if you want to, you can feel all the more comfortable staying put. This is why, perhaps, people so often approach a chair and then, before sitting on it, move the chair a few inches this way or that, finally ending up with the chair just about where it was in the first place. These moves are functional. They are a declaration of one's free will to oneself, and rather satisfying. In this one small matter you are the master of your fate.

Small moves can say things to other people. If a newcomer chooses a chair next to a couple in a crowded situation, he may make several moves with the chair. He is conveying a message: Sorry about the closeness, but it can't be helped and I am going to respect your privacy as you will mine. A reciprocal shift of a chair may signal acknowledgment.

Chair arranging by groups is a ritual worth watching. In a group of three or four women, one may be dominant and direct the sitting, including the fetching of an extra chair. More times, the members of the group work it out themselves, often with false starts and second choices. The chair arranging can take quite a bit of time on occasion – it is itself a form of recreation – but people enjoy it. Watching these exercises in civility is one of the pleasures of a good place.

Fixed individual seats deny choice. They may be good to look at, and in the form of stools, metal love seats, granite cubes, and the like, they make interesting decorative elements. That is their primary function. For sitting, however, they are inflexible and socially uncomfortable.

[. . .]

Where space is at a premium – in theatres, stadia – fixed seats are a necessity. In open spaces, however, they are uncalled for; there is so much space around them that the compression makes for awkward sitting . . . On one campus a group of metal love seats was cemented to the paving with epoxy glue; in short order they were wrenched out of position by students. The designer is unrepentant. His love seats have won several design awards.

[. . .]

A salute to grass is in order. It is a wonderfully adaptable substance, and while it is not the most comfortable seating, it is fine for napping, sunbathing, picnicking, and Frisbee throwing. Like movable chairs, it also has the great advantage of offering people the widest possible choice of sitting arrangements. There are an infinity of possible groupings, but you will note that the most frequent has people self-positioned at oblique angles from each other.

Grass offers a psychological benefit as well. A patch of green is a refreshing counter to granite and concrete, and when people are asked what they would like to see in a park, trees and grass usually are at the top of the list . . .

RELATIONSHIP TO THE STREET

Let us turn to a more difficult consideration. With the kind of amenities we have been discussing, there are second chances. If the designers have goofed on seating, more and better seating can be provided. If they have been too stingy with trees, more trees can be planted. If there is no food, a food cart can be put in – possibly a small pavilion or gazebo. If there is no water feature, a benefactor might be persuaded to donate a small pool or fountain. Thanks to such retrofitting, spaces regarded as hopeless dogs have been given new life.

What is most difficult to change, however, is what is most important: the location of the space and its relationship to the street. The real estate people are right about location, location, location. For a space to function truly well it must be central to the constituency it is to serve – and if not in physical distance, in visual accessibility . . .

The street functions as part of the plaza or square; indeed, it is often hard to tell where the street leaves off and the plaza begins. The social life of the spaces flows back and forth between them. And the most vital space of all is often the street corner. Watch one long enough and you will see how important it is to the life of the large spaces. There will be people in 100 percent conversations or prolonged goodbyes. If there is a food vendor at the corner, like Gus at Seagram, people will be clustered around him, and there will be a brisk traffic between corner and plaza.

It is a great show, and one of the best ways to make the most of it is, simply, not to wall off the plaza from it. Frederick Law Olmsted spoke of an "interior park" and an "outer park," and he argued that the latter – the surrounding streets – was vital to the enjoyment of the former. He thought it an abomination to separate the two with walls or, worse yet, with a spiked iron fence. "In expression and association," he said, "it is in the most distinct contradiction and discord with all the sentiment of a park. It belongs to a jail or to the residence of a despot who dreads assassination."

But walls are still being put up, usually in the mistaken notion that they will make the space feel safer. They do not . . . they make a space feel isolated and gloomy. Lesser defensive measures can work almost as much damage. The front rows of a space – whether ledges or steps or benches – are the best of sitting places, yet they are often modified against human use. At the General Motors Building on Fifth Avenue, the front ledges face out on one of the greatest of promenades. But you cannot sit on the ledges for more than a minute or so. There is a fussy little railing that catches you right in the small of your back. I do not think it was deliberately planned to do so. But it does and you cannot sit for more than a few moments before your back hurts. Another two inches of clearance for the railing and you would be comfortable. But day after day, year after year, one of the great front rows goes scarcely used, for want of two inches. Canted ledges, especially ones of polished marble, are another nullifying feature. You can almost sit on them if you keep pressing down on your heel hard enough.

[. . .]

A good space beckons people in, and the progression from street to interior is critical in this respect. Ideally, the transition should be such that it's hard to tell where one ends and the other begins. You shouldn't have to make a considered decision to enter; it should be almost instinctive . . .

[. . .]

ALLAN JACOBS AND DONALD APPLEYARD

"Toward an Urban Design Manifesto"

American Planning Association Journal (1987)

Editors' introduction Allan Jacobs and Donald Appleyard deplore many of the same aspects of the current urban environment that Mike Davis (p. 193), Edward Soja (p. 180), and David Harvey (p. 199) attack in actual conditions in Los Angeles, New York and other large postmodern cities: vast anonymous areas developed by giant public and private developers; dangerous, polluted, noisy, anonymous living environments; fortress-like buildings that present windowless facades to the street; cities as symbols of inequality. But "Towards an Urban Design Manifesto" moves beyond observation and critique to set out goals for urban life and advance ideas for how the urban fabric of cities might be designed to encourage a livable urban environment.

Allan Jacobs is a professor of city and regional planning at the University of California, Berkeley, where Donald Appleyard also taught until his death shortly before this selection was published. Both worked closely with Kevin Lynch, whose *The Image of the City* (p. 478) and other writings strongly influenced their work.

Jacobs has alternated between careers as a practicing city planner (in Pittsburgh, Philadelphia, New Delhi, and San Francisco) and teaching at the University of Pennsylvania and, for the last two decades, the University of California, Berkeley, Department of City and Regional Planning. While he was San Francisco's Director of City Planning, Jacobs enlisted Appleyard to work on studies of street livability in San Francisco. Jacobs called on Appleyard to help develop an award-winning citywide urban design plan reflecting Lynch's ideas.

Jacobs and Appleyard title their piece a "manifesto" and model it on the celebrated Charter of Athens adopted by the International Congress of Modern Architecture (CIAM) – the organization that advanced ideas for building contemporary cities based on Le Corbusier's principles (p. 336).

Jacobs and Appleyard do not like the vast clearance projects, highways, and high-rise buildings surrounded by enormous open space that have resulted from the CIAM's design ideology. They acknowledge that the Garden City ideas of Ebenezer Howard (p. 321) have produced some pleasant communities, but dismiss them as more like suburbs than true *cities*. Their manifesto suggests an approach far more subtle and humane than the CIAM for development and more truly urban than Howard's approach.

Jacobs and Appleyard's manifesto is grounded in both a command of academic theory and their own practical experience in city design. In this manifesto they propose urban development at densities higher than Howard proposed for garden city designs – high enough to qualify as truly urban. But they do not endorse urban densities nearly as great as the CIAM theorists do, particularly in megastructures surrounded by open space.

While Jacobs and Appleyard favor *reasonable* standards for decibel levels and street widths, they oppose excessive engineering standards that destroy the texture of urban life. Like Jane Jacobs

(p. 106), they relish some of the disorder that makes urban life enjoyable including noise, smell, and varied uses that some engineers and many modernist architects want to eliminate altogether. Jacobs and Appleyard value pedestrians and public space. And, unlike the elitist CIAM theorists, they argue that participatory planning is essential.

In *Making City Planning Work* (Chicago: Planner's Press, 1976), Jacobs alternates chapters describing the practical aspects of a city planning director's job with case studies on successful and not so successful projects he undertook during his tenure as city planning director of San Francisco. Jacobs's *Looking at Cities* (Cambridge, MA: MIT Press, 1985) grew out of a class he teaches at Berkeley in which students take him to an unfamiliar neighborhood, leave him to look carefully at it, and then compare what he finds out from observation with what they learn from examining data and city planning reports on the same area. *Looking at Cities* reminds professionals to open their eyes and experience the areas they are planning, and outlines a methodology for reading clues in the built environment that can improve urban planning. Jacobs's most recent book is *Great Streets* (Princeton, NJ: Princeton University Press, 1994), which summarizes a decade of observation and close study of notable streets throughout the world.

Donald Appleyard's *The View From the Road*, co-authored with Kevin Lynch and John Myer, (Cambridge, MA: MIT Press, 1963), and *Liveable Streets* (Berkeley: University of California Press, 1981) show how ideas expressed in this urban design manifesto can be translated into action in street design.

For an overview of urban design see Jonathon Barnett, *An Introduction to Urban Design* (New York: Harper & Row, 1982). Other works on urban design include Edumnd N. Bacon, *Design of Cities* (New York: Penguin, 1976), Vincent Scully, *American Architecture and Urbanism* (New York: Praeger, 1969), and Gordon Cullen, *Townscape* (New York: Penguin, 1976).

We think it's time for a new urban design manifesto. Almost 50 years have passed since Le Corbusier and the International Congress of Modern Architecture (CIAM) produced the Charter of Athens, and it is more than 20 years since the first Urban Design Conference, still in the CIAM tradition, was held (at Harvard in 1957). Since then the precepts of CIAM have been attacked by sociologists, planners, Jane Jacobs, and more recently by architects themselves. But it is still a strong influence, and we will take it as our starting point. Make no mistake: the charter was, simply, a manifesto – a public declaration that spelled out the ills of industrial cities as they existed in the 1930s and laid down physical requirements necessary to establish healthy, humane, and beautiful urban environments for people. It could not help but deal with social, economic, and political phenomena, but its basic subject matter was the physical design of cities. Its authors were (mostly) socially concerned architects, determined that their art and craft be responsive to social realities as well as to improving the lot of man. It would be a mistake to write them off as simply elitist designers and physical determinists.

So the charter decried the medium-size (up to six storys) high-density buildings with high land coverage that were associated so closely with slums. Similarly, buildings that faced streets were found to be detrimental to healthy living. The seemingly limitless horizontal expansion of urban areas devoured the countryside, and suburbs were viewed as symbols of terrible waste. Solutions could be found in the demolition of unsanitary housing, the provision of green areas in every residential district, and new high-rise, high-density buildings set in open space. Housing was to be removed from its traditional relationship facing streets, and the whole circulation system was to be revised to meet the needs of emerging mechanization (the automobile). Work areas should be close to but separate from residential areas. To achieve the new city, large land holdings, preferably owned by the public, should replace multiple small parcels (so that projects could be properly designed and developed).

Now thousands of housing estates and redevelopment projects in socialist and capitalist countries the world over, whether built on previously undeveloped land or developed as replacements for old urban areas, attest to the acceptance of the charter's dictums. The design notions it embraced have become part of a world design language, not just the intellectual property of an enlightened few, even though the principles have been devalued in many developments.

Of course, the Charter of Athens has not been the only major urban philosophy of this century to influence the development of urban areas. Ebenezer Howard, too, was responding to the ills of the nineteenth-century industrial city, and the Garden City movement has been at least as powerful as the Charter of Athens. New towns policies, where they exist, are rooted in Howard's thought. But you don't have to look to new towns to see the influence of Howard, Olmsted, Wright, and Stein. The superblock notion, if nothing else, pervades large housing projects around the world, in central cities as well as suburbs. The notion of buildings in a park is as common to garden city designs as it is to charter-inspired development. Indeed, the two movements have a great deal in common: superblocks, separate paths for people and cars, interior common spaces, housing divorced from streets, and central ownership of land. The garden city-inspired communities place greater emphasis on private outdoor space. The most significant difference, at least as they have evolved, is in density and building type: the garden city people preferred to accommodate people in row houses, garden apartments, and maisonettes, while Corbusier and the CIAM designers went for high-rise buildings and, inevitably, people living in flats and at significantly higher densities.

We are less than enthralled with what either the Charter of Athens or the Garden City movement has produced in the way of urban environments. The emphasis of CIAM was on buildings and what goes on within buildings that happen to sit in space, not on the public life that takes place constantly in public spaces. The orientation is often inward. Buildings tend to be islands, big or small. They could be placed anywhere. From the outside perspective, the building, like the work of art it was intended to be, sits where it can be seen and admired in full. And because it is large it is best seen from a distance (at a scale consistent with a moving auto). Diversity, spontaneity, and surprise are absent, at least for the person on foot. On the other hand, we find little joy or magic or spirit in the charter cities. They are not urban, to us, except according to some definition one might find in a census. Most garden cities, safe and healthy and even gracious as they may be, remind us more of suburbs than of cities. But they weren't trying to be cities. The emphasis has always been on "garden" as much as or more than on "city."

Both movements represent overly strong design reactions to the physical decay and social inequities of industrial cities. In responding so strongly, albeit understandably, to crowded, lightless, airless, "utilitiless," congested buildings and cities that housed so many people, the utopians did not inquire what was good about those places, either socially or physically. Did not those physical environments reflect (and maybe even foster) values that were likely to be meaningful to people individually and collectively, such as publicness and community? Without knowing it, maybe these strong reactions to urban ills ended up by throwing the baby out with the bathwater.

In the meantime we have had a lot of experience with city building and rebuilding. New spokespeople with new urban visions have emerged. As more CIAM-style buildings were built people became more disenchanted. Many began to look through picturesque lenses back to the old preindustrial cities. From a concentration on the city as a kind of sculpture garden, the townscape movement, led by the *Architectural Review*, emphasized "urban experience." This phenomenological view of the city was espoused by Rasmussen, Kepes, and ultimately Kevin Lynch and Jane Jacobs. It identified a whole new vocabulary of urban form – one that depended on the sights, sounds, feels, and smells of the city, its materials and textures, floor surfaces, facades, style, signs, lights, seating, trees, sun, and shade all potential amenities for the attentive observer and user. This has permanently humanized the vocabulary of urban design, and we enthusiastically subscribe to most of its tenets, though some in the townscape movement ignored the social meanings and implications of what they were doing.

The 1960s saw the birth of community design and an active concern for the social groups affected, usually negatively, by urban design. Designers were the "soft cops," and many professionals left the design field for social or planning vocations, finding the physical environment to have no redeeming social value. But at the beginning of the 1980s the mood in the design professions is conservative. There is a withdrawal from social engagement back to formalism. Supported by semiology and other abstract themes, much of architecture has become a dilettantish and narcissistic pursuit, a chic component of the high art consumer culture, increasingly remote from most people's everyday lives, finding its ultimate manifestation in the art gallery and the art book. City planning is too immersed in the administration and survival of housing, environmental, and energy programs and in responding to budget cuts and community demands to have any clear sense of direction with regard to city form.

While all these professional ideologies have been working themselves out, massive economic, technological, and social changes have taken place in our cities. The scale of capitalism has continued to increase, as has the scale of bureaucracy, and the automobile has virtually destroyed cities as they once were.

In formulating a new manifesto, we react against other phenomena than did the leaders of CIAM 50 years ago. The automobile cities of California and the Southwest present utterly different problems from those of nineteenth-century European cities, as do the CIAM-influenced housing developments around European, Latin American, and Russian cities and the rash of squatter settlements around the fast-growing cities of the Third World. What are these problems?

PROBLEMS FOR MODERN URBAN DESIGN

Poor living environments

While housing conditions in most advanced countries have improved in terms of such fundamentals as light, air, and space, the surroundings of homes are still frequently dangerous, polluted, noisy, anonymous wastelands. Travel around such cities has become more and more fatiguing and stressful.

Giantism and loss of control

The urban environment is increasingly in the hands of the large-scale developers and public agencies. The elements of the city grow inexorably in size, massive transportation systems are segregated for single travel modes, and vast districts and complexes are created that make people feel irrelevant.

People, therefore, have less sense of control over their homes, neighborhoods, and cities than when they lived in slower-growing locally based communities. Such giantism can be found as readily in the housing projects of socialist cities as in the office buildings and commercial developments of capitalist cities.

Large-scale privatization and the loss of public life

Cities, especially American cities, have become privatized, partly because of the consumer society's emphasis on the individual and the private sector, creating Galbraith's "private affluence and public squalor," but escalated greatly by the spread of the automobile. Crime in the streets is both a cause and a consequence of this trend, which has resulted in a new form of city: one of closed, defended islands with blank and windowless facades surrounded by wastelands of parking lots and fast-moving traffic. As public transit systems have declined, the number of places in American cities where people of different social groups actually meet each other has dwindled. The public environment of many American cities has become an empty desert, leaving public life dependent for its survival solely on planned formal occasions, mostly in protected internal locations.

Centrifugal fragmentation

Advanced industrial societies took work out of the home, and then out of the neighborhood, while the automobile and the growing scale of commerce have taken shopping out of the local community. Fear has led social groups to flee from each other into homogeneous social enclaves. Communities themselves have become lower in density and increasingly homogeneous. Thus the city has spread out and separated to form extensive monocultures and specialized destinations reachable often only by long journeys – a fragile and extravagant urban system dependent on cheap, available gasoline, and an effective contributor to the isolation of social groups from each other.

Destruction of valued places

The quest for profit and prestige and the relentless exploitation of places that attract the public have led to the destruction of much of our heritage, of historic places that no longer turn a profit, of natural amenities that become overused. In many cases, as in San Francisco, the very value of the place threatens its destruction as hungry tourists and entrepreneurs flock to see and profit from it.

Placelessness

Cities are becoming meaningless places beyond their citizens' grasp. We no longer know the origins of the world around us. We rarely know where the materials and products come from, who owns what, who is behind what, what was intended. We live in cities where things happen without warning and without our participation. It is an alien world for most people. It is little surprise that most withdraw from community involvement to enjoy their own private and limited worlds.

Injustice

Cities are symbols of inequality. In most cities the discrepancy between the environments of the rich and the environments of the poor is striking. In many instances the environments of the rich, by occupying and dominating the prevailing patterns of transportation and access, make the environments of the poor relatively worse. This discrepancy may be less visible in the low-density modern city, where the display of affluence is more hidden than in the old city; but the discrepancy remains.

Rootless professionalism

Finally, design professionals today are often part of the problem. In too many cases, we design for places and people we do not know and grant them very little power or acknowledgment. Too many professionals are more part of a universal professional culture than part of the local cultures for whom we produce our plans and products. We carry our "bag of tricks" around the world and bring them out wherever we land. This floating professional culture has only the most superficial conception of particular place. Rootless, it is more susceptible to changes in professional fashion and theory than to local events. There is too little inquiry, too much proposing. Quick surveys are made, instant solutions devised, and the rest of the time is spent persuading the clients. Limits on time and budgets drive us on, but so do lack of understanding and the placeless culture. Moreover, we designers are often unconscious of our own roots, which influence our preferences in hidden ways.

At the same time, the planning profession's retreat into trendism, under the positivist influence of social science, has left it virtually unable to resist the social pressures of capitalist economy and consumer sovereignty. Planners have lost their beliefs. Although we believe citizen participation is essential to urban planning, professionals also must have a sense of what we believe is right, even though we may be vetoed.

GOALS FOR URBAN LIFE

We propose, therefore, a number of goals that we deem essential for the future of a good urban environment: livability; identity and control; access to opportunity, imagination, and joy; authenticity and meaning; open communities and public life; self-reliance; and justice.

Livability

A city should be a place where everyone can live in relative comfort. Most people want a kind of sanctuary for their living environment, a place where they can bring up children, have privacy, sleep, eat, relax, and restore themselves. This means a well-managed environment relatively devoid of nuisance, overcrowding, noise, danger, air pollution, dirt, trash, and other unwelcome intrusions.

Identity and control

People should feel that some part of the environment belongs to them, individually and collectively, some part for which they care and are responsible, whether they own it or not. The urban environment should be an environment that encourages people to express themselves, to become involved, to decide what they want and act on it. Like a seminar where everybody has something to contribute to communal discussion, the urban environment should encourage participation. Urbanites may not always want this. Many like the anonymity of the city, but we are not convinced that the freedom of anonymity is a desirable freedom. It would be much better if people were sure enough of themselves to stand up and be counted. Environments should therefore be designed for those who use them or are affected by them, rather than for those who own them. This should reduce alienation and anonymity (even if people want them); it should increase people's sense of identity and rootedness and encourage more care and responsibility for the physical environment of cities.

Respect for the existing environment, both nature and city, is one fundamental difference we have with the CIAM movement. Urban design has too often assumed that new is better than old. But the new is justified only if it is better than what exists. Conservation encourages identity and control and, usually, a better sense of community, since old environments are more usually part of a common heritage.

Access to opportunity, imagination, and joy

People should find the city a place where they can break from traditional molds, extend their experience, meet new people, learn other viewpoints, have fun. At a functional level, people should have access to alternative housing and job choices; at another level, they should find the city an enlightening cultural experience. A city should have magical places where fantasy is possible, a counter to and an escape from the mundaneness of everyday work and living. Architects and planners take cities and themselves too seriously; the result too often is deadliness and boredom, no imagination, no humor, alienating places. But people need an escape from the seriousness and meaning of the everyday. The city has always been a place of excitement; it is theater, a stage upon which citizens can display themselves and see others. It has magic, or should have, and that depends on a certain sensuous, hedonistic mood, on signs, on night lights, on fantasy, color, and other imagery. There can be parts of the city where belief can be suspended, just as in the experience of fiction. It may be that such places have to be framed so that people know how to act. Until now such fantasy and experiment have been attempted mostly by commercial facilities, at rather low levels of quality and aspiration, seldom deeply experimental. One should not have to travel as far as the Himalayas or the South Sea Islands to stretch one's experience. Such challenges could be nearer home. There should be a place for community utopias; for historic, natural, and anthropological evocations of the modern city, for encounters with the truly exotic.

Authenticity and meaning

People should be able to understand their city (or other people's cities), its basic layout, public functions, and institutions; they should be aware of its opportunities. An authentic city is one where the origins of things and places are clear. All this means an urban environment should reveal its significant meanings; it should not be dominated only by one type of group, the powerful; neither should publicly important places be hidden. The city should symbolize the moral issues of society and educate its citizens to an awareness of them.

That does not mean everything has to be laid out as on a supermarket shelf. A city should present itself as a readable story, in an engaging and, if necessary, provocative way, for people are indifferent to the obvious, overwhelmed by complexity. A city's offerings should be revealed or they will be missed. This can affect the forms of the city, its signage, and other public information and education programs.

Livability, identity, authenticity, and opportunity are characteristics of the urban environment that should serve the individual and small social unit, but the city has to serve some higher social goals as well. It is these we especially wish to emphasize here.

Community and public life

Cities should encourage participation of their citizens in community and public life. In the face of giantism and fragmentation, public life, especially life in public places, has been seriously eroded. The neighborhood movement, by bringing thousands, probably millions of people out of their closed private lives into active participation in their local communities, has begun to counter that trend, but this movement has had its limitations. It can be purely defensive, parochial, and self-serving. A city should be more than a warring collection of interest groups, classes, and neighborhoods; it should breed a commitment to a larger whole, to tolerance, justice, law, and democracy. The structure of the city should invite and encourage public life, not only through its institutions, but directly and symbolically through its public spaces. The public environment, unlike the neighborhood, by definition should be open to all members of the community. It is where people of different kinds meet. No one should be excluded unless they threaten the balance of that life.

Urban self-reliance

Increasingly cities will have to become more self-sustaining in their uses of energy and other scarce resources. "Soft energy paths" in particular not only will reduce dependence and exploitation across regions and countries but also will help reestablish a stronger sense of local and regional identity, authenticity, and meaning.

An environment for all

Good environments should be accessible to all. Every citizen is entitled to some minimal level of environmental livability and minimal levels of identity, control, and opportunity. Good urban design must be for the poor as well as the rich. Indeed, it is more needed by the poor.

We look toward a society that is truly pluralistic, one where power is more evenly distributed among social groups than it is today in virtually any country, but where the different values and cultures of interest- and place-based groups are acknowledged and negotiated in a just public arena.

These goals for the urban environment are both individual and collective, and as such they are frequently in conflict. The more a city promises for the individual, the less it seems to have a public life; the more the city is built for public entities, the less the individual seems to count. The good urban environment is one that somehow balances these goals, allowing individual and group identity while maintaining a public concern, encouraging pleasure while maintaining responsibility, remaining open to outsiders while sustaining a strong sense of localism.

AN URBAN FABRIC FOR AN URBAN LIFE

We have some ideas, at least, for how the fabric or texture of cities might be conserved or created to encourage a livable urban environment. We emphasize the structural qualities of the good urban environment – qualities we hope will be successful in creating urban experiences that are consonant with our goals.

Do not misread this. We are not describing all the qualities of a city. We are not dealing with major transportation systems, open space, the natural environment, the structure of the large-scale city, or even the structure of neighborhoods, but only the grain of the good city.

There are five physical characteristics that must be present if there is to be a positive response to the goals and values we believe are central to urban life. They must be designed, they must exist, as prerequisites of a sound urban environment. All five must be present, not just one or two. There are other physical characteristics that are important, but these five are essential: livable streets and neighborhoods; some minimum density of residential development as well as intensity of land use; an integration of activities – living, working, shopping – in some reasonable proximity to each other; a manmade environment, particularly buildings, that defines public space (as opposed to buildings that, for the most part, sit in space); and many, many separate, distinct buildings with complex arrangements and relationships (as opposed to few, large buildings).

Let us explain, keeping in mind that all five of the characteristics must be present. People, we have said, should be able to live in reasonable (though not excessive) safety, cleanliness, and security. That means livable streets and neighborhoods: with adequate sunlight, clean air, trees, vegetation, gardens, open space, pleasantly scaled and designed buildings; without offensive noise; with cleanliness and physical safety. Many of these characteristics can be designed into the physical fabric of the city.

The reader will say, "Well of course, but what does that mean?" Usually it has meant specific standards and requirements, such as sun angles, decibel levels, lane widths, and distances between buildings. Many researchers have been trying to define the qualities of a livable environment. It depends on a wide array of attributes, some structural, some quite small details. There is no single right answer. We applaud these efforts and have participated in them ourselves. Nevertheless, desires for livability and individual comfort by themselves have led to fragmentation of the city. Livability standards, whether for urban or for suburban developments, have often been excessive.

Our approach to the details of this inclusive physical characteristic would center on the words "reasonable, though not excessive . . ." Too often, for example, the requirement of adequate sunlight has resulted in buildings and people inordinately far from each other, beyond what demonstrable need for light would dictate. Safety concerns have been the justifications for ever wider streets and wide, sweeping curves rather than narrow ways and sharp corners. Buildings are removed from streets because of noise considerations when there might be other ways to deal with this concern. So although livable streets and neighborhoods are a primary requirement for any good urban fabric – whether for existing, denser cities or for new development – the quest for livable neighborhoods, if pursued obsessively, can destroy the urban qualities we seek to achieve.

A *minimum density* is needed. By density we mean the number of people (sometimes expressed in terms of housing units) living on an area of land, or the number of people using an area of land.

Cities are not farms. A city is people living and working and doing the things they do in relatively close proximity to each other.

We are impressed with the importance of density as a perceived phenomenon and therefore relative to the beholder and agree that, for many purposes, perceived density is more important than an "objective" measurement of people per unit of land. We agree, too, that physical phenomena can be manipulated so as to render perceptions of greater or lesser density. Nevertheless, a narrow, winding street, with a lot of signs and a small enclosed open space at the end, with no people, does not make a city. Cities are more than stage sets. Some minimum number of people living and using a given area of land is required if there is to be human exchange, public life and action, diversity and community.

Density of people alone will account for the presence or absence of certain uses and services we find important to urban life. We suspect, for example, that the number and diversity of small stores and services – for instance, groceries, bars, bakeries, laundries and cleaners, coffee shops, secondhand stores, and the like – to be found in a city or area is in part a function of density. That is, that such businesses are more likely to exist, and in greater variety, in an area where people live in greater proximity to each other ("higher" density). The viability of mass transit, we know, depends partly on the density of residential areas and partly on the size and intensity of activity at commercial and service destinations. And more use of transit, in turn, reduces parking demands and permits increases in density. There must be a critical mass of people, and they must spend a lot of their time in reasonably close proximity to each other, including when they are at home, if there is to be an urban life. The goal of local control and community identity is associated with density as well. The notion of an optimum density is elusive and is easily confused with the health and livability of urban areas, with lifestyles, with housing types, with the size of area being considered (the building site or the neighborhood or the city), and with the economics of development. A density that might be best for child rearing might be less than adequate to support public transit. Most recently, energy efficiency has emerged as a concern associated with density, the notion being that conservation will demand more compact living arrangements.

Our conclusion, based largely on our experience and on the literature, is that a minimum net density (people or living units divided by the size of the building site, excluding public streets) of about 15 dwelling units (30–60 people) per acre of land is necessary to support city life. By way of illustration, that is the density produced with generous town houses (or row houses). It would permit parcel sizes up to 25 feet wide by about 115 feet deep. But other building types and lot sizes also would produce that density. Some areas could be developed with lower densities, but not very many. We don't think you get cities at 6 dwellings to the acre, let alone on half-acre lots. On the other hand, it is possible to go as high as 48 dwelling units per acre (96 to 192 people) for a very large part of the city and still provide for a spacious and gracious urban life. Much of San Francisco, for example, is developed with three-story buildings (one unit per floor) above a parking story, on parcels that measure 25 feet by 100 or 125 feet. At those densities, with that kind of housing, there can be private or shared gardens for most people, no common hallways are required, and people can have direct access to the ground. Public streets and walks adequate to handle pedestrian and vehicular traffic generated by these densities can be accommodated in rights-of-way that are 50 feet wide or less. Higher densities, for parts of the city, to suit particular needs and lifestyles, would be both possible and desirable. We are not sure what the upper limits would be but suspect that as the numbers get much higher than 200 people per net residential acre, for larger parts of the city, the concessions to less desirable living environments mount rapidly.

Beyond residential density, there must be a minimum intensity of people using an area for it to be urban, as we are defining that word. We aren't sure what the numbers are or even how best to measure this kind of intensity. We are speaking here, particularly, of the public or "meeting" areas of our city. We are confident that our lowest residential densities will provide most meeting areas with life and human exchange, but are not sure if they will generate enough activity for the most intense central districts.

There must be an *integration of activities* – living, working, and shopping as well as public, spiritual, and recreational activities – reasonably near each other.

The best urban places have some mixtures of uses. The mixture responds to the values of publicness and diversity that encourage local community identity. Excitement, spirit, sense, stimulation, and exchange are more likely when there is a mixture of activities than when there is not. There are many examples that we all know. It is the mix, not just the density of people and uses, that brings life to an area, the life of people going about a full range of normal activities without having to get into an automobile.

We are not saying that every area of the city should have a full mix of all uses. That would be impossible. The ultimate in mixture would be for each building to have a range of uses from living, to working, to shopping, to recreation. We are not calling for a return to the medieval city. There is a lot to be said for the notion of "living sanctuaries," which consist almost wholly of housing. But we think these should be relatively small, of a few blocks, and they should be close and easily accessible (by foot) to areas where people meet to shop or work or recreate or do public business. And except for a few of the most intensely developed office blocks of a central business district or a heavy industrial area, the meeting areas should have housing within them. Stores should be mixed with offices. If we envision the urban landscape as a fabric, then it would be a salt-and-pepper fabric of many colors, each color for a separate use or a combination. Of course, some areas would be much more heavily one color than another, and some would be an even mix of colors. Some areas, if you squinted your eyes, or if you got so close as to see only a small part of the fabric, would read as one color, a red or a brown or a green. But by and large there would be few if any distinct patterns, where one color stopped and another started. It would not be patchwork quilt, or an even-colored fabric. The fabric would be mixed.

In an urban environment, *buildings* (and other objects that people place in the environment) *should be arranged in such a way as to define and even enclose public space, rather than sit in space.* It is not enough to have high densities and an integration of activities to have cities. A tall enough building with enough people living (or even working) in it, sited on a large parcel, can easily produce the densities we have talked about and can have internally mixed uses, like most "mixed use" projects. But that building and its neighbors will be unrelated objects sitting in space if they are far enough apart, and the mixed uses might be only privately available. In large measure that is what the Charter of Athens, the garden cities, and standard suburban development produce.

Buildings close to each other along a street, regardless of whether the street is straight, or curved, or angled, tend to define space if the street is not too wide in relation to the buildings. The same is true of a plaza or a square. As the spaces between buildings become larger (in relation to the size of the buildings, up to a point), the buildings tend more and more to sit in space. They become focal points for few or many people, depending on their size and activity. Except where they are monuments or centers for public activities (a stadium or meeting hall), where they represent public gathering spots, buildings in space tend to be private and inwardly oriented. People come to them and go from them in any direction. That is not so for the defined outdoor environment. Avoiding the temptation to ascribe all kinds of psychological values to defined spaces (such as intimacy, belonging, protection – values that are difficult to prove and that may differ for different people), it is enough to observe that spaces surrounded by buildings are more likely to bring people together and thereby promote public interaction. The space can be linear (like streets) or in the form of plazas of myriad shapes. Moreover, interest and interplay among uses is enhanced. To be sure, such arrangements direct people and limit their freedom – they cannot move in just any direction from any point – but presumably there are enough choices (even avenues of escape) left open, and the gain is in greater potential for sense stimulation, excitement, surprise, and focus. Over and over again we seek out and return to defined ways and spaces as symbolic of urban life emphasizing the public space more than the private building.

It is important for us to emphasize *public places* and a *public* way system. We have observed that the central value of urban life is that of publicness, of people from different groups meeting each other and of people acting in concert, albeit with debate. The most important public places must be for *pedestrians*, for no public life can take place between people in automobiles. Most public space has been taken over by the automobile, for travel or parking. We must fight to restore more for the pedestrian. Pedestrian malls are not simply to benefit the local merchants. They have an essential public value. People of different kinds meet each other directly. The level of communication may be only visual, but that itself is educational and can encourage tolerance. The revival of street activities, street vending, and street theater in American cities may be the precursor of a more flourishing public environment, if the automobile can be held back.

There also must be symbolic, public meeting places, accessible to all and publicly controlled. Further, in order to communicate, to get from place to place, to interact, to exchange ideas and goods, there must be a healthy public circulation system. It cannot be privately controlled. Public circulation systems should be seen as significant cultural settings where the city's finest products and artifacts can be displayed, as in the piazzas of medieval and renaissance cities.

Finally, *many different buildings and spaces with complex arrangements and relationships* are required. The often elusive notion of human scale is associated with this requirement – a notion that is not just an architect's concept but one that other people understand as well.

Diversity, the possibility of intimacy and confrontation with the unexpected, stimulation, are all more likely with many buildings than with few taking up the same ground areas.

For a long time we have been led to believe that large land holdings were necessary to design healthy, efficient, aesthetically pleasing urban environments. The slums of the industrial city were associated, at least in part, with all those small, overbuilt parcels. Socialist and capitalist ideologies alike called for land assembly to permit integrated, socially and economically useful developments. What the socialist countries would do via public ownership the capitalists would achieve through redevelopment and new fiscal mechanisms that rewarded large holdings.

Architects of both ideological persuasions promulgated or were easily convinced of the wisdom of land assembly. It's not hard to figure out why. The results, whether by big business or big government, are more often than not inward-oriented, easily controlled or controllable, sterile, large-building projects, with fewer entrances, fewer windows, less diversity, less innovation, and less individual expression than the urban fabric that existed previously or that can be achieved with many actors and many buildings. Attempts to break up facades or otherwise to articulate separate activities in large buildings are seldom as successful as when smaller properties are developed singly.

Health, safety, and efficiency can be achieved with many smaller buildings, individually designed and developed. Reasonable public controls can see to that. And, of course, smaller buildings are a lot more likely if parcel sizes are small than if they are large. With smaller buildings and parcels, more entrances must be located on the public spaces, more windows and a finer scale of design diversity emerge. A more public, lively city is produced. It implies more, smaller groups getting pieces of the public action, taking part, having a stake. Other stipu lations may be necessary to keep public frontages alive, free from the deadening effects of offices and banks, but small buildings will help this more than large ones. There need to be large buildings, too, covering large areas of land, but they will be the exception, not the rule, and should not be in the centers of public activity.

ALL THESE QUALITIES . . . AND OTHERS

A good city must have all those qualities. Density without livability could return us to the slums of the nineteenth century. Public places without small-scale, fine-grain development would give us vast, overscale cities. As an urban fabric, however, those qualities stand a good chance of meeting many of the goals we outlined. They directly attend to the issue of livability though they are aimed especially at encouraging public places and a public life. Their effects on personal and group identity are less clear, though the small-scale city is more likely to support identity than the large-scale city. Opportunity and imagination should be encouraged by a diverse and densely settled urban structure. This structure also should create a setting that is more meaningful to the individual inhabitant and small group than the giant environments now being produced. There is no guarantee that this urban structure will be a more just one than those presently existing. In supporting the small against the large, however, more justice for the powerless may be encouraged.

Still, an urban fabric of this kind cannot by itself meet all these goals. Other physical characteristics are important to the design of urban environments. Open space, to provide access to nature as well as relief from the built environment, is one. So are definitions, boundaries if you will, that give location and identity to neighborhoods (or districts) and to the city itself. There are other characteristics as well: public buildings, educational environments, places set aside for nurturing the spirit, and more. We still have work to do.

MANY PARTICIPANTS

While we have concentrated on defining physical characteristics of a good city fabric, the process of creating it is crucial. As important as many buildings and spaces are many participants in the building process. It is through this involvement in the creation and management of their city that citizens are most likely to identify with it and, conversely, to enhance their own sense of identity and control.

AN ESSENTIAL BEGINNING

The five characteristics we have noted are essential to achieving the values central to urban life. They need much further definition and testing. We have to know more about what configurations create public space: about maximum densities, about how small a community can be and still be urban (some very small Swiss villages fit the bill, and everyone knows some favorite examples), about what is perceived as big and what small under different circumstances, about landscape material as a space definer, and a lot more. When we know more we will be still further along toward a new urban design manifesto.

We know that any ideal community, including the kind that can come from this manifesto, will not always be comfortable for every person. Some people don't like cities and aren't about to. Those who do will not be enthralled with all of what we propose.

Our urban vision is rooted partly in the realities of earlier, older urban places that many people, including many utopian designers, have rejected, often for good reasons. So our utopia will not satisfy all people. That's all right. We like cities. Given a choice of the kind of community we would *like* to live in – the sort of choice earlier city dwellers seldom had – we would choose to live in an urban, public community that embraces the goals and displays the physical characteristics we have outlined. Moreover, we think it responds to what people want and that it will promote the good urban life.

DOLORES HAYDEN

"What Would a Non-sexist City Be Like? Speculations on Housing, Urban Design, and Human Work"

from Catharine R. Stimpson *et al.* (eds.), *Women and the American City* (1981) (first published 1980)

Editors' introduction Dolores Hayden is a professor of architecture, urbanism, and American studies at Yale University where she teaches courses on American urban history and urban design. A social historian and architect, Hayden has proposed ways to make the built environment and society more responsive to women's needs, particularly the needs of working women with children. She is also committed to new ways of building cities for a more egalitarian society, with more community interaction, which are more responsive to both social and environmental concerns, and reflect a concern for public history, cultural diversity, and urban preservation.

This and other of Hayden's writings have struck a deeply responsive chord with a spectrum of architects and planners who want to design new forms of housing to better fit contemporary family structure and needs as well as with feminists favoring changes in gender and work roles. Like other of Hayden's writings it is grounded in the *ideas* of earlier feminist writers, provides evidence that changes can work based on *actual projects* which have been tried, and forcefully states Hayden's own feminist and egalitarian *values*.

Dolores Hayden's work has inspired many young architects and planners to create buildings, neighborhoods, and cities better adapted to families that do not consist of an employed husband, non-working wife, and children; where men and women can more easily share both child rearing and work outside the home; and where diverse households can interact in mutually helpful and personally satisfying ways.

Leonie Sandercock and Ann Forsyth share many of Hayden's concerns and criticisms (p. 446). They advocate incorporating feminist theory and feminist values into urban planning. Contrast Hayden's scathing critique of what is wrong with suburbia with Herbert Gans's generally positive assessment of Levittown (p. 63), the quintessential suburb of the 1950s.

Dolores Hayden's books include *The Power of Place: Urban Landscapes as Public History* (Cambridge, MA: MIT Press, 1995), *Redesigning the American Dream: The Future of Housing, Work, and Family Life* (New York: W.W. Norton, 1984) , *The Grand Domestic Revolution: A History of Feminist Designs for American Houses, Neighborhoods and Cities* (Cambridge, MA: MIT Press, 1981), and *Seven American Utopias: The Architecture of Communitarian Socialism: 1780–1975* (Cambridge, MA: MIT Press, 1976).

"Co-housing" is a current movement to build housing with more shared space than is common in subdivisions of single-family detached homes. Co-housing developments typically consist of both privately owned units and commonly owned spaces such as a common kitchen, daycare center, and perhaps a bicycle repair shop. Members of a co-housing project might pool use of appliances just as Hayden suggests. The idea of mutual support in cooking, child-rearing, gardening, and social activities is fundamental to the co-housing philosophy. The movement proposes much more interaction among members living in a co-housing project than is common in typical suburban subdivisions of single-family detached homes, but a less visionary restructuring of gender and work roles than Hayden envisages in the HOMES groups she proposes. A number of co-housing projects have been built and more are underway. The best source of information on this movement is Kathryn McCamant and Charles Durrett, *Cohousing*, 2nd edn. (Berkeley, CA: Ten Speed Press, 1992).

Writings about women and cities include Catharine Stimpson (eds.), *Women and the American City* (Chicago: University of Chicago Press, 1981) from which this selection in reprinted, Caroline Andrew and Beth Moore Milroy (eds.), *Life Spaces: Gender, Household, Employment* (Vancouver: University of British Columbia Press, 1988), Leslie Kanes Weisman, *Discrimination by Design* (Urbana and Chicago: University of Illinois Press, 1992), Daphne Spain, *Gendered Space* (Chapel Hill: University of North Carolina Press, 1992), and Clara Greed, *Women and Planning: Creating Gendered Realities* (London: Routledge, 1993). Leonie Sandercock and Ann Forsyth's, bibliography on gender issues in planning theory (p. 446) contains further bibliography of related writings.

"A woman's place is in the home" has been one of the most important principles of architectural design and urban planning in the United States for the last century. An implicit rather than explicit principle for the conservative and male-dominated design professions, it will not be found stated in large type in textbooks on land use. It has generated much less debate than the other organizing principles of the contemporary American city in an era of monopoly capitalism, which include the ravaging pressure of private land development, the fetishistic dependence on millions of private automobiles, and the wasteful use of energy.[1] However, women have rejected this dogma and entered the paid labor force in larger and larger numbers. Dwellings, neighborhoods, and cities designed for homebound women constrain women physically, socially, and economically. Acute frustration occurs when women defy these constraints to spend all or part of the work day in the paid labor force. I contend that the only remedy for this situation is to develop a new paradigm of the home, the neighborhood, and the city; to begin to describe the physical, social, and economic design of a human settlement that would support, rather than restrict, the activities of employed women and their families. It is essential to recognize such needs in order to begin both the rehabilitation of the existing housing stock and the construction of new housing to meet the needs of a new and growing majority of Americans – working women and their families.

When speaking of the American city in the last quarter of the twentieth century, a false distinction between "city" and "suburb" must be avoided. The urban region, organized to separate homes and workplaces, must be seen as a whole. In such urban regions, more than half of the population resides in the sprawling suburban areas, or "bedroom communities." The greatest part of the built environment in the United States consists of "suburban sprawl": single-family homes grouped in class-segregated areas, crisscrossed by freeways and served by shopping malls and commercial strip developments. Over 50 million small homes are on the ground. About two-thirds of American families "own" their homes on long mortgages; this includes over 77 percent of all AFL-CIO members.[2] White, male skilled workers are far more likely to be homeowners than members of minority groups and women, long denied equal credit or equal access to housing. Workers commute to jobs either in the center or elsewhere in the suburban ring. In metropolitan areas studied in 1975 and 1976, the journey to work, by public transit or private car, averaged about nine miles each way. Over 100 million privately owned cars filled two- and three-car garages (which would be considered magnificent housing by themselves in many developing countries).

The United States, with 13 percent of the world's population, uses 41 percent of the world's passenger cars in support of the housing and transportation patterns described.[3]

The roots of this American settlement form lie in the environmental and economic policies of the past. In the late nineteenth century, millions of immigrant families lived in the crowded, filthy slums of American industrial cities and despaired of achieving reasonable living conditions. However, many militant strikes and demonstrations between the 1890s and 1920s made some employers reconsider plant locations and housing issues in their search for industrial order.[4] "Good homes make contented workers" was the slogan of the Industrial Housing Associates in 1919. These consultants and many others helped major corporations plan better housing for white, male skilled workers and their families in order to eliminate industrial conflict. "Happy workers invariably mean bigger profits, while unhappy workers are never a good investment," they chirruped.[5] Men were to receive "family wages," and become home "owners" responsible for regular mortgage payments, while their wives became home "managers" taking care of spouse and children. The male worker would return from his day in the factory or office to a private domestic environment, secluded from the tense world of work in an industrial city characterized by environmental pollution, social degradation, and personal alienation. He would enter a serene dwelling whose physical and emotional maintenance would be the duty of his wife. Thus the private suburban house was the stage set for the effective sexual division of labor. It was the commodity par excellence, a spur for male paid labor and a container for female unpaid labor. It made gender appear a more important self-definition than class, and consumption more involving than production. In a brilliant discussion of the "patriarch as wage slave," Stuart Ewen has shown how capitalism and anti-feminism fused in campaigns for homeownership and mass consumption: the patriarch whose home was his "castle" was to work year in and year out to provide the wages to support this private environment.[6]

Although this strategy was first boosted by corporations interested in a docile labor force, it soon appealed to corporations that wished to move from World War I defense industries into peacetime production of domestic appliances for millions of families. The development of the advertising industry, documented by Ewen, supported this ideal of mass consumption and promoted the private suburban dwelling, which maximized appliance purchases.[7] The occupants of the isolated household were suggestible. They bought the house itself, a car, stove, refrigerator, vacuum cleaner, washer, carpets. Christine Frederick, explaining it in 1929 as Selling Mrs. Consumer, promoted homeownership and easier consumer credit and advised marketing managers on how to manipulate American women.[8] By 1931 the Hoover Commission on Home Ownership and Home Building established the private, single-family home as a national goal, but a decade and a half of depression and war postponed its achievement. Architects designed houses for Mr. and Mrs. Bliss in a competition sponsored by General Electric in 1935; winners accommodated dozens of electrical appliances in their designs with no critique of the energy costs involved.[9] In the late 1940s the single-family home was boosted by FHA and VA mortgages, and the construction of isolated, overprivatized, energy-consuming dwellings became commonplace. "I'll Buy That Dream" made the postwar hit parade.[10]

Mrs. Consumer moved the economy to new heights in the 1950s. Women who stayed at home experienced what Betty Friedan called the "feminine mystique" and Peter Filene renamed the "domestic mystique."[11] While the family occupied its private physical space, the mass media and social science experts invaded its psychological space more effectively than ever before.[12] With the increase in spatial privacy came pressure for conformity in consumption. Consumption was expensive. More and more married women joined the paid labor force, as the suggestible housewife needed to be both a frantic consumer and a paid worker to keep up with the family's bills. Just as the mass of white male workers had achieved the "dream houses" in suburbia where fantasies of patriarchal authority and consumption could be acted out, their spouses entered the world of paid employment. By 1975, the two-worker family accounted for 39 percent of American households. Another 13 percent were single-parent families, usually headed by women. Seven out of ten employed women were in the work force because of financial need. Over 50 percent of all children between the ages of one and seventeen had employed mothers.[13]

How does a conventional home serve the employed woman and her family? Badly. Whether it is in a suburban, exurban, or inner-city neighborhood, whether it is a split-level ranch house, a modern masterpiece of concrete and glass, or an old brick tenement, the house or apartment is almost invariably organized around the same set of spaces: kitchen, dining room, living room, bedrooms, garage or parking area. These spaces require someone to undertake private cooking, cleaning, child care, and usually private transportation if adults and children are to exist within it. Because of residential zoning practices, the typical dwelling will usually be physically removed from any shared community space – no commercial or communal day-care facilities, or laundry facilities, for example, are likely to be part of the dwelling's spatial domain. In many cases these facilities would be illegal if placed across property lines. They could also be illegal if located on residentially zoned sites. In some cases sharing such a private dwelling with other individuals (either relatives or those unrelated by blood) is also against the law.[14]

Within the private spaces of the dwelling, material culture works against the needs of the employed woman as much as zoning does, because the home is a box to be filled with commodities. Appliances are usually single-purpose, and often inefficient, energy-consuming machines, lined up in a room where the domestic work is done in isolation from the rest of the family. Rugs and carpets that need vacuuming, curtains that need laundering, and miscellaneous goods that need maintenance fill up the domestic spaces, often decorated in "colonial," "Mediterranean," "French Provincial," or other eclectic styles purveyed by discount and department stores to cheer up that bare box of an isolated house. Employed mothers usually are expected to, and almost invariably do, spend more time in private housework and child care than employed men; often they are expected to, and usually do, spend more time on commuting per mile traveled than men, because of their reliance on public transportation. One study found that 70 percent of adults without access to cars are female.[15] Their residential neighborhoods are not likely to provide much support for their work activities. A "good" neighborhood is usually defined in terms of conventional shopping, schools, and perhaps public transit, rather than additional social services for the working parent, such as day care or evening clinics.

While two-worker families with both parents energetically cooperating can overcome some of the problems of existing housing patterns, households in crisis, such as subjects of wife and child battering, for example, are particularly vulnerable to its inadequacies. According to Colleen McGrath, every thirty seconds a woman is being battered somewhere in the United States. Most of these batterings occur in kitchens and bedrooms. The relationship between household isolation and battering, or between unpaid domestic labor and battering, can only be guessed, at this time, but there is no doubt that America's houses and households are literally shaking with domestic violence.[16] In addition, millions of angry and upset women are treated with tranquilizers in the private home – one drug company advertises to doctors: "You can't change her environment but you can change her mood."[17]

The woman who does leave the isolated, single-family house or apartment finds very few real housing alternatives available to her.[18] The typical divorced or battered woman currently seeks housing, employment, and child care simultaneously. She finds that matching her complex family requirements with the various available offerings by landlords, employers, and social services is impossible. One environment that unites housing, services, and jobs could resolve many difficulties, but the existing system of government services, intended to stabilize households and neighborhoods by ensuring the minimum conditions for a decent home life to all Americans, almost always assumes that the traditional household with a male worker and an unpaid homemaker is the goal to be achieved or simulated. In the face of massive demographic changes, programs such as public housing, AFDC, and food stamps still attempt to support an ideal family living in an isolated house or apartment, with a full-time homemaker cooking meals and minding children many hours of the day.

By recognizing the need for a different kind of environment, far more efficient use can be made of funds now used for subsidies to individual households. Even for women with greater financial resources, the need for better housing and services is obvious. Currently, more affluent women's problems as workers have been considered "private" problems – the lack of good day care, their lack of time. The aids to overcome an environment without child care, public transportation, or food service have been "private," commercially profitable solutions: maids and baby-sitters by the hour; franchise day care or extended television viewing; fast food service; easier credit for purchasing an automobile, a washer, or a microwave oven. Not only do these commercial solutions obscure the failure of American housing policies; they also generate bad conditions for other working women. Commercial day-care and fast-food franchises are the source of low-paying nonunion jobs without security. In this respect they resemble the use of private household workers by bourgeois women, who may never ask how their private maid or child-care worker arranges care for her own children. They also resemble the insidious effects of the use of television in the home as a substitute for developmental child care in the neighborhood. The logistical problems which all employed women face are not private problems, and they do not succumb to market solutions.

The problem is paradoxical: women cannot improve their status in the home unless their overall economic position in society is altered; women cannot improve their status in the paid labor force unless their domestic responsibilities are altered. Therefore, a program to achieve economic and environmental justice for women requires, by definition, a solution that overcomes the traditional divisions between the household and the market economy, the private dwelling and the workplace. One must transform the economic situation of the traditional home-maker, whose skilled labor has been unpaid but economically and socially necessary to society; one must also transform the domestic situation of the employed woman. If architects and urban designers were to recognize all employed women and their families as a constituency for new approaches to planning and design and were to reject all previous assumptions about "woman's place" in the home, what could we do? Is it possible to build non-sexist neighborhoods and design non-sexist cities? What would they be like?

Some countries have begun to develop new approaches to the needs of employed women. The Cuban Family Code of 1974 requires men to share housework and child care within the private home. The degree of its enforcement is uncertain, but in principle it aims at men's sharing what was formerly "women's work," which is essential to equality. The Family Code, however, does not remove work from the house, and relies upon private negotiation between husband and wife for its day-to-day enforcement. Men feign incompetence, especially in the area of cooking, with tactics familiar to any reader of Patricia Mainardi's essay, "The Politics of Housework," and the sexual stereotyping of paid jobs for women outside the home, in day-care centers for example, has not been successfully challenged.[19]

Another experimental approach involves the development of special housing facilities for employed women and their families. The builder Otto Fick first introduced such a program in Copenhagen in 1903. In later years it was encouraged in Sweden by Alva Myrdal and by the architects Sven Ivar Lind and Sven Markelius. Called "service houses" or "collective houses," such projects (Figures 1 and 2) provide child care and cooked food along with housing for employed women and their families.[20] Like a few similar projects in the USSR in the 1920s, they aim at offering services, either on a commercial basis or subsidized by the state, to replace formerly private "women's work" performed in the household. The Scandinavian solution does not sufficiently challenge male exclusion from domestic work, nor does it deal with households' changing needs over the life cycle, but it recognizes that it is important for environmental design to change.

Some additional projects in Europe extend the scope of the service house to include the provision of services for the larger community or society. In the Steilshoop Project, in Hamburg, Germany, in the early 1970s, a group of parents and single people designed public housing with supporting services (Figure 3).[21] The project included a number of former mental patients as residents and therefore served as a halfway house for them, in addition to providing support services for the public-housing tenants who

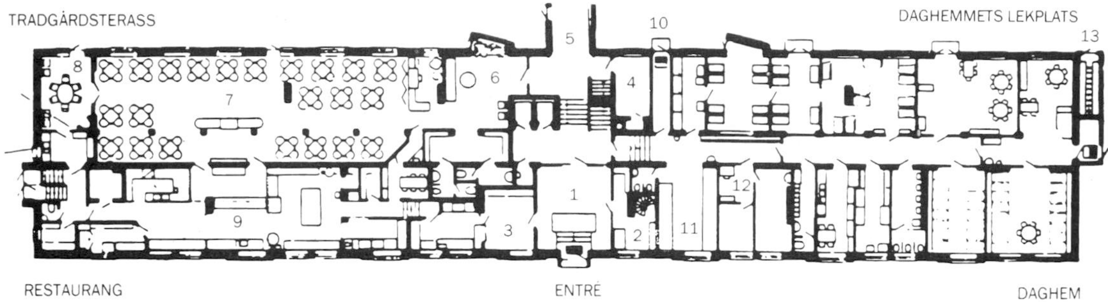

Figure 1 Sven Ivar Lind, Marieberg collective house, Stockholm, 1944, plan of entrance (entré), restaurant (restaurang), and day nursery (deghem): (1) entrance hall, (2) doorman's office, (3) restaurant delivery room, (4) real estate office, (5) connecting walkway to Swedberg House, (6) restaurant anteroom, (7) main dining room, (8) small dining room, (9) restaurant kitchen, (10) to day nursery's baby carriage room, (11) day nursery's baby changing room, (12) office for day nursery's directress, (13) to Wennerberg House's cycle garage

organized it. It suggests the extent to which current American residential stereotypes can be broken down – the sick, the aged, the unmarried can be integrated into new types of households and housing complexes, rather than segregated in separate projects.

Another recent project was created in London by Nina West Homes, a development group established in 1972, which has built or renovated over sixty-three units of housing on six sites for single parents. Children's play areas or day-care centers are integrated with the dwellings; in their Fiona House project the housing is designed to facilitate shared baby-sitting, and the day-care center is open to the neighborhood residents for a fee (Figure 4). Thus the single parents can find jobs as day-care workers and help the neighborhood's working parents as well.[22] What is most exciting here is the hint that home and work can be reunited on one site for some of the residents, and home and child-care services are reunited on one site for all of them.

In the United States, we have an even longer history of agitation for housing to reflect women's needs. In the late nineteenth century and early twentieth century there were dozens of projects by feminists, domestic scientists, and architects attempting to develop community services for private homes. By the late 1920s, few such experiments were still functioning.[23] In general, feminists of that era failed to recognize the problem of exploiting other women workers when providing services for those who could afford them. They also often failed to see men as responsible parents and workers in their attempts to socialize "women's" work. But feminist leaders had a very strong sense of the possibilities of neighborly cooperation among families and of the economic importance of "women's" work.

In addition, the United States has a long tradition of experimental utopian socialist communities building model towns, as well as the example of many communes and collectives established in the 1960s and 1970s which attempted to broaden conventional definitions of household and family.[24] While some communal groups, especially religious ones, have often demanded acceptance of a traditional sexual division of labor, others have attempted to make nurturing activities a responsibility of both women and men. It is important to draw on the examples of successful projects of all kinds, in seeking an image of a non-sexist settlement. Most employed women are not interested in taking themselves and their families to live in communal families, nor are they interested in having state bureaucracies run family life. They desire, not an end to private life altogether, but community services to support the private household. They also desire solutions that re inforce their economic independence and maximize their personal choices about child rearing and sociability.

What, then, would be the outline of a program for change in the United States? The task of reorganizing both home and work can only be accomplished by organizations of homemakers, women and men dedicated to making changes in

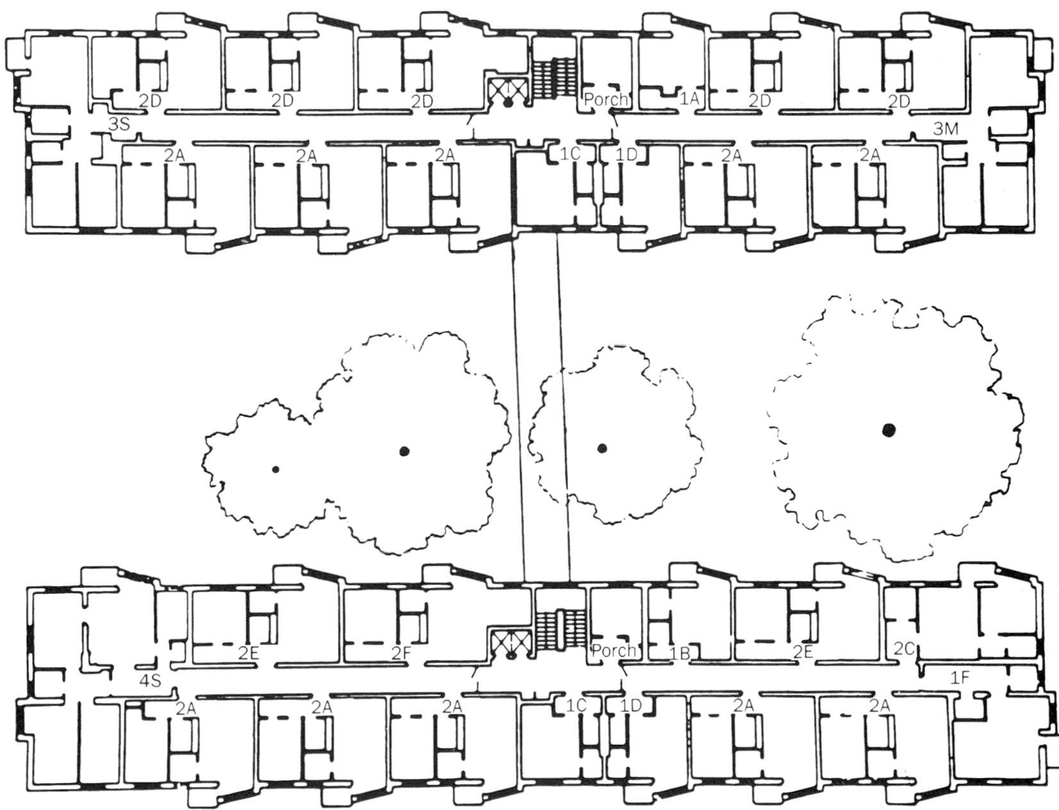

Figure 2 Plan of residential floors. Type 2A contains two rooms, bath, and kitchenette. Types 1C and 4D are efficiency units with bath and kitchenette. Type 4S includes four rooms with bath and full kitchen

the ways that Americans deal with private life and public responsibilities. They must be small, participatory organizations with members who can work together effectively. I propose calling such groups HOMES (Homemakers Organization for a More Egalitarian Society). Existing feminist groups, especially those providing shelters for battered wives and children, may wish to form HOMES to take over existing housing projects and develop services for residents as an extension of those offered by feminist counselors in the shelter. Existing organizations supporting cooperative ownership of housing may wish to form HOMES to extend their housing efforts in a feminist direction. A program broad enough to transform housework, housing, and residential neighborhoods must: (1) involve both men and women in the unpaid labor associated with housekeeping and child care on an equal basis; (2) involve both men and women in the paid labor force on an equal basis; (3) eliminate residential segregation by class, race, and age; (4) eliminate all federal, state, and local programs and laws that offer implicit or explicit reinforcement of the unpaid role of the female homemaker; (5) minimize unpaid domestic labor and wasteful energy consumption; (6) maximize real choices for households concerning recreation and sociability. While many partial reforms can support these goals, an incremental strategy cannot achieve them. I believe that the establishment of experimental residential centers, which in their architectural design and economic organization transcend traditional definitions of home, neighborhood, city, and workplace, will be necessary to make changes on this scale. These centers could be created through renovation of existing neighborhoods or through new construction.

Suppose forty households in a U.S. metropolitan area formed a HOMES group and that those households, in their composition, represented the social structure of the American population as a whole. Those forty households

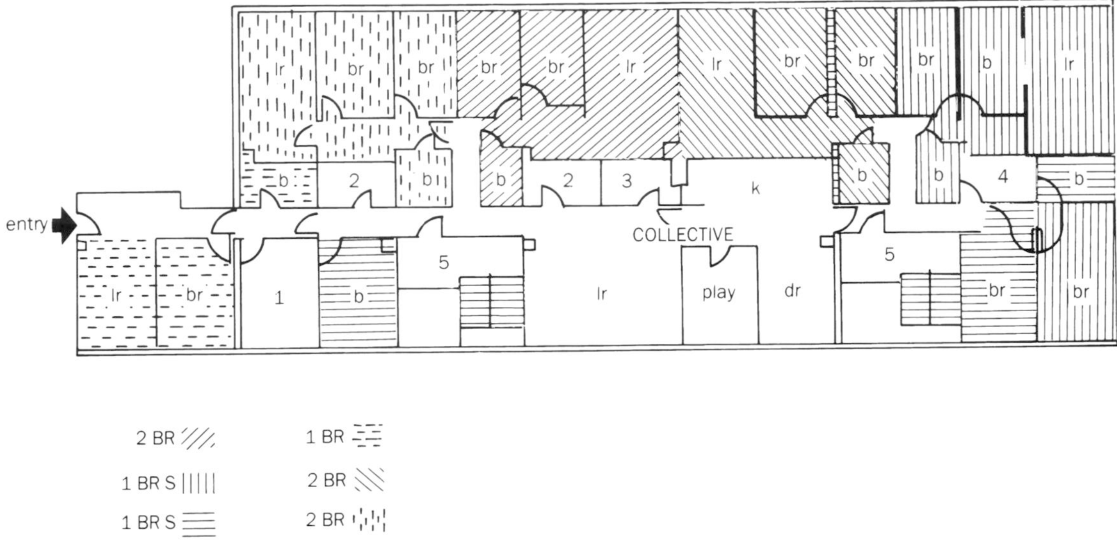

Figure 3 "Urbanes Wohnen" (urban living) Steilshoop, north of Hamburg, public housing for 206 tenants, designed by the tenant association in collaboration with Rolf Spille, 1970–3. Instead of 72 conventional units, they built 20 multifamily units and 2 studios. Twenty-six mental patients were included in the project, of whom 24 recovered. Partial floor plan. Units include private bedrooms (br), living rooms (lr), and some studios (s). They share a collective living room, kitchen, dining room, and playroom. Each private apartment can be closed off from the collective space and each is different. Key: (1) storage room, (2) closets, (3) wine cellar, (4) buanderie, (5) fire stairs

would include: seven single parents and their fourteen children (15 percent); sixteen two-worker couples and their twenty-four children (40 percent); thirteen one-worker couples and their twenty-six children (35 percent); and four single residents, some of them "displaced homemakers" (10 percent). The residents would include sixty-nine adults and sixty-four children. There would need to be forty private dwelling units, ranging in size from efficiency to three bedrooms, all with private, fenced outdoor space. In addition to the private housing, the group would provide the following collective spaces and activities: (1) a day-care center with landscaped outdoor space, providing day care for forty children and after-school activities for sixty-four children; (2) a laundromat providing laundry service; (3) a kitchen providing lunches for the day-care center, take-out evening meals, and "meals-on-wheels" for elderly people in the neighborhood; (4) a grocery depot, connected to a local food cooperative; (5) a garage with two vans providing dial-a-ride service and meals-on-wheels; (6) a garden (or allotments) where some food can be grown; (7) a home help office providing helpers for the elderly, the sick, and employed parents whose children are sick. The use of all these collective services should be voluntary; they would exist in addition to private dwelling units and private gardens.

To provide all of the above services, thirty-seven workers would be necessary: twenty day-care workers; three food-service workers; one grocery-depot worker; five home helpers; two drivers of service vehicles; two laundry workers; one maintenance worker; one gardener; two administrative staff. Some of these may be part-time workers, some full-time. Day care, food services, and elderly services could be organized as producers' cooperatives, and other workers could be employed by the housing cooperative as discussed below.

Because HOMES is not intended as an experiment in isolated community buildings but as an experiment in meeting employed women's needs in an urban area, its services should be available to the neighborhood in which the experiment is located. This will increase demand for the services and insure that the jobs are real ones. In addition, although residents of HOMES

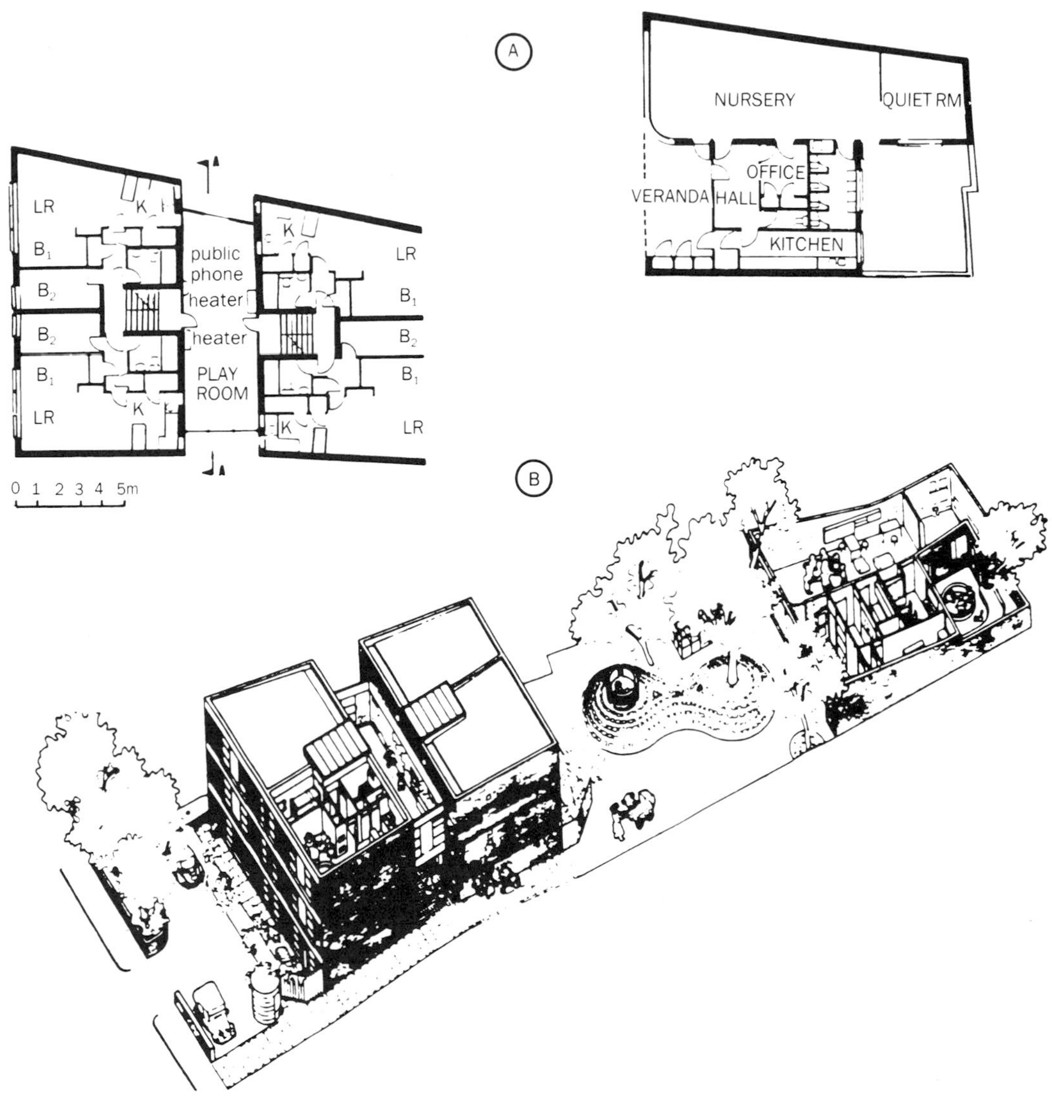

Figure 4 A, Fiona House, second-floor plan, main building, showing corridor used as a playroom, with kitchen windows opening into it; first-floor plan, rear building, showing nursery school. B, Axonometric drawing, Fiona House, Nina West Homes, London, 1972, designed by Sylvester Bone. Twelve two-bedroom units for divorced or separated mothers with additional outdoor play space and neighborhood nursery school facility. Flats can be linked by intercom system to provide an audio substitute for babysitting

should have priority for the jobs, there will be many who choose outside work. So some local residents may take jobs within the experiment.

In creating and filling these jobs it will be important to avoid traditional sex stereotyping that would result from hiring only men as drivers, for example, or only women as food-service workers. Every effort should be made to break down separate categories of paid work for women and men, just as efforts should be made to recruit men who accept an equal share of domestic responsibilities as residents. A version of the Cuban Family Code should become part of the organization's platform.

Similarly, HOMES must not create a two-class society with residents outside the project making more money than residents in HOMES jobs that utilize some of their existing domestic skills. The HOMES jobs should be paid according to egalitarian rather than sex-stereotyped attitudes about skills and hours. These jobs must be all classified as skilled work rather than as unskilled or semiskilled at present, and offer full social security and health benefits, including adequate maternity leave, whether workers are part-time or full-time.

Many federal Housing and Urban Development programs support the construction of nonprofit, low- and moderate-cost housing, including section 106b, section 202, and section 8. In addition, HUD section 213 funds are available to provide mortgage insurance for the conversion of existing housing of five or more units to housing cooperatives. HEW programs also fund special facilities such as day-care centers or meals-on-wheels for the elderly. In addition, HUD and HEW offer funds for demonstration projects which meet community needs in new ways.[25] Many trade unions, churches, and tenant cooperative organizations are active as nonprofit housing developers. A limited-equity housing cooperative offers the best basis for economic organization and control of both physical design and social policy by the residents.

Many knowledgeable nonprofit developers could aid community groups wishing to organize such projects, as could architects experienced in the design of housing cooperatives. What has not been attempted is the reintegration of work activities and collective services into housing cooperatives on a large enough scale to make a real difference to employed women. Feminists in trade unions where a majority of members are women may wish to consider building co-operative housing with services for their members. Other trade unions may wish to consider investing in such projects. Feminists in the co-op movement must make strong, clear demands to get such services from existing housing cooperatives, rather than simply go along with plans for conventional housing organized on a cooperative economic basis. Feminists outside the cooperative movement will find that cooperative organizational forms offer many possibilities for supporting their housing activities and other services to women. In addition, the recently established national Consumer Cooperative Bank has funds to support projects of all kinds that can be tied to cooperative housing.

In many areas, the rehabilitation of existing housing may be more desirable than new construction. The suburban housing stock in the United States must be dealt with effectively. A little bit of it is of architectural quality sufficient to deserve preservation; most of it can be aesthetically improved by the physical evidence of more intense social activity. To replace empty front lawns without sidewalks, neighbors can create blocks where single units are converted to multiple units; interior land is pooled to create a parklike setting at the center of the block; front and side lawns are fenced to make private outdoor spaces; pedestrian paths and sidewalks are created to link all units with the central open space; and some private porches, garages, tool sheds, utility rooms, and family rooms are converted to community facilities such as children's play areas, dial-a-ride garages, and laundries.

Figure 5A shows a typical bleak suburban block of thirteen houses, constructed by speculators at different times, where about four acres are divided into plots of one-fourth to one-half acre each. Thirteen driveways are used by twenty-six cars; ten garden sheds, ten swings, thirteen lawn mowers, thirteen outdoor dining tables, begin to suggest the wasteful duplication of existing amenities. Yet despite the available land there are no transitions between public streets and these private homes. Space is either strictly private or strictly public. Figure 6A shows a typical one-family house of 1,400 square feet on this block. With three bedrooms and den, two-and-a-half baths, laundry room, two porches, and a two-car garage, it was constructed in the 1950s at the height of the "feminine mystique."

To convert this whole block and the housing on it to more efficient and sociable uses, one has to define a zone of greater activity at the heart of the block, taking a total of one and one half to two acres for collective use (Figure 5B). Essentially, this means turning the block inside out. The Radburn plan, developed by Henry Wright and Clarence Stein in the late 1920s, delineated this principle very clearly as correct land use in "the motor age," with cars segregated from residents' green spaces, especially spaces for children. In Radburn, New Jersey, and in the Baldwin Hills district of Los Angeles, California, Wright and Stein achieved remarkably luxurious

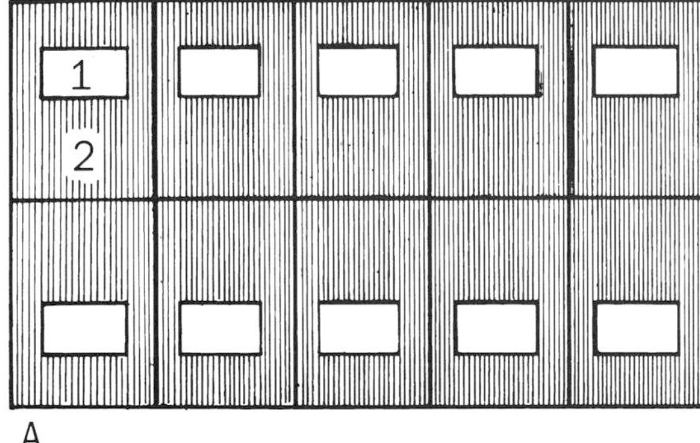

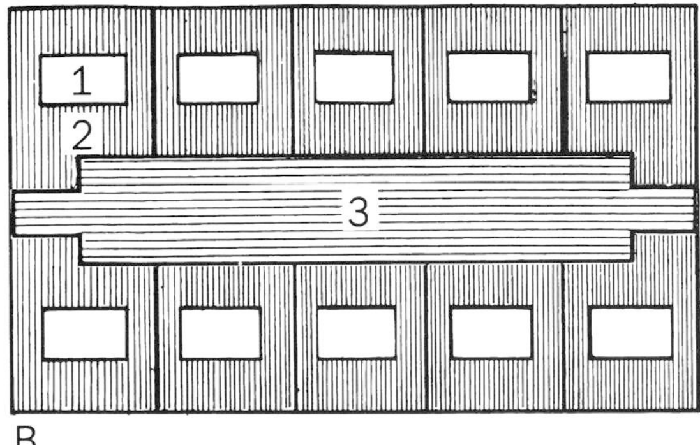

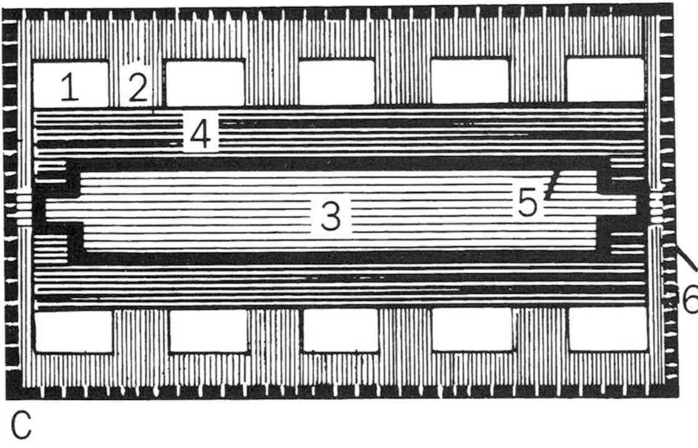

Figure 5 Diagram showing some of the possibilities of reorganizing a typical suburban block through rezoning, rebuilding, and landscaping. A, Ten single-family houses (1) on ten private lots (2); B, the same houses (1) with smaller private lots (2) after a backyard rehabilitation program has created a new village green (3) at the heart of the block; C, the same houses (1) and many small private gardens (2) with a new village green (3) surrounded by a zone for new services and accessory apartments (4) connected by a new sidewalk or arcade (5) and surrounded by a new border of street trees (6). In C, (4) can include space for such activities as day care, elderly care, laundry, and food service as well as housing, while (3) can accommodate a children's play area, vegetable or flower gardens, and outdoor seating. (5) may be a sidewalk, a vine-covered trellis, or a formal arcade. The narrow ends of the block can be emphasized as collective entrances with gates (to which residents have keys), leading to new accessory apartments entered from the arcade or sidewalk. In the densest possible situtions, (3) may be alley and parking lot, if existing street parking and public transit are not adequate

results (at a density of about seven units to the acre) by this method, since their multiple-unit housing always bordered a lush parkland without any automobile traffic. The Baldwin Hills project demonstrates this success most dramatically, but a revitalized suburban block with lots as small as one-fourth acre can be reorganized to yield something of this same effect.[26] In this case, social amenities are added to aesthetic ones as the interior park is designed to accommodate community day care, a garden for growing vegetables, some picnic tables, a playground where swings and slides are grouped, a grocery depot connected to a larger neighborhood food cooperative, and a dial-a-ride garage.

Large single-family houses can be remodeled quite easily to become duplexes and triplexes, despite the "open plans" of the 1950s and 1960s popularized by many developers. The house in Figure 6A becomes, in Figure 6B, a triplex, with a two-bedroom unit (linked to a community garage); a one-bedroom unit; and an efficiency unit (for a single person or elderly person). All three units are shown with private enclosed gardens. The three units share a front porch and entry hall. There is still enough land to give about

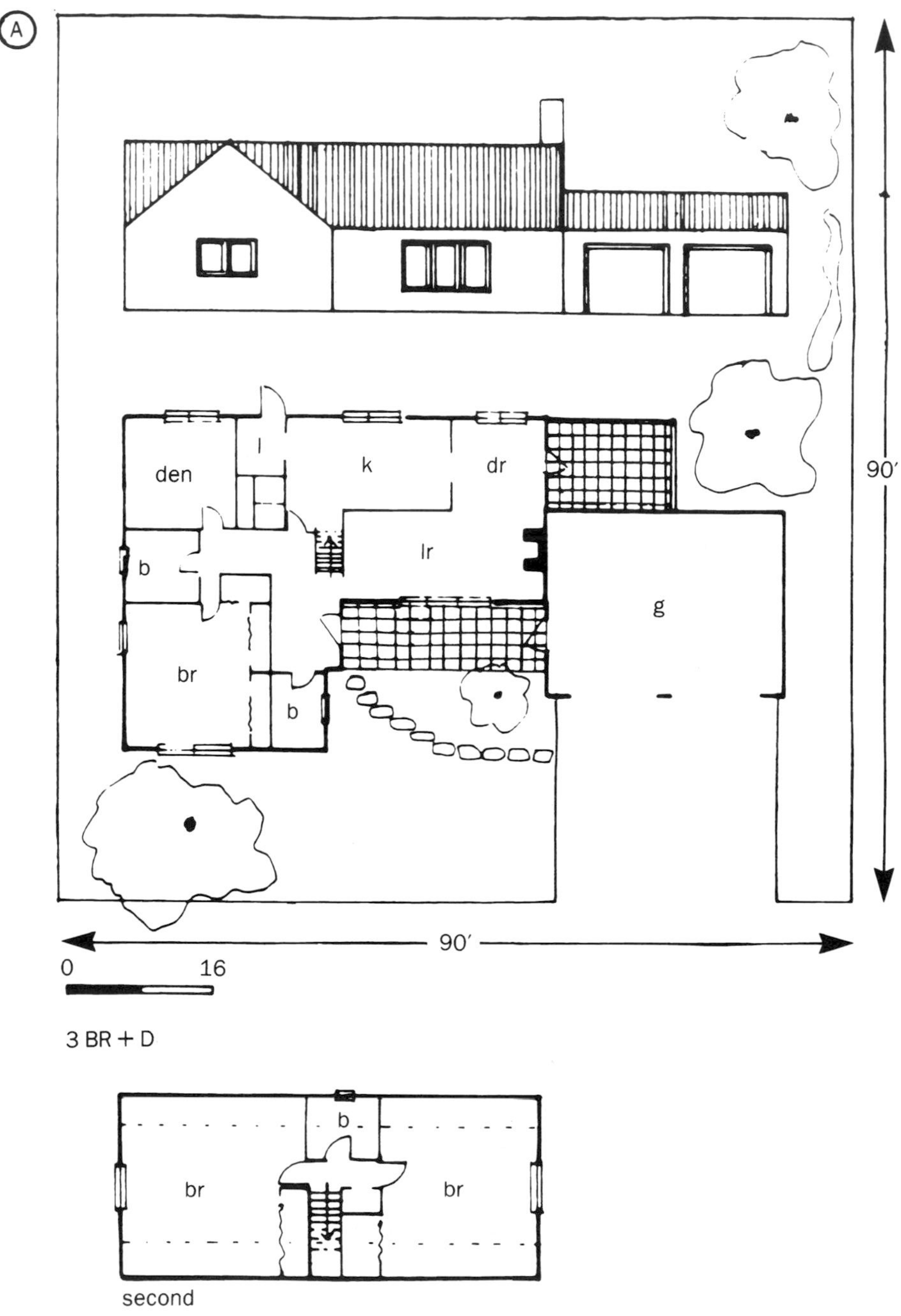

Figure 6A Suburban single-family house, plan, three bedrooms plus den

two-fifths of the original lot to the community. Particularly striking is the way in which existing spaces such as back porches or garages lend themselves to conversion to social areas or community services. Three former private garages out of thirteen might be given over to collective uses – one as a central office for the whole block, one as a grocery depot, and one as a dial-a-ride

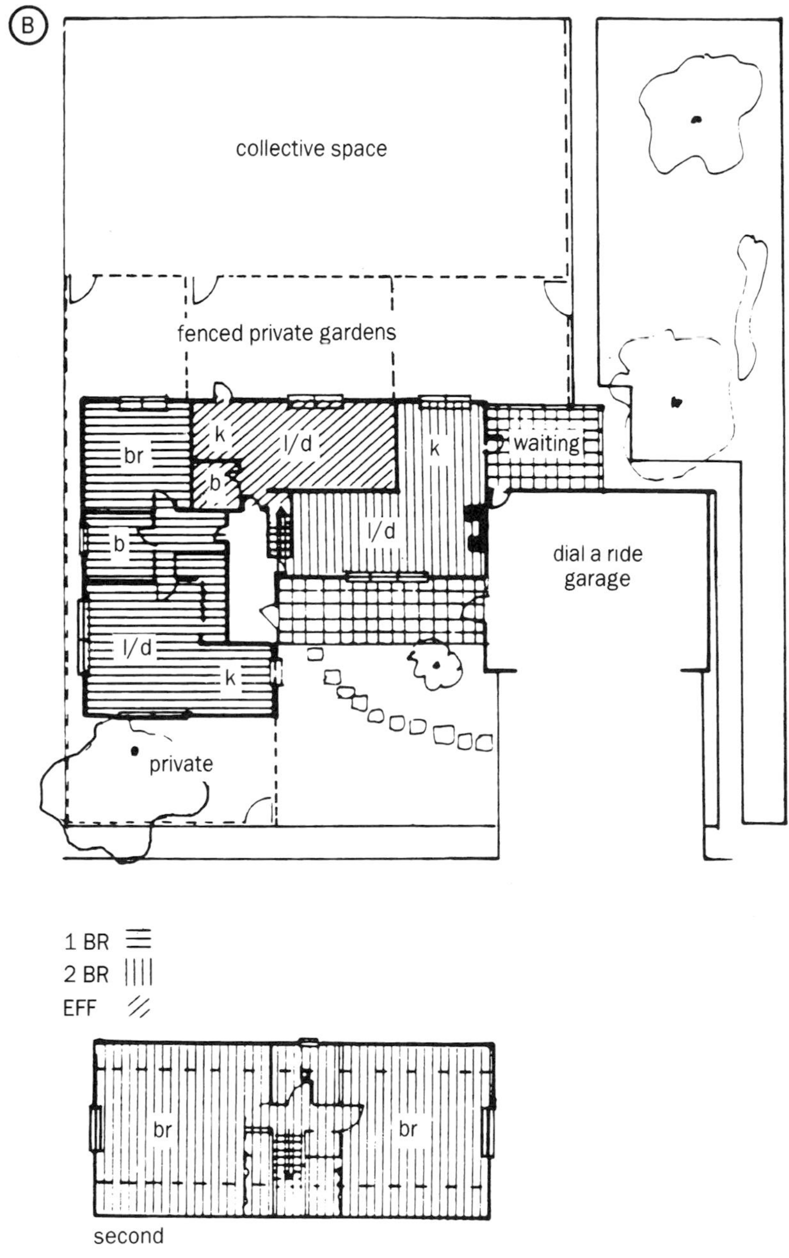

Figure 6B Proposed HOMES revitalization, same house converted to three units (two bedroom, one bedroom, and efficiency), plus dial-a-ride garage and collective outdoor space

garage. Is it possible to have only twenty cars (in ten garages) and two vans for twenty-six units in a rehabilitated block? Assuming that some residents switch from outside employment to working within the block, and that for all residents, neighborhood shopping trips are cut in half by the presence of day care, groceries, laundry, and cooked food on the block, as well as aided by the presence of some new collective transportation, this might be done.

What about neighbors who are not interested in such a scheme? Depending on the configuration of lots, it is possible to begin such a plan with as few as three or four houses. In Berkeley, California, where neighbors on Derby Street joined their backyards and created a cooperative day-care center, one absentee landlord refused to join – his entire property is fenced in and the community space flows around it without difficulty. Of course, present zoning laws must be changed, or variances obtained, for the conversion of single-family houses into duplexes and triplexes and the introduction of any sort of commercial activities into a residential block. However, a community group that is able to organize or acquire at least five units could become a HUD housing cooperative, with a nonprofit corporation owning all land and with producers' cooperatives running the small community services. With a coherent plan for an entire block, variances could be obtained much more easily than on a lot-by-lot basis. One can also imagine organizations that run halfway houses – for ex-mental patients, or runaway teenagers, or battered women – integrating their activities into such a block plan, with an entire building for their activities. Such groups often find it difficult to achieve the supportive neighborhood context such a block organization would offer.

I believe that attacking the conventional division between public and private space should become a socialist and feminist priority in the 1980s. Women must transform the sexual division of domestic labor, the privatized economic basis of domestic work, and the spatial separation of homes and workplaces in the built environment if they are to be equal members of society. The experiments I propose are an attempt to unite the best features of past and present reforms in our own society and others, with some of the existing social services available in the United States today. I would like to see several demonstration HOMES begun, some involving new construction following the program I have laid out, others involving the rehabilitation of suburban blocks. If the first few experimental projects are successful, homemakers across the United States will want to obtain day-care, food, and laundry services at a reasonable price, as well as better wages, more flexible working conditions, and more suitable housing. When all homemakers recognize that they are struggling against both gender stereotypes and wage discrimination, when they see that social, economic, and environmental changes are necessary to overcome these conditions, they will no longer tolerate housing and cities, designed around the principles of another era, that proclaim that "a woman's place is in the home."

NOTES

This paper comprised part of the text of a talk for the conference "Planning and Designing a Non-Sexist Society," University of California, Los Angeles, April 21, 1979. I would like to thank Catharine Stimpson, Peter Marris, S. M. Miller, Kevin Lynch, Jeremy Brecher, and David Thompson for extensive written comments on drafts of this paper.

1. There is an extensive Marxist literature on the importance of spatial design to the economic development of the capitalist city, including Henri Lefebre, *La Production de l'espace* (Paris: Editions Anthropos, 1974); Manuel Castells, *The Urban Question* (Cambridge, Mass.: M.I.T. Press, 1977); David Harvey, *Social Justice and the City* (London: Edward Arnold, 1974); and David Gordon, "Capitalist Development and the History of American Cities," in *Marxism and the Metropolis*, ed. William K. Tabb and Larry Sawyers (New York: Oxford University Press, 1978). None of this work deals adequately with the situation of women as workers and homemakers, nor with the unique spatial inequalities they experience. Nevertheless, it is important to combine the economic and historical analysis of these scholars with the empirical research of non-Marxist feminist urban critics and sociologists who have examined women's experience of conventional housing, such as Gerda Wekerle, "A Woman's Place Is in the City" (paper for the Lincoln Institute of Land Policy, Cambridge, Mass., 1978); and Suzanne Keller, "Women in a Planned Community" (paper for the Lincoln Institute of Land Policy, Cambridge, Mass., 1978). Only then can one begin to provide a socialist-feminist critique of the spatial design of the American city. It is also essential to develop research on housing similar to Sheila B. Kamerman, "Work and Family in Industrialized Societies," *Signs: Journal of Women in Culture and Society* 4, no. 4 (Summer 1979): 632–50, which reviews patterns of women's employment, maternity provisions, and child-care policies in Hungary, East Germany, West Germany, France, Sweden, and the United States. A comparable study of housing and related services for employed women could be the basis for more elaborate proposals for change.

Many attempts to refine socialist and feminist economic theory concerning housework are discussed in an excellent article by Ellen Malos, "Housework and the Politics of Women's Liberation," *Socialist Review* 37 (January–February 1978): 41–47. A most significant theoretical piece is Movimento di Lotta Femminile, "Programmatic Manifesto for the Struggle of Housewives in the Neighborhood," *Socialist Revolution* 9 (May–June 1972): 85–90.

2. *Survey of AFL-CIO Members Housing 1975* (Washington, D.C.: AFL-CIO, 1975), p. 16. I am indebted to Allan Heskin for this reference.
3. *Transit Fact Book*, 1977–78 ed. (Washington, D.C.: American Public Transit Association, 1978), p. 29); *Motor Vehicle Facts and Figures* (Detroit, Mich.: Motor Vehicle Manufacturers Association, 1977), pp. 29, 31, 53.
4. Gordon, pp. 48–50, discusses suburban relocation of plants and housing.
5. Industrial Housing Associates, "Good Homes Make Contented Workers," 1919, Edith Elmer Wood Papers, Avery Library, Columbia University. Also see Barbara Ehrenreich and Deirdre English, "The Manufacture of Housework," *Socialist Revolution* 5 (1975): 16. They quote an unidentified corporate official (ca. 1920): "Get them to invest their savings in homes and own them. Then they won't leave and they won't strike. It ties them down so they have a stake in our prosperity."
6. Stuart Ewen, *Captains of Consciousness: Advertising and the Social Roots of the Consumer Culture* (New York: McGraw-Hill Book Co., 1976).
7. Richard Walker, "Suburbanization in Passage," unpublished draft paper (Berkeley: University of California, Berkeley, Department of Geography, 1977).
8. Christine Frederick, *Selling Mrs. Consumer* (New York: Business Bourse, 1929).
9. Carol Barkin, "Home, Mom, and Pie-in-the-Sky" (M. Arch. thesis, University of California, Los Angeles, 1979), pp. 120–24, gives the details of this competition; Ruth Schwartz Cowan, in an unpublished lecture at M.I.T. in 1977, explained GE's choice of an energy-consuming design for its refrigerator in the 1920s, because this would increase demand for its generating equipment by municipalities.
10. Peter Filene, *Him/Her/Self: Sex Roles in Modern America* (New York: Harcourt Brace Jovanovich, 1974), p. 189.
11. Betty Friedan, *The Feminine Mystique* (1963; New York: W. W. Norton & Co., 1974), p. 307, somewhat hysterically calls the home a "comfortable concentration camp"; Filene, p. 194, suggests that men are victimized by ideal homes too, thus "domestic" mystique.
12. Eli Zaretsky, *Capitalism, the Family, and Personal Life* (New York: Harper & Row, 1976), develops Friedman's earlier argument in a more systematic way. This phenomenon is misunderstood by Christopher Lasch, *Haven in a Heartless World* (New York: Alfred A. Knopf, 1977), who seems to favor a return to the sanctity of the patriarchal home.
13. Rosalyn Baxandall, Linda Gordon, and Susan Reverby, eds., *America's Working Women: A Documentary History, 1600 to the Present* (New York: Vintage Books, 1976). For more detail, see Louise Kapp Howe, *Pink Collar Workers: Inside the World of Woman's Work* (New York: Avon Books, 1977).
14. Recent zoning fights on the commune issue have occurred in Santa Monica, Calif.; Wendy Schuman, "The Return of Togetherness," *New York Times* (March 20, 1977), reports frequent illegal down zoning by two-family groups in one-family residences in the New York area.
15. Study by D. Foley, cited in Wekerle (see n. 1 above).
16. Colleen McGrath, "The Crisis of Domestic Order," *Socialist Review* 9 (January–February 1979): 12, 23.
17. Research by Malcolm MacEwen, cited in *Associate Collegiate Schools of Architecture Newsletter* (March 1973), p. 6.
18. See, for example, Carol A. Brown, "Spatial Inequalities and Divorced Mothers" (paper delivered at the annual meeting of the American Sociological Association, San Francisco, 1978); Susan Anderson-Khleif, research report for HUD on single-parent families and their housing, summarized in "Housing for Single Parents," *Research Report, MIT-Harvard Joint Center for Urban Studies* (April 1979), pp. 3–4.
19. Patricia Mainardi, "The Politics of Housework," in *Sisterhood Is Powerful*, ed. Robin Morgan (New York: Vintage Books, 1970). My discussion of the Cuban Family Code is based on a visit to Cuba in 1978; a general review is Carollee Bengelsdorf and Alice Hageman, "Emerging from Underdevelopment: Women and Work in Cuba," in *Capitalist Patriarchy and the Case for Socialist Feminism*, ed. Z. Eisenstein (New York: Monthly Review Press, 1979). Also see Geoffrey E. Fox, "Honor, Shame and Women's Liberation in Cuba: Views of Working-Class Emigr´e Men," in *Female and Male in Latin America*, ed. A. Pescatello (Pittsburgh: University of Pittsburgh Press, 1973).
20. Erwin Muhlestein, "Kollektives Wohnen gestern und heute," *Architese* 14 (1975): 3–23.
21. This project relies on the "support structures" concept of John Habraken to provide flexible interior partitions and fixed mechanical core and structure.
22. "Bridge over Troubled Water," *Architects' Journal* (September 27, 1972), pp. 680–84; personal interview with Nina West, 1978.
23. Dolores Hayden, *A "Grand Domestic Revolution":*

Feminism, Socialism and the American Home, 1870–1930 (Cambridge, Mass.: M.I.T. Press, 1980); "Two Utopian Feminists and Their Compaigns for Kitchenless Houses," *Signs: Journal of Women in Culture and Society* 4, no. 2 (Winter 1979): 274–90; "Melusina Fay Peirce and Cooperative Housekeeping," *International Journal of Urban and Regional Research* 2 (1978): 401–20; "Challenging the American Domestic Ideal," and "Catharine Beecher and the Politics of Housework," in *Women in American Architecture*, ed. S. Torre (New York: Whitney Library of Design, 1977), pp. 22–39, 40–49; "Charlotte Perkins Gilman: Domestic Evolution or Domestic Revolution," *Radical History Review*, vol. 21 (Winter 1979–80), in press.

24. Dolores Hayden, *Seven American Utopias: The Architecture of Communitarian Socialism, 1790–1975* (Cambridge, Mass.: M.I.T. Press, 1976), discusses historical examples and includes a discussion of communes of the 1960s and 1970s, "Edge City, Heart City, Drop City: Communal Building Today," pp. 320–47.
25. I am indebted to Geraldine Kennedy and Sally Kratz, whose unpublished papers, "Toward Financing Cooperative Housing," and "Social Assistance Programs Whose Funds Could Be Redirected to Collective Services," were prepared for my UCLA graduate seminar in spring 1979.
26. See also the successful experience of Zurich, described in Hans Wirz, "Back Yard Rehab: Urban Microcosm Rediscovered," *Urban Innovation Abroad* 3 (July 1979): 2–3.

SIM VAN DER RYN and STUART COWAN

"An Introduction to Ecological Design"

from *Ecological Design* (1995)

Editors' introduction While the term "ecological design" is new, wisdom about how to build in ways that respect the natural environment is not. For millennia many cultures have built human settlements that respect the earth. Agriculture can be environmentally disruptive, but people living close to the soil often observe and learn from nature and walk gently on the land. Humans can live in harmony with forest and field, fish and fowl.

Van der Ryn and Cowan argue that the "designed mess" we have made of our neighborhoods, cities, and ecosystems is largely due to the failure to understand ecology and to use its principles as the basis for a coherent philosophy, vision, and practice of design. They are imbued with awe at the lessons nature can teach and urge designers to look to ecological processes themselves for inspiration. They urge designers, architects, and planners to practice ecological design, which they define as any form of design that minimizes environmentally destructive impacts by integrating itself with living processes. Ecological design is the effective adaptation to and integration with nature's processes.

Van Der Ryn and Cowan draw an analogy between the environment and a stack of coins. Right now we are using up our "natural capital" – lakes, forests, wetlands, and fisheries – at an alarming rate. In contrast, earlier cultures used only the natural "interest" on natural capital, leaving the capital itself intact to continue to generate natural interest forever. One way of thinking about ecological design is to envisage using only natural "interest" in the form of carefully managed yields of lumber, fish, and crops, but leaving natural capital untouched.

Van der Ryn and Cowan offer three critical strategies for the preservation of irreplaceable natural resources: conservation, regeneration, and stewardship. *Conservation* involves greater sensitivity to using up natural resources, paying attention to using less, and minimizing waste. *Regeneration* goes further. It involves repairing and renewing "a world deeply wounded by environmentally insensitive design." Finally, Van der Ryn and Cowan argue, we must learn to practice *stewardship* over a regenerated world to weave together the natural world and the built environment. Then ecological design will make possible the kind of sustainable urban development David Clark describes (p. 579) and Stephen Wheeler advocates (p. 434).

Taking a holistic and comprehensive view of design, Van der Ryn and Cowan urge the reader to reevaluate building materials and as much as possible shift away from environmentally destructive materials to ecologically sensitive ones. If, for example, producing conventional cement has unacceptable environmental consequences, they challenge designers to consider not using concrete as a "design constraint." Designers will have to learn to use new less environmentally destructive types of concrete or no concrete at all.

"An Introduction to Ecological Design" reviews the history of the ecological design movement and describes major currents in ecological design today. The authors point to ecologically sound practices of cultures as varied as Balinese rice farmers, Australian aborigines, and Berkeley, California, visionaries who built an "integral urban house" in the 1970s to demonstrate energy-efficient design. They review movements for ecologically sound town planning led by visionaries such as Ebenezer Howard (p. 321), Patrick Geddes (p. 330), Lewis Mumford (p. 92), and Frank Lloyd Wright (p. 344) and successor movements of the 1960s and 1970s.

According to Van der Ryn and Cowan these early visionary projects are maturing. Today the environmental movement is metamorphosing into a broad-based sustainability movement. Wind and solar energy technology have improved to the point that these sources offer real alternatives to use of fossil fuels. Scientists and practitioners have developed sustainable agriculture systems that work. City planners like Peter Calthorpe (p. 350) are designing compact pedestrian-oriented urban designs. Serious efforts at co-generation, recycling industrial wastes, toxic clean-up, and energy-efficient urbanism are underway throughout the world.

Review the writings of early twentieth-century visionary urban planners Howard, Geddes, Olmsted (p. 314), and Wright in light of Van Der Ryn and Cowan's description of current approaches to ecological design. Compare what Wheeler and Clark say about sustainable urban development.

Two classic writings on ecological design are Patrick Geddes, *Cities in Evolution* (London: Williams & Norgate, 1915), reprinted in Richard LeGates and Frederic Stout, *Early Urban Planning 1870–1940* (London: Routledge/Thoemmes, 1998) and Ian McHarg, *Design With Nature* (New York: John Wiley, 1992).

Other books about ecological design include Fritz Schumacher, *Small is Beautiful* (New York: HarperCollins, 1989), Amory Lovins, *Soft Energy Paths* (San Francisco: Friends of the Earth, 1977), John and Nancy Todd, *Eco-Cities to Living Machines: Principles of Ecological Design*, 2nd edn., (Berkeley, CA: North Atlantic Books, 1994), John Tillman Lyle, *Regenerative Design for Sustainable Development* (New York: Wiley, 1994), and Robert L. Thayer, *Gray World, Green Heart: Technology, Nature and the Sustainable Landscape* (New York: John Wiley, 1996). An account of a demonstration house built to illustrate ecological design principles is Helga Olkowski *et al.*, *The Integral Urban House* (San Francisco: Sierra Club Books, 1979).

OVERVIEW

We live in two interpenetrating worlds. The first is the living world, which has been forged in an evolutionary crucible over a period of four billion years. The second is the world of roads and cities, farms and artifacts, that people have been designing for themselves over the last few millennia. The condition that threatens both worlds – unsustainability – results from a lack of integration between them.

Now imagine the natural world and the humanly designed world bound together in intersecting layers, the warp and woof that make up the fabric of our lives. Instead of a simple fabric of two layers, it is made up of dozens of layers with vastly different characteristics. How these layers are woven together determines whether the result will be a coherent fabric or a dysfunctional tangle.

We need to acquire the skills to effectively interweave human and natural design. The designed mess we have made of our neighborhoods, cities, and ecosystems owes much to the lack of a coherent philosophy, vision, and practice of design that is grounded in a rich understanding of ecology. Unfortunately, the guiding metaphors of those who shape the built environment still reflect a nineteenth-century epistemology. Until our everyday activities preserve ecological integrity by *design*, their cumulative impact will continue to be devastating.

Thinking ecologically about design is a way of strengthening the weave that links nature and culture. Just as architecture has traditionally concerned itself with problems of structure, form, and aesthetics, or as engineering has with safety and efficiency, we need to consciously cultivate an ecologically sound form of design that is consonant with the long-term survival of all species. We define *ecological design* as "any form of design that minimizes environmentally destructive impacts by integrating itself with living processes." This integration implies that the design respects species diversity, minimizes resource depletion, preserves nutrient and water cycles, maintains habitat quality, and attends to all the other preconditions of human and ecosystem health.

Ecological design explicitly addresses the design dimension of the environmental crisis. It is not a style. It is a form of engagement and partnership with nature that is not bound to a particular design profession. Its scope is rich enough to embrace the work of architects rethinking their choices of building materials, the Army Corps of Engineers reformulating its flood-control strategy, and industrial designers curtailing their use of toxic compounds. Ecological design provides a coherent framework for redesigning our landscapes, buildings, cities, and systems of energy, water, food, manufacturing, and waste.

Ecological design is simply the effective *adaptation to* and *integration with* nature's processes. It proceeds from considerations of health and wholeness, and tests its solutions with a careful accounting of their full environmental impacts. It compels us to ask new questions of each design: Does it enhance and heal the living world, or does it diminish it? Does it preserve relevant ecological structure and process, or does it degrade it?

We are just beginning to make a transition from conventional forms of design, with the destructive environmental impacts they entail, to ecologically sound forms of design. There are now sewage treatment plants that use constructed marshes to simultaneously purify water, reclaim nutrients, and provide habitat. There are agricultural systems that mimic natural ecosystems and merge with their surrounding landscapes. There are new kinds of industrial systems in which the waste streams from one process are *designed to* be useful inputs to the next, thus minimizing pollution. There are new kinds of nontoxic paints, glues, and finishes. Such examples are now rapidly multiplying . . .

We have already made dramatic progress in many areas by substituting design intelligence for the extravagant use of energy and materials. Computing power that fifty years ago would fill a house full of vacuum tubes and wires can now be held in the palm of your hand. The old steel mills whose blast furnaces, slag heaps, and towering smokestacks dominated the industrial landscape have been replaced with efficient scaled-down facilities and processes. Drafty, polluting fireplaces have been replaced with compact, highly efficient ones that burn pelletized wood wastes. Many products and processes have been miniaturized, with the flow of energy and materials required to make and operate them often dramatically reduced.

These examples show that when we think differently about design, new solutions are often quick to emerge. By explicitly taking ecology as the basis of design, we can vastly diminish the environmental impacts of everything we make and build. While we've often done well in applying design intelligence to narrowly circumscribed problems, we now need to integrate ecologically sound technologies, planning methods, and policies across scales and professional boundaries.

The nutrients, energy, and information essential to life flow smoothly across scales ranging from microorganisms to continents; in contrast, design has become fragmented into dozens of separate technical disciplines, each with its own specialized language and tools. As the inventor Buckminster Fuller once noted, "Nature did not call a department heads' meeting when I threw a green apple into the pond, with the department heads having to make a decision about how to handle this biological encounter with chemistry's water and the unauthorized use of the physics department's waves." No amount of regulation, intervention, or standalone brilliance will bring us a healthier world until we begin to deliberately join design decisions into coherent patterns that are congruent with nature's own.

In a sense, evolution is nature's ongoing

design process. The wonderful thing about this process is that it is happening continuously throughout the entire biosphere. A typical organism has undergone at least a million years of intensive research and development, and none of our own designs can match that standard. Evolution has endowed individual organisms with a wide range of abilities, from harvesting sunshine to perceiving the world. Further, it has enabled communities of organisms to collectively recycle nutrients, regulate water cycles, and maintain both structure and diversity. Evolution has patiently worked on the living world to create a nested series of coherent levels, from organism to planet, each manifesting its own design integrities.

A few years ago, two Norwegian researchers set out to determine the bacterial diversity of a pinch of beech-forest soil and a pinch of shallow coastal sediment. They found well over four thousand species in each sample, which more than equaled the number listed in the standard catalogue of bacterial diversity. Even more remarkably, the species present in the two samples were almost completely distinct. This extraordinary diversity pervades the Earth's manifold habitats, from deep-sea volcanic vents to mangrove swamps, from Arctic tundra to redwood forests. It is a diversity predicated on precise adaptation to underlying conditions. Within this diversity, within a hawk's wings or a nitrogen-fixing bacterium's enzymes, lies a rich kind of design competence. In nature, there is a careful choreography of function and form bridging many scales. It is this dance that provides the wider context for our own designs. In the attempt to minimize environmental impacts, we are inevitably drawn to nature's own design strategies.

These strategies form a rich resource for design guidance and inspiration. Contemplating the patterns that sustain life, we are given crucial design clues. We learn that spider plants are particularly good at removing pollutants from the air and might serve as an effective component of a living system for purifying the confined air of office buildings. We discover that wetlands can remove vast quantities of nutrients, detoxify compounds, and neutralize pathogens, and therefore can play a role in an ecological wastewater treatment system. The sum of these simple lessons from nature's own exquisite design catalog is nothing less than a blueprint for our own survival.

Suppose we represent our working "natural capital" – forests, lakes, wetlands, salmon, and so on – with a stack of coins. This natural capital provides renewable interest in the form of sustainable fish and timber yields, crops, and clean air, water, and soil. At present, we are simply spending this capital, drawing it down to dangerously low levels, decreasing the ability of remaining ecosystems to assimilate ever-increasing quantities of waste. Such an approach cannot help but deplete natural capital.

Ecological design offers three critical strategies for addressing this loss: conservation, regeneration, and stewardship. Conservation slows the rate at which things are getting worse by allowing scarce resources to be stretched further. Typical conservation measures include recycling materials, building denser communities to preserve agricultural land, adding insulation, and designing fuel-efficient cars. Unfortunately, conservation implicitly assumes that damage must be done and that the only recourse is to somehow minimize this damage. Conservation alone cannot lead to sustainability since it still implies an annual natural resource deficit.

In the years before his death, Robert Rodale, editor of Rodale Press, was very concerned with what he termed *regeneration*. In a literal sense, regeneration is the repair and renewal of living tissue. Ecological design works to regenerate a world deeply wounded by environmentally insensitive design. This may involve restoring an eroded stream to biological productivity, re-creating habitat, or renewing soil. Regeneration is an expansion of natural capital through the active restoration of degraded ecosystems and communities. It is a form of healing and renewal that embodies the richest possibilities of culture to harmonize with nature. Regeneration not only preserves and protects: It restores a lost plenitude.

Stewardship is a particular quality of care in our relations with other living creatures and with the landscape. It is a process of steady commitment informed by constant feedback – for example, the gully is eroding, or Joe's doing poorly in math. It requires the careful maintenance and continual reinvestment that a good gardener might practice through weeding, watering, watching for pests, enriching the soil with compost, or adding new varieties. Stewardship maintains natural capital by spending frugally and investing wisely.

Ecological design embraces conservation, regeneration, and stewardship alike. If conservation involves spending natural capital more slowly, and regeneration is the expansion of natural capital, then stewardship is the wisdom to live on renewable interest rather than eating into natural capital. Conservation is already well established in the engineering and resource-management professions, but regeneration is just beginning to be explored by restoration ecologists, organic farmers, and others. Stewardship is a quality that all of us already have to some degree. Together, conservation, regeneration, and stewardship remind us of both the technical and personal dimensions of sustainability. They open up new kinds of creative endeavor even as they reaffirm the need for limits.

Careful ecological design permits such a great reduction in energy and material flows that human communities can once again be deeply integrated into their surrounding ecological communities. By carefully tailoring the scale and composition of wastes to the ability of ecosystems to assimilate them, we may begin to re-create a symbiotic relationship between nature and culture. By letting nature do the work, we allow ecosystems to flourish even as they purify and reclaim wastes, ameliorate the climate, provide food, or control flooding. It is clear that the world is a vast repository of unknown biological strategies that could have immense relevance should we develop a science of integrating the stories embedded in nature into the systems we design to sustain us. Ecological design begins to integrate these biological strategies by gently improvising upon life's own chemical vocabulary, geometry, flows, and patterns of community.

For example, if we wish to buttress a badly eroding hill, a conventional design might call for a concrete retaining wall many inches thick to hold the earth in place. Such a wall makes ostentatious use of matter and energy and does little to heal the land. In looking for an ecological design solution, we seek natural processes that perform this same work of holding the earth in place. We are led to trees, and a useful solution in practice has been to seed the hill with hundreds of willow branches.

Within months the branches sprout, providing effective sod stabilization. The willows' articulated roots are far more adapted to keeping the soil in place than concrete, which bears only a superficial relationship to the soil.

Ecological design occurs in the context of specific places. It grows out of place the way the oak grows from an acorn. It responds to the particularities of place: the sods, vegetation, animals, climate, topography, water flows, and people lending it coherence. It seeks locally adapted solutions that can replace matter, energy, and waste with design intelligence. Such an approach matches biological diversity with cultural diversity rather than compromising both the way conventional solutions do.

To design with this kind of care, we need to rigorously assess a design's set of environmental impacts. To take a simple example, consider just a few of the impacts of a typical house. Carbon dioxide emissions from the manufacture of the cement in its foundation contribute to global warming. The production of the electricity used to heat the house may contribute to acid rain in the region. Altered topography and drainage on-site may cause erosion, impacting the immediate watershed. The house might displace existing wildlife habitat. Inside the house, the health of the occupants may be compromised by emissions from the various glues, resins, and finishes used during construction. The lumber may have hastened the destruction of distant ancient forests. We are left with a somewhat disheartening picture of the wider ecological costs of a single building.

Ecological design converts these impacts from invisible side effects into explicitly incorporated design constraints. If ordinary cement's contribution to global warming renders its large-scale use undesirable, this imposes a critical design constraint. Perhaps the house can be sited in a way that minimizes cement use, or alternative, less-destructive cements can be used. If heating the house requires excessive quantities of electricity or natural gas, it may be possible to use passive solar heating through careful orientation of the building and the proper choice of building materials. In a similar way, each of the impacts can be turned into a stimulus for ecological design innovation.

Ecological design brings natural flows to the foreground. It celebrates the flow of water on the landscape, the rushing wind, the fertility of the earth, the plurality of species, and the rhythms of the sun, moon, and tides. It renders the invisible visible, allowing us to speak of it and carry it in our lives. It brings us back home. As the elements of our survival – the provenance of our food and energy, the veins of our watershed, the contours of our mountains – become vivid and present once again, they ground us in our place. We are given news of our region and the comings and goings of our fellow species. Ultimately, ecological design deepens our sense of place, our knowledge of both its true abundance and its unsuspected fragility.

Ecological design is a way of integrating human purpose with nature's own flows, cycles, and patterns. It begins with the richest possible understanding of the ecological context of a given design problem and develops solutions that are consistent with the cultural context. Such design cannot be the work of experts only. It is ultimately the work of a sustainable culture, one skilled in reweaving the multiple layers of natural and human design. Ecological designers are facilitators and catalysts in the cultural processes underlying sustainability.

We are beginning to get the pieces right, from highly efficient appliances to organic farms. However, until the pieces constitute the *texture* of everyday life, they will remain insufficient. This book is about the design wherewithal necessary to create a sustainable world. It provides a new way of seeing and thinking about design. It suggests a new set of questions and themes to order the design process. It proposes a form of design that is able to translate the vision of sustainability into the everyday objects, buildings, and landscapes around us. It embraces the best of the new ecological technologies but also inquires into the cultural foundations of sustainability. In short, it is an exploration of practical harmonies between nature and culture.

Table 1 compares conventional and ecological design in relation to a number of relevant issues.

HISTORY AND BACKGROUND

Ecological design is not a new idea. By necessity, it has been brought to a high level of excellence by many different cultures faced with widely varying conditions. The Yanomami, living with a refined knowledge of the Amazon rainforest, deliberately propagate hundreds of plant species, thereby enhancing biological diversity. Balinese aquaculture and rice terracing maintain soil fertility and pure water while feeding large numbers of people. Australian aborigines use stories and rituals to preserve an exquisitely detailed ecological map of their lands. The design rules embedded in each of these cultures have enabled them to persist for millennia.

Even during the most uncritical growth eras of the industrialized nations, there have been strong movements for ecologically sound town planning, healthy building, organic agriculture, appropriate technology, renewable energy, and interdisciplinary approaches to design. William Morris's Arts and Crafts Movement, Rudolph Steiner's biodynamic agriculture, Ebenezer Howard's garden cities, Patrick Geddes's and Lewis Mumford's regional planning, and Frank Lloyd Wright's organic architecture – each celebrated design at a human scale firmly situated in a wider ecological context. Buckminster Fuller, in an enormously productive five decades of work, tested the limits of ephemeralization – decreasing the use of materials and energy – while designing Dymaxion houses that could process their own wastes and be recycled at the end of their useful lives.

By the 1960s, various streams of stubborn ethical and aesthetic opposition to unfettered industrialization coalesced into the first modern generation of ecological design. Designer Sean Wellesley-Miller and physicist Day Chahroudi designed building "skins" based on biological metaphors and principles, but using available materials. John and Nancy Todd and their associates at the New Alchemy designed solar Arks that grew their own food, provided their own energy, and recycled their own wastes. Other experimental houses and habitats were built all over the world, including the Ouroboros House in Minneapolis, the Autonomous House at Cambridge University, and the Farallones Institute's Integral Urban House in Berkeley, California. While different in form and purpose, all of these projects shared a similar vision: Biology and ecology are the key sciences in rethinking the design of habitat. Within these projects, the metaphor of a living organism or ecosystem replaced Le Corbusier's old image of a dwelling as a "machine for living."

The house, the habitat we are most familiar with, seemed to be a good place to start this first generation of ecological design. The rural or village homestead was once the center of a largely self-sufficient system that produced a family's livelihood, its food and fiber, and its tools and toys. Over a period of several hundred years, this homestead has become an anonymous mass-produced dwelling unit, its inhabitants members of a faceless consumershed, the house itself totally dependent on outside resources to sustain its inhabitants. Rethinking home metabolism became the mission of the first generation of ecological design.

The Integral Urban House, conceived by biologists Bill and Helga Olkowski and sponsored by the Farallones Institute, started in 1973 in a ramshackle Victorian house in Berkeley, California. The oil embargo had made many people aware for the first time of their almost total dependence on an oil economy. Designers were challenged to work with the sun, turning this house from a consumer of oil for heating, cooling, electricity, and food into a producer of thermal energy, food, and electricity.

The Integral Urban House was intended to restore its inhabitants to a measure of control over the basics of their life support, reduce the outflow of money to pay for resources and services that the home and local environment could provide, and encourage interaction with local ecosystems. The idea was to integrate energy and food production and waste and water recycling directly into the home design. The Integral Urban House featured composting toilets, an aquaculture pond, organic gardens, and advanced recycling. The guiding vision was a new synthesis of architecture and biology.

[. . .]

In the 1980s, the environmental movement grew into a broad-based sustainability movement. Great technical advances were made in solar and wind energy. Lovins's Rocky Mountain Institute helped transform energy policy in many nations. Bill Mollison's "permaculture" approach to organic agriculture and healthy building gained a worldwide following from its modest start in Tasmania. Fundamental research on sustainable agriculture was performed at the University of California, Santa Cruz, and the Land Institute in Salina, Kansas. Work in landscape ecology and conservation biology provided a new set of tools for preserving biodiversity that have been effectively used by Project Wild. Peter Calthorpe, Andres Duany, and Elizabeth Plater-Zyberk created renewed interest in pedestrian-oriented town planning.

The 1990s have seen the emergence of the international ecocities movement, which is working to create healthier, more resource-efficient cities. Constructed ecosystems – wetlands and contained microcosms – are rapidly becoming an important alternative to conventional wastewater treatment systems. Industrial ecology and life-cycle analysis are already key tools for minimizing pollution. New approaches to ecological restoration and toxic decontamination show great promise. Recent attempts to integrate ecology and economics are also beginning to bear fruit, including Pliny Fisk's approach to bioregional design at the Center for Maximum Potential Building Systems in Austin, Texas. Artists like Andy Goldsworthy and Mierle Ukeles are creating works that demonstrate a deep commitment to ecological ideas.

[. . .]

The first generation of ecological design was based on small-scale experiments with living lightly in place. Many of the technologies and ideas of this generation, such as alternative building materials, renewable energy, organic foods, conservation, and recycling have been widely adopted in piecemeal fashion. We now stand at the threshold of a second generation of ecological design. This second generation is not an alternative to dominant technology and design; it is the best path for their necessary evolution.

The second generation of ecological design must effectively weave the insights of literally dozens of disciplines. It must create a viable ecological design craft within a genuine culture of sustainability rather than getting entangled in interdisciplinary disputes and turf wars. It is time to bring forth new ecologies of design that are rich with cultural and epistemological diversity.

Table 1 Characteristics of conventional and ecological design

Issue	*Conventional design*	*Ecological design*
Energy source	Usually nonrenewable and destructive, relying on fossil fuels or nuclear power; the design consumes natural capital	Whenever feasible, renewable: solar, wind, small-scale hydro, or biomass; the design lives off solar income
Materials use	High-quality materials are used clumsily, and resulting toxic and low-quality materials are discarded in soil, air, and water	Restorative materials cycles in which waste for one process becomes food for the next; designed-in reuse, recycling, flexibility, ease of repair, and durability
Pollution	Copious and endemic	Minimized; scale and composition of wastes conform to the ability of ecosystems to absorb them
Toxic substances	Common and destructive, ranging from pesticides to paints	Used extremely sparingly in very special circumstances
Ecological accounting	Limited to compliance with mandatory requirements like environmental-impact reports	Sophisticated and built in; covers a wide range of ecological impacts over the entire life-cycle of the project, from extraction of materials to final recycling of components
Ecology and economics	Perceived as in opposition; short-run view	Perceived as compatible; long-run view
Design criteria	Economics, custom, and convenience	Human and ecosystem health, ecological economics
Sensitivity to ecological context	Standard templates are replicated all over the planet with little regard to culture or place; skyscrapers look the same from New York to Cairo	Responds to bioregion: the design is integrated with local soils, vegetation, materials, culture, climate, topography; the solutions grow from place
Sensitivity to cultural context	Tends to build a homogeneous global culture; destroys local commons	Respects and nurtures traditional knowledge of place and local materials and technologies; fosters commons
Biological, cultural, and economic diversity	Employs standardized designs with high energy and materials throughput, thereby eroding biological, cultural and economic diversity	Maintains biodiversity and the locally adapted cultures and economies that support it
Knowledge base	Narrow disciplinary focus	Integrates multiple design disciplines and wide range of sciences; comprehensive
Spatial scales	Tends to work at one scale at a time	Integrates design across multiple scales, reflecting the influence of larger scales on smaller scales and smaller on larger
Whole systems	Divides systems along boundaries that do not reflect the underlying natural processes	Works with whole systems; produces designs that provide the greatest possible degree of internal integrity and coherence
Role of nature	Design must be imposed on nature to provide control and predictability and meet narrowly defined human needs	Includes nature as a partner: whenever possible, substitutes nature's own design intelligence for a heavy reliance on materials and energy

Underlying metaphors	Machine, product, part	Cell, organism, ecosystem
Level of participation	Reliance on jargon and experts who are unwilling to communicate with public limits community involvement in critical design decisions	A commitment to clear discussion and debate; everyone is empowered to join the design process
Types of learning	Nature and technology are hidden; the design does not teach us over time	Nature and technology are made visible; the design draws us closer to the systems that ultimately sustain us
Response to sustainability crisis	Views culture and nature as inimical, tries to slow the rate at which things are getting worse by implementing mild conservation efforts without questioning underlying assumptions	Views culture and nature as potentially symbiotic; moves beyond triage to a search for practices that actively regenerate human and ecosystem health

PART 8

The Future of the City

PART EIGHT

INTRODUCTION

The desire to peer into the future is a human trait as old as the biblical prophets and the oracle at Delphi. And the desire to project *urban* futures is at least as old as Plato's description of the ideal city-state in *The Republic*. But the pace of futurist predictions seems to quicken at times when great cultural and historic shifts are taking place. Such was the case during the Industrial Revolution of the nineteenth century when fantasists like Jules Verne and political idealists like Edward Bellamy (author of *Looking Backward, 2000–1887*) captured the popular imagination and when the utopian visionaries discussed in Part 5 of this volume helped to establish the theoretical basis of urban planning practice. And such is the case today as the realization becomes clearer every day that the advanced economies of the world are entering a new global, information-based, post-industrial stage of development that promises to reveal new forms of urban civilization and human community.

Clearly, a new urban paradigm is now emerging. But in order to predict the shape of the emerging postmodern city, one must have a clear sense of the probable direction of the world urbanization process described by Kingsley Davis in "The Urbanization of the Human Population," the essay with which this volume begins. Will the percentage of the world's total population living in cities, now at about 50 percent, continue to increase in the decades and centuries to come? Will the "townward drift," as Frederick Law Olmsted called it, continue? To what eventual level? 80 percent? 90? 100 percent? Or has urbanization reached its peak, ready to stabilize at more or less the present level? Perhaps the S-curve on the chart representing world urbanization will prove to be a bell curve, with the percentage of the human population living in cities going into a long, gradual decline until only a small number of the total population remains urbanized. This last possibility has spawned an intriguing, if highly conjectural, body of literature. Some, pondering the possible effects of modern transportation and telecommunications technologies, see a gradual withering away of cities. Others, particularly environmentalists, talk about urbanization reaching its natural limits and beginning to reverse in an age of ecological and economic constraints, or suggest that unrestrained growth is sure to result in widespread economic and environmental collapse.

Today, the variety of possible futures from which to choose is extraordinarily diverse. Each option mirrors some of our deepest hopes and fears. Marshall McLuhan, the 1960s guru of communications theory, suggested that the whole world would one day become a "global village," with every member of humanity interacting with every other in a real-time simulacrum of the neolithic community. Some of the more radical members of the environmentalist movement have gone "back to nature" by establishing rural communes along the fringes of urbanized civilization, while other equally radical social activists have established "urban kibbutzim" in the very hearts of the inner cities. Techno-optimists see earth's future in space colonization projects while techno-pessimists envision post-apocalypse cities like the ones depicted in *Robocop*, *Terminator*, *Blade Runner*, and other popular science-fiction films.

Some sense of the wide variety of possible urban futures is suggested by comparing the visions of Melvin Webber and Paolo Soleri. Webber, author of "The Post-City Age" (1968), argues that certain technological developments will result in an end to traditional cities and the emergence of a post-urban period of human development. Since he wrote his prophetic essay in the 1960s, the technologies he referred to were air travel and telephones. Today, the advent of computers and telecommunications make his predictions even more plausible. Even if we reject the basic idea of a reversal of the course of urbanization, Webber's insight into the way the new technological world will leave some people informationally rich and others informationally poor is a troubling reality that cannot be ignored.

Architect Paolo Soleri, on the other hand, suggests that the next stage of human evolution will be intensely dense and hyper-urban. In "Arcology: The City in the Image of Man" (1969), he proposes constructing giant megastructures, each containing populations of millions, and leaving the rest of the planet to wilderness. The Soleri sketch of an arcology that features on the opening page of this section includes a tiny Empire State Building at the same scale to make sure the viewer understands just how huge the megastructures Soleri envisions really are. In Soleri's view, nature would develop on its own terms while a new stage of human community, inside the arcologies, would thrive in the man-made "neonature."

In their own way, both Webber and Soleri are utopian visionaries engaged in brilliant, if somewhat mystical, speculation about humankind's future. It is perhaps more useful, when considering the future of the city, to come down to earth, to shorten the futurist perspective to the near term, and to project immediate futures based on present and observable trends. Among the most important such trends today are the parallel emergence of (1) a global, post-industrial, telecommunications-based economy, (2) a new, far-flung suburban ring of development that journalist Joel Garreau has dubbed "Edge City" and that Robert Fishman has analyzed as "Technoburbia" (p. 77), (3) transnational environmental concerns such as global warming and atmospheric pollution, and (4) the persistence, even intensification, of the classic social conflicts along racial and class lines that have been a feature of urban life for hundreds of years. These influences are likely to determine the course of early twenty-first-century urban development.

Anthony Downs and Manuel Castells are urban futurists who have carefully analyzed the near-term futures of urban society – Downs for the United States, Castells for Western Europe – and suggested possible directions for urban development strategies. Downs has long been concerned with issues of social justice, particularly for ghettoized inner-city African-Americans. He recognizes that the fate of all urban communities lies with new economic forces developing in the Edge City technoburbs and that the poorest communities are likely to be left behind if fundamental changes are not made in the way that Americans plan their cities of the future. In "The Need for a New Vision for the Development of Large U.S. Areas" (1989), Downs carefully lays out a new model for regional development that will, he hopes, bridge the gulf between the suburban middle class and the inner-city poor in a way that will lead to a better communal life for all.

No one, least of all Downs, thinks that solving the problem of inner-city poverty and race- and class-based segregation will be an easy task, nor that one solution is the only one worth considering. Clearly, there are many contentious issues of politics and urban governance, and even deeper concerns at the level of social structure and cultural development that will affect the future of cities. But as the neo-suburban technoburbs and the pedestrian pockets beckon, and as local economies and national sovereignties give way to global interdependence, the ongoing and sometimes explosive violence of racial and class discrimination is still there at the core of many cities, threatening to destroy urban civilization and to undermine the human community. Defusing those possible sources of conflict remains part of the unfinished business of the city of the future.

In "European Cities, the Informational Society, and the Global Economy" (1993), Manuel Castells surveys the urban future of Western Europe. Like Downs, he recognizes that racial and class conflicts – in this case exacerbated by Third World immigration – will be a troubling feature of future urban development, but his larger concerns are about the effects of economic globalism and telecommunications technologies. Castells argues that technology-based cities represent "a new industrial space," located in centers throughout Europe, Asia, and America and tied to a global "informational" economy, and are fundamentally different from anything that has ever come before. A two-tier economy and a widening gulf between the educated elites and the ghettoized, marginalized urban populations is intensifying rather than diminishing in the cities of the information economy and the global marketplace. The new urban order, abetted by new economic realities, seems cut off from its past, from history and cultural tradition as sources of communal meaning and individual identity.

In *The Rise of the Network Society* (1998), Castells observes that the implicit tendency of the work-styles of the post-industrial economy is to become detached from traditional cultures, values, and communities. Information flows through networks and across vast distances, and "the historical emergence of the space of flows," he writes, supersedes "the meaning of the space of places." For Castells, then, the real challenge of the new informational cities of Western Europe is to reconcile the "new techno-economic paradigm" and "place-based social meaning" in a way that will avoid what he calls "urban schizophrenia." "At the cultural level," he writes, "local societies, territorially defined, must preserve their identities, and build upon their historical roots, regardless of their economic and functional dependence upon the space of flows."

If Castells is one of the great theoreticians of the techno-urbanism of the future, Stephen Graham and Simon Marvin are its chroniclers and systematizers. In *Telecommunications and the City* (1996), they exhaustively catalog the multiple effects of the new technologies on economics, infrastructural development, governance, social life, cultural norms . . . and the future of cities as well. In the end, they agree fundamentally with Castells and predict a new kind of urban world – a "planetary metropolitan system" which combines traditional, place-based loyalties with new online community interactions – not the post-urban world Webber predicted.

The new telecommunications technologies are just another aspect of the human-made infrastructures of production, management, and control that have always been a part of urban life. Increasingly, however, another factor will greatly influence the future of cities: the natural environment, now seen not just as a regional rural hinterland but as a global context in which cities can either flourish or die. To use the popular sports metaphor: Nature bats last!

David Clark addresses these global environmental concerns in "The Future Urban World" from *Urban World/Global City* (1996). Clark, an urban geographer, argues that an increasingly overcrowded earth is fast becoming a global urban society of which we are all residents and that the pollution created by cities is progressively destroying the global environmental systems upon which life on the planet depends. Over the next twenty or thirty years, he predicts, most of the new urbanization will come from the developing world, and he expects the coming changes to be staggering, with the number of people who presently live in urban places likely to double by the year 2025.

Although Clark cautions against alarmist thinking and notes that urbanization "is too sensitive to economic, social and environmental change to predict more than one or two decades into the future," he nonetheless is certain that the urban world of 2025 will bear little resemblance to the urban present. Given the rapid pace of change and the great paradigm shifts of urban history, this prediction seems almost certainly true. The world of the twenty-first century is the world that those of us who are studying cities and urban planning today will inhabit. And it is how we will adapt to the challenges of the future – the new globalism, the emerging technologies, the environmental constraints, and the ongoing struggles for social justice – that will determine how cities and citizens in the years to come will achieve, or fail to achieve, a meaningful sense of personal identity and social community.

MELVIN M. WEBBER

"The Post-City Age"

Daedalus (1968)

Editors' introduction In the introduction to Kingsley Davis's "The Urbanization of the Human Population" (p. 3), we asked what the future of urbanization might be. Would the urban population of the earth stabilize at 50 percent or 70 percent? Might it increase to 100 percent? Or might the S-curve of historic urbanization become a standard bell curve in the future, charting a steady decline in the proportion of the total human population that lives in cities?

As long ago as 1968, Melvin Webber thought through the possibility of a gradual decline in urbanization in his brilliant and prophetic essay "The Post-City Age." "Prophetic?" some will ask. Major urban centers have not decreased in density and population during the last thirty years. To the contrary, major global cities – as well as newly emerging mega-cities in Asia, Africa, and Latin America – have grown tremendously in the decades since Webber made his bold and counter-intuitive prediction. Still, Webber was at least prescient in pointing to two major technological developments – air transportation and the telephone – that even in 1968 were eliminating the traditional space–time constraints on human interaction and bringing about stunning cultural changes in the nature of an increasingly global human civilization.

Webber argues that the widespread availability of commercial air transportation and global telephonic communication permits an educated and affluent class of people to live anywhere – in suburbs, in rural districts, on mountain-tops, for that matter – and still be thoroughly "urban," participating fully in intellectual, professional, and economic life. At the same time, it is precisely the poor populations that are trapped in inner cities and deprived of access to technology that are becoming increasingly "rural" in the sense that they are non-participants in global community affairs.

Are these insights still applicable, perhaps even more applicable, now that networks of instantaneous global telecommunications exist to serve the needs of technologically advanced individuals and populations? Is the society that Webber foresaw in 1968 something like the contemporary "technoburbs" (p. 77) that Robert Fishman describes? Might not computers and modern telecommunications make the Broadacre City that Frank Lloyd Wright prophesied (p. 344) even more possible today and in the future?

Melvin Webber is professor emeritus of planning at the University of California, Berkeley. He is the author of *Explorations Into Urban Structure* (Philadelphia: University of Pennsylvania Press, 1964) and has written extensively on issues of urban transit and cyberspace citizenship.

We are passing through a revolution that is unhitching the social processes of urbanization from the locationally fixed city and region. Reflecting the current explosion in science and technology, employment is shifting from the production of goods to services; increasing ease of transportation and communication is dissolving the spatial barriers to social intercourse; and Americans are forming social communities comprised of spatially dispersed members. A new kind of large scale urban society is emerging that is increasingly independent of the city. In turn, the problems of the city place generated by early industrialization are being supplanted by a new array different in kind. With but a few remaining exceptions (the new air pollution is a notable one), the recent difficulties are not place type problems at all. Rather, they are the transitional problems of a rapidly developing society-economy-and-polity whose turf is the nation. Paradoxically, just at the time in history when policy-makers and the world press are discovering the city, "the age of the city seems to be at an end."

Our failure to draw the rather simple conceptual distinction between the spatially defined city or metropolitan area and the social systems that are localized there clouds current discussions about the "crisis of our cities." The confusion stems largely from the deficiencies of our language and from the anachronistic thoughtways we have carried over from the passing era. We still have no adequate descriptive terms for the emerging social order, and so we use, perforce, old labels that are no longer fitting. Because we have named them so, we suppose that the problems manifested inside cities are, therefore and somehow, "city problems." Because societies in the past had been spatially and locally structured, and because urban societies used to be exclusively city-based, we seem still to assume that territorial is a necessary attribute of social systems.

The error has been a serious one, leading us to seek local solutions to problems whose causes are not of local origin and hence are not susceptible to municipal treatment. We have been tempted to apply city-building instruments to correct social disorders, and we have then been surprised to find that they do not work. (Our experience with therapeutic public housing, which was supposed to cure "social pathologies," and urban renewal, which was supposed to improve the lives of the poor, may be our most spectacular failures.) We have lavished large investments on public facilities, but neglected the quality and the distribution of the social services. And we have defended and reinforced home-rule prerogatives of local and state governments with elaborate rhetoric and protective legislation.

Neither crime-in-the-streets, poverty, unemployment, broken families, race riots, drug addiction, mental illness, juvenile delinquency, nor any of the commonly noted "social pathologies" marking the contemporary city can find its causes or its cure there. We cannot hope to invent local treatments for conditions whose origins are not local in character, nor can we expect territorially defined governments to deal effectively with problems whose causes are unrelated to territory or geography. The concepts and methods of civil engineering and city planning suited to the design of unitary physical facilities cannot be used to serve the design of social change in a pluralistic and mobile society. In the novel society now emerging – with its sophisticated and rapidly advancing science and technology, its complex social organization, and its internally integrated societal processes – the influence and significance of geographic distance and geographic place are declining rapidly.

This is, of course, a most remarkable change. Throughout virtually all of human history, social organization coincided with spatial organization. In preindustrial society, men interacted almost exclusively with geographic neighbors. Social communities, economies, and polities were structured about the place in which interaction was least constrained by the frictions of space. With the coming of large-scale industrialization during the latter half of the nineteenth century, the strictures of space were rapidly eroded, abetted by the new ease of travel and communication that the industrialization itself brought.

The initial counterparts of industrialization in the United States were, first, the concentration of the nation's population into large settlements and, then, the cultural urbanization of the population. Although these changes were causally linked, they had opposite spatial effects. After coming together at a common place, people entered larger societies tied to no specific place. Farming and village peoples from throughout the continent and the world migrated to the expanding cities, where they learned urban ways, acquired the occupational skills that industrialization demanded, and became integrated into the contemporary society.

In recent years, rising societal scale and improvements in transportation and communications systems have loosed a chain of effects robbing the city of its once unique function as an urbanizing instrument of society. Farmers and small-town residents, scattered throughout the continent, were once effectively removed from the cultural life of the nation. City folks visiting the rural areas used to be treated as strangers, whose styles of living and thinking were unfamiliar. News of the rest of the world was hard to get and then had little meaning for those who lived the local life. Country folk surely knew there was another world out there somewhere, but little understood it and were affected by it only indirectly. The powerful anti-urban traditions in early American thought and politics made the immigrant city dweller a suspicious character whose crude ways marked him as un-Christian (which he sometimes was) and certainly un-American. The more sophisticated urban upper classes – merchants, landowners, and professional men – were similarly suspect and hence rejected. In contrast, the small-town merchant and the farmer who lived closer to nature were the genuine Americans of pure heart who lived the simple, natural life. Because the contrasts between the rural and the urban ways-of-life were indeed sharp, antagonisms were real, and the differences became institutionalized in the conduct of politics. America was marked by a diversity of regional and class cultures whose followers interacted infrequently, if ever.

By now this is nearly gone. The vaudeville hick-town and hayseed characters have left the scene with the vaudeville act. Today's urbane farmer watches television documentaries, reads the national news magazines, and manages his acres from an office (maybe located in a downtown office building), as his hired hands ride their tractors while listening to the current world news broadcast from a transistor. Farming has long since ceased to be a handicraft art; it is among the most highly technologized industries and is tightly integrated into the international industrial complex.

During the latter half of the nineteenth century and the first third of the twentieth, the traditional territorial conception that distinguished urbanites and ruralites was probably valid: The typical rural folk lived outside the cities, and the typical urbanites lived inside. By now this pattern is nearly *reversed*. Urbanites no longer reside exclusively in metropolitan settlements, nor do ruralites live exclusively in the hinterlands. Increasingly, those who are least integrated into modern society – those who exhibit most of the attributes of rural folk – are concentrating within the highest-density portions of the large metropolitan centers. This profoundly important development is only now coming to our consciousness, yet it points up one of the major policy issues of the next decades.

Cultural diffusion is integrating immigrants, city residents, and hinterland peoples into a national urban society, but it has not touched all Americans evenly. At one extreme are the intellectual and business elites, whose habitat is the planet; at the other are the lower-class residents of city and farm who live in spatially and cognitively constrained worlds. Most of the rest of us, who comprise the large middle class, lie somewhere in-between, but in some facets of our lives we all seem to be moving from our ancestral localism toward the unbounded realms of the cosmopolites.

High educational attainments and highly specialized occupations mark the new cosmopolites. As frequent patrons of the airlines and the long-distance telephone lines, they are intimately involved in the communications networks that tie them to their spatially dispersed associates. They contribute to and consume the specialized journals of science, government, and industry, thus maintaining contact with information resources of relevance to their activities, whatever the geographic sources or their own locations. Even though some may be employed by corporations primarily engaged in manufacturing physical products, these men trade in information and ideas. They are the producers of the information and ideas that fuel the engines of societal development. For those who are tuned into the international communications circuits, cities have utility precisely because they are rich in information. The way such men use the city reveals its essential character most clearly, for to them the city is essentially a massive communications switchboard through which human interaction takes place.

Indeed, cities exist *only* because spatial agglomeration permits reduced costs of interaction. Men originally elected to locate in high-density settlements precisely because space was so costly to overcome. It is still cheaper to interact with persons who are nearby, and so men continue to locate in such settlements. Because there *are* concentrations of associates in city places, the new cosmopolites establish their offices there and then move about from city to city conducting their affairs. The biggest settlements attract the most long-distance telephone and airline traffic and have undergone the most dramatic growth during the era of city-building.

The recent expansion of Washington, D. C. is the most spectacular evidence of the changing character of metropolitan development. Unlike the older settlements whose growth was generated by expanding manufacturing activities during the nineteenth and early-twentieth centuries, Washington produces almost no goods whatsoever. Its primary products are information and intelligence, and its fantastic growth is a direct measure of the predominant roles that information and the national government have come to play in contemporary society.

This terribly important change has been subtly evolving for a long time, so gradually that it seems to have gone unnoticed. The preindustrial towns that served their adjacent farming hinterlands were essentially alike. Each supplied a standardized array of goods and services to its neighboring market area. The industrial cities that grew after the Civil War and during the early decades of this century were oriented to serving larger markets with the manufacturing products they were created to produce. As their market areas widened, as product specialization increased, and as the information content of goods expanded, establishments located in individual cities became integrated into the spatially extensive economies. By now, the large metropolitan centers that used to be primarily goods-producing loci have become interchange junctions within the international communications networks. Only in the limited geographical, physical sense is any modern metropolis a discrete, unitary, identifiable phenomenon. At most, it is a localized node within the integrating international networks, finding its significant identity as contributor to the workings of that larger system. As a result, the new cosmopolites belong to none of the world's metropolitan areas, although they use them. They belong, rather, to the national and international communities that merely maintain information exchanges at these metropolitan junctions.

Their capacity to interact intimately with others who are spatially removed depends, of course, upon a level of wealth adequate to cover the dollar costs of long-distance intercourse, as well as upon the cognitive capacities associated with highly skilled professional occupations. The intellectual and business elites are able to maintain continuing and close contact with their associates throughout the world because they are rich not only in information, but also in dollar income.

As the costs of long-distance interaction fall in proportion to the rise in incomes, more and more people are able and willing to pay the transportation and communication bills. As expense-account privileges are expanded, those costs are being reduced to zero for ever larger numbers of people. As levels of education and skill rise, more and more people are being tied into the spatially extensive communities that used to engage only a few.

Thus, the glue that once held the spatial settlement together is now dissolving, and the settlement is dispersing over ever widening terrains. At the same time, the pattern of settlement upon the continent is also shifting (moving toward long strips along the coasts, the Gulf, and the Great Lakes). These trends are likely to be accelerated dramatically by cost-reducing improvements in transportation and communications technologies now in the research-and-development stages. (The SST, COMSAT communications, high-speed ground transportation with speeds up to 500 m.p.h., TV and computer-aided educational systems, no-toll long-distance telephone service, and real-time access to national computer-based information systems are likely to be powerful ones.) Technological improvements in transport and communications reduce the frictions of space and thereby ease long-distance intercourse. Our compact, physical city layouts directly mirror the more primitive technologies in use at the time these were built. In a similar way, the locational pattern of cities upon the continent reflects the technologies available at the time the settlements grew. If currently anticipated technological improvements prove workable, each of the metropolitan settlements will spread out in low-density patterns over far more extensive areas than even the most frightened future-mongers have yet predicted. The new settlement-form will little resemble the nineteenth-century city so firmly fixed in our images and ideologies. We can also expect the large junction points will no longer have the communications advantage they now enjoy, and smaller settlements will undergo a major spurt of growth in all sorts of now isolated places where the natural amenities are attractive.

Moreover, as ever larger percentages of the nation's youth go to college and thus enter the national and international cultures, attachments to places of residence will decline dramatically. This prospect, rather than the spatial dispersion of metropolitan areas, portends the functional demise of the city. The signs are already patently clear among those groups whose worlds are widest and least bounded by parochial constraints.

Consider the extreme cosmopolite, if only for purposes of illustrative cartooning. He might be engaged in scientific research, news reporting, or international business, professions exhibiting critical common traits. The astronomer, for example, maintains instantaneous contact with his colleagues around the world; indeed, he is a day-to-day collaborator with astronomers in all countries. His work demands that he share information and that he and his colleagues monitor stellar events jointly, as the earth's rotation brings men at different locales into prime viewing position. Because he is personally committed to their common enterprise, his social reference group is the society of astronomers. He assigns his loyalties to the community of astronomers, since their work and welfare matter most to him.

To be sure, as he plays out other roles – say, as citizen, parent, laboratory director, or grocery shopper – he is a member of many other communities, both interest-based and place-defined ones. But the striking thing about our astronomer, and the millions of people like him engaged in other professions, is how little of his attention and energy he devotes to the concerns of place-defined communities. Surely, as compared to his grandfather, whose life was largely bound up in the affairs of his locality, the astronomer, playwright, newsman, steel broker, or wheat dealer lives in a life-space that is not defined by territory and deals with problems that are not local in nature. For him, the city is but a convenient setting for the conduct of his professional work; it is not the basis for the social communities that he cares most about.

PAOLO SOLERI

"Arcology: The City in the Image of Man" and "The Characteristics of Arcology"

from *Arcology: The City in the Image of Man* (1969)

Editors' introduction One need only glance at a Paolo Soleri architectural drawing (such as the one that appears on the opening page of this section) to know that one is in the presence of either an inspired visionary or a mad prophet. His "arcologies" – a combination of architecture and ecology – are megastructures so immense that they dwarf the outline of the Empire State Building that Soleri often includes as a scale reference. His techno-cities suggest populations of millions living a beehive existence in massive multi-cell units resembling the cooling towers of nuclear power plants. Is the Solerian vision a science-fiction fantasy, an Orwellian nightmare, or a new evolutionary stage in the progress of the human spirit?

Soleri was born in Italy and studied architecture at both the Turin Architectural Institute and the studio of Frank Lloyd Wright at Taliesen West in Scottsdale, Arizona – ironic, given the fact that his arcologies are the diametrical antithesis of Broadacre City (p. 344). He later founded his own Cosanti Foundation where he began building, and doggedly continues to build, Arcosanti, the world's first arcology, in the desert north of Phoenix.

During the 1960s and 1970s, Soleri became the focus of a cult-like following, and students from around the world flocked to the Arcosanti site in Arizona's Paradise Valley to sit at the feet of the master and help build the dream. In addition, many adherents of the most radical tendencies of environmentalism adopted Soleri as their architectural guru because the arcology concept appeared to quarantine (destructive) humanity from (life-giving) nature. Soleri, however, proved to be full of surprises. As a deeply humanistic artist and as a personality of charm, gentleness, and wry humor, he was no cult leader and eschewed the status of guru. As an environmentalist, his interest in nature was overwhelmingly of the human variety. In one analogy, Soleri compared his arcologies to ocean liners – campus-like temporary societies of unrelated people – and the implication that the natural world surrounding the arcology was as empty and uninviting as a vast ocean was unavoidable. Further, Soleri was quite explicit that the new arrangement of human life inside the arcology manifested a revolutionary new kind of environment, a "neonature," that was his true interest.

It has been said that Paolo Soleri takes the technological tendencies of Le Corbusier's "machine age" cities to an even higher plane. Others see his arcologies as models for cities in outer space or for earth-bound cities in an age following nuclear war or global environmental disaster. Soleri himself sees no need to rely on catastrophe to usher in the new age. For him, the evolution of arcological neonature is merely an inevitable, and welcome, next step in the history of human progress.

This selection is from *Arcology: The City in the Image of Man*, (Cambridge, MA: MIT Press, 1969). Other books by Soleri include *Arcosanti: An Urban Laboratory?*, 3rd edn. (Scottsdale, AZ: Cosanti Press, 1993), *Technology and Cosmogenesis* (New York: Paragon House, 1985), *The Omega Seed* (Garden

City, NY: Anchor, 1981), *The Bridge Between Matter and Spirit: Is Matter Becoming Spirit* (Garden City, NY: Anchor, 1973), and *The Sketchbooks of Paolo Soleri*, (Cambridge, MA: MIT Press, 1971).

ARCOLOGY: THE CITY IN THE IMAGE OF MAN

The concept is that of a structure called an arcology, or ecological architecture . . . Such a structure would take the place of the natural landscape inasmuch as it would constitute the new topography to be dealt with. This man-made topography would differ from natural topography in the following ways:

1 It would not be a one-surface configuration but a multi-level one.
2 It would be conceived in such a way as to be the carrier of all the elements that make the physical life of the city possible – places and inlets for people, freight, water, power, climate, mail, telephone; place and outlets for people, freight, waste, mail, products, and so forth.
3 It would be a large-dimensioned sheltering device, fractioning three-dimensional space in large and small subspaces, making its own weather and its own cityscape.
4 It would be the major vessel for massive flow of people and things within and toward the outside of the city.
5 It would be the organizing pattern and anchorage for private and public institutions of the city.
6 It would be the focal structure for the complex and ever-changing life of the city.
7 It would be the unmistakable expression of man the maker and man the creator. It would be diverse and singular in all of its realizations. Arcology would be surrounded by uncluttered and open landscape.

The concept of a one-structure system is not incidental to the organization of the city but central to it. It is the wholeness of a biological organism that is sought in the making of the city, as many and stringent are the analogies between the functioning of an organism and the vitality of a metropolitan structure. Fundamental to both is the element of flow. Life is there where the flow of matter and energy is abundant and uninterrupted. With a great flow gradient the city acquires a cybernetic character. The interacting of its components erases the space and time gaps that outphase the action–reaction cycles and ultimately break down the vitality of the system.

These are mechanical but fundamental premises for a functioning metropolitan life. In reality the idea is that of a very comprehensive "plumbing system" for the social animal, which the city is. The plumbing system consists of the previously mentioned man-made topography. Social, ethical, political, and aesthetic implications are left out, as they are valid and final only if and when physical conditions are realistically organized.

To dispel the aura of cerebralism or utopianism from the concept presented, there is another way to see the central problem of the city: The degree of fullness in each individual life depends on the reaching power unequivocally available to each person. In turn this reaching power is in direct proportion to the richness and variety of information coming to and going from the person. Information means not only sounds, sight, and so forth, but all the sensorial data, all the physical intermediaries that make any sensitivity possible; all kinds of inorganic, organic, organized, or man-made matter or material or instruments, from foodstuff to wireless, from toilets to television, from mothers' reprimands to theater. This wholeness of information must include packaged and remote information such as television, radio, telephone, and the communications media, as well as environmental information. Environmental information calls in the technology of transportation, distribution, and transfer, and calls for the no less fundamental quality of the environment itself.

This combination of remote and synthetic information and environmental information is indispensable to the nature of metropolitan life. In physical terms it means that the distances, the time, and the obstacles separating the person from all civilized institutions have to be scaled down to the supply of energy available to the person himself.

If we inject into the picture the sheer bulk of products and devices wanted by and forced upon each man, we can see the dimension and the absolute priority of the logistical problem. The burden of matter, part of the environmental information weighing on every man, is impressive and also irrational. This matter has to be transformed, manipulated, moved, serviced, stored, exchanged, rejected, and substituted – the warehouses of arcology will have to be enormous. One thing nomadism has not been able to teach us is frugality. What is the mechanism by which the rich and complex life of society can flow back into the structure of the city?

In a society where production is a successful and physically gigantic fact, the coordination and congruence of information, communication, transportation, distribution, and transference are the mechanics by which that society operates. It is not accidental that these are also dynamic aspects of another phenomenon, the most dynamic of all: life.

In every dynamic event of physical nature the elements of time and space, and this acceleration, speed, and deceleration, are crucial. The speed of light, a space and time shrinker, well serves the communication of information of the packaged kind – television, radio, telephone – the synthetic information. Thus a good supply of synthetic information can reach even the scattered suburbanite (for him environmental information is and remains monotone, bone stripped).

The picture is totally altered when we come to transportation and distribution. Unless the feeding in and feeding out of these two is highly centered and axialized, the laws of matter and energy will see that sluggishness and possibly stillness prevail. *Swiftness and efficiency are inversely proportional to dispersion. Scattered life is by definition deprived or parasitic.* This can be verified by approaching the problem from the opposite end: the environment is vital and living information; it is the bulk of information available to man.

Blighted environment is blighted information. The cause of blighted environment is the breakdown of environmental information occurring when there is no follow-through from synthetic information to transportation, distribution, and reach. When this occurs, the energies of the individual are exhausted in the struggle to keep the avenues of environmental information open, to keep the flow of things going. Man's mechanically low-grade energies are absorbed, not euphemistically, by cement, asphalt, steel, pollution, and all sorts of mechanical, static, and dynamic barriers in an ever-enlarging frame of space and time. The flow becomes sluggish, if it does not come to a standstill. This blighted environment is a direct consequence of sluggish or dying flow.

Impaired flow is ultimately the disproportion between the validity of the individual reach and the amount of energy that is expended to make the reach possible.

One may thus say that because of the biophysical make-up of our world, rich flow – that is, rich potentially – is the direct consequence of minimal separation between components. *Minimal separation between components cannot be achieved by using only two or three coordinates of space. Minimal separation between components is structured three-dimensionally, or it is not feasible.* The solid and not the surface is the environment where adequate flow is possible, thus where environmental information is rich and where life can flourish.

The surface of the earth, for all practical purposes, is by definition a two-dimensional configuration. *The natural landscape is thus not the apt frame for the complex life of society.* Man must make the metropolitan landscape in his own image: a physically compact, dense, three-dimensional, energetic bundle, not a tenuous film of organic matter. The man-made landscape has to be a multilevel landscape, a solid of three congruous dimensions. The only realistic direction toward a physically free community of man is toward the construction of truly three-dimensional cities. *Physical freedom, that is to say, true reaching power, is wrapped around vertical vectors . . .*

There is a further and reinforcing reason for verticality. As individuals we act horizontally and need horizontal dimensions up to six to ten times the vertical dimensions. Thus, the compactness and richness of social collective life can be found only vertically. *Around vertical vectors, megalopoly and suburbia can contract, moving from flat gigantism toward human and solid scale . . .*

If this concept is valid, as it seems to be in view of the nature of the physical and energetic world, then a dense urban structure is mandatory, regardless of the what, how, where, or when. A few generations of men reared and grown in an environment badly stripped of cultural and aesthetic scope may be sufficient for the brutalization of society. Signs that such brutalization is already at work are abundant and impressive. If man is quality against quantity, then the priority is clear. It is much too late for our present generation, bound to the spell of arrogance and license. It may even be quite late for the just born, but there is hope for the children of our children. The when is now, for lack of any reachable yesterday.

THE CHARACTERISTICS OF ARCOLOGY

A passenger liner is the closest ancestor of arcology. The common characteristics are compactness and definite boundary; the functional fullness of an organism designed for the care of many, if not most, of man's needs; a definite and unmistakable three-dimensionality. Three main characteristics on the other hand are not common: the liner is structurally and functionally designed for motion within fluid; the liner is a shell for a temporary society of unrelated people; the liner is a sealed package connected to the outside only by way of synthetic information. Relieved of these three tyrannies, the liner, the concept of it, can open up and, retaining its organizational suppleness, become truly a "machine for living," that is to say, a physical configuration that makes man physically free.

We have then architecture as the materialization of the human environment and ecology as the physical, biological, and psychological balance of conditions that account for the specific site and its participation in the whole.

Arcology becomes the cleavage of the human in the body of matter and life, probing for the ever-changing condition of the present in a manner congruous to the aestheto-compassionate nature of man.

Arcology is then that architecture so complex in scope, so sound in structure, so infrastructurally subtle and pliable, so comprehensive, and of such miniaturizing force as to alter substantially the local ecology in the human direction.

Arcology is then, *morphologically*, that of the man-made (I will call it neonature), which parallels in one an ecology and an organism – an ecology in scale, pervasiveness, and balance, an organism in complexity and dynamism; *skeletally*, a structure of such dimension, scale, and organization as to be favorable to the interplay of the forces by which man and society grow; *functionally*, a compact, dense, and efficient organization caring for the intake, processing, storing, consuming, expelling, recycling of the elements needed by the complex life of man and society; *humanly*, an apt shelter for the multiple expressions and longing of man as an individual and man as a society; *formally*, a foundation for the aesthetogenesis of nature (into neonature).

If, for the sake of clarity, one separates the not-too-separable instrumentality from the scope (ends), one may say that the instrumental purpose of arcology is the definition of a well-rounded service system which, cutting into the waste of time and space, presents man with a few extra years of "positive" time, time to use to his personal, social advantage if he so pleases. That this may be invaluable lies in the assumption that life is precious enough and unique enough to demand rightly the best environmental conditions for its flowering and that coercion and frustration are inimical to life. Life is coerced by the environment man has produced and lives in. It is basically coerced by the very fact of the physical conditions he himself has compounded.

The time-waste brought about by space-waste (functional and structural) by force of physical laws, including fatigue, results in cultural pauperism; thus, a waste of life at the level where such waste is unwarranted and unreasonable.

The achievement of the instrumental purpose of arcology coincides by force of physical laws with nature's conservation. The coincidence, which is also a reinforcing element in the qualitative scope of the life developing within the arcology itself, is a direct consequence of the identity of efficiency (in its full meaning, frugality) with axiality of life. Then instrumentality in its over-all power is vital to efficiency and thus, indirectly, is itself vital. *But the fruition of growth is in things that are not commensurable to that which would seem to be their cause.* While an apple is the fruition of a tree seeking continuity in the next apple tree, and in a sense all that is the new tree was already in the parent tree, the fruition of man is the creation of the "never been before and never to be again."

At the same time the observation that life has never been so rich is invalid in the two directions of ratio and history. The ratio fulfillment–wealth seems to dwindle constantly. This indicates an ultimate exhaustion of human values submerged by and in a mechanism of ever-powerfulness lost to man's purpose. Historically, the stage of affluence seems more a leveling of, rather than a stronger stimulus to, growth, as if affluence were at the same time cause and effect of a weakening in the thrust of evolution. Totalitarianism recognizes, or instinctively senses, this and capitalizes on it by putting ideals before affluence and in so doing, though possibly for the wrong reason, injects new purposefulness into individual motivations.

The mechanisms channeling life positively may consist of the replacement of comfort and security by joy. In joy, motivations are carried, uplifted, while in comfort and security they seem to be drugged, sinking into naught. Possibly then wealth would instrumentalize a joyful state instead of a security at any price, the negative side of conservatism. Joy comes from plenitude. Plenitude, though basically an inner condition, can be invited by an inspiring and stimulating environment and the feeling of working toward achievements that overreach one's own limitation and embrace not just oneness but otherness as well: therefrom, the fruitions of creativity.

The pertinence of arcology to the condition of man, the condition of joy or indifference, is direct and immediate. *Joy is then the burst of liveliness that comes with the fitness and coherence of a process that is developing under one's eyes; it is the opposite of senselessness and squandering.*

ANTHONY DOWNS

"The Need for a New Vision for the Development of Large U.S. Metropolitan Areas"

***Saloman Brothers* (1989)**

Editors' information As cities have evolved since the rise of capitalist industrialism, planners have had to cope with spatial and class-based divisions between rich and poor, and between suburban and inner-city populations. Beginning with the nineteenth-century settlement house movement, programs have been developed by government and private philanthropy to deal with the problems of the poor, minority slum areas of cities. Sometimes government programs try to enrich the quality of life in the poorest urban areas through job training, public health programs, housing rehabilitation, educational enrichment, crime prevention, and substance abuse programs. Other programs seek to disperse ghettoized inner-city residents through the push of urban renewal and the pull of portable housing vouchers.

Since the time of the American ghetto riots of the mid-1960s, Anthony Downs has been working on a series of alternative ghetto enrichment and dispersal strategies that combine his commitment to social justice and his realistic view of how major planning decisions are actually made, especially in the areas of transit and regional economic development. In "The Need for a New Vision for the development of Large U.S. Metropolitan Areas," he proposes a radical overhaul of the metropolitan planning process, one which abandons the earlier vision of single-family homes, an over-reliance on the automobile, and a governance structure that separates the needs of inner cities and suburban communities. In place of this, Downs argues for a more integrated system that provides suburban homes and apartments for low-wage workers and creates "governance structures that preserve substantial local authority – but within a framework that compels local governments to act responsibly to meet area-wide needs."

Downs is a senior fellow at the Brookings Institution, a Washington, DC, think-tank. He has written extensively over the past three decades on poverty, race, housing, urban sprawl, traffic congestion, metropolitan planning, and other urban issues. He often takes the logic of strategies, which have been suggested or tried on a modest scale, to their conceptual limit: What would it really take to completely disperse the population of every poor Black ghetto in America? How much would it really cost to completely eliminate traffic congestion? How much housing at what cost would really be required to provide every American household a decent, safe, and sanitary affordable housing unit?

Often, as in this selection, Downs concludes that strategies to achieve broad social goals would require massive government action and significantly increased levels of funding. This emphasis on government programs and spending differentiates Downs from conservatives like Charles Murray (p. 122) who oppose government spending to solve urban problems. But Downs is no bleeding-heart liberal. He advocates triage strategies for urban neighborhoods – leaving the worst neighborhoods to deteriorate while directing scarce resources to poor but salvageable communities. This puts him at odds with many liberals. So does the argument that Downs once put forward that the dispersal of Black ghettos would be accepted by the white majority in America only if Blacks were dispersed into white areas in such a way that the white majority would remain numerically and culturally dominant.

The selection reprinted here argues strongly for a sweeping planning strategy that eliminates the legal/governmental distinctions between impacted inner-city ghettos and comfortable suburban communities. Originally published by the Saloman Brothers brokerage house, "The Need for a New Vision for the Development of Large U.S. Metropolitan Areas" is a visionary, yet hard-headed, approach to solving what has long been an intractable problem and one that threatens the stability of any urban future in a diverse society separated by race and class divisions.

Other works by Anthony Downs include *Urban Problems and Prospects* (Chicago: Markham, 1970), *Opening Up the Suburbs: An Urban Strategy for America* (New Haven, CT: Yale University Press, 1973), *Stuck in Traffic* (Washington, DC and Cambridge, MA: Brookings Institution and Lincoln Institute of Land Policy, 1992), and *New Visions for Metropolitan America* (Washington, DC and Cambridge, MA: Brookings Institution and Lincoln Institute of Land Policy, 1992).

Many of Downs's essays on urban issues are reprinted in *The Selected Essays Of Anthony Downs*, vol. 1, *Political Theory and Public Choice* and vol. 2, *Urban Affairs And Urban Policy* (Cheltenham, UK and Northampton, MA: Edward Elgar, 1998).

INTRODUCTION

The very success of the long-dominant American ideal vision of how our metropolitan areas should develop now threatens its continuance, as that vision contains several major inner inconsistencies. This paradox is manifest in the rising protest across the nation against worsening suburban traffic congestion and declines in urban environmental quality. Up to now, these protests have been concentrated in our largest and most prosperous metropolitan areas, where the inherent inconsistencies of our dominant ideal vision have become most apparent. But some of that vision's failings will soon be visible in many more metropolitan areas.

Therefore, American society needs a new ideal vision to guide the future development of its large metropolitan areas. For us to formulate and accept such a new vision would be almost equivalent to a paradigm shift concerning urban development. Formulating such a new vision is vitally important to all participants in real estate in large metropolitan areas, including commercial developers, institutional investors, homebuilders, realtors, local officials, and homeowners. This article examines the nature and failings of the still-dominant ideal vision, and proposes an approach to a new alternative.

THE CURRENTLY DOMINANT IDEAL VISION OF DESIRABLE METROPOLITAN DEVELOPMENT

For the past few decades, one major vision about how U.S. metropolitan areas ought to be developed has become totally dominant. This ideal vision reflects the views of the vast majority of American households – especially the more than 45% who live in the suburban portions of U.S. metropolitan areas. The same vision is held by almost all suburban government officials. Hence, it has immensely influenced nearly all their policies concerning land use, transportation, etc.

This dominant ideal vision is built upon four pillars. Each is a key desire or aspiration shared by nearly all American households:

The first pillar is ownership of detached, single-family homes on spacious lots. Repeated polls show that over 90% of all American households would like to own their own homes, and the vast majority want single-family detached units. Hence, ownership of such a unit has become the heart of "the American dream" – the prevailing image of how a household "makes it" in contemporary society. Realization of this aspiration implies the dominance of very low-density settlement patterns.

The second pillar is ownership and use of a personal, private automotive vehicle. Every American wants to be able to leap into his or her own car and zoom off on an uncongested road, to wherever he or she wants to go, in total privacy and great comfort – and to arrive there in not more than 20 minutes. This factor has rapidly escalated total vehicle ownership and use. In 1983, over 87% of all U.S. households owned at least one car or truck, and 53% owned two or more such vehicles. Even among households with 1983 incomes under $10,000, over 60% owned at least one car or truck; among those with incomes of $40,000 and over, 98.9% owned at least one vehicle and 86.7% owned two or more. In 1987, there were 162 million licensed drivers in the United States, but 167 million cars and trucks in use.

During the five years 1983–1987, the United States added 9.2 million persons to its total human population, but more than twice as many cars and trucks – 20.1 million – to its total number of vehicles in use.

The third pillar of the dominant ideal vision involves the structure of suburban workplaces. They are visualized as consisting predominantly of low-rise office or industrial buildings or shopping centers, in attractively landscaped, park-like settings. Each such structure ought to be surrounded by a large supply of its own free parking so that those who work in it or patronize it can conveniently drive and park there without cost.

The fourth pillar for this ideal vision concerns governance. Most Americans want to live in small communities with strong local self-governments. They want those governments to control land use, public schools, and other key elements affecting what they perceive as the quality of neighborhood life. This institutional structure permits existing residents to have a strong voice in controlling their local environments.

The above four elements define the prevailing ideal vision of "the American dream" as it is conceived of by the vast majority of suburbanites and by many city-dwellers. All four elements express what might be termed unconstrained individualism. They represent the pursuit of an environment that maximizes one's own well being, without regard to the collective results of such behavior. This ideal vision has been reinforced over the past few decades by promotional efforts made by parts of the real estate industry and by suburban communities. Such promoters include homebuilders selling new dwellings, realtors reselling them, advertisers highlighting suburban lifestyles, politicians running for election, and planning officials trying to carry out the policies of local politicians and voters. The dominant ideal vision has become so strongly entrenched that it has become almost political suicide to challenge openly any of these four pillars of "the American dream."

THE FIRST FLAW WITHIN THE IDEAL VISION: EXCESSIVE TRAVEL

Unfortunately, the prevailing ideal vision of how metropolitan areas ought to be developed contains four major flaws. They involve basic inconsistencies between its key elements and the real preferences of the citizenry concerning how suburban life should be lived.

The first flaw is that the low-density settlement patterns required by single-family housing and low-density workplaces generate immense travel requirements. Low density inherently spreads both homes and jobs widely across the landscape. That forces people to travel long distances between where they live and where they work, shop, or play. Moreover, low-density settlement patterns cannot efficiently support mass transit; hence they can only be served by private automotive vehicles. Mass transit, whether buses or fixed-rail, can be effective only if at least one end of most journeys is relatively concentrated in a few points.

The resulting requirement for massive daily use of private vehicles is consistent with the second element of the dominant vision but has several negative consequences. The most obvious is traffic congestion. The more fully Americans achieve their ideal vision of widespread low-density homeownership and private vehicle ownership, the more congested their commuting journeys become. In addition, these heavy travel requirements cause severe air pollution, costly demands for road and other infrastructure construction, and high-level consumption of energy.

Most people who believe in this ideal vision have not yet realized that their success in achieving it is causing these adverse outcomes. Moreover, they are reluctant to come to this conclusion. Doing so would force them to admit that their own ideals and behavior are the main causes of their principal problems.

Consequently, most suburbanites suffering from excessive traffic congestion, air pollution, and other results of low-density settlement patterns blame both real estate developers and the latest-arriving residents in their suburban communities and adopt antigrowth or growth-limiting policies. But these policies do not attack the fundamental causes of the problems concerned; hence, they cannot solve such problems.

THE SECOND FLAW: NO HOUSING PROVISION FOR LOW-WAGE WORKERS

The second flaw in the dominant ideal vision is that it contains only relatively high-cost housing, providing no residences where low- and moderate-income households can afford to live. Yet such households are an integral part of American life, because they provide workers for relatively low-wage jobs. Those jobs are vital to the efficient operation of every community, including the wealthiest exurban enclaves. Low-wage workers are essential to many service firms, including but not limited to fast-food establishments, gas stations, laundries, dry cleaners, hospitals, retail stores, shopping centers, construction firms and subcontractors, and gardening and lawn-care firms. And moderate-wage workers are often the backbone of local government services, including police, firefighters, teachers, and administrative personnel.

These households cannot afford to live in the spacious, single-family homes that populate the ideal vision. In fact, many cannot afford to live in any type of single-family home. Among the 88.4 million U.S. households in 1985, 62.2% lived in detached single-family units, and 4.6% lived in attached single-family units. But 5.4% lived in mobile homes, 11.6% lived in structures containing two and four units, and 16.1% lived in structures containing five or more units. Thus, about one-third of all American households did not live in single-family homes, and over one-third were not homeowners, but renters. The ideal vision of metropolitan development does not contain any dwelling units appropriate to this "forgotten fraction."

True, U.S. suburbs – including the newest ones – contain a lot of housing other than single-family homes. Such housing is in fact being built in our metropolitan areas. Nevertheless, it does not form part of the ideal vision of how those areas should be developed. Therefore, its construction is often strongly opposed by local residents whose views are dominated by that ideal vision.

Moreover, many low- and moderate-income households cannot afford to live in brand new housing units since such units have their highest relative prices when they are first built. At that moment, they contain the most modern and up-to-date amenities and design available, and they have not yet been subjected to wear and tear. Moreover, they are most often located at the edge of the built-up territory within their metropolitan area, which is then usually among the most fashionable neighborhoods in which to live.

As time passes, these units become relatively less desirable. They no longer contain the most up-to-date amenities; they have experienced some normal wear and tear and deterioration; and their neighborhoods are no longer at the edge of the built-up territory, which has now expanded beyond them.

Hence, their relative prices decline, even though their absolute prices may rise along with inflation. If this life cycle continues, these units eventually fall in both relative prices and absolute prices enough so that low- and moderate-income households can afford to live in them. This life-cycle movement of relative prices is part of the "filtering" or "trickle-down" process through which older housing is made available to those households that cannot afford brand new units.

Most American suburbs have been built since World War II. The nation's total suburban population rose from 40.9 million in 1950 to 108.6 million in 1986 – an increase of 165.5%, compared to a 58.7% rise in total population. By 1986, the fraction of all U.S. residents living in suburbs had risen to 44.9% from 27% in 1950. In very large metropolitan areas, the newest suburbs are now located quite far from the older portions of both the central city and older suburbs.

This is particularly true in those metropolitan areas that have experienced the fastest growth in total population, such as many in California and Florida.

Historically, employers in newer suburbs have relied upon the more affordable housing inventory in older neighborhoods to provide dwellings for their low- and moderate-income workers. This arrangement permitted the newest communities to avoid providing any housing directly affordable to all the persons working within their boundaries. Hence, these communities could develop themselves in full accord with the ideal vision of universal homeownership, involving mainly detached single-family units.

However, such reliance upon the "trickle-down" process to house low-wage workers has recently become much less effective, for several reasons. First, jobs have spread out into the suburbs much more than in the past. In fact, many more new jobs are being created in the suburbs than in central cities. They include jobs located in relatively far-out suburbs, as well as close in ones. Second, the sheer size of the suburban portions of large, fast-growing metropolitan areas means that much of the older, less costly housing is many miles from new-growth regions where most new jobs are located. Low- and moderate-income workers who live in older, more central neighborhoods are far from those new jobs. Hence, they have a hard time finding job openings, or commuting to jobs they find (often because of lack of public transportation). Third, soaring land prices in many such metropolitan areas have helped raise housing prices there to extremely high levels. Few middle-income households can afford most new single-family units there. Even the few multifamily units being built there have rents far above the payment ability of most low- or moderate-income households. Fourth, the long economic expansion that began in 1982 has soaked up most available workers. This has forced the unemployment rate well below 6% nationally, and below 2% in some booming suburban areas. In addition, the changes in U.S. age distribution are creating a nationwide shortage of entry-level workers, tightening the market further. The higher wages resulting from this prosperity have become embedded in housing costs, increasing prices beyond the reach of many households.

As a result, far-out suburban portions of many large, fast-growth metropolitan areas are experiencing acute shortages of low-wage and even moderate-wage labor. Workers willing to take such jobs cannot afford to live anywhere within reasonable commuting distance of those jobs. Thus, the ideal vision's exclusive focus upon relatively new single-family housing, which is quite expensive, is inconsistent with the actual need for low- and moderate-wage workers that arises in every community.

THE THIRD FLAW: NO CONSENSUS ON HOW TO FINANCE INFRASTRUCTURES FAIRLY

A third major flaw springs from the absence of any consensus in the dominant vision about how best to finance new infrastructures, more roads, sewage systems, water systems, and school systems. They also include costs of increasing the capacity of existing arterial facilities in established areas through which new residents will pass on their way to and from work or shopping. The absence of a widespread consensus about how to pay for such added facilities "fairly" has two negative consequences: It creates severe political conflicts in many suburban areas, and it often results in gross underfunding – and therefore under-provision – of needed facilities and services. Residents already living in fast-growth areas want all the added facilities but expect services required by newcomers to be paid for by the latter. Existing residents consider it justifiable to load those marginal costs entirely onto new developments through various impact fees, exactions, proffers, and permit fees. They regard this as merely requiring the newcomers to pay "their fair share" of such costs, since newcomers are the main beneficiaries of those facilities. Hence, they vote against any increases in general taxes or general bonding powers to pay for such facilities.

In contrast, developers and potential newcomers to growth areas believe the entire community should share in paying for these added infrastructures. They cite three reasons for such sharing: First, because many existing residents benefited from past general financing of similar infrastructures when their subdivisions were being built, it is unfair of them to change the rules of the game once they have received such benefits. Second, growth creates many gains for society in general – including existing residents. Its main benefit is greater economic prosperity, which aids everyone. Therefore, society should pay for some of the marginal costs of growth. Third, loading all the marginal costs of growth onto new developments raises housing costs. That unfairly reduces the homeownership and rental opportunities of households with low and moderate incomes. From this perspective, existing residents also ought to pay "a fair share" of the costs of added growth.

Political resolution of this controversy is inherently biased in favor of existing residents because potential newcomers do not yet reside in the areas concerned. Consequently, they cannot vote on local government policies that directly affect their welfare, whereas existing residents can. The deck is thus stacked in favor of loading most marginal costs of development onto newcomers, rather than sharing those costs throughout the community.

However, realistically it is quite difficult to force newcomers to pay all the marginal costs of growth through impact fees or other exactions. Some costs of growth spring from more intensive use of existing facilities, such as expressways, arterial streets, sewage treatment plants, and school systems. Those facilities and services are shared between newcomers and previous residents. Moreover, many infrastructures can only be built in certain minimum sizes that have substantial capacity, thus presenting communities with sizable costs. For example, the addition of 200 households to an area where the sewage treatment plant has reached its absolute capacity volume of 20,000 households may require construction of a whole new treatment plant with a minimum capacity of 10,000 households. It is not possible to load the entire cost of this new plant onto those 200 newcomers. Instead, existing residents must bear some of that initial cost.

Faced by this situation, existing residents often choose to provide inadequate facilities for newcomers to hold down taxes. They make such choices without fear of being outvoted by potential newcomers, as the latter are not yet present to vote on their own behalf.

As a result, new housing and commercial subdivisions often are built without adequate infrastructures and other supporting facilities. Existing roads, streets, schools, parks, and utility systems become grossly overloaded and congested. This imposes heavy costs not only upon newcomers, but also upon the existing residents themselves. Precisely this sequence of events has been happening in Florida for some time. The Florida State government has been unable to agree on some means of financing the infrastructure to accommodate Florida's rapid population growth.

Such developments aggravate inherent political tensions between residents trying to minimize their taxes and newcomers and developers trying to accommodate new growth. Many existing residents would like to prohibit all additional growth completely. But doing so is impossible in practice because it violates basic freedoms of movement, private property rights, and contract guaranteed by the U.S. Constitution. The result in many fast growth suburban areas is an undesirable combination of high political tensions and facility overload because of underfunding and congestion. The dominant vision of how metropolitan areas ought to develop contains no means or even guidance concerning how to resolve these issues.

THE FOURTH FLAW: INABILITY TO ACCOMMODATE LULUS

The dominant ideal vision's fourth major flaw is that it contains no effective political mechanisms for resolving inevitable conflicts between the welfare of society as a whole and the welfare of geographically small parts of society. This results from extreme spatial fragmentation of government decision-making powers. Yet such fragmentation is required by the governance arrangements that most Americans prefer. They want to live in geographically small communities, each of which fully controls the land use, public schools, and other environmental aspects within its own boundaries. Such "local sovereignty" is one of the pillars of the ideal vision. But it means that each government has a highly parochial viewpoint. Its officials are concerned only with the welfare of the local residents who elect them. As a result, no public officials are primarily concerned with the welfare of society as a whole, or even of the metropolitan area as a whole.

However, every modern society must contain those facilities that benefit society as a whole, but have negative "spillover effects" upon their immediate surroundings (e.g. airports, expressways, jails, garbage incinerators, and landfills). These essential facilities are known as Locally Undesirable Land Uses (LULUs). Because of their undesirable impacts upon their immediate neighbors, no residents want to permit a LULU to be located near them. This attitude has become known as the Not In My Back Yard (NIMBY) syndrome.

Whenever a LULU has to be built, there is no effective way to find a location for it in a society designed in accordance with the predominant ideal vision. Residents near every potential site pressure their local governments to oppose locating the LULU there. Those governments have the power to reject the LULU because controls over land uses have been divided up among myriad local entities. And officials within all those entities are motivated to reject the LULU because they are politically responsible only to their own residents. Hence, it becomes almost impossible to find politically acceptable locations for facilities that every metropolitan area needs to operate efficiently. The resulting paralysis has virtually halted both major airport construction and prison expansion in most of the United States, and blocked creation of thousands of other badly needed LULUs.

Furthermore, there is a strong single-family-home bias built into the dominant ideal vision. As a result, nearly all new high-density real estate developments – especially those involving any multifamily housing – are usually considered LULUs, and are therefore extremely difficult to build. But low- and moderate-income households cannot afford to occupy new single-family homes. So the NIMBY syndrome reinforces the second flaw in the dominant ideal vision: Its failure to provide housing that low- and moderate-income households can afford.

In fact, low- and moderate-income households themselves are regarded as LULUs by many middle- and upper-income households. The latter consequently adopt a NIMBY attitude toward all housing affordable to low- and moderate-income households. They believe that such "undesirable" people near them will reduce the market values of their costly homes. About three-fourths of the financial net worth of the average American household consists of its equity in the home it owns and lives in. Hence, its members are quite sensitive about any changes in their neighborhoods that they believe would reduce the value of those equities, and very little such housing gets built in most U.S. suburbs.

SOCIAL INCONSISTENCIES RESULTING FROM THE IDEAL VISION'S FLAWS

The four major flaws in the ideal vision described above have caused the emergence and persistence of certain negative conditions in suburban America, the very heartland of that ideal vision. Hence, we must conclude that the dominant vision itself contains inherent characteristics that render it less ideal than we previously believed it to be – or than most Americans still believe it to be. These characteristics can be considered "social inconsistencies," as they are inconsistent with the high quality of life promised by the ideal vision. (Note: This situation illustrates a fundamental problem in democratic societies. They have great difficulty solving problems that arise as the long-run result of a majority's pursuit of policies that produce short-run benefits. Once a majority of citizens has begun receiving those short-run benefits, its members do not want to give them up. But they often fail to agree about how to distribute the long-run costs necessary to sustain those short-run benefits. Each subgroup among the beneficiaries tries to shift as much of its share of the costs as possible onto other subgroups.

No consensus arises about how to pay for these costs. In some cases, the costs are not paid at all. That is precisely why the United States has been unable to reduce its huge recurrent Federal deficits.)

THE ALTERNATIVE VISION OFFERED BY THE URBAN PLANNING PROFESSION

The American urban planning profession has long been aware of some of the internal inconsistencies in the dominant ideal vision described above. In particular, urban planners have repeatedly attacked the heavy reliance upon automotive vehicle travel required by very low-density settlement patterns. In fact, many urban planners have advocated an alternative vision that rejects such reliance. But their alternative has never gained substantial public acceptance in America.

This alternative has been based upon the European pattern of urban development. It featured high-density residential settlements, high-density workplaces, tightly circumscribed land use patterns that prevent peripheral sprawl, and massive use of heavily subsidized public transit systems for movement. These traits are only possible under a governance system that centralizes power over the land use patterns in each metropolitan area in a single governing body with authority over the entire area. Accompanying such centralization in Europe is use of large amounts of publicly subsidized rental housing. Such housing accommodates a sizable fraction of the entire population in subsidized multifamily units scattered throughout each metropolitan area.

For decades, U.S. urban planners have been advocating an alternative vision of future metropolitan development based largely upon this pattern. But they have been unable to persuade either the public or local government officials to accept it because this alternative directly opposes the dominant vision described earlier that most Americans cherish. It rejects too many elements of that dominant vision, including primary reliance upon owner-occupied single-family housing, widespread use of private automotive vehicles, and decentralized local government.

Furthermore, some of the key elements in the European pattern do not appeal to Americans. For example, placing large clusters of high-rise apartments right around fixed-rail transit stops repels most suburban residents, who do not like high-rise housing. In addition, the immense cost of fixed-rail transit systems, and their lack of flexibility, are not acceptable to many Americans. The only large-scale fixed-rail transit system built in the U.S. since 1945 without major Federal subsidies is the Bay Area Rapid Transit System. All the others probably never would have been created if local residents had been compelled to pay their full costs without massive Federal subsidies.

Moreover, the urban planning profession has been guilty of largely ignoring the second major flaw in the dominant vision (its inability to provide housing for low- and moderate-income households). Few urban planners have ever developed comprehensive plans that specifically indicated where to accommodate poor households. Most comprehensive plans ignore the need for low-rent housing, because explicitly locating it within the plan would arouse immense opposition from nonpoor residents.

Yet the urban planning profession has been correct in claiming that the dominant vision would not work well in the long run. Most suburbanites in fast-growth metropolitan areas are now discovering this fact, to their own dismay.

CRITERIA FOR A NEW IDEAL VISION OF FUTURE METROPOLITAN AREA DEVELOPMENT

The preceding analysis shows that American society needs a new ideal vision of how future development ought to occur in our large metropolitan areas. The current dominant vision contains too many serious inconsistencies and undesirable outcomes. The new ideal vision should conform to several specific criteria. Such conformance would enable this vision both to (1) meet the future needs of American society; and (2) avoid the problems generated by the currently dominant vision.

As noted earlier, development and widespread acceptance of a new vision would be similar to a paradigm shift in science – that is, the replacement of one fundamentally governing perception by another one. Such replacements may appear to occur suddenly, but they actually build up gradually over time.

The new ideal vision of future metropolitan growth will not require the complete rejection of the pillars on which the existing vision has been based. Rather, it will modify those pillars. This new perspective will be much more conscious of the collective impacts of individual decisions than the currently dominant perspective. Hence, it will embody individualism sensitive to collective behavior patterns, rather than unconstrained individualism. For example, we need not deny the desirability of homeownership or of single-family homes in order to recognize that a balanced society also needs rental housing and much higher average densities than we have considered ideal in the past. Similarly, we will certainly not abandon widespread use of individual vehicles. But those vehicles may be powered differently, and more people will be traveling together in shared rides or transit than at present.

Specific criteria of this new ideal vision are as follows:

First, the new vision must contain sizable areas of at least moderately high-density development, especially of housing, but also of workplaces. This is necessary to achieve three key goals: the creating of housing that is less costly than single-family homes; the reduction of the area required to accommodate both jobs and workers, thereby cutting travel requirements and reducing both congestion and air pollution; the development of more efficient spatial interrelations among buildings within suburban commercial and industrial developments. This would reduce travel requirements within such developments and promote more face-to-face contacts there. The third of these goals is far less important than the first two.

In order to make such higher-density areas politically acceptable to suburban residents, owners of single-family homes must be convinced that they can tolerate multifamily housing nearby. Such housing need not consist of big high-rise towers, but can be a mixture of low-rise and mid-rise structures, with an occasional high-rise. Owners of existing single-family homes must become confident that such housing will neither jeopardize the investments they have made in their own homes, nor expose their families to crime and violence. That will probably require careful studies of the impact of well-designed existing multifamily projects upon the market values of surrounding single-family homes. Owners of single-family homes must also be convinced that moderate-density housing can be designed to look good and therefore not reduce the aesthetic quality of their areas.

Second, the new ideal vision must encourage people to live nearer to where they work. This means that each subarea within a metropolitan area must contain a "balanced blend" of different types and prices of housing, all close to the jobs within the same subarea. Then it will be possible for workers holding jobs at all wage levels to live close to those jobs and still occupy housing they can afford.

True, people cannot be forced to live nearby their workplaces. Today, many people could live closer to where they work, but they prefer traveling longer distances in order to live in what they deem an ideal environment. But this preference is changing as congestion increases the time required to move any specific distance during peak commuting periods. Eventually, more people will decide that suffering the time losses and frustration of such worsened congestion is not worthwhile and will move either their jobs or their homes to reduce the distance between them. But this is possible only if they are compelled to live near their jobs by the appropriate cost of the housing close to those jobs. That, in turn, is possible only if each sizable subregion of the entire metropolitan area contains a mixture of different types of housing available at widely varying occupancy costs.

Achieving such "balanced blends" of housing in all parts of a large metropolitan area will not be easy. The logical spatial units within which such "balancing" should occur will rarely coincide with existing governmental boundaries. Moreover, at least one job cluster in each (i.e. downtown) contains so many jobs that not all of its workers could possibly live nearby. In order to reduce commuting travel, it may be necessary in the long run to "decant" some existing downtowns by shifting some of their existing jobs elsewhere. That is not a happy thought for owners and managers of downtown office buildings and other properties. But the goal of this article is not to determine how these criteria might be met, but rather to formulate them in the hope that others might determine how they might be implemented.

Third, the new ideal vision must contain governance structures that preserve substantial local authority – but within a framework that compels local governments to act responsibly to meet area-wide needs. That probably requires a comprehensive planning framework imposed by state governments. In such a framework, every local government would be responsible for drawing up comprehensive plans for its own future development. But all such plans would have to meet certain general criteria laid down by the state government. Such criteria might include provision of land zoned for low- and moderate-income housing, and arrangements that pushed the choosing of sites for certain region-affecting facilities up to regional or state levels to avoid complete paralysis. Moreover, the state government would have to create mechanisms to review and modify local government plans to insure that they met statewide criteria. Oregon has already put such arrangements into practice, and several other states are doing so. Thus, in considering possible future changes in existing governance within metropolitan areas, one should expect neither true metropolitan government, nor any total abandonment of local sovereignty. Simply, such sovereignty must be placed within a broader framework that enforces wider social responsibility. While this probably can occur only at the state level, where the constitutional power over local government resides, the Federal government might provide financial incentives to get more states to engage in this type of restructuring.

Fourth, the new vision should contain incentive arrangements that encourage individuals and households to take a more realistic account of the collective costs of their behavioral choices. For example, individual auto drivers are not required to pay more for taking trips alone during rush hours than during off-peak hours, because auto drivers are never charged directly for using roads. For decades, economists have pointed out that each driver who enters a well-traveled highway during peak hours creates a substantial social cost. By adding to the prevailing congestion, that driver slows down all other drivers to some degree, thereby imposing a time loss upon them. In contrast, a driver who takes the same trip during nonpeak hours does not generate any such social cost. Therefore, it would be economically efficient for society if drivers were charged a direct fee for driving during peak hours, but not during off-peak hours. Such charges would encourage more drivers to shift to off-peak periods, thereby reducing peak-hour congestion. And those charges could raise money useful in improving the nation's road system.

This type of incentive arrangement is especially important in view of the fact that most peak-hour auto trips made in the nation's largest metropolitan areas are nonwork trips, many of which could be shifted to other times relatively easily.

Unfortunately, politicians in most democracies have consistently refused to adopt differential road-use pricing. They argue that such pricing would penalize low-income drivers and favor higher-income ones, who could more easily afford to pay peak-hour charges. In a democracy where 87% of all households have cars, but most do not have high incomes, differential road pricing has always seemed to be a politically losing proposition.

This situation will change only when major freeways become so overcrowded that traffic literally slows to a crawl for several hours per day. Many southern California freeways are close to such prolonged gridlock today, and are likely to reach it soon. When they do, even low-income auto-driving voters will become frustrated enough to accept almost any proposed solution likely to be effective. Peak-hour road pricing will then be perceived in a much more favorable light because of the realistic incentives it presents to drivers.

So-called "exclusionary zoning" arrangements or "linkage fees" concerning housing could be another form of such incentives. Such exclusionary arrangements are created and sustained by political pressures on local governments from middle- and upper-income homeowners. They believe socio-economic exclusion helps maximize the market values of their homes and improve the quality of their local environments.

In many suburban areas, exclusionary zoning has made the cost of housing units too great for most low- and moderate-income households. The typical application of this type of zoning has been the "down zoning" of lots where minimum sizes are often two acres and up. In the end, this imposes a cost upon society as a whole by requiring many such households to live far from their suburban jobs and commute long distances each day. The result is more overall traffic congestion and air pollution, plus the imposition of very high travel costs upon low- and moderate-income commuters.

Therefore, those who impose exclusionary arrangements on their suburban areas should be required to offset the social costs they are generating through various arrangements that charge them for engaging in exclusionary practices. Such charges would also discourage exclusion by raising the price of housing to those who engage in it.

Several forms of such incentive arrangements concerning housing have been tried in different areas. Local or state governments could add a fee onto all market-priced housing units built or sold in exclusionary regions, and use the proceeds to subsidize units available at costs affordable to low- and moderate-income households. Another tactic would be to add a transfer tax onto the sale of single-family homes and then put the proceeds into a housing trust fund used to help finance construction of low-rent units in otherwise exclusionary communities. Another approach would be to require each developer to allocate a certain percentage of the units he or she builds for low- and moderate-income occupancy. The resulting higher costs for market-priced units could be offset to some extent by permitting the developer to build at higher-than-usual average densities in the projects concerned. Yet another tactic would be to charge developers of commercial and industrial properties in exclusionary areas "linkage fees" that are used to subsidize housing for low- and moderate-income households.

Not all of these arrangements would have the same negative incentive impact upon exclusionary zoning practices that peak-hour-driving charges would have upon rush-hour trips. But they all would compel persons engaging in exclusionary zoning practices to bear some of the collective costs their individual behavioral choices are creating for society.

Fifth, the new vision should incorporate stable and predictable strategies that would adequately finance infrastructures built to accommodate growth. Such strategies need to be stable and predictable so that all concerned can efficiently plan how to carry out new developments, both public and private. These strategies also must be fair and politically acceptable both to existing residents on the one hand, and to developers and potential newcomers on the other. Financing strategies appropriate to slow-growth areas may be different from those appropriate to fast-growth areas. In all areas, such strategies will probably have to combine some impact fees and exactions upon newcomers, with some increases in general taxes upon all residents.

Another key criterion is that the new ideal vision should not depend upon construction of costly fixed-rail mass transit systems that are politically acceptable in each area only when heavily subsidized with funds from other parts of the nation. Many urban planners will reject this criterion. But this author believes the American people will accept it – as they have for many decades. Perhaps some forms of light rail transit can be created inexpensively enough to be incorporated within viable metropolitan area transportation systems. But most such systems must be based upon freely moving automotive vehicles, whether they are cars, trucks, vans, big buses, or mini-buses. Thus, designing urban transportation systems that will reduce congestion and pollution poses a tremendous challenge to transportation planners.

Further, the ideal vision should incorporate workplace designs that combine efficient interchange among individual workplaces with big enough critical masses of jobs to create productive interchange possibilities. These characteristics are now found within traditional downtowns and within large regional shopping malls. In both places, office buildings or retail shops are located right next to each other along attractive pedestrian walkways, thereby encouraging maximum efficiency of movement among different facilities. Parking is not placed around each structure, but in external zones that are nearby but do not interfere with movements among individual units. Yet enough facilities are massed together to permit efficient comparison shopping or interchanges of services and meetings among persons working in different enterprises.

These traits are now almost totally absent from all suburban office and industrial parks. Each such structure is separated from others nearby by a sea of parking or landscaping. Therefore, meeting this criterion might require forcing owners to build new structures between existing ones, as near Tyson's Corner, in Fairfax County, Virginia, so as to recreate the interchange efficiency of downtown streets. It cannot be determined how this could be done with present fragmented private site ownership. But considering the ingenuity with which Wall Street investment bankers have reorganized existing companies into multiple entities, or created new securities, we should be equally ingenious about land-use controls and forms of ownership and management.

Finally, the new ideal vision should be internally consistent, in the dual sense that (1) the amount of travel it requires does not lead to the levels of traffic congestion or air pollution we are now encountering; and (2) that some homes in the vision are affordable to the low- and moderate-income households needed to do much of the work. We need complex computerized transportation and land-use models to test such relationships. Designing both the models and the transportation elements of the new ideal vision will not be easy. But few aspects of urban life are more important.

CONCLUSION

Is it really possible to create an alternative ideal vision of future large metropolitan area development – a new paradigm – that meets all these criteria? Our society desperately needs to make an attempt to find out. We have overwhelming current evidence that the currently dominant ideal vision does not work effectively. But we cannot displace that vision unless we have some plausible, attractive, and persuasive alternative to offer. The option traditionally offered by urban planners is never going to win widespread acceptance among the American people.

MANUEL CASTELLS

"European Cities, the Informational Society, and the Global Economy"

Journal of Economic and Social Geography (1993)

Editors' introduction The information revolution sweeping the world today has profound implications for the future of cities. The power of computers is increasing and the cost of computing is dropping at astonishing rates. Fax machines, modems, fiber-optic cable, communication satellites, a global electronic (e-mail) system, videoconferencing, the Internet, information highways, virtual reality, and multi-media – all are transforming traditional urban space and communities in ways that are still only partially understood.

Manuel Castells was born in Spain, educated as a sociologist in France, and is currently professor of city and regional planning at the University of California, Berkeley. As a young man, Castells fled Franco's authoritarian and intellectually stifling Spain for the freedom and intellectual excitement of Paris. He became a neo-Marxian and crafted sophisticated theories on the role of the capitalist state and grassroots urban protest movements. More recently, Castells has turned his attention to the implications of high technology and the information revolution for community life and urban development.

Castells sees information technologies as the fundamental instrument of the new organizational logic transforming the world today. Accordingly he uses the adjective "informational" as a type of city as "industrial" or "colonial" city might have been used in the nineteenth century. In his magisterial three-volume *The Rise of the Network Society* (1998) Castells describes how information technology will restructure relationships between rich and poor regions, labor and capital, centralization and decentralization of services, governments and nongovernmental entities, the individual and society.

Castells argues that what he calls the "space of flows" will increasingly govern the actions of power-holding organizations rather than territorially based institutions operating in the "space of places." Industry and services will be organized worldwide around the operation of their information-generating units. He envisions powerful, secretive, multinational institutions not tied to any particular place as the dominant institutions of the future.

In "European Cities, the Informational Society, and the Global Economy," Castells predicts how the basic elements of the new informational society will affect the cities of Western Europe. Among the main influences, he lists the inevitability of European Union political and economic integration keeping pace with the global economy, the cultural pressures of Third World immigration and social conflicts arising out of gentrification and the "polarized occupational structure," and the increasing importance of environmentalism and gender-based social restructuring. Thus, he argues that the "fundamental urban dualism of our time . . . opposed the cosmopolitanism of the elite, living on a daily connection to the whole world . . . to the tribalism of local communities, retrenched in their spaces that they try to control as their last stand against the macro forces that shape their lives out of their reach."

Castells' analysis raises many questions. Will local and even national governments wither in importance with the information dominance of powerful multinational institutions? Will the information revolution lead to greater social inequality between rich and poor nations? Are "dual cities" inevitable where rich and powerful professional managers manipulate impoverished masses, both blocks and continents away? May his vision of cosmopolitan elites versus retrenched local communities be too dualistic, too either-or, when the broad mass of middle-class urban residents may feel torn between both lifestyles? Will mass urban social movements emerge to reassert popular power against such a techno-Orwellian future? And can they succeed against so powerful and subtle a global system? In the late 1960s, radicals could identify and work to overthrow a territorially based dictator like Francisco Franco. But who and where is the enemy in the emerging space of flows?

Many of the visionary writers in this anthology predicted that the information (and transportation) technology of their times would undermine old, and make possible new, city patterns. Ebenezer Howard argued that garden cities were possible because people could communicate by telegraph and send goods by railroad (p. 321). Frank Lloyd Wright foresaw instant communication and rapid transportation in Broadacre City by pneumatic tube and helicopter (p. 344). And Melvin Webber predicted a "post-city age" as a result of telephones and air transportation (p. 535).

Other books on information technology and cities by Castells are *The Rise of the Network Society*, 3 vols. (Malden, MA: Blackwell, 1998), *The Informational City: Information Technology, Economic Restructuring, and the Urban-Regional Process* (Oxford and Cambridge, MA: Blackwell, 1991), an edited anthology *High Tech, Space, and Society* (Beverley Hills, CA: Sage, 1985), and, jointly authored with Sir Peter Hall, *Technopoles of the World* (London: Routledge, 1994).

Other books by Castells include *The Power of Identity* (Oxford: Blackwell, 1997), *Dual City: Restructuring New York* (New York: Russell Sage Foundation, 1991), edited with John Hull Mollenkopf, *The City and the Grassroots* (Berkeley: University of California Press, 1983) and *The Urban Question* (London: Edward Arnold, 1977).

An old axiom in urban sociology considers space as a reflection of society. Yet life, and cities, are always too complex to be captured in axioms. Thus the close relationship between space and society, between cities and history, is more a matter of expression than of reflection. The social matrix expresses itself into the spatial pattern through a dialectical interaction that opposes social contradictions and conflicts as trends fighting each other in an endless supersession. The result is not the coherent spatial form of an overwhelming social logic be it the capitalist city, the preindustrial city or the ahistorical utopia but the tortured and disorderly, yet beautiful patchwork of human creation and suffering.

Cities are socially determined in their forms and in their processes. Some of their determinants are structural, linked to deep trends of social evolution that transcend geographic or social singularity. Others are historically and culturally specific. And all are played out, and twisted, by social actors that impose their interests and their values, to project the city of their dreams and to fight the space of their nightmares.

Sociological analysis of urban evolution must start from the theoretical standpoint of considering the complexity of these interacting trends in a given time–space context. The last twenty years of urban sociology have witnessed an evolution of thinking (including my own) from structuralism to subjectivism, then to an attempt, however imperfect, at integrating both perspectives into a structural theory of urban change that, if a label rooted in an intellectual tradition is necessary, I would call Marxian, once history has freed the Marxian theoretical tradition from the terrorist tyranny of Marxism-Leninism.

I intend to apply this theoretical perspective to the understanding of the fundamental transformations that are taking place in West European cities at the end of the second millennium. In order to understand such transformations we have to refer to major social trends that are shaking up the foundations of our existence: the coming of a technological revolution centered on information technologies, the formation of a global economy, and transition to a new society, the informational society, that without ceasing to be capitalist or statist replaces the industrial society as the framework of social institutions.

But this analysis has to be at the same time general and structural (if we accept that a historical transformation is under way) and specific to a given social and cultural context, such as Western Europe (with all due acknowledgement to its internal differentiation).

In recent years, a new trademark has become popular in urban theory: capitalist restructuring. Indeed it is most relevant to pinpoint the fundamental shift in policies that both governments and corporations have introduced in the 1980s to steer capitalist economies out of their 1970s crises. Yet more often than not, the theory of capitalist restructuring has missed the specificity of the process of transformation in each area of the world, as well as the variation of the cultural and political factors that shape the process of economic restructuring, and ultimately determine its outcome.

Thus the deindustrialization processes of New York and London take place at the same time that a wave of industrialization of historic proportions occurs in China and in the Asian Pacific. The rise of the informal economy and of urban dualism takes place in Los Angeles, as well as in Madrid, Miami, Moscow, Bogota and Kuala Lumpur, but the social paths and social consequences of such similarly structural processes are so different as to induce a fundamental variegation of each resulting urban structure.

I will try therefore to analyze some structural trends underlying the current transformation of European cities, while accounting for the historical and social specificity of the processes emerging from such structural transformation.

THE THREAD OF THE NEW HISTORY

Urban life muddles through the pace of history. When this pace accelerates, cities and their people become confused, spaces turn threatening, and meaning escapes from experience. In such disconcerting yet magnificent times, knowledge becomes the only source to restore meaning, and thus meaningful action.

At the risk of schematism, and for the sake of clarity, I will summarize what seem to be the main trends that, together and in their interaction, provide the framework of social, economic and political life for European cities in this particular historical period.

First of all, we live in the midst of a fundamental technological revolution that is characterized by two features:

> As all major technological revolutions in history, the effects are pervasive. They are not limited to industry, or to the media, or to telecommunications or transportation. New technologies, that have emerged in their applications in full strength since the mid 1970s, are transforming production and consumption, management and work, life and death, culture and warfare, communication and education, space and time. We have entered a new technological paradigm.
>
> As the industrial revolution was based on energy (although it embraced many other technological fields) the current revolution is based upon information technologies, in the broadest sense of the concept, which includes genetic engineering (the decoding and reprogramming of the codes of living matter).

This technological informational revolution is the backbone (although not the determinant) of all other major structural transformations:

> It provides the basic infrastructure for the formation of a functionally interrelated world economic system.
>
> It becomes a crucial factor in competitiveness and productivity for countries throughout the world, ushering in a new international division of labour.
>
> It allows for the simultaneous process of centralization of messages and decentralization of their reception, creating a new communications world made up at the same time of the global village and of the incommunicability of those communities that are switched off from

the global network. Thus an asymmetrical space of communication flows emerges from the uneven appropriation of a global communication system.

It creates a new, intimate linkage between the productive forces of the economy and the cultural capacity of society. Because knowledge generation and information processing are at the roots of the new productivity, the ability of a society to accumulate knowledge and manipulate symbols translates into economic productivity and political military might, anchoring the sources of wealth and power in the informational capacity of each society.

While this technological revolution does not determine per se the emergence of a social system, it is an essential component of the new social structure that characterizes our world: the informational society. By this concept, I understand a social structure where the sources of economic productivity, cultural hegemony and political military power depend, fundamentally, on the capacity to retrieve, store, process and generate information and knowledge. Although information and knowledge have been critical for economic accumulation and political power throughout history, it is only under the current technological, social, and cultural parameters that they become directly productive forces. In other words, because of the interconnection of the whole world and the potential automation of most standard production and management functions, the generation and control of knowledge, information and technology is a necessary and sufficient condition to organize the overall social structure around the interests of the information holders. Information becomes the critical raw material of which all social processes and social organizations are made. Material production, as well as services, become subordinate to the handling of information in the system of production and in the organization of society. Empirically speaking, an ever growing majority of employment in Western European cities relates to information processing jobs. The growing proportion of employment in service activities is not the truly distinctive feature, because of the ambiguity of the notion of 'services' (e.g. in Third World cities a majority of the population also works in 'services', although these are indeed very different kinds of activities). What is truly fundamental is the growing quantitative size and qualitative importance of information processing activities in both goods production and services delivery. The contradictory but ineluctable emergence of the informational society shapes European cities as the onset of the industrial era marked forever the urban and rural spaces of the nineteenth century.

A third major structural trend of our epoch is the formation of a global economy. The global economy concept must be distinguished from the notion of a world economy, which reflects a very old historical reality for most European nations, and particularly for the Netherlands, which emerged as a nation through its role as one of the nodal centres of the sixteenth century's world economy. Capitalism has accumulated, since its beginnings, on a worldwide scale. This is not to say that the capitalist economy was a global economy. It is only now becoming such.

By global economy we mean an economy that works as a unit in real time on a planetary scale. It is an economy where capital flows, labour markets, commodity markets, information, raw materials, management, and organization are internationalized and fully interdependent throughout the planet, although in an asymmetrical form, characterized by the uneven integration into the global system of different areas of the planet. Major functions of the economic system are fully internationalized and interdependent on a daily basis. But many others are segmented and unevenly structured depending upon functions, countries, and regions. Thus the global economy embraces the whole planet, but not all regions and all people on the planet. In fact, only a minority of people are truly integrated into the global economy, although all the dominant economic and political centres on which people depend are indeed integrated into the global economic networks (with the possible exception of Bhutan . . .). With the disintegration of the Soviet empire, the last area of the planet that was not truly integrated into the global economy, it is restructuring itself in the most dramatic conditions to be able to reach out to the perceived avenues of prosperity of our economic model. (China already started its integration in the global capitalist economy in December 1979, while trying to preserve its statist political regime.)

This global economy increasingly concentrates wealth, technology, and power in 'the North', a vague geopolitical notion that replaces the obsolete West–East differentiation, and that roughly corresponds to the OECD countries.

The East has disintegrated and is quickly becoming an economic appendage of the North. Or at least such is the avowed project of its new leaders. The 'South' is increasingly differentiated. East Asia is quickly escaping from the lands of poverty and underdevelopment to link up, in fact, with the rising sun of Japan, in a model of development that Japanese writers love to describe as 'the flying geese pattern', with Japan of course leading the way, and the other Asian nations taking off harmoniously under the technological guidance and economic support of Japan. China is at the crossroads of a potential process of substantial economic growth at a terrible human cost as hundreds of millions of peasants are uprooted without structures able to integrate them into the new urban industrial world. South and Southeast Asia struggle to survive the process of change, looking for a subordinate yet livable position in the new world order. Most of Africa, on the contrary, finds itself increasingly disconnected from the new global economy, reduced to piecemeal, secondary functions that see the continent deteriorate, with the world only waking up from time to time to the structural genocide taking place there when television images strike the moral consciousness of public opinion and affect the political interests of otherwise indifferent policy makers. Latin America, and many regions and cities around the world, struggle in the in-between land of being only partially integrated into the global economy, and then submitted to the tensions between the promise of full integration and the daily reality of a marginal existence.

In this troubled world, Western Europe has, in fact, become a fragile island of prosperity, peace, democracy, culture, science, welfare and civil rights. However, the selfish reflex of trying to preserve this heaven by erecting walls against the rest of the world may undermine the very fundamentals of European culture and democratic civilization, since the exclusion of the other is not separable from the suppression of civil liberties and a mobilization against alien cultures. Major European cities have become nodal centres of the new global economy, but they have also seen themselves transformed into magnets of attraction for millions of human beings from all around the world who want to share the peace, democracy, and prosperity of Europe in exchange for their hard labour and their commitment to a promised land. But the overcrowded and aged Western Europe of the late twentieth century does not seem as open to the world as was the young, mostly empty America of the beginning of the century. Immigrants are not welcome as Europe tries to embark on a new stage of its common history, building the supranational Europe without renouncing its national identities. Yet the cultural isolation of the pan-European construction is inseparable from the affirmation of ethnic nationalism that will eventually turn not only against the 'alien immigrants' but against European foreigners as well. European cities will have to cope with this new global economic role while accommodating to a multi-ethnic society that emerges from the same roots that sustain the global economy.

The fourth fundamental process under way in European cities is the process of European integration, into what will amount in the twenty-first century to some form of confederation of the present nation states. This is an ineluctable process for at least fifteen countries (the current twelve EC countries plus Sweden, Austria and Switzerland) regardless of the fate of the symbolic Maastricht Treaty. If, as is generally accepted, the European Community is heading toward a common market, a common resident status for all its citizens, a common technology policy, a common currency, a common defense and a common foreign policy, all the basic prerogatives of the national state will be shifted to the European institutions by the end of the century. This will certainly be a tortuous path, with the nostalgics of the past, neofascists, neocommunists and fundamentalists of all kinds fighting the tide of European solidarity, fueling the fears of ignorance among people, building upon demagogy and opportunism. Yet however difficult the process, and with whatever substantial modifications to the current technocratic blueprints, Europe will come into existence: there are too many interests and too much political will at stake to see the project destroyed after having come this far.

The process of European integration will cause the internationalization of major political decision-making processes, and thus will trigger the fear of subordination of specific social interests to supranational institutions. But most of these specific interests express themselves on a regional or local basis rather than at the national level. Thus we are witnessing the renewal of the role of regions and cities as locuses of autonomy and political decision. In particular, major cities throughout Europe constitute the nervous system of both the economy and political system of the continent. The more national states fade in their role, the more cities emerge as a driving force in the making of the new European society.

IDENTITY AND SOCIAL MOVEMENTS

The process of historical transition experienced by Europe's cities leads to an identity crisis in its cultures and among its people, that becomes another major element of the new urban experience. This identity crisis is the result of two above mentioned processes that, however contradictory among themselves, jointly contribute to shake up the foundations of European national and local cultures. On the one hand, the march to supranationality blurs national identities and makes people uncertain about the power holders of their destiny, thus pushing them into withdrawal, either individualistic (neolibertarianism) or collective (neonationalism). On the other hand, the arrival of millions of immigrants and the consolidation of multi-ethnic, multicultural societies in most West European countries, confronts Europe head on with the reality of a nonhomogeneous culture, precisely at the moment when national identity is most threatened. There follows a crisis of cultural identity (with the corollary of collective alienation) that will mark urban processes in Europe for years to come.

More to the point: major cities will concentrate the overwhelming proportion of immigrants and ethnic minority citizens (the immigrants' sons and daughters). Thus they will also be at the forefront of the waves of racism and xenophobia that will shake up the institutions of the new Europe even before they come into existence. As a reaction to the national identity crisis we observe the emergence of territorially defined identities at the level of the region, of the city, of the neighbourhood. European cities will be increasingly oriented toward their local culture, while increasingly distrustful of higher order cultural identities. The issue then is to know if cities can reach out to the whole world without surrendering to a localistic, quasi-tribal reaction that will create a fundamental divide between local cultures, European institutions, and the global economy.

European cities are also affected by the rise of the social movements of the informational society, and in particular by the two central movements of the informational society: the environmental movement, and the women's movement.

The environmental movement is at the origin of the rise of the ecological consciousness that has substantially affected urban policies and politics. The issue of sustainable development is indeed a fundamental theme of our civilization and a dominant topic on today's political agendas. Because major cities in Europe are at the same time nodal centres of economic growth and the living places for the most environmentally conscious segment of the population, the battles for integration will be fought in the streets and institutions of these major cities.

The structural process of transformation of women's condition, in dialectical interaction with the rise of the feminist movement, has completely changed the social fabric of cities. Labour markets have been massively feminized, resulting in a change in the conditions of work and management, of struggle and negotiation, and ultimately in the weakening of a labour movement that could not overcome its sexist tradition. This also points to the possibility of a new informational labour movement that because it will have to be based on women's rights and concerns, as well as on those of men, will be historically different from its predecessor.

At the same time, the transformation of households and the domestic division of labour is fundamentally changing the demands on collective consumption, and thus urban policy. For instance, childcare is becoming as important an issue as housing in today's cities. Transportation networks have to accommodate the demands of two workers in the family, instead of relying on the free driving service provided by the suburban housewife in the not so distant past.

Some of the new social movements, the most defensive, the most reactive, have taken and will be taking the form of territorially based countercultures, occupying a given space to cut themselves off from the outside world, hopeless of being able to transform the society they refuse. Because such movements are likely to occur in major cities that concentrate a young, educated population, as well as marginal cultures that accommodate themselves in the cracks of the institutions, we will be witnessing a constant struggle over the occupation of meaningful space in the main European cities, with business corporations trying to appropriate the beauty and tradition for their noble quarters, and urban countercultures making a stand on the use value of the city, while local residents try to get on with their living, refusing to be bent by the alien wind of the new history.

Beyond the territorial battles between social movements and elite interests, the new marginality, unrelated to such social movements, is spreading over the urban space. Drug addicts, drug dealers and drug victims populate the back alleys of European cities, creating the unpredictable, awakening our own psychic terrors, and tarnishing the shine of civilized prosperity at the daily coming of darkness. The 'black holes' of our society, those social conditions from where there is no return, take their territory too, making cities tremble at the fear of their unavowed misery.

The occupation of urban space by the new poverty and the new marginality takes two forms: the tolerated ghettoes where marginalized people are permitted to stay, out of sight of the mainstream society; the open presence in the core area of cities of 'street people', a risky strategy, but at the same time a survival technique since only there do they exist, and thus only there can they relate to society, either looking for a chance or provoking a final blow.

Because the informational society concentrates wealth and power, while polarizing social groups according to their skills, unless deliberate policies correct the structural tendencies we are also witnessing the emergence of a social dualism that could ultimately lead to the formation of a dual city, a fundamental concept that I will characterize below, when considering the spatial consequences of the structural trends and social processes that I have proposed as constituting the framework that underlies the new historical dynamics of European cities.

THE SPATIAL TRANSFORMATION OF MAJOR CITIES

From the trends we have described stem a number of spatial phenomena that characterize the current structure of major metropolitan centres in Western Europe. These centres are formed by the uneasy articulation of various sociospatial forms and processes that I find useful to specify in their singularity, although it is obvious that they cannot be understood without reference to each other.

First of all, the national-international business centre is the economic engine of the city in the informational global economy. Without it, there is no wealth to be appropriated in a given urban space, and the crisis overwhelms any other project in the city, as survival becomes the obvious priority. The business centre is made up of an infrastructure of telecommunications, communications, urban services and office space, based upon technology and educational institutions. It thrives through information processing and control functions. It is sometimes complemented by tourism and travel facilities. It is the node of the space of flows that characterizes the dominant space of informational societies. That is, the abstract space constituted in the networks of exchange of capital flows, information flows, and decisions that link directional centres among themselves throughout the planet. Because the space of flows needs nodal points to organize its exchange, business centres and their ancillary functions constitute the localities of the space of flows. Such localities do not exist by themselves but by their connection to other similar localities organized in a network that forms the actual unit of management, innovation, and power.

Secondly, the informational society is not disincarnated. New elites make it work, although they do not necessarily base their power and wealth on majority ownership of the corporations. The new managerial technocratic political elite does however create exclusive spaces, as segregated and removed from the city at large as the bourgeois quarters of the industrial society. In European cities, unlike in America, the truly exclusive residential areas tend to appropriate urban culture and history, by locating in rehabilitated areas of the central city, emphasizing the basic fact that when domination is clearly established and enforced, the elite does not need to go into a suburban exile, as the weak and fearful American elite did to escape from the control of the urban population (with the significant exceptions of New York, San Francisco, and Boston).

Indeed, the suburban world of European cities is a socially diversified space, which is segmented in different peripheries around the central city. There are the traditional working class suburbs (either blue collar or white collar) of the well kept subsidized housing estates in home ownership. There are the new towns, inhabited by a young cohort of lower middle class, whose age made it difficult for them to penetrate the expensive housing market of the central city. And there are also the peripheral ghettoes of the older public housing estates where new immigrant populations and poor working families experience their exclusion from the city.

Suburbs are also the locus of industrial production in European cities, both for traditional manufacturing and for the new high technology industries that locate in new peripheries of the major metropolitan areas, close enough to the communication centres but removed from older industrial districts.

Central cities are still shaped by their history. Thus traditional working class neighbourhoods, increasingly populated by service workers rather than by industrial workers, constitute a distinctive space, a space that, because it is the most vulnerable, becomes the battleground between the redevelopment efforts of business and the upper middle class, and the invasion attempts of the countercultures trying to reappropriate the use value of the city. They often become defensive spaces for workers who have only their home to fight for, while at the same time meaningful popular neighbourhoods and likely bastions of xenophobia and localism.

The new professional middle class is torn between attraction to the peaceful comfort of the boring suburbs and the excitement of a hectic, and often too expensive, urban life. The structure of the household generally determines the spatial choice. The larger the role women play in the household, the more the proximity to jobs and urban services in the city makes central urban space attractive to the new middle class, triggering the process of gentrification of the central city. On the contrary, the more patriarchal is the middle class family, the more we are likely to observe the withdrawal to the suburb to raise children, all economic conditions being equal.

The central city is also the locus for the ghettoes of the new immigrants, linked to the underground economy, and to the networks of support and help needed to survive in a hostile society. Concentration of immigrants in some dilapidated urban areas in European cities is not the equivalent however to the experience of the American ghettoes, because the overwhelming majority of European ethnic minorities are workers, earning their living and raising their families, thus counting on a support structure that makes their ghettoes strong, family oriented communities, unlikely to be taken over by street crime.

It is in the core administrative and entertainment districts of European cities that urban marginality makes itself present. Its pervasive occupation of the busiest streets and public transport nodal points is a survival strategy deliberately designed so they can receive public attention or private business, be it welfare assistance, a drug transaction, a prostitution deal, or the customary police care.

Major European metropolitan centres present some variation around the structure of urban space we have outlined, depending upon their differential role in the European economy. The lower their position in the new informational network, the greater the difficulty of their transition from the industrial stage, and the more traditional will be their urban structure, with old established neighbourhoods and commercial quarters playing the determinant role in the dynamics of the city. On the other hand, the higher their position in the competitive structure of the new European economy, the greater the role of their advanced services in the business district, and the more intense will be the restructuring of the urban space. At the same time, in those cities where the new European society reallocates functions and people throughout the space, immigration, marginality and countercultures will be the most present, fighting over control of the territory, as identities become increasingly defined by the appropriation of space.

The critical factor in the new urban processes is, however, the fact that urban space is increasingly differentiated in social terms, while being functionally interrelated beyond physical contiguity. There follows a separation between symbolic meaning, location of functions, and the social appropriation of space in the metropolitan area. The transformation of European cities is inseparable from a deeper, structural transformation that affects urban forms and processes in advanced societies: the coming of the Informational City.

THE INFORMATIONAL CITY

The spatial evolution of European cities is a historically specific expression of a broader structural transformation of urban forms and processes that expresses the major social trends that I have presented as characterizing our historical epoch: the rise of the Informational City. By this concept I do not refer to the urban form resulting from the direct impact of information technologies on space. The Informational City is the urban expression of the whole matrix of determinations of the Informational Society, as the Industrial City was the spatial expression of the Industrial Society. The processes constituting the form and dynamics of this new urban structure, the Informational City, will be better understood by referring to the actual social and economic trends that are restructuring the territory. Thus the new international and interregional division of labour ushered in by the informational society leads, at the world level, to three simultaneous processes:

> The reinforcement of the metropolitan hierarchy exercised throughout the world by the main existing nodal centres, which use their informational potential and the new communication technologies to extend and deepen their global reach.
>
> The decline of the old dominant industrial regions that were not able to make a successful transition to the informational economy. This does not imply however that all traditional manufacturing cities are forced to decline: the examples of Dortmund or Barcelona show the possibility to rebound from the industrial past into an advanced producer services economy and high technology manufacturing.
>
> The emergence of new regions (such as the French Midi or Andalusia) or of new countries (e.g. the Asian Pacific) as dynamic economic centres, attracting capital, people, and commodities, thus recreating a new economic geography.

In the new economy, the productivity and competitiveness of regions and cities is determined by their ability to combine informational capacity, quality of life, and connectivity to the network of major metropolitan centres at the national and international levels.

The new spatial logic, characteristic of the Informational City, is determined by the preeminence of the space of flows over the space of places. By space of flows I refer to the system of exchanges of information, capital, and power that structures the basic processes of societies, economies and states between different localities, regardless of localization. I call it 'space' because it does have a spatial materiality: the directional centres located in a few selected areas of a few selected localities; the telecommunication system, dependent upon telecommunication facilities and services that are unevenly distributed in the space, thus marking a telecommunicated space; the advanced transportation system, that makes such nodal points dependent on major airports and airlines services, on freeway systems, on high speed trains; the security systems necessary to the protection of such directional spaces, surrounded by a potentially hostile world; and the symbolic marking of such spaces by the new monumentality of abstraction, making the locales of the space of flows meaningfully meaningless, both in their internal arrangement and in their architectural form. The space of flows, superseding the space of places, epitomizes the increasing differentiation between power and experience, the separation between meaning and function.

The Informational City is at the same time the Global City, as it articulates the directional functions of the global economy in a network of decision-making and information processing centres. Such globalization of urban forms and processes goes beyond the functional and the political, to influence consumption patterns, lifestyles, and formal symbolism.

Finally, the Informational City is also the Dual City. This is because the informational economy has a structural tendency to generate a polarized occupational structure, according to the informational capabilities of different social groups. Informational productivity at the top may incite structural unemployment at the bottom or downgrading of the social conditions of manual labour, particularly if the control of labour unions is weakened in the process and if the institutions of the welfare state are undermined by the concerted assault of conservative politics and libertarian ideology.

The filling of downgraded jobs by immigrant workers tends to reinforce the dualization of the urban social structure. In a parallel movement the age differential between an increasingly older native population in European cities and a younger population of newcomers and immigrants forms two extreme segments of citizens polarized simultaneously along lines of education, ethnicity, and age. There follows a potential surge of social tensions.

The necessary mixing of functions in the same metropolitan area leads to the attempt to preserve social segregation and functional differentiation through planning of the spatial layout of activities and residence, sometimes by public agencies, sometimes by the influence of real estate prices. There follows a formation of cities made up of spatially coexisting, socially exclusive groups and functions, that live in an increasingly uneasy tension vis-à-vis each other. Defensive spaces emerge as a result of the tension.

This leads to the fundamental urban dualism of our time. It opposes the cosmopolitanism of the elite, living on a daily connection to the whole world (functionally, socially, culturally), to the tribalism of local communities, retrenched in their spaces that they try to control as their last stand against the macro forces that shape their lives out of their reach. The fundamental dividing line in our cities is the inclusion of the cosmopolitans in the making of the new history while excluding the locals from the control of the global city to which ultimately their neighbourhoods belong.

Thus the Informational City, the Global City, and the Dual City are closely interrelated, forming the background of urban processes in Europe's major metropolitan centres. The fundamental issue at stake is the increasing lack of communication between the directional functions of the economy and the informational elite that performs such functions, on the one hand, and the locally oriented population that experiences an ever deeper identity crisis, on the other. The separation between function and meaning, translated into the tension between the space of flows and the space of places, could become a major destabilizing force in European cities, potentially ushering in a new type of urban crisis.

BACK TO THE FUTURE?

The most important challenge to be met in European cities, as well as in major cities throughout the world, is the articulation of the globally oriented economic functions of the city with the locally rooted society and culture. The separation between these two levels of our new reality leads to a structural urban schizophrenia that threatens our social equilibrium and our quality of life. Furthermore, the process of European integration forces a dramatic restructuring of political institutions, as national states see their functions gradually voided of relevance, pulled from the top toward supranational institutions and from the bottom toward increasing regional and local autonomy. Paradoxically, in an increasingly global economy and with the rise of the supranational state, local governments appear to be at the forefront of the process of management of the new urban contradictions and conflicts. National states are increasingly powerless to control the global economy, and at the same time they are not flexible enough to deal specifically with the problems generated in a given local society. Local governments seem to be equally powerless vis-à-vis the global trends but much more adaptable to the changing social, economic, and functional environment of cities.

The effectiveness of the political institutions of the new Europe will depend more on their capacity for negotiation and adaptation than on the amount of power that they command, since such power will be fragmented and shared across a variety of decision-making processes and organizations. Thus instead of trying to master the whole complexity of the new European society, governments will have to deal with specific sets of problems and goals in specific local circumstances. This is why local governments, in spite of their limited power, could be in fact the most adequate instances of management of these cities, working in the world economy and living in the local cultures. The strengthening of local governments is thus a precondition for the management of European cities. But local governments can only exercise such management potential if they engage in at least three fundamental policies:

The fostering of citizen participation, on the basis of strong local communities that feed local government with information, present their demands, and lay the ground for the legitimacy of local government so that they can become respected partners of the global forces operating in their territory.

The interconnection and cooperation between local governments throughout Europe, making it difficult for the global economic forces to play one government against the other, thus forcing the cooperation of global economy and local societies in a fruitful new social contract.

New information technologies should make possible a qualitative upgrading of the co-operation between local governments. A European Municipal Data Bank, and a network of instant communication between local leaders, could allow the formation of a true association of interests of the democratic representatives of the local populations. An electronically connected federation of quasi-free communes could pave the way for restoring social and political control over global powers in the informational age.

Managing the new urban contradictions at the local level by acting on the social trends that underlie such contradictions requires a vision of the new city and new society we have entered into, including the establishment of cooperative mechanisms with national governments and European institutions, beyond natural and healthy competition of parties. The local governments of the new Europe will have to do their homework in understanding their cities, if they are to assume the historical role that the surprising evolution of society could offer them.

Thus the historical specificity of European cities may be a fundamental asset in creating the conditions for managing the contradictions between the global and the local in the new context of the informational society. Because European cities have strong civil societies, rooted in an old history and a rich, diversified culture, they could stimulate citizen participation as a fundamental antidote against tribalism and alienation. And because the tradition of European cities as city states leading the pace to the modern age in much of Europe is engraved in the collective memory of their people, the revival of the city state could be the necessary complement to the expansion of a global economy and the creation of a European state. The old urban tradition of Amsterdam as a centre of politics, trade, culture and innovation, suddenly becomes more strategically important for the next stage of urban civilization than the meaningless suburban sprawl of high technology complexes that characterize the informational space in other areas of the world.

European cities, because they are cities and not just locales, could manage the articulation between the space of flows and the space of places, between function and experience, between power and culture, thus recreating the city of the future by building on the foundations of their past.

STEPHEN GRAHAM and SIMON MARVIN

"The Transformation of Cities: Towards Planetary Urban Networks" and "Telecommunications and Urban Futures"

from *Telecommunications and the City: Electronic Spaces Urban Places* (1996)

Editors' introduction Technology has always affected the development of cities, both physically and socially. Today, however, the new technologies of computing and telecommunications are bringing about truly extraordinary changes, which will unquestionably play a major part in shaping the urban future. According to Stephen Graham and Simon Marvin, both lecturers at the Centre for Urban Technology in the University of Newcastle Department of Town and Country Planning, the new telemediated city will bring about "a new type of urban world, not a post-urban world" in which the new urban form will be "an amalgam of urban places and electronic spaces."

In *Telecommunications and the City*, Graham and Marvin argue that telecommunications will affect urban development in a number of ways. Economically, they foresee the steady advance of globalized markets and new information-based means of production. "Investment in telematics," they write, already "surpasses investment in other industrial machinery." Socially and culturally, the new information-based economies will have destabilizing effects. Global marketing means intense global competition and may lead to sharp divisions between the information-rich and the information-poor trapped in "information ghettos."

Environmentally, telematics will contribute greatly to the efficient management of all kinds of infrastructural networks, but Graham and Marvin are skeptical about the beneficial effects of telecommuting as an alternative to traffic congestion. More broadly, they see telecommunications contributing greatly to the development of more sprawl in the "Edge City" technoburbs described by Robert Fishman (p. 77). "Core cities," they write, "are being turned into extended urban regions; these themselves now blend into wider megalopoli; at the final level, megalopoli merge into the planetary metropolitan system."

The impact of technology, especially telecommunications, on urban form and urban society has spawned an enormous body of literature. Foundational sources include Lewis Mumford's *Technics and Civilization* (New York: Harcourt Brace, 1934) and Marshall McLuhan's *Understanding Media: The Extensions of Man* (London: Sphere, 1964). Other seminal works include Manuel Castells' writings cited above (p. 557), William Mitchell, *City of Bits: Space, Place, and the Infobahn* (Cambridge, MA: MIT Press, 1995); Howard Rheingold, *The Virtual Community* (London: Secker and Warburg, 1994); and John E. Young, *Global Networks – Computers in a Sustainable Society* (Washington, DC: Worldwatch Paper 115, 1993).

Many more recent studies have been compiled in a number of excellent anthologies: J. Bird *et al.*

(eds), *Mapping the Futures: Local Cultures, Global Change* (London: Routledge, 1993), J. Brotchie *et al.* (eds.), *The Future of Urban Form: The Impact of New Technology* (London: Croom Helm and Nichols, 1985), *Cities of the Twenty-first Century* (London: Halsted, 1991), and J. Schmandt *et al.* (eds.), *The New Urban Infrastructure: Cities and Telecommunications* (London: Praeger, 1990). Graham and Marvin's own extensive bibliography in *Telecommunications and the City* is an invaluable resource.

THE TRANSFORMATION OF CITIES: TOWARDS PLANETARY URBAN NETWORKS

Economic restructuring, telecommunications and cities

The end of the long post-war boom in western capitalist society has triggered a massive restructuring which has radically altered cities. Globalisation and the intensification of global competition have torn away the traditional industrial fabric of many western cities through 'deindustrialisation'. Huge transfers of manufacturing activity have focused on less developed and newly industrialised countries (LDCs and NICs) creating a global division of labour. The vertically integrated manufacturing giants of the so-called 'Fordist' era are everywhere being replaced by more responsive and flexible networked corporations operating across these global distances and tending to buy in goods and services from small firms. In western cities, information, high-tech manufacturing, service and leisure industries have grown (albeit patchily), forcing great changes in urban labour markets and urban socioeconomic dynamics. Political shifts towards liberalisation and the growth of investment markets have led to a remarkable boom in financial services. This has fuelled the growth of the larger cities which are placed at the hubs of the global electronic and financial services networks.

The result of these shifts is that economic activity involving processing and adding value to knowledge and information now dominate the economics of western cities as never before. Even commodity-based industries such as retailing and manufacturing are increasingly information rich. With the unprecedented turbulence, competitiveness and volatility of markets, higher inputs of knowledge and information are being used to reduce uncertainty and improve responsiveness. Because information has such a central place in production in all sectors, it, too, is emerging as a key commodity to be bought, sold, traded and exchanged in markets. This is made possible by the capabilities that telematics bring for processing, storing and controlling vast flows of electronic information on a continuous and real-time basis. In short, as a result of all these shifts, industrial cities have been transformed into 'post industrial' or 'information' cities, dominated by consumption industries and the processing and circulation of knowledge and symbolic goods rather than physical goods. A corresponding shift has gone on in the labour markets of cities, with 60–70 per cent of new jobs typically now concerned with some form of information processing, distribution or production. Investment in telematics – the basic information infrastructures of cities – now surpasses investment in other industrial machinery. Because telematics make information so easy to move around, these shifts increasingly take a global complexion, tied in with the wider shifts towards globalisation and the growing power of TNCs. The result of the pervasive application of telematics across all economic sectors is that markets for computers and communications equipment are growing rapidly. This growth is dominated overwhelmingly by the three urban-industrial blocks which are the powerhouses of global capitalism: North America, Western Europe and Japan.

This economic transformation in cities, however, has been associated with the growth of structural unemployment. The growth of well-paid, knowledge-intensive jobs has been dwarfed by the loss of manufacturing jobs and the growth of poor quality jobs in retailing, leisure and tourism. Globalisation and the application of telematics also seem to be associated with the fracturing and disintegration of city economics, as they become 'exposed' to global telematics networks that invisibly and silently cross-cut them. The result is that cities are being restructured from internally integrated wholes to collections of units which operate as nodes on international, and, increasingly, global economic networks. It is increasingly impossible to understand the forces that are shaping the restructuring of cities from a purely local perspective; contemporary city economics can only be understood through their relations to global economic, political and technological changes. The instantaneity of telematics networks is a key facilitator of this linking of the 'local' into the 'global', largely through the construction of corporate telematics networks and global transport networks. As a result, cities are now being tied together with a new level simultaneity. The interaction between and within cities now approaches real time – or at least operates with an unprecedented velocity. The global urban world now operates as a vast set of international systems based on electronic telecommunications-based flows of information, money, services, labour power, commodities and images as well as by advanced transport networks . . .

But the fortunes of cities are very uneven within these shifts. The world financial capitals have emerged as key command and control centres where the best jobs are located. Certain smaller, usually non-industrial cities have managed to specialise in advanced manufacturing, research and development or high technology services. Others have emerged as key tourist centres. At the same time, however, many older industrial cities have had to compete even for low order service jobs such as those in back offices, branch plant manufacturing and shopping centres. In this context, because of the speed of these systems and the erosion of the attachment to place, city economics are more turbulent and face very uncertain futures.

Urban social and cultural change

Such economic restructuring and the telematics-based globalisation of cities has been associated with profound urban social and cultural change. On the social front, the urban economic crisis of the 1980s precipitated the end of the post-war welfare–Keynesian consensus and the ascendancy of neo-liberal approaches to urban management. In concert with the effects of economic restructuring, these changes have forced social and geographical polarisation within cities.

The ways in which new telecommunications and telematics innovations are involved in the social life of cities tends to both reflect and support this polarisation. While affluent and elite groups are beginning to orient themselves to the Internet and home informatics and telematics systems, other groups are excluded by price, lack of skills or threaten to be exploited at home by such new technologies. Advanced telecommunications and transport networks open up the world to be experienced as a single global system for some. But others remain physically trapped in 'information ghettos' where even the basic telephone connection is far from a universal luxury.

The growing divisions between affluent and poor areas can lead to rising fear of crime, the 'fortressing' of neighbourhoods through electronics surveillance systems, and an increasingly home-based urban culture where people's working, shopping, access to services and social interaction may become mediated more via telematics than by social interaction in the public spaces of cities. The parallel shift towards market-based telecommunications regimes has added further momentum to this polarisation and growing unevenness in the social landscapes of cities. It is clear that while many speak of 'time–space compression' through new technologies, it is important to differentiate between the diverse social experiences of these processes of change for different groups of people.

A major area of debate currently centres on the degree to which telematies can be used to support socially liberating and progressive changes within cities – by overcoming the isolation of disabled groups for example. Whilst there are certainly some examples of such liberating applications, for example with 'virtual communities' for marginalised and housebound groups, critical commentators stress that, on the whole, such technologies may in fact be a basis for exacerbating further the social and geographical polarisation within urban places. Current evidence certainly suggests that the future of those in cities who are at the margins of the information society seems grim. The hyped-up promises of technological fixes in the new telematics era threaten to have a distinctly hollow ring when considering the position of the most disadvantaged groups within the contemporary city.

These changes in the social dynamics of western cities are in turn interwoven with the cultural dimensions of globalisation, a process which is closely tied into wider shifts towards a global, 'post-modern' urban culture. For example, advances in telecommunications and telematics are, along with the liberalisation of media regulations, helping to support the emergence of truly global cultural and media industries. These feed in to support social and cultural change in international systems of cities. There is little doubt that the human experience of place and the social construction of cultural identities by groups and individuals are being radically altered in these 'global times'. This is because new advanced telecommunications act as conduits for flows of images, knowledge, information and symbols which integrate places and people into the global cultural system in 'real time'. Thus, traditional national mass broadcasting systems are giving way to a broadening range of global systems of mass communications (cable and satellite TV) and interactive personal communications (such as electronic mail and the World Wide Web on the Internet). Through these systems a growing proportion of social interaction and cultural flow take place.

But interpreting the implications and effects of these shifts remains highly contested. To some, shifts towards a more participative and interactive media culture through the explosion of the Internet and 'cyberspace' represent the empowerment and liberation of individuals and groups who were simply passive consumers of media in the past. To critics, though, the globalisation of urban culture is little more than the ruthless commodification of all information by fast-growing media conglomerates such as Rupert Murdoch's News International corporation. It is argued that this erases local differences; it superimposes western cultures over non-western ones; it polarises social access to information; and it leads to a cacophony of signs which often alienates urban inhabitants within an 'uprooted' and bewildering global culture.

Whether one stresses the positive or critical interpretations, or some blend of the two, there is little doubt that, in this movement-dominated post-modern world, the whole cultural idea of cities is being redefined. This transformation is creating a 'global sense of place' which challenges all previous ideas about what it means to live in a region, nation or city. But this does not simply erase the attachment of people to urban places. Instead, a complex interaction between telemediated cultural exchanges in electronic spaces and place-based ones in urban places is emerging. Networked communities of interest now span the globe, based on specialised foci of interest or various combinations of ethnicity, gender, sexuality, profession, etc. These mesh and interact with place-based communities, but in ways which we have only begun to understand.

Urban environments

These economic and social shifts have led to a growing concern amongst urban planners and policy-makers to address the *environmental* dimensions of their cities. Planners are trying to address the legacies of pollution and dereliction from the industrial era as well as the side-effects of burgeoning traffic congestion. The need to compete as an attractive business environment is joining with wider social awareness to force environmental issues to the top of the urban agenda. Concern now centres on the need for environmentally sustainable urban futures. Once again much attention here has turned to the potential role of telecommunications and telematics for contributing towards the development of more sustainable cities. Most often, however, this analysis has been hampered by the assumption that telematics-based flows of electronic information can be 'used directly to substitute for the environmentally damaging flows of physical transportation'. More critical scrutiny of such ideas casts doubt on such claims, however, because they rely on oversimplified conceptions of the relationships between urban environments and telecommunications.

Rather than simply substituting for transportation, telecommunications have a wide range of contradictory linkages. Telecommunications can help to stimulate more travel as cheaper and more accessible forms of communication generate new demands for the physical movement of people and goods. At the same time new services such as road information systems and auto route guidance can help drivers to overcome the uncertainties of traffic congestion, thereby improving the attractiveness of the road network. Although telecommuting and teleworking initiatives may be able to help reduce levels of peak-time congestion they have a multitude of second-order effects. The teleworker has to heat and power the home during the day, the road space created by the teleworker can quickly induce new traffic on to the road network while weakening the need to live in close physical proximity to work can encourage urban sprawl. Similar contradictions also occur in the energy, water and waste sectors. Telecommunications do not necessarily lead to reductions in material flows through cities. While the new control capabilities of telematics do have the potential to shape the level and/or time of resource consumption the current supply-oriented logic of network management does not necessarily provide much incentive to infrastructure to reduce total flows.

Urban transport and infrastructure

Telecommunications are radically transforming the management and provision of urban transportation and infrastructure networks. Old ideas about the role of monopolistic, standardised and universally available infrastructure networks available to meet all demands for movement, mobility, heat, power and water supply are being challenged. Increasingly telecommunications are supporting the splintering of infrastructure services to facilitate competition on what were previously considered to be monopolistic networks. Telecommunications enable infrastructure providers more effectively to control their networks by identifying the costs of servicing different types of customers. At the same time these capabilities provide the opportunity of providing premium, enhanced and value added services to particular groups of customers.

The application of telecommunications technologies in the management of networks is developing in tandem with shifts towards the liberalisation and privatisation of infrastructure networks. Privatisation has transformed service provision from monopolistic, universally available, standardised systems into complex new patterns of service provision. Increasing levels of choice for large users mean that incumbent operators are forced to remove cross subsidies from large to small users to compete with new entrants who cherry pick their most valuable customers. Telematics play a central role in these new logics of infrastructure management. For instance, Geographical Information Systems (GISs) provide the tools for allocating costs to different types of customer, allowing new entrants to target the most lucrative and profitable customers. The new control capabilities of telematics networks being fitted over what were previously 'dumb' infrastructure networks provide much more sophisticated control and operational systems. These can significantly improve the efficiency and profitability of networks by more effectively balancing demand and supply of services. New smart metering technologies enable premium customers to have increasing levels of choice while prepayment metering technology based on smart cards allows utilities to socially dump expensive, marginal and poor customers.

Telecommunications are also having profound implications within the transportation sector. Historically, there have always been close linkages between communication technologies and transportation. The telegraph and later the telephone enabled railway companies to standardise time while monitoring and controlling the movement of trains across their rail networks. In the contemporary city these linkages have significantly strengthened as the bundle of applications within the road transportation informatics technologies lead to important changes in the use and management of transportation networks. Electronic road pricing systems have been demonstrated in cities across Europe allowing transport authorities potentially to charge for road space according to levels of congestion in real time. The development of freight logistic systems linking together production and distribution sites electronically is enabling companies drastically to cut warehouse stocks and centralise warehouse and distribution functions, thereby more closely utilising the transport network as part of the production process. These companies are also relying more heavily on transportation informatics to maintain real time contact with drivers to route deliveries according to congestion levels and the demands of retailers. Public transport operators are also utilising telematics to provide real time information to users and to make efficient use of expensive transport capacity by more efficiently meeting demands.

Consequently, telematics are facilitating radical transformation in all types of infrastructure networks. New information and telematics technologies are helping to create new markets in infrastructure services, introduce competition onto networks, differentiate between particular types of customers and provide a wide range of value added services. But these new logics of network management hardly figure in the urban policy literature. With the exception of telecommunications networks themselves, there has been relatively little academic or policy interest in how the new capabilities of telecommunications are transforming the management and provision of urban networked services.

Urban physical form

Reflecting the economic, social and political restructuring of western cities, the urban *landscape and physical urban forms* of advanced industrial cities are in turn being radically reshaped. Global economic forces are taking over local property markets. Derelict or decaying old industrial spaces are being transformed into post-modern urban developments as foci of global consumption and culture. The sprouting of new telecommunications equipment, and the infusion of many new telecommunications infrastructures into the old fabric of the city, have been an essential part of this transformation. In addition the deconcentration of many cities, and the emergence of the multicentred urban area – what Jean Gottman (1983) called 'megalopolis' – has been, at least in part, facilitated by the new capabilities of telecommunications and telematics for supporting dispersed economic activities away from urban cores. Suburban office complexes, business and technology parks, out-of-town shopping malls and, increasingly, whole 'Edge Cities' are reshaping the physical layout of urban areas, using the combined decentralising power of automobiles and telematics. Core cities are being turned into extended urban regions; these themselves now blend into wider megalopoli; at the final level megalopoli merge into the planetary metropolitan system. Instead of single centres linked into some single central place hierarchy, then, very complex networks between cities are emerging based on complex complementary relationships. This trend is particularly advanced in North America. But this decentralisation does not represent the end of cities as we know it. Complex combinations of both decentralisation and centralisation are occurring simultaneously, with the world cities in particular the focus of new pressures for more centralisation because of telematics.

Urban planning, policy and governance

The overwhelming importance of the *economic* imperative in cities means that the increasing emphasis of urban governance is on public–private partnerships oriented towards an explicit economic development agenda rather than the social, redistributional one that characterised the post-war period. City authorities have been plunged into a new competitive era in which they act as 'urban entrepreneurs' in increasingly global 'marketplaces' for investment from multinational corporations, public agencies, media, sport and leisure corporations and tourists.

Because city economics today operate as fragmented collections of nodes on global networks, this fight for an improved nodal status is intense and very competitive. Increasingly, elected local governments work in corporatist ways with non-elected public agencies and local firms, utilities and property interests to fight for the regeneration of their cities. Telecommunications companies of all types are, once again, involved here as growing players in such local coalitions because they are dependent upon long-term revenues from the infrastructure they have, quite literally, sunk into the physical fabric of their home cities. Therefore they have much to gain from supporting the local growth of information-intensive economic sectors. The importance of telecommunications and telematics to the image and competitiveness of urban areas mean that they are a key focus of such entrepreneurial policy. These new approaches to urban government reflect the emergence of truly international systems of interlinked cities, where urban policy-makers need to consider the role of their city as a node on urban networks, mediated by advanced telecommunications and global transport systems. To parallel the globalisation of their markets, telecommunications companies are increasingly being set up on a global basis so as to meet the needs of multinational corporations for private, global networks.

Urban governance and public services are also being transformed through the application of telematics. The wider shift from Keynesian to individualistic and conservative welfare regimes is driving cost cutting, privatisation and restructuring in urban welfare services at the same time as increased polarisation is increasing demands for these services. In these circumstances, the talk is now to reinvent government more along business lines and to use telematics innovations as the new mechanism for delivering services with minimum costs and maximum flexibility. As a result, telematics-based restructuring in urban governments is burgeoning, often with the result of replacing physical, staffed service delivery offices with virtual and electronic ones. Many routine functions are also being outsourced to distant, even less developed country, locations.

[. . .]

TELECOMMUNICATIONS AND URBAN FUTURES

[. . .]

Our long journey through the complex terrain of city-telecommunications relations is now complete. We have built up a picture in unprecedented detail of how all aspects of the development, management and planning of cities increasingly relate to the pervasive application of telecommunications and telematics. Our perspective has been broad and comprehensive – some might say over-ambitious. We have critically reviewed the theoretical approaches that can be taken to city–telecommunications relations. Using this, we have constructed a basic framework blending the insights of political economy and social constructivism to build up the argument that contemporary cities can only be understood as parallel constructions within both urban place and electronic space. Without understanding both, and the many interaction points between them, we believe that we will never be able to approach or understand the totality of the current transformation underway in advanced capitalist cities . . .

This book suggests that the understanding of the contemporary city requires that one should grasp the complex interactions between urban places – as fixed sites which 'hold down' social, economic and cultural life – and electronic spaces, with their diverse flows of information, capital, services, labour and media which flit through urban places on their instantaneous paths across geographical space.

A new type of urban world, not a post-urban world . . .

Clearly, the growth of electronic spaces is not somehow leading to the dissolution of cities, as so often argued by futurists and utopianists. Urban functions are not being completely substituted by demateralised activities operating entirely within electronic spaces. In fact, this view is naive, shortsighted and dangerous; it perpetuates simplistic ideas about cities and telecommunications and undermines the potential for critical and sophisticated policy debates. We live in a fundamentally urban civilisation: cities as places still matter and will continue to matter. Urban places remain the unique arenas which bring together the webs of relations and 'externalities' that sustain global capitalism. They are of fundamental importance as the terrain for social and cultural life; they house the vast majority of our population; and they seem likely to remain the key economic, social, physical, cultural and political concentrations of advanced capitalist society. Much of what goes on in cities cannot be telemediated . . . Urban places and electronic spaces can be seen to influence and shape each other, to be recursively linked; it is this recursive interaction which will define the future of cities.

In fact, this interaction is shifting us to a new type of urban world rather than a post-urban world and there will be complex terrains of winners and losers in this process. To discredit further the utopian dreams of some new technological rural idyll, what is emerging is a *more totally urbanised* world, where rural spaces and lifestyles are being drawn into an urban realm because of the time–space transcending capabilities of telecommunications and fast transportation networks. This represents not the death of cities and the renaissance of genuinely rural ways of life, but the emergence of a 'super-urban' and 'super-industrial' capitalist society operating via global networks . . .

There is also little evidence that transport demand and physical flows of people and goods are actually slackening to any significant extent as a result of the use of telecommunications as substitution – even though telecommunications demand is burgeoning. Demand for both transportation and telecommunications within and between cities is growing rapidly. This suggests that the trend is towards both a movement and communications intensive society based on growing flows of goods, people, services, information, data and images. Both, in fact, tend to feed off each other in positive ways. The main focus of innovations in teleworking is around the major cities, not in widely dispersed electronic cottages. Most teleworkers are still reliant on transport links to and from work at some stage during their work routines. What is happening is the use of combinations of transport and telematics innovations to make work processes more flexible in time and space. While this may reconfigure transport patterns within and between cities, there seems little sign yet that it will begin to undermine the practice of travel.

On the other hand, though, we do not argue that cities are unaffected by the remarkable extension of electronic spaces; their pervasive growth critically affects all aspects of urban development. New conceptual treatments of the 'city' and the 'urban' are required, raising many questions as to how cities can sustain the transition to telemediation in a progressive way. Of key importance here is the inherent logic of polarisation, which seems to be locked into current processes of economic and social development in cities. This polarisation is both reflected in, and supported and reinforced by, the development of electronic spaces. Fewer city economies seem set to do well; patterns of economic health become more starkly uneven at all spatial scales; and processes of change seem to reinforce the privilege and power of social elites while marginalising, excluding and controlling larger and larger proportions of the population of cities.

Clearly these shifts represent a transformation of society as much as a transformation of cities. Important questions are raised about what words like 'city' and 'urban' can actually mean in a world where urbanity seems almost total; where globalisation and global networks seem to undermine the notion that what goes on in cities is radically different from the processes elsewhere; and where electronic spaces provide the 'sites' for a growing proportion of urban life.

The city as an amalgam of urban places and electronic spaces

This book has shown how the contemporary city is, more than ever before, an amalgam whereby the fixed and tangible aspects of familiar urban life interact continuously with the electronic and the intangible. We have explored many examples across all aspects of urban life whereby superimposed systems combining presence in urban places with interactions in electronic spaces are being constructed – witness the electronic financial markets, the back-office networks, the global media flows, the 'intelligent building', 'intelligent city' and 'smart home' debates, the surveillance networks, the transport–telecom interactions, the virtual communities and the civic electronic spaces. Fixed constructions of places and buildings in urban places linked into electronic networks and 'spaces' seem to define contemporary urbanism. Telematics – the supports for electronic spaces – are woven increasingly into the built environments of cities. They are also filling the corridors between them as key infrastructures to underpin the shift to global urban and infrastructural networks. The social and cultural aspects of urban life also operate increasingly through constructions which meld the built environments of cities with uses and applications of telecommunications.

Together, the diverse electronic spaces that we have explored amount to a hidden and parallel universe of buzzing electronic networks. Largely free of space and time constraints, these interact with and impinge on the tangible and familiar dynamics of urban life on a twenty-four-hour-a-day basis and at all geographical scales. Thus, a car, rail, plane or bus journey, and the physical flows of water, commodities, manufactured goods and energy are supported by a parallel electronic 'networld'. These monitor, shape and control the physical flows underway on a real-time basis. Traffic snarl-ups are now the launch pads for countless electronic conversations and interactions. The apparently lifeless world of the office block hides an 'intelligent building' – a hub in an electronic universe of twenty-four-hour-a-day global flows of capital, services and labour power. The daily life of an urban resident leaves a continuous set of 'digital images' as it is mapped out by a wide array of surveillance systems – closed circuit TV cameras, electronic transaction systems, road transport informatics and the like. The fortressing of affluent neighbourhoods relies on old-fashioned walls and gates linked into sophisticated electronic surveillance systems. The most ordinary suburbs of most cities now act as hubs in the growing electronic cacophony of global image and media flows and the ongoing participation of people in virtual communities, often on a global basis. Urban policies and strategies are increasingly directed to try to shape both urban places and electronic spaces. This shadowy world of electronic spaces exists through the instantaneous flows of electrons and photons within cities and across planetary metropolitan networks, which, unseen, underpin virtually all that we experience in our daily lives.

Not surprisingly, this world of electronic spaces is as diverse and complex as the landscapes and life of the actual cities. Like the geographical landscapes of cities, there are many segmentations, divisions and social struggles underway over the definition and shaping of electronic space. There are 'information black holes' and 'electronic ghettos' where the poor remain confined to the traditional marginalised life of the physically confined, and there are intense concentrations of infrastructure in city centres and elite suburbs supporting the corporate classes and transnational corporations. As with geographical landscapes, the results can be 'read' as reflections of complex processes whereby social, ethnic, gender and power relations play out against the backdrop of the globalising political economy of capitalism.

The proliferation of electronic spaces seems to be heavily involved as we move from the standardised, rigid, hierarchical and rhythmic world of the industrial age to new and much more fluid societal processes. Telecommunications and electronic spaces bring profoundly new relationships between cities and the fundamental matrices of space, time and power within a globalising capitalist society. We feel that they are implicated as central underpinnings to the changing nature of capitalism and the capitalist city – not as technologies 'impacting' autonomously on cities but as social constructions that are being used to explore new ways of controlling and organising space, time and social processes in a crisis-ridden urban world.

These complex interactions between urban places and electronic spaces challenge the simplicity of many widely held assumptions about telecommunications, notably those stemming from technological determinism and futurism, which are widely infused into research on telecommunications and cities. In much urban telecommunications research, we diagnose the use of many over-simplified and 'autonomous' notions of technology, unjustified assumptions and a dearth of real empirical analysis. As we have seen, neither the urban studies nor the urban policy communities have made much progress in understanding these dynamics with the sophistication they demand. Most approaches still only accommodate urban places and the tangible social and economic dynamics of cities. Mechanical, Euclidean and Cartesian notions of urban development, distance, time and space are often still privileged over 'electronic' ones, which accommodate the interrelationships between electronic spaces and other forms of space and time. Remarkably few urban commentators even mention telecommunications and telematics; when they do, the treatment tends to be rather simplistic. Old-fashioned ideas of the development of cities that are separated in 'Euclidian' space and shaped by the physical 'friction of distance' surrounding them often still implicitly dominate and are used to try and accommodate the effects of telecommunications. Concepts that capture the free and flexible meshing of times and spaces together into global networks, and the often bewildering processes of time–space compression which are associated with telematics, are still poorly developed.

Overcoming the myths of determinism: contingency with bias

To improve our understanding of the recursive interactions between cities and telecommunications, we must leave behind the myths of determinism, both technological and social. The assumption that telecommunications impact in some simple, universal and linear way on cities is still common; the stress that they simply reflect some abstract capitalist political economy is still prevalent in some critical literatures. Both, we argue, are unhelpful. An integrated perspective of cities and telecommunications teaches us that social action shapes telecommunications applications in cities in diverse and contingent ways, even if this goes on against the backcloth of broader political economic trends. New telecommunications technologies bring new options and capabilities within which urban processes can be shaped. Much of what we have seen in this book represents a range of efforts to address problems and crises in this way. But the complex interactions between the social and technological lead to diverse effects, some intended, some unintended, and to the emergence of new problems. Social conflict and struggle between unevenly equipped groups and organisations is a key feature of the processes at work . . .

These struggles inevitably bias the design and application of telecommunications in cities. Transnational corporations gain access to optic fibre and private networks; people in disadvantaged ghettos are lucky to access a pay phone. But this bias does not shape and determine all the applications and developments of the technology in all cities. Once technologies are available, political and social struggle and actions can redirect their application and change their effects – just as political and social influences can redirect the shaping of urban politics and the built environments of the urban places. Thus, similar technologies can be used in very different ways and have very different effects. The effects of telecommunications on cities also seem to vary considerably across time and space. They do not conform to the simplistic and often naive models of technological forecasters where cities were assumed to be 'impacted' universally, linearly and directly by telecommunications. In other words, the implications of telecommunications for cities are indeterminate. They are not predefined either through some abstract technological 'logic' or by some mechanistic political economy imprinting itself on to society. These effects tend to be evolutionary rather than revolutionary. Like all technologies, telecommunications and telematics have influences only through their involvement in the wider process of social, political-economic and geographical transformation . . .

This leads to complex and apparently contradictory situations. The same technologies can be applied to empower and assist disadvantaged groups as well as to disenfranchise or exploit them. There are many examples of disabled people, women and ethnic minorities benefiting substantially from telematics. Telematics can be used to strengthen the public, local and civic dimensions of cities as well as to support social fragmentation and atomisation. They can help improve urban public transport or further the domination of cars in cities. They can assist in the search for sustainable models of urban development or help maintain the growth of highly unsustainable cities. And profoundly different political styles of urban government can all develop, each using telematics in particular ways to support their approach. The clear lesson is that telecommunications and telematics are nothing if not flexible.

The effects of telecommunications within cities are therefore complex and ambivalent. They allow urban infrastructures and transport systems actually to extend their capacities, so removing barriers to further urban growth and concentration. Telecommunications are supporting environmentally damaging increases in transport flows as well as promising assistance in the drive towards sustainable cities through telecommuting and reduced transport flows. Telecommunications assist the globalisation of the economy in which city economics are being fragmented as nodes on the global telecommunications networks of transnational corporations. They also provide new policy tools of urban management at the local level through which the public, civic face of cities can be strengthened and local economic and social cohesion supported. On the social front, telecommunications can help overcome isolation, disadvantage and disability, as well as furthering the degree to which the 'information poor' are marginalised and exploited. On the one hand, new telematics services promise a global playground to the affluent elites who are 'switched into' broadband cable TV and telematics systems and whose lives are saturated with these technologies at home and work. A short distance away, however, there are invariably ghettos of the so-called 'information poor' who fail to benefit even from the supposedly 'universal' basic telephone service – that most rudimentary point of access to the so-called 'information society'. Many households in inner cities without the basic phone are actually surrounded in the physical sense by lattices of sophisticated optic fibre networks supporting corporate telecommunications flows from which they are totally excluded. The nature of telematics means that physical proximity has very little to do with electronic proximity or access.

Telecommunications and urban futures

This leads us to the future of urban places and electronic spaces. Debates about the future of cities and urban society, while gaining ground, are currently somewhat stifled. Futurism is often discredited; the fall-out from the failures of many futuristic and modernist urban plans of the 1960s and 1970s continues. Faith in science and technology as redeeming forces has long since withered with the collapse of modernist assumptions about progress, knowledge and technical rationality. The rate and complexity of change in contemporary cities makes even more daunting the task of the extrapolation of these processes and the prediction of urban futures. Given the crisis of urban social polarisation, the many questions over the future of urban economics, the continuing environmental crisis in cities, and the radical shifts underway in urban policy and planning, it is not surprising that looking into the future often seems something of a luxury from current standpoints. The 'paradigm challenge' brought by the proliferation of electronic spaces further compounds these problems. The inertia and dominance of anachronistic ideas about cities are deeply rooted; the barriers facing sophisticated analysis and intervention in urban telecommunications are daunting. And these problems make it even more difficult to be normative and suggest what kinds of cities and electronic spaces we want and how these urban ideals may be brought to fruition. Of course the answers to such questions presume that there is such a group which can be easily identified from the many disparate and conflicting bodies that make up cities. Perhaps debate should start here by exploring how this 'we' might first be identified.

Thankfully, these issues lie beyond the scope of this book. While aiming to be comprehensive, this book represents only a small part of this much wider project: to reconfigure urban studies and urban policy-making in ways which directly reflect the increasingly tele-mediated nature of contemporary cities and so help reinvigorate debates about urban futures. The overwhelming conclusion of this book is that a concerted research drive is urgently needed if telecommunications are to take their appropriate place at the centre of current conceptions and understanding about the development, planning and management of cities. Critical urban research now needs to turn to the complex interactions between electronic spaces and urban places.

DAVID CLARK

"The Future Urban World"

from *Urban World/Global City* (1996)

Editors' introduction Ever since the earliest urban concentrations were located along major rivers, allowing for the construction of extensive irrigation systems for the production of a food supply, surrounding environmental conditions have influenced the development of cities.

Today, global environmental conditions are an increasingly important constraint on the size and direction of future urban development. Many urban theorists are now increasingly persuaded that cities are imbedded in and dependent on the surrounding rural environments and on land, water, and atmospheric resources that are regional, continental, and even global in scope.

David Clark, head of the Department of Geography at the University of Coventry, argues that urban life in the twenty-first century "can only be achieved and maintained through careful management and planning of resources at both global and local scales." In "The Future Urban World" he surveys the full range of environmental challenges to future urbanization and calls for sweeping international agreements, such as those originally put forward at the Rio de Janeiro "Earth Summit" in 1992, if "a sustainable urban future for all is to be secured." "There is ample evidence," he writes, "that the global physical environment is being degraded and that many major cities are near to exhausting their abilities to cope with their exploding populations."

Clark is a careful scientist, not an alarmist. He recognizes that population predictions are highly speculative and that many doomsday scenarios of the recent past – such as predictions that there would be global mass starvation by the 1980s! – have proven to be false. Further, he admits that much of the scientific evidence for the dire effects of global warming are "equivocal" and that there are grounds for believing that some of the more pessimistic arguments found in the environmental and sustainable planning literature "have been overstated." Nonetheless, he suggests that certain real concerns – unchecked sprawl and the disappearance of green space, the "sheer pressure of numbers" on urban infrastructures and basic services, and the probable increase of greenhouses gases, sulfur dioxide, and particulate matter emissions – are issues that will require efficient regulation and "transfrontier responsibility." "If urban life is to be sustained much beyond the present century, then steps must be taken now so as to prevent further damage to the environment and to bequeath resources to succeeding generations."

For other views on sustainable planning, see Sym Van der Ryn and Stuart Cowan's "An Introduction to Ecological Design" (p. 519) and Stephen Wheeler's "Planning Sustainable and Livable Cities" (p. 434). Many of the most important sources of information on the influence of the global environment on future urban development are included in the references at the end of this selection from Clark's *Urban World/Global City* (London and New York: Routledge, 1996), and that book also contains an excellent and thorough bibliography of recent writings on other aspects of global urbanism – population growth and urbanization, the growth of world cities, social patterns, and cultural changes. Other useful sources include Mark Sagoff's *The Economy of Earth* (Cambridge: Cambridge University Press, 1988), and Joel Cohen's *How Many People can the Earth Support?* (New York: Norton, 1995).

It is appropriate in a book which adopts a broadly historical approach to conclude with a brief consideration of how the urban world may evolve in the near future. Forecasting population levels, distributions and socioeconomic conditions in a single country is a difficult enough task, and attempts to undertake this sort of exercise at the global scale can only yield predictions which are little more than guesstimates. This chapter is therefore grounded in speculation rather than in detailed analysis. What is clear, however, because they are products of long-term and deep-seated processes which have yet to run their course, is that urban growth and urbanisation will lead to further significant urban development. Most will be in those parts of the world which are presently classified as developing. Even over the next quarter of a century the changes which are expected are staggering. Extrapolation of current trends suggests that the number of people who presently (1996) live in urban places is likely to double by the year 2025.

The scale of urban development which these figures imply raises important questions as to whether such a geographical pattern can be supported. It is difficult to imagine a world with twice as many urban residents as today, and it is important to focus on issues of maintenance and sustainability. Cities are elements in global economic and environmental systems which are both vulnerable and fragile. Although they represent a highly efficient use of space and provide unrivaled opportunities for production and social interaction, they consume prodigious amounts of finite resources, far more than a rural population of equivalent size. There are grave doubts as to whether future cities can be sustained in economic terms, how the population can be fed and how they can generate and distribute sufficient wealth to support their residents at acceptable standards of living. A parallel concern is with the ecological implications of further urban development. Cities are widely seen as being parasitic in that they draw air and water from the natural environment and generate large quantities of pollution and waste. Little is reused and recycled. Many would argue that their emissions are progressively destroying the global environmental systems upon which life on the planet, and hence their own viability, depends. The prospect for further massive urban development necessarily focuses attention upon the implications for the environment and whether the urban future is sustainable.

Such questions raise a further set of issues concerning the need for regulation today so as to ensure the continuation of cities into the future. If urban life is to be sustained much beyond the present century, then steps must be taken now so as to prevent further damage to the environment and to bequeath adequate resources to succeeding generations. Action is required at the global scale to cut harmful emissions and to prevent indiscriminate and unnecessary exploitation of scarce resources. Within national boundaries, there is a need to deal with local sustainability issues including urban servicing and waste disposal. Agendas for such intervention are presently emerging although these are more high-level statements of intent than examples of concerted and effective action. The urban future is likely to depend as much upon the success of international agencies and governments in shaping urban development as it is on the unregulated growth and redistribution of the population.

THE URBAN FUTURE

The direction and scale of contemporary urban growth and urbanisation point to the emergence by 2025 of an urban world that will bear little resemblance to the urban present. This much is certain, but filling in the detail by country and by region is problematical because of the deficiencies of the data. A related difficulty is methodological and concerns the interpretation of past trends and the ways in which they are used in forecasting. Urban development is too sensitive to economic, social and environmental change to predict more than one or two decades into the future (Hardoy and Satterthwaite, 1990). An example of the very different pictures which can emerge is provided by the projections for city populations in the year 2000 that were published by the United Nations between 1973 and 1984. The projected population of Mexico City in the year 2000 was 31.6 million in the Population Division's 1973–5 assessment, but was much lower at 25.8 million in the 1984–5 forecast. Similarly, the 1973–5 projection for the population of Beijing in 2000 was 19.1 million, but this had been revised downwards to 11.3 million in 1984–5. The scale of revision by the leading statistical agency underlines the innate difficulties which are involved in long-term urban forecasting.

The most recent estimate by the United Nations (1991) is that by the year 2025 there will be some 5.5 billion people, out of a world population of 8.5 billion, living in urban places. This future urban population is roughly the same as the total population of the world today. Some 4.4 billion will be living in towns and cities in what are presently classified as developing countries. The population of China's urban places will be close to 1 billion and India's is expected to be some 740 million.

Urban growth will be accompanied by increased urbanisation. Some 65 per cent of the world's population is expected to be urban by the year 2025. It follows . . . that this increase will occur principally because of the urbanisation of the population across large parts of Africa and Asia. These regions will be most radically affected by urban development, both urban growth and urbanisation, in the next quarter century.

The most striking feature of the predicted urban geography of the year 2025 is the uniformly high level of urban development in the Americas. The population of all of the principal countries of North, Central and South America is expected to be over 60 per cent urban and in most it will be in excess of 80 per cent. The Americas are presently highly urbanised so this change represents a consolidation of existing patterns. Similarly high levels of urban development are anticipated in Australia, Japan, parts of the Middle East, North Africa and most of Europe. Urbanisation levels in excess of 60 per cent are expected across the whole of Asia north of the Himalayas.

The United Nations forecast suggests that levels of urban development in Africa and southern Asia will be very much higher than today, but will vary considerably from country to country. Although the population in most countries will be more urban than rural, the proportion living in towns and cities in Burundi, Malawi, Rwanda, Ethiopia, Uganda, Burkina Faso, Afghanistan, Nepal, Bhutan, Cambodia, East Timor and Papua New Guinea is expected to be less than 40 per cent. Such countries have yet to go through the phase of rapid urbanisation that is a characteristic feature of the cycle of urban development. They will be the world's last remaining rural territories. At 51 per cent, India is expected to be only marginally more urban than rural in 2025.

If migration continues as it has in the past, and there is presently no suggestion that it will not, many who move from rural areas to urban places will go to the largest cities. The number of mega-cities is expected to rise to 28 in 2000 and to be more than 60 in 2025 (United Nations, 1991). Strong patterns of primacy already exist in many African countries and these are likely to be reinforced rather than reduced by urbanisation and urban growth. An important consequence of continued primacy is the further polarisation of the settlement hierarchy . . . A widening of the gap between town and country is expected to be one of the major consequences of the urban transition in Africa over the next thirty years.

The urban geography of the developed world is likely to be very different. Here, urban populations are presently high and the principal shifts will take place among cities. Rather than a concentration in a small number of large cities, which is the current pattern, the population is expected to be more evenly spread across many smaller centres. Cities of about 200,000 are likely to be the most attractive, as they are large enough to sustain an acceptable range of services without the congestion and pollution that are associated with life in the mega-city. As the benefits of centrality and agglomeration lessen further, many people will be drawn to places which may be too small, in future terms, to merit designation as urban. The spatial structure of large cities is likely to be transformed by shifts in population and economic activity. Central area densities are likely to fall significantly as people and businesses move to the suburbs and beyond. Many of the areas which are vacated will become parkland and open space, so that cities will take on a 'doughnut' form. In the longer term, however, it is possible that central area populations will rise as cities begin to reurbanise. Decentralisation of population at the local scale, and deconcentration at the national level will significantly reduce urban/rural differences, so producing a 'rurban' arrangement. The expected trends in different parts of the world will lead to a progressive inversion of the contemporary urban pattern at the global scale. A small-city and rural orientation will increasingly characterise the landscape in developed economies but strongly urban-centred mega-city societies will emerge and predominate in the developing world.

ISSUES OF SUSTAINABILITY

The preceding forecasts paint a disturbing picture of major population increase and further massive urban development in the first half of the next century. They point to a future in which the population in most parts of the world will live in urban places and in many cases in mega-cities. Such is the scale of urban development which is predicted, however, that it raises questions as to whether urban development of this magnitude can be sustained. It is difficult, given the current pressures on resources and the environment, to see how urban populations can double over the next twenty-five years without some form of economic or ecological breakdown. How will the urban population be fed, and what effects will mass concentration of population have upon the global environment and upon local ecosystems? Posing such questions is easy but arriving at plausible answers is much more difficult. The debate on sustainability is comparatively new and is strongly infused with speculation and conjecture. Hard information is lacking and the significance of such scientific evidence as is available, as for example on the scale of ozone depletion and global warming, is much disputed. Particular difficulties surround the availability of data for developing countries where environmental monitoring is in its infancy. Such problems do not of themselves negate discussion but they mean that it must be conducted at a very general level. It follows that few clear conclusions can be reached that might assist urban planners and managers.

A general concern with sustainable development emerged as a key issue with the heightened environmental awareness of the late 1980s. It gave rise to the emergence of 'green' parties in many developed countries in which the membership was committed to forcing environmental issues to the centre of the political debate. Evidence for global warming, the depletion of the ozone layer, and the detrimental effect of acid rain highlighted the need for urgent action to prevent further environmental degradation. Sustainable development can be interpreted in a number of different ways; indeed Pearce, Markandya and Barbier (1989) in an appendix to their book *Blueprint for a Green Economy* quote twenty-four separate definitions. The most widely accepted is that of the report of the World Commission on Environment and Development (WCED, 1987), also known as the Brundtland Commission, which defined sustainable development as 'development which meets the needs of the present without compromising the ability of future generations to meet their own needs' (p. 43). It further elevated the notion of sustainable development to the level of an operational concept which embodied the principles and values which it saw as desirable and necessary so as to deal effectively with the crisis of the environment and the development process. Emphasis was placed on the need for action today so as to provide for economic and ecological viability tomorrow.

Haughton and Hunter (1994) argue that the concept of sustainability involves three major principles. The first is that of 'intergenerational equity' and concerns the legacy which is left to future generations. It argues that the success of cities in the future depends to a large extent upon the assets and resources that are available and it is therefore incumbent upon the current generation not to indulge in indiscriminate and wasteful consumption. A sustainable future requires that national capital assets of at least equal value to those of the present are passed on to succeeding generations. The second is that a fair and equitable use of present resources is clearly necessary and this is enshrined in the principle of 'social justice'. Some form of central control over access to and use of resources is implied. The fact that both resources and consumption are widely distributed and are interdependent means that such management must be at a broad scale. A third precondition for sustainable development is that of 'transfrontier responsibility' insofar as key issues such as pollution, waste disposal and climatic warming are not constrained by national or regional boundaries but are essentially global in cause and consequence.

Against this background, Blowers (1993) identifies five fundamental goals that should guide all decisions concerning future development so as to ensure sustainability. The first concerns conservation and involves the need to ensure the supply of natural resources for present and future generations through the efficient use of land, less wasteful use of non-renewable resources, their replacement by renewable resources wherever possible, and the maintenance of biological diversity. The second concerns the use of physical resources and their impact on the land. It seeks to ensure that development and the use of the built environment is in harmony with the natural environment and that the relationship between the two is one of balance and mutual enhancement. A third goal is to prevent or reduce the processes that downgrade or pollute the environment and to promote the regenerative capacity of ecosystems. The final two goals are social and political in character. The aim of goal four is to prevent any development that increases the gap between rich and poor and to encourage development that reduces social inequality. The final goal is to change attitudes, values and behaviour by encouraging increased participation in political decision-making and in initiating environmental improvements at all levels from the local community upwards.

[. . .]

Some indication of the highly detrimental effect of cities is provided by measures of the levels of suspended particulate matter such as soil, soot, smoke, metals and acids which are found over cities as opposed to adjacent rural areas. Goudie (1989) reports the average concentration of suspended particulate matter found in the commercial areas of a number of cities as 400 micrograms per cubic metre in Calcutta, 170 in Madrid and Prague, 147 in Zagreb, 43 in Tokyo, and 24 in London and Brussels. These values compare with concentrations of less than 10 micrograms per cubic metre in rural areas.

An estimated 1.4 billion urban residents world-wide are exposed to averages for suspended particulate matter or sulphur dioxide (or both) that are higher than the levels recommended by the World Health Organisation. Research reported by Hardoy, Mitlin and Satterthwaite (1992: p. 79) underlines the severity of the pollution which hangs over many Third World cities. For example, in Shanghai there are seven power stations, eight steel works, 8,000 industrial boilers, 1,000 kilns, 15,000 restaurant stoves and one million cooking stoves, most using coal with a high sulphur content. In 1991 the annual average concentration of sulphur dioxide in the urban core was more than twice the recommended level. The annual average for total suspended solids was more than four times that recommended. Similar situations exist in Sao Paulo and Bangkok, where suspended particulate matter routinely exceeds recommended levels at all the monitoring stations in the city. The effects of emission may be compounded locally by physical geography. Santiago has one of the highest levels of air pollution, because the surrounding mountains impede natural ventilation.

[. . .]

The concern for the sustainability of cities has been expressed at two levels. The first is global and involves a wide range of issues surrounding the long-term stability of the earth's environment and the implication for cities. It is clear that the world's cities cannot remain prosperous if the aggregate impact of their economies' production and their inhabitants' consumption draws on global resources at unsustainable rates and deposits wastes in global sinks at levels which lead to detrimental climatic change. The second is local and involves the possibility that urban life could be undermined from within because of congestion, pollution and waste generation and their accompanying social and economic consequences. These different concerns focus attention upon the need for intervention at an international scale by governments working together on agreed programmes, and at the domestic level by city authorities addressing the local sustainability issues over which they can exercise some control.

There is growing evidence from climatological research that the earth's atmosphere is being degraded to an unacceptable extent, with serious implications for life on the planet. There are particular concerns for the well-being of the global climate which it is believed is being threatened by the depletion of the ozone layer and by atmospheric warming. The layer of ozone which exists in the upper atmosphere is of vital importance in the global energy balance because it reduces the amount of harmful solar radiation which is received at the earth's surface. Ozone occurs when oxygen reacts with ultra-violet light to give a molecule of three oxygen atoms, and its concentration and distribution within the lower stratosphere remain roughly constant under normal conditions. There is growing evidence, however, that the natural cycle of creation and breakdown of ozone in the upper atmosphere has been seriously interrupted by certain compounds, especially chlorofluorocarbons (CFCs). These chemicals are widely used in refrigeration, aerosols, packaging and cleaning, and levels of production have increased significantly in recent years as these applications have grown. CFCs live for a long time and have accumulated in large concentrations in the lower stratosphere where they are thought to have caused a general thinning of the ozone layer and the appearance of ozone holes. The most extensive hole is that over the Antarctic and there are indications that it is increasing in size and may now cover parts of Australasia and South America. Severe ozone depletion has also been observed during winter months in middle and high latitudes in the northern hemisphere (Tolba and El-Kholy, 1992).

It is widely believed that ozone depletion has led to a rise in ultraviolet radiation, which in turn is affecting human health and is threatening many natural and semi-natural ecosystems. The magnitudes are difficult to establish but a conservative estimate is that a 1 per cent reduction in stratospheric ozone leads to a 3–4 per cent increase in non-melanoma skin cancers (Turner, Pearce and Bateman, 1994). Sunbathers in areas where the ozone layer is thinnest are at greatest risk. Ozone depletion is also thought to cause eye damage and to suppress people's immune systems. The yield of some commercial food crops may be reduced, as may fish stocks (Tolba and El-Kholy, 1992).

A second set of pollution-related changes is believed to be raising average temperatures across the world, with far-reaching implications for climate, global sea levels and the functioning of local ecosystems. Global temperatures are regulated by a layer of natural greenhouse gases in the atmosphere, including water vapour, carbon dioxide, methane and nitrous oxide. These trap longwave radiation emitted by the earth and reflect some of it back to the surface in the form of heat. There is now compelling evidence that a build-up of pollution in the atmosphere has compounded the greenhouse effect and has caused long-term global warming (O'Riordan, 1989). The principal pollutant is carbon dioxide, which is produced during the burning of fossil fuels, but CFCs are important absorbers of longwave radiation as well. Atmospheric levels of carbon dioxide have risen by around one quarter over the past two centuries, with about half of the increase occurring in the last forty years (Kelly and Karas, 1990). Historically, emissions were higher in the developed world but the fastest growth today is occurring in developing countries in association with coal-fired heating, inefficient power stations, and the rise in the number of motor vehicles.

Although the scientific evidence on the scale of change is equivocal, global warming is widely seen as an increasing long-term threat. The environment is finely balanced and a rise in average global temperatures of as little as 1 per cent could melt polar icecaps sufficiently to raise sea levels across the world by half a metre. Such a change would threaten densely populated areas on deltas and coastal plains. Many major cities, including New Orleans, Amsterdam, Shanghai, Dhaka and Cairo, are wholly or partly below present sea level and any rise would significantly increase the risk of flooding. Increases in temperature may also contribute to desertification and reduce agricultural production, especially in areas which are presently semi-arid (McMichael, 1993). Global warming is also predicted to lead to greater climatic fluctuations, accentuating summer temperatures and depressing those in winter. Such large-scale shifts could have particular consequences for cities in semi-arid environments where climatic conditions are presently marginal.

The threats posed to cities by a degradation of the global environment are potentially serious, but they are likely to accrue only in the long term. A more immediate possibility is that cities could be seriously undermined from within because of the sheer pressure of numbers on infrastructure and basic services. There are many concerns involved, including fears over the ways in which the built environment is evolving and the implications for the effective functioning of cities as economic and social systems (Haughton and Hunter, 1994). Many cities, especially in the developed world, are suburbanising to such an extent that their coherence and integration is being compromised. Others, particularly in the developing world, are severely stretched to provide public services today, and large numbers of their residents lack basic utilities and amenities. The principal implications for sustainability can be illustrated by an examination of the very different problems of urban sprawl, water supply and waste disposal.

Land use is a focus of growing concern because of fears that cities are in danger of consuming too much of this locally finite resource and are evolving as spatial forms which are not sustainable. The area of cities increases as populations rise and this, combined with greater locational flexibility for individuals and industries, has led to urban sprawl in place of the high-density compact urban forms that are associated with the pre-automobile age. In some cities the space demands of the car account for almost one-third of the urban land area, rising to two-thirds in inner Los Angeles (McMichael, 1993). Areal expansion is widely seen as inefficient, as it is associated with high energy consumption and increased pollution. Evidence is provided by McGlynn, Newman and Kenworthy (1991), who ranked a number of cities from across the world into five groups, from large sprawling US-type cities with high automobile dependence, to compact cities where there is little reliance upon the car. A high positive correlation was found between urban type and environmental impact. Compact cities had fewest adverse consequences for the environment, but sprawling cities had major detrimental effects. Sprawl is further condemned because of its adverse effects on the countryside and because some observers believe that it leads to cities which lack social cohesion, dynamism and vibrancy (Unwin and Searle, 1991). Others point out that in the long term, urban sprawl is counterproductive. Many of the benefits of the car are short-lived, as rising levels of ownership and use soon lead to congestion and paralysis, undermining the urban structures which the car helped to create.

Although low-density living has many supporters, not least among those who enjoy the environmental attractions of suburbia, there is a widespread view that the physical expansion of cities needs to be checked. There is a limit, probably already exceeded in some countries, to the extent to which the built environment can be allowed to encroach on green land. Such restraint does not, however, necessarily imply a return to compact cities, as high-density living has many disadvantages including congestion, noise, and lack of open space. An alternative way forward is to build new ecological communities based on notions such as permaculture where the population would be self-reliant and self-sufficient (Orrskog and Snickars, 1992). Such proposals for environmentally efficient settlements are attractive to many, though it is uncertain how they could generate the wealth to support large numbers of people at standards of living which would be acceptable in the twenty-first century. Under present circumstances they seem far-fetched but they may gain currency if the functioning of cities is seriously undermined by urban sprawl.

Although there are many cities that have large areas, the principal sustainability issues in the developing world are those of service provision rather than urban sprawl. Urban growth has placed undue stress upon municipal authorities and many cities are deficient in even the most basic public services. The lack of a piped water supply is especially significant, as many health problems are linked to water. Millions of residents of Third World cities have no alternative but to use contaminated water, or at least supplies whose price is high and whose quality is not guaranteed ... For example, although most of the population in Accra has access to piped water, the system is often not operational. In Bangkok, Dar es Salaam and Kinshasa, over half of households are connected and the occupants of the remainder must either use standpipes or buy water from vendors. The situation in Dakar is worse, as only 28 per cent of households have private water connections, while 68 per cent rely on public standpipes and 4 per cent buy from water carriers.

A similar pattern of low but variable provision characterises sewage disposal and garbage collection. Hardoy, Mitlin and Satterthwaite (1992) estimate that around two-thirds of the urban population in the Third World have no hygienic means of disposing of sewage and an even greater number lack an adequate means of disposing of waste water. Most cities in Africa and many in Asia have no sewers at all, and human waste and waste water end up untreated in canals, rivers and ditches. Where sewage systems exist, they rarely serve more than the population that lives in the richer residential areas. Some 70 per cent of the population of Mexico City live in housing served by sewers, but this leaves some three million people who do not. In Buenos Aires it is estimated that the habitations of 6 million of the 11.3 million inhabitants are not connected to the sewer system.

[. . .]

Despite the largely pessimistic tone of the sustainability literature there are, however, some grounds for believing that the arguments have been overstated. The debate on sustainability is in its infancy and many of the points that are made lack detailed empirical evidence and need to be evaluated in the light of experience. Sceptics are keen to point out that the present concerns are merely the most recent in a string of doomsday predictions for the city, none of which have materialised. Arguments that the population grows more rapidly than food production, so leading eventually to widespread famine and social breakdown, have been expounded by a succession of writers from Malthus in the late eighteenth century to Meadows *et al.* (1972) in their highly influential work on the *Limits to Growth*. Although differences in growth rates exist, the evidence is that such analysts have consistently underestimated the capacity of farmers to raise outputs. There are famines in several parts of the world today, but these are distributional problems as there is no shortage of food overall. The Malthusian paradox is that the agricultural sector in many developed countries is contracting, under strong government pressure, in order to deal with overproduction. A crisis of food supply is likely at some stage in the future, but it seems some way off.

A different basis for some optimism about the future is the success of past attempts to deal with urban environmental problems. Most cities in the developed world are cleaner today than they were twenty years ago, and their residents enjoy higher levels of health and amenity as a result. Air quality is one area in which improvements have been dramatic. The reasons are both technological, involving the switch from coal to oil and gas as energy sources, and political, as governments have introduced and enforced clean air legislation. The drive for clean air continues as the burning of oil and gas create different types of pollutant, but these can now more easily be tackled at source by improving the efficiency of combustion at power stations. The example of improvements in air quality suggests, however, that there is some justification for believing that with appropriate intervention and direction, a sustainable future for cities is a realistic possibility.

MANAGING THE URBAN FUTURE

The overwhelming message in the sustainability literature is that the urban populations which are envisaged in the first section of this chapter can only be achieved and maintained through careful planning and management of resources at both global and local scales. Agreement is required among states to work together with other countries on a common agenda to rehabilitate and to protect the global environment. Action is also necessary to regulate urban development within national boundaries. A start has been made, in that international protocols on environment and development have recently been signed in which most nation states have entered into commitments to promote sustainable human settlement. The expectation is that governments will set environmental goals for cities and take appropriate steps to reduce resource use and pollution. Some countries and interest groups within countries have already embarked on this course of action.

Attempts to provide for sustainable development by tackling environmental issues at the global level have been led by the United Nations. Examples of past initiatives include the UNESCO Man and the Biosphere programme, which was launched in 1971, and the Centre for Urban Settlements' HABITAT programme for sustainable cities (HABITAT, 1987). Agreements to phase out CFCs were entered into under the Montreal Protocol which was introduced in 1987. The most comprehensive initiative to date, however, was the 'Earth Summit' Conference on Environment and Development held in Rio de Janeiro in 1992 (Johnson, 1993). With around 175 nations represented, with over 100 heads of state and government, and with over 1,500 officially accredited non-governmental organisations, the summit is thought to have been the largest international gathering ever held. The conference debated a wide range of issues, grouped under the eight headings of atmosphere, biodiversity/biotechnology, institutions, legal instruments, finance, technology transfer, freshwater resources and forests. Two international treaties, on climatic change and biodiversity, were opened for signature and were ratified by over 150 nations. A number of declarations and commitments to enter into further discussions were also made. In addition, a massive 600-page tome detailing a comprehensive agenda for the 21st century was adopted as a framework for future national and international steps in the fields of environment and development.

[. . .]

The fact that an Earth Summit was held at all and that it led to several important agreements and declarations underlines the growing international concern for a sustainable future and a recognition that coordinated global action is required. Critics have argued, however, that the summit achieved little because it failed to strike the necessary 'global bargain' between the developed and the developing worlds (Johnson, 1993). As envisaged, this bargain involves commitments from the developing countries over greenhouse gases, forests and sustainable development in return for concessions from the developed countries on finance, technology transfer and implementation. For Johnson (1993), the fact that greater agreement was not reached was because of the unwillingness of the developed countries to tackle their profligate lifestyles. This was coupled with the refusal of the developing countries to agree to limit the exploitation of their own natural resources so as to address the imbalance caused by the developed countries who have already used up so much of the earth's environmental capital and generated so much waste. The failure to reach agreements on a target date for stabilising emissions of greenhouse gases and on the protection of forests were the two principal failures of the conference.

It is far too early to identify and evaluate the practical outcomes of the Earth Summit. Such consequences will be slow to emerge and any effects will only be measurable in the long term. Particular difficulties surround Agenda 21, since its status is that of a framework for national action. No mandatory rules are specified and there are no clear targets against which progress can be measured. While it includes many practical suggestions to assist in achieving sustainable development, it does not have the power or resources to ensure implementation. The governments which attended the summit did not agree to transfer the necessary authority to any international institution. The onus of responsibility is upon individual nations, each of which is likely to respond differently. Few countries have systems of planning and management through which the necessary regulation can be introduced immediately, and many of those that do are committed to minimal intervention, in the belief that market mechanisms should decide. Many states have a tradition of acting independently and out of self-interest. The Earth Summit recognised sustainable urban development as an important goal for the next century, but further international agreement and commitment, and concentrated action, are required before rhetoric is likely to be translated into reality.

CONCLUSION

This book has attempted to analyse and account for the salient characteristics of the urban world and the global city. It has adopted a broad historical and geographical sweep, viewing urban development as a long-term process and seeing the world as being progressively, but as yet incompletely, interlinked and interconnected as an urban place in economic and social terms. Neither set of trends has run its course and though many urban changes have happened, especially in the last thirty years, more are in progress and are likely to occur. 'Urban world' and 'global city' are convenient catch-phrases which summarise dominant themes, but they describe conceptual ideals rather than the present situation.

[. . .]

It is only comparatively recently that the concept of the urban world has begun to have any meaning. Urban development up to mid-century was largely restricted to the core countries of the world economy. It was most advanced in those parts of north western Europe and North America which had been industrialised the longest and had dominated extensive political and economic empires. Elsewhere, urban development was embryonic, reflecting the widespread inability of pre-industrial economies to raise productivity and surpluses to levels necessary for significant and sustainable urban growth. When measured at the global scale the wholesale switch of population from rural to urban places is a phenomenon of the last thirty years. It is principally a product of changes in the distribution of population in developing countries. The factors involved are identified and accounted for by interdependency theory, which sees urban growth and urbanisation as consequences of the evolution of capitalism and its changing spatial relations. Of particular importance is the recent globalisation of the economy, a development which is reflected in and achieved through the rise of transnational corporations and global finance and producer services institutions, most of which are concentrated in a small number of world cities. The key development is the emergence of a new international division of labour in which production is dispersed to and so accelerates urban development in the peripheral areas within the world-economy.

As well as the shift into towns and cities, the world is progressively becoming urbanised in a social and behavioural sense. Traditionally, urban patterns of association and behaviour, though they themselves were highly varied, were a function of or related to place, being restricted to those who actually lived in cities. Today, the lifestyles and values of urbanites are being extended across the globe, both as a direct corollary of urban growth and urbanisation, and because they can be observed, copied and adopted in rural areas via telecommunications and the media. Urban images and messages, once largely Western in origin, are becoming more diverse as the producers of media products increase in number. The ability to participate in an urban way of life is becoming increasingly independent of location. The world is fast becoming a global urban society of which we are all residents.

The key issue which surrounds the urban world is whether it can continue in its present form. Urban patterns are now well established in most countries, but whether they can absorb a predicted doubling in the urban population over the next quarter century seems highly questionable. Doomsday scenarios have been invoked before and have come to nothing, but the sheer scale of likely growth suggests that they must be taken seriously this time. There is ample evidence that the global physical environment is being degraded and that many major cities are near to exhausting their abilities to cope with their exploding populations. Urgent and decisive action is required by governments if a sustainable urban future for all is to be secured.

[. . .]

REFERENCES

Blowers, A. (1993) *Planning for Sustainable Development* (London: Earthscan).

Goudie, A. (1989) *The Nature of the Environment* (Oxford: Blackwell).

Hardoy, J. E., Mitlin D., and Satterthwaite, D. (1992) *Environmental Problems in Third World Cities* (London: Earthscan).

Hardoy, J. E. and Satterthwaite, D. (1990) "Urban change in the third world. Are recent trends a useful pointer to the urban future?" in D. Cadman and G. Payne (eds.) *The Living City* (London: Routledge), 75–110.

Haughton, G. and Hunter, C. (1994) *Sustainable Cities* (London: Regional Studies Association).
Johnson, S. P. (1993) *The Earth Summit: The United Nations Conference on Environment and Development* (London: Graham & Trotman).
Kelly, P. M. and Karas, J. H. W. (1990) "The greenhouse effect," *Capital and Class* 38: 17–27.
McGlynn, G., Newman, P. and Kenworthy, J. (1991) "Land use and transport: the missing link in urban consolidation," *Urban Futures* (special issue 1, July), 9–18.
McMichael, M. (1993) *Planetary Overload: Global Environmental Change and the Health of the Human Species* (Cambridge: Cambridge University Press).
Meadows, D. H., Meadows, D. L., Randers, J. and Behrens, W. W. (1972) *The Limits to Growth* (New York: University Books).
O'Riordan, T. (1989) "The challenge for environmentalism" in R. Peet, and N. Thrift (eds.) *New Models in Geography: Volume 1* (London: Arnold), 77–104.
Orrskog, L. and Snickars, F. (1992) "On the sustainability of urban and regional structures" in M. Breheny (ed.) *Sustainable Development and Urban Form* (London: Pion).
Pearce, D., Markandya, A. and Barbier, E. B. (1989) *Blueprint for a Green Economy* (London: Earthscan).
Tolba, M. and El-Kholy, O. A. (1992) *The World Environment, 1972–92: Two Decades of Challenge* (London: Chapman and Hall).
Turner, R. K., Pearce, D. and Bateman, I. (1994) *Environmental Economics: An Elementary Introduction* (Hemel Hempstead: Harvester Wheatsheaf).
Unwin, N. and Searle, G. (1991) "Ecologically sustainable development and urban development," *Urban Futures* (special issue 4, November), 1–12.
WCED (1987) [The Brundtland Report] *Our Common Future* (Oxford: Oxford University Press).

ILLUSTRATION CREDITS

COVER PHOTO

Centre Pompidou Plaza, Paris, France, from Algimantas Kezys, *Cityscapes* (Chicago: Loyola University Press, 1988) © Algimantas Kezys. Reproduced with permission of Algimantas Kezys and Loyola University Press.

PLATES

1 Palace of Sargon II of Khorsabad. Jim Harter (ed.), *Images of World Architecture* (New York: Bonanza Books, 1997).
2 Theater of Dionysus, Athens, Greece, fifth century BCE. Jim Harter (ed.), *Images of World Architecture* (New York: Bonanza Books, 1997).
3 Medieval marketplace, Monpazier, France (© Jay Vance).
4 The nineteenth-century industrial city. Augustus Welby Pugin, *Contrasts: Or A Paralled Between the Noble Edifices of the Middle Ages and Corresponding Buildings of the Present Day: Showing the Present Decay of Taste* (London: Charles Dolman, 1841).
5 Birds-eye view of La Crosse, Wisconsin, showing the principal business street. *Frank Leslie's Illustrated Newspaper* (April 30, 1887).
6 Levittown, New York, 1947 (© Levittown, New York Public Library).
7 The auto-centered metropolis. Security Pacific Collection, Los Angeles Public Library (© Los Angeles Public Library).
8 Technoburbia. Microsoft Corporate Headquarters, Redmond, Washington. Used with permission by Microsoft.
9 Unknown artist, "Music in the Street, Music in the Parlor" (1868, *The Illustrated News*). Public domain.
10 Sol Eytinge, Jr., "The Hearth-stone of the Poor" (1876, *Harper's Weekly*). Public domain.
11 C. A. Barry, "City Sketches" (1855, *Ballou's Pictorial Drawing-Room Companion*). Public domain.
12 Unknown artist, "Bicycles and Tricycles – How They Are Made" (1887, *Frank Leslie's Illustrated Newspaper*). Public domain.
13 George N. Barnard, "Burning Mills, Oswego, New York" (1853, *Oswego Daily Times*) (Courtesy George Eastman House).
14 Edward Anthony, "A Rainy Day on Broadway" (1859) (Courtesy George Eastman House).
15 Jacob Riis, "Bandits' Roost, 39 1/2 Mulberry Street" New York (1889). The Jacob A. Riis collection # 101 (© Museum of the City of New York).
16 Lewis Hine, Ellis Island immigrant portraits (ca. 1910). In Grove S. Dow, *Social Problems of Today* (New York: Thomas Y. Crowell, 1925). Public domain.
17 Alfred Stieglitz, "The Steerage" (1907) (Courtesy of the Museum of Modern Art, New York).
18 Dorothea Lange, untitled (roadside sign) (1939). Farm Security Administration, US Department of Agriculture. Printed in C. E. Lively and Conrad Tauber, *Rural Migration in the United States* (Washington, DC: Works Progress Administration, 1939). Public domain.
19 Charles Sheeler, "Ford Plant, Detroit"

(1927) (Courtesy of the Museum of Modern Art, New York).

20 Weegee (Arthur Fellig), "The Critic" (1943) (© 1994, International Center of Photography, New York. Bequest of Wilma Wilcox).

21 King Vidor, still from *The Crowd* (1926) (Photograph courtesy of the Museum of Modern Art, New York, Film Stills Archive).

22 Fritz Lang, still from *Metropolis* (1929) (Photograph courtesy of the Museum of Modern Art, New York, Film Stills Archive).

23 Central Park, New York 1863 (© Museum of the City of New York).

24 Arturo Soria y Mata's plan for a linear city around Madrid, 1894. From Arturo Soria y Mata, *The Linear City* in Richard LeGates and Frederic Stout (eds.), *Early Urban Planning 1870–1940* (London: Routledge/Thoemmes, 1998).

25 Ebenezer Howard's plan for a Garden City, 1898. Ebenezer Howard, *Garden Cities of To-morrow* in Richard LeGates and Frederic Stout (eds.) *Early Urban Planning 1870–1940* (London: Routledge/Thoemmes, 1998).

26 Plan for Welwyn Garden City, 1909. Royal Town Planning Institute. Public domain.

27 Le Corbusier's "Plan Voisin" for a hypothetical city of three million people, 1925. Image Le Corbusier, *Urbanisme* (London: John Rodher, 1929).

28 Plan for Radburn, New Jersey, 1929. Clarence Stein and Henry Wright architects.

29 Frank Lloyd Wright's plan for Broadacre City, 1935. Frank Lloyd Wright, *When Democracy Builds* (Chicago: University of Chicago Press, 1945).

30 Paseo del Rio, San Antonio, Texas (© Alexander Garvin).

31 Quincy Market, Boston, Massachusetts (© Alexander Garvin).

32 Peter Calthorpe's plan for a pedestrian pocket (© Peter Calthorpe Associates).

PART PAGE ILLUSTRATIONS

Prologue Lisa Ryan, 1998.

PART 1 Hartman *Schedel*, WeltChronik (Nuremberg, 1493).

PART 2 Gustav Dore, "London Traffic," 1872.

PART 3 Frederic Stout and Lisa Ryan, Geography collage, 1999.

PART 4 Thomas Nast, "Who Stole The People's Money," *Harpers Weekly* August 19, 1871.

PART 5 Robert Owen, *The Crisis* (London: J. Eamonson, 1832).

PART 6 Postwar British cartoon.

PART 7 © Peter Calthorpe Associates.

PART 8 "Arcology" by Paolo Soleri. Courtesy of the Cosanti Foundation.

COPYRIGHT INFORMATION

PROLOGUE

K. DAVIS Davis, Kingsley, "The Urbanization of the Human Population," *Scientific American* (September 1965). Copyright © 1965 by Scientific American, Inc. All rights reserved. Reprinted with permission.

1 THE EVOLUTION OF CITIES

CHILDE Childe, V. Gordon, "The Urban Revolution," *Town Planning Review*, 21 (April 1950), 3-17. Reprinted by permission.

KITTO Kitto, H. D. F., "The Polis" from *The Greeks* (London: Penguin Books, 1951; revised edition, 1957). Copyright © 1951, 1957, by H. D. F. Kitto. Reprinted with permission of Penguin Books Ltd.

PIRENNE Pirenne, Henri, "City Origins" and "Cities and European Civilization" from *Medieval Cities*, copyright © 1925 (renewed 1952, paperback edition 1969) by Princeton University Press. Reprinted by permission.

ENGELS Public domain

DU BOIS Public domain

GANS Gans, Herbert J., "Levittown and America" from *The Levittowners* Copyright © 1967 by Herbert J. Gans. Reprinted by permission of Pantheon Books, a division of Random House, Inc.

WARNER Warner, Sam Bass, Jr. "The Megalopolis: 1920–" from *The Urban Wilderness: A History of the American City* (Berkeley: University of California Press, 1995). Originally published by Harper & Row. Copyright © 1972 by Sam Bass Warner, Jr. Reprinted by permission.

FISHMAN Fishman, Robert, "Beyond Suburbia: The Rise of the Technoburb" from *Bourgeois Utopias: The Rise and Fall of Suburbia.* Copyright © 1987 by Basic Books, Inc. Reprinted by permission of Basic Books, a division of HarperCollins Publishers, Inc.

2 URBAN CULTURE AND SOCIETY

MUMFORD Mumford, Lewis, "What Is a City?," *Architectural Record*, LXXXII (November 1937), copyright 1937 by McGraw-Hill, Inc. All rights reserved. Reproduced with the permission of the publisher.

WIRTH Wirth, Louis, "Urbanism as a Way of Life," *American Journal of Sociology*, XLIV, 1 (July 1938). Copyright 1938 by the University of Chicago Press. Reprinted by permission.

JACOBS Jacobs, Jane, "The Uses of Sidewalks: Safety" from *The Death and Life of Great American Cities.* Copyright © 1961 by Jane Jacobs. Reprinted by permission of Random House, Inc. and Jonathan Cape.

MURRAY Murray, Charles, "Choosing A Future," pp. 219–237 from *Losing Ground: American Social Policy 1950–1980.* Copyright

© 1984 by Charles Murray. Reprinted by permission of the author and Basic Books, a division of HarperCollins Publishers, Inc.

WILSON Wilson, William Julius, "From Institutional to Jobless Ghettos" from *When Work Disappears: The World of the New Urban Poor* (New York: Knopf, 1996). Copyright © 1996 by William Julius Wilson. Reprinted by permission.

ZUKIN Zukin, Sharon, "Whose Culture? Whose City?" from *The Cultures of Cities* (Oxford: Blackwell, 1995). Copyright © 1995 by Blackwell Publishers. Reprinted by permission.

STOUT Stout, Frederic, "Visions of a New Reality: The City and the Emergence of Modern Visual Culture." Copyright © 1999 by Frederic Stout.

3 URBAN SPACE

BURGESS Burgess, Ernest W., "The Growth of the City: An Introduction to a Research Project" from Robert E. Park, Ernest W. Burgess, and Roderick D. McKenzie, *The City* (Chicago: University of Chicago Press, 1925). Copyright © 1925 by the University of Chicago. Reprinted by permission of the University of Chicago Press.

JACKSON Jackson, John Brinckerhoff, "The Almost Perfect Town," *Landscape*, Vol. 2, No. 1 (Autumn 1952). Reprinted with permission of the author.

RYBCZYNSKI Rybczynski, Witold, "The New Downtown" from *City Life: Urban Expectations in a New World*. Published by HarperCollins Publishers Ltd. Copyright © 1995 by Witold Rybczynski. Reprinted by permission of Brandt & Brandt Literary Agents, Inc.

SOJA Soja, Edward, "Taking Los Angeles Apart: Towards a Postmodern Geography" from *Postmodern Geographies: The Reassertion of Space in Critical Social Theory* (London: Verso, 1989). Copyright © Edward Soja. Reprinted by permission of the author.

M. DAVIS Davis, Mike, "Fortress L.A." from *City of Quartz: Excavating the Future in Los Angeles* (London: Verso, 1990). Copyright © 1990 by Verso. Reprinted by permission.

HARVEY Harvey, David, "Social Justice, Postmodernism, and the City," *International Journal of Urban and Regional Research*, Vol. 16, No. 4 (1992). Copyright © 1992 by Blackwell Publishers/Joint Editors. Reprinted by permission.

SASSEN Sassen, Saskia, "A New Geography of Centers and Margins: Summary and Implications" from *Cities in a World Economy*, pp. 1–8, 119–24, copyright © 1994 by Pine Forge Press. Reprinted by permission.

4 URBAN POLITICS, GOVERNANCE, AND ECONOMICS

MOLLENKOPF Mollenkopf, John, "How to Study Urban Political Power" from *A Phoenix in the Ashes: The Rise and Fall of the Koch Coalition in New York City Politics*. Copyright © 1992 by Princeton University Press. Reprinted by permission.

MAYER Mayer, Margit, "Post-Fordist City Politics" from Ash Amin (ed.), *Post-Fordism: A Reader* (Oxford and Cambridge, Mass.: Blackwell, 1994). © Margit Mayer. Reprinted by permission of the author. This selection is based on a paper prepared for the "Challenges in Urban Management Conference" held at the University of Newcastle upon Tyne, UK, in 1993. An earlier version of the selection also appeared in Patsy Healey *et al.* (eds.), *Managing Cities: The New Urban Context* (Chichester: Wiley, 1995).

ARNSTEIN Arnstein, Sherry, "A Ladder of Citizen Participation," *Journal of the American Institute of Planners*, Vol. 8, No. 3 (July 1969). © *American Planning Association Journal*. Reprinted by permission.

WILSON and KELLING Wilson, James Q. and Kelling, George L., "Broken Windows," *Atlantic Monthly*, Vol. X, No. X (March 1982). Copyright © 1982 by James Q. Wilson. Reprinted by permission.

SAVAGE and WARDE Savage, Mike and Alan Warde, "Cities and Uneven Economic Development" from *Urban Sociology, Capitalism, and Modernity*. Copyright © 1993 Macmillan Press Ltd. and Continuum Publishing Group.

PORTER Porter, Michael, "The Competitive Advantage of the Inner City," reprinted by permission of *Harvard Business Review*, Vol. X, No. X (1995). Copyright © 1995 by the President and Fellows of Harvard College; all rights reserved.

5 URBAN PLANNING HISTORY AND VISIONS

LEGATES and STOUT LeGates, Richard T. and Stout, Frederic, "Modernism and Modern Urban Planning, 1870–1940," a version of "Editors' Introduction" from *Selected Essays*, volume 1 of LeGates, Richard T. and Stout, Frederic (eds.), *Early Urban Planning, 1870–1940* (London: Routledge/Thoemmes, 1998). © Richard LeGates and Frederic Stout. Reprinted by permission of Routledge/Thoemmes and the authors.

OLMSTED Public domain

HOWARD Public domain

GEDDES Public domain

LE CORBUSIER Le Corbusier, "A Contemporary City" from *The City of Tomorrow and its Planning* (translated from the 8th French edition of *Urbanisme* by Frederick Etchells, copyright © 1929). Copyright © 1971 by MIT Press. Reprinted by permission.

WRIGHT Wright, Frank Lloyd, "Broadacre City: A New Community Plan," *Architectural Record*, LXXVII (April, 1935), copyright © 1935 by McGraw-Hill, Inc. All rights reserved. Reproduced with the permission of the publisher and the Frank Lloyd Wright Foundation, Scottsdale, Ariz.

CALTHORPE Calthorpe, Peter, "The Pedestrian Pocket" from Doug Kelbaugh (ed.), *The Pedestrian Pocket Book* (New York: Princeton Architectural Press, 1989). Reprinted by permission of Princeton Architectural Press.

6 URBAN PLANNING THEORY AND PRACTICE

HALL Hall, Peter, "The City of Theory" from *Cities of Tomorrow: An Intellectual History of Urban Planning and Design in the Twentieth Century* (updated edition) (Oxford: Basil Blackwell, 1996). Copyright © 1996 by Peter Hall. Reprinted by permission of Basil Blackwell and Peter Hall.

KAISER and GODSCHALK Kaiser, Edward J. and Godschalk, David R., "Twentieth Century Land Use Planning: A Stalwart Family Tree," *American Planning Association Journal*, Vol. 61, No. 3, 365–85 (Summer, 1995). Reprinted by permission.

GARVIN Garvin, Alexander, "A Realistic Approach to City and Suburban Planning" and "Ingredients of Success" from *The American City: What Works, What Doesn't* (New York: McGraw-Hill, 1996). Reproduced with the permission of the McGraw-Hill Companies.

FORESTER Forester, John, "Planning in the Face of Conflict," *American Planning Association Journal*, Vol. 53, No. 3 (Summer, 1987). © *Journal of the American Planning Association*. Reprinted by permission.

DAVIDOFF Davidoff, Paul, "Advocacy and Pluralism in Planning," *Journal of the American Institute of Planners*, Vol. XXI, No. 4 (November 1965). © *Journal of the American Planning Association*. Reprinted by permission.

WHEELER Wheeler, Stephen, "Planning Sustainable and Livable Cities." Copyright © 1998 by Stephen Wheeler.

SANDERCOCK and FORSYTH Sandercock, Leonie, and Forsyth, Ann, "A Gender Agenda: New Directions for Planning Theory," *American Planning Association Journal*, Vol. 58, No. 1 (Winter 1992). © *Journal of the American Planning Association*. Reprinted by permission.

7 PERSPECTIVES ON URBAN DESIGN

SITTE Public domain.

LYNCH Lynch, Kevin, "The City Image and its Elements" from *The Image of the City* (Cambridge, Mass.: MIT Press, 1961). Copyright © 1960 by the Massachusetts Institute of Technology and the President and Fellows of Harvard College. Reprinted by permission of MIT Press.

WHYTE Whyte, William H., "The Design of Spaces" from *City: Rediscovering the Center.* Copyright © 1988 by William Whyte. Used by permission of Doubleday, a division of Bantam Doubleday Dell Publishing Group, Inc.

JACOBS and APPLEYARD Jacobs, Allan and Appleyard, Donald, "Toward an Urban Design Manifesto," *American Planning Association Journal*, Vol. 53, No. 1 (Winter 1987). © *Journal of the American Planning Association.* Reprinted by permission.

HAYDEN Hayden, Dolores, "What Would a Non-sexist City Be Like? Speculations on Housing, Urban Design, and Human Work," from Catherine R. Stimpson, Elsa Dixler, Martha J. Nelson, and Kathryn B. Yatrakis (eds.), *Women and the American City* (Chicago: University of Chicago Press, 1981; first published 1980). Copyright © 1980, 1981 by the University of Chicago Press. Reprinted by permission of the University of Chicago Press. Originally published as a supplement to *Signs*, Vol. 5, © 1980 University of Chicago Press.

VAN DER RYN and COWAN Van Der Ryn, Sim and Cowan, Stuart, "An Introduction to Ecological Design" from *Ecological Design* (Washington, DC: Island Press, 1995). © Island Press.

8 THE FUTURE OF THE CITY

WEBBER Webber, Melvin M., "The Post-City Age" © *Daedalus*. Reprinted by permission of *Daedalus*, the journal of the American Academy of Arts and Sciences, from the issue entitled "The Conscience of the City," Fall 1968, Vol. 97, No. 4.

SOLERI Soleri, Paolo, "Arcology: The City in the Image of Man" and "The Characteristics of Arcology" from *Arcology: The City in the Image of Man*. Copyright © 1969 by the Massachusetts Institute of Technology. Reprinted by permission of MIT Press and the Cosanti Foundation.

DOWNS Downs, Anthony, "The Need for a New Vision for the Development of Large U.S. Metropolitan Areas." Copyright © 1989 Salomon Brothers Inc. This information is derived from an article written by Anthony Downs, a former Salomon Brothers Inc employee. Although the information in the article was obtained from sources that Saloman Brothers Inc believed to be reliable, Salomon does not guarantee its accuracy, and such information may be incomplete or condensed. All figures included constitute Salomon's judgment as of the original publication date.

CASTELLS Castells, Manuel, "European Cities, the Informational Society, and the Global Economy." © Manuel Castells. This selection was first given as a lecture at the Centrum voor Grootstedelijk Onderzock, University of Amsterdam. It was published in the *Journal of Economic and Social Geography*, lxxxiv, 4 (1993), and republished in the March/April edition of *New Left Review*, No. 240.

GRAHAM and MARVIN Graham, Stephen and Marvin, Simon, "The Transformation of Cities: Towards Planetary Urban Networks" and "Telecommunications and Urban Futures" from *Telecommunications and the City: Electronic Spaces, Urban Places* (London: Routledge, 1996). Reprinted by permission of Steven Graham and Simon Marvin and Routledge.

CLARK Clark, David, "The Future Urban World" from *Urban World/Global City* (London and New York: Routledge, 1996). Reprinted by permission.

INDEX